HEAT TRANSFER
IN INDUSTRIAL
COMBUSTION

Charles E. Baukal, Jr.

CRC Press
Taylor & Francis Group
Boca Raton London New York

CRC Press is an imprint of the
Taylor & Francis Group, an **informa** business

CRC Press
Taylor & Francis Group
6000 Broken Sound Parkway NW, Suite 300
Boca Raton, FL 33487-2742

First issued in paperback 2019

© 2000 by Taylor & Francis Group, LLC
CRC Press is an imprint of Taylor & Francis Group, an Informa business

No claim to original U.S. Government works

ISBN-13: 978-0-8493-1699-9 (hbk)
ISBN-13: 978-0-367-39859-0 (pbk)

Library of Congress Card Number 99-088045

Library of Congress Cataloging-in-Publication Data

Baukal, Charles E.
    Heat transfer in industrial combustion / Charles E. Baukal, Jr.
        p. cm.
    Includes bibliographical references and index.
    ISBN 0-8493-1699-5 (alk. paper)
    1. Heat--Transmission.   2. Combustion engineering.   I. Title

TJ260.B359 2000
621.402′2--dc21
                                                      99-088045

Visit the Taylor & Francis Web site at
http://www.taylorandfrancis.com

and the CRC Press Web site at
http://www.crcpress.com

Preface

This book is intended to fill a gap in the literature for books on heat transfer in industrial combustion, written primarily for the practicing engineer. Many textbooks have been written on both heat transfer and combustion, but both types of book generally have only a limited amount of information concerning the combination of heat transfer and industrial combustion. One of the purposes of this book is to codify the many relevant books, papers, and reports that have been written on this subject into a single, coherent reference source.

The key difference for this book compared to others is that it looks at each topic from a somewhat narrow scope to see how that topic affects heat transfer in industrial combustion. For example, in Chapter 2, the basics of combustion are considered, but from the limited perspective as to how combustion influences the heat transfer. There is very little discussion of combustion kinetics because in the overall combustion system, the kinetics of the chemical reactions in the flame only significantly impact the heat transfer in somewhat limited circumstances. Therefore, this book does not attempt to go over subjects that have been more than adequately covered in other books, but rather attempts to look at those subjects through the narrow lens of how they influence the heat transfer in the system.

The book is basically organized in three parts. The first part deals with the basics of heat transfer in combustion and includes chapters on the modes of heat transfer, computer modeling, and experimental techniques. The middle part of the book deals with general concepts of heat transfer in industrial combustion systems and includes chapters on heat transfer from flame impingement, from burners, and in furnaces. The last part of the book deals with specific applications of heat transfer in industrial combustion and includes chapters on lower and higher temperature applications and some advanced applications. The book has discussions on the use of oxygen to enhance combustion and on flame impingement, both of particular interest to the author. These subjects have received very little, if any, coverage in previous books on heat transfer in industrial combustion.

As with any book of this type, there are many topics that are not covered. The book does not address other aspects of heat transfer in combustion such as power generation (stationary turbines or boilers) and propulsion (internal combustion, gas turbine or rocket engines), which are not normally considered to be industrial applications. It also does not treat packed bed combustion, material synthesis in flames, or flare applications, which are all fairly narrow in scope. Because the vast majority of industrial applications use gaseous fuels, that is the focus of this book, with only a cursory discussion of solid and liquid fuels. This book basically concerns atmospheric combustion, which is the predominant type used in industry. There are also many topics that are discussed in the book, but with a very limited treatment. One example is optical diagnostics. The reason for the limited discussion is that there has been very little application of such techniques to industrial combustors because of the difficulties in making them work on a large scale in sometimes hostile environments.

This book attempts to focus on those topics that are of interest to the practicing engineer. It does not profess to be exhaustively comprehensive, but does attempt to provide references for the interested reader who would like more information on a particular subject. As most authors know, it is always a struggle about what to include and what not to include in a book. Here, the guideline that has been used is to minimize the theory and maximize the applications, while at the same time trying to at least touch on the relevant topics for heat transfer in industrial combustion.

About the Author

Charles E. Baukal, Jr., Ph.D., P.E., is the Director of the John Zink Company LLC R & D Test Center in Tulsa, OK. He has 20 years of experience in the fields of heat transfer and industrial combustion and has authored more than 50 publications in those fields, including editing the book *Oxygen-Enhanced Combustion* (CRC Press, Boca Raton, FL, 1998). He has a Ph.D. in mechanical engineering from the University of Pennsylvania, is a licensed Professional Engineer in the state of Pennsylvania, has been an adjunct instructor at several colleges, and has eight U.S. patents.

Acknowledgment

This book is dedicated to my wife Beth, to my children Christine, Caitlyn, and Courtney, and to my mother Elaine. This book is also dedicated to the memories of my father, Charles, Sr. and my brother, Jim, who have both gone on to be with their maker.

The author would like to thank Tom Smith of Marsden, Inc. (Pennsauken, NJ) and Buddy Eleazer of Air Products (Allentown, PA) for the opportunities to learn firsthand about heat transfer in industrial combustion. The author would also like to thank David Koch and Dr. Roberto Ruiz of John Zink Company LLC (Tulsa, OK) for their support in the writing of this book. Last but not least, the author would like to thank the good Lord above, without whom this would not have been possible.

Charles E. Baukal, Jr., Ph.D., P.E.

Nomenclature

Symbol	Description	Units
A	Area	ft^2 or m^2
c	Speed of light	ft/sec or m/sec
c_p	Specific heat	Btu/lb-°F or J/kg-K
C_p	Pitot-Static probe calibration constant	dimensionless
d	Diameter	in. or mm
D	Dimensionless diameter $= d/d_n$	dimensionless
Da	Damköhler number (see Eq. 2.19)	dimensionless
e	Hemispherical emissive power	Btu/hr-ft^2 or kW/m^2
E	Error	dimensionless
E	Hemispherical emissive power	Btu/hr or kW
$F_{1\text{-}2}$	Radiation view factor from surface 1 to surface 2	dimensionless
Gr	Grashoff number (see Eq. 3.12)	dimensionless
h	Convection heat transfer coefficient	Btu/hr-ft^2-°F or W/m^2-K
h^C	Chemical enthalpy	Btu/lb or J/kg
h_{fusion}	Heat of fusion	Btu/lb or J/kg
h^S	Sensible enthalpy $= \int c_p dt$	Btu/lb or J/kg
h^T	Total enthalpy $= h^C + h^S$	Btu/lb or J/kg
H	Fuel heat content	Btu/lb or kJ/kg
I	Radiation intensity	Btu/hr-ft^2-μm or W/m^2-μm
k	Thermal conductivity	Btu/hr-ft-°F or W/m-K
K	Non-absorption factor for radiation	dimensionless
K_a	Absorption coefficient for radiation	dimensionless
K_s	Scattering coefficient for radiation	dimensionless
l	Length	in. or mm
l_v	Potential core length for velocity	in. or mm
L	Distance between the burner and the target $= l_j/d_n$	dimensionless
L	Radiation path length through a gas	ft or m
L_m	Mean beam length	ft or m
Le	Lewis number $= \rho c_p D_{i-mix}/k$	dimensionless
$\dot{m}$	Mass flow rate	lb/hr or kg/hr
Ma	Mach number $= v/c$	dimensionless
MW	Molecular weight	lb/lb-mole or g/g-mole
Nu	Nusselt number (see Eq. 3.3)	dimensionless
p_{st}	Static pressure	psig or Pa
p_t	Total pressure	psig or Pa
Pr	Prandtl number $= c_p \mu/k$ (see Eq. 3.2)	dimensionless
q	Heat flow	Btu/hr or kW
q''	Heat flux	Btu/hr-ft^2 or kW/m^2
q_f	Burner firing rate	Btu/hr or kW
q_i	Heat absorbed by calorimeter i	Btu/hr or kW
Q	Gas flow rate	ft^3/hr or m^3/hr
r	Radial distance from the burner centerline	in. or mm
R	Dimensionless radius $= r/d_n$	dimensionless
Ra	Rayleigh number (see Eq. 3.11)	dimensionless
Re	Reynolds number $= \rho vd/\mu$ (see Eq. 3.1)	dimensionless
Ri	Richardson number (see Eq. 3.13)	dimensionless
S	Stoichiometry (see Eqs. 2.3 and 2.5)	dimensionless

S_L	Laminar flame speed	ft/s or m/s
t	Temperature	°F or K
T	Absolute temperature	°R or K
Tu	Turbulence intensity	dimensionless
v	Velocity	ft/s or m/s
V	Volume	ft³ or m³
x	Axial distance from the burner to the target stagnation point	in. or mm
X	Distance from the burner to the target stagnation point = x/d_n	dimensionless
X_v	Potential core length for velocity = l_v/d_n	dimensionless

Greek Symbols

α	Absorptivity	dimensionless
β	Velocity gradient	s^{-1}
$\tilde{\beta}$	Volume coefficient of expansion	°R⁻¹ or K⁻¹
δ	Boundary layer thickness	in. or mm
ε	Emissivity	dimensionless
η	Thermal efficiency	dimensionless
$\varkappa$	Absorption coefficient of a luminous gas	ft⁻¹ or m⁻¹
μ	Absolute or dynamic viscosity	lb/ft-s or kg/m-s
γ	Turbulence enhancement factor (see Eq. 7.25)	dimensionless
Ω	Oxidizer composition = $\dfrac{O_2 \text{ volume in the oxidizer}}{O_2 + N_2 \text{ volume in the oxidizer}}$	dimensionless
ϕ	Equivalence ratio = $\dfrac{\text{Stoichiometric oxygen/Fuel volume ratio}}{\text{Actual oxygen/Fuel volume ratio}}$	dimensionless
δ	Soot radiation index (see Eq. 3.53)	dimensionless
λ	Fuel mixture ratio (see Eq. 2.8)	dimensionless
λ	Wavelength	μm
ν	Frequency	s^{-1}
ν	Kinematic viscosity	ft²/s or m²/s
ρ	Density	lb/ft³ or kg/m³
ρ	Reflectivity	dimensionless
σ	Stefan-Boltzmann constant	Btu/hr-ft2-°R⁴ or W/m²-K⁴
Φ	Surface catalytic efficiency (see Eq. 7.26)	dimensionless
θ_{radm}	Radiometer field of view	degrees
τ	Optical density	s
τ	Transmissivity	dimensionless
τ	Time	s

Subscripts

b	Stagnation body or target		n	Burner nozzle
conv	Convective heat transfer		NG	Natural gas
e	Edge of boundary layer		p	Probe
eff	Effective diameter		r	Radial direction
f	Fluid		rad	Thermal radiation
f	Film temperature (see Eq. 4.7)		radm	Radiometer
g	Gas		rec	Recovery temperature (see Eq. 7.6)
∞	Ambient conditions		ref	Reference temperature (see Eq. 7.5)
j	Jet		s	Stagnation point
j	Thermocouple junction		T	Turbulent
l	Load		T/C	Thermocouple
K	Kolmogorov		w	Wall (target surface)
m	Medium		υ	Volumetric
max	Maximum			

Table of Contents

Chapter 1   Introduction

1.1   Importance of Heat Transfer in Industrial Combustion 1
   1.1.1   Energy Consumption..1
   1.1.2   Research Needs 1
1.2   Literature Discussion 6
   1.2.1   Heat Transfer..6
   1.2.2   Combustion 7
   1.2.3   Heat Transfer and Combustion 7
1.3   Combustion System Components 8
   1.3.1   Burners 8
      1.3.1.1   Competing Priorities 9
      1.3.1.2   Design Factors...10
         1.3.1.2.1   Fuel...11
         1.3.1.2.2   Oxidizer..12
         1.3.1.2.3   Gas Recirculation 13
      1.3.1.3   General Burner Types 13
         1.3.1.3.1   Mixing Type..13
         1.3.1.3.2   Oxidizer Type...14
         1.3.1.3.3   Draft Type...16
         1.3.1.3.4   Heating Type...18
   1.3.2   Combustors...18
      1.3.2.1   Design Considerations 18
         1.3.2.1.1   Load Handling...18
         1.3.2.1.2   Temperature..19
         1.3.2.1.3   Heat Recovery...19
      1.3.2.2   General Classifications...19
         1.3.2.2.1   Load Processing Method 19
         1.3.2.2.2   Heating Type...20
         1.3.2.2.3   Geometry..20
         1.3.2.2.4   Heat Recuperation..21
   1.3.3   Heat Load..21
      1.3.3.1   Process Tubes...21
      1.3.3.2   Moving Substrate 21
      1.3.3.3   Opaque Materials 22
      1.3.3.4   Transparent Materials...22
   1.3.4   Heat Recovery Devices...23
      1.3.4.1   Recuperators..23
      1.3.4.2   Regenerators..23
References 24

Chapter 2   Some Fundamentals of Combustion

2.1   Combustion Chemistry 29
   2.1.1   Fuel Properties 29
   2.1.2   Oxidizer Composition 30

2.1.3   Mixture Ratio 30

2.1.4   Operating Regimes 33

2.2   Combustion Properties 34

     2.2.1   Combustion Products 34

          2.2.1.1   Oxidizer Composition 34

          2.2.1.2   Mixture Ratio 37

          2.2.1.3   Air and Fuel Preheat Temperature 38

          2.2.1.4   Fuel Composition 40

     2.2.2   Flame Temperature 40

          2.2.2.1   Oxidizer and Fuel Composition 40

          2.2.2.2   Mixture Ratio 41

          2.2.2.3   Oxidizer and Fuel Preheat Temperature 43

     2.2.3   Available Heat 43

     2.2.4   Flue Gas Volume 46

2.3   Exhaust Product Transport Properties 48

     2.3.1   Density 49

     2.3.2   Specific Heat 51

     2.3.3   Thermal Conductivity 53

     2.3.4   Viscosity 55

     2.3.5   Prandtl Number 58

     2.3.6   Lewis Number 60

References 64

Chapter 3   Heat Transfer Modes

3.1   Introduction 65

3.2   Convection 65

     3.2.1   Forced Convection 66

          3.2.1.1   Forced Convection from Flames...................66

          3.2.1.2   Forced Convection from Outside Combustor Wall 68

          3.2.1.3   Forced Convection from Hot Gases to Tubes 68

     3.2.2   Natural Convection 68

          3.2.2.1   Natural Convection from Flames...................68

          3.2.2.2   Natural Convection from Outside Combustor Wall 69

3.3   Radiation 69

     3.3.1   Surface Radiation 72

     3.3.2   Nonluminous Radiation 82

          3.3.2.1   Theory 82

          3.3.2.2   Combustion Studies 90

               3.3.2.2.1   Total Radiation 90

               3.3.2.2.2   Spectral Radiation 98

     3.3.3   Luminous Radiation 102

          3.3.3.1   Theory 102

          3.3.3.2   Combustion Studies 105

               3.3.3.2.1   Total Radiation 106

               3.3.3.2.2   Spectral Radiation 108

3.4   Conduction 109

     3.4.1   Steady-State Conduction..110

     3.4.2   Transient Conduction 113

3.5   Phase Change 114

     3.5.1   Melting 114

     3.5.2   Boiling 114

        3.5.2.1   Internal Boiling 116
        3.5.2.2   External Boiling 116
   3.5.3   Condensation 117
References 117

Chapter 4   Heat Sources and Sinks

4.1   Heat Sources 123
   4.1.1   Combustibles 123
        4.1.1.1   Fuel Combustion 123
        4.1.1.2   Volatile Combustion 123
   4.1.2   Thermochemical Heat Release 124
        4.1.2.1   Equilibrium TCHR 126
        4.1.2.2   Catalytic TCHR 127
        4.1.2.3   Mixed TCHR 127
4.2   Heat Sinks 127
   4.2.1   Load 128
        4.2.1.1   Tubes 128
        4.2.1.2   Substrate 129
        4.2.1.3   Granular Solid 129
        4.2.1.4   Molten Liquid 131
        4.2.1.5   Surface Conditions 132
                4.2.1.5.1   Radiation 133
                4.2.1.5.2   Catalyticity 134
   4.2.2   Wall Losses 140
   4.2.3   Openings 143
        4.2.3.1   Radiation 143
        4.2.3.2   Gas Flow Through Openings 143
   4.2.4   Material Transport 145
References 145

Chapter 5   Computer Modeling

5.1   Combustion Modeling 149
5.2   Modeling Approaches 150
   5.2.1   Fluid Dynamics 151
        5.2.1.1   Moment Averaging 151
        5.2.1.2   Vortex Methods 152
        5.2.1.3   Spectral Methods 152
        5.2.1.4   Direct Numerical Simulation 153
   5.2.2   Geometry 153
        5.2.2.1   Zero-Dimensional Modeling 153
        5.2.2.2   One-Dimensional Modeling 153
        5.2.2.3   Multi-dimensional Modeling 154
   5.2.3   Reaction Chemistry 154
        5.2.3.1   Nonreacting Flows 155
        5.2.3.2   Simplified Chemistry 155
        5.2.3.3   Complex Chemistry 156
   5.2.4   Radiation 156
        5.2.4.1   Nonradiating 156
        5.2.4.2   Participating Media 157
   5.2.5   Time Dependence 158

	5.2.5.1	Steady State	158
	5.2.5.2	Transient	158
5.3	Simplified Models		159
5.4	Computational Fluid Dynamic Modeling		159
	5.4.1	Increasing Popularity of CFD	159
	5.4.2	Potential Problems of CFD	161
	5.4.3	Equations	162
		5.4.3.1 Fluid Dynamics	162
		5.4.3.2 Heat Transfer	164
		5.4.3.3 Chemistry	169
		5.4.3.4 Multiple Phases	170
	5.4.4	Boundary and Initial Conditions	171
		5.4.4.1 Inlets and Outlets	172
		5.4.4.2 Surfaces	172
		5.4.4.3 Symmetry	172
	5.4.5	Discretization	173
		5.4.5.1 Finite Difference Technique	173
		5.4.5.2 Finite Volume Technique	174
		5.4.5.3 Finite Element Technique	175
		5.4.5.4 Mixed	175
		5.4.5.5 None	175
	5.4.6	Solution Methods	175
	5.4.7	Model Validation	176
	5.4.8	Industrial Combustion Examples	177
		5.4.8.1 Modeling Burners	177
		5.4.8.2 Modeling Combustors	178
References			181

Chapter 6   Experimental Techniques

6.1	Introduction		195
6.2	Heat Flux		195
	6.2.1	Total Heat Flux	195
		6.2.1.1 Steady-State Uncooled Solids	196
		6.2.1.2 Steady-State Cooled Solids	196
		6.2.1.2.1 Single Cooling Circuit	196
		6.2.1.2.2 Multiple Cooling Circuits	196
		6.2.1.2.3 Surface Probe	197
		6.2.1.3 Steady-State Cooled Gages	197
		6.2.1.3.1 Gradient Through a Thin Solid Rod	197
		6.2.1.3.2 Thin Disk Calorimeter	197
		6.2.1.3.3 Heat Flux Transducer	197
		6.2.1.4 Transient Uncooled Targets	198
		6.2.1.5 Transient Uncooled Gages	198
		6.2.1.5.1 Slug Calorimeter	199
		6.2.1.5.2 Heat Flux Transducer	200
	6.2.2	Radiant Heat Flux	200
		6.2.2.1 Heat Flux Gage	200
		6.2.2.2 Ellipsoidal Radiometer	203
		6.2.2.3 Spectral Radiometer	204
		6.2.2.4 Other Techniques	204
	6.2.3	Convective Heat Flux	206

6.3 Temperature...207
 6.3.1 Gas Temperature 207
  6.3.1.1 Suction Pyrometer...207
  6.3.1.2 Optical Techniques..209
  6.3.1.3 Fine Wire Thermocouples...209
  6.3.1.4 Line Reversal 213
 6.3.2 Surface Temperature 213
  6.3.2.1 Embedded Thermocouple 213
  6.3.2.2 Infrared Detectors 214
6.4 Gas Flow 216
 6.4.1 Gas Velocity 216
  6.4.1.1 Pitot Tubes...216
  6.4.1.2 Laser Doppler Velocimetry 218
  6.4.1.3 Other Techniques 219
 6.4.2 Static Pressure Distribution 219
  6.4.2.1 Stagnation Velocity Gradient 219
  6.4.2.2 Stagnation Zone 220
6.5 Gas Species 221
6.6 Other Measurements 221
6.7 Physical Modeling..223
References 223

Chapter 7 Flame Impingement
7.1 Introduction 231
7.2 Experimental Conditions..233
 7.2.1 Configurations 234
  7.2.1.1 Flame Normal to a Cylinder in Crossflow 235
  7.2.1.2 Flame Normal to a Hemispherically Nosed Cylinder...236
  7.2.1.3 Flame Normal to a Plane Surface 236
  7.2.1.4 Flame Parallel to a Plane Surface.................................238
 7.2.2 Operating Conditions 238
  7.2.2.1 Oxidizers 238
  7.2.2.2 Fuels 240
  7.2.2.3 Equivalence Ratios 240
  7.2.2.4 Firing Rates 243
  7.2.2.5 Reynolds Number 243
  7.2.2.6 Burners 243
  7.2.2.7 Nozzle Diameter 246
  7.2.2.8 Location 246
 7.2.3 Stagnation Targets 246
  7.2.3.1 Size..247
  7.2.3.2 Target Materials 248
  7.2.3.3 Surface Preparation 248
  7.2.3.4 Surface Temperatures..250
 7.2.4 Measurements...250
7.3 Semianalytical Heat Transfer Solutions 250
 7.3.1 Equation Parameters 257
  7.3.1.1 Thermophysical Properties 258
  7.3.1.2 Stagnation Velocity Gradient 261
   7.3.1.2.1 Analytical Solutions 261
   7.3.1.2.2 Empirical Correlations 262

7.3.2   Equations...263
         7.3.2.1   Sibulkin Results 263
         7.3.2.2   Fay and Riddell Results 263
         7.3.2.3   Rosner Results..264
7.3.3   Comparisons With Experiments 264
         7.3.3.1   Forced Convection (Negligible TCHR)................................264
                 7.3.3.1.1   Laminar Flow 265
                 7.3.3.1.2   Turbulent Flows 265
         7.3.3.2   Forced Convection with TCHR 266
                 7.3.3.2.1   Laminar Flow 266
                 7.3.3.2.2   Turbulent Flow 268
7.3.4   Sample Calculations..268
         7.3.4.1   Laminar Flames Without TCHR..268
         7.3.4.2   Turbulent Flames Without TCHR...269
         7.3.4.3   Laminar Flames with TCHR 270
7.3.5   Summary 270
7.4   Empirical Heat Transfer Correlations...271
7.4.1   Thermophysical Properties 272
7.4.2   Flames Impinging Normal to a Cylinder 272
         7.4.2.1   Local Convection Heat Transfer..273
                 7.4.2.1.1   Laminar and Turbulent Flows....................273
                 7.4.2.1.2   Turbulent Flows 273
         7.4.2.2   Average Convection Heat Transfer.......................................273
                 7.4.2.2.1   Laminar Flows 273
                 7.4.2.2.2   Laminar and Turbulent Flows....................274
                 7.4.2.2.3   Flow Type Unspecified 274
         7.4.2.3   Average Convection Heat Transfer with TCHR....................274
                 7.4.2.3.1   Flow Type Unspecified 274
         7.4.2.4   Average Radiation Heat Transfer...274
                 7.4.2.4.1   Laminar and Turbulent Flows....................274
         7.4.2.5   Maximum Convection and Radiation Heat Transfer.............275
                 7.4.2.5.1   Turbulent Flows 275
7.4.3   Flames Impinging Normal to a Hemi-Nosed Cylinder.........................275
         7.4.3.1   Local Convection Heat Transfer..275
                 7.4.3.1.1   Laminar and Turbulent Flows....................275
                 7.4.3.1.2   Turbulent Flows 276
         7.4.3.2   Local Convection Heat Transfer with TCHR 276
                 7.4.3.2.1   Turbulent Flows 276
7.4.4   Flames Impinging Normal to a Plane Surface 276
         7.4.4.1   Local Convection Heat Transfer..276
                 7.4.4.1.1   Laminar Flows 276
                 7.4.4.1.2   Turbulent Flows 277
         7.4.4.2   Local Convection Heat Transfer with TCHR 279
                 7.4.4.2.1   Laminar Flows 279
                 7.4.4.2.2   Turbulent Flows 279
         7.4.4.3   Average Convection Heat Transfer.......................................279
                 7.4.4.3.1   Laminar Flows 279
                 7.4.4.3.2   Turbulent Flows 279
7.4.5   Flames Parallel to a Plane Surface 280
         7.4.5.1   Local Convection Heat Transfer With TCHR 280

            7.4.5.1.1   Laminar Flows 280
            7.4.5.1.2   Turbulent Flows 280
     7.4.5.2   Local Convection and Radiation Heat Transfer 281
            7.4.5.2.1   Turbulent Flows 281
References 281

Chapter 8   Heat Transfer from Burners

8.1   Introduction 285
8.2   Open-Flame Burners...285
     8.2.1   Momentum Effects...285
     8.2.2   Flame Luminosity 285
     8.2.3   Firing Rate Effects 287
     8.2.4   Flame Shape Effects 292
8.3   Radiant Burners..296
     8.3.1   Perforated Ceramic or Wire Mesh Radiant Burners 300
     8.3.2   Flame Impingement Radiant Burners...301
     8.3.3   Porous Refractory Radiant Burners..302
     8.3.4   Advanced Ceramic Radiant Burners 307
     8.3.5   Radiant Wall Burners 311
     8.3.6   Radiant Tube Burners 311
8.4   Effects on Heat Transfer 316
     8.4.1   Fuel Effects 316
          8.4.1.1   Solid Fuels...316
          8.4.1.2   Liquid Fuels 316
          8.4.1.3   Gaseous Fuels 316
          8.4.1.4   Fuel Temperature 317
     8.4.2   Oxidizer Effects 318
          8.4.2.1   Oxidizer Composition 318
          8.4.2.2   Oxidizer Temperature..318
     8.4.3   Staging Effects 320
          8.4.3.1   Fuel Staging 321
          8.4.3.2   Oxidizer Staging 322
     8.4.4   Burner Orientation 322
          8.4.4.1   Hearth-Fired Burners 323
          8.4.4.2   Wall-Fired Burners..324
          8.4.4.3   Roof-Fired Burners 325
          8.4.4.4   Side-Fired Burners 326
     8.4.5   Heat Recuperation 326
          8.4.5.1   Regenerative Burners 327
          8.4.5.2   Recuperative Burners 329
          8.4.5.3   Furnace or Flue Gas Recirculation.............................331
     8.4.6   Pulse Combustion 331
8.5   In-Flame Treatment..335
References 337

Chapter 9   Heat Transfer in Furnaces

9.1   Introduction 345
9.2   Furnaces 346
     9.2.1   Firing Method 347

          9.2.1.1 Direct Firing 347
          9.2.1.2 Indirect Firing 347
          9.2.1.3 Heat Distribution 349
     9.2.2 Load Processing Method 351
          9.2.2.1 Batch Processing 351
          9.2.2.2 Continuous Processing 351
          9.2.2.3 Hybrid Processing 352
     9.2.3 Heat Transfer Medium 352
          9.2.3.1 Gaseous Medium 352
          9.2.3.2 Vacuum 353
          9.2.3.3 Liquid Medium 353
          9.2.3.4 Solid Medium 354
     9.2.4 Geometry 354
          9.2.4.1 Rotary Geometry 355
          9.2.4.2 Rectangular Geometry 358
          9.2.4.3 Ladle Geometry 358
          9.2.4.4 Vertical Cylindrical Geometry 360
     9.2.5 Furnace Types 360
          9.2.5.1 Reverberatory Furnace 360
          9.2.5.2 Shaft Kiln 362
          9.2.5.3 Rotary Furnace 362
9.3 Heat Recovery 363
     9.3.1 Recuperators 364
     9.3.2 Regenerators 364
     9.3.3 Gas Recirculation 366
          9.3.3.1 Flue Gas Recirculation 366
          9.3.3.2 Furnace Gas Recirculation 366
References 367

Chapter 10   Lower Temperature Applications

10.1 Introduction 369
10.2 Ovens and Dryers 369
     10.2.1 Predryer 369
     10.2.2 Dryer 372
10.3 Fired Heaters 375
     10.3.1 Reformer 376
     10.3.2 Process Heater 378
10.4 Heat Treating 384
     10.4.1 Standard Atmosphere 387
     10.4.2 Special Atmosphere 387
References 391

Chapter 11   Higher Temperature Applications

11.1 Introduction 395
     11.1.1 Furnaces 395
     11.1.2 Industries 395
11.2 Metals Industry 396
     11.2.1 Ferrous Metal Production 396
          11.2.1.1 Electric Arc Furnace 396
          11.2.1.2 Smelting 399

11.2.1.3 Ladle Preheating 399
11.2.1.4 Reheating Furnace...402
11.2.1.5 Forging 407
11.2.2 Aluminum Metal Production 408
11.3 Minerals Industry 410
11.3.1 Glass 411
11.3.1.1 Types of Traditional Glass-Melting Furnaces 412
11.3.1.2 Unit Melter 412
11.3.1.3 Recuperative Melter 412
11.3.1.4 Regenerative or Siemens Furnace 412
11.3.1.4.1 End-Port Regenerative Furnace 413
11.3.1.4.2 Side-Port Regenerative Furnace................413
11.3.1.5 Oxygen-Enhanced Combustion for Glass Production.............414
11.3.1.6 Advanced Techniques for Glass Production.....................416
11.3.2 Cement and Lime 417
11.3.3 Bricks, Refractories, and Ceramics 419
11.4 Waste Incineration..419
11.4.1 Types of Incinerators 422
11.4.1.1 Municipal Waste Incinerators 422
11.4.1.2 Sludge Incinerators 423
11.4.1.3 Mobile Incinerators 424
11.4.1.4 Transportable Incinerators 425
11.4.1.5 Fixed Hazardous Waste Incinerators 425
11.4.2 Heat Transfer in Waste Incineration 426
References 428

Chapter 12   Advanced Combustion Systems
12.1 Introduction 433
12.2 Oxygen-Enhanced Combustion 433
12.2.1 Typical Use Methods 434
12.2.1.1 Air Enrichment...434
12.2.1.2 O_2 Lancing 435
12.2.1.3 Oxy/Fuel...435
12.2.1.4 Air-Oxy/Fuel 436
12.2.2 Operating Regimes..437
12.2.3 Heat Transfer Benefits 437
12.2.3.1 Increased Productivity...437
12.2.3.2 Higher Thermal Efficiencies 438
12.2.3.3 Higher Heat Transfer Efficiency 438
12.2.3.4 Increased Flexibility...438
12.2.4 Potential Heat Transfer Problems 439
12.2.4.1 Refractory Damage 439
12.2.4.2 Nonuniform Heating 440
12.2.4.2.1 Hotspots 440
12.2.4.2.2 Reduction in Convection..........................440
12.2.5 Industrial Heating Applications 440
12.2.5.1 Metals 440
12.2.5.2 Minerals..440
12.2.5.3 Incineration 441
12.2.5.4 Other...441

12.3 Submerged Combustion 441
    12.3.1 Metals Production 441
    12.3.2 Minerals Production..443
    12.3.3 Liquid Heating 444
12.4 Miscellaneous...444
    12.4.1 Surface Combustor-Heater..445
    12.4.2 Direct-Fired Cylinder Dryer 445
References 446

Appendices

Appendix A: Reference Sources for Further Information.......................................449
Appendix B: Common Conversions 451
Appendix C: Methods of Expressing Mixture Ratios for CH_4, C_3H_8, and H_2 453
Appendix D: Properties for CH_4, C_3H_8, and H_2 Flames.....................................459
Appendix E: Fluid Dynamics Equations 497
Appendix F: Material Properties 501

Author Index...521

Subject Index..535

1 Introduction

1.1 IMPORTANCE OF HEAT TRANSFER IN INDUSTRIAL COMBUSTION

This chapter section briefly attempts to establish the importance of heat transfer in industrial combustion by first looking at how much energy is consumed by industry and then by the number of recommendations for continued research into heat transfer for industrial combustion applications.

1.1.1 ENERGY CONSUMPTION

Industry relies heavily on the combustion process as shown in Table 1.1. The major uses for combustion in industry are shown in Table 1.2. Hewitt et al. (1994) have listed some of the common heating applications used in industry, as shown in Table 1.3.[1] Typical industrial combustion applications can also be characterized by their temperature ranges, as shown in Figure 1.1. As can be seen in Figure 1.2, the demand for energy is expected to continue to rapidly increase. Most of the energy (88%) is produced by the combustion of fossil fuels like oil, natural gas, and coal. According to the U.S. Department of Energy, the demand in the industrial sector is projected to increase by 0.8% per year to the year 2020.[2]

The objective in nearly all industrial combustion applications is to transfer that energy to some type of load for thermal processing of that load.[3] Examples of the many types of thermal processes include heating solids to the softening point for forming, drying for moisture removal, and chemical processing in calcining. Depending on the application, the heat may be transferred directly from the flame to the load, or indirectly from the flame to a heat transfer medium like a ceramic tube. The sheer amount of energy used by industry makes heat transfer in industrial combustion an important subject.

1.1.2 RESEARCH NEEDS

Many studies have recommended further research into heat transfer in industrial combustion for a wide range of reasons. One obvious reason includes increasing fuel efficiency in the light of the substantial energy consumption in industry in the combustion of fossil fuels. Another reason is to optimize existing processes to increase throughput or productivity for a given size combustion system. Further research is needed to develop new processes as new materials and products need to be heated during processing. Research is needed in both the computer simulation of the combustion processes and in making experimental measurements in those processes.

Knowles (1986) described future combustion systems using natural gas.[4] The systems involving both combustion and heat transfer included higher temperature air preheaters and higher heat flux furnaces. Pohl et al. (1986) gave 15 recommendations for research to improve energy efficiency in the process industries.[5] The major ones that included heat transfer were:

- Furnace design, heat and mixing patterns
- Heat recovery and energy efficiency of flares
- Recuperative burner for low heating value gases
- Impinging (pulsed) heat transfer
- Rich flames for higher thermal radiation

TABLE 1.1
The Importance of Combustion to Industry

Industry	% Total Energy from (at the point of use)		
	Steam	Heat	Combustion
Petroleum refining	29.6	62.6	92.2
Forest products	84.4	6.0	90.4
Steel	22.6	67.0	89.6
Chemicals	49.9	32.7	82.6
Glass	4.8	75.2	80.0
Metal casting	2.4	67.2	69.6
Aluminum	1.3	17.6	18.9

Source: From U.S. Dept. of Energy, Energy Information Administration as quoted in the *Industrial Combustion Vision,* prepared by the U.S. Dept. of Energy, May 1998.

TABLE 1.2
Major Process Heating Operations

Metal melting
- Steel making
- Iron and steel melting
- Nonferrous melting

Metal heating
- Steel soaking, reheat, ladle preheating
- Forging
- Nonferrous heating

Metal heat treating
- Annealing
- Stress relief
- Tempering
- Solution heat treating
- Aging
- Precipitation hardening

Curing and forming
- Glass annealing, tempering, forming
- Plastics fabrication
- Gypsum production

Fluid heating
- Oil and natural gas production
- Chemical/petroleum feedstock preheating
- Distillation, visbreaking, hydrotreating, hydrocracking, delayed coking

Bonding
- Sintering, brazing

Drying
- Surface film drying
- Rubber, plastic, wood, glass products drying
- Coal drying
- Food processing
- Animal food processing

Calcining
- Cement, lime, soda ash
- Alumina, gypsum

Clay firing
- Structural products
- Refractories

Agglomeration
- Iron, lead, zinc

Smelting
- Iron, copper, lead

Non-metallic materials melting
- Glass

Other heating
- Ore roasting
- Textile manufacturing
- Food production
- Aluminum anode baking

Source: From *Industrial Combustion Vision,* U.S. Dept. of Energy, May 1998.

TABLE 1.3
Examples of Processes in the Process Industries Requiring Industrial Combustion

Process Industry	Examples of Processes Using Heat
Steel making	Smelting of ores, melting, annealing
Chemicals	Chemical reactions, pyrolysis, drying
Nonmetallic minerals (bricks, glass, cement and other refractories)	Firing, kilning, drying, calcining, melting, forming
Metal manufacture (iron and steel, and nonferrous metals)	Blast furnaces and cupolas, soaking and heat treatment, melting, sintering, annealing
Paper and printing	Drying

Source: Adapted from G.F. Hewitt, G.L. Shires, and T.L. Bott, Eds., *Process Heat Transfer,* CRC Press, Boca Raton, FL, 1994, 2.

Wilde (1986) made recommendations for solving some combustion problems in the steel industry.[6] The recommendations that specifically concerned heat transfer included:

- Tighter furnace enclosures to reduce heat loss and air infiltration
- Improved flame quality to optimize furnace operation
- Improved methods for heat recuperation for preheating air

Drake (1986) stated that the biggest problem in the glass industry is how to improve the thermal efficiency of glass melting furnaces whose designs still resemble the 1860 Siemens design.[7] A specific need that was identified was a better understanding of the heat transfer to the glass melt to optimize the number and placement of fuel injectors, development of new burners, and regulation of the flames corresponding to given loads.

Gupta and Lilley (1987) discussed research needs in practical combustion systems.[8] The recommendations were divided into the areas of experimentation and simulation. The following areas in experimentation need more research:

- Fluid dynamic problems, including particle and velocity measurements in two-phase flows, turbulence measurements, and turbulence-particle interactions
- Diagnostic tool development, including diagnostics for two-phase flows, flow visualization, and improvements in velocity measurement, laser probes, and laser-based spectroscopic measurements
- Investigation into specific types of combustion problems where some of the latest diagnostic techniques have not been applied yet

In the area of computer simulations, further advances are needed in two-phase combustion modeling, general model development is needed for interacting phenomena, more efficient and accurate solution schemes are needed, and specifically, turbulent chemically reacting flows need further investigation.

Chace et al. (1989) noted a number of research needs for applications in the following industries: primary metals; chemical; petroleum and coal products; stone, clay, and glass; paper; and food.[9] The needs recommended that involved heat transfer in combustion included:

- Heat transfer enhancement in steel furnaces
- Heat treatment in aluminum furnaces and ovens

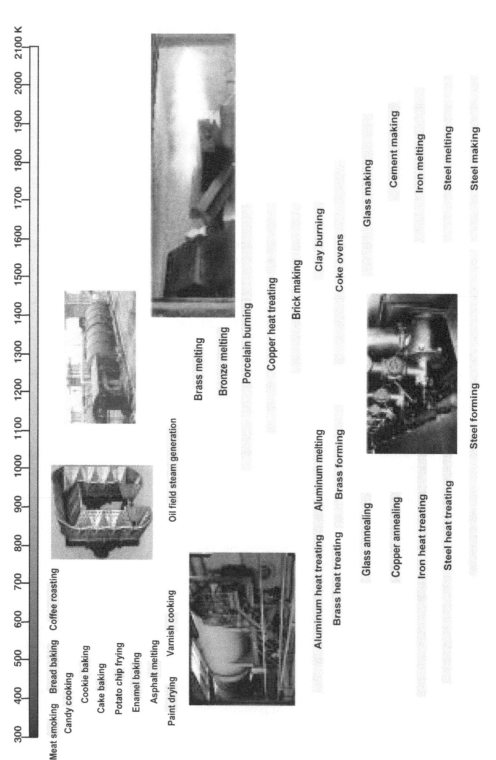

FIGURE 1.1 Temperature ranges of common industrial combustion applications. (Courtesy of Werner Dahm, 1998. With permission.)

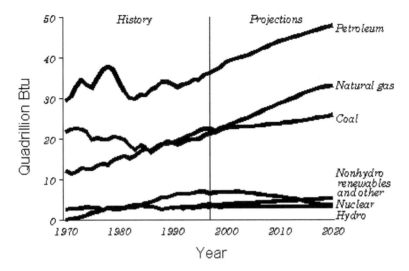

FIGURE 1.2 Historical and projected world energy consumption. (U.S. Dept. of Energy, Energy Information Administration, Annual Energy Outlook 1999, Report DOE/EIA-0383(99) Washington, D.C.)

- Enhanced heat transfer in glass furnaces
- Heat transfer enhancement in the chemical industry
- Fluidized-bed heating system for the petroleum industry
- Enhanced heat transfer rates in thermal processing in the chemical industry
- Enhanced heat transfer in the pulp and paper industry
- Accelerated drying in the food industry

Viskanta (1991) discussed some selected techniques for enhancing the heat transfer in fossil-fuel-fired industrial furnaces.[10] Such improvements could lead to higher productivity and efficiency and, in some cases, reduced equipment size. Further research was recommended to improve models for radiative transfer in industrial furnaces that have fluctuating temperature and concentrations fields.

A study by the U.S. Department of Energy (DOE) identified many research opportunities involving heat transfer in industrial combustion.[11] In the petroleum refining industry, high-temperature furnace efficiency improvements through flame radiation enhancement were recommended. Some recommendations were given for steel industry research, specifically in reheat furnaces: improve the uniformity of the generation and application of heat to the steel and conduct fundamental flame research to verify the actual heat transfer achieved by various types of burners. For the metal casting industry, improvements were recommended for waste heat recovery and for general heat transfer in the heating process. Development of enhanced heat transfer mechanisms was recommended for the chemical industry. Optimizing the heat transfer to molten glass, improving waste heat recovery, and improving computer models of the heating and melting process were recommended as research needs in the glass industry. In the aluminum industry, furnace efficiency and productivity improvements through flame radiation for secondary aluminum melters and treating furnaces were recommended. Improved computer models were recommended for simulating the black liquor combustion process in the pulp and paper industry.

Kurek et al. (1998) noted that recent research in industrial combustion has been directed at improving the energy utilization in the process industries.[12] Five general areas of research were listed and three of those concern improving heat transfer in some way: (1) improving process productivity (increasing heat flux rates), (2) improving process temperature uniformity (improving the heat flux distribution), and (3) improving thermal efficiency (improving the energy transfer from the heat source to the load).

The U.S. DOE developed a "technology roadmap" for industrial combustion with the help of representatives from users and manufacturers in industry and from academia.[13] A number of research needs in industrial combustion, directly or indirectly concerning heat transfer, were identified:

- New furnace designs (heat transfer needed for the analysis)
- Cost-effective heat recovery processes
- Optimization of the emissivity of materials used in furnaces or burners
- Increased combustion intensity (heat release per unit of furnace volume)
- Adaptation of computational fluid dynamics models to design burners
- Development of new equipment and methods for heating and transferring heat
- Development of hybrid or other methods to increase heat transfer to loads

1.2 LITERATURE DISCUSSION

The subject of this book is heat transfer — specifically in industrial combustion systems. This chapter section briefly considers some of the relevant literature on the subjects of heat transfer, combustion, and the combination of combustion with heat transfer. Many textbooks have been written on both heat transfer and combustion, but both types of book generally have only a limited amount of information concerning the combination of heat transfer and industrial combustion. Most of these books were written at a highly technical level for use in upper-level undergraduate or graduate-level courses. The books typically have broad coverage, with less emphasis on practical applications due to the nature of their target audience. This chapter section briefly surveys books related to heat transfer, combustion, and heat transfer in combustion. A list of relevant journals and trade magazines on this subject is given in Appendix A.

1.2.1 HEAT TRANSFER

Numerous excellent books have been written on the subject of heat transfer. However, almost none of them have any significant discussion of combustion. This is not surprising as the field of heat transfer is very broad, which makes it very difficult to be exhaustively comprehensive. Many of the heat transfer textbooks have no specific discussion of heat transfer in industrial combustion but do treat gaseous radiation heat transfer.[14-19,19a]

The heat transfer books written specifically about radiation often have sections covering heat transfer from luminous and nonluminous flames. Those topics are also discussed in this book in Chapter 4. Hottel and Sarofim's (1967) book has a good blend of theory and practice regarding radiation.[20] It also has a chapter specifically devoted to applications in furnaces. Love's (1968) book on radiation has short theoretical discussions of radiative heat transfer in flames and measuring flame parameters, but no other significant discussions of flames and combustion.[21] Özisik's (1973) book focuses more on interactions between radiation and conduction and convection, with no specific treatment of combustion or flames.[22] A short book by Gray and Müller (1974) is aimed toward more practical applications of radiation.[23] Sparrow and Cess (1978) have a brief chapter on nonluminous gaseous radiation, in which the various band models are discussed.[24]

Some of the older books on heat transfer are more practically oriented, with less emphasis on theory. Kern's classic book *Process Heat Transfer* has a chapter specifically on heat transfer in furnaces, primarily boilers and petroleum refinery furnaces.[25] Hutchinson (1952) gives many graphical solutions of conduction, radiation and convection heat transfer problems, but nothing specifically for flames or combustion.[26] Hsu (1963) has helpful discussions on nonluminous gaseous radiation and luminous radiation from flames.[27] Welty (1974) discusses heat exchangers, but not combustors or flames.[28] Karlekar and Desmond (1977) give a brief presentation on nonluminous gaseous radiation, but no discussion of flames or combustion.[29] Ganapathy's (1982) book on applied heat transfer is one of the better ones concerning heat transfer in industrial combustion and includes a chapter on fired-heater design.[30] Blokh's (1988) book is also a good reference for heat transfer in industrial combustion although it is aimed at power boilers and does not specifically address industrial combustion processes.[31] It has much information on flame radiation from a wide range of fuels, including pulverized coal, oils, and gases.

A few handbooks on heat transfer have been written, but these also tend to have little if anything on industrial combustion systems.[32-34]

1.2.2 COMBUSTION

Many theoretical books have been written on the subject of combustion but have little if anything on the heat transfer from the combustion process.[35-39] Barnard and Bradley (1985) have a brief chapter on industrial applications, but little on heat transfer in those processes or from flames.[40] A recent book by Turns (1996), designed for undergraduate- and graduate-level combustion courses, contains more discussions of practical combustion equipment and of heat transfer than most similar books.[41]

There have also been many books written on the more practical aspects of combustion. Griswold's (1946) book has a substantial treatment of the theory of combustion, but is also very practically oriented and includes chapters on gas burners, oil burners, stokers and pulverized-coal burners, heat transfer (although brief), furnace refractories, tube heaters, process furnaces, and kilns.[42] Stambuleanu's (1976) book on industrial combustion has information on actual furnaces and on aerospace applications, particularly rockets.[43] There is much data in the book on flame lengths, flame shapes, velocity profiles, species concentrations, liquid and solid fuel combustion, with a limited amount of information on heat transfer. A book on industrial combustion has significant discussions on flame chemistry, but only a total of about one page on heat transfer from flames.[44] Keating's (1993) book on applied combustion is aimed at engines but has no treatment of industrial combustion processes.[45] A recent book by Borman and Ragland (1998) attempts to bridge the gap between the theoretical and practical books on combustion.[46] However, the book has little discussion about the types of industrial applications considered here and no discussion about heat transfer in those applications. Even handbooks on combustion applications have little if anything on industrial combustion systems.[47-51]

1.2.3 HEAT TRANSFER AND COMBUSTION

Only a few books have been written with any significant coverage of heat transfer in industrial combustion.[52-56a] However, most of these are fairly old and are somewhat outdated. Reed's (1981) book has a chapter on heat transfer but does not give a single equation in that chapter.[48] There have been a number of conferences sponsored by, among others, the American Society of Mechanical Engineers on the subject of heat transfer in combustion.[57-76] As with any conference proceedings, the coverage and quality varies widely. Very few of the papers in those proceedings concerned heat transfer in industrial combustion. The present book attempts to codify the relevant papers from those conferences, as well as from numerous other sources, into a single coherent reference source.

Churchill and Lior (1982) have written a review paper on heat transfer in combustors, primarily limited to papers from 1977–1981.[77] The scope of that paper was broader than that of the present book. It included heat transfer in power generation, internal combustion engines, and fluidized beds, which are not included here. The paper also considered, for example, heat transfer in a blast furnace, a topic that has been specifically excluded here as it does not fit the narrow definition of industrial combustion used here. While only brief discussions are given due to length restrictions, 176 references have been given that may be useful for the researcher.

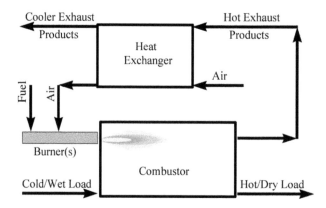

FIGURE 1.3   Schematic of the major components in a combustion system.

1.3   COMBUSTION SYSTEM COMPONENTS

There are four components that are important in the transfer of thermal energy from a combustion process to some type of heat load (see Figure 1.3). One component is the burner, which combusts the fuel with an oxidizer to release heat. Another component is the load itself, which can greatly affect how the heat is transferred from the flame. In most cases, the flame and the load are located inside of a combustor, which may be a furnace, heater, or dryer (which is the third component in the system). In some cases, there may be some type of heat recovery device to increase the thermal efficiency of the overall combustion system, which is the fourth component of the system. Each of these components is briefly considered in this chapter section. Various aspects of these components are discussed in more detail in other chapters of the book.

Although there are other important components in a combustion system (e.g., the flow control system), they do not normally have a significant impact on the heat transfer from the flame to the load. An exception would be the flow controls for a pulsed combustion system where the cycling of either the fuel or oxidizer supply valves can cause the pulsing, which can significantly increase the heat transfer from the flame to the load. Pulse combustion is discussed in more detail in Chapter 12. In general, however, the other components in a combustion system do not usually influence the heat transfer, which is the subject of this book.

1.3.1   BURNERS

The burner is the device used to combust the fuel, with an oxidizer to convert the chemical energy in the fuel into thermal energy. A given combustion system may have a single burner or many burners, depending on the size and type of the application. For example, in a rotary kiln, a single burner is located in the center of the wall on one end of a cylindrically shaped furnace (see Figure 1.4). The heat from the burner radiates in all directions and is efficiently absorbed by the load. However, the cylindrical geometry has some limitations concerning size and load type that make its use limited to certain applications such as melting scrap aluminum or producing cement clinker. A more common combustion system has multiple burners in a rectangular geometry (see Figure 1.5). This type of system is generally more difficult to analyze because of the multiplicity of heat sources and because of the interactions between the flames and their associated products of combustion.

There are many factors that go into the design of a burner. This chapter section briefly considers some of the important factors to be taken into account for a particular type of burner, with specific emphasis on how those factors impact heat transfer. These factors also affect other things (e.g.,

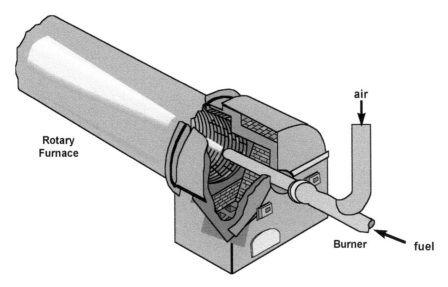

FIGURE 1.4 Single burner in a rotary kiln.

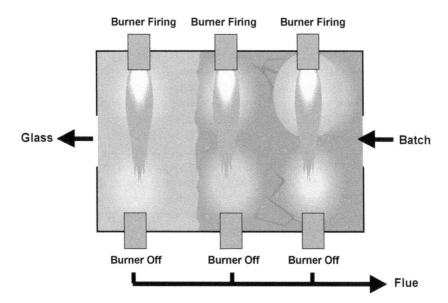

FIGURE 1.5 Multiple burners in a side-fired regenerative glass furnace.

pollutant emissions), which will only briefly be discussed since they normally only influence the heat transfer characteristics for a given burner design under fairly limited and special conditions.

1.3.1.1 Competing Priorities

There have many changes in the traditional designs that have been used in burners, primarily because of the recent interest in reducing pollutant emissions. In the past, the burner designer was primarily concerned with efficiently combusting the fuel and transferring the energy to a heat load. New and increasingly more stringent environmental regulations have added the need to consider the pollutant emissions produced by the burner. In many cases, reducing pollutant emissions and

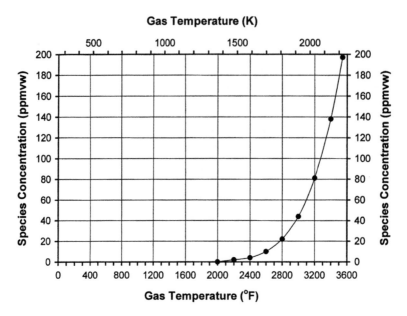

FIGURE 1.6 Dependence of NO (in parts per million on a wet basis, or ppmvw) on gas temperature for adiabatic equilibrium combustion of air/CH_4.

maximizing combustion efficiency are at odds with each other. For example, a well-accepted technique for reducing NOx emissions is known as staging, where the primary flame zone is deficient of either fuel or oxidizer.[78] The balance of the fuel or oxidizer may be injected into the burner in a secondary flame zone or, in a more extreme case, may be injected somewhere else in the combustion chamber. Staging reduces the peak temperatures in the primary flame zone and also alters the chemistry in a way that reduces NOx emissions because fuel-rich or fuel-lean zones are less conducive to NOx formation than near-stoichiometric zones. Figure 1.6 shows how the NOx emissions are affected by the exhaust product temperature. Since thermal NOx is exponentially dependent on the gas temperature, even small reductions in the peak flame temperature can dramatically reduce NOx emissions. However, lower flame temperatures often reduce the radiant heat transfer from the flame since radiation is dependent on the fourth power of the absolute temperature of the gases. Another potential problem with staging is that it may increase CO emissions, which is an indication of incomplete combustion and reduced combustion efficiency. However, it is also possible that staged combustion may produce soot in the flame, which can increase flame radiation. The actual impact of staging on the heat transfer from the flame is highly dependent on the actual burner design.

In the past, the challenge for the burner designer was often to maximize the mixing between the fuel and the oxidizer to ensure complete combustion, especially if the fuel was difficult to burn, as in the case of low heating value fuels such as waste liquid fuels or process gases from chemicals production. Now the burner designer must balance the mixing of the fuel and the oxidizer to maximize combustion efficiency while simultaneously minimizing all types of pollutant emissions. This is no easy task as, for example, NOx and CO emissions often go in opposite directions, as shown in Figure 1.7. When CO is low, NOx may be high, and vice versa. Modern burners must be environmentally friendly, while simultaneously efficiently transferring heat to the load.

1.3.1.2 Design Factors

There are many types of burner designs that exist due to the wide variety of fuels, oxidizers, combustion chamber geometries, environmental regulations, thermal input sizes, and heat transfer

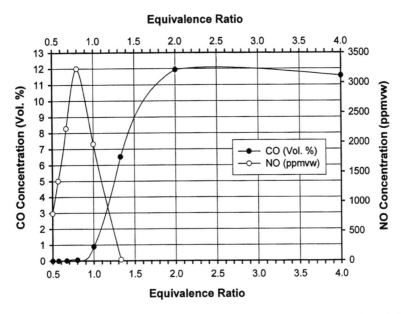

FIGURE 1.7   Dependence of NO and CO on equivalence ratio for adiabatic equilibrium air/CH₄ flames.

requirements, which includes things like flame temperature, flame momentum, and heat distribution. Some of these design factors are briefly considered here.

1.3.1.2.1   Fuel

Depending on many factors, certain types of fuels are preferred for certain geographic locations due to cost and availability considerations. Gaseous fuels — particularly natural gas — are commonly used in most industrial heating applications in the U.S. In Europe, natural gas is also commonly used, along with light fuel oil. In Asia and South America, heavy fuel oils are generally preferred although the use of gaseous fuels is on the rise. Fuels also vary depending on the application. For example, in incineration processes, waste fuels are commonly used either by themselves or with other fuels like natural gas. In the petrochemical industry, fuel gases often consist of a blend of several fuels, including gases like hydrogen, methane, propane, butane, and propylene.

Fuel choice has an important influence on the heat transfer from a flame. In general, solid fuels like coal and liquid fuels like oil produce very luminous flames that contain soot particles that radiate like blackbodies to the heat load. Gaseous fuels like natural gas often produce nonluminous flames because they burn so cleanly and completely, without producing soot particles. A fuel like hydrogen is completely nonluminous, as there is no carbon available to produce soot. In cases where highly radiant flames are required, a luminous flame is preferred. In cases where convection heat transfer is preferred, a nonluminous flame may be preferred in order to minimize the possibility of contaminating the heat load with soot particles from a luminous flame. Where natural gas is the preferred fuel and highly radiant flames are desired, new technologies are being developed to produce more luminous flames. These include things like pyrolyzing the fuel in a partial oxidation process,[79] using a plasma to produce soot in the fuel,[80] and generally controlling the mixing of the fuel and oxidizer to produce fuel-rich flame zones that generate soot particles.[81] Therefore, the fuel itself has a significant impact on the heat transfer mechanisms between the flame and the load. In most cases, the fuel choice is dictated by the customer as part of the specifications for the system and is not chosen by the burner designer. The designer must make the best of whatever fuel has been selected. In most cases, the burner design is optimized based on the choice for the fuel.

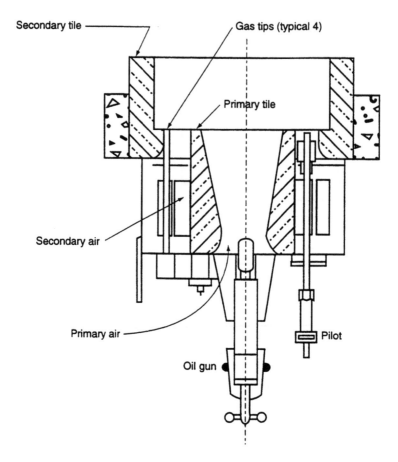

FIGURE 1.8 Typical combination oil and gas burner. (Courtesy of American Petroleum Institute, Washington, D.C.[82] With permission.)

In some cases, the burner may have more than one type of fuel. An example is shown in Figure 1.8.[82] Dual-fuel burners are typically designed to operate on either gaseous or liquid fuels. These burners are used where the customer may need to switch between a gaseous fuel like natural gas and a liquid fuel like oil, usually for economic reasons. These burners normally operate on one fuel or the other, and occasionally on both fuels. Another application where multiple fuels may be used is in waste incineration. One method of disposing of waste liquids contaminated with hydrocarbons is to combust them by direct injection through a burner. The waste liquids are fed through the burner, which is powered by a traditional fuel such as natural gas or oil. The waste liquids often have very low heating values and are difficult to combust without auxiliary fuel. This further complicates the burner design where the waste liquid must be vaporized and combusted concurrently with the normal fuel used in the burner.

1.3.1.2.2   Oxidizer

The predominant oxidizer used in most industrial heating processes is atmospheric air. This can present challenges in some applications where highly accurate control is required due to the daily variations in the barometric pressure and humidity of ambient air. The combustion air is sometimes preheated and sometimes blended with some of the products of combustion, which is usually referred to as flue gas recirculation (FlGR). In certain cases, preheated air is used to increase the overall thermal efficiency of a process. FlGR is often used to both increase thermal efficiency and reduce NOx emissions. The thermal efficiency is increased by capturing some of the energy in the exhaust gases that are used to preheat the incoming combustion oxidizer. NOx emissions can also be reduced because the peak flame temperatures are reduced, which can reduce the NOx emissions, which are highly temperature dependent. There are also many high-temperature combustion processes that use an oxidizer containing a higher proportion of oxygen than the 21% (by volume) found in normal atmospheric air. This is referred to as oxygen-enhanced combustion (OEC) and has many benefits, which include increased productivity and thermal efficiency while reducing the exhaust gas volume and pollutant emissions.[83] A simplified global chemical reaction for the stoichiometric combustion of methane with air is given as follows:

$$CH_4 + 2O_2 + 7.52N_2 \rightarrow CO_2 + 2H_2O + 7.52N_2 + \text{Trace species} \tag{1.1}$$

This compares to the same reaction where the oxidizer is pure O_2 instead of air:

$$CH_4 + 2O_2 \rightarrow CO_2 + 2H_2O + \text{Trace species} \tag{1.2}$$

The volume of exhaust gases is significantly reduced by the elimination of N_2. In general, a stoichiometric oxygen-enhanced methane combustion process can be represented by:

$$CH_4 + 2O_2 + xN_2 \rightarrow CO_2 + 2H_2O + xN_2 + \text{Trace species} \tag{1.3}$$

where $0 \leq x \leq 7.52$, depending on the oxidizer. The N_2 contained in air acts as a ballast that may inhibit the combustion process and have negative consequences. The benefits of using oxygen-enhanced combustion must be weighed against the added cost of the oxidizer, which in the case of air is essentially free except for the minor cost of the air-handling equipment and power for the blower. The use of a higher purity oxidizer has many consequences with regard to heat transfer from the flame, which are considered elsewhere in this book. Oxygen-enhanced combustion is considered in more detail in Chapter 12.

1.3.1.2.3   Gas recirculation

A common technique used in combustion systems is to design the burner to induce furnace gases to be drawn into the burner to dilute the flame, usually referred to as furnace gas recirculation (FuGR). Although the furnace gases are hot, they are still much cooler than the flame itself. This dilution may accomplish several purposes. One is to minimize NOx emissions by reducing the peak temperatures in the flame, as in FlGR. However, furnace gas recirculation may be preferred to FlGR because no external high-temperature ductwork or fans are needed to bring the product gases into the flame zone. Another reason to use furnace gas recirculation may be to increase the convective heating from the flame because of the added gas volume and momentum. An example of flue gas recirculation into the burner is shown in Figure 1.9.[84]

1.3.1.3   General Burner Types

There are numerous ways to classify burners. Some of the common ones are discussed in this chapter section, with a brief consideration as to how the heat transfer is impacted.

1.3.1.3.1   Mixing type

A common method for classifying burners is according to how the fuel and the oxidizer are mixed. In premixed burners, shown in a cartoon in Figure 1.10 and schematically in Figure 1.11, the fuel and the oxidizer are completely mixed before combustion begins. Porous radiant burners are usually of the premixed type (see Chapter 8). Premixed burners often produce shorter and more intense flames, as compared to diffusion flames. This can produce high-temperature regions in the flame,

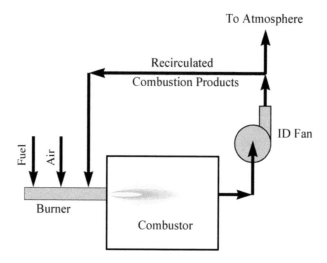

FIGURE 1.9   Schematic of flue gas recirculation.

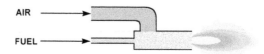

FIGURE 1.10   Premix burner.

leading to nonuniform heating of the load and higher NOx emissions. However, in flame impingement heating, premixed burners are useful because the higher temperatures and shorter flames can
enhance the heating rates (see Chapter 7).

In diffusion-mixed burners, shown schematically in Figure 1.12, the fuel and the oxidizer are
separated and unmixed prior to combustion, which begins where the oxidizer/fuel mixture is within
the flammability range. Oxygen/fuel burners (see Chapter 12) are usually diffusion burners, primarily for safety reasons, to prevent flashback and explosion in a potentially dangerous system.
Diffusion gas burners are sometimes referred to as "raw gas" burners as the fuel gas exits the burner
essentially intact with no air mixed with it. Diffusion burners typically have longer flames than
premixed burners, do not have as high temperature a hotspot, and usually have a more uniform
temperature and heat flux distribution.

It is also possible to have partially premixed burners, as shown schematically in Figures 1.11
and 1.13, where a portion of the fuel is mixed with the oxidizer. This is often done for stability
and safety reasons where the partial premixing helps anchor the flame, but not fully premixing
lessens the chance for flashback. This type of burner often has a flame length and temperature and
heat flux distribution that is between the fully premixed and diffusion flames.

Another burner classification based on mixing is known as staging: staged air and staged fuel.
A staged-air burner is shown in a cartoon in Figure 1.14 and schematically in Figure 1.15. A staged-
fuel burner is shown in a cartoon in Figure 1.16 and schematically in Figure 1.17. Secondary and
sometimes tertiary injectors in the burner are used to inject a portion of the fuel and/or the oxidizer
into the flame, downstream of the root of the flame. Staging is often done to reduce NOx emissions
and produce longer flames. These longer flames typically have a lower peak flame temperature and
more uniform heat flux distribution than nonstaged flames.

1.3.1.3.2   Oxidizer type

Burners and flames are often classified according to the type of oxidizer used. The majority of
industrial burners use air for combustion. In many of the higher temperature heating and melting

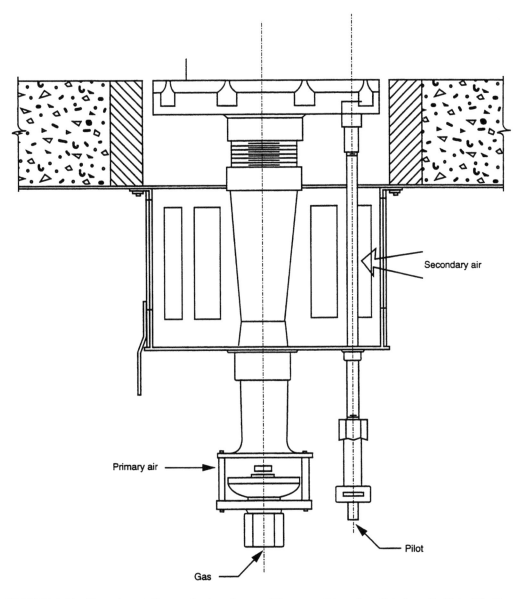

FIGURE 1.11   Typical partially premixed gas burner. (Courtesy of American Petroleum Institute, Washington, D.C.[82] With permission.)

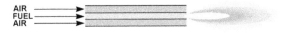

FIGURE 1.12   Diffusion burner.

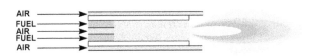

FIGURE 1.13   Partially premixed burner.

FIGURE 1.14   Staged-air burner.

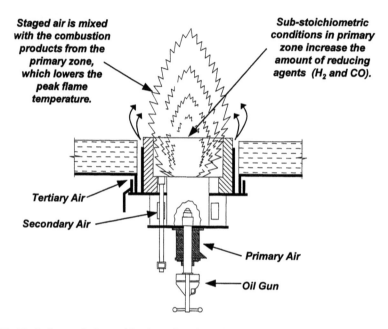

FIGURE 1.15   Typical staged-air combination oil and gas burner. (Courtesy of John Zink Co. LLC, Tulsa, OK. With permission.)

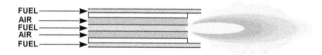

FIGURE 1.16   Staged-fuel burner.

applications, such as glass production, the oxidizer is pure oxygen. In other applications, the oxidizer is a combination of air and oxygen, often referred to as oxygen-enriched air combustion. The latter two types of oxidizers are discussed in more detail in Chapter 12.

Another way to classify the oxidizer is by its temperature. It is common in many industrial applications to recover heat from the exhaust gases by preheating the incoming combustion air — either with a recuperator or a regenerator (discussed below). Such a burner is often referred to as a preheated air burner.

1.3.1.3.3   Draft type

Most industrial burners are known as forced-draft burners. This means that the oxidizer is supplied to the burner under pressure. For example, in a forced-draft air burner, the air used for combustion is supplied to the burner by a blower. In natural-draft burners, the air used for combustion is induced into the burner by the negative draft produced in the combustor. A schematic is shown in Figure 1.18 and an example is shown in Figure 1.19. In this type of burner, the pressure drop and combustor stack height are critical in producing enough suction to induce enough combustion air into the burners. This type of burner is commonly used in the chemical and petrochemical industries in

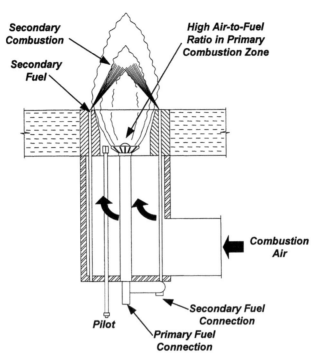

FIGURE 1.17 Typical staged-fuel gas burner. (Courtesy of John Zink Co. LLC, Tulsa, OK. With permission.)

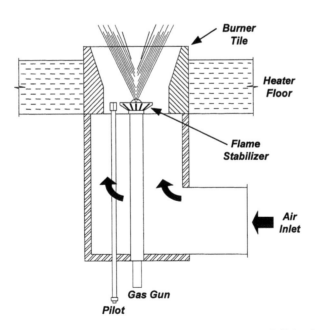

FIGURE 1.18 Typical natural draft gas burner. (Courtesy of John Zink Co. LLC, Tulsa, OK. With permission.)

fluid heaters. The main consequence of the draft type on heat transfer is that the natural-draft flames are usually longer than the forced-draft flames so that the heat flux from the flame is distributed over a longer distance and the peak temperature in the flame is often lower.

FIGURE 1.19   Natural draft burner. (Courtesy of John Zink Co. LLC, Tulsa, OK. With permission.)

1.3.1.3.4   Heating type

Burners are often classified as to whether they are direct or indirect heating. In direct heating, there is no intermediate heat exchange surface between the flame and the load. In indirect heating, such as radiant tube burners (see Chapter 8), there is an intermediate surface between the flame and the load. This is usually done because the combustion products cannot come in contact with the load because of possible contamination.

1.3.2   COMBUSTORS

This chapter section briefly introduces the combustors that are commonly used in industrial heating and melting applications. These combustors are discussed in more detail in Chapters 9, 10, and 11.

1.3.2.1   Design Considerations

There are many important factors that must be considered when designing a combustor. This chapter section only briefly considers a few of those factors and how they may influence the heat transfer in the system.

1.3.2.1.1   Load handling

A primary consideration for any combustor is the type of material that will be processed. The various types of loads are considered later in this chapter and also in more detail in Chapter 4. One obvious factor of importance in handling the load and transporting it through the combustor is its physical state — whether it is a solid, liquid, or gas. Another factor is the transport properties of the load. For example, the solid might be granular or it might be in the form of a sheet (web). Related to this is how the solid will be fed into the combustor. A granular solid could be fed continuously into a combustor with a screw conveyor or it could be fed in with discrete charges from a front-end loader. The shape of the furnace will vary according to how the material will be transported through it. For example, limestone is fed continuously into a rotating and slightly downward-inclined cylinder.

1.3.2.1.2   Temperature

In this book, industrial heating applications have been divided into two categories: higher and lower temperatures. The division between the two is somewhat arbitrary but mainly concerns the different types of applications used in each. For example, most of the metal- and glass-melting applications fall into the higher temperature categories, as the furnace temperatures are often well over 2000°F (1400K). They use technologies like air preheating and oxygen enrichment (see Chapter 12) to achieve those higher temperatures. Lower temperature applications include dryers, process heaters, and heat treating and are typically below about 2000°F (1400K). Although many of these processes may use air preheating, it is primarily to improve the thermal efficiency and not to get higher flame temperatures. Those processes rarely use oxygen enrichment, which usually only works economically for higher temperature processes. Obviously, the combustors are designed differently for higher and lower temperature processes. The heat transfer mechanisms are often different as well. In higher temperature processes, the primary mode is often radiation, while in lower temperature applications, convection often plays a significant role.

1.3.2.1.3   Heat recovery

When heat recovery is used in an industrial combustion process, it is an integral part of the system. The two most popular methods are regenerative and recuperative, which are discussed briefly below and also in Chapter 8. The heat recovery system is important in the design of the combustor as it determines the thermal efficiency of the process and the flame temperatures in the system. It also influences the heat transfer modes as it may increase both the radiation and convection because of higher flame temperatures. Another type of heat recovery used in some processes is furnace or flue gas recirculation, where the exhaust products are recirculated back through the flame. This also influences the heat transfer and furnace design as it can moderate the flame temperature but increase the volume flow of gases through the combustion chamber.

1.3.2.2   General Classifications

There are several ways that a combustor can be classified; these are briefly discussed in this chapter section. Each type has an impact on the heat transfer mechanisms in the furnace.

1.3.2.2.1   Load processing method

Furnaces are often classified as to whether they are batch or continuous. In a batch furnace, the load is charged into the furnace at discrete intervals where it is heated. There may be multiple load charges, depending on the application. Normally, the firing rate of the burners is reduced or turned off during the charging cycle. On some furnaces, a door may also need to be opened during charging. These significantly impact the heat transfer in the system as the heat losses during the charge cycle are very large. The radiation losses through open doors are high and the reduced firing rate may not be enough to maintain the furnace temperature. In some cases, the temperature on the inside of the refractory wall, closest to the load, may actually be lower than the temperature of the refractory at some distance from the inside, due to the heat losses during charging. The heating process and heat transfer are dynamic and constantly changing as a result of the cyclical nature of the load charging. This makes analysis of these systems more complicated because of the need to include time in the computations.

In a continuous furnace, the load is constantly fed into and out of the combustor. The feed rate may change, sometimes due to conditions upstream or downstream of the combustor or due to the production needs of the plant, but the process is nearly steady state. This makes continuous processes simpler to analyze as there is no need to include time in the computations. It is often easier to make meaningful measurements in continuous processes due to their steady-state nature.

There are some furnaces that are semicontinuous, where the load may be charged in a nearly continuous fashion, but the finished product may be removed from the furnace at discrete intervals. An example is an aluminum reverberatory furnace that is charged using an automatic side-well feed mechanism (see Chapter 11). In that process, shredded scrap is continuously added to a circulating bath of molten aluminum. When the correct alloy composition has been reached and the furnace has a full load, some or all of that load is then tapped out of the furnace. The effect on heat transfer is somewhere between that for batch and continuous furnaces.

1.3.2.2.2   Heating type

As described above for burners, combustors are often classified as indirect or direct heating. In indirect heating, there is some type of intermediate heat transfer medium between the flames and the load that keeps the combustion products separate from the load. One example is a muffle furnace where there is a high-temperature ceramic muffle between the flames and the load. The flames transfer their heat to the muffle, which then radiates to the load which is usually some type of metal. The limitation of indirect heating processes is the temperature limit of the intermediate material. Although ceramic materials have fairly high temperature limits, other issues such as structural integrity over long distance spans and thermal cycling can still reduce the recommended operating temperatures. Another example of indirect heating is in process heaters where fluids are transported through metal tubes that are heated by flames. Indirect heating processes often have fairly uniform heat flux distributions because the heat exchange medium tends to homogenize the energy distribution from the flames to the load. The heat transfer from the heat exchange surface to the load is often fairly simple and straightforward to compute because of the absence of chemical reactions in-between. However, the heat transfer from the flames to the heat exchange surface and the subsequent thermal conduction through that surface are as complicated as if the flame was radiating directly to the load.

As a result of the temperature limits of the heat exchange materials, most higher temperature processes are of the direct heating type where the flames can directly radiate heat to the load.

1.3.2.2.3   Geometry

Another common way of classifying combustors is according to their geometry, which includes their shape and orientation. The two most common shapes are rectangular and cylindrical. The two most common orientations are horizontal and vertical, although inclined furnaces are commonly used in certain applications (e.g., rotary cement furnaces). An example of using the shape and orientation of the furnace as a means of classification would be a vertical cylindrical heater (sometimes referred to as a VC) used to heat fluids in the petrochemical industry. Both the furnace shape and orientation have important effects on the heat transfer in the system. They also determine the type of analysis that will be used. For example, in a VC heater, it is often possible to model only a slice of the heater due to its angular symmetry, in which case cylindrical coordinates would be used. On the other hand, it is usually not reasonable to model a horizontal rectangular furnace using cylindrical coordinates, especially if buoyancy effects are important.

Some furnaces are classified by what they look like. One example is a shaft furnace used to make iron. The raw materials are loaded into the top of a tall, thin, vertically oriented cylinder. Hot combustion gases generated at the bottom through the combustion of coke flow up through the raw materials which get heated. The melted final product is tapped out of the bottom. The furnace looks and acts almost like a shaft because of the way the raw materials are fed in through the top and exit at the bottom. A transfer chamber used to move molten metal around in a steel mill is often referred to as a ladle because of its function and appearance. These ladles are preheated using burners before the molten metal is poured into them to prevent the refractory-lined vessels from thermally shocking.

Another aspect of the geometry that is important in some applications is whether the furnace is moving or not. For example, in a rotary furnace for melting scrap aluminum, the furnace rotates to enhance mixing and heat transfer distribution. This again affects the type of analysis that would be appropriate for that system and can add some complexity to the computations.

The burner orientation with respect to the combustor is also sometimes used to classify the combustor. For example, a wall-fired furnace has burners located in and firing along the wall.

1.3.2.2.4   Heat recuperation

In many heat processing systems, energy recuperation is an integral part of the combustion system. Often, the heat recuperation equipment is a separate component of the system and not part of the burners themselves. Depending on the method used to recover the energy, the combustors are commonly referred to as either recuperative or regenerative (see discussion below). The heat transfer in these systems is a function of the energy recovery system. For example, the higher the combustion air preheat temperature, the hotter the flame and the more radiant heat that can be produced by that flame. The convective heat transfer may also be increased due to the higher gas temperature and also due to the higher thermal expansion of the gases which increases the average gas velocity through the combustor.

1.3.3   Heat Load

This chapter section is a brief introduction to some of the important issues concerning the heat load in a furnace or combustor. A more detailed treatment is given in Chapter 4.

1.3.3.1   Process Tubes

In petrochemical production processes, process heaters are used to heat petroleum products to operating temperatures. The fluids are transported through the process heaters in process tubes. These heaters often have a radiant section and a convection section. In the radiant section, radiation from burners heats the process tubes. In the convection section, the combustion products heat the tubes by flowing over the tubes. The design of the radiant section is especially important because flame impingement on the tubes can cause premature failure of the tubes or cause the hydrocarbon fluids to coke inside the tubes, which reduces the heat transfer to the fluids.

1.3.3.2   Moving Substrate

In some applications, heaters and burners are used to heat or dry moving substrates or webs. An example is shown in Figure 1.20. One common application is the use of gas-fired infrared (IR) burners to remove moisture from paper during the forming process.[85] These paper webs can travel at speeds over 300 m/s (1000 ft/s) and are normally dried by traveling over and contacting steam-heated cylinders. IR heaters are often used to selectively dry certain portions of the web that may be wetter than others. For example, if the target moisture content for the paper is 5%, then the entire width of the paper must have no more than 5% moisture. Streaks of higher moisture areas

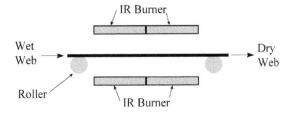

FIGURE 1.20   Gas-fired infrared burners heating a moving substrate.

often occur in sections along the width of the paper. Without selectively drying those areas, those streaks would be dried to the target moisture level, which means that the rest of the sheet would be dried to even lower moisture levels. This creates at least two important problems. The first is lost revenue because paper is usually sold on a weight basis; any water unnecessarily removed from the paper decreases its weight and therefore results in lost income. Another problem is a reduction in the quality of the paper. If areas of the paper are too dry, they do not handle as well in devices like copiers and printers and are not nearly as desirable as paper of uniform moisture content. Therefore, selective drying of the paper only removes the minimum amount of water from the substrate. The challenge of this application is to measure the moisture content profile across the width of a sheet that may be several meters wide and moving at hundreds of meters per second. That information must then be fed to the control system for the IR heaters, which then must be able to react almost instantaneously. This is possible today because of advances in measurement and controls systems.

Another example of a moving substrate application is using IR burners to remove water during the production of fabrics in textile manufacturing.[86] Moving substrates present unique challenges for burners. Often, the material being heated can easily be set on fire if there is a line stoppage and the burner is not turned off quickly enough. This means that the burner control system must be interlocked with the web-handling equipment so that the burners can be turned off immediately in the event of a line stoppage. If the burners have substantial thermal mass, then the burners may need to be retracted away from the substrate during a stoppage, or heat shields may need to be inserted between the burners and the substrate to prevent overheating.

Convection dryers are also used to heat and dry substrates. Typically, high-velocity heated air is blown at the substrate from both sides so that the substrate is elevated between the nozzles. In many cases, the heated air is used for both heat and mass transfer, to volatilize any liquids on or in the substrate such as water, and then carry the vapor away from the substrate.

An important aspect of heating webs is how the energy is transferred into the material. For example, dry paper is known to be a good insulator. When steam cylinders are used to heat and dry paper, they become less and less effective as the paper becomes drier because the heat from the cylinder cannot conduct through the paper as well as when it is moist since the thermal conductivity of the paper increases with moisture content. IR burners are effective for drying paper because the radiant energy transfers into the paper and is absorbed by the water. The radiant penetration into the paper actually increases as the paper becomes drier, unlike with steam cylinders which become less effective.

1.3.3.3 Opaque Materials

This type of load encompasses a wide range of materials, including granular solids like limestone and liquids like molten metal. For this type of load, the heat transfers to the surface of the load and must conduct down into the material. This process can be enhanced by proper mixing of the materials so that new material is constantly exposed to the surface, as in rotary kilns or in aluminum reverberatory furnaces which have molten metal pumps to continuously recirculate the metal through the heating zone. The potential problems with this method include overheating the surface materials or having lower thermal efficiencies by limiting the heat transfer to the surface to prevent overheating.

1.3.3.4 Transparent Materials

The primary example of this type of load is glass, which has selective radiant transmission properties. In glass-melting processes (see Chapter 11), the primary mode of heat transfer is by radiation. As shown in Chapter 3, flames have specific types of radiant outputs that vary as a function of wavelength. If the flame is nonluminous, the flame usually has higher radiant outputs in the preferred wavelengths for water and carbon dioxide bands. If the flame is luminous, it has a broader, more graybody-type spectral radiant profile. Chapter 4 shows that luminous flames are preferred in melting glass because of the selective transmission properties of molten glass. This allows a significant portion of the radiation received at the surface of the glass to penetrate into the glass, which enhances heat transfer rates and reduces the chances of overheating the surface, which would reduce product quality.

1.3.4   HEAT RECOVERY DEVICES

Heat recovery devices are often used to improve the efficiency of combustion systems. Some of these devices are incorporated into the burners, but more commonly they are another component in the combustion system, separate from the burners. These heat recovery devices incorporate some type of heat exchanger, depending on the application. The two most common types have been recuperators and regenerators, which are briefly discussed next. A fuller treatment of this topic is contained in Chapter 9. Reed (1987) predicts an increasing importance for heat recovery devices in industrial combustion systems for increasing heat transfer and thermal efficiencies.[87]

1.3.4.1   Recuperators

A recuperator is a low- to medium-temperature (up to about 1300°F, or 700°C), continuous heat exchanger that uses the sensible energy from hot combustion products to preheat the incoming combustion air. These heat exchangers are commonly counterflow, where the highest temperatures for both the combustion products and the combustion air are at one end of the exchanger with the coldest temperatures at the other end. Lower temperature recuperators are normally made of metal, while higher temperature recuperators may be made of ceramics. Recuperators are typically used in lower temperature applications because of the limitations of the metals used to construct these heat exchangers. These are discussed in more detail in section 9.3.1.

1.3.4.2   Regenerators

A regenerator is a higher temperature, transient heat exchanger used to improve the energy efficiency of high-temperature heating and melting processes, particularly in the high-temperature processing industries like glass production. In a regenerator, energy from the hot combustion products is temporarily stored in a unit constructed of firebricks. This energy is then used to heat the incoming combustion air during a given part of the firing cycle up to temperatures in excess of 2000°F (1000°C).

Regenerators are normally operated in pairs. During one part of the cycle, the hot combustion gases are flowing through one of the regenerators and heating up the refractory bricks, while the combustion air is flowing through and cooling down the refractory bricks in the second regenerator. Both the exhaust gases and the combustion air directly contact the bricks in the regenerators, although not both at the same time since each is in a different regenerator at any given time. After a sufficient amount of time (usually from 5 to 30 min), the cycle is reversed so that the cooler bricks in the second regenerator are then reheated while the hotter bricks in the first regenerator exchange their heat with the incoming combustion air. A reversing valve is used to change the flow from one gas to another in each regenerator. These are discussed further in section 9.3.2.

Davies (1986) noted that some of the following questions should be considered concerning regenerative burners:[88]

- What will the fuel savings be (compared to no heat recuperation)?
- What is the maximum allowable flue gas temperature?
- How big is the heat exchanger?
- What is the air pressure drop through the exchanger?
- How long will the exchanger run without plugging?
- Is the flue gas path through the furnace altered?

REFERENCES

1. G.F. Hewitt, G.L. Shires, and T.R. Bott, *Process Heat Transfer*, CRC Press, Boca Raton, FL, 1994.
2. U.S. Department of Energy, Energy Information Administration, *Annual Energy Outlook 1999*, Report DOE/EIA-0383(99) Washington, D.C.
3. R.J. Reid, What You Should Know About Combustion, *Glass Industry*, 70(7), 24-35, 1989.
4. D.F. Knowles, Combustion system requirements for advanced gas-fired processes, in *Industrial Combustion Technologies*, M.A. Lukasiewicz, Ed., American Society of Metals, Warren, PA, 1986, 323-325.
5. J.H. Pohl, W.S. Lanier, J. Keller, J. Patton, and R. Jain, Technologies to increase the efficient use of energy in the process industries, in *Industrial Combustion Technologies*, M.A. Lukasiewicz, Ed., American Society of Metals, Warren, PA, 1986, 313-321.
6. J.D. Wilde, Combustion problems in the steel industry, in *Industrial Combustion Technologies*, M.A. Lukasiewicz, Ed., American Society of Metals, Warren, PA, 1986, 13-22.
7. R.A. Drake, Combustion progress, problems, needs in the glass industry, in *Industrial Combustion Technologies*, M.A. Lukasiewicz, Ed., American Society of Metals, Warren, PA, 1986, 23-25.
8. A. Gupta and D.G. Lilley, An overview of research needs in practical combustion systems, in *Heat Transfer in Furnaces*, C. Presser and D.G. Lilley, Eds., New York, ASME HTD-Vol. 74, 1-10, 1987.
9. A.S. Chace, H.R. Hazard, A. Levy, A.C. Thekdi, and E.W. Ungar, Combustion Research Opportunities for Industrial Applications, U.S. Department of Energy Report DOE/ID-10204-2, 1989.
10. R. Viskanta, Enhancement of heat transfer in industrial combustion systems: problems and future challenges, in *Thermal Engineering 1991, Proceedings of the ASME/JSME Joint Conference,* J.R. Lloyd and Y. Kurosaki, Eds., ASME, 5, 161-173, 1991.
11. J.G. Keller, N.R. Soelberg, and G.F. Kessinger, Industry-Identified Combustion Research Needs, Idaho National Engineering Lab report INEL-95/0578, Idaho Falls, ID, 1995.
12. H.S. Kurek, M. Khinkis, W. Kunc, A. Touzet, A. de La Faire, T. Landais, A. Yerinov, and O. Semernin, Flat radiant panels for improving temperature uniformity and product quality in indirect-fired furnaces, *Proc. of 1998 International Gas Research Conf.*, Vol. V: Industrial Utilization, D.A. Dolenc, Ed., Gas Research Institute, Chicago, 1998, 15-23.
13. U.S. Department of Energy, Industrial Combustion Technology Roadmap, Washington, D.C., April 1999.
14. B. Gebhart, *Heat Transfer*, 2nd edition, McGraw-Hill, New York, 1971.
15. F. Kreith and M.S. Bohn, *Principles of Heat Transfer*, Harper & Row, New York, 1986.
16. J.P. Holman, *Heat Transfer*, 7th edition, McGraw-Hill, New York, 1990.
17. A. Bejan, *Heat Transfer*, Wiley, New York, 1993.
18. F.P. Incropera and D.P. Dewitt, *Introduction to Heat Transfer*, 3rd edition, Wiley, New York, 1996.
19. A.F. Mills, *Heat Transfer*, 2nd edition, Prentice-Hall, Englewood Cliffs, NJ, 1998.
19a. W.S. Janna, *Engineering Heat Transfer,* 2nd edition, CRC Press, Boca Raton, FL, 2000.
20. H.C. Hottel and A.F. Sarofim, *Radiative Transfer*, McGraw-Hill, New York, 1967.
21. T.J. Love, *Radiative Heat Transfer*, Merrill Publishing, Columbus, OH, 1968.
22. M. Özisik, *Radiative Transfer and Interactions with Conduction and Convection*, Wiley, New York, 1973.
23. W.A. Gray and R. Müller, *Engineering Calculations in Radiative Heat Transfer*, Pergamon, Oxford, U.K., 1974.
24. E.M. Sparrow and R.D. Cess, *Radiation Heat Transfer*, augmented edition, Hemisphere, Washington, D.C., 1978.
25. J.B. Dwyer, Furnace calculations, in *Process Heat Transfer*, D.Q. Kern, Ed., McGraw-Hill, New York, 1950.
26. F.W. Hutchinson, *Industrial Heat Transfer*, Industrial Press, New York, 1952.
27. S.T. Hsu, *Engineering Heat Transfer*, D. Van Nostrand Co., Princeton, NJ, 1963.
28. J.R. Welty, *Engineering Heat Transfer*, Wiley, New York, 1974.
29. B.V. Karlekar and R.M. Desmond, *Engineering Heat Transfer*, West Publ. Co., St. Paul, MN, 1977.
30. V. Ganapathy, *Applied Heat Transfer*, PennWell Books, Tulsa, OK, 1982.
31. A.G. Blokh, *Heat Transfer in Steam Boiler Furnaces*, Hemisphere, Washington, D.C., 1988.
32. W.M. Rohsenow, J.P., Hartnett, and E.N., Ganic, *Handbook of Heat Transfer Applications*, McGraw-Hill Book Company, New York, 1985.

33. N.P. Cheremisinoff, Ed., *Handbook of Heat and Mass Transfer*, 4 volumes, Gulf Pub. Co., Houston, TX, 1986, Vol. 1: Heat Transfer Operations, Vol. 2: Mass Transfer and Reactor Design, Vol. 3: Catalysis, Kinetics, and Reactor Engineering, and Vol. 4: Advances in Reactor Design and Combustion Science.

34. F. Kreith, Ed., *The CRC Handbook of Thermal Engineering*, CRC Press, Boca Raton, FL, 2000.

35. R.A. Strehlow, *Fundamentals of Combustion*, Inter. Textbook Co., Scranton, PA, 1968.

36. F.A. Williams, *Combustion Theory*, Benjamin/Cummings Publishing, Menlo Park, CA, 1985.

37. B. Lewis and G. von Elbe, *Combustion, Flames and Explosions of Gases*, 3rd edition, Academic Press, New York, 1987.

38. W. Bartok and A.F. Sarofim, Eds., *Fossil Fuel Combustion*, Wiley, New York, 1991.

38a. R.M. Fristrom, *Flame Structure and Processes,* Oxford University Press, New York, 1995.

39. I. Glassman, *Combustion*, 3rd edition, Academic Press, New York, 1996.

40. J.A. Barnard and J.N. Bradley, *Flame and Combustion*, 2nd edition, Chapman and Hall, London, 1985.

41. S.R. Turns, *An Introduction to Combustion*, McGraw-Hill, New York, 1996.

42. J. Griswold, *Fuels, Combustion and Furnaces*, McGraw-Hill, New York, 1946.

43. A. Stambuleanu, *Flame Combustion Processes in Industry,* Abacus Press, Tunbridge Wells, U.K., 1976.

44. E. Perthuis, *La Combustion Industrielle*, Éditions Technip, Paris, 1983.

45. E.L. Keating, *Applied Combustion*, Marcel Dekker, New York, 1993.

46. G. Borman and K. Ragland, *Combustion Engineering*, McGraw-Hill, New York, 1998.

47. C.G. Segeler, Ed., *Gas Engineers Handbook*, Industrial Press, New York, 1965.

48. R.D. Reed, *Furnace Operations*, 3rd edition, Gulf Publishing, Houston, 1976.

49. R. Pritchard, J.J., Guy, and N.E., Connor, *Handbook of Industrial Gas Utilization*, Van Nostrand Reinhold, New York, 1977.

50. R.J. Reed, *North American Combustion Handbook*, Volume I, 3rd edition, North American Mfg. Co., Cleveland, OH, 1986.

51. IHEA, *Combustion Technology Manual*, 5th edition, Industrial Heating Equipment Assoc., Arlington, VA, 1994.

52. W. Trinks and M.H., Mawhinney, *Industrial Furnaces*, Vol. I, 5th edition, John Wiley & Sons, New York, 1961.

53. A.E. Aldersley, *Fuels, Combustion and Heat Transfer*, Brick Development Association, London, 1964.

54. W.A. Gray, J.K. Kilham, and R. Muller, *Heat Transfer from Flames*, Elek, London, 1976

55. E. E. Khalil, *Modelling of Furnaces and Combustors*, Abacus Press, Tunbridge Wells, Kent, 1982.

56. J.M. Rhine, and R.J. Tucker, *Modelling of Gas-Fired Furnaces and Boilers*, McGraw-Hill, London, 1991.

56a. J.R. Cornforth, Ed., *Combustion Engineering and Gas Utilisation*, E&FN Spon, London, 1992.

57. K.H. Khalil, F.M. El-Mahallawy, and E.E. Khalil, Eds., *Flow, Mixing, and Heat Transfer in Furnaces*, Conference on Mechanical Power Engineering (1st: 1977, Cairo University), Pergamon Press, Oxford, U.K., 1978.

58. R. Lazzeretti and S.N.B. Murthy, Eds., *Proceedings of the United States-Italy Joint Workshop on Heat Transfer and Combustion*, U.S.A.-Italy Joint Workshop on Heat Transfer and Combustion (1982: Pisa, Italy), sponsored by the National Science Foundation and Consiglio nazionale della ricerche, s.n., 3 volumes, 1982.

59. C.K. Law, Ed., Heat transfer in fire and combustion systems, presented at the *23rd National Heat Transfer Conference,* Denver, CO, August 4-7, 1985, sponsored by the K-11 Committee on Heat Transfer in Fire and Combustion Systems, the Heat Transfer Division, ASME, New York, HTD, Vol. 45, 1985.

60. C. Presser and D.G. Lilley, Eds., Heat transfer in furnaces, presented at the *24th National Heat Transfer Conference and Exhibition,* Pittsburgh, PA, August 9-12, 1987, sponsored by the Heat Transfer Division, ASME, New York, HTD Vol. 74, 1987.

61. R.K. Shah, Ed., *Heat transfer phenomena in radiation, combustion, and fires*, presented at the 1989 National Heat Transfer Conference, Philadelphia, PA, August 6-9, 1989, sponsored by the Heat Transfer Division, ASME, New York, HTD Vol. 106, 1989.

62. N. Ashgriz, Ed., Heat transfer in combustion systems, *Proceedings of the Winter Annual Meeting of the American Society of Mechanical Engineers,* San Francisco, CA, December 10-15, 1989, sponsored by the Heat Transfer Division, ASME, New York, HTD Vol. 122, 1989.

63. B. Farouk, Ed., Heat transfer in combustion systems, presented at *AIAA/ASME Thermophysics and Heat Transfer Conference,* June 18-20, 1990, Seattle, WA, sponsored by the Heat Transfer Division, ASME, New York, HTD Vol. 142, 1990.

64. W.L. Grosshandler and H.G. Semerjian, Eds., Heat and mass transfer in fires and combustion systems, presented at the *Winter Annual Meeting of the American Society of Mechanical Engineers,* Dallas, TX, November 25-30, 1990, sponsored by the Heat Transfer Division, ASME, New York, HTD Vol. 148, 1990.

65. M.G. Carvalho, F. Lockwood, and J. Taine, Eds., Heat transfer in radiating and combusting systems, *Proceedings of Eurotherm Seminar no. 17,* 8-10 October 1990, Cascais, Portugal, Springer-Verlag, Berlin, 1991.

66. R.J. Santoro and J.D. Felske, Eds., Heat transfer in fire and combustion systems, 1991, presented at the *28th National Heat Transfer Conference,* Minneapolis, MN, July 28-31, 1991, sponsored by the Heat Transfer Division, ASME, New York, HTD Vol. 166, 1991.

67. M.F. Modest, T.W. Simon, and M. Ali Ebadian, Eds., Experimental/numerical heat transfer in combustion and phase change, presented at the *28th National Heat Transfer Conference,* Minneapolis, MN, July 28-31, 1991, sponsored by the Heat Transfer Division, ASME, New York, HTD, Vol. 170, 1991.

68. S.C. Yao and J.N. Chung, Eds., Heat and mass transfer in fires and combustion systems, presented at the *Winter Annual Meeting of the American Society of Mechanical Engineers,* Atlanta, GA, December 1-6, 1991, sponsored by the Heat Transfer Division, ASME, New York, HTD Vol. 176, 1991.

69. A.M. Kanury and M.Q. Brewster, Eds., Heat transfer in fire and combustion systems, presented at the *28th National Heat Transfer Conference and Exhibition,* San Diego, CA, August 9-12, 1992, sponsored by the Heat Transfer Division, ASME, New York, HTD Vol. 199, 1992.

70. P. Cho and J. Quintiere, Eds., Heat and mass transfer in fire and combustion systems, 1992: presented at the *Winter Annual Meeting of the American Society of Mech. Engineers,* Anaheim, CA, November 8-13, 1992, sponsored by the Heat Transfer Division, ASME, New York, HTD Vol. 223, 1992.

71. B. Farouk, M. P. Menguc, R. Viskanta, C. Presser, and S. Chellaiah Eds., Heat transfer in fire and combustion systems, presented at the *29th National Heat Transfer Conference,* Atlanta, GA, August 8-11, 1993, sponsored by the Heat Transfer Division, ASME, New York, HTD Vol. 250, 1993.

72. W.W. Yuen, and K.S. Ball, Eds., Heat transfer in fire and combustion systems, presented at the *6th AIAA/ASME Thermophysics and Heat Transfer Conference,* Colorado Springs, CO, June 20-23, 1994, sponsored by the Heat Transfer Division, ASME, New York, HTD Vol. 272, 1994.

73. R.B. Peterson, O.A. Ezekoye, and T. Simon, Combustion and fire research, *Proceedings of the 1995 National Heat Transfer Conference,* August 6-8 1995, Portland, OR, sponsored by the Heat Transfer Division, ASME, Vol. 2, 1995.

74. M.C. McQuay, K. Annamalai, W. Schreiber, D. Choudhury, E. Bigzadeh, and A. Runchal, Eds., Heat transfer in combustion systems, *Proceedings of the 1996 31st National Heat Transfer Conference,* Houston, TX, August 3-6 1996, sponsored by the Heat Transfer Division, ASME, New York, HTD Vol. 328, 1996.

75. K. Annamalai, S. Acharya, A.K. Gupta, R. Altenkirch, and C. Presser, Eds., *Symposium on Fire and Combustion, Proc. of 32nd National Heat Transfer Conference*, August 10-12, 1997, Baltimore, MD, sponsored by the Heat Transfer Division, ASME HTD-Vol. 352, New York, 1997.

76. R.A. Nelson, K.S. Ball, and Z.M. Zhang, Eds., Combustion and radiation heat transfer, *Proceedings of the ASME Heat Transfer Division – 1998,* Vol. 2, ASME HTD-Vol. 361-2, ASME, New York, 1998.

77. S.W. Churchill and N. Lior, Heat transfer in combustion chambers, *Heat Transfer 1982,* Vol. 1: Review and Keynote Papers, U. Grigull, E. Hahne, K. Stephan, and J. Straub, Eds., Hemisphere, Washington, D.C., 1982, 289-299.

78. J.L. Reese, G.L. Moilanen, R. Borkowicz, C. Baukal, D. Czerniak, and R. Batten, State-of-the-Art of NOx Emission Control Technology, ASME paper 94-JPGC-EC-15, *Proceedings of Int. Joint Power Generation Conf.,* Phoenix, AZ, 3-5 October, 1994.

79. M.L. Joshi, M.E. Tester, G.C. Neff, and S.K. Panahi, Flame particle seeding with oxygen enrichment for NOx reduction and increased efficiency, *Glass,* 68(6), 212-213, 1990.

80. R. Ruiz and J.C. Hilliard, Luminosity enhancement of natural gas flames, *Proc. of 1989 International Gas Research Conf.,* T.L. Cramer, Ed., Govt. Institutes, Rockville, MD, 1990, 1345-1353.

81. A.G. Slavejkov, T.M. Gosling, and R.E. Knorr, Low-NOx Staged Combustion Device for Controlled Radiative Heating in High Temperature Furnaces, U.S. patent 5,611,682, March 18, 1997.

82. API Publication 535: Burner for Fired Heaters in General Refinery Services, 1st edition, American Petroleum Institute, Washington, D.C., July 1995.

83. C.E. Baukal, Ed., *Oxygen-Enhanced Combustion*, CRC Press, Boca Raton, FL, 1998.

84. K.J. Fioravanti, L.S. Zelson, and C.E. Baukal, Flame Stabilized Oxy-Fuel Recirculating Burner, U.S. Patent 4,954,076 issued 04 September 1990.

85. S. Longacre, Using infrared to dry paper and its coatings, *Process Heating*, 4(2), 45-49, 1997.

86. T.M. Smith and C.E. Baukal, Space-age refractory fibers improve gas-fired infrared generators for heat processing textile webs, *Journal of Coated Fabrics*, 12(3), 160-173, 1983.

87. R.J. Reed, Future consequences of compact, highly effective heat recovery devices, in *Heat Transfer in Furnaces*, C. Presser and D.G. Lilley, Eds., ASME HTD, New York, 74, 23-28, 1987.

88. T. Davies, Regenerative burners for radiant tubes — field test experience, in *Industrial Combustion Technologies*, M.A. Lukasiewicz, Ed., American Society of Metals, Warren, PA, 1986, 65-70.

2 Some Fundamentals of Combustion

2.1 COMBUSTION CHEMISTRY

Combustion is usually considered to be the controlled release of heat and energy from the chemical reaction between a fuel and an oxidizer. This is in contrast to a fire or explosion which are usually uncontrolled and undesirable. Virtually all of the combustion in industrial processes uses a hydrocarbon fuel. A generalized combustion reaction for a typical hydrocarbon fuel can be written as follows:

$$\text{fuel} + \text{oxidizer} \rightarrow CO_2 + H_2O + \text{Other species} \tag{2.1}$$

The "other species" depends on what oxidizer is used and what is the ratio of the fuel to oxidizer. The most commonly used oxidizer is air, which consists of nearly 79% N_2 by volume and is normally carried through in the combustion process. If the combustion is fuel rich, meaning there is not enough oxygen to fully combust the fuel, then there will be unburned hydrocarbons in the exhaust products and little if any excess O_2. If the combustion is fuel lean, meaning there is more oxygen than required to fully combust the fuel, then there will be excess O_2 in the exhaust products. The exhaust gas composition is very important in determining the heat transfer in the system. Unburned hydrocarbons in the exhaust indicate that the fuel was not fully combusted and therefore not all of the available heat was released. High excess O_2 levels in the exhaust usually indicate that too much oxidizer was supplied. The excess oxidizer carries sensible energy out the exhaust, which again means that some of the available heat of the fuel was not fully utilized to heat the load. If the oxidizer is air, then a large proportion of the available energy in the fuel will normally be carried out the flue with the exhaust products. This is discussed in more detail in Chapter 12.

2.1.1 FUEL PROPERTIES

Table E.2 in Appendix E gives some of the properties for gaseous fuels commonly used in industrial combustion systems. As expected, the fuel has a significant influence on the heat transfer in a combustion system. One of the most important properties is the heating value of the fuel. This is used to determine how much fuel must be combusted to process the desired production rate of material that is being heated. The heating value is specified as either the higher heating value (HHV) or the lower heating value (LHV). The LHV excludes the heat of vaporization, which is the energy required to convert liquid water to steam. This means that the LHV assumes all of the products of combustion are gaseous, which is generally the case for nearly all industrial combustion applications. If the combustion products were to exit the process at a temperature low enough that all of the water is converted from a gas to a liquid, then the heat of condensation would be released into the process as an additional source of energy. The HHV of a fuel includes that energy.

The composition of the fuel is important in determining the composition of the products of combustion and the amount of oxidizer that will be needed to combust the fuel, both of which are discussed below. It is also important for determining the soot-producing tendency of the fuel which is discussed in Chapter 3. The density of the fuel is needed to determine flow rates through the fuel delivery system and the associated pipe sizes.

2.1.2   OXIDIZER COMPOSITION

There are two common types of oxidizers used in industrial combustion processes. The majority of those processes use air as the oxidizer. However, many of the higher temperature processes use an oxidizer containing a higher concentration of oxygen than found in air (approximately 21% by volume). This type of combustion is referred to as oxygen-enhanced combustion and is discussed further in Chapter 12.[1] In many cases, the production rate in a heating process can be significantly increased with only relatively small amounts of oxygen enrichment. In most cases, air/fuel burners can successfully operate with an oxidizer containing up to about 30% O_2 with little or no modifications.[2] At greater oxygen concentrations, the flame may become unstable or the flame temperature may become too high for a burner designed to operate under air/fuel conditions. In higher temperature applications where the benefits of higher purity oxygen justify the added costs, higher purity oxidizers may be used (>90% O_2). The heating process is greatly intensified by the high purity oxygen. It has only been in the last decade that a significant number of combustion systems have been operated in the intermediate oxygen purities (30 – 90% O_2), primarily for economic reasons as the cost of high-purity oxidizers is not justified. The oxidizer purity has a significant influence on the heat transfer in a combustion system, as will be shown throughout this book.

2.1.3   MIXTURE RATIO

An important consideration when studying a combustion system is the ratio of the fuel to the oxidizer. There are many ways that this can be specified. These are briefly considered here.

A global combustion reaction using CH_4 as the fuel can be written as:

$$CH_4 + (xO_2 + yN_2) \rightarrow CO, CO_2, H_2, H_2O, N_2, NOx, O_2, \text{Trace species} \quad (2.2)$$

The stoichiometry of a reaction indicates the ratio of oxygen to fuel for a given system. One method of quantifying the stoichiometry is to only consider the O_2 in the oxidizer, since the inerts in the oxidizer are not needed for the reaction:

$$S_1 = \frac{\text{Volume flow rate of } O_2 \text{ in the oxidizer}}{\text{Volume flow rate of fuel}} \quad (2.3)$$

If CH_4 is again used as an example, a global simplified stoichiometric reaction with air can be written as:

$$CH_4 + (2O_2 + 7.52N_2) \rightarrow CO_2 + 2H_2O + 7.52N_2 \quad (2.4)$$

where air is represented as $2O_2 + 7.52N_2$. In that case,

$$S_1 = \frac{2}{1} = 2$$

This method of specifying stoichiometry is more commonly used for combustion systems incorporating oxygen enrichment. This is because the oxidizer composition can vary widely, but the amount of oxygen supplied to the combustion system is what is of importance.

The most common way of defining the stoichiometry or mixture ratio in industry in the U.S. is as follows:

$$S_2 = \frac{\text{Volume flow rate of oxidizer}}{\text{Volume flow rate of fuel}} \quad (2.5)$$

For the example reaction given in Equation 2.4, this stoichiometry would be calculated as follows:

$$S_2 = \frac{2 + 7.52}{1} = 9.52$$

The problem with the stoichiometry definition commonly used in industry (S_2) is that the stoichiometry must be recalculated whenever the oxidizer composition changes and stoichiometric conditions change for each oxidizer composition. This is not a concern if air is always used as the oxidizer, which is the case for the vast majority of combustion processes. For example, consider the reaction of methane and pure oxygen:

$$CH_4 + 2O_2 \rightarrow CO_2 + 2H_2O + 7.52N_2 \tag{2.6}$$

The two stoichiometric ratios would be calculated as follows:

$$S_1 = \frac{2}{1} = 2, \; S_2 = \frac{2}{1} = 2$$

For this limited case where the oxidizer is pure O_2, $S_1 = S_2$; but in the example above using air as the oxidizer, $S_1 \neq S_2$. For both cases, S_1 is the same, while S_2 is different. The benefit of using S_1 is that the stoichiometry is independent of the oxidizer composition, so stoichiometric conditions are the same for any oxidizer composition. In Equation 2.1, $S_1 = x/1 = x$. Theoretically, for the complete combustion of CH_4 with no excess O_2, $S_1 = 2.0$, no matter how much N_2 is in the oxidizer, while S_2 varies depending on the actual N_2 content in the oxidizer, although the same amount of O_2 may be present. One can define the stoichiometric ratio for theoretically perfect combustion as S_1^P, which is equal to 2.0 for methane, no matter what the quantity of other gases besides O_2 in the oxidizer. Then, the stoichiometric ratio for theoretical combustion for S_2 would be S_2^P which, unlike S_1^P, would vary depending on the actual oxidizer composition.

Actual flames generally require some excess O_2 for complete combustion of the fuel. This is due to incomplete mixing between the fuel and oxidant. For the fuel-rich combustion of CH_4, $S_1 < 2.0$. For the fuel-lean combustion of CH_4, $S_1 > 2.0$. These generalizations are not possible for S_2 without specifying the oxidizer composition first.

A common way of specifying the oxidizer composition is by calculating the O_2 mole fraction in the oxidizer, which can be defined as:

$$\Omega = \frac{\text{Volume flow rate of O}_2 \text{ in the oxidizer}}{\text{Total volume flow rate of oxidizer}} \tag{2.7}$$

Using Equation 2.1, $\Omega = x/(x + y)$. If the oxidizer is air that contains approximately 21% O_2 by volume, then $\Omega = 0.21$. If the oxidizer is pure O_2, then $\Omega = 1.0$. The O_2 enrichment level is sometimes used. This refers to the incremental O_2 volume above that found in air. For example, if $\Omega = 0.35$, then the O_2 enrichment would be 14% (35% − 21% = 14%).

Another way of specifying the fuel mixture ratio is:

$$\lambda_1 = \frac{\text{Actual volumetric ratio of oxygen:fuel}}{\text{Stoichiometric volumetric ratio of oxygen:fuel}} \tag{2.8}$$

The way it is commonly specified in industry in Europe is as follows:

$$\lambda_2 = \frac{\text{Actual volumetric ratio of oxidant:fuel}}{\text{Stoichiometric volumetric ratio of oxidant:fuel}} \tag{2.9}$$

or

$$\lambda_i = \frac{S_i}{S_i^P} \tag{2.10}$$

Using this definition for the mixture ratio, $\lambda_i < 1.0$ for fuel-rich flames and $\lambda_i > 1.0$ for fuel-lean flames. Many industrial combustion processes run with approximately 3% more O_2 than is theoretically needed for perfect combustion. That is often the amount of excess O_2 required to minimize the emissions of unburned hydrocarbons and ensure the complete combustion of the fuel. This may be due to mixing limitations between the fuel and the oxidizer, especially in non-premixed systems. Too much excess O_2 means that energy is being wasted heating up excess combustion air, instead of the load. Therefore, it is desirable to only use just enough excess O_2 to get low CO emissions. An example of a simplified global reaction for methane with 3% excess O_2 is the reaction:

$$CH_4 + (2.06O_2 + 7.75N_2) \rightarrow CO_2 + 2H_2O + 0.06O_2 + 7.75N_2 \tag{2.11}$$

For this reaction,

$$S_1 = \frac{2.06}{1} = 2.06, \ S_2 = \frac{2.06 + 7.75}{1} = 9.81$$

with $S_1^P = 2.0$ and $S_2^P = 9.52$. Then, for the reaction in Equation 2.11,

$$\lambda_1 = \frac{S_1}{S_1^P} = \frac{2.06}{2.0} = 1.03, \ \lambda_2 = \frac{S_1}{S_1^P} = \frac{9.81}{9.52} = 1.03$$

As can be seen, these two ratios are the same, so a single ratio λ can be given, using either Equation 2.8 or 2.9. This is an advantage of using this method to specify the mixture ratio.

Another way of specifying the mixture ratio in industry is by the amount of either excess O_2 (XO_2), or the amount of excess air (X_A), on a volume basis. For the example in Equation 2.11, $X_{O_2} = 3\%$ and $X_A = 14.4\%$.

In the scientific community, it is common to use the equivalence ratio (ϕ) to specify the mixture ratio, which is the inverse of λ. Again, there are two possible ways this could be defined:

$$\phi_1 = \frac{\text{Stoichiometric volumetric ratio of oxygen:fuel}}{\text{Actual volumetric ratio of oxygen:fuel}} \tag{2.12}$$

or

$$\phi_2 = \frac{\text{Stoichiometric volumetric ratio of oxidant:fuel}}{\text{Actual volumetric ratio of oxidant:fuel}} \tag{2.13}$$

It can again be shown that these are the same, so one obtains

$$\phi = \frac{S_i^P}{S_i} \tag{2.14}$$

or

$$\phi = \frac{1}{\lambda} \qquad (2.15)$$

A comparison of these various methods of specifying the mixture ratio is shown in the Appendices. Tables C1-C3 in Appendix C show various methods of specifying the mixture ratios for methane, propane, and hydrogen, respectively, as example fuels. As will be shown in the next chapter section, this mixture is very important in determining the heat release and transport properties for the exhaust products.

2.1.4 OPERATING REGIMES

Most industrial flames are turbulent. A turbulent Reynolds number (Re_T) can be defined as:

$$Re_T = \frac{v'l_0}{v_0} \qquad (2.16)$$

One can define a turbulent characteristic length scale, commonly called the Kolmogorov length, l_K:

$$l_K = \frac{l_0}{\left(\dfrac{vl_0}{v_0}\right)^{3/4}} \qquad (2.17)$$

where l_0 is a characteristic length scale usually associated with large eddies, v is a characteristic velocity, and v_0 is the characteristic kinematic viscosity. The Kolmogorov length is representative of the dimension where dissipation occurs. The Taylor length scale can be defined as the ratio of the strain rate to the viscous forces:

$$l_T = \frac{l_0^2}{Re_T} \qquad (2.18)$$

where v' is a characteristic velocity fluctuation from the mean velocity v, which is an indicator of the turbulence level. The flame thickness, l_L, is a characteristic length scale of the flame. The various lengths can be used to characterize the flame[3]:

$l_L < l_K$ Wrinkled flame
$l_K < l_L < l_T$ Severely wrinkled flame
$l_T < l_L < l_0$ Flamelets in eddies
$l_0 < l_L$ Distributed reaction front

A nondimensional Damköhler number (Da) can be defined that indicates the type of reaction time which is significant for the specific type of combustion reaction:

$$Da = \frac{l_0 S_L}{v' l_L} \qquad (2.19)$$

where S_L is the laminar flame speed. This number is the ratio of the reaction time to the flow rate. If this number is high, then the assumption is often made in modeling that the reactions are infinite rate or equilibrium, which greatly simplifies modeling. If this ratio is low, then finite rate chemistry must be used, which increases the complexity and computation time of model simulations.

2.2 COMBUSTION PROPERTIES

This chapter section briefly considers the combustion product composition, flame temperature, available heat, and flue gas volume for combustion as commonly used in industrial applications. These are important in calculating the heat transfer from the flame and exhaust gases to the furnace and to the load.

2.2.1 COMBUSTION PRODUCTS

There are a number of variables that can have a significant impact on the products of combustion. Some of the important variables include the oxidizer composition, mixture ratio, air and fuel preheat temperatures, and fuel composition. These are briefly discussed here.

2.2.1.1 Oxidizer Composition

The stoichiometric combustion of CH_4 with air can be represented by the following global equation:

$$CH_4 + 2O_2 + 7.52N_2 \rightarrow CO_2, 2H_2O, 7.52N_2, \text{Trace species} \qquad (2.20)$$

It can be seen that over 70 vol% of the exhaust gases is N_2. Similarly, a stoichiometric O_2/CH_4 combustion process can be represented by:

$$CH_4 + 2O_2 \rightarrow CO_2, 2H_2O, \text{Trace species} \qquad (2.21)$$

The volume of exhaust gases is significantly reduced by the elimination of N_2. In general, a stoichiometric oxygen-enhanced methane combustion process can be represented by:

$$CH_4 + 2O_2 + xN_2 \rightarrow CO_2 + 2H_2O + xN_2 + \text{Trace species} \qquad (2.22)$$

where $0 \leq x \leq 7.52$, depending on the oxidizer.

The actual composition of the exhaust products from the combustion reaction depends on several factors, including the oxidizer composition, the temperature of the gases, and the equivalence ratio. A cartoon showing an adiabatic equilibrium combustion reaction is shown in Figure 2.1. An adiabatic

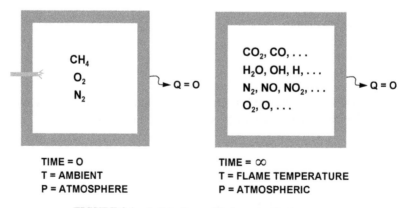

FIGURE 2.1   Adiabatic equilibrium reaction process.

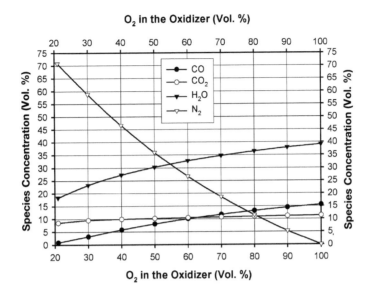

FIGURE 2.2 Major species concentrations vs. oxidizer (O_2 + N_2) composition for an adiabatic equilibrium stoichiometric CH_4 flame.

process means that no heat is lost during the reaction, or that the reaction occurs in a perfectly insulated chamber. This is not the case in an actual combustion process where heat is lost from the flame by radiation. Figure 2.2 shows the predicted major species for the adiabatic equilibrium combustion of CH_4 as a function of the oxidizer composition. The calculations were made using a NASA computer program that minimizes the Gibbs free energy of a gaseous system.[4] An equilibrium process means that there is an infinite amount of time for the chemical reactions to take place, or the reaction products are not limited by chemical kinetics. In actuality, the combustion reactions are completed in fractions of a second. As expected, Figure 2.2 shows that as N_2 is removed from the oxidizer, the concentration of N_2 in the exhaust products decreases correspondingly. Likewise, there is an increase in the concentrations of CO, CO_2 and H_2O. For this adiabatic process, there is a significant amount of CO at higher levels of O_2 in the oxidizer. Figure 2.3 shows the predicted minor species for the same conditions as Figure 2.2. Note that trace species have been excluded from this figure. The radical species H, O, and OH all increase with the O_2 in the oxidizer. NO initially increases and then decreases after about 60% O_2 in the oxidizer as more N_2 is removed from the system. When the oxidizer is pure O_2, no NO is formed because no N_2 is available. Unburned fuel in the form of H_2 and unreacted oxidizer in the form of O_2 also increase with the O_2 concentration in the oxidizer. This increase in radical concentrations, unburned fuel in the form of CO and H_2, and unreacted O_2 are all due to chemical dissociation that occurs at high temperatures.

The actual flame temperature is lower than the adiabatic equilibrium flame temperature due to imperfect combustion and radiation from the flame. The actual flame temperature is determined by how well the flame radiates its heat and how well the combustion system, including the load and the refractory walls, absorbs that radiation. A highly luminous flame generally has a lower flame temperature than a highly nonluminous flame. The actual flame temperature will also be lower when the load and the walls are more radiatively absorptive. This occurs when the load and walls are at lower temperatures and have higher radiant absorptivities. These effects are discussed in more detail in Chapter 4. As the gaseous combustion products exit the flame, they typically lose more heat by convection and radiation as they travel through the combustion chamber. The objective of a combustion process is to transfer the chemical energy contained in the fuel to the load, or in some cases to the combustion chamber. The more thermally efficient the combustion process, the more heat that is transferred from the combustion products to the load and to the combustion chamber. Therefore, the

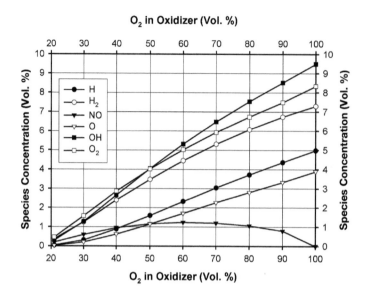

FIGURE 2.3 Minor species concentrations vs. oxidizer ($O_2 + N_2$) composition for an adiabatic equilibrium stoichiometric CH_4 flame.

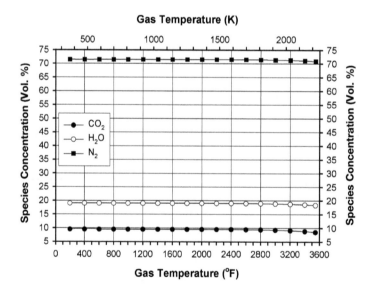

FIGURE 2.4 Equilibrium calculations for the predicted gas composition of the major species as a function of the combustion product temperature for stoichiometric air/CH_4 flames.

gas temperature in the exhaust stack is desirably much lower than in the flame in a thermally efficient heating process. The composition of the combustion products then changes with gas temperature.

Figure 2.4 shows the predicted major species for the equilibrium combustion of CH_4 with "air" (21% O_2, 79% N_2) as a function of the gas temperature. The highest possible temperature for the air/CH_4 reaction is the adiabatic equilibrium temperature of 3537°F (2220K). For the air/CH_4 reaction, there is very little change in the predicted gas composition as a function of temperature. Figure 2.5 shows the predicted minor species for the same conditions as in Figure 2.4. For the air/CH_4, none of the minor species exceed 1% by volume. As the gas temperature increases, chemical dissociation increases.

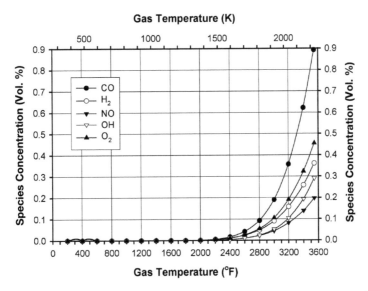

FIGURE 2.5 Equilibrium calculations for the predicted gas composition of the minor species as a function of the combustion product temperature for stoichiometric air/CH$_4$ flames.

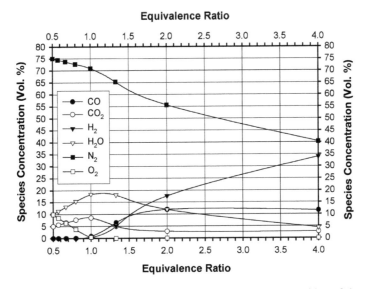

FIGURE 2.6 Adiabatic equilibrium calculations for the predicted gas composition of the major species as a function of the equivalence ratio for air/CH$_4$ flames.

2.2.1.2 Mixture Ratio

Figure 2.6 shows the predicted gas composition for the adiabatic equilibrium combustion of air/CH$_4$ as a function of the equivalence ratio. The O$_2$ and N$_2$ concentrations in the exhaust gases strictly decrease with the equivalence ratio. The H$_2$O and CO$_2$ concentrations peak at stoichiometric conditions ($\phi = 1.0$). This is important as both of these gases produce nonluminous gaseous radiation (see Chapter 3). As expected, the unburned fuels in the form of H$_2$ and CO both increase with equivalence ratio. This will be reflected in the available heat (discussed below) as not all of the fuel is fully combusted.

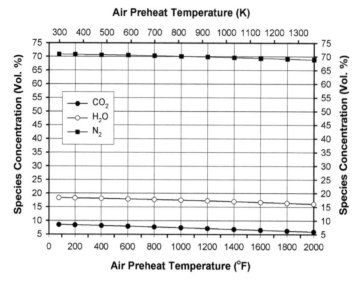

FIGURE 2.7 Adiabatic equilibrium stoichiometric calculations for the predicted gas composition of the major species as a function of the air preheat temperature for air/CH_4 flames.

2.2.1.3 Air and Fuel Preheat Temperature

In many industrial combustion processes, heat is recovered to improve the overall thermal efficiency of the process to reduce operating costs. The recovered heat is most commonly used to preheat the incoming combustion air and is sometimes used to preheat the incoming fuel. Preheating either the air or the fuel affects the composition of the combustion products. Figure 2.7 shows the major species predicted for the combustion of air and CH_4 where the air is preheated up to as high as 2000°F (1366K). CO_2, H_2O, and N_2 all decrease with air preheat, due to chemical dissociation. Figure 2.8 shows that the minor species increase with air preheat. Figures 2.9 and 2.10 show the

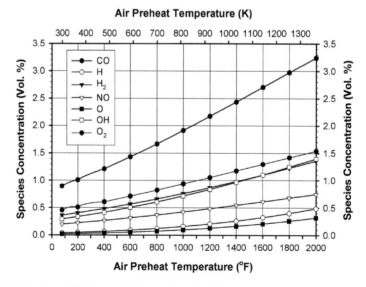

FIGURE 2.8 Adiabatic equilibrium stoichiometric calculations for the predicted gas composition of the minor species as a function of the air preheat temperature for air/CH_4 flames.

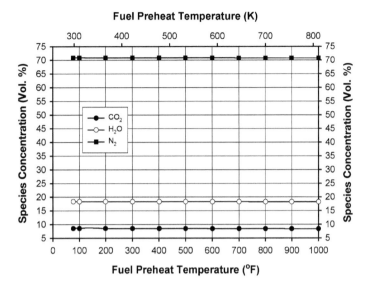

FIGURE 2.9 Adiabatic equilibrium stoichiometric calculations for the predicted gas composition of the major species as a function of the fuel preheat for air/CH_4 flames.

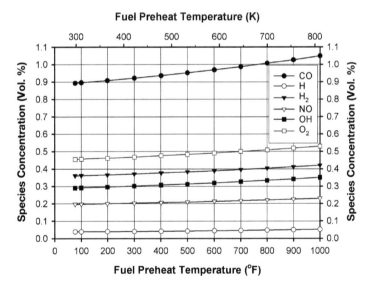

FIGURE 2.10 Adiabatic equilibrium stoichiometric calculations for the predicted gas composition of the minor species as a function of the fuel preheat for air/CH_4 flames.

major and minor species, respectively, for air/CH_4 flames as a function of the CH_4 preheat temperature up to temperatures as high as 1000°F (811K). Due to safety considerations and the possibility of sooting up the fuel supply piping, higher fuel preheat temperatures are not practical or recommended under most conditions. The figures show that there is only a slight decrease in the concentrations of the major species and a slight increase in the concentrations of the minor species. This is due to the fact that the mass of fuel is relatively small compared to the mass of combustion air supplied to the system. This means that preheating the combustion air has a much more significant impact than preheating the fuel for a given preheat temperature.

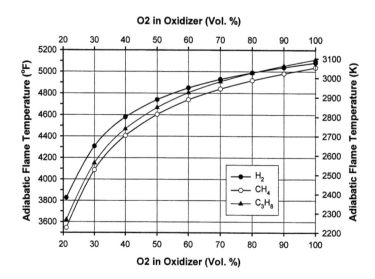

FIGURE 2.11 Adiabatic flame temperature vs. oxidizer composition for adiabatic equilibrium stoichiometric air/H_2, air/CH_4, and air/C_3H_8 flames.

2.2.1.4 Fuel Composition

Combustion products have been calculated for the following four fuels: H_2, CH_4, C_3H_8, and blends of H_2 and CH_4. These are intended to be representative of fuels commonly used in industrial applications. In terms of luminosity, H_2 produces nonluminous flames, CH_4 produces low-luminosity flames, and C_3H_8 produces higher luminosity flames. The predicted combustion product compositions for each fuel under a variety of operating conditions are given in Appendix D.

2.2.2 FLAME TEMPERATURE

The flame temperature is a critical variable in determining the heat transfer from the flame to the load, as will be shown in Chapter 3. This chapter section shows how the adiabatic flame temperature is affected by the oxidizer and fuel compositions, the mixture ratio, and the air and fuel preheat temperatures. As previously mentioned, real flame temperatures are not as high as the adiabatic flame temperature, but the trends are comparable and representative of actual conditions.

2.2.2.1 Oxidizer and Fuel Composition

The flame temperature increases significantly when air is replaced with oxygen because N_2 acts as a diluent that reduces the flame temperature. Figue 2.11 is a plot of the adiabatic equilibrium flame temperature for CH_4 combustion, as a function of the oxidizer composition, for a stoichiometric methane combustion process. The flame temperature varies from 3600°F to 5000°F (2300K to 3000K) for air and pure oxygen, respectively. The graph shows a rapid rise in the flame temperature from air up to about 60% O_2 in the oxidizer. The flame temperature increases at a slower rate for higher O_2 concentrations.

Table 2.1 lists the adiabatic flame temperatures for a number of fuels where the oxidizer is either air or pure O_2. From this table, it can be seen that the fuel composition has a strong impact on the flame temperature. Figure 2.12 shows how the flame temperature varies for a fuel blend of H_2 and CH_4. The temperature increases as the H_2 content in the blend increases. It is important to note that the increase is not linear, with a more rapid increase at higher levels of H_2. Because of the relatively high cost of H_2 compared to CH_4 and C_3H_8, it is not used in many industrial applications. However, high H_2 fuels are often used in many of the hydrocarbon and petrochemical

TABLE 2.1
Adiabatic Flame Temperatures

Fuel	Air		O_2	
	°F	K	°F	K
H_2	3807	2370	5082	3079
CH_4	3542	2223	5036	3053
C_2H_2	4104	2535	5556	3342
C_2H_4	3790	2361	5256	3175
C_2H_6	3607	2259	5095	3086
C_3H_6	4725	2334	5203	3138
C_3H_8	3610	2261	5112	3095
C_4H_{10}	3583	2246	5121	3100
CO	3826	2381	4901	2978

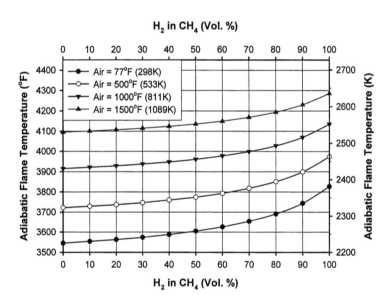

FIGURE 2.12 Adiabatic equilibrium flame temperature vs. fuel blend composition for stoichiometric air/fuel flames at different air preheat temperatures.

applications for fluid heating. Those fuels are by-products of the chemical manufacturing process and therefore much less expensive than purchasing H_2 from an industrial gas supplier and more cost effective than using other purchased fuels.

2.2.2.2 Mixture Ratio

Figure 2.13 is a similar plot of the adiabatic equilibrium flame temperature for CH_4 flames as a function of the stoichiometry for four different oxidizer compositions ranging from air to pure O_2. The peak flame temperatures occur at stoichiometric conditions. The lower the O_2 concentration in the oxidizer, the more the flame temperature is reduced by operating at nonstoichiometric conditions (either fuel rich or fuel lean). This is due to the higher concentration of N_2, which absorbs heat and lowers the overall temperature. Figure 2.14 shows the adiabatic flame temperature as a function of the equivalence ratio for three fuels: H_2, CH_4, and C_3H_8. The peak temperature occurs at stoichiometric conditions ($\phi = 1.0$). In that case, there is just enough oxidizer to fully

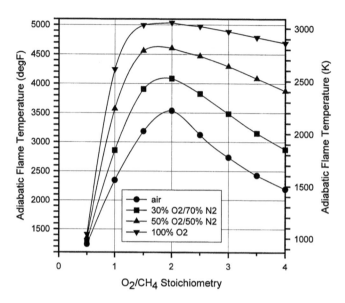

FIGURE 2.13 Adiabatic equilibirium flame temperature vs. stoichiometry for a CH_4 flame and various oxidizers.

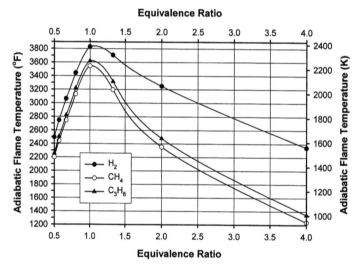

FIGURE 2.14 Adiabatic equilibrium flame temperature vs. equivalence ratio for air/H_2, air/CH_4, and air/C_3H_8 flames.

combust all the fuel. Any additional oxidizer absorbs sensible energy from the flame and reduces the flame temperature. In most real flames, the peak flame temperature often occurs at slightly fuel-lean conditions ($\phi < 1.0$). This is due to imperfect mixing where slightly more O_2 is needed to fully combust all the fuel. Nearly all industrial combustion applications are run at fuel-lean conditions to ensure that the CO emissions are low. Therefore, depending on the actual burner design, the flame temperature may be close to its peak, which is often desirable for maximizing heat transfer. One problem often encountered when maximizing the flame temperature is that the NOx emissions are also maximized since NOx increases approximately exponentially with gas temperature. This has led to many design concepts for reducing the peak flame temperature in the flame to minimize NOx emissions.[5] This also affects the heat transfer from the flame and is discussed in Chapters 3 and 8.

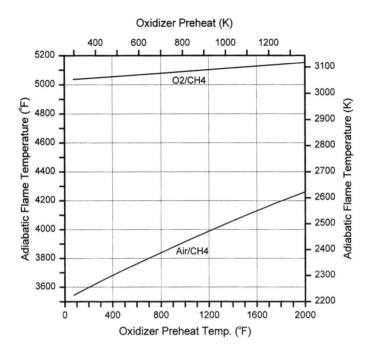

FIGURE 2.15 Adiabatic equilibrium flame temperature vs. oxidizer preheat temperature for stoichiometric air/CH$_4$ and O$_2$/CH$_4$ flames.

2.2.2.3 Oxidizer and Fuel Preheat Temperature

Figure 2.15 shows how the adiabatic flame temperature varies as a function of the oxidizer preheat temperature for air/CH$_4$ and O$_2$/CH$_4$ flames. The increase in flame temperature is relatively small for the O$_2$/CH$_4$ flame because the increased sensible heat of the O$_2$ is only a fraction of the chemical energy contained in the fuel. For air/CH$_4$ flames, preheating the air has a more dramatic impact because the increase in sensible heat is very significant due to the large mass of air in the combustion reaction. Figure 2.16 shows that the adiabatic flame temperature increases rapidly for air/fuel flames for each of the three fuels shown. Figure 2.12 also shows how preheating the air in the combustion of a blended fuel affects the flame temperature. Again, the higher the air preheat, the higher the temperature of the combustion products. Figure 2.17 shows that preheating the fuel has less impact on the flame temperature than preheating the oxidizer because of the mass flow rate differences, as previously discussed.

2.2.3 AVAILABLE HEAT

The available heat in a combustion system is important in determining the overall thermal efficiency and is therefore a factor when calculating the heat transfer in the process. It would be less effective to try to maximize the heat transfer in a system that inherently has a low available heat. Available heat is defined as the gross heating value of the fuel, less the energy carried out of the combustion process by the hot exhaust gases. The heat lost from a process through openings in the furnace, through the furnace walls, or by air infiltration are not considered in calculating the theoretical available heat as they are dependent on the process. The theoretical available heat should be proportional to the amount of energy actually absorbed by the load in an actual process, which is directly related to the thermal efficiency of the system. Therefore, the theoretical available heat is used here to show the thermal efficiency trends as functions of exhaust gas temperature, oxidizer and fuel compositions, mixture ratio, and air and fuel preheat temperatures.

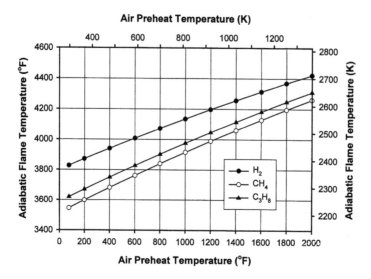

FIGURE 2.16 Adiabatic equilibrium flame temperature vs. air preheat temperature for stoichiometric air/H₂, air/CH₄, and air/C₃H₈ flames.

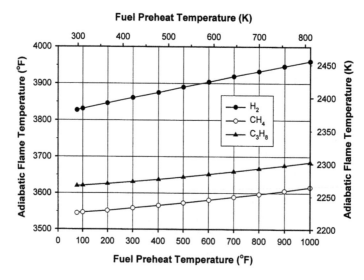

FIGURE 2.17 Adiabatic equilibrium flame temperature vs. fuel preheat temperature for stoichiometric air/H₂, air/CH₄, and air/C₃H₈ flames.

Figure 2.18 shows how the available heat decreases rapidly with the exhaust gas temperature and is relatively independent of the fuel composition for the three fuels shown. Then, to maximize the thermal efficiency of a process, it is desirable to minimize the exhaust gas temperature. This is usually done by maximizing the heat transfer from the exhaust gases to the load (and furnace walls) and by recovering some of the heat in the exhaust gases by preheating the fuel and/or the oxidizer. Figure 2.19 shows how the available heat, for stoichiometric air/CH₄ and O₂/CH₄ flames, varies as a function of the exhaust gas temperature. As the exhaust temperature increases, more energy is carried out of the combustion system and less remains in the system. The available heat decreases to zero at the adiabatic equilibrium flame temperature where no heat is lost from the gases. The figure shows that even at gas temperatures as high as 3500°F (2200K), the available

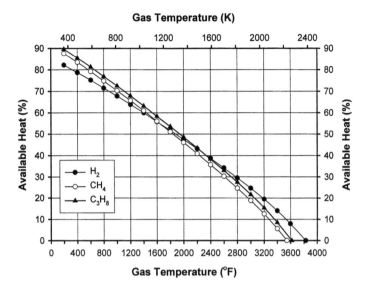

FIGURE 2.18   Available heat vs. gas temperature for stoichiometric air/H$_2$, air/CH$_4$, and air/C$_3$H$_8$ flames.

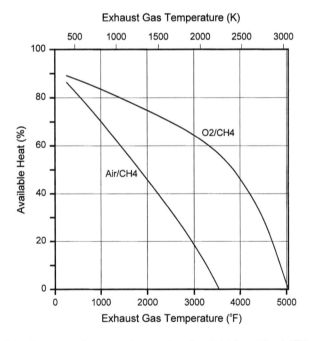

FIGURE 2.19   Available heat vs. exhaust gas temperature for stoichiometric air/CH$_4$ and O$_2$/CH$_4$ flames.

heat of an O$_2$/CH$_4$ system is still as high as 57%. The figure also shows that it is usually not very economical to use air/CH$_4$ systems for high-temperature heating and melting processes. At an exhaust temperature of 2500°F (1600K), the available heat for the air/CH$_4$ system is only a little over 30%. Heat recovery in the form of preheated air is commonly used for higher temperature heating processes to increase the thermal efficiencies.

Figure 2.20 is a graph of the available heat for the combustion of CH$_4$ as a function of the O$_2$ concentration in the oxidizer, for three different exhaust gas temperatures. As the exhaust gas

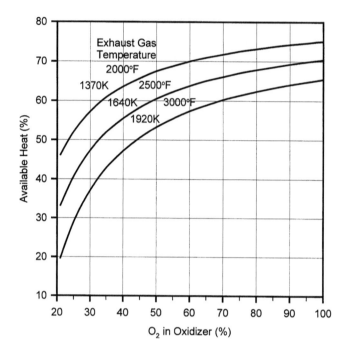

FIGURE 2.20   Available heat vs. oxidizer composition for a stoichiometric CH_4 flame, at exhaust temperatures of 2000°F, 2500°F, and 3000°F.

temperature increases, the available heat decreases because more energy is carried out the exhaust stack. There is an initial rapid increase in available heat as the O_2 concentration in the oxidizer increases from the 21% found in air. That is one reason why O_2 enrichment has been a popular technique for using OEC because the incremental increase in efficiency is very significant.

Figure 2.21 shows how the available heat increases with the oxidizer preheat temperature. The thermal efficiency of the air/CH_4 doubles by preheating the air to 2000°F (1400K). For the O_2/CH_4 flames, the increase in efficiency is much less dramatic by preheating the O_2. This is because the initial efficiency with no preheat is already 70% and because the mass of the O_2 is not nearly as significant in the combustion reaction as compared to the mass of air in an air/fuel flame. There are also safety concerns when flowing hot O_2 through piping, heat recuperation equipment, and a burner. Figure 2.22 shows that the available heat increases rapidly for any of the three fuels shown as the combustion air preheat temperature increases.

The fuel savings for a given technology can be calculated using the available heat curves:

$$\text{Fuel savings }(\%) = \left(1 - \frac{AH_2}{AH_1}\right) \times 100 \qquad (2.23)$$

where AH_1 is the available heat of the base case process and AH_2 is the available heat using a new technology. For example, if the base case process has an available heat of 30% and the available heat using the new technology is 45%, then the fuel savings equal $(1 - 45/30) \times 100 = -50\%$, which means that 50% less fuel is needed for process 2 compared to process 1.

2.2.4   FLUE GAS VOLUME

The flow rate of gases through a combustion chamber is proportional to the convective heat transfer to the load. There are several factors that influence this flow rate. One is the gas temperature since

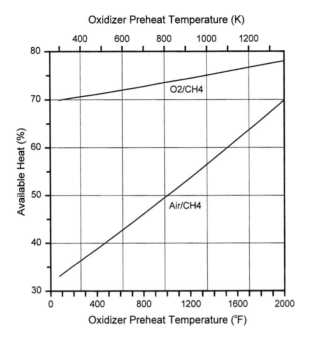

FIGURE 2.21   Available heat vs. oxidizer preheat temperature for equilibrium stoichiometric air/CH$_4$ and O$_2$/CH$_4$ flames at an exhaust gas temperature of 2500°F (1644K).

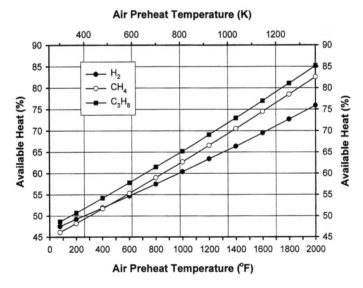

FIGURE 2.22   Available heat vs. air preheat temperature for equilibrium stoichiometric air/H$_2$, air/CH$_4$, and air/C$_3$H$_8$ flames at an exhaust gas temperature of 2000°F (1366K).

higher temperature gases have higher actual flow rates (e.g., actual cubic feet per hour or ACFH) due to the thermal expansion of the gases. This means that preheating the fuel or the oxidizer, which both normally increase the flame temperature, would produce higher actual flow rates. However, the flow rate of the gases is the same when corrected to standard temperature and pressure conditions (STP). Another factor that has a very strong influence on the gas flow rate through the combustion system is the oxidizer composition. Oxygen-enhanced combustion basically involves

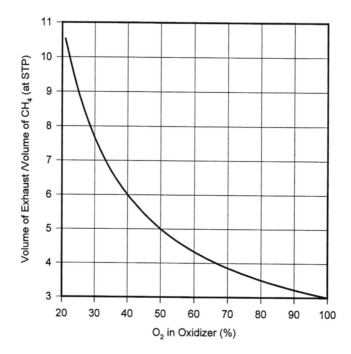

FIGURE 2.23   Normalized flue gas volume vs. oxidizer composition for a stoichiometric CH_4 flame.

removing N_2 from the oxidizer. A major change compared to air/fuel combustion is the reduction in the flue gas volume. Figure 2.23 shows the exhaust gas flow rate, normalized to the fuel flow rate at standard temperature and pressure conditions (e.g., standard cubic feet per hour or SCFH), for the stoichiometric combustion of CH_4 where it has been assumed that all the combustion products are CO_2, H_2O, and N_2 (except when the oxidizer is pure O_2 when there is no N_2). This means that for each unit volume of fuel, 3 normalized volumes of gas are produced for oxy/fuel compared to 10.5 volumes for air/fuel. This reduction can have both positive and negative effects, but the effect on convective heat transfer is a reduction in the average gas velocity through a given combustor and a resulting reduction in convection to the load.

2.3   EXHAUST PRODUCT TRANSPORT PROPERTIES

The transport properties of the gas species in the combustor are important for determining the heat transfer and fluid dynamics. The properties are highly temperature and species dependent. The properties here have been calculated using the software program originally developed for NASA.[6] Sandia Labs (Livermore, CA) has also developed a computer package that can be used to compute these properties.[7] The methods used to calculate the gas properties are not considered here because they are adequately treated elsewhere[8-12] and because software is available for making those calculations. The purpose of this chapter section is to show how the important gas properties for heat transfer in industrial combustors vary as functions of the fuel and oxidizer composition, mixture ratio, and air preheat temperatures. Note that the property variation as a function of the fuel preheat temperature has not been included here as the changes were minimal (see Appendix C).

The gas composition and gas temperature are needed to calculate nonluminous gaseous radiation (see Chapter 3). The gas transport properties are needed to calculate the convection heat transfer coefficient, which is often of the form:

$$Nu = a\, Pr^{b}\, Re^{c} \tag{2.24}$$

where the Nu is the Nusselt number, Pr is the Prandtl number, Re is the Reynolds number, and a, b, and c are constants. The constants depend on the flow conditions, but often $b \approx 1/3$ and $c \approx 2/3$. The convection heat transfer coefficient h is then calculated from the Nusselt number using:

$$\text{Nu} = \frac{hd}{k} \tag{2.25}$$

where d is a characteristic dimension for the flow system and k is the fluid thermal conductivity. The gas properties are needed to calculate the Nusselt, Prandtl, and Reynolds numbers, as will be shown next.

2.3.1 DENSITY

The gas density (ρ) can be used to calculate the Reynolds number ($\text{Re} = \rho v \, d/\mu$), which is usually needed to compute the convective heat transfer coefficient (h), which is discussed in the Chapter 3. The density is also used to calculate the average gas velocity through the combustor, which is also usually needed to compute the convection coefficient:

$$\bar{v} = \frac{\dot{m}}{\rho A} \tag{2.26}$$

where $\dot{m}$ is the mass flow rate of gases (which is known and equal to the sum of the incoming fuel + oxidizer flow rates) and A is the cross-sectional area of the combustor. The gas density is inversely proportional to the gas temperature so that as the temperature increases, the density decreases.

Figure 2.24 shows that the gas density declines approximately proportional to the inverse of the absolute gas temperature. Figure 2.25 shows how the gas density decreases rapidly as the O_2 content in the oxidizer increases, for all three fuels shown. This is a consequence of the increased flame temperatures, as shown in Figure 2.11. A lower gas density means a lower Reynolds number and therefore reduced convective heat transfer, if all other variables remain the same. However, the mass flow rate of gases is also decreasing, as shown in Figure 2.23. Therefore, the average gas

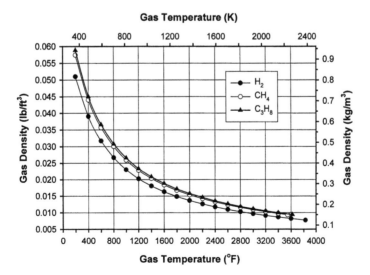

FIGURE 2.24 Calculated gas density for equilibrium stoichiometric flames as a function of the combustion product temperature and fuel composition.

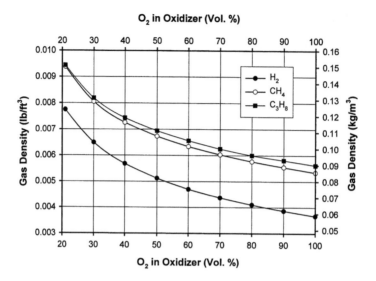

FIGURE 2.25   Calculated gas density for adiabatic equilibrium stoichiometric flames as a function of the oxidizer (O_2 + N_2) and fuel compositions.

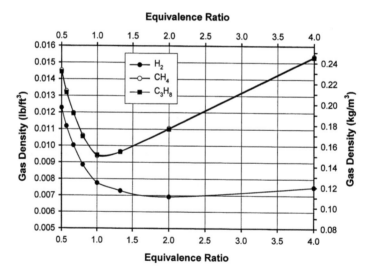

FIGURE 2.26   Calculated gas density for adiabatic equilibrium air/fuel flames as a function of the equivalence ratio and fuel composition.

velocity is not significantly impacted as a result of the combined effect of lower density and lower mass flow rate so that the impact on convection due to gas velocity is minimal. Figure 2.26 shows that the gas density reaches a minimum at intermediate equivalence ratios. This again can be attributed to the adiabatic equilibrium flame temperature as shown in Figure 2.14. Figure 2.27 shows that the gas density decreases nearly linearly as the air preheat temperature increases, which correlates inversely with the curves for the flame temperature in Figure 2.16. Figure 2.28 shows that the gas density does not decrease linearly as a function of the gas blend composition, as might be expected intuitively. Again, the density corresponds inversely to the adiabatic flame temperatures shown in Figure 2.12.

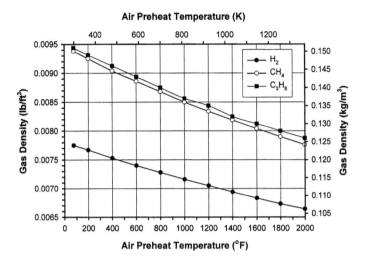

FIGURE 2.27 Calculated gas density for adiabatic equilibrium stoichiometric air/fuel flames as a function of the air preheat temperature and fuel composition.

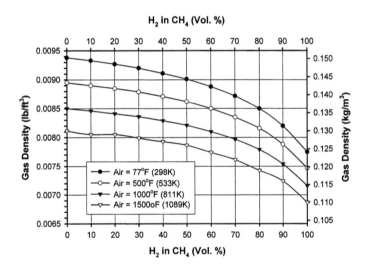

FIGURE 2.28 Calculated gas density for adiabatic equilibrium stoichiometric air/fuel flames as a function of the fuel blend composition and air preheat temperature.

2.3.2 Specific Heat

The gas specific heat (c_p), sometimes referred to as the gas heat capacity, is another transport property that has an impact on the convective heat transfer in a system. It is used to calculate the Prandtl number ($\mathrm{Pr} = c_p\mu/k$), which is often used to calculate the convective heat transfer coefficient (h).

Figure 2.29 shows a nonlinear increase in the gas specific heat as a function of the exhaust product temperature. The specific heat increases more rapidly at higher temperatures. Figure 2.30 shows that the exhaust gas specific heat increases almost linearly as the O_2 purity in the oxidizer increases for all three fuels considered. All other things being the same, this would enhance the convective heat transfer from the combustion product gases to the load. Figure 2.31 shows a much more complicated relationship between the specific heat and the equivalence ratio, including a strong fuel dependence as well. All three fuels show an initial increase in the specific heat as the

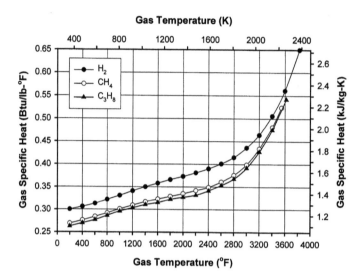

FIGURE 2.29 Calculated gas specific heat for equilibirium stoichiometric flames as a function of the combustion product temperature and fuel composition.

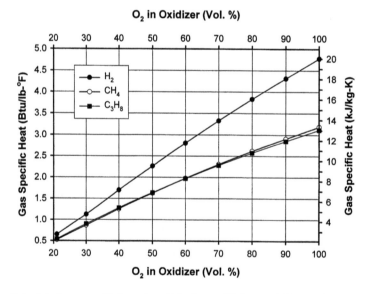

FIGURE 2.30 Calculated gas specific heat for adiabatic equilibrium stoichiometric flames as a function of the oxidizer ($O_2 + N_2$) and fuel compositions.

equivalence ratio increases, reaching a local maximum at stoichiometric conditions ($\phi = 1.0$). Beyond stoichiometric conditions, the specific heat then decreases, plateaus, and increases again. In the case of CH_4, the specific heat increases very rapidly at high equivalence values. Although the relationship between specific heat and equivalence ratio is fairly complicated, the reality is that most industrial combustion processes are operated at slightly fuel-lean conditions where there is a strong but more linear relationship between the equivalence ratio and the specific heat. Figure 2.32 shows an almost linear dependence of specific heat on air preheat temperature for all three fuels. Figure 2.33 shows that the specific heat increases rapidly at high H_2 contents in H_2/CH_4 fuel blends. Intuitively, one might expect a linear relationship; but as can be seen in, the flame temperature shows a very similar relationship to the H_2 content in the blend.

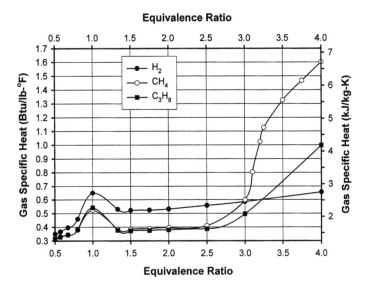

FIGURE 2.31 Calculated gas specific heat for adiabatic equilibrium air/fuel flames as a function of the equivalence ratio and fuel composition.

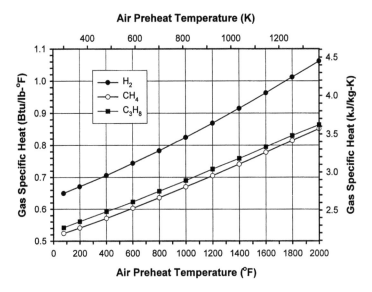

FIGURE 2.32 Calculated gas specific heat for adiabatic equilibrium stoichiometric air/fuel flames as a function of the air preheat temperature and fuel composition.

2.3.3 THERMAL CONDUCTIVITY

Like the specific heat, the gas thermal conductivity (k) affects the Prandtl number, which in turn affects the convective heat transfer coefficient. In this case, there is an inverse relationship between the thermal conductivity and the Prandtl number. As the thermal conductivity increases (decreases), the Prandtl number decreases (increases) along with the convection coefficient, assuming all other variables remain constant. The thermal conductivity of a gas is approximately dependent on $\sqrt{T}$, where T is the absolute temperature.[13]

Figure 2.34 shows a similar nonlinear increase in thermal conductivity with gas temperature as for the specific heat. Figure 2.35 shows that the thermal conductivity increases rapidly as the

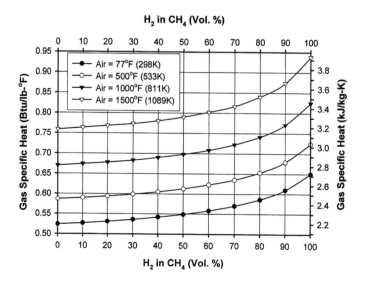

FIGURE 2.33 Calculated gas specific heat for adiabatic equilibrium stoichiometric air/fuel flames as a function of the fuel blend composition and air preheat temperature.

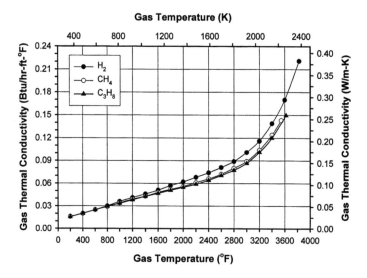

FIGURE 2.34 Calculated gas thermal conductivity for equilibrium stoichiometric flames as a function of the combustion product temperature and fuel composition.

O_2 content in the oxidizer increases. The relationship is almost linear although there is a faster increase at lower O_2 contents compared to that at higher O_2 oxidizers. Figure 2.36 again shows a complicated relationship between a transport property and the equivalence ratio. There is a local maxima at stoichiometric conditions ($\phi = 1.0$). For H_2, the local maxima also is the overall maximum for the range of equivalence ratios computed. For CH_4, there is a rapid increase in the thermal conductivity at very fuel-rich conditions (high equivalence ratios), with the conductivity exceeding the local maxima at stoichiometric conditions. Although not as dramatic, there is a similar phenomenon for C_3H_8. Although most industrial processes are run at slightly fuel-lean conditions, there is still a rapid change in thermal conductivity on the fuel-lean side of stoichiometric conditions. Figure 2.37 shows a much simpler relationship between conductivity and combustion air preheat temperature. The conductivity increases slightly faster than linearly as the preheat temperature

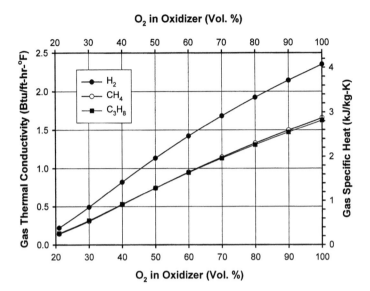

FIGURE 2.35 Calculated gas thermal conductivity for adiabatic equilibrium stoichiometric flames as a function of the oxidizer ($O_2 + N_2$) and fuel compositions.

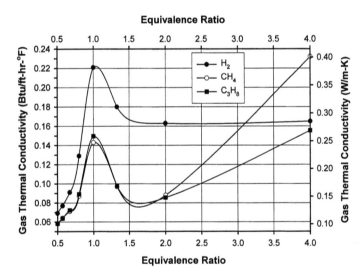

FIGURE 2.36 Calculated gas thermal conductivity for adiabatic equilibrium air/fuel flames as a function of the equivalence ratio and fuel composition.

increases. As previously shown for other transport properties, Figure 2.38 shows that the thermal conductivity increases much more rapidly as the H_2 content in the H_2/CH_4 fuel blend increases.

2.3.4 VISCOSITY

The absolute or dynamic viscosity μ is a measure of momentum diffusion. Gas viscosity has a similar relationship to $\sqrt{T}$ as does thermal conductivity.[13] The viscosity is important in calculating both the Prandtl and Reynolds numbers, but in opposite ways. As the gas viscosity increases (decreases), the Prandtl number increases (decreases) and the Reynolds number decreases (increases), assuming all other variables are constant. The kinematic viscosity ν is related to the dynamic viscosity as follows:

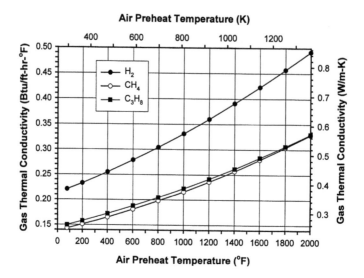

FIGURE 2.37 Calculated gas thermal conductivity for adiabatic equilibrium stoichiometric air/fuel flames as a function of the air preheat temperature and fuel composition.

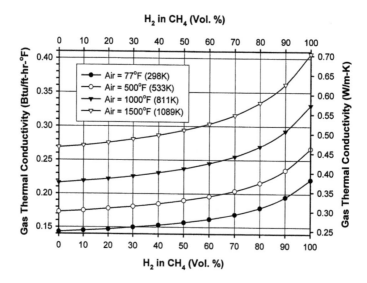

FIGURE 2.38 Calculated gas thermal conductivity for adiabatic equilibrium stoichiometric air/fuel flames as a function of the fuel blend composition and air preheat temperature.

$$\nu = \frac{\mu}{\rho} \tag{2.27}$$

and is sometimes used to define the Reynolds number:

$$Re = \frac{\nu d}{\nu} \tag{2.28}$$

where ν is the fluid velocity and d is some characteristic dimension for the system under investigation.

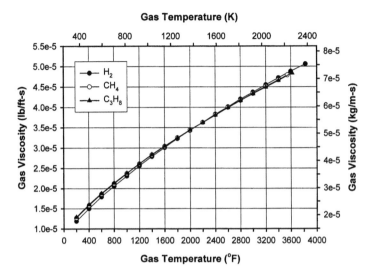

FIGURE 2.39 Calculated gas viscosity for equilibrium stoichiometric flames as a function of the combustion product temperature and fuel composition.

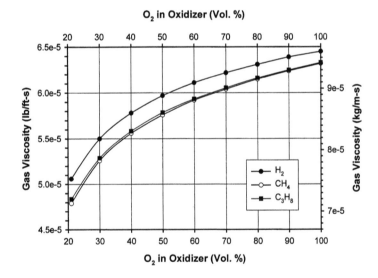

FIGURE 2.40 Calculated gas viscosity for adiabatic equilibrium stoichiometric flames as a function of the oxidizer (O_2 + N_2) and fuel compositions.

Figure 2.39 shows a nearly linear increase in gas viscosity as a function of the exhaust product temperature. Figure 2.40 shows that the gas viscosity increases as the O_2 content in the oxidizer increases, similar to the adiabatic flame temperature as shown in Figure 2.11. Figure 2.41 shows that the gas viscosity peaks at an equivalence ratio of 1.0 (stoichiometric conditions) and declines as the mixture becomes either more fuel rich or more fuel lean. Again, this compares to the adiabatic flame temperature as a function of equivalence ratio as shown in Figure 2.14. As shown in Figure 2.42, the gas viscosity increases with the air preheat temperature, comparable to the flame temperature as shown in Figure 2.16. Figure 2.43 shows that the viscosity increases as the H_2 content increases in an H_2/CH_4 fuel blend. Just as for the adiabatic flame temperature shown in, the viscosity increases more rapidly at higher H_2 contents.

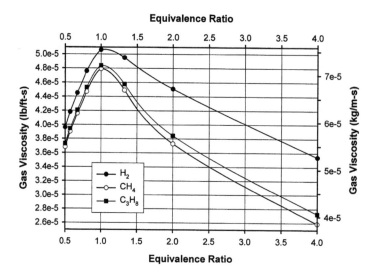

FIGURE 2.41 Calculated gas viscosity for adiabatic equilibrium air/fuel flames as a function of the equivalence ratio and fuel composition.

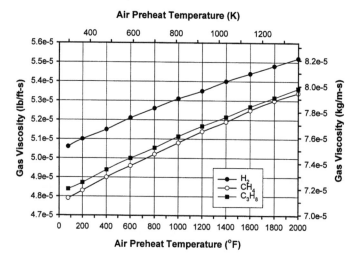

FIGURE 2.42 Calculated gas viscosity for adiabatic equilibrium stoichiometric air/fuel flames as a function of the air preheat temperature and fuel composition.

2.3.5 PRANDTL NUMBER

As was previously shown, the Prandtl number is often used to calculate the convection heat transfer coefficient. The components of Pr, including the specific heat, viscosity, and thermal conductivity, were discussed above. This chapter section shows how the combination of these variables which forms the Prandtl number changes as functions of the fuel and oxidizer compositions, the mixture ratio, and the air preheat temperature. Since there is little change in Pr as a function of the fuel preheat temperature, that is not included here but the data are given in Appendix D.

Figure 2.44 shows that the Prandtl number decreases as a function of temperature, but in a nonuniform way. Initially, it decreases moderately quickly, then decreases more slowly, and finally

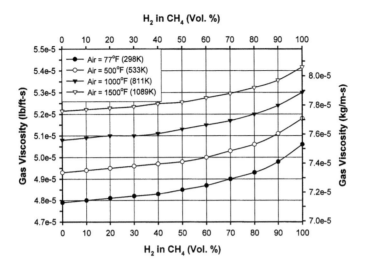

FIGURE 2.43 Calculated gas viscosity for adiabatic equilibrium stoichiometric air/fuel flames as a function of the fuel blend composition and air preheat temperature.

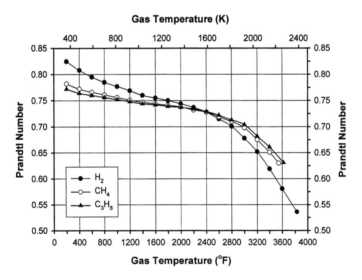

FIGURE 2.44 Calculated Prandtl number for equilibrium stoichiometric flames as a function of the combustion product temperature and fuel composition.

decreases rapidly at higher temperatures. Figure 2.45 also shows a highly nonlinear relationship between the Prandtl number and the oxidizer composition. For CH_4 and C_3H_8, the Prandtl number decreases rapidly at first and then levels off at higher O_2 contents. For H_2, the Prandtl number actually has a minima at about 50% O_2 content. Figure 2.46 shows a highly nonlinear relationship between the Prandtl number and the equivalence ratio. All three fuels show local maxima and minima at $1.0 < \phi < 2.0$. In Figure 2.47, the Prandtl number declines almost linearly with the air preheat temperature. Figure 2.48 shows that the Prandtl number declines as the H_2 content in an H_2/CH_4 fuel blend decreases, and decreases rapidly at high H_2 contents.

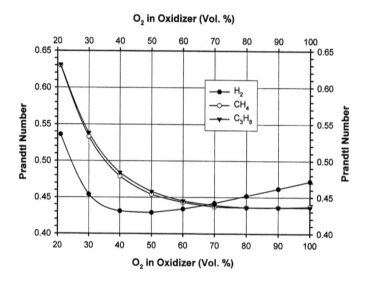

FIGURE 2.45 Calculated Prandtl number for adiabatic equilibrium stoichiometric flames as a function of the oxidizer ($O_2 + N_2$) and fuel compositions.

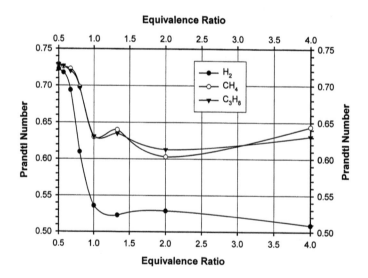

FIGURE 2.46 Calculated Prandtl number for adiabatic equilibrium air/fuel flames as a function of the equivalence ratio and fuel composition.

2.3.6 LEWIS NUMBER

The Lewis number is the ratio of the thermal diffusivity to the molecular (mass) diffusivity and is defined as:

$$Le = \frac{k/\rho c_p}{D} \tag{2.29}$$

where D is the mass diffusivity for the fluid. The importance of the Lewis number to heat transfer in combustion systems is considered in more detail in Chapter 4. In general, for $Le > 1$, there are some enhancements in convective heat transfer due to chemical recombination reactions.

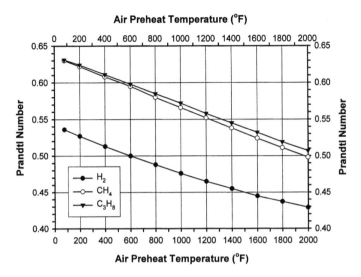

FIGURE 2.47 Calculated Prandtl number for adiabatic equilibrium stoichiometric air/fuel flames as a function of the air preheat temperature and fuel composition.

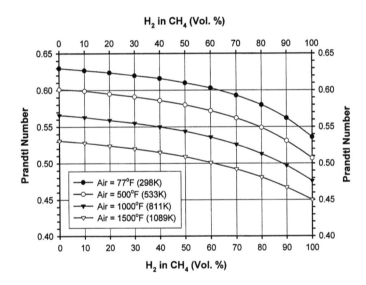

FIGURE 2.48 Calculated Prandtl number for adiabatic equilibrium stoichiometric air/fuel flames as a function of the fuel blend composition and air preheat temperature.

Figure 2.49 shows that the Lewis number is 1 for temperatures below about 2300°F (1500K), depending on the fuel, and then rises fairly rapidly at higher temperatures. Figure 2.50 shows that the Lewis number is greater than unity for all oxidizer compositions under adiabatic equilibrium conditions, which equates to the highest flame temperature possible for those conditions. The shape of the curves is interesting in that Le peaks at intermediate oxidizer compositions and actually declines at higher O_2 contents. Figure 2.51 shows a dramatic peak in the Lewis number at stoichiometric conditions, with the Lewis number actually going below 1.0 at higher equivalence ratios. Figure 2.52 shows that the Lewis number increases almost linearly with the air preheat temperature for adiabatic equilibrium conditions. The Lewis number increases more rapidly as the H_2 content in a fuel blend of H_2/CH_4 increases, as shown in Figure 2.53.

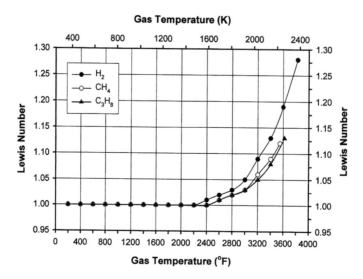

FIGURE 2.49   Calculated Lewis number for equilibrium stoichiometric flames as a function of the combustion product temperature and fuel composition.

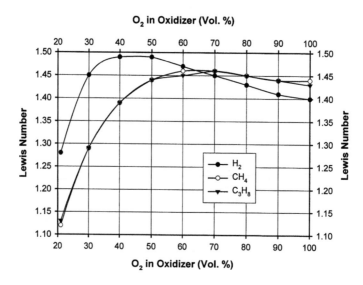

FIGURE 2.50   Calculated Lewis number for adiabatic equilibrium stoichiometric flames as a function of the oxidizer ($O_2 + N_2$) and fuel compositions.

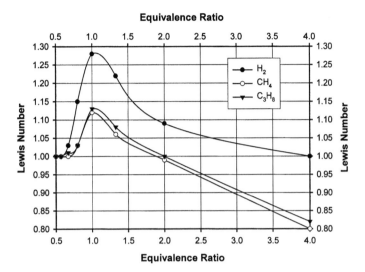

FIGURE 2.51 Calculated Lewis number for adiabatic equilibrium air/fuel flames as a function of the equivalence ratio and fuel composition.

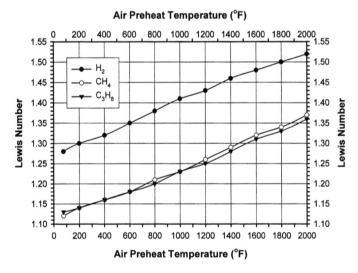

FIGURE 2.52 Calculated Lewis number for adiabatic equilibrium stoichiometric air/fuel flames as a function of the air preheat temperature and fuel composition.

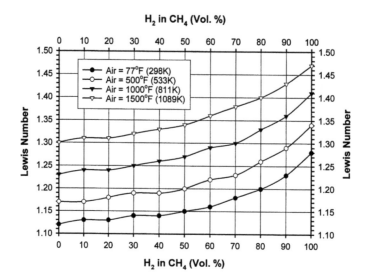

FIGURE 2.53 Calculated Lewis number for adiabatic equilibrium stoichiometric air/fuel flames as a function of the fuel blend composition and air preheat temperature.

REFERENCES

1. C. Baukal, Ed., *Oxygen-Enhanced Combustion*, CRC Press, Boca Raton, FL, 1998.
2. A.I. Dalton and D.W. Tyndall, Oxygen Enriched Air/Natural Gas Burner System Development, NTIS Report PB91-167510, Springfield, VA, 1989.
3. I. Glassman, *Combustion*, 3rd edition, Academic Press, New York, 1996.
4. S. Gordon and B.J. McBride, Computer Program for Calculation of Complex Chemical Equilibrium Compositions, Rocket Performance, Incident and Reflected Shocks, and Chapman-Jouguet Detonations, NASA Report SP-273, 1971.
5. J.L. Reese, G.L. Moilanen, R. Borkowicz, C. Baukal, D. Czerniak, and R. Batten, State-of-the-Art of NO_x emission control technology, ASME Paper 94-JPGC-EC-15, *Proceedings of Int. Joint Power Generation Conf.*, Phoenix, AZ, 3-5 October, 1994.
6. S. Gordon, B.J. McBride, and F.J. Zeleznik, Computer Program for Calculation of Complex Chemical Equilibrium Compositions and Applications. Supplement I. Transport Properties, NASA Technical Memorandum 86885, Washington, D.C., 1984.
7. R.J. Kee, G. Dixon-Lewis, J. Warnatz, M.E. Coltrin, and J.A. Miller, A Fortran Computer Package for the Evaluation of Gas-Phase, Multicomponent Transport Properties, Sandia National Laboratory Report SAND86-8246, Livermore, CA, 1986.
8. R.B. Bird, W.E. Stewart, and E.N. Lightfoot, *Transport Phenomena*, Wiley, New York, 1960.
9. N.V. Tsederberg, *Thermal Conductivity of Gases and Liquids*, MIT Press, Cambridge, MA, 1965.
10. P.E. Liley, R.C. Reid, and E. Buck, Physical and Chemical Data, Section 3 in *Perry's Chemical Engineers' Handbook*, 6th edition, R.H. Perry and D. Green, Eds., McGraw-Hill, New York, 1984.
11. R.C. Reid, J.M. Prausnitz, and B.E. Poling, *The Properties of Gases and Liquids*, 4th edition, McGraw-Hill, New York, 1987.
12. D.R. Lide, Ed., *CRC Handbook of Chemistry and Physics*, 79th edition, CRC Press, Boca Raton, FL, 1998.
13. B. Gebhart, *Heat Transfer*, 2nd edition, McGraw-Hill, New York, 1971.

3   Heat Transfer Modes

3.1   INTRODUCTION

There are some nondimensional numbers that are commonly used in heat transfer analysis. The Reynolds number (Re) is the ratio of the inertial forces to the viscous forces in a flow:

$$\text{Re} = \frac{\rho v l}{\mu} = \frac{vl}{\nu} \tag{3.1}$$

where ρ is the fluid density, v is the fluid velocity, l is the characteristic length scale (e.g., diameter for a pipe for flow through a pipe), μ is the absolute viscosity, and ν is the kinematic viscosity. The Reynolds number is low for laminar flows and high for turbulent flows, with transition flow at values in between. The actual range for each type of flow depends on the flow geometry. The Prandtl number (Pr) is the ratio of momentum diffusivity to thermal diffusivity and is defined as:

$$\text{Pr} = \frac{c_p \mu}{k} \tag{3.2}$$

where c_p is the fluid constant pressure specific heat and k is the fluid thermal conductivity. For many gases, $\text{Pr} \approx 0.7$. The Nusselt number (Nu) is the ratio of the convective and conductive heat transfer rates:

$$\text{Nu} = \frac{hl}{k} \tag{3.3}$$

where h is the convective heat transfer coefficient. In forced convection flows, Nu is commonly a function of Pr and Re and is used to determine the convection coefficient.

3.2   CONVECTION

Convection heat transfer is caused by fluid motion past a material, where the fluid is either at a higher or lower temperature than the material. In industrial combustion applications, the fluid is usually at a higher temperature than the medium it is heating. At least one person has argued that convection is not actually a separate mode of heat transfer, but that it is a subset of conduction because the energy must still conduct from the fluid to the material.[1] While that may true on a microscopic scale in the boundary layer next to the material, convection is a fundamentally different process from conduction and is treated here as such, which is the convention in standard heat transfer texts. Forced convection is often a very important mode of heat transfer in industrial combustion systems. In limited applications, natural convection may also be important because of high-temperature gradients that may exist. Forced and natural convection may be important in industrial combustion and are briefly considered next. A cartoon of forced and natural convection is shown in Figure 7.3. A number of books are available specifically on convection heat transfer.[2-10]

3.2.1 FORCED CONVECTION

Forced convection heat transfer occurs when a fluid is forcefully directed at or over a medium (liquid or solid) and may be simply calculated using:

$$q = hA\left(t_f - t_m\right) \tag{3.4}$$

where q is the heat flux to the medium (Btu/hr or kW), h is the convective heat transfer coefficient (Btu/ft²-hr-°F or W/m²-K), A is the surface area of the medium in contact with the moving fluid, t_f is the fluid temperature (°F or K), and t_m is the temperature (°F or K) of the medium. A simple example is given below.

Example 3.1

Given: $h = 10$ Btu/ft²-hr-°F, $A = 10$ ft², $t_f = 300°F$, and $t_m = 70°F$.

Find: q

Solution: $q = h\,A\,(t_m - t_f) = (10$ Btu/ft²-hr-°F$)(10$ ft²$)(300°F - 70°F) = 23{,}000$ Btu/hr

This is an overly simplified example because the convection heat transfer coefficient, h, has been given. In most problems, this value must be calculated from an appropriate correlation which may be fairly complicated, or worse, may not exist for the exact configuration under study. In the latter case, either experiments need to be done to determine this value or an approximate value must be used. The correlations for the convection coefficient are often complicated functions that normally depend on the flow geometry, the fluid velocity, and the fluid properties. In addition, for most real problems, the fluid and medium temperatures vary by location and are not constant over the entire medium surface. Forced convection heat transfer from a flame to the load and inner combustor walls and from the outer combustor walls to the ambient are both potentially important mechanisms. Forced convection heat transfer in industrial combustion is often more difficult to analyze and calculate because of the large temperature difference between the combustion products and the surfaces (furnace walls and heat load) and because the gas properties vary widely as a function of both temperature and composition (see Chapter 2).

3.2.1.1   Forced Convection from Flames

In many conventional furnace heating processes, forced convection is only a small fraction of the total heat transfer to the product. Most of the heating emanates from the radiation from the hot refractory walls. However, in flame impingement, with no furnace enclosure, forced convection may be 70% to 90% of the total heat flux.[11,12] For flame temperatures up to about 2600°F (1700K), forced convection is the dominant mechanism in flame impingement heat transfer.[13]

For low-temperature flames, as is common in air/fuel combustion systems, forced convection has generally been the only mechanism considered. In highly dissociated oxygen/fuel flames, a large fraction of the heat release is from exothermic reactions. However, even for those flames, forced convection is still an important contributor to the overall heat transfer to the target.

The turbulence level directly affects the importance of forced convection. The flow regime is determined by the Reynolds number, defined in Equation 3.1. For example, for flame jet impingement (see Figure 3.1) there are many possible choices for the length l. One is the burner outlet nozzle diameter, d_n; another is the axial distance from the nozzle exit to the surface being heated, l_j; a third possibility is the width of the jet at the edge of the stagnation zone, d_j; still another is some dimension of the material being heated. For a disk or cylinder, it may be the diameter, d_b. For a plane surface, it may be the radial distance from the stagnation point, r. In one case, the width of the water cooling channel in a target used in a flame impingement study was used.[13]

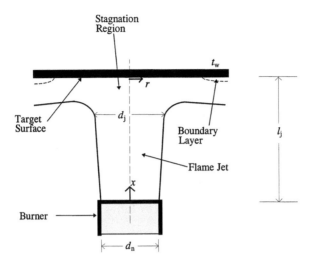

FIGURE 3.1 Flame jet impingement.

Laminar flames have been used in many flame impingement studies.[13-23] Sibulkin developed a semianalytical solution for the heat transfer for laminar flow, normal to the stagnation point of an axisymmetric, blunt-nosed target:[24]

$$q_s'' = 0.763 \left(\beta_s \rho_e \mu_e \right)^{0.5} \Pr_e^{-0.6} c_{p_e} \left(t_e - t_w \right) \tag{3.5}$$

This result has been the basis for all other semianalytical flame impingement heat transfer solutions.[25]

Turbulent flames have also been commonly used.[12,26-38] A typical example of an empirical equation, incorporating the turbulence intensity Tu, was given by Hustad[37] as

$$q_s'' = \frac{k_e}{d_b} \left\{ 0.41 \, \mathrm{Re}_{b,e}^{0.6} \Pr_e^{0.35} \mathrm{Tu}^{0.15} \left(\frac{\Pr_e}{\Pr_w} \right)^{0.25} \right\} \left(t_e - t_w \right) \tag{3.6}$$

This was developed for flames produced by jets of CH_4 and C_3H_8, into ambient air. These are known as pure diffusion ($\phi = \infty$) flames. The flames impinged normal to uncooled steel pipes. These experiments were done to simulate fires caused by ruptured fuel pipes in the petrochemical industry. Similar empirical correlations are discussed further in Chapter 7.

Babiy (1974) presented a correlation for the convective heat transfer from combustion gases to carbon particles in pulverized coal combustion processes.[39] A commonly used equation for the convective heat transfer between a gas and a sphere is given by:

$$\mathrm{Nu}_d = 2 + 0.17 \, \mathrm{Re}_d^{0.66} \tag{3.7}$$

where d is the diameter of the sphere. For pulverized coal combustion, this equation is modified as follows:

$$\mathrm{Nu}_d^* = \mathrm{Nu}_d \left[145 \exp \left(\frac{-5000}{T_g} \right) \right] \tag{3.8}$$

where T_g is the absolute temperature of the gas (K). The equation applies for T_g = 1200–1600K (1700–2400°F), O_2 = 5–21%, d = 150–1000 μm, and $\mathrm{Re}_d < 1$.

3.2.1.2   Forced Convection from Outside Combustor Wall

The second place where forced convection is often important is where heat is transferred from the hot shell of a combustor to the cooler ambient air. Griswold (1946)[40] has recommended the following empirical correlation for the natural convection coefficient for wind flowing against a surface:

$$h_c = \frac{1 + 0.225v}{\left(t_w - t_\infty\right)} \tag{3.9}$$

where h_c = convection coefficient (Btu/hr-ft^2-°F), v = air velocity, direction unspecified (ft/sec), t_w = outside wall temperature (°F), and t_∞ = ambient air temperature (°F).

3.2.1.3   Forced Convection from Hot Gases to Tubes

In the convection section of a heater, Monrad (1932)[41] recommended the following empirical correlation for the convective heat transfer from hot exhaust products to a bank of staggered tubes:

$$h_c = \frac{1.6G^{2/3}\,T_g^{0.3}}{d^{1/3}} \tag{3.10}$$

where h_c = convection coefficient (Btu/hr-ft^2-°F), G = gas mass velocity at minimum free cross-sectional area (lb/ft^2-sec), T_g = absolute gas temperature (°R), and d = tube diameter (in.).

3.2.2   NATURAL CONVECTION

Natural convection is sometimes referred to as buoyancy-induced flow.[42] There are two common situations where natural convection heat transfer may be important in industrial combustion systems. The first is transferring heat from the flame in a system where the gas velocities are very low. The second is transferring heat from the outside shell of a combustor to the environment. Each is briefly considered next.

3.2.2.1   Natural Convection from Flames

One measure of the intensity of natural convection is the Rayleigh number:

$$\mathrm{Ra} = \frac{g\tilde{\beta}_e q_f l_j^2}{\rho_e c_{p_e} v_e^3} \tag{3.11}$$

which is analogous to the Reynolds number for forced convection. Higher Rayleigh numbers indicate more natural convection. Another measure of intensity is the Grashoff number:

$$\mathrm{Gr} = \frac{g\tilde{\beta}\left(t_w - t_\infty\right)l^3}{v^2} \tag{3.12}$$

where g is the gravity constant, $\tilde{\beta}$ is the volume coefficient of expansion (= $1/T$ for an ideal gas), and v is the kinematic viscosity. The Richardson number, Ri, is one measure of the importance of buoyancy compared to forced convection. It is defined as:

$$\mathrm{Ri} = \frac{\mathrm{Gr}}{\mathrm{Re}_n^2} \tag{3.13}$$

which is the ratio of the buoyant force to the inertial force. Conolly and Davies (1972) studied stoichiometric, laminar flames impinging on a hemi-nosed cylinder.[19] The flame was parallel to the cylinder and impinging on the nose. A variety of fuels and oxidizers were used. Buoyancy effects were negligible. The criterion was that buoyancy may be neglected, compared to forced convection, for Ri < 0.05. Wang (1993) numerically modeled a nonreacting jet of ambient air, impinging on an infinite flat plate.[43] It was concluded that natural convection is important only when Re_j is low, and the temperature difference between the jet and the stagnation surface is large. The critical Ri was estimated to be approximately 0.02. You (1985)[23] determined the heat transfer from a buoyant flame to a flat plate in terms of the Rayleigh number as

$$q_b'' = 31.2\left(q_f/l_j^2\right)\mathrm{Ra}_e^{-1/6}\,\mathrm{Pr}_e^{-3/5} \qquad (3.14)$$

The flames were produced by upward jets of pure natural gas ($\phi = \infty$) into ambient air. The jets impinged on a horizontal surface to simulate a fire spreading over the ceiling of a room. Natural convection is more important in low-velocity flames. Both Beér (1968)[11] and Vizioz (1971)[27] stated that the effects of buoyancy were negligible in their studies, due to the high burner exit velocities.

3.2.2.2 Natural Convection from Outside Combustor Wall

The second place where natural convection is often important is where heat is transferred from the hot shell of a combustor to the cooler ambient air. Griswold[40] has recommended the following empirical correlation for the natural convection coefficient for quiet air:

$$h_c = \frac{0.53C\left(T_w - T_\infty\right)^{0.27}}{\left[(T_w + T_\infty)/2\right]^{0.18}} \qquad (3.15)$$

where h_c = convection coefficient (Btu/hr-ft^2-°F), C = shape constant (1.39 for a vertical wall or 1.79 for a furnace arch), T_w = outside wall temperature (°R), and T_∞ = ambient air temperature (°R).

Example 3.2

Given: $t_w = 150°F$, $t_\infty = 70°F$.

Find: Average heat loss per unit area by natural convection from a vertical surface using the above equation.

Solution: $q'' = h_c(t_w - t_\infty) = \dfrac{0.53(1.39)\left[(150+460)-(70+460)^{0.27}\right]}{\left\{[(150+460)+(70+460)]/2\right\}^{0.18}}(150-70) = 83$ Btu/hr-ft^2

3.3 RADIATION

Thermal radiation is one of the most important heat transfer mechanisms in industrial furnaces.[44] Radiation is a unique method of heat transfer as no medium is required for energy transport — it can be transmitted through a vacuum or through a medium. Radiation is simply the transmission of energy by electromagnetic waves, which are characterized by their wavelength or frequency and are related as follows:

$$\lambda = \frac{c}{v} \qquad (3.16)$$

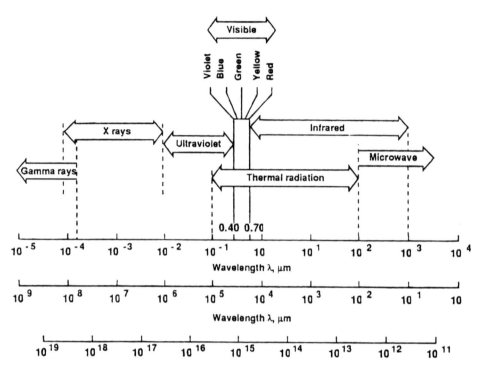

FIGURE 3.2 Electromagnetic spectrum. (Courtesy of CRC Press, Boca Raton, FL. From *International Encyclopedia of Heat & Mass Transfer,* Hewitt et al., 1997, 914.)

where λ is the wavelength, c is the speed of light, and ν is the frequency. In the fields of heat transfer and combustion, wavelength is more commonly used. The classification of the various types of radiation is shown in Figure 3.2. In industrial combustion heating, the most important type of radiation is infrared. The human optic nerves are sensitive to radiation in the wavelengths from 0.38 to 0.76 μm, which means that one can see radiation in that band. In practical terms, one cannot see thermal radiation from bodies at temperatures below about 900°F (500°C).[45]

There are four possible things that can happen to radiation incident on a medium (solid, liquid, or gas). The radiation can be absorbed, reflected, transmitted, or some combination of these three which is most often the case, as shown in Figure 3.3. In general,

$$\alpha + \rho + \tau = 1 \tag{3.17}$$

where α is the absorptivity of the medium, ρ is the reflectivity, and τ is the transmissivity, which are defined as:

$$\alpha \equiv \frac{\text{Absorbed part of incoming radiation}}{\text{Total incoming radiation}}$$

$$\rho \equiv \frac{\text{Reflected part of incoming radiation}}{\text{Total incoming radiation}}$$

$$\tau \equiv \frac{\text{Transmitted part of incoming radiation}}{\text{Total incoming radiation}}$$

For most solid materials, the transmissivity is low except for materials like glass and plastics. The reflectivity of most solids is low, unless they are highly polished (e.g., new stainless steel). For liquids, the transmissivity may be significant, especially for fluids with high water contents. For

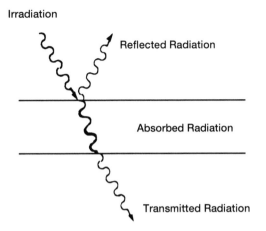

Irradiation

Reflected Radiation

Absorbed Radiation

Transmitted Radiation

FIGURE 3.3 Radiant energy absorbed, reflected and transmitted through a material. (Courtesy of CRC Press, Boca Raton, FL. From *International Encyclopedia of Heat & Mass Transfer,* Hewitt et al., 1997, 917.)

most gases, the transmissivity is generally very high with negligible absorptance and reflectance. These radiative properties are extremely important in determining how much radiation will be transferred to and from a medium. This is further complicated by the fact that these radiative properties may be functions of wavelength, angle of incidence, surface condition, and thickness. An example of wavelength dependence is solid glass, which transmits shorter wavelength UV radiation but absorbs longer wavelength IR radiation. At high angles of incidence, a surface may be more reflective; while at normal incident angles, a surface may be more absorptive. Highly oxidized metals have high absorptivities and low reflectivities, while highly polished metals have lower absorptivities and high reflectivities. A very thin layer of a solid material can have significant transmissivity, while a thick layer of the same solid may have no transmissivity. The radiative properties can also change over time, such as when an initially reflective metal surface becomes less reflective as it oxidizes. Therefore, an important challenge in computing radiative heat transfer in combustion systems is determining radiative properties.

There are three common forms of radiation heat transfer in industrial heating applications: (1) radiation from a solid surface, (2) radiation from a gaseous medium (usually referred to as nonluminous radiation), and (3) radiation from particles in a gaseous medium (usually referred to as luminous radiation). These are briefly discussed next and are also considered in other pertinent chapters elsewhere in this book. A cartoon of the three types is shown in Figure 7.5. Each of these three types is considerably different and must be treated accordingly. Also, it is not uncommon for two or all three types to be important in industrial heating. In industrial furnaces, as much as 90% of the heat transfer to the load can be by radiation.[46]

The "Total Schmidt Method" has been used to determine the amount of radiation coming from the walls and that coming from the flame, which are received by the load.[47] The flame radiation is measured by sighting the flame with a radiation pyrometer where the background behind the flame is water-cooled. The radiation from the water-cooled wall is negligible compared to that from the flame and is ignored:

$$q_1 = \sigma A \varepsilon_f T_f^{\,4} \qquad (3.18)$$

where σ is the Stefan–Boltzmann constant (see Eq. 3.28), ε_f is the average flame emissivity, A is the area radiating, and T_f is the average absolute flame temperature. Then, the radiation from the flame with a hot blackbody background (furnace wall) is measured:

$$q_2 = \sigma A \varepsilon_f T_f^{\,4} + \sigma A \left(1 - \alpha_f\right) T_w^{\,4} \qquad (3.19)$$

where α_f is the average flame absorptivity and T_w is the average wall temperature. Finally, the radiation from the hot blackbody background (furnace wall) is measured immediately after the flame is extinguished:

$$q_3 = \sigma A T_w^4 \qquad (3.20)$$

The flame is assumed to be a graybody so that:

$$\alpha_f = \varepsilon_f \qquad (3.21)$$

Solving the above four equations for the flame emissivity ε_f:

$$\varepsilon_f = 1 - \left(\frac{q_2 - q_1}{q_3} \right) \qquad (3.22)$$

and then solving for flame temperature T_f:

$$T_f = \left(\frac{q_1}{\sigma A \varepsilon_f} \right)^{1/4} \qquad (3.23)$$

Lowes and Newall (1974) note that there are some errors that arise with this method:[48]

1. The absorptivity of the flame is not equal to the emissivity unless the blackbody source is at the flame temperature.
2. The emissivities obtained assume a graybody, which is normally not strictly true.
3. Soot particles in the flame can cause radiation scattering, which may give high absorptivity values.

They showed that the error due to the blackbody not being at the flame temperature can be 42% if the blackbody temperature was 1000K and the flame temperature was 2000K. Leblanc et al. (1974) used a variation of this technique to distinguish between the radiation from the walls and from the flame, to the load.[49] An ellipsoidal radiometer was used to measure the total radiation received by the load. Then the flame was turned off and the radiation was again measured, which was the contribution by the walls. The radiation from the flame was then the difference between that total and the wall:

$$q_{rad,total} = K \, q_{rad,walls} + q_{rad,flame} \qquad (3.24)$$

where the flame radiation from the walls has been corrected by a non-absorption factor K, which is obtained experimentally. Lihou (1977) reviewed analytical techniques for calculating nonspectral radiation for use in furnace design.[50] Some general books on radiation heat transfer are available.[45,51-58]

3.3.1 Surface Radiation

The spectral or monochromatic emissive power of a blackbody is dependent on the wavelength of the radiation and on the absolute temperature of the blackbody and is calculated as follows:

$$E_{b\lambda} = \frac{C_1}{\lambda^5 \left(e^{C_2/\lambda T} - 1 \right)} \tag{3.25}$$

where $E_{b\lambda}(T)$ is the monochromatic emissive power of a blackbody in Btu/hr-ft²-µm (W/m³), $C_1 = 1.1870 \times 10^8$ Btu/µm⁴-ft²-hr (3.7415×10^{-16} W-m²), $C_2 = 2.5896 \times 10^4$ µm-°R (1.4388×10^{-2} m-K), T is the absolute temperature of the body in °R (K), and λ is the wavelength in µm (m). Figure 3.4 shows some graphs of this relation as a function of wavelength and temperature. As can be seen, the peak radiation shifts to shorter wavelengths as the body temperature increases. That wavelength can be computed from Wein's displacement law as follows:

$$\lambda_{max} = \frac{c}{T} \tag{3.26}$$

where $c = 5215.6$ µm-°R (2897.6 µm-K).

Example 3.3

Given: Blackbody temperature = 2000°F.

Find: Wavelength for peak radiation.

Solution: $\lambda_{max} = \dfrac{c}{T} = \dfrac{5215.6\mu - °R}{(2000 + 460)°R} = 2.12\ \mu m$

This wavelength may be important for certain types of heat loads that selectively absorb radiation at preferred wavelengths. In those cases, to maximize the radiant efficiency, the heat source temperature should be coupled to the preferred wavelength of the load. Note that this will not necessarily maximize the radiation to the load as higher temperatures produce more radiation and more energy may be absorbed by the load even if the spectral absorptivity is lower because of the higher radiant power density, as will be shown in a later example.

The integrated hemispherical emissive power of a blackbody, computed over all wavelengths, is a function only of its absolute temperature and is described by:

$$e_b = \sigma T^4 \tag{3.27a}$$

$$E_b = \sigma A T^4 \tag{3.27b}$$

where A is the area (ft² or m²), T is the absolute temperature of the body (°R or K), and σ is the Stefan-Boltzmann constant, which has the value of:

$$\sigma = 0.1714 \times 10^{-8}\ \text{Btu}/\text{hr-ft}^2\text{-}°R^4 \left(5.669 \times 10^{-8}\ \text{W}/\text{m}^2\text{-}K^4 \right) \tag{3.28}$$

Example 3.4

Given: Surface temperature = 2000°F.

Find: Blackbody emissive power.

Solution: $e_b = \sigma T^4 = (0.1714 \times 10^{-8}\ \text{Btu/hr-ft}^2\text{-}°R^4)(2000 + 460°R)^4 = 62,800$ Btu/hr-ft²

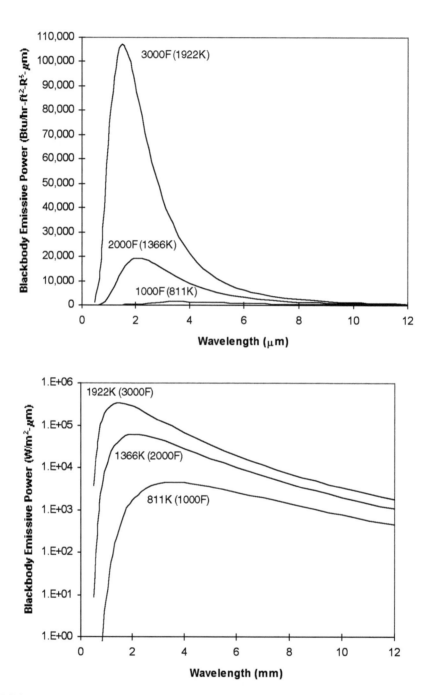

FIGURE 3.4 Blackbody emissive power in A: English units, linear scale, and B: metric units, logarithmic scale.

If all the radiation received at a solid surface is absorbed by that surface, the body is referred to as a *blackbody* and has an absorptivity of one ($\alpha = 1$). In addition, a blackbody not only absorbs all incident radiation, but it also emits the maximum amount of energy possible for the given temperature. The emissivity (ε) of a medium is the fraction of energy a body emits (E) for a given temperature, compared to the amount it could emit (E_b):

$$\varepsilon = \frac{E}{E_b} \tag{3.29}$$

Then, for a blackbody, $E = E_b$ and $\varepsilon = 1$. Real bodies are typically not perfect blackbodies and do not absorb all the incident energy received on their surfaces or emit the maximum amount of energy possible. The radiant heat transfer absorbed by real surfaces is a function of the absorptivity of the surface and the total radiant energy emitted by a body can be calculated from:

$$E = \varepsilon \sigma A T^4 \tag{3.30}$$

Example 3.5

Given: Surface temperature = 2000°F, $\varepsilon = 0.4$, $A = 10$ ft².

Find: Total radiant energy emitted by the body.

Solution: $E_b = \varepsilon \sigma A T^4 = (0.4)(0.1714 \times 10^{-8}$ Btu/hr-ft²-°R⁴$)(10$ ft²$)(2000 + 460°$R$)^4 =$ 251,000 Btu/hr

Thus, a surface with an emissivity of 0.4 will emit 40% of the energy that a blackbody would emit at the same temperature.

The above equation for E_b assumes that the emissivity is a constant value. The absorptivity and emissivity of some surfaces are a function of the temperature and the wavelength of radiation. Kirchoff's law states that:

$$\varepsilon(\lambda, T) = \alpha(\lambda, T) \tag{3.31}$$

so that at equilibrium conditions, a surface absorbs and emits the same amount of radiation. This assumes that the surface condition remains the same. If the absorptivity and emissivity are independent of wavelength, the surface is said to be a *graybody*. In engineering calculations, most surfaces can be treated as graybodies. The spectral emissive power can be written as:

$$e_\lambda = \varepsilon_\lambda \sigma T^4 \tag{3.32}$$

The following example shows a comparison of the radiation absorbed as a function of wavelength for a load whose emissivity varies with wavelength.

Example 3.6

Given: Surface emissivities: $\varepsilon_{2\mu m} = 0.4$, $\varepsilon_{3\mu m} = 0.8$.

Find: Radiant energy absorbed for heat sources matched to these wavelengths.

Solution: Find temperature whose blackbody curve peaks at the given wavelengths:

$$T_1 = \frac{5215.6 \ \mu m \cdot °R}{2 \ \mu m} = 2608°R, \text{ and } T_2 = \frac{5215.6 \ \mu m \cdot °R}{3 \ \mu m} = 1739°R$$

Then calculate the radiation that would be absorbed at those temperatures at the given wavelength:

$$e_{2\mu m} = \varepsilon_{2\mu m}\, \sigma T_1^4 = (0.4)\big(0.1714 \times 10^{-8}\ \text{Btu/hr-ft}^2\text{-}^\circ\text{R}^4\big)(2608^\circ\text{R})^4$$

$$= 31,700\ \text{Btu/hr-ft}^2$$

$$e_{3\mu m} = \varepsilon_{3\mu m}\, \sigma T_2^4 = (0.8)\big(0.1714 \times 10^{-8}\ \text{Btu/hr-ft}^2\text{-}^\circ\text{R}^4\big)(1739^\circ\text{R})^4$$

$$= 12,500\ \text{Btu/hr-ft}^2$$

As can be seen in this example, the spectral radiant efficiency is higher at 3 μm with a source temperature of 1739°R because 80% of the radiant energy is absorbed ($\varepsilon_{3\,\mu m} = 0.8$). However, much more energy is absorbed at the higher source temperature, although it is less efficient. Figure 3.5 shows an example of the spectral emissivity of some materials. Figure 3.6 shows that some materials may have a wide a range of total emissivities; and Figure 3.7 shows how the total normal emissivity varies with temperature. Table 3.1 gives a list of normal total emissivities for various surfaces as a function of temperature. Table 3.2 shows total emissivities for refractory materials commonly used in furnaces, as a function of temperature.[59] Glinkov (1974) gave several examples of how the emissivity varies with temperature and wavelength for different solids, including refractories, copper, brass, steel and steel alloys, and open-hearth slags.[60]

The emissivity of the refractory used in a combustor is important for determining the surface radiation heat transfer between the walls, the load, and the flame. The heat transfer in many industrial combustion processes is dominated by radiation from the hot refractory walls. Docherty and Tucker (1986) numerically studied the influence of wall emissivity on furnace performance.[61] The predictions showed that fuel consumption decreases as furnace wall emissivity increases, although there is no effect when the furnace atmosphere is gray. Transient furnace operation or poor wall insulation reduces the beneficial effects of high wall emissivity. Elliston et al. (1987) analytically and numerically showed that wall surface emissivity has only a negligible impact on heat transfer to a load inside a furnace.[62] In a study sponsored by the Gas Research Institute (Chicago, IL), it was determined that the total normal emittances of common, commercially available refractories — including dense insulating firebrick and porous ceramic fiber — ranged from 0.3 to 0.7 at 1800°F (1300K).[63] The emissivity of so-called high-emittance coatings was also measured and it was found

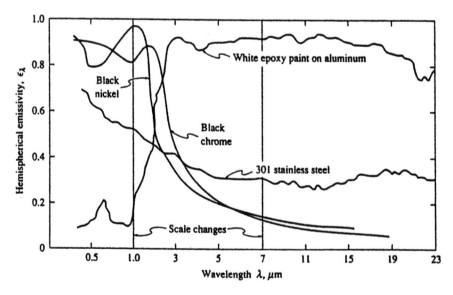

FIGURE 3.5   Spectral, hemispherical emissivities of several spectrally selective surfaces. (Courtesy of CRC Press, Boca Raton, FL. From *The CRC Handbook of Mechanical Engineering*, F. Kreith, Ed., 1998, 4-61.)

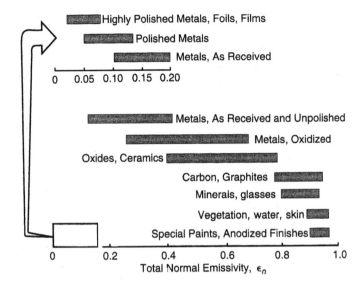

FIGURE 3.6 Total emissivity ranges for various materials. (Courtesy of CRC Press, Boca Raton, FL.)

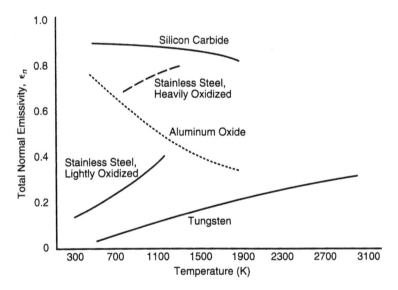

FIGURE 3.7 Total emissivity ranges vor various materials as a function of temperature. (Courtesy of CRC Press, Boca Raton, FL.)

that few of the coatings actually had high emissivities. Those coatings could extend the range of refractory emittances from 0.3 to 0.9 at 1800°F (1300K). Another part of that study investigated the effect of using high-emittance coatings in large industrial furnaces.[64] Testing was done on a 50-ton/hour (45 m-ton/hr) reheat furnace. Experimental results showed no significant change in furnace efficiency (at most, a 6% increase) using the highest available emissivity coatings on the refractory walls of the furnace. Another study showed that high-emissivity coatings applied to hundreds of furnaces in metallurgical, petrochemical, ceramic, mechanical, and other industries in the Peoples Republic of China showed energy savings of 5% to 10%, up to a maximum of 28%.[65] The reported emissivity of the coating ranged from 0.80 to 0.92, depending on the surface temperature and wavelength.

TABLE 3.1
Normal Total Emissivities of Various Surfaces as a Function of Temperature

Surface	T (°C)	Emissivity
A. Metals and their oxides		
Aluminum		
Polished	100	0.095
Commercial sheet	100	0.09
Heavily oxidized	95–505	0.20–0.31
Brass		
Highly polished		
73.2 Cu, 26.7 Zn	245–355	0.028–0.031
62.4 Cu, 36.8 Zn. 0.4 Pb, 0.3 Al	255–375	0.033–0.037
82.9 Cu, 17.0 Zn	275	0.030
Rolled plate, natural surface	20	0.06
Dull plate	50–350	0.22
Oxidized by heating at 600°C	200–600	0.61–0.59
Copper		
Carefully polished electrolytic copper	80	0.018
Commercial emeried, polished, but pits remaining	20	0.030
Commercial, scraped shiny, but not mirrorlike	20	0.072
Plate, heated for a long time, covered with thick oxide layer	25	0.78
Molten copper	1075–1275	0.16–0.13
Gold		
Pure, highly polished	225–625	0.018–0.035
Iron and steel (not including stainless)		
Electrolytic iron, highly polished	175–225	0.052–0.064
Steel, polished	100	0.066
Iron, polished	425–1025	0.14–0.38
Cast iron, polished	200	0.21
Cast iron, newly turned	20	0.44
Cast iron, turned and heated	880–988	0.60–0.70
Mild steel, cleaned with organic solvents	25	0.12, 0.15
Oxidized surfaces		
Iron plate, pickled, then rusted red	20	0.61
Iron plate, completely rusted	20	0.69
Rolled sheet steel	20	0.66
Cast iron, oxidized at 600°C	200–600	0.64–0.78
Steel, oxidized at 600°C	200–600	0.79
Sheet steel		
With rough oxide layer	25	0.80
With shiny oxide layer	25	0.82
Steel plate, rough	40–370	0.94–0.97
Molten surfaces		
Cast iron	1300–1400	0.29
Mild steel	1600–1800	0.28
Lead		
Pure (99.6%) unoxidized	125–225	0.057–0.075
Oxidized at 150°C	200	0.63
Mercury	0–100	0.09–0.12
Nickel		
Electroplated, polished	25	0.045
Polished	100	0.072

TABLE 3.1 (continued)
Normal Total Emissivities of Various Surfaces as a Function of Temperature

Surface	T (°C)	Emissivity
Electroplated, not polished	20	0.11
Plate, oxidized by heating at 600°C	200–600	0.37–0.48
Nickel alloys		
Copper-nickel, polished	100	0.059
Nichrome wire, bright	50–1000	0.65–0.79
Platinum		
Pure polished plate	225–625	0.054–0.104
Silver		
Polished, pure	225–625	0.020–0.032
Stainless steels		
Type 304 (8 Cr; 18 Ni)		
Polished	100	0.074
Light silvery, rough, brown, after heating	215–490	0.44–0.36
After 42 h at 525°C	215–525	0.62–0.73
Thorium oxide	275–500	0.58–0.36
Thorium oxide	500–825	0.36–0.21
Tin		
Bright	50	0.06
Commercial tin-plated steel iron	100	0.07, 0.08
Zinc		
Commercial 99.1% pure, polished	225–325	0.045–0.053
Oxidized by heating at 400°C	400	0.11
Galvanized sheet iron, fairly bright	30	0.23
Galvanized sheet iron, gray oxidized	25	0.28

B. Refractories, Building Materials, Paints, and Miscellaneous Materials

Surface	T (°C)	Emissivity
Alumina (99.5–85 Al_2O_3; 0–12 SiO_2; 0–1 Fe_2O_3)		
Effect of mean grain size		
10 μm	1010–1565	0.30–0.18
50 μm		0.39–0.28
100 μm		0.50–0.40
Alumina-silica (showing effect of Fe)		
80–58 Al_2O_3; 16–38 SiO_2; 0.4 Fe_2O_3	1010–1570	0.61–0.43
36–26 Al_2O_3; 50–60 SiO_2; 1.7 Fe_2O_3		0.73–0.62
61 Al_2O_3; 35 SiO; 2.9 Fe_2O_3		0.78–0.68
Asbestos		
Board	25	0.96
Paper	40–370	0.93–0.94
Brick		
Red, rough, but no gross irregularities	20	0.93
Building	1000	0.45
Fireclay	1000	0.75
Carbon		
Filament	1040–1405	0.526
Graphitized	100–320	0.76–0.75
Graphitized	320–500	0.75–0.71
Thin layer on iron plate	20	0.927
Thick coat	20	0.967

TABLE 3.1 (continued)
Normal Total Emissivities of Various Surfaces as a Function of Temperature

Surface	T (°C)	Emissivity
Glass		
Smooth	20	0.94
Magnesite refractory brick	1000	0.38
Paints, lacquers, varnishes		
Snow-white enamel varnish on rough iron plate		
Black shiny lacquer, sprayed on iron	25	0.875
Radiator paint; white, cream, bleach	100	0.79, 0.77, 0.84
Plaster, rough lime	10–90	0.91
Quartz		
Rough, fused	20	0.93
Glass, 1.98 mm thick	280–840	0.90–0.41
Glass, 6.88 mm thick	280–840	0.93–0.47
Silica (98 SiO_2; Fe-free), effect of grain size		
10 μm	1010–1565	0.42–0.33
70–600 μm		0.62–0.46
Water	0–100	0.95–0.963

Source: Courtesy of CRC Press, Boca Raton, FL.

TABLE 3.2
Normal Total Emissivities of Refractories as a Function of Temperature

Material	Temperature, °F (°C)						
	200 (93)	400 (200)	800 (430)	1600 (870)	2000 (1090)	2400 (1320)	2800 (1540)
Fireclay brick	0.90	0.90	0.90	0.81	0.76	0.72	0.68
Silica brick	0.90	—	—	0.82–0.65	0.78–0.60	0.74–0.57	0.67–0.52
Chrome-magnesite brick	—	—	—	0.87	0.82	0.75	0.67
Chrome brick	0.90	—	—	0.97	0.98	—	—
High-alumina brick	0.90	0.85	0.79	0.50	0.44	—	—
Mullite brick	—	—	—	0.53	0.53	0.62	0.63
Silicon carbide brick	—	—	—	0.92	0.89	0.87	0.86

Source: From D.H. Hubble, *The Making, Shaping and Treating of Steel,* 11th edition, Steelmaking and Refining Volume, R.J. Fruehan, Ed., AISE Steel Foundation, Pittsburgh, PA, 1998, 159-290.

In some industrial combustion processes, the surface absorptivity can change over time, which affects the performance of the system. This is particularly true in coal-fired processes where the ash may be deposited on tube surfaces. Wall et al. (1993) showed that the normal emittance of particulate ash is highly dependent on the particle size and surface temperature.[66] In a subsequent study, Wall et al. (1995) studied the effects of ash deposits on the heat transfer in a coal-fired furnace.[67] The heat transfer from the combustion products to the tubes is primarily by radiation, with a lesser amount by convection heating. The heat must conduct through the ash deposits and the tube wall, before heating the fluid inside the tubes primarily by convection. As the ash deposit melts to form a slag, the absorptance increases dramatically to values approaching 0.9. The thermal conductivity of the deposit is highly dependent on its physical state, especially its porosity. It is interesting to note that the heat flux through the deposits decreases during the initial phases of its growth and then actually increases before reaching a steady-state value when the deposits have reached maturity. The measured thermal conduction coefficient ranged from 0.3 to 0.5 kW/m^2-K (53 to 88 Btu/ft^2-hr-°F). The important conclusion of these studies is that the radiative and conductive properties of ash deposits depend on the physical and chemical character of the deposits.

An important analysis in combustion processes is often radiation heat transfer in an enclosure. The furnace walls are normally at a higher temperature than the heat load and radiate energy to that load. When the space inside an enclosure is either a vacuum or contains a gas like air, which is essentially transparent to radiation, then the medium in the combustion space is referred to as *non-participating*. This means that it does not absorb any of the radiation passing through it. If the combustion space contains a radiatively absorbing gas like CO_2, H_2O, or CO, then the medium is referred to as *participating* because it does absorb some of the radiation passing through it. In the combustion of fossil fuels, the products of combustion usually contain significant quantities of CO_2 and H_2O, so the combustion space contains participating media. An assumption that is often made for a first-order analysis is that the concentrations of participating gases are low enough due to dilution by N_2 that the combustion space can be treated as nonparticipating to simplify the analysis. The net radiant heat transfer from one surface to another can be calculated from:

$$q_{1 \Leftrightarrow 2} = \sigma A_1 \, F_{1 \to 2} \left(T_1^4 - T_2^4 \right) = \sigma A_2 \, F_{2 \to 1} \left(T_1^4 - T_2^4 \right) \tag{3.33}$$

where $q_{1 \Leftrightarrow 2}$ is the net energy transferred between surfaces 1 and 2, $F_{i \to j}$ is the view factor or radiation shape factor, which is the diffuse radiation leaving surface i and received at surface j, and

$$A_1 \, F_{1 \to 2} = A_2 \, F_{2 \to 1} \tag{3.34}$$

due to the reciprocity theory. Equations, charts, tables, and graphs of radiation view factors are available elsewhere.[51,55] An example is given to illustrate this type of analysis (see Example 3.7). The view factor between two identical, parallel, directly opposed rectangles (see Figure 3.8) is given by:

$$F_{1 \to 2} = F_{2 \to 1} = \frac{2}{\pi X Y} \left\{ \begin{array}{l} \ln \sqrt{\dfrac{(1+X^2)(1+Y^2)}{1+X^2+Y^2}} + X\sqrt{1+Y^2} \, \tan^{-1} \dfrac{X}{\sqrt{1+Y^2}} \\[2ex] + Y\sqrt{1+X^2} \, \tan^{-1} \dfrac{Y}{\sqrt{1+X^2}} - X \tan^{-1} X - Y \tan^{-1} Y \end{array} \right\} \tag{3.35}$$

where $X = a/c$, $Y = b/c$, a = length of the rectangles, b = width of the rectangles, and c = spacing between the rectangles. As can be seen, this is a fairly complicated relationship for a relatively simple geometric configuration. For more complicated and for general surface orientations, computer analysis becomes necessary.

Example 3.7

Given: Furnace roof temperature of 2000°F, average load temperature = 300°F, combustion space length = 20 ft, width = 10 ft, height = 8 ft.

Find: Net radiant heat transfer from the roof to the floor, assuming both are blackbodies.

Solution: Find view factor from floor to ceiling:

$a = 20$ ft, $b = 10$ ft, $c = 8$ ft, $X = a/c = 20/8 = 2.5$, $Y = b/c = 10/8 = 1.25$

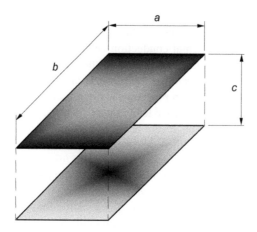

FIGURE 3.8   Geometry for radiation between parallel plates.

Using the formula above for the view factor between parallel rectangles,

$F_{1 \rightarrow 2}$ = 0.356

T_1 = 2000°F = 2460°R, T_2 = 300°F = 760°R

$q_{1 \rightarrow 2}$ = (0.1714 × 10^{-8} Btu/hr-ft²-°R⁴)(0.356)(20 ft × 10 ft)[(2460°R)⁴ − (760°R)⁴]

        = 4.43 × 10^6 Btu/hr

Beér studied flame impingement inside a furnace and experimentally determined that radiation was at least 10% of the total heat flux.[11] It was not specified how much of that radiation came from the hot walls and how much came from the flame. Vizioz[27] and Smith[29] studied flame impingement on a flat plate in a hot furnace. It was determined that radiation and convection were of comparable magnitude. Vizioz measured radiation to be 4% to 100% of the total heat flux. Smith calculated the surface radiant emission to the target, using Hottel's zone method.[51] It was 30% to 43% and 10% to 17% of the total heat transfer to water-cooled and air-cooled flat plates, respectively. Matsuo studied turbulent, preheated air/coke oven gas flames, impinging on a metal slab, inside a hot furnace.[30] The top of the slab was exposed to the impinging flame. The rest was exposed to the radiation from the furnace walls. Furnace radiation was the dominant mechanism: (1) for large L, (2) for high t_w, and (3) for large R. Ivernel calculated the radiation from hot furnace walls to a hemi-nosed cylinder.[32] It was up to 42% of the total heat flux for impinging O_2/natural gas flames. You measured the convective and total heat flux, using gages plated with gold and black foils, respectively.[23] By subtracting the convection from the total heat flux, radiation was calculated to be up to 35% of the convective flux. Van der Meer estimated that the radiation, from the hot inner refractory wall of a tunnel burner, was up to 15% of the total heat flux to the target.[38]

3.3.2   NONLUMINOUS GASEOUS RADIATION

3.3.2.1   Theory

The equation of radiative transfer will not be considered here as it is discussed in many radiation textbooks and is not commonly used as such to solve industrial combustion problems. The complete combustion of hydrocarbon fuels produces, among other things, CO_2 and H_2O. These gaseous products generate nonluminous radiation, which has been extensively studied.[52] This heat transfer mode depends on the gas temperature level, the partial pressure and concentration of each species, and the molecular path length through the gas. Based on the study of CO flames, Garner (1965) concluded that most of the radiant energy from the flames was primarily chemiluminescence,[68] but this was later proven to be wrong.[69] Ludwig et al. (1973) gave a very detailed discussion of the theory of infrared radiation from combustion gases, including the terms and their definitions, calculation techniques for both homogeneous and inhomogeneous gases, models for specific molecules (both diatomic and polyatomic), actual computed radiation data, a discussion of the accuracy of the models, and some predictive techniques for calculating rocket exhausts.[70] Examples of the spectral emissivity of H_2O and CO_2 as functions of wavelength and path length are shown in Figures 3.9 and 3.10, respectively.

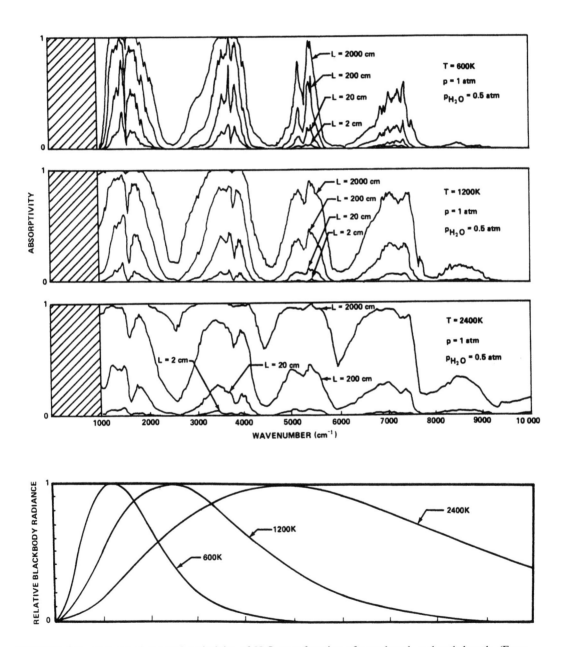

FIGURE 3.9   Calculated spectral emissivity of H_2O as a function of wavelength and path length. (From *Handbook of Infrared Radiation from Combustion Gases*, Ludwig et al., Eds., NASA, 1973, 241.)

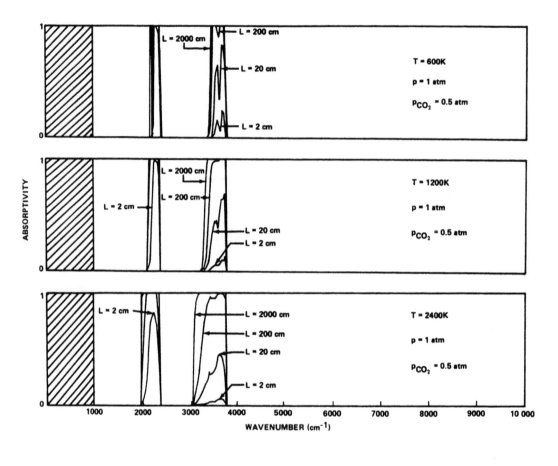

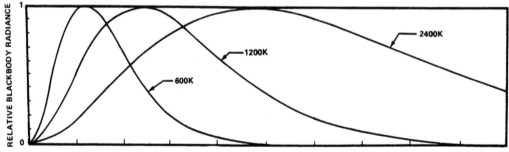

FIGURE 3.10   Calculated spectral emissivity of CO_2 as a function of wavelength and path length. (From Ludwig et al., *Handbook of Infrared Radiation from Combustion Gases*, NASA, 1973, 244.)

The total emissivity can be calculated from Leckner (1972).[71] The individual emissivity of either CO_2 or H_2O is given by:

$$\varepsilon_i\left(p_a L, p, T_g\right) = \varepsilon_0\left(p_a L, T_g\right)\left(\frac{\varepsilon}{\varepsilon_0}\right)\left(p_a L, p, T_g\right) \tag{3.36}$$

where ε_i = emissivity of the individual gas, p_a = partial pressure of the gas, L = path length through the gas, T_g = absolute temperature of the gas, and ε_0 = emissivity of the individual gas at a reference state (atmospheric pressure and $p_a \rightarrow 0$ but $p_a L > 0$). The first term in Equation 3.36 is calculated using:

TABLE 3.3
Correlation Constants for Gas Emissivity Equations

Gas	Water Vapor			Carbon Dioxide		
M, N	2,2			2,3		
$c_{00} \quad \cdots \quad c_{N1}$	−2.2118	−1.1987	0.035596	−3.9893	2.7669	−2.1081	0.39163
$\vdots \quad \ddots \quad \vdots$	0.85667	0.93048	−0.14391	1.2710	−1.1090	1.0195	−0.21897
$c_{0M} \quad \cdots \quad c_{NM}$	−0.10838	−0.17156	0.045915	−0.23678	0.19731	−0.19544	0.044644

	Water Vapor	Carbon Dioxide
P_E	$(p + 2.56 p_a / \sqrt{t})/p_0$	$(p + 0.28 p_a)/p_0$
$(p_a L)_m/(p_a L)_0$	$13.2 t^2$	$0.054/t^2, \quad t < 0.7$ $0.225 t^2, \quad t > 0.7$
a	$2.144, \qquad\qquad t < 0.75$ $1.88 - 2.053 \log_{10} t, \quad t > 0.75$	$1 + 0.1/t^{1.45}$
b	$1.10/t^{1.4}$	0.23
c	0.5	1.47

Note: $T_0 = 1000$ K, $p_0 = 1$ bar, $t = T/T_0$, $(p_a L)_0 = 1$ bar cm.

Source: From F. Kreith, Ed., *CRC Handbook of Mechanical Engineering*, CRC Press, Boca Raton, FL, 1998, 4-73.

$$\varepsilon_0 \left(p_a L, T_g \right) = \exp \left[\sum_{i=0}^{M} \sum_{j=0}^{N} c_{ij} \left(\frac{T_g}{T_0} \right)^j \left(\log_{10} \frac{p_a L}{(p_a L)_0} \right)^i \right] \tag{3.37}$$

where T_0 = absolute reference temperature of the gas (1000K) and c_{ij} are constants. The second term in Equation 3.36 is calculated from:

$$\left(\frac{\varepsilon}{\varepsilon_0} \right) \left(p_a L, p, T_g \right) = \left\{ 1 - \frac{(a-1)(1-P_E)}{a+b-1+P_E} \exp \left[-c \left(\log_{10} \frac{(p_a L)_m}{p_a L} \right)^2 \right] \right\} \tag{3.38}$$

where a, b, c, P_E, and $(p_a L)_m/p_a L$ are given in Table 3.3. Graphical results for H_2O and CO_2 are shown in Figures 3.11 and 3.12, respectively. The total emissivity is then calculated using:

$$\varepsilon_{CO_2 + H_2O} = \varepsilon_{CO_2} + \varepsilon_{H_2O} - \Delta\varepsilon \tag{3.39}$$

where the $\Delta\varepsilon$ accounts for the overlap between the H_2O and CO_2 bands and is calculated from:

$$\Delta\varepsilon = \left(\frac{\xi}{10.7 + 101\xi} - 0.0089\xi^{10.4} \right) \left(\log_{10} \frac{\left(\pi_{H_2O} + P_{CO_2} \right) L}{(p_a L)_0} \right)^{2.76} \tag{3.40}$$

and

$$\xi = \frac{P_{H_2O}}{P_{H_2O} + P_{CO_2}} \tag{3.41}$$

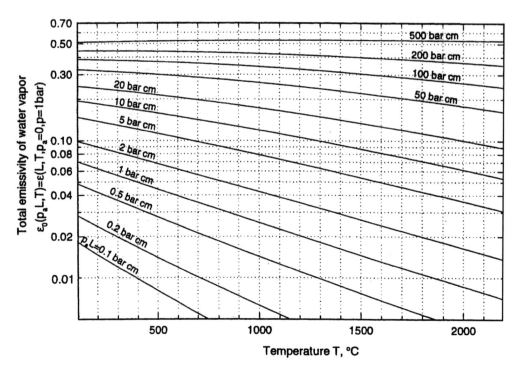

FIGURE 3.11 Total emissivity of water vapor at the reference state of a total gas pressure $p = 1$ bar and a partial pressure of H_2O $p_a \to 0$. (Courtesy of CRC Press, Boca Raton, FL. From *The CRC Handbook of Mechanical Engineering,* F. Kreith, Ed., 1998, 4-74.)

Example 3.8

Given: Combustion products containing 9% CO_2, 18% H_2O, and the balance N_2, at a temperature of 1500°F, with a mean beam length of 10 ft, and at atmospheric pressure.

Find: Gas emissivity.

Solution: Calculate p_aL for CO_2 and H_2O to use graphs:

CO_2: $p_aL = (0.09$ bar$)(305$ cm$) = 27$ bar-cm

H_2O: $p_aL = (0.18$ bar$)(305$ cm$) = 55$ bar-cm

Look up on graphs the ε_0 for CO_2 and H_2O at a temperature of 820°C:

CO_2: $\varepsilon_0 \approx 0.12$

H_2O: $\varepsilon_0 \approx 0.29$

Calculate correction factors $\varepsilon/\varepsilon_0$ using Equation 3.38 and Table 3.3:

$t = T/T_0$ (1090K/1000K) = 1.09

CO_2: $P_E = (1.0 + 0.28(0.09))/1.0 = 1.03$

H_2O: $P_E = (1.0 + 2.56/\sqrt{1.09})/1.0 = 3.45$

CO_2: $\dfrac{(p_aL)_m}{p_aL} = 0.225t^2 = 0.225(1.09)^2 = 0.267$

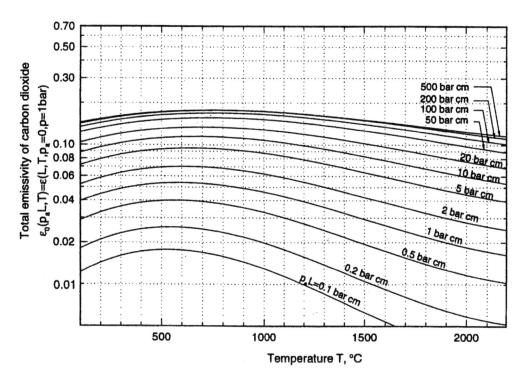

FIGURE 3.12 Total emissivity of carbon dioxide at the reference state of a total gas pressure $p = 1$ bar and a partial pressure of CO_2 $p_a \rightarrow 0$. (Courtesy of CRC Press, Boca Raton, FL. From *The CRC Handbook of Mechanical Engineering*, F. Kreith, Ed., 1998, 4-75.)

H_2O: $\dfrac{(p_a L)_m}{p_a L} = 13.2 t^2 = 13.2(1.09)^2 = 15.7$

CO_2: $a = 1 + 0.1/(1.09)^{1.45} = 1.09$; $b = 0.23$; $c = 1.47$

H_2O: $a = 1.88 - 2.053 \log_{10}(1.09) = 1.80$; $b = 1.10/(1.09)^{1.4} = 0.97$; $c = 0.5$

CO_2: $= \left(\dfrac{\varepsilon}{\varepsilon_0}\right) = \left\{1 - \dfrac{(1.09-1)(1-1.03)}{1.09+0.23-1+1.03}\exp\left[-1.47\left(\log_{10} 0.267\right)^2\right]\right\} = 1.001$

H_2O: $= \left(\dfrac{\varepsilon}{\varepsilon_0}\right) = \left\{1 - \dfrac{(1.80-1)(1-3.45)}{1.80+0.07-1+3.45}\exp\left[-0.5\left(\log_{10} 15.7\right)^2\right]\right\} = 1.184$

CO_2: $\varepsilon = (\varepsilon/\varepsilon_0)\,\varepsilon_0 = (1.001)(0.12) = 0.120$

H_2O: $\varepsilon = (\varepsilon/\varepsilon_0)\,\varepsilon_0 = (1.184)(0.29) = 0.343$

Calculate ξ: $\xi = \dfrac{0.18}{0.18+0.09} = 0.67$

Calculate $\Delta\varepsilon$:

$$\Delta\varepsilon = \left(\dfrac{0.67}{10.7+101(0.67)} - 0.0089(0.67)^{10.4}\right)\left(\log_{10}\dfrac{(0.18+0.09)(305\text{ cm})}{1\text{ bar-cm}}\right)$$

$\Delta\varepsilon = 0.051$

$\varepsilon = 0.120 + 0.343 - 0.051 = 0.412$

The absorptivity of H_2O and CO_2 can be estimated using:

$$\alpha\left(p_aL, p, T_g, T_s\right) = \left(\frac{T_g}{T_s}\right)^{1/2} \varepsilon\left(p_aL\frac{T_s}{T_g}, p, T_s\right) \tag{3.42}$$

where T_s is the surface temperature such as a furnace wall. The correction for the band overlap between H_2O and CO_2 is calculated using:

$$\alpha_{CO_2+H_2O} = \alpha_{CO_2} + \alpha_{H_2O} - \Delta\varepsilon \tag{3.43}$$

where $\Delta\varepsilon$ is estimated with a pressure path length of p_aLT_s/T_g.

Example 3.9

Given: Using the data from the previous example and a wall temperature of 1000°F.

Find: Gas absorptivity.

Solution: Calculate p_aLT_s/T_g for CO_2 and H_2O to use graphs:

CO_2: p_aLT_s/T_g = (0.09 bar)(305 cm)(1000 + 460)/(1500 + 460) = 20 bar-cm

H_2O: p_aLT_s/T_g = (0.18 bar)(305 cm)(1000 + 460)/(1500 + 460) = 41 bar-cm

Look up on graphs the ε_0 for CO_2 and H_2O at a temperature of 540°C (1000°F):

CO_2: $\varepsilon_0 \approx 0.12$, H_2O: $\varepsilon_0 \approx 0.28$

Calculate correction factors $\varepsilon/\varepsilon_0$ using Equation 3.38 and Table 3.3:

$t = (811\text{K}/1000\text{K}) = 0.811$

CO_2: $P_E = (1.0 + 0.28(0.09))/1.0 = 1.03$

H_2O: $P_E = (1.0 + 2.56/\sqrt{0.811})/1.0 = 3.84$

CO_2: $\dfrac{(p_aL)_m}{p_aL} = 0.225t^2 = 0.225(0.811)^2 = 0.148$

H_2O: $\dfrac{(p_aL)_m}{p_aL} = 13.2t^2 = 13.2(0.811)^2 = 8.68$

CO_2: $a = 1 + 0.1/(0.811)^{1.45} = 1.14$; $b = 0.23$; $c = 1.47$

H_2O: $a = 1.88 - 2.053 \log_{10}(0.811) = 2.07$; $b = 1.10/(0.811)^{1.4} = 1.47$; $c = 0.5$

CO_2: $\left(\dfrac{\varepsilon}{\varepsilon_0}\right) = \left\{1 - \dfrac{(1.14-1)(1-1.03)}{1.14+0.23-1+1.03} \exp\left[-1.47\left(\log_{10} 0.148\right)^2\right]\right\} = 1.001$

H_2O: $\left(\dfrac{\varepsilon}{\varepsilon_0}\right) = \left\{1 - \dfrac{(2.07-1)(1-3.84)}{2.07+1.47-1+3.84} \exp\left[-0.5\left(\log_{10} 8.68\right)^2\right]\right\} = 1.307$

CO_2: $\varepsilon = (\varepsilon/\varepsilon_0)\, \varepsilon_0 = (1.001)(0.12) = 0.120$

H_2O: $\varepsilon = (\varepsilon/\varepsilon_0)\, \varepsilon_0 = (1.307)(0.28) = 0.366$

CO_2: $\alpha = \sqrt{\dfrac{1500+460}{1000+460}}(0.120) = 0.139$

$$H_2O: \quad \alpha = \sqrt{\frac{1500 + 460}{1000 + 460}} \, 0.366 = 0.424$$

$$\alpha = 0.139 + 0.424 - 0.051 = 0.512$$

Relatively simple but accurate calculations for isothermal gases, which are absorbing-emitting but not scattering, in an isothermal black-walled enclosure can be computed using:

$$q = \sigma \left\{ \left[1 - \alpha\left(L_m\right) \right] T_w^4 - \varepsilon\left(L_m\right) T_g^4 \right\} \qquad (3.44)$$

where L_m is the average mean beam length and T_w is the absolute temperature of the isothermal wall. The mean beam length is a function of the geometry of the enclosure and can be calculated using the formulas in Table 3.4. For geometries not listed in that table, the following formula can be used:

$$L_m = 3.6 \ V/A \qquad (3.45)$$

where V is the volume of the enclosure and A is the surface area inside the enclosure. Gulic (1974) gave a modified form to calculate the mean beam length:[72]

$$L_m = 3.6 \frac{V}{A} \left(\frac{\tau e^{-\tau}}{\tau - 1 + e^{-\tau} + \tau e^{-\tau}} \right) \qquad (3.46)$$

where τ is the optical density of the gas.

Cess (1974) has given a relatively simple procedure for calculating infrared gaseous radiation that uses an analytical approximate band model.[73] The results showed good agreement with a numerical solution. Greif (1974) presented some experimental and theoretical results for infrared radiation from a turbulent flow of air, CO_2, and steam.[74] Kovotny (1974) showed the importance of pressure in calculating gaseous radiation heat transfer.[75] In a review paper, Edwards (1976) cautioned about the use of gray-gas models and analysis to represent the truly spectrally-dependent phenomena of gaseous radiation.[76] Trout (1977) gave some simple methods for estimating both nonluminous and luminous radiation from industrial flames.[77] Tien and co-workers (1968, 1982) reviewed radiation from flames.[78,79] They extensively reviewed the available models for both nonluminous and luminous gaseous radiation. Edwards and Balakrishnan (1973) presented some correlations for calculating the radiative properties of H_2O, CO_2, CO, NO, SO_2, and CH_4 for use in solving gas radiation problems.[80] Taylor and Foster (1974) presented empirical curve-fitted equations for the total emissivities of nonluminous and luminous flames for CO_2–H_2O and CO_2–H_2O–soot mixtures arising in oil and gas combustion.[81] Coppalle and Vervisch (1983) calculated the total emissivities of high-temperature (2000–3000K) gases (CO_2–H_2O mixtures).[82] Edwards and Matavosian (1984) developed some scaling rules for calculating the total absorptivity and emissivity of gases and tabulated and graphed results for various combinations of gas composition, product of partial pressure times the path length, and gas temperature.[83] Howell (1988) reviewed thermal radiation in participating media and gave numerous references for further information.[84] The P-N, two-flux, discrete ordinate, finite element, zoning, and Monte Carlo methods of analysis were briefly discussed. Wieringa et al. (1990) modeled the spectral radiation produced by natural gas flames in a regenerative glass furnace.[85] The spectral radiation from H_2O and CO_2 were modeled using 15 spectral bands. The numerical results showed that the furnace wall emissivity had only minimal impact on the furnace efficiency.

TABLE 3.4

Mean Beam Lengths for Radiation Calculations

Geometry of Gas Volume	Characterizing Dimension L	Geometric Mean Beam Length L_0/L	Average Mean Beam Length L_m/L	L_m/L_0
Sphere radiating to its surface	Diameter, L = D	0.67	0.65	0.97
Infinite circular cylinder to bounding surface	Diameter, L = D	1.00	0.94	0.94
Semi-infinite circular cylinder to:	Diameter, L = D			
Element at center of base		1.00	0.90	0.90
Entire base		0.81	0.65	0.80
Circular cylinder (height/diameter = 1) to:	Diameter, L = D			
Element at center of base		0.76	0.71	0.92
Entire surface		0.67	0.60	0.90
Circular cylinder (height/diameter = 2) to:	Diameter, L = D			
Plane base		0.73	0.60	0.82
Concave surface		0.82	0.76	0.93
Entire surface		0.80	0.73	0.91
Circular cylinder (height/diameter = 0.5) to:	Diameter, L = D			
Plane base		0.48	0.43	0.90
Concave surface		0.53	0.46	0.88
Entire surface		0.50	0.45	0.90
Infinite semicircular cylinder to center of plane rectangular face	Radius, L = R	—	1.26	—
Infinite slab to its surface	Slab thickness, L	2.00	1.76	0.88
Cube to a face	Edge L	0.67	0.6	0.90
Rectangular $1 \times 1 \times 4$ parallelepipeds:	Shortest edge, L			
To 1×4 face		0.90	0.82	0.91
To 1×1 face		0.86	0.71	0.83
To all faces		0.89	0.81	0.91

Source: From *CRC Handbook of Chemistry and Physics,* 79th edition, CRC Press, Boca Raton, FL, 1998, 12-191–12-192.

3.3.2.2 Combustion Studies

Two types of nonluminous radiation measurements have been made in previous combustion studies: total and spectral. The total radiation measurements were typically made with some type of detector that measured the overall radiation received, with no wavelength dependence. More recent measurements have been made that gave the radiation as a function of wavelength. These two methods of measurements applied to nonluminous radiation are discussed briefly here.

3.3.2.2.1   Total radiation

In some previous combustion experimental studies, nonluminous gaseous radiation has been significant. Kilham tested nearly stoichiometric, laminar air/CO flames, impinging normal to an uncooled refractory cylinder.[14] Flame radiation was 5% to 16% of the total heat flux. Jackson also tested laminar flames, impinging normal to refractory cylinders. A variety of fuels, oxidizers, and stoichiometries were tested. The measured flame radiation was up to 5% of the total heat flux. Dunham tested nearly stoichiometric, laminar air/CO flames, impinging normal to the nose of a hemi-nosed cylinder.[86] The estimated nonluminous radiation was up to 13% of the total heat flux.

Ivernel tested nearly stoichiometric natural gas flames, impinging normal to the nose of a hemi-nosed cylinder.[32] The calculated nonluminous flame radiation was up to 34% of the total heat flux.

In other combustion studies, nonluminous radiation has not been significant. Giedt measured and calculated nonluminous radiation to be less than 2% of the total heat flux.[87] Woodruff used the same configuration as Giedt, except that the flames were turbulent.[16] The measured nonluminous radiation was negligible compared to the total heat flux. Shorin and Pechurkin tested a wide variety of flames impinging normal to a flat plate.[26] Radiation was experimentally determined to be negligible. Purvis studied O_2/CH_4 and O_2/C_3H_8 flames impinging normal to a flat plate.[88] The calculated nonluminous flame radiation was negligible compared to the total heat flux. Although the C_3H_8 flames were very fuel rich, there was no discussion of luminous flame radiation. Hoo-gendoorn tested stoichiometric, laminar air/natural gas flames, impinging normal to a water-cooled flat plate.[21] Flame radiation effects did not exceed 5% of the total heat flux. No radiation measure-ments or calculations were given. Davies tested natural gas flames, impinging normal to a water-cooled cylinder.[89] A range of stoichiometries and oxidizers was tested. Using an estimated flame emissivity of 0.01, the calculated nonluminous radiation was only 2% of the total heat flux. Van der Meer tested stoichiometric, laminar and turbulent, air/natural gas flames, impinging normal to a water-cooled plate.[90] The flame radiation was said to be negligible because of the very low emissivity of a thin hot gas layer. No supporting calculations were cited.

Baukal and Gebhart (1997) studied total flame radiation from oxygen-enhanced natural gas flames.[91] A total, narrow-angle radiometer was used to measure the flame radiation as a function of position from the burner, as shown in Figure 3.13. Five parameters were investigated, including the burner firing rate (q_f), the oxidizer composition (Ω), the equivalence ratio (ϕ), the axial position along the flame (L_r), and the radial position from the flame (D_r). The parameters of primary interest were Ω, q_f, and L_r. Most of the tests were done at $\phi = 1.0$ and $D_r = 0.5$. The equivalence ratio was only varied through a narrow range for two reasons. The first is that the equivalence ratio is operated in a narrow band around stoichiometric conditions in nearly all industrial heating and melting processes. The second reason is that only nonluminous radiation was studied in this particular study. If the burner operated at very fuel-rich conditions ($\phi \rightarrow \infty$), luminous radiation would have become important. Figure 3.14 shows the effects of both the firing rate and the axial distance along the flame length. The minimum L_r was 0.5. For $L_r < 0.5$, the radiometer would have viewed the

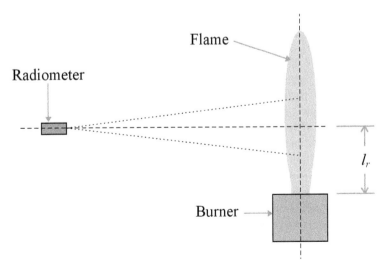

FIGURE 3.13 Experimental setup for measuring total nonluminous gaseous radiation from an open-flame diffusion burner as a function of the distance from the burner outlet. (From C.E. Baukal and B. Gebhart, *Int. J. Heat Mass Transfer,* 44(11), 2539-2547, 1997.)

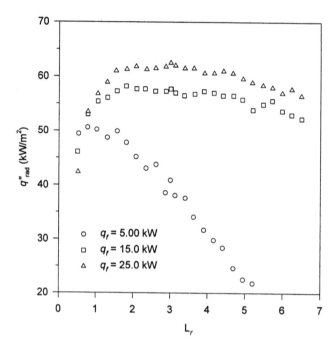

FIGURE 3.14 Total nonluminous gaseous radiation (q''_{rad}) as a function of distance from the burner (L_r) and firing rate (q_f = 5.0, 15.0, 25.0 kW) for Ω = 1.00, ϕ = 1.00, D_r = 0.5. (From C.E. Baukal and B. Gebhart, *Int. J. Heat Mass Transfer,* 44(11), 2539-2547, 1997.)

flame and part of the burner itself. At low firing rates, the flame radiation decreased rapidly with the distance from the burner exit. At intermediate and higher firing rates, the flame radiation was relatively constant for L_r from about 2 to 5. Figure 3.15 is a similar graph except that the oxidizer composition, instead of the firing rate, was varied. The flame radiation was relatively constant over a wide range of L_r, for Ω = 1.0. At Ω = 0.35, the peak flame radiation occurred at about L_r = 1.5 and then slowly decreased with L_r. Figure 3.16 shows how the radial location of the radiometer affected the measurements. At D_r = 1.0, the radiometer viewed the entire width of the flame. The effective path length decreased as D_r increased because the flame had a circular cross-section. This reduced the average flame radiation as shown in the plot. However, the reduction was relatively small. Figure 3.17 shows how the equivalence ratio and the oxidizer compositions affected the flame radiation. For the smaller Ω, the highest flame radiation occurred at slightly fuel-lean conditions. For a pure O_2 oxidizer, the highest flame radiation occurred in a band around stoichiometric conditions (ϕ = 1.0). Figure 3.18 shows how the flame radiation varied as a function of the oxidizer composition for a fixed firing rate and equivalence ratio. The only parameter that was varied was the amount of N_2 in the oxidizer. The fuel and O_2 flow rates were constant. Re_n decreased from 6800 to 2400 as Ω increased from 0.28 to 1.00. The plot shows that the flame radiation increased by more than 2.5 times by removing N_2 from the oxidizer. This is a result of higher flame temperatures and partial pressures of CO_2 and H_2O. Figure 3.19 shows how the flame radiation increased as the firing rate increased and as the equivalence ratio increased. Figure 3.20 is a plot of flame radiation as a function of the equivalence ratio and the firing rate for a constant Reynolds number of 4500. As N_2 was removed from the oxidizer (Ω increasing), the firing rate had to be increased to maintain a fixed Re_n. This shows that a higher heat release density may be achieved by increasing the O_2 concentration in the oxidizer, for a given nozzle diameter. It also implies that Re_n by itself is not a sufficient parameter to indicate the performance of oxygen-enhanced flames. The oxidizer composition Ω must also be specified. Figure 3.21 shows the peak flame radiation measured for a given firing rate and oxidizer composition. For Ω = 1.00, the peak radiation increased

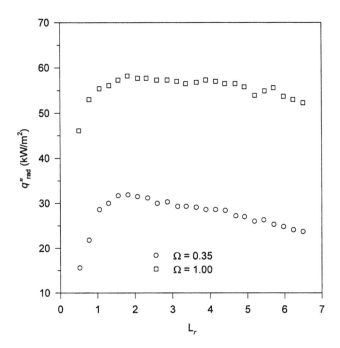

FIGURE 3.15 Total nonluminous gaseous radiation (q''_{rad}) as a function of distance from the burner outlet (L_r) and oxidizer composition (Ω = 0.35, 1.00) for q_f = 15.0 kW, ϕ = 1.00, and D_r = 0.5. (From C.E. Baukal and B. Gebhart, *Int. J. Heat Mass Transfer,* 44(11), 2539-2547, 1997.)

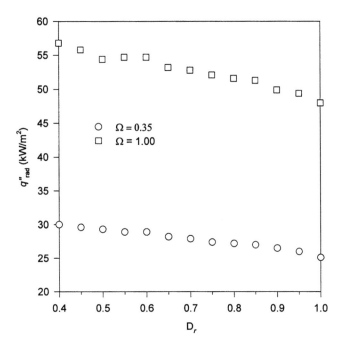

FIGURE 3.16 Total nonluminous gaseous radiation (q''_{rad}) as a function of distance from the burner outlet (D_r) and oxidizer composition (Ω = 0.35, 1.00) for q_f = 15.0 kW, ϕ = 1.00, and L_r = 3.0. (From C.E. Baukal and B. Gebhart, *Int. J. Heat Mass Transfer,* 44(11), 2539-2547, 1997.)

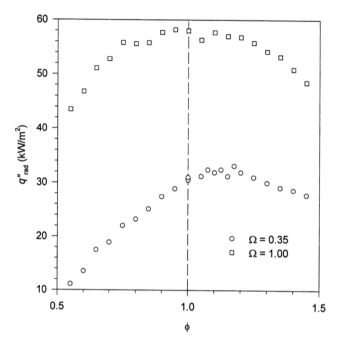

FIGURE 3.17 Total nonluminous gaseous radiation (q''_{rad}) as a function of the equivalence ratio (ϕ) and oxidizer composition (Ω = 0.35, 1.00) for q_f = 15.0 kW, L_r = 3.0, and D_r = 0.5. (From C.E. Baukal and B. Gebhart, *Int. J. Heat Mass Transfer,* 44(11), 2539-2547, 1997.)

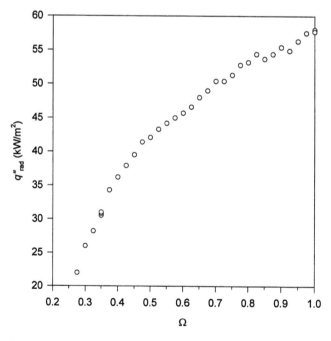

FIGURE 3.18 Total nonluminous gaseous radiation (q''_{rad}) as a function of the oxidizer composition (Ω) for q_f = 15.0 kW, ϕ = 1.00, L_r = 3.0, and D_r = 0.5. (From C.E. Baukal and B. Gebhart, *Int. J. Heat Mass Transfer,* 44(11), 2539-2547, 1997.)

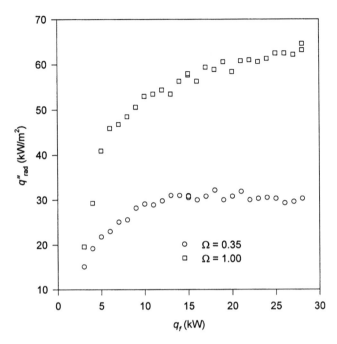

FIGURE 3.19 Total nonluminous gaseous radiation (q''_{rad}) as a function of the firing rate (q_f) and oxidizer composition ($\Omega = 0.35$, 1.00) for $\phi = 1.00$, $L_r = 3.0$, and and $D_r = 0.5$. (From C.E. Baukal and B. Gebhart, *Int. J. Heat Mass Transfer*, 44(11), 2539-2547, 1997.)

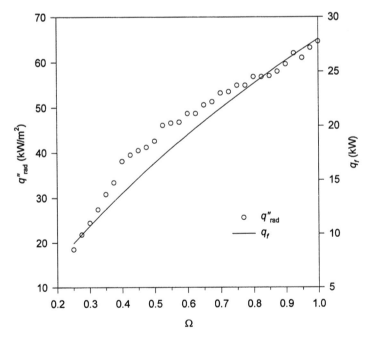

FIGURE 3.20 Total nonluminous gaseous radiation (q''_{rad}) as a function of the oxidizer composition (Ω) for $Re_n = 4500$, $\phi = 1.00$, $L_r = 3.0$, and $D_r = 0.5$. (From C.E. Baukal and B. Gebhart, *Int. J. Heat Mass Transfer*, 44(11), 2539-2547, 1997.)

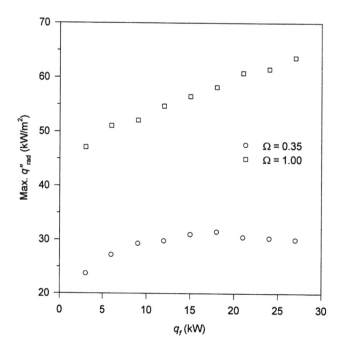

FIGURE 3.21 Maximum total nonluminous gaseous radiation (q''_{rad}) as a function of the firing rate (q_f) and oxidizer composition ($\Omega = 0.35$, 1.00) for $\phi = 1.00$ and $D_r = 0.5$ with L_r variable. (From C.E. Baukal and B. Gebhart, *Int. J. Heat Mass Transfer*, 44(11), 2539-2547, 1997.)

with the firing rate. For $\Omega = 0.35$, the peak radiation was relatively constant over a wide range of firing rates. Figure 3.22 shows the approximate axial location for the peak flame radiation. Initially, the peak flame locations were closer to the burner for $\Omega = 1.00$, compared to $\Omega = 0.35$. At higher firing rates, this trend reversed. For $\Omega = 1.00$, these locations were fairly well-defined. For $\Omega = 0.35$, there was a range of positions along the flame length where the maximum flame radiation values were obtained. The locations given in the figure are approximately in the center of the range. Figure 3.23 is a plot of the location for the peak flame radiation that has been normalized to the visible length of the flame. This shows that for high Ω flames, the axial location of the peak flame radiation was at about 14% of the visible flame length. For the lower Ω flames, this location was between about 9% and 15% of the flame length.

The important conclusions of the study were:

1. Thermal radiation increased dramatically by removing N_2 from the oxidizer
2. The flame radiation increased with the firing rate with a more dramatic increase at higher O_2 concentrations in the oxidizer
3. Higher flame radiation was measured at or near stoichiometric conditions where typical industrial heating processes are operated
4. For higher firing rates, the radiation was nearly constant over a wide range of axial locations in the flame
5. At lower firing rates, the flame radiation decreased with the axial distance from the burner outlet

For higher O_2 concentrations, the peak flame radiation increased with the firing rate. For lower O_2 concentrations, the peak flame radiation was nearly constant over a wide range of firing rates. The location of the peak flame radiation varied from 9% to 15% of the visible flame length. This location was more defined for the higher O_2 flames in terms of both its absolute position and of the position normalized by the flame length.

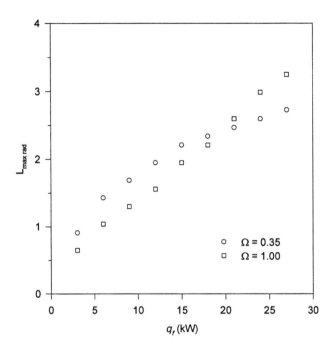

FIGURE 3.22 Axial distance of maximum radiation ($L_{max\,rad}$) as a function of the firing rate (q_f) and oxidizer composition (Ω = 0.35, 1.00) for ϕ = 1.00, and D_r = 0.5. (From C.E. Baukal and B. Gebhart, *Int. J. Heat Mass Transfer*, 44(11), 2539-2547, 1997.)

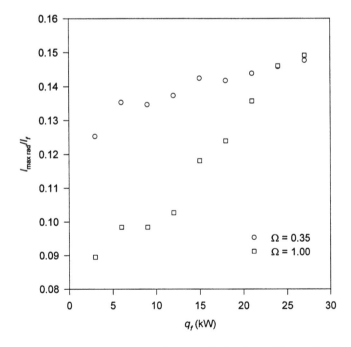

FIGURE 3.23 Normalized axial location of maximum radiation ($l_{max\,rad}/l_f$) as a function of firing rate (q_f) and oxidizer composition (Ω = 0.35, 1.00) for ϕ = 1.00, and D_r = 0.5. (From C.E. Baukal and B. Gebhart, *Int. J. Heat Mass Transfer*, 44(11), 2539-2547, 1997.)

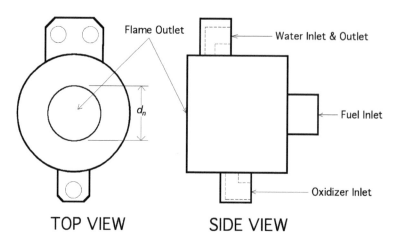

FIGURE 3.24   Burner used for flame radiation study using oxygen-enhanced burners. (From C.E. Baukal, Heat Transfer from Flames Impinging Normal to a Plane Surface, Ph.D. thesis, University of Pennsylvania, 1996.)

3.3.2.2.2   Spectral radiation

Ji and Baukal (1998) did the first systematic experimental study of the nonluminous and luminous spectral radiation from oxygen-enhanced flames.[92] Two common quantities are measured in spectral radiometry: spectral radiance and spectral irradiance. Spectral radiance quantifies the energy flux in a unit wavelength interval that is emitting from a source surface per unit area and unit solid angle. The spectral irradiance is a measure of the energy flux in a unit wavelength interval that is incident onto a target surface per unit area. For an optical source with a well-defined uniform emitting surface, spectral radiance can be measured accurately and converted into spectral irradiance. The lack of such well-defined and uniform radiating surfaces for flames renders the measurement of spectral radiance impractical. However, spectral irradiance is still a valid quantity to measure in flames and has more practical meaning in heat transfer to targets. For this reason, spectral irradiance was measured, despite the fact that spectral radiance has been used in some previous theoretical studies on flame radiation.[44]

A custom-built version of the commercial ICSM burner from Nordsea Gas Technology Ltd. (Cheshire, England) was used for the study, as shown in Figure 3.24, where $d_n = 38.5$ mm. It was a round form of the rectangular burners[93] used in industrial heating and melting processes. Details of this burner and its operating characteristics, as well as fuel and oxidizer flow specifications, are given elsewhere.[94] A key feature of the burner was the wide variety of oxidizers that could be used, ranging from air to pure oxygen, for a range of firing rates. Another key feature was that the burner produced a uniform, nearly one-dimensional flame, so that the flame radiation was more uniform compared to diffusion flames. A firing rate of 17,000 Btu/hr (5.0 kW) was used for most of the measurements, so that there was no significant radiation at a height 1.6 ft (0.5 m) above the burner surface.

As shown in Figure 3.25, three spectral monochromators were set at 11 ft (3.5 m) away from the burner centerline, and 0.82 ft (0.25 m) above the burner surface so that their field of view covered the entire radiation length of the flames. Because the flames were in open air, without any furnace enclosure and without impinging on any targets or nearby walls, ambient emission and reflection from the surroundings were negligible. Three parameters were varied to study their effects on the spectral radiation from oxygen-enhanced/natural gas flames: oxygen enrichment ratio (Ω), fuel equivalence ratio (ϕ), and firing rate or fuel input (q_f). The oxygen enrichment ratio ranged from $\Omega = 0.21$ (air) to $\Omega = 1.00$ (pure O_2). For the specific natural gas composition used, the stoichiometric ratio of oxygen:fuel was 2.08. Fuel-rich and fuel-lean flames had fuel equivalence ratios >1.0 and <1.0, respectively.

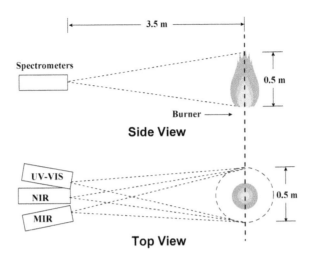

FIGURE 3.25 Experimental setup of the Monolight spectral radiometer (UV-VIS = UV and visible, NIR = near-IR, and MIR = mid-IR). (From B. Ji and C.E. Baukal, *Proc. 1998 Int. Gas Research Conf.,* Nov. 8-11, 1998, San Diego, CA, D. Dolenc, Ed., 5, 422-433, 1998.)

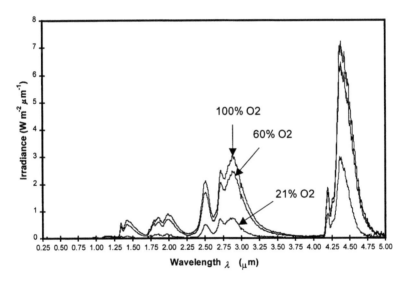

FIGURE 3.26 Overall spectra for flame radiation dependence on oxygen enrichment ratio Ω. (From B. Ji and C.E. Baukal, *Proc. 1998 Int. Gas Research Conf.,* Nov. 8-11, 1998, San Diego, CA, D. Dolenc, Ed., 5, 422-433, 1998.)

As will be shown in the flame radiation spectra, most of the flame radiation fell within the infrared region. Ultraviolet (0.25–0.38 μm) radiation, while important in some flame diagnostics, contributed little to the total radiation intensity. Visible (0.38–0.78 μm) radiation also made only a small contribution to the total radiation intensity. However, as shown later (Chapter 4), the visible radiation is a direct indicator of the penetrating radiation that is most effective for glass heating and melting. For a relative comparison among the flames studied, the radiation spectra over the entire wavelength region (0.25–5.0 μm) was numerically integrated and then normalized to the integrated radiation intensity of the flame with $q_f = 5.0$ kW, $\Omega = 1.0$, and $\phi = 1.0$. This normalization eliminated possible errors due to the uncertainties in the spectral radiometer calibration. Normalization also helped to reveal the general trends as the flame parameters changed. The results are summarized in Figure 3.26, which shows three flame radiation spectra with $q_f = 5.0$ kW, $\phi = 1.0$,

TABLE 3.5
Flame Radiation Intensity Measurements

W	ϕ	$I_{nrm(total)}$	$I_{nrm(pen.)}$	$I_{nrm(BK7)}$	γ
1.00	0.67	0.90 ± 0.06	0.89 ± 0.06	0.82 ± 0.06	0.15 ± 0.01
1.00	0.80	0.99 ± 0.07	0.98 ± 0.07	0.95 ± 0.07	0.13 ± 0.01
1.00	1.00	$1.00*$	$1.00*$	$1.00*$	0.14 ± 0.01
1.00	1.33	0.93 ± 0.07	0.98 ± 0.07	0.99 ± 0.07	0.17 ± 0.01
1.00	1.50	0.91 ± 0.06	0.93 ± 0.07	0.87 ± 0.06	0.20 ± 0.01
1.00	1.60	0.84 ± 0.06	0.83 ± 0.06	0.88 ± 0.06	0.21 ± 0.01
1.00	1.79	0.80 ± 0.06	0.83 ± 0.06	0.83 ± 0.06	0.37 ± 0.03
1.00	1.89	0.75 ± 0.05	0.87 ± 0.06	0.96 ± 0.07	0.55 ± 0.04
1.00	2.00	0.80 ± 0.06	1.12 ± 0.08	1.10 ± 0.08	1.02 ± 0.07
0.60	1.00	0.73 ± 0.05	0.70 ± 0.05	0.75 ± 0.05	0.12 ± 0.01
0.21	1.00	0.26 ± 0.02	0.19 ± 0.01	0.19 ± 0.01	0.005 ± 0.001

* The $\Omega = 1.0$ and $\phi = 1.0$ intensities were used as the reference in the normalization.

Source: From B. Ji and C.E. Baukal, *Proc. 1998 Int. Gas Research Conf.,* Nov. 8–11, 1998, San Diego, CA, D. Dolenc, Ed., 5, 422-433, 1998.

and $\Omega = 1.00, 0.60,$ and 0.21, respectively. The integrated results are listed in Table 3.5. The radiation intensity from the pure oxygen/natural gas flame ($\Omega = 1.00$) was about 4 times that from the air/natural gas flame ($\Omega = 0.21$) for the same fuel consumption. This is because nitrogen in the air absorbs energy from the combustion reaction, but N_2 does not radiate. Figure 3.27 is the UV portion of the spectra plotted on a logarithmic scale. Despite the relatively small contribution of UV radiation to the total intensity, the presence of UV emissions (predominantly from the electronically

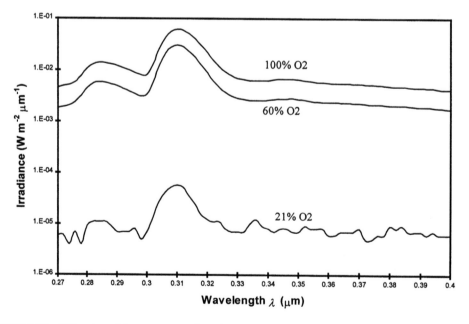

FIGURE 3.27 UV region of the spectra on a logarithmic scale for the flame radiation dependence on oxygen enrichment ratio Ω. (From B. Ji and C.E. Baukal, *Proc. 1998 Int. Gas Research Conf.,* Nov. 8-11, 1998, San Diego, CA, D. Dolenc, Ed., 5, 422-433, 1998.)

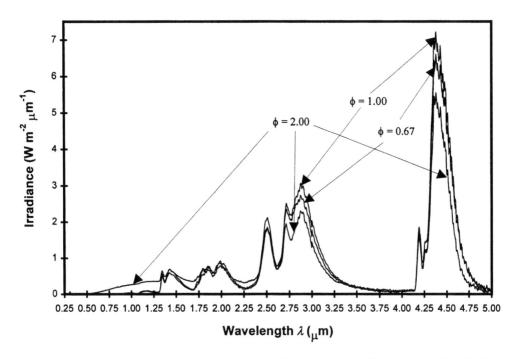

FIGURE 3.28 Flame radiation dependence on fuel equivalence ratio ϕ. Spectra were plotted with $q_f =$ 5.0 kW, $\Omega = 1.0$, and $\phi = 0.67$, 1.00, and 2.00, respectively. (From B. Ji and C.E. Baukal, *Proc. 1998 Int. Gas Research Conf.,* Nov. 8-11, 1998, San Diego, CA, D. Dolenc, Ed., 5, 422-433, 1998.)

excited OH radicals) can be explored as a way to visualize nonluminous flames in hot and highly luminous furnaces. The UV radiation from the pure oxygen flame was about 3 orders of magnitude stronger than that from the air/fuel flame. Such a dramatic difference cannot be simply accounted for by thermodynamic equilibrium at different flame temperatures. Rather, this was an indication that significantly more nonequilibrium chemiluminescence of OH radicals exists in oxygen-enhanced flames. It can be inferred that the OH emission-based band-reversal method, while successfully performed in CH_4/N_2O flames,[95] will not yield the correct flame temperature in oxygen-enhanced flames. Figure 3.28 shows three flame radiation spectra with $q_f = 5.0$ kW, $\Omega = 1.0$, and $\phi = 0.67$, 1.00, and 2.00, respectively. The normalized total radiation intensities for ϕ in the range of 0.67 to 2.00 are summarized in Table 3.5. The reduced radiation intensities in the fuel lean cases ($\phi < 1.0$) are due to the fact that excessive oxygen carries away heat, but does not radiate. Although the radiation intensities in the fuel-rich cases ($\phi > 1.0$) are apparently less than that in the stoichiometric case ($\phi = 1.00$), it should be noted that not all of the fuel was consumed due to lack of available oxygen. In a more realistic combustion system, oxygen would be introduced in a later stage to complete the combustion. With this in mind, the fuel-rich cases would actually provide more radiation for the same fuel consumption. Since the flames were in open air, actual equivalence ratios were different from the control settings, due to air entrainment. Nevertheless, the general trends remain the same.

The integrated spectral radiation intensities were nearly the same for stoichiometric flames with $\Omega = 1.00$, $q_f = 5$ kW and $\Omega = 0.60$, $q_f = 7$ kW. For a given O_2 enrichment level, the radiation intensity increased more slowly than the increase in firing rate. Figure 3.29 shows radiation spectra from stoichiometric flames with different oxygen enrichment ratios and firing rates. The radiation intensity did not increase for air-fired flames ($\Omega = 0.21$) when the firing rate increased from 5 kW to 10 kW. These trends for the change in radiation as a function of the firing rate are consistent with the findings in another study.[94] While a quantitative comparison may depend on the specific

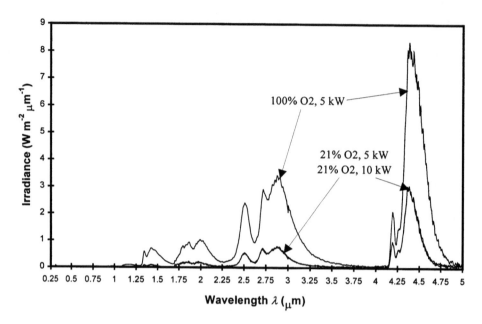

FIGURE 3.29   Flame radiation dependence on firing rate and oxygen enrichment ratio. (From B. Ji and C.E. Baukal, *Proc. 1998 Int. Gas Research Conf.,* Nov. 8-11, 1998, San Diego, CA, D. Dolenc, Ed., 5, 422-433, 1998.)

burner design, the qualitative trend is generally true. Nitrogen in the air reduces radiation intensity. Since increasing the firing rate in air/natural gas flames is accompanied by increasing nitrogen flow, the increase in combustion heat cannot be as efficiently released by radiation as in the oxygen-enriched flames.

3.3.3   Luminous Radiation

3.3.3.1   Theory

Luminous flames are produced by the continuous radiant emission of particles in the flame, such as soot, which radiate approximately as blackbodies. Yagi and Iino (1961) studied both the luminous (q_{rs} = radiation from soot) and nonluminous (q_{rg} = gaseous radiation) from turbulent diffusion flames.[96] The comparison of the two types of radiation is shown in Figure 3.30, which shows that the soot radiation is greater than the gaseous radiation. Echigo et al. (1967) studied luminous radiation from flames.[97] They noted that a rigorous definition of luminous and nonluminous flames does not exit. It had long been assumed that soot remains in the solid phase during the combustion process and emits a continuous spectrum of visible and infrared radiation. Based on experiments, they hypothesized that the dehydrogenation and polymerization of hydrocarbon fuels occurs in the liquid phase, and that the decomposed and polymerized compound ("pre-soot substance") emits banded spectra; then, the soot particles agglomerate after dehydrogenation and polymerization are completed. Gray et al. (1976) noted three size categories for particles: large, small, and intermediate.[98] Particles absorb, diffract, or attenuate radiation to varying degrees depending on their size. The soot generated in a flame is highly dependent, among other things, on the fuel composition. Luminous radiation is usually important when liquid and solid fuels, like oil and coal, are used. It is usually not significant for gaseous fuels, like natural gas. Fuels with higher carbon-to-hydrogen weight ratios (see Table 3.6) tend to produce sootier flames. In a survey paper, Wagner (1978) reviewed soot formation in combustion.[99] It was noted that soot forms at temperatures ranging from 1000 to 2500°C (1800 to 4500°F). Soot consists primarily of carbon, formed into long chains. The

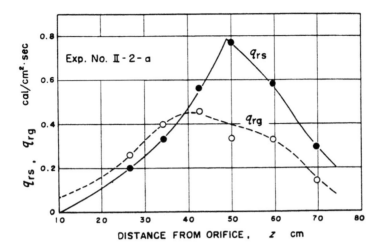

FIGURE 3.30 Soot (q_{rs}) and gaseous (q_{rg}) radiation from a turbulent diffusion flame with a fuel consisting of 11.9% CH_4, 7.5% C_3H_8, 22.5% C_3H_6, 2.8% heavy hydrocarbons, 8.4% CO, 18.9% H_2, 3.5% O_2, 2.1% CO_2, 22.4% N_2, and a nozzle diameter of 7 mm. (Courtesy of The Combustion Institute, Pittsburgh, PA.[96] With permission.)

TABLE 3.6
Sooting Tendency of Common Gaseous Fuels

Fuel	Formula	C	H	C/H Mass Ratio
Hydrogen	H_2	0	2	0.00
Methane	CH_4	1	4	2.97
Ethane	C_2H_6	2	6	3.96
Propane	C_3H_8	3	8	4.46
Butane	C_4H_{10}	4	10	4.76
Propylene	C_3H_6	3	6	5.95
Acetylene	C_2H_2	2	2	11.89
Carbon monoxide	CO	1	0	∞

total amount of soot formed is usually small in comparison to the amount of available carbon. Tien and Lee (1982) reviewed some of the various models available for calculating the emissivity of luminous flames, including the homogeneous non-gray model, the homogeneous gray model, and the nonhomogeneous non-gray model.[79] Glassman (1988) discussed the detailed chemistry of soot formation as it relates to fuel composition.[100] The graph in Figure 3.31 shows how the flame emissivity varies with fuel type.[101] For fuel gases with C:H weight ratios between 3.5 and 5.0, the data was correlated by either of two correlations:

$$\varepsilon = 0.20\sqrt{\frac{LHV}{900}} \qquad (3.47)$$

where LHV is the lower heating value of the fuel in Btu/ft³, or

$$\varepsilon = 0.048\sqrt{MW_{fuel}} \qquad (3.48)$$

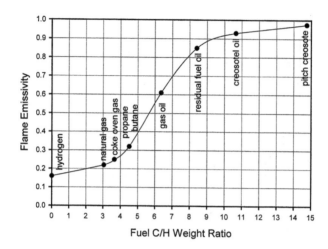

FIGURE 3.31 Emissivities of industrial-scale, turbulent diffusion flames of various fuels. (Courtesy of Gulf Publishing, Houston, TX.[101] With permission.)

where MW_{fuel} is the molecular weight of the fuel. For liquid fuels with C:H ratios between 5 and 15, the following correlation was determined:

$$\varepsilon = 1 - 68.2e^{-2.1\sqrt{C/H}} \qquad (3.49)$$

where C/H is the weight ratio of carbon to hydrogen for the fuel. Haynes (1991) reviewed soot formation in flames but did not specifically discuss the effects on heat transfer.[102]

The laminar smoke point for a fuel is another indicator of flame luminosity.[103] Fuels with lower smoke point heights produce more luminous flames than high smoke point fuels. The laminar smoke point is determined by measuring the distance from the burner outlet where a vertical, laminar jet of burning pure fuel just begins to produce smoke. The lower the height where smoke starts to form, the higher tendency the fuel has to produce soot.

Lahaye and Prado (1983) have edited an extensive book on soot generated in combustion.[104] The focus of the book is on the chemistry of soot formation and destruction, with no real consideration of the heat transfer from luminous flames. Propane and acetylene have been two commonly used fuels to study soot. Longwell reported that soot in the form of polycyclic aromatic hydrocarbons (PAHs) can come from the fuel, fuel non-flame pyrolysis, or from quenching a fuel-rich mixture region in the flame.[105] Howard and Bittner (1983) noted the formation of high-molecular-weight species prior to and during the appearance and growth of soot.[106] Bittner et al. (1983) noted that soot formation is strongly related to fuel composition and that aromatic fuels have a much higher propensity to soot.[107] Prado et al. (1983) discussed soot inception and growth through nucleation and agglomeration.[108] Calcote (1983) noted the following steps involved in soot formation:[109]

1. Formation of precursors
2. Nucleation, which transforms molecules to particles
3. Soot particle growth, which increases the molecular weight
4. Coagulation of soot particles to form single larger particles where the identity of the original colliding particles has been lost
5. Agglomeration of particles that adhere to each other to form a chain
6. Aggregation of colliding particles to form a cluster of individual particles that are still individually distinguishable
7. Oxidation of particles in any of the above steps to reduce the particle size and H:C ratio

Blokh (1988) has provided a lengthy treatment of luminous flame radiation produced by the burning of pulverized coal, oil, natural gas, and combinations of these.[110]

Chemiluminesence is light emission produced when an atom or molecule is elevated to an electronically excited state in a chemical reaction. Stambuleanu (1976) noted that visible luminosity in flames is often actually chemiluminesence and not luminosity at all.[111]

Blokh (1974) discussed theoretical and experimental research on radiative properties of soot particles in luminous flames and coal particles in pulverized coal flames.[112] Kunitomo (1974) developed a method for calculating luminous flame radiation from a liquid fuel at pressures up to 20 atm (20 bar).[113] Hammond and Beér (1974) made spectroradiometric measurements in sooty oil flames to determine spectral attenuation coefficients.[114] Ku and Shim (1990) developed models to predict the radiative properties of soot particles in flames.[115] To calculate luminous radiation from soot particles, the extinction coefficient, single-scattering albedo, and phase function are needed in the visible and near-infrared wavelengths. The formation processes and physical properties of soot, light scattering, and extinction by small particles, the effect of complex refractive index, the effect of size distribution, the effects of particle size distribution, and the effect of agglomeration were all discussed. In addition to the discussions of models available and suggested refinements for soot particle properties, diagnostic techniques for making these measurements were also discussed. Bockhorn (1994) edited a book on soot formation in combustion, including several sections on modeling, although nothing specifically on industrial combustion.[116] Fridman et al. (1997) showed that flame luminosity can be enhanced by adding pyrene ($C_{16}H_{10}$).[117] Farias et al. (1998) computed the soot radiation properties of phase function, albedo, extinction coefficient, and emissivity using the integral equation formulation for scattering (IEFS).[118]

The properties of a luminous gas can be expressed as:[119,120]

$$\varepsilon_\lambda = 1 - \exp(-\kappa_L L) \tag{3.50}$$

where ε_L is the monochromatic emittance of the luminous gas, κ_L is the absorption coefficient of the luminous gas, and L is the equivalent length of the radiating system. Hottel has recommended the following forms for the absorption coefficient:[121]

$$\kappa_L = \frac{Ck_1}{\lambda^{0.95}} \tag{3.51a}$$

in the infrared region (0.8 to 10 μm), or

$$\kappa_L = \frac{Ck_2}{\lambda^{1.39}} \tag{3.51b}$$

in the visible region (0.3 to 0.8 μm), where C is the soot concentration and k_1 and k_2 are constants specific to the flame under investigation. The need for empirically determined constants is undesirable as each flame must be tested prior to analysis of a given system.[51]

3.3.3.2 Combustion Studies

This chapter section gives a sampling of combustion studies concerning luminous radiation. Leblanc and Goracci (1973) noted the inverse relationship between flame luminosity and flame temperatures.[46] That is, if the flame is very luminous, it radiates its energy more efficiently and therefore has a lower temperature. If it is very nonluminous, then the flame temperature will generally be much higher because it does not as efficiently release its energy. Leblanc and Goracci reported flame emissivities ranging from 0.2 to 0.7. Example 3.10 shows a comparison of how much the flame temperature can change, depending on the emissivity (luminosity) of the flame.

Example 3.10

Given: Flame 1: $\varepsilon_1 = 0.2$, $t_1 = 3000°F$; flame 2: $\varepsilon_2 = 0.7$; load: $t_L = 300°F$.

Find: Find temperature of flame 2 (t_2), assuming same total radiant outputs.

Solution: Since total radiant outputs are the same,

$$\varepsilon_1\left(T_1^4 - T_L^4\right) = \varepsilon_2\left(T_2^4 - T_L^4\right), \text{ or}$$

$$T_2 = \left[\frac{\varepsilon_1}{\varepsilon_2}\left(T_1^4 - T_L^4\right) + T_L^4\right]^{1/4}$$

$$T_2 = 2533°R, \text{ or } t_2 = 2073°F.$$

The flame temperature of the luminous flame is nearly 1000°F (560°C) lower than for the nonluminous flame. Although the assumptions made in Example 3.10 would not be strictly true for most flames, it shows how the flame emissivity affects the flame temperature.

3.3.3.2.1   Total radiation

Beér and Howarth (1968) presented radiation measurements from industrial flames produced by fuels having a variable composition of oil and coke oven gas as shown in Figure 3.32.[122] As expected, the more oil in the flames, the more radiation from the flame. Maesawa et al. (1968) reported radiation measurements from industrial residual oil flames in a vertical cylindrical furnace.[123] As shown in Figure 3.33, the peak emissivity and radiant heat flux occurred at about 1 m (3 ft) from

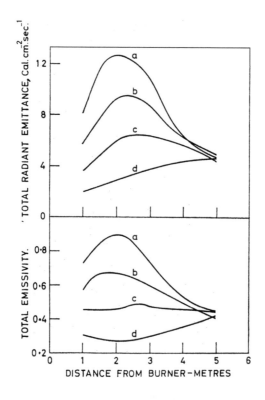

FIGURE 3.32   Effect of fuel composition and distance from an industrial burner for the following fuels: (a) 100% oil, (b) 40% oil/60% coke oven gas, (c) 20% oil/80% coke oven gas, and (d) 100% coke oven gas. (Courtesy of Gulf Publishing, Houston, TX.[122] With permission.)

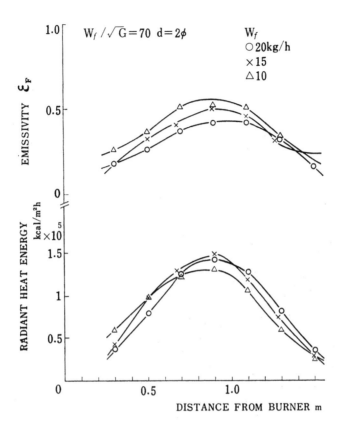

FIGURE 3.33 Measured radiant heat flux and calculated flame emissivity as a function of the distance from the burner for a residual oil flame in a vertical cylindrical furnace. (Courtesy of The Combustion Institute, Pittsburgh, PA.[123] With permission.)

the burner in a 3-m (10 ft) long furnace. Gill et al. (1968) measured and predicted the radiant heat transfer from pulverized coal flames in a 6 m (20 ft) long × 1 m (3 ft) wide rectangular, water-cooled furnace.[124] The peak heat flux occurred at approximately the midpoint of the furnace length. You (1985) measured the radiation from pure diffusion ($\phi = \infty$), natural gas flames.[23] Much of this radiation may have been luminous for these very fuel-rich flames. Radiation to the stagnation point was 13% to 26% of the total heat flux. Radiation was negligible at the edge of the plate (R = 7.3). Hustad calculated the radiant flux to be 7% to 14% and 20% to 40% of the total flux for CH_4 and C_3H_8 pure diffusion flames ($\phi = \infty$), respectively.[37] Hustad assumed that these long flames were optically thick radiating cylinders. The convective flux was calculated using correlations for flow over a cylinder. This was subtracted from the measured total flux to determine a calculated radiant flux. This calculated radiation compared favorably with the measured radiation. The following empirical correlation was derived:

$$q''_{b,\text{rad}} = 5 + 3.9\left(l_j - x_{\text{liftoff}}\right) \tag{3.52}$$

where $q''_{b,rad}$ is in kW/m² and x_{liftoff} is the axial distance from the nozzle exit to the start of the flame. Luminous radiation was not considered in other studies, where the flames were very fuel rich (e.g., Reference 28), although it may have been important.

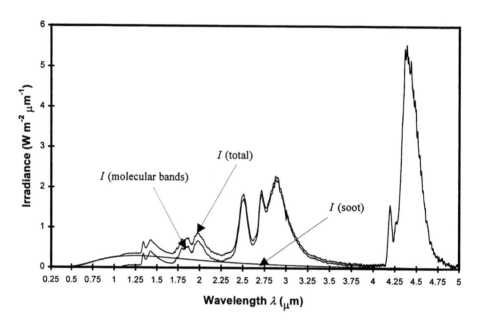

FIGURE 3.34 Soot radiation from a fuel-rich flame (ϕ = 2.00). (From B. Ji and C.E. Baukal, *Proc. 1998 Int. Gas Research Conf.,* Nov. 8-11, 1998, San Diego, CA, D. Dolenc, Ed., 5, 422-433, 1998.)

3.3.3.2.2 Spectral radiation

The Ji and Baukal (1998) study discussed above also investigated luminous flame radiation.[92] The increased penetrating radiation in the extreme fuel-rich case (ϕ = 2.00) reported above comes from the soot radiation. Incomplete combustion causes the fuel to crack, which forms soot particles at high temperatures. These hot soot particles emit blackbody-type radiation. The total radiation in this case is composed of discrete molecular band spectra (from hot combustion products) super-imposed onto a blackbody continuum (from soot particles). The continuum radiation in the ϕ = 2.00 flame spectra can be accounted for by a blackbody at 2318K (3713°F), which has peak radiation at 1.25 μm. Figure 3.34 decomposes the ϕ = 2.00 flame spectra into its soot blackbody continuum and its molecular band spectra components. The discrete band spectra are dominated by the hot combustion products: H_2O bands at 1.14, 1.38, 1.87, and 2.7 μm, and CO_2 bands at 2.7 and 4.3 μm. The CH_4 fuel molecules (2.3 and 3.3 μm bands) and combustion intermediate CO molecules (4.7 μm band) do not make a significant contribution to the flame radiation.

For quantitative characterization of the relative contributions from soot and molecular radiation, a soot radiation index γ was defined as:

$$\gamma = \frac{I_{(\text{soot at } 1.250\,\mu m)}}{I_{(\text{molecule at } 1.346\,\mu m)}} \tag{3.53}$$

Since the soot blackbody radiation peaks at 1.250 μm and there is little molecular radiation at this wavelength, it was therefore assumed that:

$$I_{(\text{soot at } 1.250\,\mu m)} = I_{(\text{total at } 1.250\,\mu m)} \tag{3.54}$$

The H_2O molecular radiation has a sharp rising (band head) peak at 1.346 μm, therefore it can be approximated that:

$$I_{(\text{molecule at } 1.346\,\mu m)} = I_{(\text{total at } 1.346\,\mu m)} - I_{(\text{soot at } 1.250\,\mu m)} \tag{3.55}$$

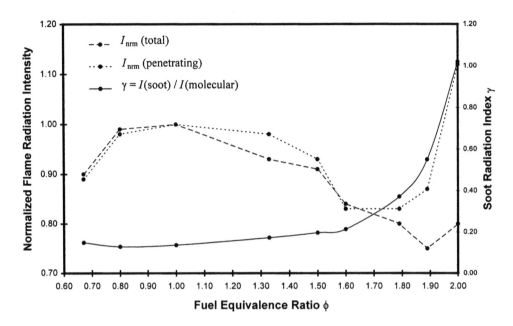

FIGURE 3.35 Normalized total radiation intensity $I_{nrm(total)}$, normalized penetrating radiation intensity $I_{nrm(pen.)}$, and soot radiation index γ as a function of fuel equivalence ratio ϕ for $\Omega = 1.00$ flames. (From B. Ji and C.E. Baukal, *Proc. 1998 Int. Gas Research Conf.*, Nov. 8-11, 1998, San Diego, CA, D. Dolenc, Ed., 5, 422-433, 1998.)

Because the Monolight spectrometer had a nearly flat response sensitivity over the narrow wavelength range from 1.250 to 1.346 µm, the value of soot radiation index γ reported here is independent of the spectrometer calibration. Table 3.5 lists the soot radiation index γ for $q_f = 5.0$ kW flames.

Figure 3.35 shows the normalized total radiation intensity $I_{nrm(total)}$, the normalized penetrating radiation intensity $I_{nrm(pen.)}$, and the soot radiation index γ as a function of the fuel equivalence ratio (ϕ) for $\Omega = 1.00$, $q_f = 5.0$ kW flames. As ϕ increased to fuel rich, both $I_{nrm(total)}$ and $I_{nrm(pen.)}$ decreased first due to incomplete combustion. At $\phi = 1.79$, $I_{nrm(pen.)}$ started to increase due to increased soot radiation. This increase in $I_{nrm(pen.)}$ corresponds to a rapid increase in the soot radiation index γ. Finally, at $\phi = 2.00$, $I_{nrm(total)}$ also increased since the much increased soot radiation became a significant part of the total radiation. The soot radiation index γ correlates closely to the soot concentration in the flame. A new commercial high-radiation burner[125] has a soot index of close to 4, compared to the highest value in this study, which was slightly greater than 1 (see Table 3.5).

3.4 CONDUCTION

Thermal conduction is often overlooked when considering heat transfer in combustion systems. Although it is not an important heat transfer mode in the combustion space, it is important in determining the heat loss through the refractory walls. Conduction is important in the design of furnaces because of the thermal expansion that occurs as the furnace heats up, especially considering the difference in expansion between refractories and the metal shell.[126]

Thermal conduction has played an important role in many flame impingement heating applications. In some processes, high thermal conduction rates are desired. An example is a rapid reheat furnace. There, the goal is to raise the temperature of metal parts. Because metals generally have high thermal conductivities, heat can be quickly conducted through the part. This reduces the temperature gradient between the surface and the interior of the part. High gradients may cause the part to warp or deform. In other applications, low thermal conduction is desired. An example is thermal spallation. In this process, a high-intensity flame impinges directly on a solid that has a low thermal conductivity. The heat transfers slowly into the solid, due to its low conductivity. The surface is very hot. Just below the surface, the solid is near ambient temperature. This results in very large internal temperature gradients, which produce high thermal stresses. These stresses cause the solid to fracture. Thermal spallation has been shown to be a cost-effective method to "drill" through rocks.[35]

In addition to transferring heat through the target, conduction has been used to measure the heat flux in flame impingement experimental studies. Both steady-state and transient methods have been used (see Figure 7.4). These are briefly discussed here. More detailed information on thermal conduction heat transfer is available in books specifically written on that subject.[127-132]

3.4.1 Steady-State Conduction

Thermal conduction through refractory walls is important in determining the heat losses for a given process. The heat loss through a composite wall made of different refractory materials, shown in Figure 3.36, can be calculated using:

$$q = \frac{t_h - t_c}{l_a/k_a A + l_b/k_b A + l_c/k_c A} \tag{3.56}$$

where t_h is the hotface temperature, t_c is the coldface temperature, l_i is the thickness of layer i, k_i is the thermal conductivity of layer i, and A is the cross-sectional area for conduction. This equation assumes that the thermal conductivity of each layer does not change with temperature, which makes the analysis more complicated. It also assumes that there is no contact resistance between each layer. For example, if there is an air gap between each layer, this would give a large resistance to heat flow that should be included. Equation 3.56 is for a three-layer system, but any number of layers can be used by adding or subtracting the appropriate number of thermal conductances ($l_i/k_i A$). This equation can be used to solve for the hot- or coldface temperatures, for the heat flux through the wall, or for a temperature inside the wall. It is assumed that the thicknesses (l_i), thermal conductivities (k_i), and cross-sectional area are known. Table 3.7 shows the thermal conductivity for various ceramics and other insulating materials.[133]

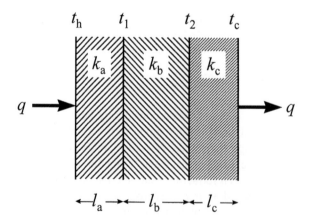

FIGURE 3.36 Thermal conduction through a composite wall.

TABLE 3.7

Thermal Conductivity of Ceramics and Other Insulating Materials

Material	Dens. g/cm³	t °C	Ther. cond. W/m K	Material	Dens. g/cm³	t °C	Ther. cond. W/m K
Alumina (Al_2O_3)	3.8	100	30	Cotton wool	0.08	30	0.04
		400	13	Diatomite	0.2	0	0.05
		1300	6			400	0.09
		1800	7.4		0.5	0	0.09
	3.5	100	17			400	0.16
		800	7.6	Ebonite	1.2	0	0.16
Al_2O_3 + MgO		100	15	Felt, flax	0.2	30	0.05
		400	10		0.3	30	0.04
		1000	5.6	Fuller's earth	0.53	30	0.1
Asbestos	0.4	-100	0.07	Glass wool	0.2	−200 to 20	0.005
		100	1			50	0.04
		0	0.09			100	0.05
Asbestos + 85% MgO	0.3	30	0.08			300	0.08
Asphalt	2.1	20	0.06	Graphite			
Beryllia (BeO)	2.8	100	210	100 mesh	0.48	40	0.18
		400	90	20-40 mesh	0.7	40	1.29
		1000	20	Linoleum cork	0.54	20	0.08
		1800	15	Magnesia (MgO)		100	36
	1.85	50	64			400	18
		200	40			1200	5.8
		600	23			1700	9.2
Brick, dry	1.54	0	0.04	MgO + SiO_2		100	5.3
Brick, refractory:						400	3.5
alosite		1000	1.3			1500	2.3
aluminous	1.99	400	1.2	Mica:			
		1000	1.3	muscovite		100	0.72
diatomaceous	0.77	100	0.2			300	0.65
		500	0.24			600	0.69
	0.4	100	0.08	phlogopite		100	0.66
		500	0.1	Canadian		300	0.19
fireclay	2	400	1			600	0.2
		1000	1.2	Micanite		30	0.3
silicon carbide	2	200	2	Mineral wool	0.15	30	0.04
		600	2.4	Perlite, expanded	0.1	−200 to 20	0.002
vermiculite	0.77	200	0.26	Plastics:			
		600	0.31	bakelite	1.3	20	1.4
Calcium oxide		100	16	celluloid	1.4	30	0.02
		400	9	polystyrenc foam	0.05	−200 to 20	0.033
		1000	7.5	mylar foil	0.05	−200 to 20	0.0001
Cement mortar	2	90	0.55	nylon		−253	0.10
Charcoal	0.2	20	0.055			−193	0.23
Coal	1.35	20	0.26			25	0.30
Concrete	1.6	0	0.8	polytetrafluoroethylene		−253	0.13
Cork	0.05	0	0.03			−193	0.16
		100	0.04			25	0.26
	0.35	0	0.06			230	2.5
		100	0.08	urethane foam	0.07	20	0.06

TABLE 3.7 (continued)
Thermal Conductivity of Ceramics and Other Insulating Materials

Material	Dens. g/cm³	t °C	Ther. cond. W/m K	Material	Dens. g/cm³	t °C	Ther. cond. W/m K
Porcelain		90	1	Uranium dioxide		100	9.8
Rock:						400	5.5
basalt		20	2			1000	3.4
chalk		20	0.92	Wood:			
granite	2.8	20	2.2	balsa, ^	0.11	30	0.04
limestone	2	20	1	fir, ^	0.54	20	0.14
sandstone	2.2	20	1.3	fir, ‖	0.54	20	0.35
slate, ^		95	1.4	oak		20	0.16
slate, ‖		95	2.5	plywood		20	0.11
Rubber:				pine, ^	0.45	60	0.11
sponge	0.2	20	0.05	pine, ‖	0.45	60	0.26
92 percent		25	0.16	walnut, ^	0.65	20	0.14
Sand, dry	1.5	20	0.33	Wool	0.09	30	0.04
Sawdust	0.2	30	0.06	Zinc oxide		200	17
Shellac		20	0.23			800	5.3
Silica aerogel	0.1	−200 to 20	0.003	Zirconia (ZrO₂)		100	2
Snow	0.25	0	0.16			400	2
Steel wool	0.1	55	0.09			1500	2.5
Thoria (ThO₂)		100	10	Zirconia + silica		200	5.6
		400	5.8			600	4.6
		1500	2.4			1500	3.7
Titanium dioxide		100	6.5				
		400	3.8				
		1200	3.3				

Thermal conductivity values for ceramics, refractory oxides, and miscellaneous insulating materials are given here. The thermal conductivity refers to samples with density indicated in the second column. Since most of these materials are highly variable, the values should only be considered as a rough guide.

Source: From *CRC Handbook of Chemistry and Physics,* 79th edition, CRC Press, Boca Raton, FL, 1998, 12-191–12-192.

Example 3.11

Given: Cold face temperature of 140°F, heat flux to the wall of 400 Btu/hr-ft², 8 in. of fireclay brick.

Find: The hot face temperature.

Solution: Assume that the conductivity through the steel shell is negligible. Note that heat flux per unit area is given (q/A). Rearrange the equation above to solve for the hot face temperature:

$$t_h = t_c + \left(\frac{q}{A}\right)\left(\frac{l_{firebrick}}{k_{firebrick}}\right)$$

From Table 3.7, $k_{firebrick}$ = 1.1 W/m-K = 0.64 Btu/hr-ft-°F,

$l_{firebrick}$ = 8/12 ft, t_c = 140°F

solving for t_h = 557°F

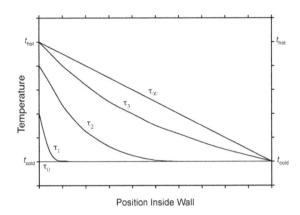

FIGURE 3.37   Transient thermal conduction through a wall.

Steady-state conduction has been important in flame impingement studies. Internally cooled targets have been used in many of the experimental studies.[15,17,19,21,26-29,31,33,34,36,89,90] The flame impinges on the hot side of the target. The coolant flows past the cold side of the target (see Figures 7.1 and 7.2). In other studies, the target has been an uncooled ceramic.[11,13,14,18,27] In both the cooleds metal and uncooled ceramic studies, the hot side and cold side temperatures have been approximately constant, producing steady-state conduction through the target.

Steady-state conduction methods have been employed in the design of heat flux gages (see Chapter 6) used in flame impingement studies. These methods involve measuring the temperature gradient through a piece of metal in the gage. The gage is heated on one side by the flame. The other side is commonly maintained at a fixed temperature, using a coolant. Long thin solid rods,[11,15,17,28,33,36] thin disks, and heat flux transducers· have been used as gages.

3.4.2   TRANSIENT CONDUCTION

Transient conduction is an important aspect of heat transfer in industrial combustion that is often overlooked. It is significant during start-up of a process where the heat distribution through the walls of a furnace has not reached the steady state. Figure 3.37 shows how the temperature distribution through a refractory wall changes as a function of time. At τ_o, the temperature through the wall is uniform (t_{cold}), when heat is then applied to the left side of the wall. After some time (τ_∞), the wall temperature distribution reaches steady-state.

Another instance where transient conduction is important is during the heating of a load in a batch heating process. The load is initially cold after it first is charged into the furnace. As heat is applied to the surface, the surface temperature rises quickly, while the temperatures inside the load are still low until the heat has time to conduct through the load. The thermal conductivity of the load determines how fast the temperature profile equilibrates inside the load. For example, aluminum conducts very quickly compared to refractory materials like bricks or ceramics. This can be an important consideration if the material conducts slowly, as the surface can be overheated before the heat has a chance to conduct inside the load. This reduces the product quality and process yield.

Uncooled metal targets have been used in previous flame impingement experimental studies (see Chapter 7). The target materials have been steel,[12,18,30,37] inconel,[32] iron,[87] molybdenum,[16] brass,[20] and copper.[35] These targets were exposed to the flame for only a short time. Kilham[18] and Nawaz[20] inserted the target into the flame for less than 1 sec. Giedt and co-workers had test durations of between 20 and 50 sec.[16,87] The targets were instrumented with an array of thermocouples. The heat flux to the surface was estimated using an inverse transient heat conduction calculation. This method was discussed by Gebhart.[131]

Transient conduction has also been used in the design of slug calorimeter heat flux gages (see Chapter 6), as used in previous studies.[18-20,32,34,36,37] Heat is received on a slug of metal inside the calorimeter. The slug is assumed to be internally isothermal at any instant in time. The heat flux is calculated from the rise in the thermal capacitance of the slug as a function of time.

3.5 PHASE CHANGE

The three primary phase changes that can be encountered in industrial combustion applications include melting, boiling, and condensation. These are briefly discussed here.

3.5.1 MELTING

Melting involves changing the phase of a solid to a liquid. This occurs in many industrial combustion applications. Examples include melting of scrap aluminum to make aluminum alloys, melting scrap steel in the production of steel alloys, and melting of batch materials like silica used in the production of glass. Melting involves heating the solid up to the melting temperature and then supplying enough energy to change the phase from a solid to a liquid (heat of fusion):

$$q_m = c_p\left(t_m - t_0\right) + h_{fusion} \tag{3.57}$$

where c_p is the specific heat of the solid, t_0 is the initial temperature of the solid, t_m is the melting temperature of the solid, and h_{fusion} is the heat of fusion of the solid. Typical melting temperatures, specific heats, and heats of fusion for various solids are given in Table 3.8.

Example 3.12

    Find: From Table 3.8, the melting temperature of aluminum is 660.32°C, the specific heat is 0.897 J/g-K, and the heat of fusion is 397 J/g.

    Given: Calculate how much energy is required to melt aluminum that is initially at 20°C.

    Solution: The energy to heat the aluminum to the melting temperature and change it from a solid to a liquid =

                $q_1 = c_p\,(t_m - t_0) = (0.897\ \text{J/g-K})\,(660.32 - 20)\ \text{K} = 574\ \text{J/g}$

                $q_2 = 397\ \text{J/g}$

                $q_m = q_1 + q_2 = 574 + 397 = 971\ \text{J/g} = 971\ \text{kJ/kg}$

3.5.2 BOILING

Boiling involves changing the phase of a liquid to a vapor. The most common occurrence of boiling is when the heat from the combustion process is used to vaporize a process fluid or fluid like water used to simulate the heat flux to a load. For example, both the internal and external boiling of water have been considered in previous flame impingement experimental studies (see Figure 7.7). In internal boiling, the coolant vaporizes inside the target. In external boiling, water, in the combustion products, vaporizes on the outside of the target. Both types are sometimes referred to as forced convection boiling.[134] This refers to boiling where the flow is through a passage or over a surface. The flow is two-phase since both liquid and vapor are present. Only internal boiling has been explicitly discussed in previous impingement studies. Although not discussed in any of those studies, external boiling may have occurred in some of the previous experiments. The effects of boiling were not quantified in any study. Further information on this topic is available in books specifically devoted to boiling.[135-142]

TABLE 3.8
Melting Temperatures and Heats of Fusion for Various Materials

Metal (symbol)	Atomic weight	t_m °C	t_b °C	ΔH_{fus} J/g	ρ_{25} g/cm³	$\alpha \times 10^6$ K⁻¹	c_p J/g K	λ W/cm K
Actinium (Ac)		1051	3198		10			
Aluminum (Al)	26.98	660.32	2519	397	2.70	23.1	0.897	2.37
Antimony (Sb)	121.76	630.63	1587	163.2	6.68	11.0	0.207	0.243
Barium (Ba)	137.33	727	1897	52	3.62	20.6	0.204	0.184
Beryllium (Be)	9.01	1287	2471	877	1.85	11.3	1.825	2.00
Bismuth (Bi)	208.98	271.40	15.64	54.1	9.79	13.4	0.122	0.0787
Cadmium (Cd)	112.41	321.07	767	55.1	8.69	30.8	0.232	0.968
Calcium (Ca)	40.08	842	1484	213	1.54	22.3	0.647	2.00
Cerium (Ce)	140.11	798	3443	39.0	6.77	6.3	0.192	0.113
Cesium (Cs)	132.91	28.44	671	15.8	1.93		0.242	0.359
Chromium (Cr)	52.00	1907	2671	404	7.15	4.9	0.449	0.937
Cobalt (Co)	58.93	1495	2927	275	8.86	13.0	0.421	1.00
Copper (Cu)	63.55	1084.62	2562	208.7	8.96	16.5	0.385	4.01
Dysprosium (Dy)	162.50	1412	2567	68.1	8.55	9.9	0.170	0.107
Erbium (Er)	167.26	1529	2868	119	9.07	12.2	0.168	0.145
Europium (Eu)	151.96	822	1529	60.6	5.24	35.0	0.182	0.139[a]
Gadolinium (Gd)	157.25	1313	3273	63.6	7.90	9.4[b]	0.236	0.105
Gallium (Ga)	69.72	29.76	2204	80.2	5.91		0.371	0.406
Gold (Au)	196.97	1064.18	2856	63.7	19.3	14.2	0.129	3.17
Hafnium (Hf)	178.49	2233	4603	152.4	13.3	5.9	0.144	0.230
Holmium (Ho)	164.93	1474	2700	103[a]	8.80	11.2	0.165	0.162
Indium (In)	114.82	156.60	2072	28.6	7.31	32.1	0.233	0.816
Iridium (Ir)	192.22	2446	4428	213.9	22.5	6.4	0.131	1.47
Iron (Fe)	55.85	1538	2861	247.3	7.87	11.8	0.449	0.802
Lanthanum (La)	138.91	918	3464	44.6	6.15	12.1	0.195	0.134
Lead (Pb)	207.20	327.46	1749	23.0	11.3	28.9	0.129	0.353
Lithium (Li)	6.94	180.5	1342	432	0.534	46	3.582	0.847
Lutetium (Lu)	174.97	1663	3402	126[a]	9.84	9.9	0.154	0.164
Magnesium (Mg)	24.30	650	1090	349	1.74	24.8	1.023	1.56
Manganese (Mn)	54.94	1246	2061	235.0	7.3	21.7	0.479	0.0782
Mercury (Hg)	200.59	−38.83	356.73	11.4	13.5336		0.140	0.0834
Molybdenum (Mo)	95.94	2623	4639	390.7	10.2	4.8	0.251	1.38
Neodymium (Nd)	144.24	1021	3074	49.5	7.01	9.6	0.190	0.165
Neptunium (Np)		644			20.2			0.063
Nickel (Ni)	58.69	1455	2913	298	8.90	13.4	0.444	0.907
Niobium (Nb)	92.91	2477	4744	323	8.57	7.3	0.265	0.537
Osmium (Os)	190.23	3033	5012	304.2	22.59	5.1	0.130	0.876
Palladium (Pd)	106.42	1554.9	2963	157.3	12.0	11.8	0.246	0.718
Platinum (Pt)	195.08	1768.4	3825	113.6	21.5	8.8	0.133	0.716
Plutonium (Pu)		640	3228		19.7	46.7		0.0674
Polonium (Po)		254	962		9.20			0.20
Potassium (K)	39.10	63.38	759	59.3	0.89		0.757	1.024
Praseodymium (Pr)	140.91	931	3520	48.9	6.77	6.7	0.193	0.125
Promethium (Pm)		1042	3000[a]	53[a]	7.26	11[a]	0.19[a]	0.15[a]
Protactinium (Pa)	231.04	1572		53.4	15.4			
Radium (Ra)		700			5			
Rhenium (Re)	186.21	3186	5596	324.5	20.8	6.2	0.137	0.479

TABLE 3.8 (continued)
Melting Temperatures and Heats of Fusion for Various Materials

Metal (symbol)	Atomic weight	t_m °C	t_b °C	ΔH_{fus} J/g	ρ_{25} g/cm^3	$\alpha \times 10^6$ K^{-1}	c_p J/g K	λ W/cm K
Rhodium (Rh)	102.91	1964	3695	258.4	12.4	8.2	0.243	1.50
Rubidium (Rb)	85.47	39.31	688	25.6	1.53		0.363	0.582
Ruthenium (Ru)	101.07	2334	4150	381.8	12.1	6.4	0.238	1.17
Samarium (Sm)	150.36	1074	1794	57.3	7.52	12.7	0.197	0.133
Scandium (Sc)	44.96	1541	2836	314	2.99	10.2	0.568	0.158
Silver (Ag)	107.87	961.78	2162	104.8	10.5	18.9	0.235	4.29
Sodium (Na)	22.99	97.72	883	113	0.97	71	1.228	1.41
Strontium (Sr)	87.62	777	1382	84.8	2.64	22.5	0.301	0.353
Tantalum (Ta)	180.95	3017	5458	202.1	16.4	6.3	0.140	0.575
Technetium (Tc)		2157	4265		11			0.506
Terbium (Tb)	158.93	1356	3230	67.9	8.23	10.3	0.182	0.111
Thallium (Tl)	204.38	304	1473	20.3	11.8	29.9	0.129	0.461
Thorium (Th)	232.04	1750	4788	59.5	11.7	11.0	0.113	0.540
Thulium (Tm)	168.93	1545	1950	99.4	9.32	13.3	0.160	0.169
Tin (Sn)	118.71	231.93	2602	59.2	7.26	22.0	0.228	0.666
Titanium (Ti)	47.88	1668	3287	295	4.51	8.6	0.523	0.219
Tungsten (W)	183.84	3422	5555	284.5	19.3	4.5	0.132	1.74
Uranium (U)	238.03	1135	4131	38.4	19.1	13.9	0.116	0.276
Vanadium (V)	50.94	1910	3407	422	6.0	8.4	0.489	0.307
Ytterbium (Yb)	173.04	819	1196	44.3	6.90	26.3	0.155	0.385
Yttrium (Y)	88.91	1522	3345	128	4.47	10.6	0.298	0.172
Zinc (Zn)	65.39	419.53	907	112	7.14	30.2	0.388	1.16
Zirconium (Zr)	91.22	1855	4409	230	6.52	5.7	0.278	0.227

[a] Estimated.
[b] At 100°C.

Source: Courtesy of CRC Press, Boca Raton, FL.[133]

3.5.2.1 Internal Boiling

One example of internal boiling has been inside targets heated by flame impingement. Kilham and co-workers used targets that were cooled by glycol to prevent boiling inside the targets.[15,17,33,36] Rajani used a water-cooled steel disk as a target.[31] The target surface temperature was estimated to be 400 to 480°F (200 to 250°C). Calculations showed that boiling inside the target was probable, based on the expected heat flux rates. To prevent boiling, internal fins were added in the cooling passages. This increased both the internal surface area and the turbulence level of the cooling water flow. Both of these factors increased the effective convective heat transfer coefficient inside the target. Then, the calculated convection coefficient was high enough to preclude boiling in the target. However, no measurements were made to determine if boiling did occur.

3.5.2.2 External Boiling

External boiling is important in many processes involving drying of a material, particularly a surface. For example, removing water from the surface of wet metal scrap material can involve external boiling to vaporize the free moisture. This occurs during the initial heating phase of scrap melting processes, such as steel or aluminum, when the scrap is wet from recent rain. This moisture vaporization can represent a large additional heat load to the process that may vary from day to day and from season to season depending on the weather. This mechanism has not been discussed in any of the previous flame impingement studies although it may have been present. Kilham and Purvis[18] and Nawaz[20] studied O_2/CH_4 flames. The combustion products should have been about 67% water vapor by volume. These flames impinged normal to the nose of an uncooled, stainless steel, hemi-nosed cylinder. The cylinder was inserted into the flame with a solenoid. The temperature rise of the cylinder, as a function of time, was recorded. The heat flux was determined by calculating the rate of temperature rise, in the range between about 63°F and 300°F (17°C and 150°C). Because of the high concentration of water in the impinging gases, some of the water may have condensed on the target when it was below 212°F (100°C). The condensed water may then have boiled off the target, as its temperature increased above 212°F (100°C). The boiling would have been in the temperature range that was used to calculate the heat flux. If boiling was important, there may have been a plateau in the temperature curves (which were not given), near 212°F (100°C). Rauenzahn studied stoichiometric, turbulent O_2/C_3H_8 flames impinging on an uncooled copper slab.[35] The measured slab surface temperature ranged from 63°F to 410°F (17°C to 210°C). The combustion products should have been about 57% water by volume. Again, that water may have condensed on the target when the surface temperature was below 212°F (100°C), and then subsequently boiled off when the surface temperature increased above 212°F (100°C).

3.5.3 CONDENSATION

Condensation involves converting a vapor to a liquid. This is not often encountered in industrial combustion applications. However, it has been used in some flame impingement studies that attempted to determine the heat flux to a simulated load (see Figure 7.8). Kilham and co-workers used targets internally cooled by ethylene glycol.[15,17,33,36] The target surface temperature was maintained above the boiling point of water (212°F [100°C]) and below that of the glycol (388°F [198°C]). This prevented the water in the combustion products from condensing on the outside of the target. It also prevented the glycol from boiling inside the target. Water-cooled targets have been used in many studies.[21,26-29,34,89,90] The target surface temperature has commonly been maintained below 212°F (100°C) to prevent the cooling water from boiling inside the target. However, the water in the combustion products may have condensed on the outside of the target. Such condensation heat transfer effects were not considered in those studies. For air/fuel flames, the water content in the combustion products is only about 18% by volume. However, in O_2/CH_4 flames, for example, the combustion products may contain up to 67% water by volume. Condensation may be important in any study (e.g., Reference 18) where O_2 is the oxidizer, and where the target surface temperature is below 212°F (100°C).

REFERENCES

1. T.M. Smith, private communication, 1982.
2. V.S. Arpaci, *Convection Heat Transfer*, Prentice-Hall, Englewood Cliffs, NJ, 1984.
3. C.S. Fang, *Convective Heat Transfer*, Gulf Publishing, Houston, TX, 1985.
4. S. Kakac, R.K. Shah, and W. Aung, Eds., *Handbook of Single-Phase Convective Heat Transfer*, Wiley, New York, 1987.
5. L.C. Burmeister, *Convective Heat Transfer*, 2nd edition, Wiley, New York, 1993.
6. W.M. Kays and M.E. Crawford, *Convective Heat and Mass Transfer*, 3rd edition, McGraw-Hill, New York, 1993.
7. A. Bejan, *Convection Heat Transfer*, 2nd edition, Wiley, New York, 1994
8. M. Kaviany, *Principles of Convective Heat Transfer*, Springer-Verlag, New York, 1994.
9. S. Kakac and Y. Yener, *Convective Heat Transfer*, 2nd edition, CRC Press, Boca Raton, FL, 1995.
10. P.H. Oosthuizen, *An Introduction to Convective Heat Transfer*, McGraw-Hill, New York, 1999.

11. J.M. Beér and N.A. Chigier, Impinging jet flames, *Comb. Flame*, 12, 575-586, 1968.
12. A. Milson and N.A. Chigier, Studies of methane-air flames impinging on a cold plate, *Comb. Flame*, 21, 295-305, 1973.
13. E.G. Jackson and J.K. Kilham, Heat transfer from combustion products by forced convection, *Ind. Eng. Chem.*, 48(11), 2077-2079, 1956.
14. J.K. Kilham, Energy transfer from flame gases to solids, *Third Symposium on Combustion and Flame and Explosion Phenomena*, The Williams and Wilkins Co., Baltimore, MD, 1949, 733-740.
15. R.A. Cookson and J.K. Kilham, Energy transfer from hydrogen-air flames, *Ninth Symposium (International) on Combustion*, Academic Press, New York, 1963, 257-263.
16. L.W. Woodruff and W.H. Giedt, Heat transfer measurements from a partially dissociated gas with high Lewis number, *J. Heat Transfer*, 88, 415-420, 1966.
17. J.K. Kilham and P.G. Dunham, Energy transfer from carbon monoxide flames, *Eleventh Symposium (International) on Combustion*, The Combustion Institute, Pittsburgh, PA, 1967, 899-905.
18. J.K. Kilham and M.R.I. Purvis, Heat transfer from hydrocarbon-oxygen flames, *Comb. Flame*, 16, 47-54, 1971.
19. R. Conolly and R.M. Davies, A study of convective heat transfer from flames, *Int. J. Heat Mass Transfer*, 15, 2155-2172, 1972.
20. S. Nawaz, Heat Transfer from Oxygen Enriched Methane Flames, Ph.D. thesis, The University of Leeds, Leeds, U.K., 1973.
21. C.J. Hoogendoorn, C.O. Popiel, and T.H. van der Meer, Turbulent heat transfer on a plane surface in impingement round premixed flame jets, *Proc. of 6th Int. Heat Trans. Conf.*, Toronto, 4, 107-112, 1978.
22. J.K. Kilham and M.R.I. Purvis, Heat transfer from normally impinging flames, *Comb. Sci. Tech.*, 18, 81-90, 1978.
23. H.-Z. You, An investigation of fire-plume impingement on a horizontal ceiling: 2. Impingement and ceiling-jet regions, *Fire & Materials*, 9(1), 46-56, 1985.
24. M. Sibulkin, Heat transfer near the forward stagnation point of a body of revolution, *J. Aero. Sci.*, 19, 570-571, 1952.
25. C.E. Baukal and B. Gebhart, A review of semi-analytical solutions for flame impingement heat transfer, *Int. J. Heat Mass Transfer*, 39(14), 2989-3002, 1996.
26. S.N. Shorin and V.A. Pechurkin, Effectivnost' teploperenosa na poverkhnost' plity ot vysokotemperaturnoi strui produktov sjoraniya razlichnykh gazov, *Teoriya i Praktika Szhiganiya Gaza*, 4, 134-143, 1968.
27. J.-P. Vizioz and T.M. Lowes, Convective Heat Transfer from Impinging Flame Jets, Int'l Flame Res. Found. Report F 35/a/6, IJmuiden, The Netherlands, 1971.
28. E. Buhr, G. Haupt, and H. Kremer, Heat transfer from impinging turbulent jet flames to plane surfaces, *Combustion Institute European Symposium 1973*, F.J. Weinberg, Ed., Academic Press, New York, 1973, 607-612.
29. R.B. Smith and T.M. Lowes, Convective Heat Transfer from Impinging Tunnel Burner Flames — A Short Report on the NG-4 Trials, Int. Flame Res. Found. Report F 35/a/9, IJmuiden, The Netherlands, 1974.
30. M. Matsuo, M. Hattori, T. Ohta and S. Kishimoto, The Experimental Results of the Heat Transfer by Flame Impingement, Int. Flame Res. Found. Report F 29/1a/1, IJmuiden, The Netherlands, 1978.
31. J.B. Rajani, R. Payne and S. Michelfelder, Convective Heat Transfer from Impinging Oxygen–Natural Gas Flames. Experimental Results from the NG5 Trials, Int. Flame Res. Found. Report F 35/a/12, IJmuiden, The Netherlands, 1978.
32. A. Ivernel and P. Vernotte, Etude expérimentale de l'amélioration des transferts convectis dans les fours par suroxygénation du comburant, Rev. Gén. Therm., Fr., Nos. 210-211, pp. 375-391, 1979.
33. G.K. Hargrave and J.K. Kilham, The Effect of Turbulence Intensity on Convective Heat Transfer From Premixed Methane-Air Flames, *Inst. Chem. Eng. Symp. Ser.*, 2(86), 1025-1034, 1984.
34. M.E. Horsley, M.R.I. Purvis, and A.S. Tariq, Convective heat transfer from laminar and turbulent premixed flames, *Heat Transfer 1982*, U. Grigull, E. Hahne, K. Stephen, and J. Straub, Eds., Hemisphere, Washington, D.C., 3, 409-415, 1982.
35. R.M. Rauenzahn, Analysis of Rock Mechanics and Gas Dynamics of Flame-Jet Thermal Spallation Drilling, Ph.D. thesis, MIT, Cambridge, MA, 1986.
36. G.K. Hargrave, M. Fairweather, and J.K. Kilham, Forced convective heat transfer from premixed flames. 2. Impingement heat transfer, *Int. J. Heat Fluid Flow*, 8(2), 132-138, 1987.

37. J.E. Hustad and O.K. Sønju, Heat transfer to pipes submerged in turbulent jet diffusion flames, in *Heat Transfer in Radiating and Combusting Systems*, Springer-Verlag, Berlin, 1991, 474-490.

38. T.H. van der Meer, Stagnation point heat transfer from turbulent low Reynolds number jets and flame jets, *Exper. Therm. Fluid Sci.*, 4, 115-126, 1991.

39. V.I. Babiy, Solid/gas phase heat exchange in combustion of powdered fuel, in *Heat Transfer in Flames*, N.H. Afgan and J.M. Beer, Eds., Scripta Book Company, Washington, D.C., 1974, chap. 7, 131-139.

40. J. Griswold, *Fuels, Combustion, and Furnaces*, McGraw-Hill, New York, 1946.

41. C.C. Monrad, Heat Transmission in Convection Sections of Pipe Stills, *Ind. Eng. Chem.*, 24, 505-509, 1932.

42. B. Gebhart, Y. Jaluria, R. Mahajan, and B. Sammakia, *Buoyancy-Induced Flows and Transport*, Hemisphere, New York, 1988.

43. Y.B. Wang, C. Chaussavoine, and F. Teyssandier, Two-dimensional modelling of a non-confined circular impinging jet reactor — fluid dynamics and heat transfer, *Int. J. Heat Mass Trans.*, 36(4), 857-873, 1993.

44. G.M. Faeth, J.P. Gore, S.G. Chuech, and S.-M. Jeng, Radiation from turbulent diffusion flames, *Annu. Rev. of Fluid Mech. Heat Transfer*, 2, 1-38, 1989.

45. W.A. Gray and R. Müller, *Engineering Calculations in Radiative Heat Transfer*, Pergamon, Oxford, U.K., 1974.

46. B. Leblanc and E. Goracci, Example of applications in the field of heat transfer in hot wall furnaces, *La Rivista dei Combustibili*, 27(4-5), 155-164, 1973.

47. H. Schmidt, Prüfung der Strahlungsgesetze der Bunsenflamme, *Annln. Phys.*, 29, 971-1028, 1909.

48. T.M. Lowes and A.J. Newall, Problems associated with the characterisation (sic) of flame radiation in glass tanks, *Glass Tech.*, 12(2), 32-35, 1971.

49. B. Leblanc, A.B. Ivernel, and J. Chedaille, New concepts on heat transfer processes in high temperature furnaces, Heat transfer, *Proceedings of International Heat Transfer Conference, 5th,* Vol. 1, Japanese Soc. Mech. Eng., Tokyo, 1974, 61-65.

50. D.A. Lihou, Review of furnace design methods, *Trans. IChemE*, 55(4), 225-242, 1977.

51. J.A. Wiebelt, *Engineering Radiation Heat Transfer*, Holt, Rinehart and Winston, New York, 1966.

52. H.C. Hottel and A.F. Sarofim, *Radiative Transfer*, McGraw-Hill, New York, 1967.

53. T.J. Love, *Radiative Heat Transfer*, Merrill Publ., Columbus, OH, 1968.

54. E.M. Sparrow and R.D. Cess, *Radiation Heat Transfer*, augmented edition, Hemisphere, Washington, D.C., 1978.

55. D.K. Edwards, *Radiation Heat Transfer Notes*, Hemisphere, Washington, D.C., 1981.

56. R. Siegel and J.R. Howell, *Thermal Radiation Heat Transfer*, Second Edition, Hemisphere, Washington, D.C., 1981.

57. M.Q. Brewster, *Thermal Radiative Transfer and Properties*, Wiley, New York, 1992.

58. M.F. Modest, *Radiative Heat Transfer*, McGraw-Hill, New York, 1993.

59. D.H. Hubble, Steel plant refractories, in *The Making, Shaping and Treating of Steel*, 11th edition, Steelmaking and Refining Volume, R.J. Fruehan, Ed., AISE Steel Foundation, Pittsburgh, PA, 1998, 159-290.

60. M.A. Glinkov, Flame as a problem of the general theory of furnaces, in *Heat Transfer in Flames*, N.H. Afgan and J.M. Beer, Eds., Scripta Book Company, Washington, D.C., 1974, chap. 9, 159-177.

61. P. Docherty and R.J. Tucker, The influence of wall emissivity on furnace performance, *J. Inst. Energy*, 59(438), 35-37, 1986.

62. D.G. Elliston, W.A. Gray, D.F. Hibberd, T.-Y. Ho, and A. Williams, The effect of surface emissivity on furnace performance, *J. Inst. Energy*, 60(445), 155-167, 1987.

63. C.L. DeBellis, Effect of refractory emittance in industrial furnaces, in *Fundamentals of Radiation Heat Transfer*, W.A. Fiveland, A.L. Crosbie, A.M. Smith, and T.F. Smith, Eds., ASME HTD-Vol. 160, New York, 1991, 105-115.

64. C.L. DeBellis, Evaluation of high-emittance coatings in a large industrial furnace, in *Heat Transfer in Fire and Combustion Systems — 1993*, B. Farouk, M.P. Menguc, R. Viskanta, C. Presser, and S. Chellaiah, Eds., New York, ASME HTD-Vol. 250, 1993, 190-198.

65. X. Xin, G. Qingchang, and J. Yi, High emissivity refractory coating improves furnace output in China, *Industrial Heating*, LXVI(6), 49-50, 1999.

66. T.F. Wall, S.P. Bhattacharya, D.K. Zhang, R.P. Gupta, and X. He, The properties and thermal effects of ash deposits in coal fired furnaces, *Prog. Energy Combust. Sci.*, 19, 487-504, 1993.

67. T.F. Wall, S.P. Bhattacharya, L.L. Baxter, G. Richards, and J.N. Harb, The character of ash deposits and the thermal performance of furnaces, *Fuel Processing Technology*, 44, 143-153, 1995.

68. W.E. Garner, Radiant energy from flames, *First Symposium (International) on Combustion*, The Combustion Institute, Pittsburgh, PA, 1965, 19-23.

69. A.F. Sarofim, Radiative heat transfer in combustion: friend or foe, Hoyt C. Hottel Plenary Lecture, *Twenty-First Symp. (Int.) on Combustion*, The Combustion Institute, Pittsburgh, PA, 1986, 1-23.

70. C.B. Ludwig, W. Malkmus, J.E. Reardon, and J.A.L. Thomson, *Handbook of Infrared Radiation from Combustion Gases*, National Aeronautics and Space Administration Report SP-3080, Washington, D.C., 1973.

71. B. Leckner, Spectral and total emissivity of water vapor and carbon dioxide, *Comb. Flame*, 19, 33-48, 1972.

72. M. Gulic, A new formula for determining the effective beam length of gas layer or flame, in *Heat Transfer in Flames*, N.H. Afgan and J.M. Beer, Eds., Scripta Book Company, Washington, D.C., 1974, chap. 12, 201-208.

73. R.D. Cess, Infrared gaseous radiation, in *Heat Transfer in Flames*, N.H. Afgan and J.M. Beer, Eds., Scripta Book Company, Washington, D.C., 1974, chap. 15, 231-248.

74. R. Grief, Experimental and theoretical results with infrared radiating gases, in *Heat Transfer in Flames*, N.H. Afgan and J.M. Beer, Eds., Scripta Book Company, Washington, D.C., 1974, chap. 16, 249-253.

75. J.L. Kovotny, The effect of pressure on heat transfer in radiating gases, in *Heat Transfer in Flames*, N.H. Afgan and J.M. Beer, Eds., Scripta Book Company, Washington, D.C., 1974, chap. 17, 255-269.

76. D.K. Edwards, Molecular gas band radiation, in *Advances in Heat Transfer*, Vol. 12, T.F. Irvine and J.P. Hartnett, Eds., Academic Press, New York, 1976, 115-193.

77. H.E. Trout, Heat transfer. IV. Radiation from gases and flames, *Industrial Heating*, Vol. XLIV, No. 8, 28-33, 1977.

78. C.L. Tien, Thermal Radiation Properties of Gases, *Advances in Heat Transfer*, T.F. Irvine and J.P. Hartnett, Eds., Academic Press, New York, 5, 253-324, 1968.

79. C.L. Tien and S.C. Lee, Flame Radiation, *Prog. Energy Combust. Sci.*, 8, 41-59, 1982.

80. D.K. Edwards and A. Balakrishnan, Thermal radiation by combustion gases, *Int. J. Heat Mass Transfer*, 16, 25-40, 1973.

81. P.B. Taylor and P.J. Foster, The total emissivities of luminous and non-luminous flames, *Int. J. Heat Mass Transfer*, 17, 1591-1605, 1974.

82. A. Coppalle and P. Vervisch, The total emissivities of high-temperature flames, *Combust. Flame*, 49, 101-108, 1983.

83. D.K. Edwards and R. Matavosian, Scaling rules for total absorptivity and emissivity of gases, *J. Heat Transfer*, 106, 684-689, 1984.

84. J.R. Howell, Thermal radiation in participating media: the past, the present, and some possible futures, *J. Heat Transf.*, 110, 1220-1229, 1988.

85. J.A. Wieringa, J.J. Ph. Elich, and C.J. Hoogendoorn, Spectral effects of radiative heat-transfer in high-temperature furnaces burning natural gas, *J. Inst. Energy*, 63(456), 101-108, 1990.

86. P.G. Dunham, Convective Heat Transfer from Carbon Monoxide Flames, Ph.D. thesis, The University of Leeds, Leeds, U.K., 1963.

87. W.H. Giedt, L.L. Cobb, and E.J. Russ, Effect of Hydrogen Recombination on Turbulent Flow Heat Transfer, ASME Paper 60-WA-256, New York, 1960.

88. M.R.I. Purvis, Heat Transfer from Normally Impinging Hydrocarbon Oxygen Flames to Surfaces at Elevated Temperatures, Ph.D. thesis, The University of Leeds, Leeds, U.K., 1974.

89. D.R. Davies, Heat Transfer from Working Flame Burners, B.S. thesis, Univ. of Salford, Salford, U.K., 1979.

90. T.H. van der Meer, Heat Transfer from Impinging Flame Jets, Ph.D. thesis, Technical University of Delft, the Netherlands, 1987.

91. C.E. Baukal and B. Gebhart, Oxygen-enhanced/natural gas flame radiation, *Int. J. Heat Mass Transfer*, 44(11), 2539-2547, 1997.

92. B. Ji and C.E. Baukal, Spectral radiation properties of oxygen-enhanced/natural gas flames, *Proc. 1998 International Gas Research Conference*, 8-11 November 1998, San Diego, CA, Dan Dolenc, Ed., 5, 422-433, 1998.

93. D.R. Davies and P.J.,Young, Strip Edge Heating Burner, U.S. Patent 4,756,685, issued July 12, 1988.
94. C.E. Baukal, Heat Transfer from Flames Impinging Normal to a Plane Surface, Ph.D. thesis, Univ. of Pennsylvania, 1996.
95. W. R. Anderson, Measurement of the Line Reversal Temperature of OH in CH_4/N_2O Flames, Technical Report ARBRL-TR-02280, 1981.
96. S. Yagi and H. Iino, Radiation from soot particles in luminous flames, *Eighth Symposium (International) on Combustion*, The Combustion Institute, Pittsburgh, PA, 1961 (reprinted 1991), 288-293.
97. R. Echigo, N. Nishiwaki and M. Hirata, A study on the radiation of luminous flames, *Eleventh Symposium (International) on Combustion*, The Combustion Institute, Pittsburgh, PA, 1967, 381-389.
98. W.A. Gray, J.K. Kilham, and R. Müller, *Heat Transfer from Flames*, Elek Science, London, 1976.
99. H.G. Wagner, Soot Formation in Combustion, *Seventeenth Symposium (International) on Combustion*, The Combustion Institute, Pittsburgh, PA, 1978, 3-19.
100. I. Glassman, Soot Formation in Combustion Processes, *Twenty-Second Symposium (International) on Combustion*, The Combustion Institute, Pittsburgh, PA, 1988, 295-311.
101. E. Talmor, *Combustion Hot Spot Analysis for Fired Process Heaters*, Gulf Publishing, Houston, TX, 1982.
102. B.S. Haynes, Soot and Hydrocarbons in Combustion, in *Fossil Fuel Combustion*, W. Bartok and A.F. Sarofim, Eds., Wiley, New York, 1991, chap. 5.
103. P.B. Sunderland, S. Mortazavi, G.M. Faeth, and D.L. Urban, Laminar smoke points of nonbuoyant jet diffusion flames, *Combust. Flame*, 96, 97-103, 1994.
104. J. Lahaye and G. Prado, Eds., *Soot in Combustion Systems and its Toxic Properties*, Plenum Press, New York, 1983.
105. J.P. Longwell, Polycyclic aromatic hydrocarbons and soot, in *Soot in Combustion Systems and its Toxic Properties*, J. Lahaye and G. Prado Eds., Plenum Press, New York, 1983, 37-56.
106. J.B. Howard and J.D. Bittner, Structure of Sooting Flames, in *Soot in Combustion Systems and its Toxic Properties*, J. Lahaye and G. Prado Eds., Plenum Press, New York, 1983, 57-93.
107. J.D. Bittner, J.B. Howard, and H.B. Palmer, Chemistry of intermediate species in the rich combustion of benzene, in *Soot in Combustion Systems and its Toxic Properties*, J. Lahaye and G. Prado Eds., Plenum Press, New York, 1983, 95-125.
108. G. Prado, J. Lahaye, and B.S. Haynes, Soot Particle Nucleation and Agglomeration, in *Soot in Combustion Systems and its Toxic Properties*, J. Lahaye and G. Prado Eds., Plenum Press, New York, 1983, 145-161.
109. H.F. Calcote, Ionic mechanisms of soot formation, in *Soot in Combustion Systems and its Toxic Properties*, J. Lahaye and G. Prado Eds., Plenum Press, New York, 1983, 197-215.
110. A.G. Blokh, *Heat Transfer in Steam Boiler Furnaces*, Hemisphere, Washington, D.C., 1988.
111. A. Stambuleanu, *Flame Combustion Processes in Industry*, Abacus Press, Tunbridge Wells, U.K., 1976.
112. A. Blokh, The problem of flame as a disperse system, in *Heat Transfer in Flames*, N.H. Afgan and J.M. Beer, Eds., Scripta Book Company, Washington, D.C., 1974, chap. 6, 111-130.
113. T. Kunitomo, Luminous flame emission under pressure up to 20 atm, in *Heat Transfer in Flames*, N.H. Afgan and J.M. Beer, Eds., Scripta Book Company, Washington, D.C., 1974, chap. 18, 271-281.
114. E.G. Hammond and J.M. Beér, Spatial Distribution of Spectral Radiant Energy in a Pressure Jet Oil Flame, in *Heat Transfer in Flames*, N.H. Afgan and J.M. Beer, Eds., Scripta Book Company, Washington, D.C., 1974, chap. 19, 283-291.
115. J.C. Ku and K-H Shim, The effects of refractive indices, size distribution, and agglomeration on the diagnostics and radiative properties of flame soot particles, in *Heat and Mass Transfer in Fires and Combustion Systems*, W.L. Grosshandler and H.G. Semerjian, Eds., New York, ASME HTD-Vol. 148, 1990, 105-115.
116. H. Bockhorn, Ed., *Soot Formation in Combustion,* Springer-Verlag, Berlin, 1994.
117. A.A. Fridman, S.A. Nestor and A.V. Saveliev, Effect of pyrene addition on the luminosity of methane flames, in *ASME Proceedings of the 32nd National Heat Transfer Conf.*, Vol. 3: Fire and Combustion, L. Gritzo and J.-P. Delplanque, Eds., ASME, New York, 1997, 7-12.
118. T.L. Farias, M.G. Carvalho, and Ü.Ö. Köylü, Radiative heat transfer in soot-containing combustion systems with aggregation, *Int. J. Heat Mass Trans.*, 41(17), 2581-2587, 1998.
119. M. Jacob, *Heat Transfer*, Vol. 2, Wiley, New York, 1957.
120. W.H. McAdams, *Heat Transmission*, McGraw-Hill, New York, 1954.

121. H.C. Hottel and F.P. Broughton, Determination of true temperature and total radiation from luminous gas flames, *Ind. Eng. Chem.*, 4, 166-175, 1933.

122. J.M. Beér and C.R. Howarth, Radiation from flames in furnaces, *Twelfth Symposium (International) on Combustion*, The Combustion Institute, Pittsburgh, PA, 1968, 1205-1217.

123. M. Maesawa, Y. Tanaka, Y. Ogisu, and Y. Tsukamoto, Radiation from the luminous flames of liquid fuel jets in a combustion chamber, *Twelfth Symposium (International) on Combustion*, The Combustion Institute, Pittsburgh, PA, 1968, 1229-1237.

124. D.W. Gill, D.J. Loveridge, and G.G. Thurlow, A comparison of predicted and measured heat-transfer rates in a pulverized-coal-fired furnace, *Twelfth Symposium (International) on Combustion*, The Combustion Institute, Pittsburgh, PA, 1968, 1239-1246.

125. A.G. Slavejkov, T.M. Gosling, and R.E. Knorr, Method and Device for Low-NO_x High Efficiency Heating in High Temperature Furnace. U.S. Patent 5,575,637, 1996.

126. J.D. Gilchrist, *Fuels, Furnaces and Refractories*, Pergamon Press, Oxford, U.K., 1977.

127. V.S. Arpaci, *Conduction Heat Transfer*, Addison-Wesley, Reading, MA, 1966.

128. M.N. Özisik, *Boundary Value Problems of Heat Conduction*, Dover, New York, 1968.

129. U. Grigull and H. Sandner, *Heat Conduction*, Hemisphere, Washington, D.C., 1984.

130. G.E. Myers, *Analytical Methods in Conduction Heat Transfer*, Genium Publishing, Schenectady, NY, 1987.

131. B. Gebhart, *Heat Transfer and Mass Diffusion*, McGraw-Hill, New York, 1993.

132. D. Poulikakos, *Conduction Heat Transfer*, Prentice-Hall, Englewood Cliffs, NJ, 1994.

133. D.R. Lide, Ed., *CRC Handbook of Chemistry and Physics*, 79th edition, CRC Press, Boca Raton, FL, 1998.

134. W.M. Rohsenow, Boiling, in *Handbook of Heat Transfer Fundamentals*, second edition, W.M. Rohsenow, J.P. Hartnett, and E.N. Ganic, Eds., McGraw-Hill, New York, 1985.

135. S.V. Stralen and R. Cole, Eds., *Boiling Phenomena*, Vol. 1, Hemisphere, Washington, D.C., 1979.

136. J.R. Thome, *Enhanced Boiling Heat Transfer*, Hemisphere, Washington, D.C., 1990.

137. R.T. Lahey, Ed., *Boiling Heat Transfer: Modern Developments and Advances*, Elsevier, New York, 1992.

138. K. Stephen, *Heat Transfer in Condensation and Boiling*, Springer-Verlag, 1992.

139. J.G. Collier and J.R. Thome, *Convective Boiling and Condensation*, 3rd edition, Oxford University Press, Oxford, U.K., 1996.

140. L.S. Tong and Y.S. Tang, *Boiling Heat Transfer and Two-Phase Flow*, 2nd edition, Taylor & Frances, Bristol, PA, 1997.

141. S.G. Kandlikar, M. Shoji, and V.K. Dhir, Eds., *Handbook of Phase Change: Boiling and Condensation*, Hemisphere, Washington, D.C., 1999.

142. M. Lehner and F. Mayinger, *Convective Flow and Pool Boiling*, Taylor & Francis, Bristol, PA, 1999.

4 Heat Sources and Sinks

4.1 HEAT SOURCES

The primary objective of industrial heating and melting processes is to supply energy in the form of heat to some type of load. There are several possibilities for the source of that energy. In this book, only energy produced through combustion processes is considered.

4.1.1 COMBUSTIBLES

There are two general kinds of combustibles that are sources of energy in industrial combustion processes. The source that is present in every system considered in this book is the fuel supplied through the burner. A second source that may be present in a limited number of applications is volatiles coming out of the material being heated. These are briefly considered next.

4.1.1.1 Fuel Combustion

The primary source of energy in industrial heating and melting systems is from the combustion of a fossil fuel. Although electrical heating elements are used in some limited applications, this is a relatively small portion of the overall industrial sector. The overall fuel combustion reaction may be written as:

$$\text{fuel} + \text{oxidizer} \rightarrow \text{combustion products} + \text{heat}$$

In this book, the transfer of that generated heat to both the load and to the combustor are of interest. The fuel and the oxidizer are generally supplied through one or more burners that produce the desired flame shape and heat transfer rates to the load.

The fuels may be solids (e.g., coal or coke), liquids (e.g., no. 2 fuel oil), gaseous (e.g., natural gas or propane), or some combination of these (e.g., natural gas and no. 2 fuel oil). Table 4.1 shows a comparison of heating values for various fuels on a weight basis. Most of those shown have higher or gross heating values of approximately 20,000 Btu/lb (47 MJ/kg), even for liquid and gaseous fuels. Hydrogen has a much higher heating value by weight, while carbon monoxide has a much lower heating value. There are many factors that may be considered when selecting the fuel for a given application including: operating costs, equipment costs, availability, pollutant emissions, storage and handling, and convenience among others. As is common in the petrochemical industry (see Chapter 10), the fuel may be a byproduct of another chemical process which can significantly improve the overall economics of a given plant.

4.1.1.2 Volatile Combustion

Volatiles are narrowly defined here as hydrocarbon vapors that evaporated from the liquid form. The overall volatiles combustion reaction may be written as:

$$\text{volatiles} + \text{oxidizer} \rightarrow \text{combustion products} + \text{heat}$$

In the industrial applications considered here, the volatiles are generated by the heating of liquid or solid hydrocarbons. These volatiles then go into the combustion space above the load where they can be combusted.

TABLE 4.1
Heating Values for Different Types of Fuels

Fuel	Higher Heating Value		Lower Heating Value	
	Btu/lb	MJ/kg	Btu/lb	MJ/kg
#2 Distillate oil	18,993	44.180	17,855	41.533
#6 Residual oil	18,126	42.163	17,277	40.188
Acetylene, C_2H_2	21,502	50.014	20,769	48.309
Carbon, C	14,093	32.780	14,093	32.780
Carbon Monoxide, CO	4,347	10.11	4,347	10.11
Ethane, C_2H_6	22,323	51.923	20,418	47.492
Hydrogen, H_2	61,095	142.11	51,623	120.08
Methane, CH_4	23,875	55.533	21,495	49.997
Methanol, CH_3OH	9,700	22.56	8,400	19.54
Octane, C_8H_{18}	20,796	48.371	19,291	44.871
Propane, C_3H_8	21,669	50.402	19,937	46.373

Source: Adapted from Reed, *North American Combustion Handbook,* Cleveland, OH, 1986.[48]

In certain heating operations, such as calcining, volatile matter may evolve from the load. An example is the processing of petroleum coke in a rotary hearth calciner where volatiles from the coke are combusted in the calciner to generate heat.[1] Volatiles are also often generated in waste incineration applications (see Chapter 11) where the hydrocarbons contaminating a solid (e.g., soil) are volatilized during the decontamination process. In processing waste aluminum beverage containers, volatiles are often generated as the paints burn off the containers.

These volatiles are important and must be considered for several reasons if they are in significant quantities. They are a source of energy in the process, which may mean the heat input from the burners can be reduced. Sufficient excess air, over and above that for the burners, must be provided in the combustor to fully combust the volatiles. There must be sufficient residence time, temperature and mixing in the combustion system to ensure the volatiles are fully reacted. In fuel combustion, the flow rate of the fuel at a specific location is given. The analysis of volatile combustion may be more complicated than for fuel combustion, because the volatile generation rate involves complicated mass transfer processes. The volatile evolution may occur over a wide space in the combustor.

4.1.2 THERMOCHEMICAL HEAT RELEASE

This mechanism refers to the exothermic release of energy from reacting gases during cooling. In high-temperature flames, hot and dissociated gaseous species are produced. Upon contact with a cooler target, these products cool down and exothermically recombine into more thermodynamically stable molecules. This process has been given many names. It has commonly been called *chemical recombination* or simply *recombination.*[2-10] Some studies have referred to it as *convection vivre* or live convection.[11-13] Another name for this process is *aerothermochemistry,*[14] which is commonly used in the aerospace field. That name includes the physical effects of the chemical reactions, along with the fluid dynamics in stagnation flows. Cookson (1960) referred to the process as the *exothermic displacement of equilibrium.*[15] The cooler solid body modifies the chemical equilibrium processes. Baukal et al. have referred to the process as *thermochemical*[6,14] *heat release* or TCHR.[16-18] This name includes the aspects of thermodynamics, chemical reactions, and exothermic energy release. None of the other names explicitly indicate the importance that the exothermic heat release has on the heat transfer.

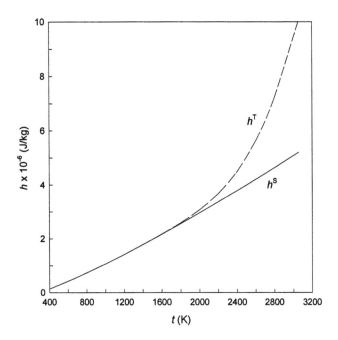

FIGURE 4.1 Sensible and total enthalpies as a function of the gas temperature, calculated for a stoichiometric equilibrium O_2/CH_4 flame. (From C.E. Baukal and B. Gebhart, *Therm. Fluid Sci.*, 15, 323-335, 1997.)

The products of many combustion processes contain dissociated species. The level of such dissociation increases with the flame temperature. When a flame impinges on a cool surface, these species diffuse in the direction of the concentration gradients, toward the lower temperature regions. As the gases cool, they exothermically recombine with other such species to form more stable molecules. These new species are thermodynamically favored at lower temperature levels. For example, when CH_4 is combusted adiabatically with pure O_2, significant amounts of unburned fuel in the form of CO (16 vol%) and H_2 (7 vol%) are produced, along with radicals like O (4 vol%), H (5 vol%), and OH (9 vol%). This composition is shown in Figures 2.2 and 2.3. As these combustion products cool to temperatures below about 2400°F (1600K), they react to form CO_2 and H_2O, while simultaneously releasing energy. However, when CH_4 is combusted with air, the final combustion products are essentially all CO_2, H_2O, and N_2. This is due to the lower flame temperature. The large concentration of N_2 acts as a heat sink and moderates the flame temperature.

The heat release from radical recombination becomes important when high-temperature dissociated gases contact cooler bodies. One example is the catalytic reaction of hydrogen atoms to form stable H_2 molecules:

$$2H + M = H_2 + 431 \text{ Joules} \qquad (4.1)$$

where M is a catalytic surface. Giedt studied fuel-rich, turbulent, O_2/C_2H_2 flames, flowing parallel to a flat plate.[7] It was estimated that H atom recombination increased the heating rate by 30% to 90%, compared to forced convection. In high-temperature flame impingement, the combustion products diffuse through the boundary layer to the colder surface. They exothermically react and form new species. Giedt found that two chemical mechanisms initiate the thermochemical heat release: equilibrium and catalytic. Nawaz (1973) referred to a third mechanism as mixed flow, which is a mixture of equilibrium and catalytic chemistries.[19] These three mechanisms are shown schematically in Figure 7.6.

Figure 4.1 shows the calculated sensible (h^S) and total (h^T) enthalpies for an equilibrium O_2/CH_4 flame, using the program developed by Gordon and co-workers.[20,21] The chemical enthalpy (h^C) is

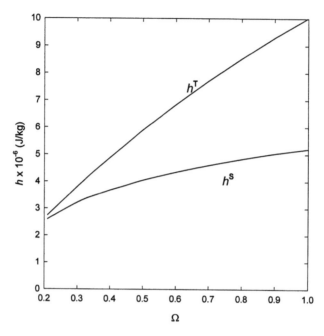

FIGURE 4.2 Calculated adiabatic equilibrium sensible (h^S) and total (h^T) enthalpies for a stoichiometric CH_4 flame with a variable oxidizer ($O_2 + N_2$) composition. (From C.E. Baukal and B. Gebhart, *Therm. Fluid Sci.*, 15, 323-335, 1997.)

the difference between the total and the sensible enthalpies, ($h^T - h^S$). The chemical enthalpy includes the chemical energy contained in the dissociated gas species. In forced convection, the driving potential is the sensible enthalpy difference between the gas and the wall, ($h^S_g - h^S_w$). The graph shows that TCHR becomes significant as the gas temperature increases above about 2800°F (1800K). At the adiabatic equilibrium flame temperature of 5037°F (3054K), the sensible energy and chemical enthalpies are nearly the same. Therefore, at high gas temperatures, TCHR may be of the same order of magnitude as forced convection. Figure 4.2 shows the difference between the total and sensible enthalpies as a function of the oxidizer composition for adiabatic equilibrium combustion. It can be seen that as the O_2 content in the oxidizer increases, the chemical enthalpy (difference between the total and sensible enthalpies) increases rapidly. This is a consequence of dissociation at the higher temperatures as shown in Figure 2.11, which produces more dissociated species as shown in Figure 2.3. This explains why TCHR is more important as the O_2 content in the oxidizer increases, which is consistent with the results of many experimental studies that are briefly discussed next.

4.1.2.1   Equilibrium TCHR

This has also been referred to as a *homogeneous* effect. The gas-phase chemical reactions occur in the boundary layer. Unstable species collide in the gas phase with other atoms or molecules. These are the third bodies that initiate the reactions. The reaction time is much less than the time required for the gases to diffuse to the surface. Free radicals enter the laminar boundary layer by molecular diffusion. The diffusion rate is small compared to the chemical reaction rate. Therefore, there is a higher probability of homogeneous free radical chemical reactions. Buhr (1973) studied fuel-rich, turbulent, air/CH_4 flames impinging normal to a water-cooled disk.[22] Chemical reaction effects were negligible. Because of higher levels of dissociation, TCHR is very significant when O_2, instead of air, is used as the oxidizer. Kilham and Purvis (1978) tested mostly fuel-rich, laminar, O_2/CH_4 and C_3H_8 flames impinging normal to a refractory plate.[23] Heat flux gages, made of silicon carbide (nearly noncatalytic) and platinum (highly catalytic), were used to try to measure TCHR effects. No difference in heating rates was found. This led to the conclusion that the TCHR process occurred in the boundary layer, before reaching the surface. Therefore, it was an equilibrium effect. In a similar study, Baukal and Gebhart (1997) came to the same conclusion.[18]

Many equations have been recommended for the combined heat transfer from forced convection and equilibrium TCHR.[17] One example is that recommended by Rosner (1961):[14]

$$q_s'' = 0.763 \left(\beta_s \rho_e \mu_e \right)^{0.5} \mathrm{Pr}_e^{-0.6} \left[1 + \left(\mathrm{Le}_e - 1 \right) \frac{h_e^C - h_w^C}{h_e^T - h_w^T} \right]^{0.6} \left(h_e^T - h_w^T \right) \tag{4.2}$$

More equations are given in Chapter 7.

4.1.2.2 Catalytic TCHR

This has also been called a *heterogeneous* effect. It involves chemical diffusion reactions at the target surface. The chemical reaction times are much greater than the transit time for the diffusing species to reach the surface. There is insufficient time for the radical species to react before reaching the surface. Some surface materials may catalytically accelerate these reactions.[24] Turbulence may also accelerate these reactions.[22]

Several equations have been recommended for the combined heat transfer from forced convection and catalytic TCHR.[17] Rosner also developed an equation for this type of TCHR:[14]

$$q_s'' = 0.763 \left(\beta_s \rho_e \mu_e \right)^{0.5} \mathrm{Pr}_e^{-0.6} \left[1 + \left(\mathrm{Le}_e^{0.6} - 1 \right) \frac{h_c^C - h_w^C}{h_e^T - h_w^T} \right] \left(h_e^T - h_w^T \right) \tag{4.3}$$

The difference between Equations 4.2 and 4.3 is the position of the exponent 0.6. Depending on the conditions, the predictions may be nearly the same using either equation.[17] More equations, including catalytic TCHR, are given in Chapter 7.

4.1.2.3 Mixed TCHR

This mechanism was suggested by Nawaz (1973).[19] It is a combination of equilibrium and catalytic TCHR. Some of the dissociated species in the flame may react within the boundary layer prior to reaching the surface. Some of the species may react catalytically upon contact with the cool surface. Some of the dissociated species may also remain unreacted. This may occur if either some of the gases do not reach the surface, or if the surface is not perfectly catalytic. No semianalytical heat transfer solutions have been recommended for this type of TCHR.

4.2 HEAT SINKS

Heat sinks are places where energy goes in a combustion system. As discussed in Chapter 2, a large heat sink is the energy carried out of a process by the combustion products. The available heat is commonly defined as the gross heating value of the fuel input to a system, less the energy carried out of the process by the combustion products. Since the available heat has already been discussed, including how various parameters affect it, this chapter section only considers the other heat sinks in a system, including the load, the walls of the combustor, losses through the openings in the combustion chamber, and energy carried out by the load transport system.

A Sankey diagram is often used to show where all the energy goes in a combustion system. A simplified example is shown in Figure 4.3. There can be as much or as little detail as needed. For example, there can be finer detail in the losses to include air infiltration, conduction losses through

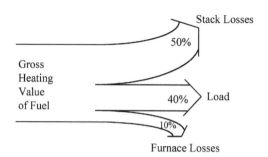

FIGURE 4.3 Example of a Sankey diagram.

the wall, and radiation losses through openings. This analysis is useful for determining what areas to focus on for improving the thermal efficiency of a system. It can also be used to determine the maximum possible system efficiency to determine what modifications to a combustion system make economic sense. For example, there is a diminishing return to recovering the sensible energy from low-temperature exhaust products. Replacing an existing heat recovery device with a more efficient and newer design may not be cost-justified if the payback is too long for the given capital investment.

4.2.1 LOAD

The primary objective in industrial heating and melting processes involving combustion is to transfer heat from the combustion of a fuel to a load. The type of load determines what type of heating process should be used. In the heating of liquids like petroleum products, there is an intermediate heat exchanger in the form of metal tubes that carry the fluids. Heat is transferred from the combustion products to the tubes by both radiation and convection, which is then transferred through the tube wall by conduction where it is then transferred by convection to the process fluid. In the heating of solids and some liquids (like molten metal or glass), the heat is transferred more directly from the combustion products to the load. However, there is a very important contribution of heat flux from hot refractory walls to the load which in some cases can be more than the heat transferred directly from the combustion products.

The properties of the load have a significant impact on how effectively heat is transferred. In the case of the heating of process tubes, the inside and outside conditions of the tubes greatly influence the heat transfer. Deposits on either side of the tube generally reduce the heat transfer efficiency. However, there are some types of coatings deliberately applied to the tubes that can improve the heating efficiency. Some loads have spectrally dependent radiant emissivity properties. For example, molten glass selectively transmits certain wavelengths of radiation. Another example is molten aluminum, which can be very reflective if there is little or no slag layer on the surface. Other loads like granular solids may be highly absorptive over a wide range of wavelengths and can be treated as gray- or nearly blackbodies. Therefore, the properties of the load itself are as important in the heating process as the heat source itself. Proper coupling of the heat source and the load maximizes the thermal efficiency of the process. Madsen et al. (1993) showed that if the emissivity of the burner is the same as the emissivity of the load, then the thermal efficiency for the conditions under investigation increased by 29%.[25]

4.2.1.1 Tubes

In most chemical and petrochemical applications, the materials are processed in heaters where fluids flow through and are heated in horizontal or vertical tubes. Therefore, heat transfer to these tubes is critical to these heating applications. In many heaters, the section closest to the burners is referred to as the radiant section because the predominant mode of heat transfer is by radiation. A downstream section of the heater is referred to as the convection section because the predominant mode of heat transfer is forced convection.

The inside and outside conditions of the tubes in the radiant section greatly affect the heat transfer from the flame to the tubes. If the outside of the tubes becomes coated with a hydrocarbon, then the radiant absorptivity of the tubes increases and the radiation from the flames to the tubes also increases. However, the heat transfer to the fluid inside the tubes may decrease as a result of a build-up on the outside of the tubes if the thermal conductivity of the coating is lower than that of the metal.

Biede et al. (1995) have shown that as slag from a pulverized coal combustor builds up on waterwall tubes, the heat transfer to the water in the tubes decreases and that the heat absorbed is a function of the slag composition.[26] Harb et al. (1994) developed a model for the heat transfer through furnace wall deposits.[27] The model calculates the fraction of particles that stick to the walls, the porosity and thermal conductivity of the deposit, and the fraction of liquid and crystalline phases in the deposit. The results showed that heat transfer to the wall rapidly decreased initially but then slowed asymptotically for longer times. The heat flux decreased by a factor of about 4 after only about 100 minutes. The model showed that the emittance of the deposit had only a relatively small effect on the heat transfer.

Hellander and Bryant (1993) described a ceramic coating that can be applied to the outside of process tubes to increase the heat transfer performance of the process heater.[28] The coating has a high radiant emissivity (>0.85) and high thermal conductivity 580 Btu-in./ft²-hr-°F (84 W/m-K) so that it absorbs more radiation than an uncoated tube and conducts that energy readily through the thin (0.00175 to 0.00225 in., 0.0445 to 0.0572 mm) coating. The desired characteristics of such a coating include:

- Minimize surface reactions to prevent deposition and oxidation
- Be non-stick to prevent the build-up of combustion products on the tubes
- Prevent carbon and oxygen diffusion, which corrode the tubes
- Be water-based and made of environmentally friendly materials
- Cure quickly to minimize downtime for repairs
- Be stable at temperatures of at least 1800°F (1300K)
- Be thermally shock resistant

Efficiency improvements of 4% to 8% were reported for the coating.

4.2.1.2 Substrate

In many drying processes, moisture is removed from a substrate that is often traveling at high speeds. Common examples of these substrates include paper, carpets, plastics, metal sheets, and textiles. Radiant heating is often used in drying substrates, especially those with some type of surface coating where contact with a hot solid like a steam-heated cylinder or with high-velocity hot gases like in convection dryers can reduce the product quality.[29] Examples are shown in Figures 1.20 and 4.4.

In radiant heating, to maximize efficiency, it is important to match the spectral output of the radiant source with the spectral radiant absorptivity of the load.[30] Van der Drift et al. (1997) studied the effects of coatings on infrared burners on the efficiency of radiant heat transfer to a paper substrate and to a metal substrate coated with three different power paints (white, blue, and black).[31] They showed that the thermal efficiency can be improved by as much as 10% by matching the radiant spectral output of the burner to the spectral absorptivity of the load. Both the paper and the paint powders had high absorptivities (nearly blackbody) above about 3 μm and very low absorptivities below about 3 μm.

4.2.1.3 Granular Solid

Many granular solids are heated in industrial furnaces. Common examples include cement and lime manufacturing. The heat transfer to granular solids has some unique challenges compared to other load materials. The surface of a granular solid often has a very high emissivity due to the many

FIGURE 4.4   Infrared burner heating a continuously moving substrate. (Courtesy of Marsden, Inc., Pennsauken, NJ).

interstices in the surface, which makes it a very good absorber for radiant energy. The irregular surface also improves forced convection heat transfer because the granules act as tiny fins, as in a heat exchanger. The only mode of heat transfer that often suffers with granular solids is the thermal conduction through the load. Granular solids often have low thermal conductivities due to the air spaces between the granules, since air has a very low thermal conductivity. Therefore, it is common practice to process granular solids in furnaces where there is high contact area between the heat source and the load. In a rotary kiln, this is accomplished by rotating the furnace, which constantly exposes new material to the heat source as the granules tumble during the rotation. A cut-away of a load in a rotating kiln is shown in Figure 4.5. In a shaft kiln, the hot exhaust products flow up through the granules, resulting in very high heat transfer rates due to the very high contact area between the hot gases and the solids. A furnace with a thick stationary load of solid granules would have a low thermal efficiency and probably overheat the load surface due to the inability of the granules to conduct the heat away from the surface. An effective thermal conductivity, which includes inter-particle radiation, is often used to model the conduction heat transfer through granular solids.[1]

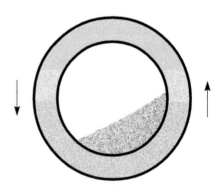

FIGURE 4.5   Cross-sectional view of a granular load in a rotary kiln.

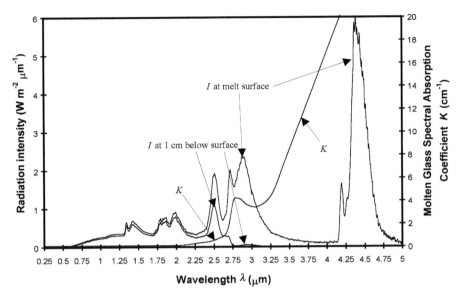

FIGURE 4.6 Interaction of flame radiation with high-temperature glass melt. (From B. Ji and C.E. Baukal, *Proc. of 1998 International Gas Research Conference,* Nov. 8-11, 1998, San Diego, CA, D. Dolenc, Ed., 5, 422-433, 1998.)

4.2.1.4 Molten Liquid

In a study by Ji and Baukal (1998), the nonluminous and luminous radiation from oxygen-enhanced flames were studied to determine the impact on melting glass.[32] This was of particular interest because of the transparent characteristics of glass. The object of the study was to determine if nonluminous or luminous flames transfer more radiant heat to the molten glass and to determine which type of radiation penetrates deeper into the glass to minimize overheating of the glass surface. In glass melting, heat is transferred primarily by emission and absorption of flame radiation during the melting and much of the forming processes in the glass industry.[33-36] The study by Ji and Baukal focused on the spectral radiation properties of natural gas flames that were enhanced with oxygen and on the absorption of flame radiation by molten glass. While there has been much theoretical research related to this subject,[34,35,37] this was the first systematic experimental study on the spectral radiation properties of oxygen-enhanced/natural gas flames.

Because one of the most important heat transfer mechanisms in a glass furnace is absorption of the flame radiation by the glass melt, one must consider how flame radiation propagates through the glass melt over the entire spectral range. Several studies on high-temperature glass melt absorption spectra[33,34,36,38] showed that the spectral absorption coefficient increases dramatically at about 2.7 μm. The glass melt essentially becomes opaque at longer wavelengths due to strong absorption from the O-H and Si-O bonds in silicate. This is a universal feature of molten glass. The spectral absorption at wavelengths shorter than 2.7 μm varies with the impurities and/or additives in the glass melt, the temperature, and the furnace atmosphere.[33,36,38] For low iron (FeO) content molten glass without other impurities or additives, the spectral absorption at wavelengths shorter than 2.7 μm is relatively small and constant.[34] The spectral absorption data of such low iron content glass melts in Reference 34 have been used for the case study in this work. Figure 4.6 shows the spectral absorption coefficient K of low iron content molten glass as reported in Reference 34. Also plotted in Figure 4.6 are the spectral radiation intensity of a flame with $\Omega = 1.0$ and $\phi = 2.0$, $q_f = 5$ kW, and the intensity after the radiation propagates through 1 cm (0.4 in) of molten glass, respectively. It is evident from Figure 4.6 that only radiation at wavelengths $\lambda < 2.7$ μm has significant penetration into molten glass. Because glass is a poor heat conductor, the radiation energy at $\lambda > 2.7$ μm only heats the top

FIGURE 4.7 Natural gas flame impinging on a water-cooled target. (From C.E. Baukal and B. Gebhart, *Therm. Fluid Sci.*, 15, 323-335, 1997.)

surface of the glass, which can cause overheating that may reduce the glass quality. Therefore, only the penetrating ($\lambda < 2.7$ μm) radiation is the effective radiation for molten glass. In order to improve heat transfer efficiency and reduce fuel consumption in glass furnaces, it is important that the penetrating radiation intensity is maximized, rather than the total radiation intensity. To compare the penetrating radiation intensity among the various flames, the spectral radiation was numerically integrated for $\lambda < 2.7$ μm. The results are summarized in Table 3.5. For the penetrating radiation, the pure oxygen/natural gas flame produced 5 times more radiation than the air-fired flame, for the same fuel consumption and stoichiometry. Taking this into account, the fuel savings using oxygen enrichment is more pronounced for glass furnaces than the savings based only in terms of the total radiation. For oxygen/natural gas flames with different fuel equivalence ratios, a fuel-rich flame ($\phi = 2.00$) apparently produced less total radiation than a stoichiometric flame ($\phi = 1.00$), due to incomplete combustion (see Table 3.5). However, the fuel-rich flame emits more penetrating radiation than a stoichiometric flame although not all the fuel was consumed.

4.2.1.5 Surface Conditions

Baukal and Gebhart (1997) studied the effects of the surface conditions on the heat transfer from impinging natural gas flames as shown in Figure 4.7.[18] Different surface coatings were used to see how the radiation and thermochemical heat release were affected. The objective of this study was to determine the relative importance of thermal radiation and to determine how the thermochemical heat release is influenced by the surface properties of the target. The study investigated the heat transfer from oxygen-enhanced/natural gas flames ($q_f = 15$ kW) impinging normal to a water-cooled metal disk ($d_b = 135$ mm) segmented into six concentric calorimetric rings as shown in Figure 4.8. The outer ring (#6) acted as a thermal guard to prevent lateral heat transfer into ring 5. A section view through a typical target cooling ring calorimeter is shown in Figure 4.9. A high-temperature, nickel-based, anti-seize lubricant was used between the layers to ensure good thermal contact. The back of the target, which contained the coolant inlet and outlet tubes, was fully insulated to thermally isolate each calorimeter. The steady-state cooled target technique[39] was used to measure the heat flux to each calorimeter. These ring calorimeters were concentrically located around the stagnation point to determine the radial heat flux profiles. The inlet and outlet coolant temperatures to each calorimeter were measured with 4-wire resistance temperature detectors (RTDs). The total heat flux (q_i''), to each calorimeter ring i, was determined using:

$$q_i'' = \frac{\dot{m}_i c_p \left(t_{\text{out}_i} - t_{\text{in}_i} \right)}{A_i}$$

(4.4)

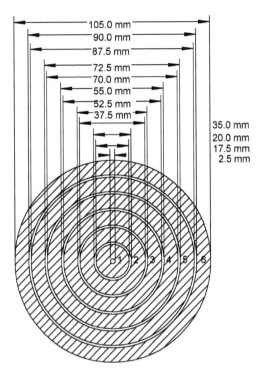

FIGURE 4.8   Plan view of target. (From C.E. Baukal and B. Gebhart, *Therm. Fluid Sci.,* 15, 323-335, 1997.)

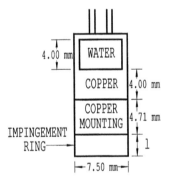

FIGURE 4.9   Section view through a typical target calorimeter ring. (From C.E. Baukal and B. Gebhart, *Therm. Fluid Sci.,* 15, 323-335, 1997.)

where $\dot{m}_i$ is the cooling water mass flow rate to ring i, c_p is the average specific heat of the cooling water, t_{out_i} is the outlet temperature of the cooling water to ring i, t_{in_i} is the inlet temperature of the cooling water to ring i, and A_i is the impingement surface area of ring i.

4.2.1.5.1   Radiation

Radiation was not directly measured, but was inferred by comparing the total heat fluxes to surfaces with various emissivities. In many of the tests, the target surface was untreated. Both the brass and the stainless steel surfaces were relatively reflective in their natural state. The natural state of the copper was not as reflective as the brass or the stainless steel. The untreated brass and copper surfaces oxidized rapidly, after less than an hour of flame impingement testing. The stainless steel surfaces oxidized much more slowly. They were reflective even after several hours of testing. In one type of test, the target surface was highly polished. This was done to make the surface reflective to minimize its radiant absorptivity. In another type of test, the target surface was coated with a highly absorptive, high-temperature flat black paint, similar to lampblack. The typical emissivities

TABLE 4.2
Total Emissivities of Various Surfaces

Material	Condition	Temp. (°C)	ε
Brass	Highly polished	250	0.028
	Rolled plate, natural surface	22	0.06
	Dull plate	50	0.06
	Oxidized by heating		0.60
Copper	Polished	100	0.052
	Plate, thick oxide layer	22	0.78
Stainless steel	Polished	100	0.074
	Type 304, after heating	220	0.44
Lampblack	Coating		0.95

Source: From B. Gebhart, *Heat Transfer,* McGraw-Hill, New York, 1971.

of the surfaces used are given in Table 4.2,[40] which shows the range of emissivities that were tested. The emissivity of lampblack is 34 times higher than that of highly polished brass.

Figure 4.10 shows how the heat flux varied with the surface radiant characteristics for an $\Omega = 0.30$ flame impinging on a stainless steel surface. R_{eff} is the dimensionless distance that divided a given calorimeter into equal impingement surface areas. Figure 4.11 is a set of plots for a pure O_2 flame impinging on a brass surface. Figure 4.12 is a similar set of plots for a pure O_2 flame impinging on a copper surface. The differences in the heat fluxes, using the various surface emissivities, were relatively small. The blackened surfaces generally had the highest heat fluxes, and the polished surfaces generally had the lowest fluxes. The difference in the heat flux, between the polished and the blackened copper surfaces, was small at smaller L. This may have been because the copper lost its reflectivity within minutes after the flame began to impinge on the surface. By the time the L = 0.5 tests were conducted, only about half of the copper surface was still highly reflective. The untreated surface had heat fluxes generally in between the blackened and polished surfaces. Figure 4.13 more clearly shows these trends for the heat flux at the stagnation point of the brass target. In general, for this series of tests, the blackened surfaces had the highest heat fluxes, the polished surfaces had the lowest heat fluxes, and the untreated surfaces had fluxes in between the blackened and the polished surfaces. The largest differences in the fluxes were at smaller R_{eff}. However, the largest difference in the heat flux between the polished and blackened surfaces was only 9.8%. For larger R_{eff}, the differences in the measured heat fluxes were negligible. These results showed that nonluminous radiation played a relatively minor role in this study and confirmed the results of previous investigators, while contradicting the results of other investigators (see Chapters 3 and 7).

4.2.1.5.2 Catalyticity

In some of the tests, the brass impingement surface was coated with platinum. As shown in Table 4.3, platinum has a very high recombination coefficient, which is a measure of the efficiency of a surface to catalyze the recombination of hydrogen atoms.[41] Previous flame impingement studies recommended that the TCHR effects could be calculated strictly based on the H atom recombination since the predicted results using H atom TCHR showed good agreement with the measured heat fluxes.[4,7,8,19,42] The platinum coating was used to determine if there were significant chemical reactions at the surface. In another type of test, the target surface was coated with a material having a low catalytic efficiency. The target rings were plasma spray-coated with pure alumina. As an oxide of aluminum with surface characteristics close to Pyrex, alumina has a very low catalytic efficiency and can be considered noncatalytic. This type of coating minimizes catalysis of the chemical reactions at the target surface.

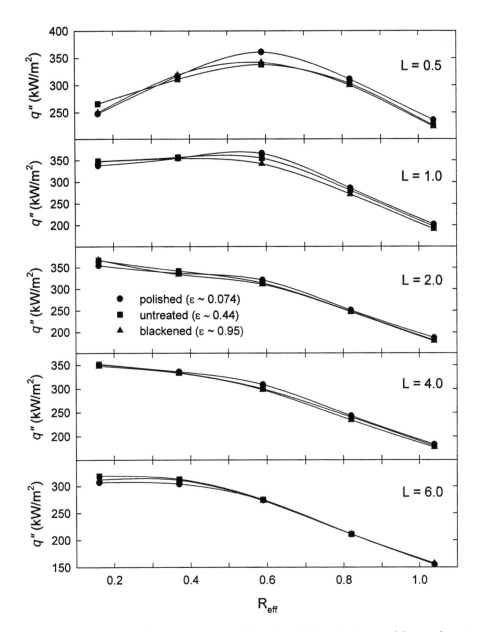

FIGURE 4.10 Total heat flux (q'') for flames ($q_f = 15.0$ kW, $\Omega = 0.30$) impinging on stainless steel targets with various surface radiant characteristics. (From C.E. Baukal and B. Gebhart, *Therm. Fluid Sci.,* 15, 323-335, 1997.)

Figure 4.14 shows a series of plots of the heat flux to the target as a function of the surface catalyticity. The oxidizer was pure O_2 where TCHR effects should have been the greatest since pure O_2 flames produce higher levels of dissociated species compared to air/fuel flames. At small L and small R_{eff}, the platinum-coated surface had a distinctively higher heat flux, which is shown more clearly in Figure 4.14. At L = 0.5 and $R_{eff} = 0.16$, the platinum-coated surface received 12% more heat flux than the alumina-coated surface. Three identical sets of tests were run to ensure the statistical significance of those results. At other locations, the differences in the heat flux, using surfaces of various catalyticities, were small. The heat fluxes to untreated surfaces were comparable to the fluxes to alumina-coated surfaces. Again, the largest differences in the fluxes were at smaller R_{eff}. However, the largest difference in the heat flux, between the platinum-coated and the alumina–coated surfaces,

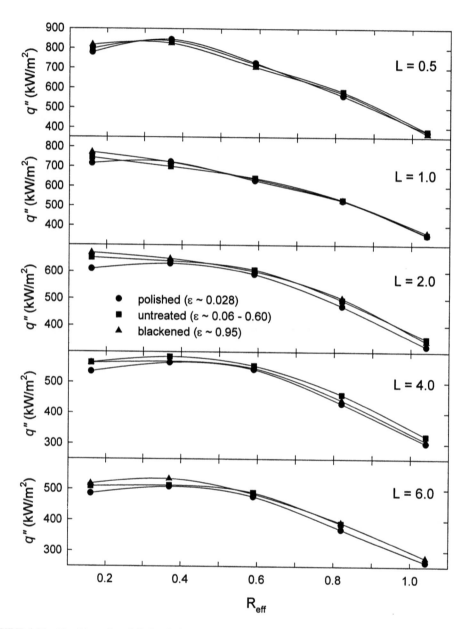

FIGURE 4.11 Total heat flux (q'') for O_2/natural gas flames ($q_f = 15.0$ kW) impinging on brass targets with various surface radiant characteristics. (From C.E. Baukal and B. Gebhart, *Therm. Fluid Sci.*, 15, 323-335, 1997.)

was only 12%. For L > 0.5, the difference in the heat flux between the platinum and alumina surfaces was negligible.

There are two important factors related to that study that determined the importance of TCHR. These include the temperature and the composition of the flame gases. As was shown in Chapter 2, these are related to each other and more dissociated species are normally produced in high Ω flames, compared to low Ω flames. It has also already been shown that TCHR is related to the amount of chemical energy contained in the combustion gases. Higher levels of dissociation increase the amount of chemical energy (h^C) contained in the gases. This chemical energy is of two types. Some of the energy is contained in the form of unburned fuels like CO and H_2. Some of the energy is also contained in the form of dissociated species like O, H, and OH. When the dissociated species

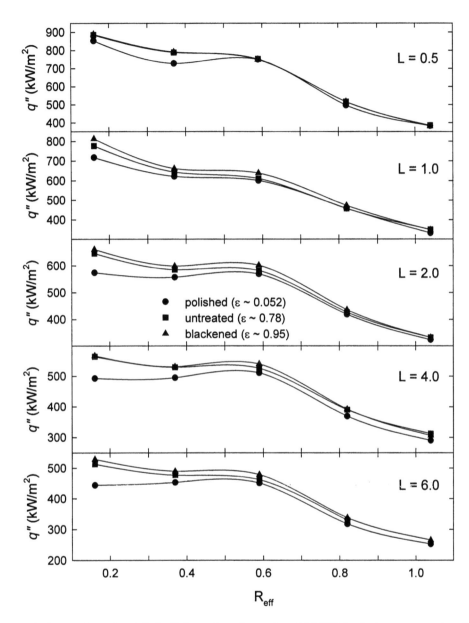

FIGURE 4.12 Total heat flux (q'') for O_2/natural gas flames ($q_f = 15.0$ kW) impinging on copper targets with various surface radiant characteristics. (From C.E. Baukal and B. Gebhart, *Therm. Fluid Sci.,* 15, 323-335, 1997.)

cool down, they exothermically recombine into more stable species like H_2O and CO_2. The second factor that has an important influence on TCHR is the gas temperature. Higher gas temperatures generally contain more dissociated species. Therefore, the temperature influences the gas composition. It was shown in Figure 4.2 that pure O_2 flame gases are expected to contain the highest levels of chemical enthalpy. However, that graph assumed adiabatic equilibrium conditions which are not present in actual flames. For a pure O_2 oxidizer, Figure 4.1 showed how the temperature of the flame gases affects the amount of chemical enthalpy. Even for pure O_2 flames, the amount of chemical enthalpy in the flame only begins to become important for temperatures above about 1800K (2800°F). Then, for real flames, the importance of TCHR is also dependent on the flame temperature. The gas temperatures for pure O_2 flames were not measured in the Baukal and Gebhart

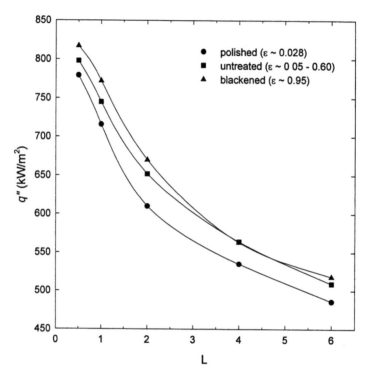

FIGURE 4.13   Total heat flux (q'') for O_2/natural gas flames ($q_f = 15.0$ kW) impinging on the stagnation point ($R_{eff} = 0.16$) of brass targets with various surface radiant characteristics. (From C.E. Baukal and B. Gebhart, *Therm. Fluid Sci.,* 15, 323-335, 1997.)

TABLE 4.3
Recombination Coefficient of Various Surfaces for Hydrogen Atoms

Surface	Ti	V	Cr	Mn	Fe	Ni	Cu	Pd	Pt	Ag	Au	Al	Pyrex
Recombination coefficient	0.10	0.15	0.16	0.20	0.17	0.18	0.19	0.20	0.25	0.13	0.10	0.001	0.00075

Source: From B.J. Wood and H. Wise, *J. Chem. Phys.,* 29, 1416-1417, 1958.

---

study,[18] but appear to have been well over 2400K (3900°F) based on the failure of the type L thermocouple at much lower oxidizer compositions. Although no quantitative conclusions could be made concerning TCHR, it appears there would have been substantial quantities of radical species in the impinging gases so that TCHR should have been an important mode of heat transfer for the $\Omega = 1.00$ flames.

No method is known for directly measuring TCHR. In the Baukal and Gebhart study,[18] one objective was to determine if catalytic TCHR was important. Three types of surfaces were tested: platinum-coated, untreated, and alumina-coated. If the catalyticity of the surface was important, then the heat fluxes to the platinum-coated surface should have been noticeably higher than the fluxes to the alumina-coated surface. The measurements showed that there was very little difference in the fluxes, except for small R and L. This confirmed the observations of Kilham and Purvis (1978), who studied O_2/CH_4 and O_2/C_3H_8 flames impinging on flat refractory blocks.[8] Both silicon carbide and platinum calorimeters were imbedded flush with the surface of the blocks. No differences in the heat fluxes were detected using these two types of calorimeters. The results of the

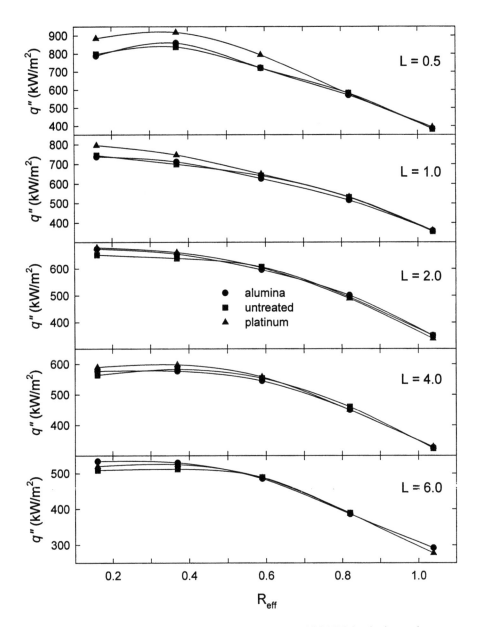

FIGURE 4.14 Total heat flux (q'') for O_2/natural gas flames ($q_f = 15.0$ kW) impinging on brass targets with various surface catalyticities. (From C.E. Baukal and B. Gebhart, *Therm. Fluid Sci.*, 15, 323-335, 1997.)

Baukal and Gebhart study also showed that a platinum-coated surface only enhanced the heat transfer at very close axial and radial target locations. Platinum is one of the most widely used catalysts in catalytic combustion.[43] It is commonly used to oxidize hydrocarbons at much lower temperatures than is possible in normal flames. These lower temperatures significantly reduce pollutant emissions, especially NOx, which is exponentially dependent on the gas temperature. At close distances between the flame and the target, the combustion reactions may not have been fully completed. The platinum may have enhanced those reactions.

The differences in the heat flux, with and without the platinum coating, were relatively small. There are several possible explanations for this. The first is the low surface-area-to-volume ratio for platinum applied to a flat surface. Normally, catalysts are applied to high surface area substrates, such as a honeycomb structure. A coating on a flat surface, as used by Baukal and Gebhart, has a very low surface-area-to-volume ratio compared to common catalyst supports. Higher surface areas enhance the effect of the catalyst.[44] Deterioration of the catalyst is another effect that may have reduced the catalyticity of the platinum.[45] After only a few test runs, the platinum surface conditions visibly changed. The initially reflective surface obtained a darker, less reflective appearance. This change in the surface may have indicated a reduction in the catalyst effectiveness. A third possible explanation is that catalytic TCHR was not an important effect for the conditions studied. It is well known that one must compensate for chemical recombination when using platinum thermocouples to measure gas temperatures in flames.[46] This is usually done by coating the thermocouple with a material having a low catalyticity, such as silica. Leah and Carpenter (1953) showed how uncoated platinum thermocouples overpredict the temperature, compared to silica-coated thermocouples.[47] However, in the Baukal and Gebhart study, no significant differences were observed using targets with either high or low catalyticity. This may be explained in several ways. Equilibrium TCHR may be a much more dominant mechanism. The much cooler gases in the stagnation region may cause the dissociated species to react before they contact the surface. The reaction times may be small compared to the diffusion times to the surface. Therefore, the concentration of dissociated species may have been very small by the time the gases reached the target surface. A second effect may have been the fluid dynamics. Many of the flame gases may never have contacted the surface, due to the boundary layer. A third explanation may be that the gas residence time, compared to the time for the species to diffuse to the surface, may have been too short. There may not have been enough time for the gases to react at the surface before being carried away by the flow. In any case, catalytic TCHR did not appear to be a significant mechanism for the flame impingement conditions studied there. Because mixed TCHR is a combination of equilibrium and catalytic TCHRs, one might also conclude that mixed TCHR was not an important mechanism. It has been shown that the heat transfer correlations may predict nearly identical heat fluxes for either equilibrium or catalytic TCHR, depending on the test conditions.[17]

Figure 4.15 shows a comparison of the effects of both the radiative and catalytic surface properties. The platinum-coated surface clearly had the highest heat flux. There was little difference in the heat flux to the target with the other surface conditions.

4.2.2 WALL LOSSES

No industrial combustor is perfectly insulated, which means that heat will flow from the inside of the combustor to the outside environment by thermal conduction. An example of calculating conduction losses through a refractory wall was given in Chapter 3. Wall losses are important for many reasons. High wall losses can significantly reduce the thermal efficiency of a process. High outside combustor temperatures due to high wall losses can be a safety hazard to personnel operating the equipment. Hotspots on the outside wall can be an indicator of a problem where refractory may have become damaged or thinned due to the process, or flames may be impinging undesirably on the wall, which may be a problem caused by the burners or improper operating conditions. The type and thickness of refractory used in a combustor depends on many factors, especially economics. Again, there is a diminishing return to increasing the thickness of the refractory, which may only give a marginal reduction in the heat loss through the wall and is typically a relatively small fraction of the overall heat losses in a system. Reed (1986) has given a good discussion of calculating wall losses and included tables and graphs, using real materials used in industrial combustors, for making these calculations more quickly.[48]

Wall losses are due to both radiation and convection. The skin temperature of the outside furnace wall, usually a metal, is at a temperature higher than the ambient surroundings, causing energy to radiate from the wall to the ambient. For analysis purposes, the ambient is considered to be a large enclosure at a constant temperature so that the view factor from the wall to the

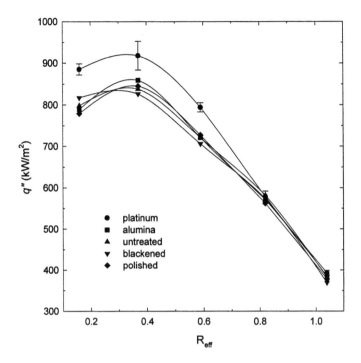

FIGURE 4.15   Total heat flux (q'') for O_2/natural gas flames (q_f = 15.0 kW) at a fixed axial position (L = 0.5), for brass targets with various surface treatments. (From C.E. Baukal and B. Gebhart, *Therm. Fluid Sci.,* 15, 323-335, 1997.)

surroundings can be considered to be one. Then, the radiant losses from the outside combustor wall to the surroundings can be simply calculated using:

$$q_{w,rad} = \varepsilon_w \sigma A \left(T_w^4 - T_\infty^4 \right) \qquad (4.5)$$

where ε_w and T_w are the emissivity and absolute temperature of the wall, respectively, A is the area of the wall, and T_∞ is the absolute temperature of the ambient.

There are two types of convection that can be important in calculating wall losses, depending on the ambient conditions. Forced convection is only significant if there is gas flow over the surface of the outside combustor wall. This can be the case for combustors located outside where the wind may be a factor. The average forced convection heat transfer coefficient for flow parallel to an isothermal plate can be calculated using:[49]

$$\overline{Nu} = 0.037 \, Re_l^{4/5} \, Pr^{1/3} \qquad (4.6)$$

where $\overline{Nu} = \overline{h}l/k$, $Re_l = \rho v l/\mu$ is the Reynolds number calculated with the length l in the direction of the flow and $Pr = c_p \mu/k$ for the air flowing over the plate.

Example 4.1

Given: Air at 81°F flowing at 20 miles/hr parallel to the outer wall (20 ft in the direction of flow and 10 ft high) of a furnace where the wall temperature is 150°F.

Find: Heat lost from the wall by forced convection.

Solution: Determine gas properties for air at 81°F (see Table F.1a):

$\rho = 0.0725$ lb/ft^3, $k = 0.0152$ Btu/ft-hr-°F, $\mu = 0.0447$ lb/ft-sec, Pr = 0.707

Calculate Re and then Nu:

$v = 20$ miles/hr = 29 ft/sec,

Re = (0.0725 lb/ft^3)(29 ft/s)(20 ft)/(1.24 × 10^{-5} lb/ft-sec) = 3.39 × 10^6,

$\overline{Nu} = 0.037(3.39 × 10^6)^{4/5} (0.707)^{1/3} = 5524$

Calculate h = Nu k/l = (5524)(0.0152 Btu/ft-hr-°F)/(20 ft) = 4.20 Btu/ft^2-hr-°F

Solve for heat loss: $q = h A (t_w - t_\infty)$

$= (4.20$ Btu/hr-ft^2-°F)(20 ft × 10 ft)(150°F − 81°F)

$= 58,000$ Btu/hr

For the relatively narrow temperature range given in this problem, there is very little variation in the properties with temperature. It may be desirable to compute the properties at the film temperature defined as:

$$t_f = \frac{t_w - t_\infty}{2}$$ (4.7)

Natural convection occurs because the furnace wall is at a higher temperature than the surroundings, which causes the air next to the wall to heat up and then rise due to buoyancy. Convective heat transfer coefficients for the free convection of air from a surface can range from 1 to 5 Btu/hr-ft^2-°F (6 to 30 W/m^2-K).[50] The actual value depends on the geometry and temperature difference between the wall and the ambient. Holman (1981) has given some simplified equations for natural convection from various surfaces to ambient air.[51] For laminar ($10^4 < Gr_f Pr_f < 10^9$) natural convection from vertical surfaces to air:

$$h = 1.42 \left(\frac{t_w - t_\infty}{l} \right)^{1/4}$$ (4.8)

where t_w is the wall temperature, t_∞ is the air temperature, and l is the vertical length of the wall. For turbulent ($Gr_f Pr_f > 10^9$) natural convection from a vertical surface to air:

$$h = 0.95 \left(t_w - t_\infty \right)^{1/3}$$ (4.9)

For laminar ($10^4 < Gr_f Pr_f < 10^9$) natural convection from a heated flat plat facing upward to air:

$$h = 1.32 \left(\frac{t_w - t_\infty}{l} \right)^{1/4}$$ (4.10)

For turbulent ($Gr_f Pr_f > 10^9$) natural convection from a heated flat plat facing upward to air:

$$h = 1.43 \left(t_w - t_\infty \right)^{1/3}$$ (4.11)

where the Grashoff number is calculated from:

$$Gr = \frac{g \tilde{\beta} \left(t_w - t_\infty \right) l^3}{v^2}$$ (4.12)

where g is the gravity constant, $\tilde{\beta}$ is the volume coefficient of expansion ($= 1/T$ for an ideal gas), and ν is the kinematic viscosity. Both the Grashoff and Prandtl numbers are calculated at the film temperature t_f.

4.2.3 OPENINGS

Energy can be lost by openings in the furnace by one of two mechanisms. Heat can be radiated out of the hot furnace to the cooler surroundings, or gases can travel through openings in the furnace (ambient air can infiltrate into the furnace or hot gases can exit through openings). Both are briefly discussed here.

4.2.3.1 Radiation

Most industrial combustion chambers can be considered to be a blackbody when calculating the radiant heat loss from relatively small openings in the furnace. The view factor from a small opening to the ambient environment can be considered to be one. Then, the radiant heat loss from a furnace opening to the surroundings can be simply calculated using:

$$q_{opening,rad} = \sigma A \left(T_w^4 - T_\infty^4 \right) \tag{4.13}$$

where A is the area of the opening.

Example 4.2

Given: Average furnace interior temperature $= 1500°F$; ambient temperature $= 70°F$; opening of 4 in. $\times$ 10 in.

Find: Heat lost from the opening by radiation.

Solution: $q_{opening,rad} = (0.1714 \times 10^{-8}$ Btu/hr-ft^2-(R^4)$(4/12$ ft $\times$ $10/12$ ft$)(1960^4 - 530^4)°$R^4
$= 14{,}000$ Btu/hr

Radiant losses through furnace openings can present several problems. They obviously reduce the efficiency of the process. They can present a hazard to personnel working around the furnace. They can also damage surrounding equipment, particularly electronics, which are not capable of handling the radiant heat load and higher temperatures.

Some furnaces are equipped with sight or view ports for operators to look into the furnace to view the operation. Although some view ports have no window and are just an opening in the furnace, accepted safety practice is to have some type of glass over the opening to protect the operators from hot gases exiting the furnaces through those ports. These windows are typically made of quartz, which can handle high temperatures. A good view port design should have a movable shutter between the inside of the furnace and the glass to minimize heat losses from the view port and to protect the glass from the high temperatures and any particles in the furnace atmosphere that might accumulate on the glass and obscure the view. The heat loss through the glass is less than through an uncovered opening because the glass only transmits part of the radiation hitting it. Glass selectively transmits radiant energy as a function of wavelength, so the actual amount of radiation transmitted through the glass is dependent on its spectral characteristics. To simply calculations, an average transmissivity is usually used.

4.2.3.2 Gas Flow Through Openings

The second major type of heat loss through a furnace opening is by gas flow through openings in the furnace. Depending on the furnace pressure conditions at the opening, cold air from outside the furnace may be drawn inside, or hot gases inside the furnace may exit through the openings. Many industrial combustion processes are operated at negative furnace pressures to prevent any of the gases in the furnace from leaking into the environment. This protects the operators and equipment in the surrounding area from being overheated. If there are toxic chemicals in the process, negative furnace pressures also protect them from toxic gases leaking into the environment. For example, waste incinerators are operated at negative furnace pressures to prevent any of the waste gases from going into the environment until they have been fully incinerated and treated in downstream scrubbing equipment. A large potential source of air infiltration for batch heating processes occurs when charge doors are opened to add new charge materials or to remove finished products. Another potentially large source for air infiltration are the openings used to transport materials into and out of continuous furnaces where there must be some type of gap for the materials to enter the furnace. When cold air leaks into a furnace, it absorbs sensible energy from the process. This reduces the thermal efficiency of the process. However, it is often possible to adjust the combustion process to account for this air leakage by reducing the amount of combustion air flow to the burners. It is still preferable to minimize air infiltration as it is by nature uncontrolled, which makes it more difficult to control the overall combustion system. Assuming that any air leaking into a furnace does not participate in the combustion process but merely absorbs sensible energy, the heat loss can be calculated using:

$$q_{\text{air infiltration}} = \dot{m}_{air} c_p \left(t_{\text{exhaust}} - t_\infty \right) \tag{4.14}$$

where $\dot{m}_{air}$ is the mass flow rate of air, c_p is the average specific heat of the air, and t_{exhaust} is the temperature of the air when it exits the flue.

Example 4.3

Given: Ambient air (70°F) is infiltrating into a furnace at a rate of 100 scfh; the exhaust gas temperature is 1800°F.

Find   Heat lost due to air infiltration.

Solution: $\dot{m}_{air}$ = (10 ft³/hr)(0.075 lb/ft³ at 70°F) = 0.75 lb/hr

$q_{\text{air infiltration}}$ = (0.75 lb/hr)(0.24 Btu/lb-°F)(1800°F − 70°F) = 311 Btu/hr

In this example, the heat loss due to air infiltration is very small. This mechanism only becomes significant for large volumes of air infiltration. Note that although the flow rate of infiltrating air is almost impossible to measure directly, it is fairly simple to calculate. It requires an accurate measure of the fuel and combustion air flow rates into the furnace and of the O_2 content in the exhaust gas products. The composition of the combustion products is estimated, without air infiltration, using the principles discussed in Chapter 2. Then, the measured O_2 in the exhaust is compared against the calculated value, which should be lower. The difference is then attributed to air infiltration. Simple chemical and mass balances are then used to estimate how much air infiltrated into the process.

The second way heat can be lost from gas flow through a furnace opening is by hot gases flowing out of the opening. Some industrial combustion processes are operated at slightly positive furnace pressures to prevent air from infiltrating into the process. One example is heat treating systems that often require very low levels of oxygen in the furnace atmosphere to produce high-quality parts. Uncontrolled air leakage into that type of process would reduce the product yields. Then, for this type of process, it is preferred to have a slight flow of gases exiting the furnace. Hot gases exiting a furnace constitute a different type of heat loss that is often more difficult to estimate. If there were no openings in the furnace except for the exhaust stack, these gases would eventually exit the furnace anyway. By leaking out before reaching the designed exit (exhaust stack), these gases do not have an opportunity to release more of their heat inside the furnace. This means that they may also leak out at a higher temperature than if they exited out of the stack. Determining the temperature and flow rate of the gases exiting out openings other than the flue and estimating how much energy they would have released if they had remained in the furnace longer are somewhat difficult to determine. This calculation requires fairly specific knowledge about a specific process and is beyond the scope of this book.

4.2.4 MATERIAL TRANSPORT

Another potential heat loss in a combustion system is the transport system used for putting materials into and taking materials out of the combustor. Some examples will serve to illustrate this type of loss. In a walking-beam steel reheat furnace, rectangular steel bars are "walked" through the furnace on beams under the bars. These beams lift the bars and carry them forward an incremental distance at discrete time intervals depending on the specific process. The beams may be water-cooled steel or may be made of a hard refractory to protect them from the heat. The water-cooled beams will obviously remove more heat from the process than refractory beams, but they will also last longer. Another example of material transport losses is in a belt furnace where parts are loaded onto a metal mesh belt so they can be carried through the furnace. If the belt remained mostly inside the furnace, except for a small section at the entrance where the parts are loaded, then the heat losses would be minimal. However, most of these heat-treating furnaces have an internal cooling zone in the furnace that is an integral part of the heat-treating process in order to get the desired metal properties. In that case, the metal mesh belt carries a significant amount of energy out of the process because the cooling zone typically blows cool gases onto the parts being heat-treated and through the belt, which cools the belt and carries away heat.

A different type of heat loss involving the material transport system is when the product being heated is being circulated out of the furnace during a portion of the cycle. An example is in a side-well aluminum reverberatory furnace (see Chapter 11) where the molten aluminum is circulated outside the combustion chamber where cold scrap material can be added into the circulating metal bath. In that case, there will be radiation and natural convection losses from the molten aluminum exposed to the ambient environment.

REFERENCES

1. H.C. Meisingset, J.G. Balchen, and R. Fernandez, Mathematical modelling of a rotary hearth calciner, *Light Metals 1996*, W. Hale, Ed., The Minerals, Metals & Materials Society, Warren, PA, 1996, 491-497.
2. R. Conolly and R.M. Davies, A study of convective heat transfer from flames, *Int. J. Heat Mass Trans.*, 15, 2155-2172, 1972.
3. M.E. Horsley, M.R.I. Purvis, and A.S. Tariq, Convective heat transfer from laminar and turbulent premixed flames, *Heat Transfer 1982*, U. Grigull, E. Hahne, K. Stephan, and J. Straub, Eds., Hemisphere, Washington, D.C., 3, 409-415, 1982.
4. J.K. Kilham and M.R.I. Purvis, Heat Transfer From Hydrocarbon-Oxygen Flames, *Comb. Flame*, 16, 47-54, 1971.
5. R. Viskanta, Heat transfer to impinging isothermal gas and flame jets, *Exper. Therm. Fluid Sci.*, 6, 111-134, 1993.
6. J.A. Fay and F.R. Riddell, Theory of stagnation point heat transfer in dissociated air, *J. Aero. Sci.*, 25, 73-85, 1958.
7. W.H. Giedt, L.L. Cobb and E.J. Russ, Effect of Hydrogen Recombination on Turbulent Flow Heat Transfer, ASME Paper 60-WA-256, New York, 1960.
8. J.K. Kilham and M.R.I. Purvis, Heat transfer from normally impinging flames, *Comb. Sci. Tech.*, 18, 81-90, 1978.
9. R.A. Cookson and J.K. Kilham, Energy transfer from hydrogen-air flames, *Ninth Symposium (International) on Combustion*, Academic Press, New York, 1963, 257-263.

10. J.K. Kilham and P.G. Dunham, Energy transfer from carbon monoxide flames, *Eleventh Symposium (International) on Combustion*, The Comb. Inst., Pittsburgh, PA, 1967, 899-905.

11. A. Ivernel and P. Vernotte, Etude expérimentale de l'amélioration des transferts convectis dans les fours par suroxygénation du comburant, *Rev. Gén. Therm., Fr.*, 210-211, 375-391, 1979.

12. J.M. Beér and N.A. Chigier, Impinging jet flames, *Comb. Flame*, 12, 575-586, 1968.

13. A. Milson and N.A. Chigier, Studies of methane-air flames impinging on a cold plate, *Comb. Flame*, 21, 295-305, 1973.

14. D.E. Rosner, Convective Heat Transfer with Chemical Reaction, Aeron. Res. Lab. Rept. ARL 99, Part 1, AD269816, 1961.

15. R.A. Cookson, An Investigation of Heat Transfer from Flames, Ph.D. thesis, The University of Leeds, Leeds, U.K., 1960.

16. C.E., Baukal, L.K. Farmer, B. Gebhart, and I. Chan, Heat transfer mechanisms in flame impingement heating, *Proceedings of the 1995 Int'l Gas Research Conference*, D.A. Dolenc, Ed., Gov't Institutes, Inc., Rockville, MD, 2, 2277-2287, 1996.

17. C.E. Baukal and B. Gebhart, A review of semi-analytic solutions for flame impingement heat transfer, *Int. J. Heat Mass Trans.*, 39(14), 2989-3002, 1996.

18. C.E. Baukal and B. Gebhart, Surface condition effects on flame impingement heat transfer, *Therm. Fluid Sci.*, 15, 323-335, 1997.

19. S. Nawaz, Heat Transfer from Oxygen Enriched Methane Flames, Ph.D. thesis, The University of Leeds, Leeds, U.K., 1973.

20. S. Gordon and B.J. McBride, Computer Program for Calculation of Complex Chemical Equilibrium Compositions, Rocket Performance, Incident and Reflected Shocks, and Chapman-Jouget Detonations, NASA Report SP-273, Washington, D.C., 1971.

21. S. Gordon, B.J. McBride, and F.J. Zeleznik, Computer Program for Calculation of Complex Chemical Equilibrium Compositions and Applications. Supplement I. Transport Properties, NASA Technical Memorandum 86885, Washington, D.C., 1984.

22. E. Buhr, G. Haupt, and H. Kremer, Heat transfer from impinging turbulent jet flames to plane surfaces, *Combustion Institute European Symposium 1973*, F.J. Weinberg, Ed., Academic Press, New York, 1973, 607-612.

23. J.K. Kilham and M.R.I. Purvis, Heat transfer from normally impinging flames, *Comb. Sci. Tech.*, 18, 81-90, 1978.

24. W.A. Hardy and J.W. Linnett, Mechanisms of atom recombination on surfaces, *Eleventh Symposium (International) on Combustion*, The Combustion Institute, Pittsburgh, PA, 1967, 167-179.

25. O.H. Madsen, M. Andersen, R.E. Peck, and M.R. Kulkarni, Infrared flux matching for improved radiant heating, *Proc. of 1992 International Gas Research Conf.*, H.A. Thompson, Ed., Govt. Institutes, Rockville, MD, 1993, 2025-2034.

26. O. Biede, J.S. Lund, R. Thiemke, O. Rathmann, and L. Wolff, Measured heat absorption in a test furnace for various coal types, *Proceedings of the 3rd European Conf. on Industrial Furnaces and Boilers*, Lisbon, Portugal, 18-21 April, 1995.

27. J.N. Harb, P.N. Slater and G.H. Richards, A mathematical model for the build-up of furnace wall deposits, *Proceedings of the Conf. on the Impact of Ash Deposition on Coal-Fired Plants*, Solihull, U.K., June 1993, Taylor & Francis, 1994, 637-644.

28. J.C. Hellander and G.B. Bryant, Use ceramic coatings to improve efficiency of gas-fired furnaces, *Hydrocarbon Processing*, 72(10), 59-60, 1993.

29. P. Colette, Surface of coated grades enhanced by high intensity infrared drying, *Pulp and Paper*, 64, 89-95, 1990.

30. O.H. Madsen, M. Andersen, R.E. Peck, and M.R. Kulkarni, Infrared flux matching for improved radiant heating, *Proc. of 1992 International Gas Research Conf.*, Gas Research Institute, 16-19 November 1992, Orlando, FL, Gov't Institutes Inc., Rockville, MD, 5, 74-83, 1992.

31. A. van der Drift, N.B.K. Rasmussen, and K. Jorgensen, Improved efficiency drying using selective emittance radiant burners, *Applied Therm. Engrg.*, 1(8-10), 911-920, 1997.

32. B. Ji and C.E. Baukal, Spectral radiation properties of oxygen-enhanced/natural gas flames, *Proceedings of 1998 International Gas Research Conference*, Dan Dolenc, Ed., 8-11 November 1998, San Diego, CA, 5, 422-433, 1998.

33. H. Franz, Infrared absorption of molten soda-lime-silica glasses containing transition metal oxides, *Int. Congr. Glass, Sci. Tech. Comm.*, 9(1), 243-260, 1971.
34. J.I. Berg, Near infrared absorption coefficient of molten glass by emission spectroscopy, *Int. J. of Thermophysics*, 2(4), 381-394, 1981.
35. F. Ammouri, C. Champinot, W. Bechara, E. Djadvan, M. Till, and B. Marie, Influence of oxy-firing on radiation transfer to the glass melt in an industrial furnace: importance of spectral radiation model, *Glastech. Ber. Glass Sci. Tech.*, 70(7), 201-206, 1997.
36. J. Endrys, F. Geotti-Bianchini, and L. De Riu, Study of the high-temperature spectral behavior of container glass. *Glastech. Ber. Glass Sci. Tech.*, 70(5), 126-136, 1997.
37. G.M. Faeth, J.P. Gore, S.G. Chuech, and S.-M. Jeng, Radiation from turbulent diffusion flames, *Annu. Rev. of Fluid Mech. & Heat Transfer*, 2, 1-38, 1989.
38. D. Banner and S. Klarsfeld, High temperature infrared spectra of silicate melts, *Phys. Non-cryst. Solids*, L.D. Pye, W.C. La Course, H.J. Stevens, Eds., London, 1992, 371-375.
39. ASTM, Standard Method for Measuring Heat Flux Using a Water-Cooled Calorimeter, American Society of Testing and Materials Standard E422, Philadelphia, PA, 1983.
40. B. Gebhart, 1971, *Heat Transfer*, McGraw-Hill, New York, 1971.
41. B.J. Wood, and H. Wise, Diffusion and heterogeneous reaction. II. Catalytic activity of solids for hydrogen-atom recombination, *J. Chem. Phys.*, 29, 1416-1417, 1958.
42. L.W. Woodruff, and W.H. Giedt, Heat transfer measurements from a partially dissociated gas with high Lewis number, *J. Heat Trans.*, 88, 415-420, 1966.
43. H. Arai, and M. Machida, Recent progress in high-temperature catalytic combustion, *Catalysis Today*, 10, 81-95, 1991.
44. H. Arai, T. Yamada, K. Eguchi, and T. Seiyama, Catalytic combustion of methane over various Perovskite-type oxides, *Applied Catalysis*, 26, 265-276, 1986.
45. R.L. Burwell, Synthesis gas formation by direct oxidation of methane over Pt monoliths, *Chemtracts-Inorganic Chemistry*, 5, 100-105, 1993.
46. R.M. Fristrom, *Flame Structure and Processes*, Oxford Univ. Press, New York, 1995.
47. A.S. Leah, and N. Carpenter, The estimation of atomic oxygen in open flames and the measurement of temperature, in *Fourth Symposium (International) on Combustion*, B. Lewis, H.C. Hottel, and A.J. Nerad, Eds., Williams and Wilkins Co., Baltimore, MD, 1953, 274-285.
48. R.J. Reed, *North American Combustion Handbook. Volume I: Combustion, Fuels, Stoichiometry, Heat Transfer, Fluid Flow*, 3rd edition, North Amer. Mfg. Co., Cleveland, OH, 1986.
49. R.H. Pletcher, External flow forced convection, in *Handbook of Single-Phase Convective Heat Transfer*, S. Kakaç, R.K. Shah, and W. Aung, Eds., Wiley, New York, 1987.
50. F. Kreith and M.S. Bohn, *Principles of Heat Transfer*, 4th edition, Harper & Row, New York, 1986.
51. J.P. Holman, *Heat Transfer*, 5th edition, McGraw-Hill, New York, 1981.

5 Computer Modeling

5.1 COMBUSTION MODELING

There are many reasons to model a combustion system. The most obvious is to gain insight into a particular configuration in order to optimize it. Optimization means different things to different people. It may mean maximizing thermal efficiency, minimizing pollutant emissions, maximizing throughput, minimizing operating costs, or some combination of these. Another reason to model is in the development of new technologies. New geometries can be tested relatively quickly compared to building an entire combustion system. Ideally, modeling is done in conjunction with experimentation to validate a particular new design. Doing modeling first can save considerably on prototype development time and costs by eliminating particular designs without having to actually test them. However, in most cases, it is not possible to completely eliminate prototype testing because of the uncertainty and limitations of combustion modeling, especially when it comes to new configurations that may never have been tried before. Another reason for modeling is to aid in scaling systems to either larger or smaller throughputs. Simple velocity or residence time scaling laws often do not apply to complicated combustion problems.[1] Modeling can be used for predictive purposes to test different scenarios that may be too risky or expensive to try in an existing operational industrial combustion system. For example, a glass producer may want to evaluate the impact of replacing an existing air preheat system with pure oxygen. Another reason to do combustion modeling is to help determine the location for instrumentation. For example, models can be used to help decide where to locate thermocouples in a furnace wall at potential hot spots in order to prevent refractory damage. Although experiments are normally used to validate modeling results, the opposite may also be true. In industrial combustion systems, large-scale probes may be necessary due to water-cooling requirements for survivability. These large probes can cause significant disturbances in the process, which can be simulated with models. The model results can then be used to determine the relevance of experimental measurements. Models can also be used to simulate potentially dangerous conditions to assess the consequences in order to design the proper safety equipment and procedures. A more recent use of computer modeling is for control of processes where the models are used to predict the results under the given conditions and then adjust the operating parameters to produce the desired results. This includes the use of artificial intelligence, where the control system has a large database of past operating conditions and the associated results so that the system can then predict and adjust itself to meet new operating conditions. Examples include making adjustments as equipment ages and deteriorates as well as for new materials being processed. In the past, these adjustments would have been based on the knowledge and experience of the operators and were often trial-and-error. Newer control systems promise more sophisticated and systematic evaluation of the given operating conditions and desired results.

One of the risks in computer modeling is that too much faith may be placed in the results. Some tend to believe anything generated by a computer. However, if the computer models have not been properly validated, then the results may be highly suspect. For the foreseeable future, it is likely that models will continue to use various approximations (e.g., turbulence) in order to get solutions in a reasonable amount of time. Therefore, the user must exercise good judgment and not try to overextend the results beyond what is warranted. For example, in many cases, models are very useful in predicting pollutant emission trends but are often very inaccurate in predicting the actual emissions. Knowledge of the model's capabilities helps one understand which results are more reliable and which ones are less reliable. The bulk fluid flow and heat transfer in a combustion system can usually be predicted with a high degree of accuracy, while the small-scale turbulence and trace species predictions may be less reliable. Therefore, it is recommended that computer modeling of combustion systems only be done by those who have been properly trained in that area.

Patankar and Spalding (1974) noted some of the important aspects of the problem statement for industrial combustion modeling problems:[2]

- Geometry of the combustion chamber
- Fuel and air input conditions
- Thermal boundary conditions
- Thermodynamic, transport, radiative, and chemical-kinetic properties, and the desired outputs of models
- Velocity, temperature, composition, etc. throughout the chamber
- Heat flux and temperature at the wall

A number of books have been written on the subject of modeling combustion processes. However, very few have specifically concerned large-scale industrial combustion systems. Khalil (1982) presented modeling results for six large-scale industrial furnaces that had published experimental data for comparison.[3] These six studies had burners with and without quarls (burner tiles); methane, natural gas, and propane fuels; firing rates ranging from 0.74 to 13 MW (2.5 to 44×10^6 Btu/hr); furnace lengths ranging from 4.5 to 11 m (15 to 36 ft); and swirl numbers ranging from 0 to 5.0. The modeling results using the k-ε turbulence model were in good agreement with the published experimental data. Oran and Boris (1991) have edited a large book on combustion modeling.[4] Part 1 of the book concerns modeling the chemistry of combustion; Part 2 contains information on flames and flame structure; Part 3 is on high-speed reacting flows; and Part 4 is humorously entitled "(Even More) Complex Combustion Systems" and has chapters on liquid and solid fuel combustion, as well as on pulse combustion. The book is more theoretical in nature and is intended for aerospace combustion. However, it does have some useful information pertinent to industrial combustion, which is referred to later in this chapter. A book edited by Larrouturou (1991) looks at the modeling of some fundamental processes in combustion science, but does not specifically consider large-scale industrial flames.[5] The papers have significant discussions of flame chemistry and fluid dynamics, but very little on heat transfer from the flames. Chung (1993) has edited a book that contains chapters on the various techniques used to model combustion processes, but without any specific applications to industrial combustion problems.[6] In a handbook on fluid dynamics, Lilley (1998) has a brief treatment of combustion modeling with only very brief discussions of industrial applications.[7] Many papers have also been written on the subject of modeling the heat transfer in combustion systems, which are discussed here.

5.2   MODELING APPROACHES

A complete combustion system may be extremely complex and can include a wide range of physical processes that are often highly interactive and interdependent. A given combustion system may include:

- Turbulent fluid dynamics in the flame with laminar fluid dynamics in the bulk of the combustor
- Multidimensional flows, which could include swirl
- Multiple phases that could include gases, liquids, and solids, depending on the fuel composition
- Very high temperature, velocity, and species gradients in the flame region with much lower gradients in the bulk of the combustor
- Large material property variations caused by the wide range of temperatures, species, and solids present in the system

- Multiple modes of heat transfer, especially radiation that is highly nonlinear and may include wavelength dependence
- Complex chemistry involving numerous reactions and many species, most of which are in trace amounts
- Porous media
- Catalytic chemical reactions in some limited applications
- Complex, nonsymmetrical furnace geometries
- Multiple flame zones produced by burners that may be operated at different conditions and whose flames interact with each other
- A heat load that may be moving and interacting with the combustion space above it in a nonlinear manner
- A heat load that may produce volatile species during the heating process
- A heat load whose properties may vary greatly with temperature, physical state, and even wavelength for radiation
- A transient heating and melting process that may include discrete material additions and withdrawals

There are many challenges caused by this complexity, including inadequate physics to properly model the problem, large numbers of grid points requiring large amounts of computer memory, and long computation times. The simulation results may be difficult to validate as many of the experimental measurements are difficult, time-consuming, and costly to make in industrial combustors. Therefore, in most combustion simulations, simplifying assumptions must be made to get cost-effective solutions in the amount of time available for a given problem. The actual simplifications depend on many factors, including the level of accuracy required, the available amount of computing power, the skill and knowledge of the modeler, the experience with the given system being simulated, and the time available to get a solution. These simplifying approaches are briefly discussed here. More detailed information on each aspect of the modeling is given later in this chapter. Spalding (1963) discussed simplifying approaches to combustion modeling and noted that the main concern is which modeling rules can be ignored to simplify the problem and then to estimate the errors in the resulting predictions.[8] He also noted the difficulty in matching all the dimensionless groups in a large-scale problem with small-scale experiments. Weber et al. (1993) classified models for designing industrial burners into three categories.[9] First-order methods give rough qualitative estimates of heat fluxes and flame shapes. Second-order methods give higher accuracy results than first-order methods for temperature, oxygen concentration, and heat flux. Third-order methods further improve accuracy over second-order methods and give detailed species predictions in the flame that are useful for pollutant formation rates. The order used will in large part depend on the information and accuracy needed.

5.2.1 FLUID DYNAMICS

There are a variety of methods available to simulate the fluid flow in a combustion system. The Navier-Stokes equations are generally accepted as providing an "exact" model for turbulent fluid flow systems.[10] Unfortunately, these equations for systems of practical interest are too complicated to solve exactly, either analytically or numerically. Therefore, different types of approximations have been suggested for solving these equations. These are very briefly discussed next, with appropriate references for the reader interested in more detail.

5.2.1.1 Moment Averaging

This has been by far the most popular method used in simulating large-scale industrial combustion problems, primarily because of the ready availability of commercial software programs like PHOEN-ICS, FLUENT, FLOW-3D, TEACH, PCGC-3, Harwell-3D, GENMIX, and others to solve these

problems. In this method, the turbulent velocity components are decomposed into average and fluctuating terms and solved using the famous k-ε closure equations.[11] Despite the well-known limitations
of this approach, it remains the most popular choice for solving practical combustion problems. This
may be due to the fact that it has been around for decades and therefore the software has been highly
developed. Finite difference,[12,13] finite element,[14,15] finite volume,[16,17] and spectral element[18] techniques
have been used to simulate fluid flows, some including turbulent combustion. The commercial codes
today are very user friendly and have excellent pre- and post-processing packages to make setting up
the problem and viewing the results relatively simple and straightforward. Because of the popularity
and widespread use of this method, it is discussed in more detail later in this chapter.

5.2.1.2 Vortex Methods

Most numerical approaches for solving fluid flow problems use an Eulerian scheme with a fixed
coordinate system that is discretized into small parts. One problem with this approach is that there
may often be areas in the flow where the gradients are very high and require very fine discretization,
while in nearby areas the gradients may be much lower and need much less discretization. To
further complicate this disparity, these areas may be moving. Finite difference solution convergence
problems result from having fine cells next to coarse cells. Therefore, the choice is to either use
finer or coarser cells for both areas. If finer cells are used, then accuracy is improved, but with a
significant penalty in solution times. If coarser cells are used, then solution times are improved,
but accuracy is sacrificed. An alternative approach is to use a Lagrangian system with a moving
coordinate system that can keep track of the finer details of high gradient areas, without the burden
of unnecessary detail in areas that do not require it. Some Lagrangian methods use grid points that
are transported along flow trajectories, while other Lagrangian methods are grid-free.[19] The Navier-
Stokes equations are set up and solved in terms of vorticity:

$$\frac{\partial \vec{\omega}}{\partial t} + \vec{u} \cdot \nabla \vec{\omega} = \vec{\omega} \cdot \nabla \vec{u} \tag{5.1}$$

where $\omega = \nabla \times \vec{u}$, $\nabla = (\partial/\partial x, \partial/\partial y, \partial/\partial z)$, $\vec{u} = (u, v, w)$, and $\vec{x} = (x, y, z)$. Velocities are then
calculated from the vorticity solutions. This method has been applied to industrial combustion
simulations.[20] Variations of this method have also been referred to as large-eddy simulations
(LES).[21,22] Dahm and co-workers have developed a method known as the Local Integral Moment
(LIM), which is based on large-eddy simulation concepts.[23-26]

5.2.1.3 Spectral Methods

This is an approximation method where the solutions for the scalar variables in the partial differential
equations are simulated as a truncated series expansion:[27]

$$C(x,t) = \sum_{k=0}^{N} c_k(t)\phi(x) \tag{5.2}$$

where $C(x,t)$ is a scalar variable such as temperature, N is the finite wavenumber truncation cutoff,
$c_k(t)$ are the expansion coefficients, and $\phi_k(x)$ are the basis functions, which are chosen to best
represent the flow. This solution approach is more global than finite difference discretization
approaches, which tend to be more local. Therefore, spectral methods can provide more accurate
approximations of the solution compared to moment methods, although this is not always the case.
Solution times may be longer and the selection of the proper basis functions is critical to the success
of this approach, which has been used in combustion problems[28,29] but has not been a popular

method for solving industrial combustion problems. This method could become more popular if the appropriate user-friendly software were developed and commercialized.

5.2.1.4 Direct Numerical Simulation

In this method, usually referred to as DNS,[30-34] no assumptions are made regarding the turbulent behavior of the flow. The exact Navier-Stokes equations are solved at small enough length and time scales that the complete physics of the problem can be captured. This approach obviously requires tremendous computing power and is not currently used for solving industrial combustion problems. However, as rapid advancements in computers continue, including parallel processing, large memories, and fast computer speeds, this method may become more prevalent in the future. This method is currently being used to solve fundamental combustion[35] and aerospace propulsion problems using supercomputers where the simulation costs are not a significant portion of the overall cost of new developments. At this time, the economics of DNS are not justified for most industrial combustion equipment manufacturers and end users where the cost of these calculations could dwarf the actual cost of the combustion system itself.

5.2.2 Geometry

There are several different levels of complexity concerning the geometry of a given combustion system, ranging from zero-dimensional up to fully three-dimensional. These levels are briefly discussed here and have been discussed in more detail by Khalil.[3]

5.2.2.1 Zero-Dimensional Modeling

Numerous modeling approaches to handling the complexity of large-scale combustors are possible and have been used. Before the advent of CFD codes, a common modeling approach was to do an overall heat and material balance on the system. This is often referred to as zero-dimensional modeling because it does not give any spatial resolution. This type of zero-dimensional modeling does not involve any analysis of the fluid dynamics. It can, however, include detailed analysis of the chemical reactions and is often referred to as a stirred reactor or stirred vessel. This type of modeling was made easier with the advent of electronic spreadsheets, but still requires numerous assumptions and simplifications. A more recent type of zero-dimensional model may include very detailed chemistry but still no fluid flow. In that type, the reactor is assumed to be typically either constant pressure or constant volume. The main variable then becomes time, which may be finite or infinite (equilibrium). Zero-dimensional models give a reasonable approximation of the overall performance of the system, but give very little information on the detailed performance, such as where potential hotspots in the furnace wall might be. Despite the obvious disadvantages, there are some advantages of zero-dimensional modeling. One is that solutions can be obtained very quickly. This is important in parameter studies where a large number of variables are to be investigated and where fast results are needed. Another advantage is that these models can be very helpful in developing an understanding of the system performance, which can sometimes be lost when detailed analyses are done. One can see the forest, before looking at the individual trees in the forest. Another advantage is that this type of modeling does not require the same level of training as complicated modeling, so it can be done by a wider group of personnel. One example is the zero-dimensional model of furnaces to rapidly heat cylindrical metal billets.[36] Another example of a zero-dimensional model is given by Kuo (1994) to simulate a batch-fed solid waste incineration process.[37]

5.2.2.2 One-Dimensional Modeling

The next level of complexity involves one-dimensional modeling. This is where only one spatial dimension is considered. Although this greatly simplifies the number of equations, these models may still be fairly complicated and provide many details into the spatial changes of a given parameter. One-dimensional modeling is often used to examine the detailed chemistry in a combustion process which may be simulated as a plug flow reactor.

Despite the limitations, there are advantages to using this type of geometrical simplification. In certain applications, these models are particularly relevant, with little or no sacrifice in accuracy and resolution. An example is in porous radiant burners and flat flames which are both essentially one-dimensional in nature. Another obvious advantage is that faster results are possible, compared to multidimensional modeling. One-dimensional models also greatly simplify the task of radiation modeling, which can become very complicated in multidimensional geometries. However, it should be noted that one-dimensional models may still be fairly complicated and may include very detailed chemistry, multiple phases, porous media, and radiation. As an example, Singh et al. (1991) reported on a one-dimensional model used to simulate ceramic radiant burners.[38] For that type of burner (see Chapter 8), the one-dimensional model is generally very adequate.

5.2.2.3 Multidimensional Modeling

The highest level of geometrical complexity involves multidimensional modeling, both two- and three-dimensional geometries. Geometry simplifications are often used to reduce the computing requirements for simulating combustion systems. Wherever possible, three-dimensional problems are simulated by two-dimensional models or by axisymmetric geometries, which are three-dimensional problems that can be solved in two spatial variables. In the early days of CFD, it was not uncommon to simulate a rectangular furnace as a cylindrical axisymmetric geometry to reduce the problem from 3D to 2D. A related simplification is modeling certain types of cylindrical problems as angular slices, instead of modeling the entire cylinder. For example, if a burner has four injectors equally spaced angularly and radially from the centerpoint, then this can be modeled as a 90° slice of a cylinder using symmetric boundary conditions.

Another type of geometric simplification in multidimensional modeling involves limiting the number of grid points due to the limitations of the computer speed and memory. It may not always be possible to model the entire combustion system, so an approach that has often been taken is to separately simulate the flame region where small-scale effects are important and the combustor where large-scale effects are predominant. The results of the flame simulation can then be used as inputs to the large-scale modeling of the combustor itself. For example, a single flame can be more accurately input as a heat source using the detailed modeling results for that flame. Another common method for minimizing the number of grid points is to model only a small portion or section of a combustor. For example, most glass furnaces have multiple burners symmetrically firing parallel to the molten glass bath. Often, only a single slice of the furnace containing one burner is modeled. Although this precludes simulating the flame-to-flame interactions, it is a reasonable assumption to make in order to get timely and cost-effective solutions of acceptable accuracy.

The obvious advantage of multidimensional modeling is that much higher spatial resolution is possible. This can provide important insight into the problem that is not possible with simpler geometrical models. This resolution is particularly important in simulating burner performance since the burner geometry is normally too complicated to model as a one-dimensional problem. However, there are some obvious disadvantages to multidimensional modeling, including longer computational times, difficulties in visualizing and interpreting the results, and more difficulty in separating the effects of individual parameters. Gillis and Smith (1990) evaluated a 3D model for industrial furnaces and compared modeling results against experimental data for two pilot-scale furnaces.[39]

5.2.3 Reaction Chemistry

The reaction chemistry is the second important aspect of most industrial combustion problems. Modeling approaches for this chemistry range from nonreacting up to multiple reactions with multiple species, and finite rate kinetics. The different approaches commonly used in modeling combustion problems are briefly discussed next.

5.2.3.1 Nonreacting Flows

When CFD codes first became commercially available, the chemistry submodels were very primitive and greatly increased the computation time, often beyond the capability of the available hardware. Therefore, a common approach to simulating combustion problems was to model them as nonreacting flows. This has sometimes been referred to as "cold-flow" modeling, which is really a misnomer because the flame was often simulated as a flow input of hot inert gases to the combustor. A variation of this approach is to use a nonreacting gas that has the thermophysical properties (such as viscosity, thermal conductivity, and specific heat) of the combustion products as a function of temperature. Those properties are separately calculated, typically using some type of equilibrium chemistry calculation. The properties are then curve-fit with temperature and included in the CFD codes. This type of nonreacting-flow model can give fairly accurate predictions in many cases for the overall energy transfer in a large-scale combustor.

Nonreacting flow modeling may grossly oversimplify a problem, but it can give considerable insight into the flow patterns inside the combustor. The flame can also be simplified to be a heat source, in order to avoid modeling the chemical reactions in the flame zone. The difficulty is how to specify the heat release profile of the flame, especially since that is something usually desired of the modeling itself. Although there are some advantages to using nonreacting chemistry, such as simplicity and speed, this approach is rarely used in most types of combustion modeling today because it is too limited and unrealistic.

5.2.3.2 Simplified Chemistry

The term "simplified" chemistry is a somewhat relative term, but generally refers to reducing the number of chemical equations used to represent a system, reducing the complexity of the reaction mechanism, or a combination of both. In the first approach, a very limited number of reactions and species are used to represent the actual combustion reaction system which may involve hundreds of reactions and dozens of species. In this approach, a greatly reduced set of reactions is used.[40] Often, the goal of this approach is to predict flow and heat transfer information, but not detailed species such as pollutant emissions like NOx. Simplified chemistry was often used in the early days of CFD modeling because of the limitations of the submodels and computer memory and because the main interests were things like the heat transfer to the load and the walls and the bulk gas flow in the system. For example, the earliest models for simulating the combustion of methane used a single-step reaction such as the following:

$$CH_4 + O_2 + N_2 \rightarrow CO_2 + H_2O + O_2 + N_2 \tag{5.3}$$

Infinite rate kinetics were used and no minor species were included. This simplified chemistry could obviously not be used to predict pollutant emissions like NOx, but was useful for simulating the flow patterns and heat transfer in the combustor. An example of a slightly more complicated reaction set is given by Westbrook and Dryer (1981):[41]

$$CH_4 + 1.5O_2 \rightarrow CO + 2H_2O \tag{5.4a}$$

$$CO + 0.5O_2 \rightarrow CO_2 \tag{5.4b}$$

A more popular approach in recent years has been to use slightly more complicated reduced sets. An example of a four-step reduced mechanism set for methane flames is given by:[42]

$$CH_4 + 2H + H_2O \rightarrow CO + 4H_2 \qquad\qquad (5.5a)$$

$$CO + H_2O \rightarrow CO_2 + H_2 \qquad\qquad (5.5b)$$

$$H + H + M \rightarrow H_2 + M \qquad\qquad (5.5c)$$

$$O_2 + 3H_2 \rightarrow 2H + 2H_2O \qquad\qquad (5.5d)$$

Another aspect of simplified chemistry models involves not only the number of equations used, but the type of chemical kinetics being simulated. This second approach to simplifying the chemistry is sometimes referred to as reaction mechanism simplification, or mechanism reduction. Infinite rate kinetics, or equilibrium chemistry, is an example of this type of approach often used in combustion modeling. This means that the chemical reactions are assumed to be infinitely fast and therefore independent of time. This is often a reasonable assumption to make but, again, is dependent on the specific problem and required level of accuracy. Another variation of this approach is an empirical correlation for the chemistry of a given system.

5.2.3.3 Complex Chemistry

Another approach to modeling combustion systems is to use very detailed chemistry. This approach is commonly used if detailed information on gas species is required, such as when, for example, NOx emissions need to be predicted. Again, "complex" chemistry is a somewhat vague and relative term, but here refers to multistep reactions with multiple species. The actual numbers of reactions and species depend on a given problem and the level of detail required. Complex chemistry also concerns finite rate kinetics where the reaction rates are time dependent.

5.2.4 RADIATION

Kocaefe et al. (1987) have given a brief review of some of the methods used for radiation modeling.[43] They conclude that the imaginary planes and discrete transfer methods have good accuracy and low computation times, while the zone method has the lowest computation time if the interchange factors are known or only calculated once. Some of the approaches used to handle radiation are discussed in this chapter section.

5.2.4.1 Nonradiating

Another type of simplification involves using known empirical relationships for the problem at hand. These empirical correlations normally apply only to a specific set of conditions and problems but can be very useful for reducing the size and complexity of the problem. For example, it may be possible to simulate the nonlinear radiation from the flame to the load and combustor walls as a type of radiation heat transfer coefficient in order to make the radiation linear with temperature and therefore much easier to solve:

$$h_{rad} = f\left(T_{source}^4 - T_{sink}^4\right) \approx f\left(T_{source} - T_{sink}\right) \qquad\qquad (5.6)$$

This approach should be used with caution only after careful examination and understanding of the system under investigation. Although this may limit the generality of the problem, this type of simplification may greatly reduce the time to get solutions. This makes it possible to do more simulations of the problem and may be especially useful for finding optimized conditions.

The key to using any simplifications is to understand the resulting inaccuracies they introduce. Therefore, it is usually prudent to have experimental data to compare against any simplified numerical simulations. It is also advisable to use the most complicated possible model for at least a base-case problem, which can then be used to compare against the simplified results. If the simplified results compare favorably to the full-blown simulation, there is some justification for using the simplifications. However, if the simplifications do not compare favorably to the comprehensive model results, then further analysis is warranted to understand the discrepancies.

As computer power continues to improve, fewer and fewer simplifications will be necessary. Eventually, it will be possible to do direct numerical simulations (DNS) so that even the turbulent fluid flow will not need to be approximated because it will be possible and practical to model the small length scales present in turbulent fluid flows.

5.2.4.2 Participating Media

Participating media includes nonluminous gaseous radiation and luminous radiation from particle-laden gaseous flows. Bhattacharjee and Grosshandler (1989) noted three factors that complicate gaseous radiation modeling:[44]

1. The spectral variation of the properties requires calculations over the spectrum.
2. The gaseous composition is not homogeneous over the entire space which means that integrations must be done over every line of sight.
3. The asymmetry of most real problems means integration over all solid angles.

Sivathanu et al. (1990) noted that there are accurate methods for calculating weakly radiating turbulent diffusion flames, but that it is much more difficult to model the strongly radiating turbulent diffusion flames that are used in many industrial combustion systems.[45] Turbulence can significantly increase the mean radiation levels from diffusion flames.[46-51]

Hoogendoorn et al. (1990) used a 15-band gas radiation model in a well-stirred furnace zone method to simulate the nonluminous radiation in a natural gas-fired regenerative glass melter.[52] The results showed that as much as 99% of the heat transfer to the melt was by radiation. They compared the axial heat flux distribution using both a simplified plug-flow model and a more complete model. The plug-flow model gave both unrealistically high fluxes in the middle of the melter and low fluxes near the ends of the melter. The addition of additives to the combustion products was shown to slightly increase the heat flux to the glass. However, this would lead to lower flue gas outlet temperatures, which reduces the performance of the regenerative air preheater so that the overall effect of the additive in this process would be minimal. Increasing the roof emissivity from 0.4 to 1 was shown to increase the heat transfer to the glass melt by 8%.

Zhenghua and Holmstedt (1997) presented a fast narrow-band computer model (FASTNB) for predicting the radiation intensity in a general nonisothermal and nonhomogeneous combustion environment.[53] The model was used to calculate the spectral absorption coefficients for CO_2, H_2O and soot. It was claimed to be as much as 20 times faster than a benchmark model called RADCAL,[54-56] with only a 1% deviation from that model. Further development was proposed for inclusion of other gases like carbon monoxide, methane, propylene, and acetylene.

Liu et al. (1998) have presented a new approximate method for non-graygas radiative heat transfer using a statistical narrow-band model that utilizes a local absorption coefficient which is calculated using local properties rather than global properties.[57] The main advantage of the proposed approximate method is considerable savings in computational time — up to a 2 orders of magnitude reduction. The method also improves the accuracy of the calculations compared to methods using global properties.

Gritzo and Strickland (1999) presented a gridless integral method for solving the radiative transport equations for use in combustion calculations using Lagrangian techniques to solve the fluid dynamics.[58] Their approach is particularly compatible with parallel computing. It is shown that this method compares favorably against other popular methods used in grid-based solution techniques, which can have significant errors when adapted to gridless solution schemes. Previous methods to solve for radiation have relied on grid-based calculations and are not optimal for transport element methods.

For some combustion processes — primarily those involving solid and liquid fuels — spectral radiation from particulates may be significant. Ahluwalia and Im (1990) presented a three-dimensional spectral radiation model that they used to model the burning of deeply cleaned coals in a pulverized coal furnace.[59] Spectroscopic data were used to calculate the absorption coefficients of the gases. The extinction and scattering efficiencies of the particulates were calculated using Mie theory. The optical properties of the char, ash, and soot were determined from reflectivity, transmissivity, and extinction measurements. The radiation from the char was as much as 30% of the nonluminous gaseous radiation. The heat transfer in the furnace ranged from 168 to 221 MW (5.73 to 7.54×10^8 Btu/hr), depending on the specific fuel used. It is noted that ashes rich in iron enhance radiative heat transfer and fine grinding of the coal improves furnace heat absorption. In a later paper, Ahluwalia and Im (1994) used a hybrid technique to solve spectral radiation involving gases and particulates in coal furnaces.[60] To optimize computational speed and accuracy, the discrete ordinate method (S_4), modified differential approximation (MDA), and P_1 approximation were combined and used in different ranges of optical thicknesses. The MDA method has been shown to be sufficiently accurate for all optical thicknesses, but computationally slow for the optically thin and thick limits.[61] There were significant discrepancies between the predicted and calculated heat fluxes. This was explained by the difficulty in making heat flux measurements in industrial furnaces. The soot, char, and ash contributions to heat transfer were approximately 15%, 3%, and up to 14%, respectively.

5.2.5   TIME DEPENDENCE

Another critical aspect of combustion modeling is whether or not the solution is time dependent. Nearly all industrial combustion processes are time dependent at small length scales, due to turbulence. However, these processes are normally modeled as steady-state systems because of the limitations of the turbulent submodels, the large increase in computer time required to simulate transient combustion, and the lack of need for such detailed information in most industrial combustion systems.

5.2.5.1   Steady State

In steady-state simulations, there is no time dependence of the solution. The problem with turbulent combustion problems is that the time scale, especially in the near-flame region, is very small. To accurately simulate an entire system using such a small time scale is normally computationally prohibitive. However, this is not usually a problem in most cases because that type of detail is not required. The standard approach has been to simulate the average properties. There is considerable debate about averaging turbulent properties, but the reality is that virtually all commercial codes have some type of turbulence averaging (discussed in more detail below). For many continuous industrial applications, the heating process is essentially steady state with fixed and usually known combustion chamber wall temperatures, fuel firing rates, and material feed rates. Processes that are truly varying with time, such as batch heating processes, may be simulated using average conditions over the entire cycle or by making a series of steady-state calculations to simulate various steps in the process.

5.2.5.2   Transient

Transient or time-varying calculations are rarely made for large-scale industrial combustion processes due to the large computational time required, which usually exceeds the amount of time allowed for the needed simulations. As computer speeds continue to increase, this type of computation will increase in popularity for those applications that have significant variations during a given cycle. A good example of such an application is scrap metal melting. Initially, a charge of cold solid scrap metal is charged into a colder furnace. Then the burners begin to heat up and melt down the metal, as well as heat up the furnace. At any given time, there may be a mixture of solid and liquid metal. When the charge is at or near fully molten, a second charge of cold scrap may be added to the first melted charge. Several more charges are possible, depending on the application. An accurate simulation of this process should include a fully transient computation.

5.3 SIMPLIFIED MODELS

Wilson (1997) has presented a new and more efficient technique for simulating the first-order dependence of a flame in a furnace.[62] The technique is called the moving boundary flame model. Dynamic state variables are created to keep track of the size of the flame in the furnace. The technique is claimed to be a considerable improvement over point reactor models, but is not intended to replace detailed multidimensional CFD models. Its accuracy depends on data from either experimental data or from more detailed combustion simulations. This reduced model for the flame makes it possible to do online dynamic simulations that may be used for burner diagnostics and controls.

Gray et al. (1993) developed a simplified model called COMBUST for use on a personal computer to solve the radiant heat transfer in gas-fired furnaces.[63] The model was capable of solving either direct- or indirect-fired furnaces and used the Hottel zone model to solve for the radiation heat transfer. Rumminger et al. (1996) used a one-dimensional model to simulate a porous radiant burner.[64] The bilayered reticulated ceramic burner consisted of an outer layer (known as the flame support) with large pores (4 pores/cm^2) that extracts heat from the postflame gases, and an inner layer (known as the diffuser) with small pores (25 pores/cm^2) that prevents the flame from flashing back. The flame was assumed to be one-dimensional, laminar, and steady state. The gas was assumed to be optically thin and the solid was taken to be spectrally gray. Detailed chemical kinetics were used with the assumption that the gases are adiabatic at the gas outlet. Radiation was simulated as a gray gas. The surface temperature of the porous radiant burner was computed, for which the radiant losses from the burner were equivalent to the convective heat loss from the gases. The following properties were used for the outer layer porous medium: radiation extinction coefficient of 115 m^{-1} (35 ft^{-1}), scattering albedo of 0.72, pore diameter of 0.22 cm (0.087 in.), porosity of 0.8, bulk thermal conductivity of 1.0 W/m-K (0.58 Btu/hr-ft-°F). For the inner layer porous medium, the following properties were used: radiation extinction coefficient of 1000 m^{-1} (300 ft^{-1}), scattering albedo of 0.77, pore diameter of 0.022 cm (0.0087 in.), porosity of 0.65, bulk thermal conductivity of 1.0 W/m-K (0.58 Btu/hr-ft-°F). The convection correlation for gas flow in reticulated ceramics was from Younis and Viskanta (1993).[65] Kendall and Sullivan (1993) presented a similar one-dimensional model to that of Rumminger et al., where the flow was simplified but the chemistry was detailed, using a code called PROF (PRemixed One-dimensional Flame).[66] A sample result is given in Figure 5.1. The figure shows that the calculated radiant output has little dependence on emissivity in the range from 0.4 to 1.0.

5.4 COMPUTATIONAL FLUID DYNAMIC MODELING

Although the term "computational fluid dynamics" (CFD) can be applied to a wide range of simulations, here CFD represents multidimensional (2D and 3D) modeling of the fluid flow, including chemistry and heat transfer. Any of these three components (fluid flow, chemistry, and heat transfer) may range from simplified to highly complex. The simulations can be computed by any number of different types of schemes (e.g., finite difference, finite volume, etc.).

5.4.1 INCREASING POPULARITY OF CFD

Computational fluid dynamics (CFD) is a numerical tool for simulating the complicated fluid flow, heat transfer, and chemical reactions in a combustor. This tool has been gaining in popularity in recent years because of a number of factors. One obvious change is the dramatic increase in

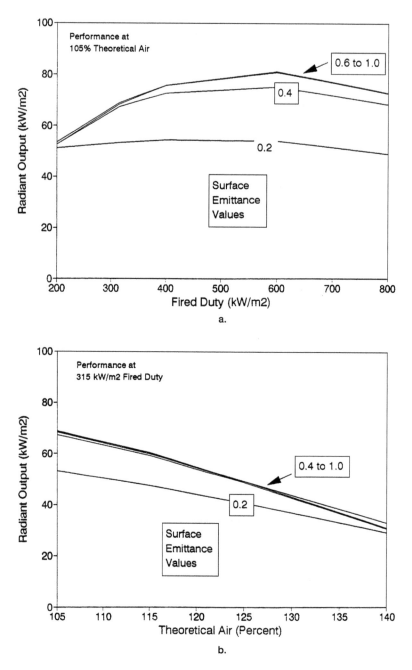

FIGURE 5.1 Calculated radiant output for a porous radiant burner using the PROF code. Performance as a function of A: surface heat release rate, and B: theoretical air. (Courtesy of Gas Research Institute, Chicago, IL.[66] With permission.)

computer power that is available at a cost that is affordable for virtually all businesses. Each new generation of computer hardware continues to have more speed and more memory at lower cost. The personal computers of today are more powerful than the workstations of only a few years ago. Another reason for the growing popularity of CFD codes is that they are now available at a reasonable cost, which usually includes some type of support to aid the user both in the general use of the code as well as the application of the code to the specific needs of the user. Another change is that CFD computer codes now have very easy-to-use front ends for setting up problems and viewing the results. Before the days of the commercial CFD codes, data had to be input in a certain format that generally required detailed knowledge of both the code and the computer. The results were generally only available in tabular form, which made it difficult and unwieldy to visualize the predictions. Now, the data can be input without any knowledge of the details inside the code and without regard for the specific computer operating system. Graphical user interfaces are much more powerful and can be used to quickly set up very complicated flow geometries, including smoothly contoured walls that had to modeled using a stairstep type of approach in the past. The results of the modeling can be quickly and easily viewed in visually appealing and useful formats. One researcher has renamed "CFD" to "colorized fluid dynamics" because of the explosion of color and pretty pictures that can be generated with the codes that are available today.[67] Another important factor in the adoption of CFD is the improvements in the physics and number of submodels that are available to the user. It is now possible to model compressible and incompressible, viscous and nonviscous, laminar and turbulent, high and low pressure, reacting and nonreacting, multiphase, and many other types of flows. The user often has a choice of several submodels for thermal radiation, turbulence, soot, and pollution chemistry, for example. Another important factor that has accelerated the use of CFD is experimental validation, wherein experimental data has been compared against numerical predictions to show the validity of the predictions. Where the correlation between the predictions and the experimental data was poor, changes have been made to the code to improve its weaknesses. As with most tools of this type, increased adoption of the codes has led to greater acceptance by the engineering community. This growth in popularity is expected to continue to increase as more and more engineers are trained in its use.

5.4.2 POTENTIAL PROBLEMS OF CFD

As with most things, CFD modeling is not a panacea that can be used to solve all problems. As the saying goes, "Garbage in = garbage out," or the results are only as good as the input data. It is incumbent upon the user to input the relevant geometry and boundary conditions into the simulation and to select the appropriate submodels for a given problem. Because of the ease-of-use of the codes available today, it is possible for someone with little or no training in fluid dynamics, heat transfer, and chemical reactions to make predictions that may or may not be credible. It is often observed that anything predicted by a computer must, by definition, be correct. Those who are skilled in this art know that this is far from the truth and tend to use the codes to predict trends, rather than to guarantee absolute numbers, unless they have a great deal of experience with a particular type of problem and have experimental data to validate the code under those conditions. It is very easy for the CFD codes to be misapplied to problems that are beyond the range of their validity. Therefore, it is appropriate to use the caveat "let the user beware."

There are still many limitations of the physical models in the codes. One example of considerable importance in most combustion problems is turbulence. The empirical k-ε model has been around for many years and has been widely used despite its many known limitations. Two other submodels of great importance in combustion modeling are radiation and chemical reactions. These are all discussed in some detail later. Suffice it to say that further research is required and is ongoing to improve those submodels.

Combustion problems are among the most complex that CFD codes are used to solve because they usually involve complicated geometries and fluid dynamics, heat transfer including nonlinear thermal radiation, and chemically reacting flows that may include many species and literally hundreds of chemicals reactions. Many industrial combustors have two very different length scales because the combustor itself may be relatively large, while the length scale required to properly simulate the individual flames in the combustor may be several orders of magnitude smaller. Therefore, large-scale problems may require hundreds of thousands, if not millions, of grid points. As the complexity of the problem increases, so does the number of iterations required to get a converged solution. The large number of grid points and complicated physics often equate to long computation times, depending on the available hardware. A calculation for a large industrial furnace with multiple burners and the full set of physics, modeled on a typical workstation, can take literally weeks to obtain a converged solution. Normally, multiple simulations of a given problem are required to find the optimum set of operating conditions. However, it is rare that weeks are available to obtain those solutions. This means that some simplification of the problem is required. This usually involves fairly intimate knowledge of the physics of the problem and preferably some prior knowledge of typical results. It is usually desirable to have some base case against which the model results can be compared, to determine the validity of the numerical predictions. Sometimes, the time, experience with related problems, and base-case data are lacking so that the modeler is left to use his or her best judgment as to how to simplify a given problem. Experienced users know the inherent dangers in blindly simplifying a problem in order to get a "solution" within the time available.

It is tempting to use only CFD to design new combustion equipment because it is often much cheaper and faster than building prototypes that are usually tested first under controlled laboratory conditions before trying them out in actual field installations. Further, it is also tempting to use CFD to guarantee the performance of combustion equipment because it may be difficult, if not impossible, to test the equipment in every conceivable type of application. Too much confidence in CFD codes is potentially dangerous without proper experimental validation. A more logical approach is to use a combination of numerical modeling in conjunction with experimental measurements. CFD modeling can be used to dramatically reduce the cycle times for developing new products by rapidly simulating a wide range of configurations, which would be both time-consuming and expensive to do with prototypes. CFD modeling can also be used to scale-up lab results or field results from one specific application to another type of application. Only proper experimental validation can assure the user of the usefulness of the modeling results. Unfortunately, this step is often overlooked or ignored and can result in spurious predictions.

5.4.3   EQUATIONS

In this chapter section, the equations are given — without derivation — for flows in rectangular coordinates. Equations for cylindrical and spherical coordinate systems are given in the Appendix E. A more complete discussion of these equations and their derivations is given in many other places (e.g., see Reference 68) and has not been repeated here for the sake of brevity.

5.4.3.1   Fluid Dynamics

The unsteady equations of motion for an incompressible Newtonian fluid with constant viscosity in rectangular coordinates (x,y,z) are given as follows:

$$\frac{\partial u}{\partial \tau} + u\frac{\partial u}{\partial x} + v\frac{\partial u}{\partial y} + w\frac{\partial u}{\partial z} = f_x - \frac{1}{\rho}\frac{\partial p}{\partial x} + v\left(\frac{\partial^2 u}{\partial x^2} + \frac{\partial^2 u}{\partial y^2} + \frac{\partial^2 u}{\partial z^2}\right) \tag{5.7}$$

$$\frac{\partial v}{\partial \tau} + u\frac{\partial v}{\partial x} + v\frac{\partial v}{\partial y} + w\frac{\partial v}{\partial z} = f_y - \frac{1}{\rho}\frac{\partial p}{\partial y} + v\left(\frac{\partial^2 v}{\partial x^2} + \frac{\partial^2 v}{\partial y^2} + \frac{\partial^2 v}{\partial z^2}\right) \tag{5.8}$$

$$\frac{\partial w}{\partial \tau} + u\frac{\partial w}{\partial x} + v\frac{\partial w}{\partial y} + w\frac{\partial w}{\partial z} = f_z - \frac{1}{\rho}\frac{\partial p}{\partial z} + v\left(\frac{\partial^2 w}{\partial x^2} + \frac{\partial^2 w}{\partial y^2} + \frac{\partial^2 w}{\partial z^2}\right) \tag{5.9}$$

where f_i is some type of body force such as buoyancy.

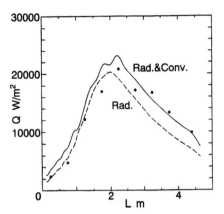

FIGURE 5.2 Calculated heat flux rates to a furnace wall compared with experimental measurements. (Courtesy of Taylor & Francis, Washington, D.C.[107] With permission.)

In industrial combustion, it is normally assumed that the flows are low Mach number, which simplifies the fluid dynamics where the flow is incompressible.[69] One of the earliest and still widely used algorithms was developed by Patankar and Spalding.[70,71] The algorithm is known by the acronym SIMPLE, which stands for semi-implicit pressure-linked equation.

A discussion of turbulence modeling has been adequately treated in many other places and is only briefly considered here.[72-96] Spalding (1976) discussed the eddy-break-up model for turbulent combustion.[97] Prudnikov (1959),[98] Spalding (1976),[99] Bray (1978, 1996),[100,101] Pope (1990),[102] and Ashurst (1994)[103] have given reviews of combustion in turbulent flames. Jones and Whitelaw (1984) compared some of the available turbulence models against experimental data.[104] Faeth (1986) reviewed the interactions between turbulence and heat and mass transfer processes in flames.[105] Arpaci (1993) discussed a method for including the interaction between turbulence and radiation.[106] Yoshimoto et al. (1996) gave a typical example of modeling a furnace using the k-ε turbulence model.[107] A comparison of the numerical results against experimental measurements is shown in Figure 5.2. Lindstedt and Váos (1999) gave a good discussion of the closure problem using the Reynolds stress equations to solve turbulent flame models.[108] Swaminathan and Bilger (1999) discussed the stationary laminar flamelet and conditional moment closure submodels used for simulating turbulent combustion.[109] Bilger (1976) modeled turbulent diffusion flames using cylindrical coordinates and the following generalized governing equation:[110]

$$\frac{\partial}{\partial x}\left(\bar{\rho}\tilde{u}\phi\right)+\frac{1}{r}\frac{\partial}{\partial r}\left(r\bar{\rho}\tilde{v}\phi\right)=\frac{1}{r}\frac{\partial}{\partial r}\left(r\mu_{\text{eff},\phi}\frac{\partial\phi}{\partial r}\right)+S_{\phi} \tag{5.10}$$

where x is the streamwise direction, u is the streamwise velocity, r is the radial direction, v is the radial velocity, ρ is the density, $\mu_{\text{eff},\phi}$ and S_{ϕ} are given in Table 5.1, and ϕ is the variable under consideration: 1, $\tilde{u}$, $\tilde{f}$ mixture fraction), k (turbulent kinetic energy), ε (eddy dissipation rate), or g (mixture fraction fluctuations). The formulation is based on Favre (mass-weighted) averaged quantities:

$$\tilde{\phi}=\frac{\overline{\rho\phi}}{\bar{\rho}} \tag{5.11}$$

where the overbar represents a conventional time average. The constants in Table 5.1 are empirical and were determined by matching predictions and measurements for constant-density round jets.

TABLE 5.1
Turbulence Model Parameters

ϕ	$\mu_{\text{eff},\phi}$	S_ϕ
1	—	0
$\tilde{u}$	$\mu + \mu_t$	$a(\bar{\rho} - \rho_\infty)$
$\tilde{f}$	$(\mu/\text{Sc}) + (\mu_t/\sigma_f)$	0
k	$\mu + (\mu_t/\sigma_k)$	$\mu_t (\partial \tilde{u}/\partial r)^2 - \rho\varepsilon$
ε	$\mu + (\mu_t/\sigma_\varepsilon)$	$[C_{\varepsilon1}\mu_t (\partial \tilde{u}/\partial r)^2 - C_{\varepsilon2}\bar{\rho}\varepsilon](\varepsilon/k)$
g	$(\mu/\text{Sc}) + (\mu_t/\sigma_g)$	$C_{g1}\mu_t(\partial \tilde{f}/\partial r)^2 - C_{g2}\bar{\rho}\varepsilon/k$

C_μ	$C_{\varepsilon1}$	C_{g1}	$C_{\varepsilon2} = C_{g2}$	σ_k	s_ε	$\sigma_f = \sigma_g$	Sc
0.09	1.44	2.8	1.87	1.0	1.3	0.7	0.7

$\mu_t = \bar{\rho}C_\mu k^2/\varepsilon$

Source: Adapted from R.W. Bilger, *Prog. Energy Combust. Sci.,* 1, 87-109, 1976.

Turbulence is very important in most industrial combustion applications that involve high-speed flows. Therefore, this phenomenon must be included in most types of models if representative results are to be expected. Some examples of modeling of swirling flows, which may be important in certain types of combustors, are given elsewhere in this chapter.

5.4.3.2   Heat Transfer

The energy equation for an incompressible fluid can be written as:

$$\rho c_p \left(\frac{\partial t}{\partial \tau} + u\frac{\partial t}{\partial x} + v\frac{\partial t}{\partial y} + w\frac{\partial t}{\partial z} \right) = \frac{\partial}{\partial x}\left(k\frac{\partial t}{\partial x} \right) + \frac{\partial}{\partial y}\left(k\frac{\partial t}{\partial y} \right) + \frac{\partial}{\partial z}\left(k\frac{\partial t}{\partial z} \right) + \dot{q} + \Phi \qquad (5.12)$$

where

$$\Phi = 2\mu \left[\left(\frac{\partial u}{\partial x}\right)^2 + \left(\frac{\partial v}{\partial y}\right)^2 + \left(\frac{\partial w}{\partial z}\right)^2 + \frac{1}{2}\left(\frac{\partial v}{\partial x} + \frac{\partial u}{\partial y}\right)^2 + \frac{1}{2}\left(\frac{\partial w}{\partial y} + \frac{\partial v}{\partial z}\right)^2 + \frac{1}{2}\left(\frac{\partial u}{\partial z} + \frac{\partial w}{\partial x}\right)^2 \right] \quad (5.13)$$

The source term $\dot{q}$ contains the terms for calculating radiation that are dependent on the absolute temperatures raised to the fourth power, which makes the system highly nonlinear and much more difficult to solve. One technique often employed in numerical solution schemes is to "turn off" the radiation terms for some of the calculations to allow the solution to stabilize and for the iterations to be completed more quickly. For example, a particular solution scheme may only solve for the radiation source term every ten iterations.

The neglect of radiation cannot be justified in combustion system modeling.[50] Nonluminous and luminous radiation can greatly complicate a problem because of the spectral dependence of the solution. Cess and Tiwari (1972)[111] and Ludwig et al. (1973)[112] gave a very extensive treatment of gaseous radiation, including the methods available at that time for analyzing those types of problems. They give different techniques used for computing nonluminous radiation and both experimental and computational data on a wide variety of gases, including CO, CO_2, H_2O, HCl, HF, NO, and OH. Beér and co-workers (1972, 1974) presented a discussion of some of the early methods used for radiation analysis, including the zone method and some of the flux methods.[113,114] Lowes et al. (1974) reviewed some of the methods used to analyze radiation in furnaces, including two flux and multi-flux models.[115] Buckius and Tien (1977) showed that computations using non-gray homogeneous and nonhomogeneous radiation models for infrared flame radiation compared favorably with experimental measurements.[116] Crosbie and Dougherty (1981) gave an extensive review of exact methods for solving the radiative transport equation.[117] However, it is noted that exact solutions are not practical for engineering problems.[50] Wall et al. (1982) used a simple zoned model using the Monte Carlo technique for the radiative heat transfer in a pilot-scale furnace for oil and gas flames.[118] They used a convective coefficient of 5.8 W/m²-K (1.0 Btu/hr-ft²-°F) for the transfer from the gas to the furnace walls. Predictions showed good agreement with experimental measurements. Hayasaka (1987) described a method called radiative heat ray (RHR), which was intended to model the actual radiation phenomenon from an atom and claimed to be more computationally efficient than either the Hottel zone or Monte Carlo methods.[119] Bhattacharjee and Grosshandler (1989, 1990) developed a model, called the effective angle model (EAM), that promised computer storage and computational time savings compared to other models.[44,120] The EAM should be effective for calculating radiation in two-dimensional or cylindrical combustors with black walls that have either a specified temperature or heat flux condition. Viskanta and Mengüç (1990) gave an extensive review of radiative heat transfer in combustion systems, including some simple examples.[121] They noted that the following characteristics are needed from a radiation model:

- Capability of handling inhomogeneous and spectrally dependent properties
- Capability of handling highly anisotropic radiation fields due to large temperature gradients and anisotropically scattering particles present in the medium
- Compatibility with finite-difference/finite-element algorithms for solving transport equations

They also compared the different techniques for modeling radiative heat transfer, as shown in Table 5.2. Komornicki and Tomeczek (1992) developed a modification of the wide-band gas model for use in calculating flame radiation.[122] Figure 5.3 shows that their model compared well against experimental data and both a narrow-band model and an unmodified wide-band model. Soufiani and Djavdan (1994) compared the weighted-sum-of-gray-gases (WSGG) and the statistical narrow-band (SNB) radiation models.[123] The WSGG model is much less computationally intensive than the SNB model. They found that the WSGG only introduced small errors when the gas mixture was nearly isothermal and surrounded by cold walls. However, significant inaccuracies were found when using the WSGG where large temperature gradients existed. Lallemant et al. (1996)[124] compared nine popular total emissivity models used in CFD modeling for H_2O-CO_2 homogenous mixtures with the exponential wide-band model (EWBM).[125,126] They recommended the use of the EWBM in conjunction with weighted-sum-of-gray-gas models. A more recent review by Carvalho and Farias (1998) presents the various models that have been used to simulate radiation in combustion systems.[127] These methods include:

- The zone method usually referred to as Hottel's zonal method[128]
- The Monte Carlo method, which is a statistical method[129]
- The Schuster-Hamaker-type flux models[2,130,131]
- The Schuster-Schwarzschild-type flux models[132-135]
- The spherical harmonic flux models (*P-N* approximations)[136]
- Discrete ordinates approximations[137-141]
- Finite volume method[142-143]
- Discrete transfer method[144-146]

A schematic of some of the popular radiation models is shown in Figure 5.4.[49] The discrete exchange factor (DEF) method has been used by Naraghi and co-workers.[147-149] Denison and Webb (1993) presented a spectral radiation approach for generating weighted-sum-of-gray gases models.[150]

TABLE 5.2
Comparison of Techniques for Modeling Radiative Heat Transfer

Method	Remarks	Advantages	Disadvantages
Mean-beam length	Approximation of radiation heat flux using concept of gas emissivity	Simple, possible to include detailed spectral information	Isothermal system; uncertain accuracy; insufficient detail; difficult to generalize
Zone	Approximation of system by finite size zones containing uniform temperature and composition gases	Nonhomogeneities in temperature and concentration of gases can be accounted for	Cumbersome; restricted to relatively simple geometries; difficult to account for scattering and spectral information of gases; not compatible with numerical algorithms for solving transport equations
Differential spherical harmonics, moment	Approximation of RTE in terms of the moments of intensity	RTE is recast into a system of differential equations; absorption and scattering can be accounted for; compatible with numerical algorithms for solving transport equations	Unknown accuracy as the relationship between RTE and the flux equations is not always explicit
Flux and discrete ordinates	Approximation of angular intensity distribution along discrete directions and solutions of these equations numerically	Flexible; higher order approximations are accurate; can account for spectral absorption by gases and scattering by particles; compatible with numerical algorithms for solving transport equations	Time consuming; requires iterative solution of finite-difference equations; simple flux approximations are not accurate
Discrete transfer, ray tracing, numerical	Solves RTE approximately along a line-of-sight	Can use spectral information; flexible; compatible with numerical algorithms	Time consuming if scattering by particles is to be accounted for; accuracy is poor if few rays are considered in scattering media
Monte-Carlo	Simulation of physical process using purely statistical techniques and following individual photons	Flexibility for application to complex geometries; absorption and scattering by particles can be accounted for	Can be time consuming; not compatible with numerical algorithms for solving transport equations
Hybrid	New procedures, relatively untested, which use a combination of two or more methods	Different methods can be developed to account for geometric effects; flexible; may be compatible with the numerical algorithm	Relatively untested; cannot be generalized to all systems

Source: From R. Viskanta and M.P. Mengüç, *Handbook of Heat and Mass Transfer,* N. Cheremisinoff, Ed., Gulf Publishing, Houston, TX, Vol. 4, 970-971, 1990. With permission.

RTE = radiation transport equation

Unlike some other methods, the absorption coefficient is the modeled radiative property that permits arbitrary solution of the radiative transfer equation.

Some examples of the application of various radiation models to industrial combustion processes are given next. Siddall and Selcuk (1974) described the application of the two-flux method to a process gas heater.[151] Docherty and Fairweather (1988) showed that their predictions using the discrete transfer method for radiation from nonhomogeneous combustion products compared favorably to narrow-band calculations, as shown in Figure 5.5.[146] Cloutman and Brookshaw (1993) described a numerical algorithm for solving radiative heat losses from an experimental burner.[152] Abdullin and Vafin (1994) modeled the radiative properties of a waterwall combustor to determine

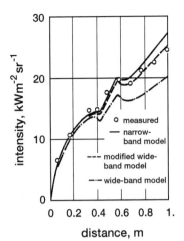

FIGURE 5.3 Calculated and experimental radiation intensity values for a natural gas flame at a distance of 0.6 m from the burner exit. (Courtesy of Elsevier.[122] With permission.)

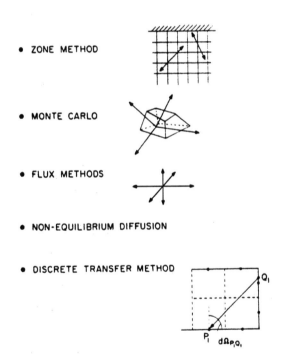

FIGURE 5.4 Common approaches to radiation modeling. (Courtesy of The Combustion Institute, Pittsburgh, PA.[49] With permission.)

their effects on the heat transfer in a tube furnace.[153] They modeled downfired burners bounded by rows of vertical tubes with the exhaust at the bottom of the furnace. The results showed peak heat fluxes at about 20% of the distance from the ceiling and the floor, with radiation far exceeding convection. As expected, the tube emissivity was an important parameter in the heat flux in the combustor. The partial pressure of the combustion products also had an interesting effect because it affected the gas radiation and absorptivity. The peak radiation to the waterwall was predicted for a gas partial pressure of the combination of CO_2 and H_2O of 0.27 atm (0.27 barg).

Ahluwalia and Im (1992) presented an improved technique for modeling the radiative heat transfer in coal furnaces.[154] Coal furnaces differ from gas-fired combustion processes because of

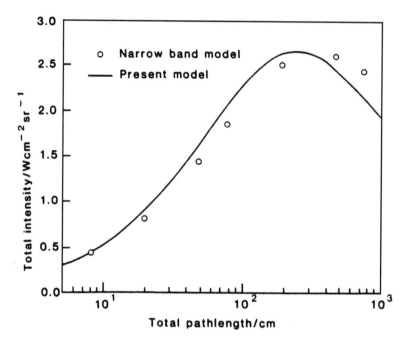

FIGURE 5.5   Predicted radiation intensity as a function of path length for narrow-band and discrete transfer radiation models. (Courtesy of The Combustion Institute, Pittsburgh, PA.[146] With permission.)

the presence of char and ash, which produce significant quantities of luminous radiation. This improved technique was developed to help solve three-dimensional spectral radiation transport equations for the case of absorbing, emitting, and anisotropically scattering media present in coal systems. The incorporation of spectral radiation can significantly increase the computational time and complexity, depending on how the spectra is discretized. The technique is a hybrid combination of the discrete ordinate method (S_4),[155] modified differential approximation (MDA),[156] and P_1 approximation for use in different ranges of optical thicknesses. It combines a char burnout model and spectroscopic data for H_2O, CO_2, CO, char, soot, and ash. It is used to determine the influence of ash composition, ash content, and coal preparation on heat absorption by the furnace. In the simulation of an 80-MW_e corner-fired coal boiler, predicted wall heat fluxes ranging from approximately 100 to 600 kW/m^2 (32,000 to 190,000 $Btu/hr\text{-}ft^2$) compared favorably with experimental measurements.

Song and Viskanta (1987) discussed the modeling of radiation and turbulence as applied to combustion.[157] Figure 5.6 shows the predicted heat flux for models with and without turbulence/radiation interactions. Köylü and Faeth (1993) discussed modeling the properties of flame-generated soot.[158] They evaluated approximate methods for calculating the following properties for both individual aggregates and polydisperse aggregate populations: the Rayleigh scattering approximation,[159] Mie scattering for an equivalent sphere,[160] and Rayleigh-Debye-Gans (R-D-G) scattering[161] for both given and fractal aggregates. Available measurements and computer simulations were not adequate to properly evaluate the approximate prediction methods. Given those limitations, Rayleigh scattering generally underestimated scattering, Mie scattering for an equivalent sphere was unreliable, and R-D-G scattering gave the most reliable results. Bressloff et al. (1996) presented a coupled strategy for predicting soot and gas species concentrations, and radiative exchange in turbulent combustion.[162] Good agreement was found with experimental data on temperature, mixture fraction, and soot volume fraction. Bai et al. (1998) discussed soot modeling in turbulent jet diffusion flames.[163] Brookes and Moss (1999) showed the intimate connection between soot production and flame radiation, which must be accurately accounted for in modeling.[164]

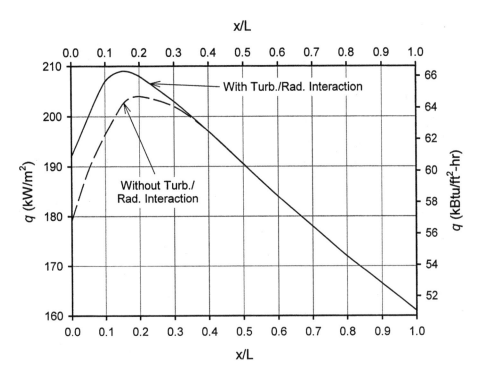

FIGURE 5.6 Heat flux distributions along the sink: comparison of the total (convective + radiative) with and without turbulence/radiation interactions. (From T.H. Song and R. Viskanta, *J. Thermophysics,* 1(1), 56-62, 1987. With permission.)

Numerous methods exist for modeling nonluminous spectral radiation from combustion products like H_2O, CO_2, and CO. Wide-band models were the first to be used because they are the simplest to implement.[165,166] More advanced models incorporated narrow-band approximations. Taine (1983)[167] and Hartmann et al. (1984)[168] have computed line-by-line calculations for single absorption bands of CO_2 and H_2O, respectively. Goody (1964) developed a statistical narrow-band model.[169] Properties used in the band models are often taken from Ludwig and co-workers (1973).[112] Song et al. (1997) used a statistical narrow-band hybrid model to calculate the gaseous radiation heat transfer in an side-port-fired glass furnace firing on natural gas.[170] This study showed the need to include spectral calculations because a gray-medium assumption overestimates the heat transfer and produces inconsistent results.[171]

5.4.3.3 Chemistry

Many schemes have been proposed for the number of equations and reaction rates that can be used to simulate combustion reactions. One that has commonly been used is known as CHEMKIN, which was developed at Sandia National Labs.[172] Another source of chemical kinetic data is a database, formed through funding by Gas Research Institute (Chicago, IL), known as GRI-Mech.[173] The National Institute of Standards and Technology or NIST (Washington, D.C.) has also assembled and maintains a very extensive database (over 37,000 separate reactions for over 11,400 distinct reactant pairs) of chemical kinetic data taken from over 11,000 papers.[174] The CEC Group on Evaluation of Kinetic Data for Combustion Modeling was established by the European Energy Research and Development Programme to compile a database of critically evaluated chemical kinetic data.[175,176] Gaz de France has sponsored research toward improving the chemical kinetic modeling of natural gas combustion.[177] Gardiner (1984),[178] Sloane (1984),[179] and Libby and Williams (1994)[180] have edited books concerning combustion chemistry. Golden (1991) reviewed the rate parameters used in combustion chemistry modeling.[181] A number of papers have been written that discuss chemistry in combustion processes.[182-191] Some papers specifically consider the interaction of turbulence and the chemical reactions.[192]

One approach that has been used to simplify the chemistry in combustion modeling is to use a statistical approach referred to as the probability-density function (PDF) approach.[193,194] This is coupled with the solution of the energy and momentum equations, along with the species equations. The PDF approach is most suited to turbulent reactive flows since the complex chemical reactions can be treated without modeling assumptions.[195] However, some simplifications are usually required because of the excess computer time requirement for complete PDF modeling.

Modeling soot formation in flames is also a challenging aspect of the simulation. Coelho and Carvalho (1994) compared different soot formation models for turbulent diffusion propane flames with 500°C (930°F) air preheat.[196] They used the soot formation models given by Khan and Greeves (1974)[197] and Stewart et al. (1991),[198] and the soot oxidation models given by Magnussen and Hjertager (1977),[199] Lee et al. (1962),[200] and Nagle and Strickland-Constable (1961).[201] By comparing with available soot data, it was found that the Stewart et al.[198] model gave better predictions than the Khan and Greeves model,[197] once the constants were properly tuned. There was not enough data to determine which soot oxidation model gave the best predictions. Delichatsios and Orloff (1988) studied the interaction between luminous flame radiation and turbulence.[202] They concluded that soot formation was determined by the straining rate of the small (Kolmogorov) scales. Boerstoel et al. (1996) found that experimental data compared favorably with several different soot formation and oxidation models for a high-temperature furnace.[203] Said et al. (1997) proposed a simple two-equation model for soot formation and oxidation in turbulent flames.[204] Xu et al. (1998) studied the soot produced by fuel-rich, oxygen/methane, atmospheric pressure laminar premixed flames.[205] Their measurements showed good agreement with the soot computational models proposed by Frenklach and Wang (1990)[206] and Leung and Lindstedt (1995).[207] In certain industrial heating processes, there may be additional chemical reactions besides those involved in the combustion. The additional reactions may come from the material processing in the combustor or they may also come from downstream processing of the exhaust gases, especially in the case of many of the post-treatment pollutant reduction technologies such as selective noncatalytic reduction (SNCR) or methane re-burn for NOx reduction. In the latter case, it may be argued that the downstream treatment technologies involving combustion are a part of the overall combustion system. Several examples will suffice to illustrate these "other" chemical reactions. In the glass melting process, there are many chemical reactions produced during the melt-in of the incoming batch materials, including the production of CO_2 and some corrosive species. In the flash smelting process used for the processing of copper, there are many chemical reactions involving copper, sulfur, and iron that are separate from the combustion reactions. In the methane re-burn NOx reduction technology, methane is injected downstream of the main combustion zone where it chemically reduces much of the NO generated in the flame region back to N_2 and O_2. It is not the purpose here to detail the noncombustion reactions, but merely to point out that they may need to be included in the model as they are directly or indirectly coupled to the combustion system.

5.4.3.4 Multiple Phases

In some combustion systems, multiple phases are present. The most notable involve the combustion of liquid and solid fuels. In the case of a liquid fuel, the fuel is atomized or vaporized into a fine spray that can then be combusted. In the case of a solid fuel, the fuel normally must be finely ground so that complete combustion can be achieved. In both cases, the modeling effort is significantly complicated. In the U.S., the vast majority of industrial combustion processes use gaseous fuels. Therefore, modeling the combustion of liquid or solid fuels will not be treated here. It is recommended that the interested reader consult some of the numerous references that are available for liquid and solid fuel combustion modeling and given throughout this chapter and in the general references cited in Chapter 1.

5.4.4 BOUNDARY AND INITIAL CONDITIONS

There are several types of boundary conditions that may be applicable to a particular problem, depending on the specific conditions. For the sake of argument, a two-dimensional rectangular coordinate system (x,y) will be used for illustration purposes. A boundary condition of the first kind is where a variable is specified as a function along a given boundary:

$$\phi = f(x,y,\tau) \tag{5.14}$$

where ϕ is the variable and τ is time. Examples include a constant temperature wall $t(x,y = 0,\tau) = 2000°F$ or a constant inlet velocity $u(x = 0,y,\tau) = 100$ ft/sec. If the scalar is zero everywhere on a given boundary, this is referred to as a homogeneous boundary condition of the first kind.

A boundary condition of the second kind is when the gradient of a variable is a function of the position and time:

$$\frac{\partial \phi}{\partial n} = f(x,y,\tau) \tag{5.15}$$

where n is the normal to a surface. An example would be a constant heat flux at a horizontal wall: $\partial t/\partial y(y = 0) = 100$ Btu/ft^2-hr. If the gradient is equal to zero everywhere on a boundary, this is known as a homogeneous boundary condition of the second kind.

A boundary condition of the third kind is a linear combination of the first and second kinds:

$$a\phi + b\frac{\partial \phi}{\partial n} = f(x,y,\tau) \tag{5.16}$$

where a and b are functions not dependent on ϕ, and n is the normal to a surface. An example would be where the convection to a surface is equal to the conduction into the surface: $k\partial t/\partial y(y = 0) + ht_1 = ht_0$, where k is the thermal conductivity of the solid and h is the convection coefficient. If the function is equal to zero everywhere on a boundary, this is known as a homogeneous boundary condition of the third kind.

These three types cover many common boundary conditions. There are also other types of boundary conditions that do not fit these types. For example, a radiative boundary condition is a function of a fourth-power temperature law, which makes it highly nonlinear. These types of boundary conditions are sometimes approximated by boundary conditions of the third kind where possible to simplify the solution of the problem.

Each type of equation (conservation of mass, momentum, species, etc.) must have a sufficient number of boundary conditions to get a unique solution. A simple example will illustrate this. The steady-state, one-dimensional heat conduction through an infinitely long slab of finite width can be written as:

$$q = -k\frac{\partial t}{\partial x} \tag{5.17}$$

Equation 5.17 can be integrated to find the heat flux through the slab:

$$\int\limits_{x=a}^{x=b} q \, dx = -\int\limits_{x=a}^{x=b} k \frac{\partial t}{\partial x} \, dx = -k\left[t(x=b) - t(x=a)\right] \qquad (5.18)$$

where it has been assumed that the thermal conductivity k is a constant. Boundary conditions of the first kind are needed at each side of the slab to get a unique solution.

Initial conditions refer to the conditions that exist at time = zero. If the problem is steady state, then no initial conditions are required. However, for transient problems, the initial conditions must be specified for all variables, prior to initiating the computations, to get a unique solution.

5.4.4.1   Inlets and Outlets

The simplest type of inlet condition is to have uniform scalar and vector properties, such as a fluid flowing into a chamber at constant temperature, pressure, and velocity, with a single species. This applies to many problems, but there are many more that are not quite so simple. Many problems have some type of velocity profile at the inlet, for example, gases flowing through a pipe entering a burner. A blended fuel with multiple constituents needs the capability of having multiple species in a single inlet stream. These nonuniform inlet conditions are available on commercial codes.

In older commercial software packages, outlet cells were specified as such, but no other conditions could be assigned to them. Newer packages now have the capability of specifying outlet conditions. For many problems where the fluids flow strictly out of the system, this is not necessary. However, for problems with recirculating flows, it is often possible that gases may be drawn into the system through some of the "outlet" cells. This can be useful for including, for example, ambient air infiltration into a furnace.

5.4.4.2   Surfaces

The primary variables used most often for surfaces are constant temperature or constant heat flux. It is also possible to put in temperature and heat flux profiles. Newer codes also have provisions to put in the thermal conductivity as a function of temperature for the wall to calculate conduction through the walls. Radiation characteristics such as emissivity can also be specified for surfaces. This is important for virtually all industrial combustion problems. Many codes also have the provision to put in porous surfaces for gases to flow through; these are especially relevant for simulating porous refractory burners (see Chapter 8).

5.4.4.3   Symmetry

There are many problems that have some type of symmetry which can be used to reduce the number of computations needed to simulate a combustion system. One type of symmetry is where, for example, the left half of a furnace is a mirror image of the right half. In that case, only one of the halves needs to be modeled, with a symmetry boundary condition along the axis of symmetry in the middle of the furnace. Another type of symmetry is slice symmetry, where only a single slice of a furnace is modeled to reduce the computation time. For example, a furnace with multiple burners symmetrically located on both sides of the furnace would be time-consuming to accurately model. A single burner in the middle of a side could be modeled with symmetry boundary conditions on both sides and on one end (the middle of the furnace). Although the symmetry is not strictly correct, as the exhaust is typically at one end of the furnace, the symmetry model may have sufficient accuracy for the problem under consideration (e.g., the heat flux rates to the wall or to the load). Another example of where a symmetry boundary condition can be used is in a cylindrical furnace that has burners symmetrically located on one end of the furnace. For example, assume the furnace has six burners located on one end of the cylinder and firing into the cylinder. Assuming these burners are located symmetrically every 60° around the wall, only one sixth of the furnace needs to be simulated where periodic symmetric boundary conditions are used to simulate a single slice. The beauty of the codes available today is that the results for the entire furnace can be displayed if desired although only a fraction was actually modeled.

5.4.5 DISCRETIZATION

As CFD initially developed, the physical space being modeled was discretized into many smaller spaces in order to convert the partial differential equations into algebraic equations. In general, the more grids used, the higher the accuracy. Unfortunately, this often meant that solutions were grid dependent. That is, different solutions were computed for different-sized grids. Therefore, it was often necessary to run many simulations with different-sized grids in order to find the relationship between the grid and the solution so that a truly grid-independent solution could be found. This was very computationally intensive. Early codes only had the capability of a uniform grid everywhere. Later codes had provisions for varying grid sizes so that only fine gridding was used in areas of high gradients, while a coarser grid could be used where variables were only changing slowly. The newest codes have adaptive grid capability, wherein the grid is automatically adjusted by the code to achieve the optimal mesh for a given problem under the constraints given by the user. Another grid improvement over early codes is known as boundary-fitted coordinates (BFC), where smooth curving surfaces can be simulated as such without having to do a stair-step approximation which was previously necessary for orthogonal gridding methods). This section briefly discusses the various approaches commonly used for discretizing CFD problems.

5.4.5.1 Finite Difference Technique

Özisik (1994) has discussed the general use of the finite difference technique in heat transfer and fluid flow problems, but without reactions.[208] Figure 5.7 shows a typical grid for the finite difference method. The finite difference technique is one of the oldest methods for numerically solving fluid flow problems and dates back to at least 1910.[13] In the finite difference method, the differential fluid flow equations are replaced by discrete difference approximations. The gridwork of points

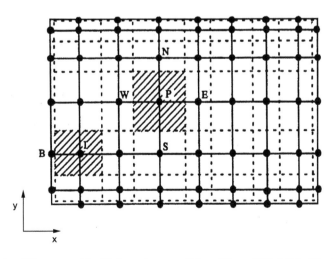

FIGURE 5.7 Finite difference grid with boundary condition. (From K.D. Mish, *The Handbook of Fluid Dynamics,* R.W. Johnson, Ed., CRC Press, Boca Raton, FL, 1998, 26-9.)

used to compute these approximations is known as the finite difference grid, which is regularly spaced for optimal problem convergence. The regular spacing is one of the biggest constraints in using this method. It is possible for some problems to convert irregularly spaced grid points into regular spacing by means of a transformation, but it is often difficult to determine the transformation for complicated problems. Forward, centered, and backward differencing schemes can be used. To illustrate, assume there is a two-dimensional curve with the function $y = f(x)$. Take three points on the curve: $(x_0, f(x_0))$, $(x_1, f(x_1))$, and $(x_2, f(x_2))$, where x_1 is spaced at Δx from x_0, and x_2 is spaced Δx from x_1. Then, the slope (m) of the line $f(x)$ at x_1 can be approximated with difference equations using a forward difference technique:

$$m_{\text{forward}} = \frac{f(x_2) - f(x_1)}{x_2 - x_1} = \frac{f(x_1 + \Delta x) - f(x_1)}{\Delta x} \tag{5.19}$$

The slope of the line can be estimated using a backward difference technique:

$$m_{\text{backward}} = \frac{f(x_1) - f(x_0)}{x_1 - x_0} = \frac{f(x_1) - f(x_1 - \Delta x)}{\Delta x} \tag{5.20}$$

Or, the slope of the line can be estimated using a centered difference technique:

$$m_{\text{centered}} = \frac{f(x_2) - f(x_0)}{x_2 - x_0} = \frac{f(x_1 + \Delta x) - f(x_1 - \Delta x)}{2\Delta x} \tag{5.21}$$

This same type of discretization is used to solve the fluid flow equations in difference form. The centered difference technique often produces the fastest rate of convergence in most problems.

5.4.5.2 Finite Volume Technique

The finite volume technique is probably the most popular in commercial codes, possibly because it has been used since the earliest days of commercial CFD codes. Figure 5.8 shows a typical two-dimensional grid for the finite volume method. This technique developed from the modification of

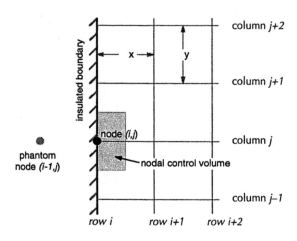

FIGURE 5.8 Finite volume grid and control volumes. (From S.V. Patankar, K.C. Karki, and K.M. Keldar, *The Handbook of Fluid Dynamics,* R.W. Johnson, Ed., CRC Press, Boca Raton, FL, 1998, 27-6.)

the finite difference technique, in order to solve more complicated flow geometries.[13] In this method, finite volumes are used around each grid point to calculate the discretized fluid flow equations, along with neighboring grid points. In the figure, the point of interest is P, which communicates with its neighboring gridpoints N, E, S, and W through the four faces of the hatched volume around P. In this method, the spacing between points does not need to be equal, nor does the size of the control volumes. The differential equations are then discretized in terms of control volumes and fluxes through the faces (see Reference 71 for the discretized finite volume equations).

5.4.5.3 Finite Element Technique

Reddy and Gartling (1994) presented the general use of the finite difference technique in heat transfer and fluid flow problems, but without reactions.[209] This technique is well-known for solving solid modeling problems to predict, for example, stresses in a deformed solid body. It has only relatively recently been applied to fluid flow problems. The fundamental difference between the finite element, and the finite difference and finite volume techniques is that the finite element technique uses the integral equations for fluid flow rather than the differential equations. The fluid flow equations are discretized in terms of flux vectors. Some functional form is assumed for the approximation that will be used to solve the integral equations, where the time- and space-independent variables are separated. Then, various solution techniques are used to solve the resulting integral equations with the assumed approximation form. A common method is to assume that the solution is in the form of a polynomial, where the coefficients in the polynomial are solved numerically to get the best approximation for the assumed form. The reader should consult the relevant references for more details on this technique.[14,15,209]

5.4.5.4 Mixed

Pember et al. (1995) used a hybrid discretization scheme to simulate an industrial burner.[69] Their grid consisted of uniform rectangular cells everywhere except at the furnace walls, for which they used finite elements to solve what they called mixed cells where the cells contained some fluid and some solid boundary. This method was termed an "embedded boundary method" and was offered as an alternative to using body-fitted coordinates.

5.4.5.5 None

As discussed above for the vortex methods of solution (see section 5.2.1.2), no discretization of the space is used to simulate the fluid flow. Instead, the properties of the vortices are tracked as they evolve and dissipate, without regard to any grid system. This is a more accurate way of simulating the space as it eliminates the problems of how to discretize the space. There are also no problems with the numerical errors associated with discretization.

5.4.6 Solution Methods

Shyy and Mittal (1998) have presented a number of common methods used to solve the Navier-Stokes equations.[210] In most cases, the objective is to reduce integro-differential equations into algebraic equations, usually by discretization, that can then be solved in a variety of ways. The choice of solution method depends on the type of problem, especially as to whether the problem can be linearized or not. Linearized equations are commonly put into matrix form and then solved using a variety of available matrix reduction techniques, such as the Gauss Elimination Method, the Thomas Algorithm, Gauss-Seidel Iteration, Successive Over-Relaxation, Red and Black Ordering Scheme, or LU Decomposition. Nonlinear systems may be solved with other techniques, such as Newton-Raphson Iteration.

5.4.7 MODEL VALIDATION

Roache (1998) has written an entire book about verification and validation of computation fluid dynamics modeling.[211] He noted the distinction between verification ("solving the equations right") and validation ("solving the right equations"). Here, no discussion will be given on verification as this is available in any good CFD book (including the book by Roache), which as Roache noted is a more mathematical issue. For the purposes of this chapter section, it is assumed that the mathematical models in a given code have been properly debugged and produce reliable results within a given accuracy range. Of more concern in combustion modeling is validation, to make sure the appropriate physics are being used for the problem under consideration and that those physics are properly simulated. Validation is a much more difficult problem than verification, especially in combustion simulations, due to the difficulties of making relevant and accurate measurements in harsh environments.

One of the seductive aspects of computer modeling is that virtually any type of problem can be simulated. How a problem is modeled depends on many things; but if enough assumptions are made, it is possible to generate computational "results." For the naïve and inexperienced, the tendency may be to believe anything that is generated by a computer — because, how can a computer be wrong? The caveat, "garbage in, garbage out" definitely applies to computer modeling of complex industrial combustion problems. Any given problem may have many assumptions that need to be made, so the results are only as good as the model and the accompanying assumptions. Paraphrasing an anonymous researcher, "Everyone believes a computer analysis except the one who did it, and no one believes experimental results except the one who made them." This is to say that most people inherently realize the difficulties of making experimental measurements in complex geometries, but most naturally believe the results generated by a computer.

As any good modeler knows, a model is only as good as its validation. Models must constantly be tested against experimental measurements when they are applied to new problems. Model validation is particularly difficult for industrial combustion problems because of the difficulty in making measurements in harsh environments and because of the cost involved in making those measurements. Most measurements made in industrial combustors are with intrusive water-cooled probes because many of the nonintrusive laser-based techniques have not yet been developed for large scales, are not rugged enough for the environments, or are too costly to use outside the lab. These intrusive probes are often larger than those used in labs because more water cooling is required in high-temperature combustors. Therefore, the flow is disturbed by the probes, which makes it more difficult to compare the measurements with the modeling results. In general, there is relatively little experimental data available for industrial combustors that has enough information to do a complete model validation. This is an important research need for the future — to generate comprehensive data sets in a wide range of industrial combustion systems that can be used for model validation. Some typical model validation cases are given next; they are representative of those available to date.

Fiveland et al. (1996) presented four validation cases for comparison against codes developed by Babcock and Wilcox.[212] Although the codes are primarily directed at large-scale boilers, the validations were done for a broader range of cases, including flow in a curved duct, nonreacting flow and natural gas combustion in a swirl stabilized flame, and swirling flow coal combustion in a one-sixth scale model of a utility boiler. Kaufman and Fiveland (1995) generated a large set of experimental data for the swirl-stabilized natural gas flame case in the Burner Engineering Research Lab at Sandia National Labs (Livermore, CA) as part of a program partially funded by the Gas Research Institute (Chicago, IL).[213] This data was used for the model validation. The model results were generally very good, except in the recirculation zones. Further work was recommended to improve the chemistry and turbulence models.

5.4.8 INDUSTRIAL COMBUSTION EXAMPLES

The two major parts of industrial combustion problems include the burners and the combustors. These are often modeled separately for a variety of reasons, as previously discussed. Examples of modeling different types of burners and combustors are given next.

5.4.8.1 Modeling Burners

There have been numerous papers on modeling industrial burners. A sampling of references is given for modeling the following industrial burners:

- Radiant tube burners[214-215]
- Swirl burners[217]
- Pulse combustion burner[218]
- Porous radiant burners[219-224]
- Industrial hydrogen sulfide burner[225]

Butler et al. (1986) gave a general discussion of modeling burners using the finite volume techniques.[226] Schmücker and Leyens (1998) described the use of CFD to design a new nozzle-mix burner referred to as the Delta Burner.[227] Schmidt et al. (1998) described the use of CFD to redesign a burner, originally firing on coal, to fire on natural gas for use in a rotary kiln.[228]

Modeling radiant burners poses the additional challenge of simulating a porous medium.[229] Perrin et al. (1986) discussed the use of a numerical model for the design of a single-ended radiant tube for immersion in and heating of a bath of molten zinc.[230] A sample result of the temperature and heat flux distribution from the tube is shown in Figure 5.9. As can be seen, most of the heat transfer is by radiation. Hackert et al. (1998) simulated the combustion and heat transfer in two-dimensional porous burners.[231] Two different porous geometries were simulated: a honeycomb consisting of parallel, nonconnecting passages and a separated plates geometry consisting of parallel

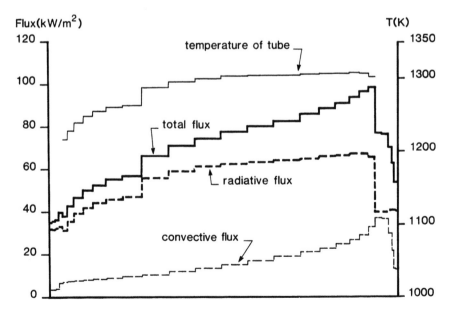

FIGURE 5.9 Predicted temperature and heat flux on the shell of a single-ended radiant tube immersed in molten zinc. (Courtesy of the American Society of Metals, Warren, PA.[230] With permission.)

but broken walls where there was no continuous solid path, which minimizes the importance of solid conduction. Spatial calculated temperatures compared favorably with measured values. The radiant efficiency in the stable region of the flame ranged from 15% to 17% and 22% for the two geometries, respectively. They determined a volumetric Nu of 5.4 ± 0.3, regardless of burning rate or pore size, for the separated plates geometry; this compares favorably with other values reported in the literature. Fu et al. (1998) used a one-dimensional model to simulate the performance of a porous radiant burner.[232] The model accounted for the interaction of convection, conduction, radiation, and chemical reaction in the burner, which consisted of two layers of reticulated ceramics having different porosities. The model showed that the radiant efficiency increased with the volumetric heat transfer coefficient and the effective thermal conductivity of the solid matrix, and decreased when the firing rate increased and the equivalence ratio decreased.

5.4.8.2   Modeling Combustors

Numerous papers have been presented over the past few decades on the modeling of industrial combustors. Some of the earliest work was done at Imperial College (U.K.).[2,233,234] A sampling of references are given for modeling industrial combustors:

- General combustors and boilers[235-249]
- Pulse combustors[250,272-274]
- Glass furnaces[251-256]
- Aluminum reverberatory furnaces[257]
- Metal reheat furnaces[36,258-260]
- Radiant tube batch furnace[261]
- Flash smelting furnaces[262,263]
- Industrial coal combustors[264-268]
- Steam reformers for liquefied petroleum gas conversion[269]
- Fluid catalytic crackers[270,271]
- Rotary hearth calciners[275]
- Rotary kilns[276,277]
- Cement rotary kiln[278]
- Vertical lime kilns[279]
- Generic oxy/fuel-fired furnace[280]
- Rotary kiln incinerator[281-283]
- Municipal solid waste incinerator ash melting furnace[284]
- Hazardous waste incineration furnace[285]
- Baking ovens[286]

Carvalho and Nogueira (1997) modeled glass-melting furnaces, cement kilns, and baking ovens.[287] Song and Viskanta (1986) did a parametric study of the thermal performance of a generic natural gas-fired furnace.[288] Some papers have specifically focused on radiation modeling in combustors,[289] while other papers have considered swirling flows in furnaces.[290,291]

Carvalho and Nogueira (1993) reviewed the modeling of industrial glass melting processes.[292] They noted that modeling glass melting is especially important because of the difficulty in making measurements due to the very high temperatures and corrosive environment. The models are useful for improving glass quality, increasing thermal efficiency, reducing pollution emissions, and improving equipment reliability. They recommended further research in simulating flow modeling, batch melting, fining and refining, foam formation/elimination, homogenizing, refractories, radiative transfer, and the mass transfer between the glass melt and the combustion chamber. Glass modeling is particularly difficult if the combustion space and molten glass are coupled together, because of the large disparity in flow types. The combustion space may have turbulent gas flow while the molten glass is a very low-speed flow of highly viscous material. The system is further complicated by the transparent radiative characteristic of the glass and by the chemical reactions occurring in both the gas space and liquid glass phase. In many cases, electrodes are located in the liquid glass and gases may be bubbled through the glass to stimulate stirring and circulation patterns in the glass. The load consists of fine solid materials at the inlet of the tank and liquid glass at the outlet of the tank. Despite this complexity, models have been successfully used to further the understanding of and improvement in the glass production process.

Viskanta and co-workers did parametric computational studies of both direct- and indirect-fired furnaces.[293-296] Chapman et al. (1989) presented the results of parametric studies of a direct-fired, continuous reheating furnace.[293] They developed a simplified mathematical model that accurately calculated the heat balance throughout the furnace. The combustion space was divided into zones that were considered to be well-stirred reactors. The load in each zone was further subdivided into smaller control volumes. Radiation was modeled using Hottel's zone method. A primary objective of the study was to compute the furnace thermal efficiency as a function of a variety of parameters, including the load velocity, load emissivity, furnace combustion space height, and refractory emissivity. The furnace efficiency increased rapidly with the initial increase in load velocity and then leveled off with further increases in velocity. The load heat flux was relatively insensitive to the load velocity — except for the lowest speed, where the flux was considerably lower. The heat flux to the load was most sensitive to the load emissivity, as shown in Figure 5.10. The heat flux to the load was relatively insensitive to the height of the combustion space. Although the model used in the study was fairly simple, with no detailed fluid flow calculations, it was useful for studying a wide range of values for different parameters — which makes it a valuable design tool for studying other configurations. In a companion study, Chapman et al. (1989) also did a parametric modeling study of direct-fired batch reheating furnaces.[294] Again it was shown that the heat flux

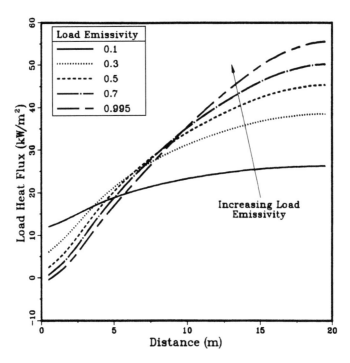

FIGURE 5.10 Load heat flux as a function of the load emissivity and distance in the furnace for a direct-fired continuous reheat furnace. (Courtesy of ASME, New York.[293] With permission.)

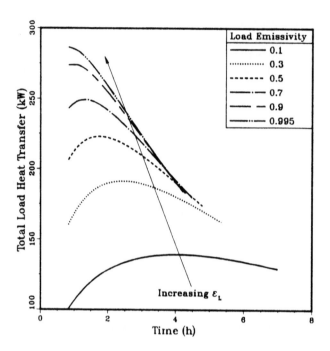

FIGURE 5.11 Load heat flux as a function of the load emissivity and distance in the furnace for a direct-fired batch reheat furnace. (Courtesy of ASME, New York.[294] With permission.)

to the load, was very sensitive to the emissivity of the load as shown in Figure 5.11. Chapman et al. (1994) modeled a direct-fired metal reheat furnace with impinging flame jets.[295] The model simulated an actual steel reheat furnace at an Inland Steel Co. plant in East Chicago, Indiana. A typical result, shown in Figure 5.12, shows how the heat transfer to the load was affected by the fuel firing rate.

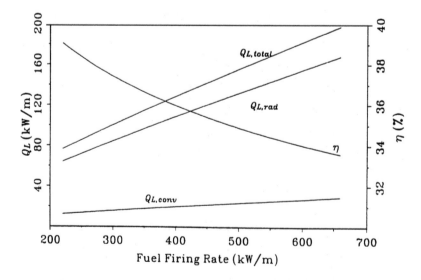

FIGURE 5.12 Load heat flux (total, radiative and convective) as a function of the fuel firing rate for a direct-fired continuous reheat furnace. (Courtesy of Gordon & Breach, London.[296] With permission.)

REFERENCES

1. R. Weber, Scaling Characteristics of aerodynamics, heat transfer, and pollutant emissions in industrial flames, *Twenty-Sixth Symposium (International) on Combustion*, The Combustion Institute, Pittsburgh, PA, 1996, 3343-3354.
2. S. Patankar and B. Spalding, Simultaneous predictions of flow patterns and radiation for three-dimensional flames, in *Heat Transfer in Flames*, N.H. Afgan and J.M. Beer, Eds., Scripta Book Co., Washington, D.C., 1974, 73-94.
3. E.E. Khalil, *Modelling of Furnaces and Combustors*, Abacus Press, Kent, U.K., 1982.
4. E.S. Oran and J.P. Boris, Eds., *Numerical Approaches to Combustion Modeling*, Vol. 135, Progress in Astronautics and Aeronautics, American Institute of Aeronautics and Astronautics, Washington, D.C., 1991.
5. B. Larrouturou, *Recent Advances in Combustion Modelling*, World Scientific, Singapore, 1991.
6. T.J. Chung, *Numerical Modeling in Combustion*, Taylor & Francis, Washington, D.C., 1993.
7. D.G. Lilley, Chemically reacting flows (combustion), in *The Handbook of Fluid Dynamics*, R.W. Johnson, Eds., CRC Press, Boca Raton, FL, 1998, chap. 16.
8. D.B. Spalding, The art of partial modeling, *Ninth Symposium (International) on Combustion*, Academic Press, New York, 1963, 833-843.
9. R. Weber, A.A. Peters, P.P. Breithaupt, and B.M.V. Visser, Mathematical modeling of swirling pulverized coal flames: What can combustion engineers expect from modeling?, *Amer. Soc. of Mech. Eng. (ASME) FACT* 17, 71-86, 1993.
10. E.S. Oran and J.P. Boris, Detailed modeling of combustion systems, *Prog. in Energy and Comb. Science*, 7(1), 1-72, 1981.
11. B.E. Launder and D.B. Spalding, The numerical computation of turbulent flows, in *Lectures in Mathematical Modeling of Turbulence*, Academic Press, London, 1972.
12. J.I. Ramos, Finite-difference methods in turbulent combustion, in *Numerical Modeling in Combustion*, T.J. Chung, Ed., Taylor & Francis, Washington, D.C., 1993, 281-373.
13. K.D. Mish, Finite difference method, in *The Handbook of Fluid Dynamics*, R.W. Johnson, Ed., CRC Press, Boca Raton, FL, 1998, chap. 26.
14. T.J. Chung, Finite element methods in turbulent combustion, in *Numerical Modeling in Combustion*, T.J. Chung, Ed., Taylor & Francis, Washington, D.C., 1993, 375-397.
15. A.J. Baker, Finite element method, in *The Handbook of Fluid Dynamics*, R.W. Johnson, Ed., CRC Press, Boca Raton, FL, 1998, chap. 28.
16. H.A. Dwyer, Finite-volume methods in turbulent combustion, in *Numerical Modeling in Combustion*, T.J. Chung, Ed., Taylor & Francis, Washington, D.C., 1993, 399-408.
17. S.V. Patankar, K.C. Karki, and K.M. Keldar, Finite volume method, in *The Handbook of Fluid Dynamics*, R.W. Johnson, Ed., CRC Press, Boca Raton, FL, 1998, chap. 27.
18. G.E. Karniadakis and R.D. Henderson, Spectral element methods for incompressible flows, in *The Handbook of Fluid Dynamics*, R.W. Johnson, Ed., CRC Press, Boca Raton, FL, 1998, chap. 29.
19. A.F. Ghoniem, Vortex simulation of reacting shear flow, in *Numerical Approaches to Combustion Modeling*, E.S. Oran and J.P. Boris, Eds., Vol. 135, Progress in Astronautics and Aeronautics, American Institute of Aeronautics and Astronautics, Washington, D.C., 1991, 305-348.
20. L.-F. Martins and A.F. Ghonien, Simulation of the nonreacting flow in a bluff-body burner — Effect of the diameter ratio, in *Heat and Mass Transfer in Fires and Combustion Systems*, W.L. Grosshandler and H.G. Semerjian, Eds., ASME HTD-Vol. 148, New York, 1990, 33-44.
21. C. Fureby, E. Lundgren, and S.-I. Möller, Large eddy simulation of combustion, in *Tranport Phenomena in Combustion*, Vol. 2, S.H. Chan, Ed., Taylor & Francis, Washington, D.C., 1996, 1083-1094.
22. H.G. Weller, G. Tabor, A.D. Gosman, and C. Fureby, Application of a flame-wrinkling LES combustion model to a turbulent mixing layer, *Twenty-Seventh Symposium (International) on Combustion*, The Combustion Institute, Pittsburgh, PA, 1998, 899-907.
23. G. Tryggvason and W.J.A. Dahm, An integral method for mixing, chemical reactions, and extinction in unsteady strained diffusion layers, *Comb. Flame*, 83(3-4), 207-220, 1990.
24. C.H.H. Chang, W.J.A. Dahm, and G. Tryggvason, Lagrangian model simulations of molecular mixing, including finite rate chemical reactions, in temporally developing shear layer, *Phys. Fluids A*, 3(5), 1300-1311, 1991.

25. W.J.A. Dahm, G. Tryggvason, and M. Zhuang, Intergral method solution of time-dependent strained diffusion-reaction layers with multistep kinetics, *SIAM J. Appl. Math.*, 56(4), 1039-1059, 1996.

26. W.J.A. Dahm, G. Tryggvason, J.A. Kezerle and R.V. Serauskas, Simulation of turbulent flow and complex chemistry by local integral moment (LIM) modeling, *Proc. of 1995 International Gas Research Conf.*, D.A. Dolenc, Ed., Govt. Institutes, Rockville, MD, 1996, 2169-2178.

27. M.Y. Hussaini and T.A. Zang, Spectral Methods in Fluid Dynamics, *Annu. Rev. of Fluid Mechanics*, 19, 339-367, 1987.

28. P.A. McMurtry and P. Givi, Spectral simulations of reacting turbulent flows, in *Numerical Approaches to Combustion Modeling*, E.S. Oran and J.P. Boris, Eds., Vol. 135, Progress in Astronautics and Aeronautics, American Institute of Aeronautics and Astronautics, Washington, D.C., 1991, 257-303.

29. P. Givi and C.K. Madnia, Spectral methods in combustion, in *Numerical Modeling in Combustion*, T.J. Chung, Ed., Taylor & Francis, Washington, D.C., 1993, 409-452.

30. V. Eswaran and S.B. Pope, Direct numerical simulations of the turbulent mixing of a passive scalar, *Physics of Fluids*, 31(3), 506-520, 1988.

31. J.H. Chen, J.M. Card, M. Day, and S. Mahalingam, Direct numerical simulation of turbulent non-premixed methane-air flames, in *Tranport Phenomena in Combustion*, Vol. 2, S.H. Chan, Ed., Taylor & Francis, Washington, D.C., 1996, 1049-1060.

32. T. Poinsot, Using direct numerical simulations to understand premixed turbulent combustion, *Twenty-Sixth Symposium (International) on Combustion*, The Combustion Institute, Pittsburgh, PA, 1996, 219-232.

33. T. Baritaud, T. Poinsot, and M. Baum, Eds., *Direct Numerical Simulation for Turbulent Reacting Flows*, Editions Technip, Paris, 1996.

34. M. Boger, D. Veynante, H. Boughanem, and A. Trouvé, Direct numerical simulation analysis of flame surface density concept for large eddy simulation of turbulent premixed combustion, *Twenty-Seventh Symposium (International) on Combustion*, The Combustion Institute, Pittsburgh, PA, 1998, 917-925.

35. P.A. McMurtry and P. Givi, Direct numerical simulations of mixing and reaction in non-premixed homogeneous turbulent flows, *Comb. Flame*, 77, 171-185, 1989.

36. R.M. Davies, D.M. Lucas, B.E. Moppett, and R.A. Galsworthy, Isothermal model studies of rapid heating furnaces, *J. Inst. Fuel*, 44, 453-461, 1971.

37. J.T. Kuo, System simulation and control of batch-fed solid waste incinerators, in *Heat Transfer in Fire and Combustion Systems*, W.W. Yuen and K.S. Ball, Eds., New York, ASME HTD-Vol. 272, 55-62, 1994.

38. S. Singh, M. Ziolkowski, J. Sultzbaugh, and R. Viskanta, Mathematical model of a ceramic burner radiant heater, in *Fossil Fuel Combustion 1991*, R. Ruiz, Ed., New York, ASME PD-Vol. 33, 111-116, 1991.

39. P.A. Gillis and P.J. Smith, An evaluation of three-dimensional computational combustion and fluid dynamics for industrial furnace geometries, *Twenty-Third Symposium (International) on Combustion*, The Combustion Institute, Pittsburgh, PA, 1990, 981-991.

40. M. Frenklach, Reduction of chemical reaction models, in *Numerical Approaches to Combustion Modeling*, E.S. Oran and J.P. Boris, Eds., Vol. 135, Progress in Astronautics and Aeronautics, American Institute of Aeronautics and Astronautics, Washington, D.C., 1991, 129-154.

41. C.K. Westbrook and F.L. Dryer, Simplified reaction mechanisms for the oxidation of hydrocarbon fuels in flames, *Comb. Sci. Tech.*, 27, 31-43, 1981.

42. N. Peters, Systematic reduction of flame kinetics: principles and details, in *Dynamics of Reactive Systems*, A.L. Kuhl, J.R. Bowen, J.-C. Leyer, and A. Borisov, Eds., Vol. 113, Progress in Astronautics and Aeronautics, American Institute of Aeronautics and Astronautics, Washington, D.C., 1988, 67-86.

43. Y.S. Kocaefe, A. Charette, and M. Munger, Comparison of the various methods for analysing the radiative heat transfer in furnaces, *Proceedings of the Combustion Institute Canadian Section Spring Technical Meeting*, Vancouver, Canada, May 1987, 15-17.

44. S. Bhattacharjee and W.L. Grosshandler, Effect of radiative heat transfer on combustion chamber flows, *Comb. Flame*, 77, 347-357, 1989.

45. Y.R. Sivathanu, J.P. Gore, and J. Dolinar, Transient scalar properties of strongly radiating flames, in *Heat and Mass Transfer in Fires and Combustion Systems*, W.L. Grosshandler and H.G. Semerjian, Eds., ASME HTD-Vol. 148, 45-56, 1990.

46. G. Cox, On radiant heat transfer in diffusion flames, *Combust. Sci. Tech.*, 17, 75-78, 1977.

47. V.P. Kabashnikov and G.I. Kmit, Influence of turbulent fluctuations on thermal radiation, *Appl. Spect.*, 31, 963-967, 1979.

48. W.L. Grosshandler and P. Joulain, The effect of large scale fluctuations on flame radiation, *Prog. Astro. and Aero.*, Vol. 105, Part II, AIAA, Washington, D.C., 1986, 123-152.

49. A.F. Sarofim, Radiative heat transfer in combustion: friend or foe, Hoyt C. Hottel Plenary Lecture, *Twenty-First Symp. (International) on Combustion*, The Combustion Institute, Pittsburgh, PA, 1986, 1-23.

50. R. Viskanta and M.P. Menguc, Radiation heat transfer in combustion systems, *Prog. Energy Combust. Sci.*, 8, 97-160, 1987.

51. G.M. Faeth, J.P. Gore, S.G. Chuech, and S.M. Jeng, Radiation from turbulent diffusion flames, *Annu. Rev. Numerical Fluid Mech. & Heat Trans.*, C.L. Tien and T.C. Chawla, Eds., Hemisphere, New York, Vol. 2, 1-38, 1989.

52. C.J. Hoogendoorn, L. Post and J.A. Wieringa, Modelling of combustion and heat transfer in glass furnaces, *Glastech. Ber.*, 63(1), 7-12, 1990.

53. Y. Zhenghua and G. Holmstedt, Fast, narrow-band computer model for radiation calculations, *Num. Heat Transfer*, Part B, 31, 61-71, 1997.

54. W.L. Grosshandler, Radiation from Nonhomogeneous Fires, Tech. Rep. FMRC, Sept. 1979.

55. W.L. Grosshandler, Radiative heat transfer in nonhomogeneous gases: a simplified approach, *Int. J. Heat Mass Transfer*, 23, 1447-1459, 1980.

56. W.L. Grosshandler, RADCAL: A Narrow-Band Model for Radiation Calculation in a Combustion Environment, NIST Tech. Note 1402, April 1993.

57. F. Liu, Ö.L. Gülder, and G.J. Smallwood, Non-grey gas radiative transfer analyses using the statistical narrow-band model, *Int. J. Heat Mass Transfer*, 41(14), 2227-2236, 1998.

58. L.A. Gritzo and J.H. Strickland, A gridless solution of the radiative transfer equation for fire and combustion calculations, *Combust. Theory Modelling*, 3, 159-175, 1999.

59. R.K. Ahluwalia and K.H. Im, Radiative Heat Transfer in PC Furnaces Burning Deeply Cleaned Coals, U.S. Dept. of Energy Report DE91 006022, Argonne National Laboratory, Argonne, IL, 1990.

60. R.K. Ahluwalia and K.H. Im, Spectral radiative heat-transfer in coal furnaces using a hybrid technique, *J. Inst. Energy*, 67, 23-29, 1994.

61. H.M. Park, R.K. Ahluwalia, and K.H. Im, Three-dimensional radiation in absorbing-emitting-scattering media using the modified differential approximation, *Int. J. Heat Mass Trans.*, 36(5), 1181-1189, 1993.

62. T.L. Wilson, A three-region, moving boundary model of a furnace flame, *Proceedings of the 5th Symp. On High Performance Computing*, Atlanta, GA, April 1997, U.S. Dept. of Energy Report DE97003334, Oak Ridge National Lab, TN, 1997.

63. W.A. Gray, E. Hampartsoumian, J.M. Taylor, M. Gover, and J. Sykes, Modelling of radiant heat transfer in gas-fired furnaces using a practical personal computer package, *Proc. of 1992 International Gas Research Conf.*, H.A. Thompson, Ed., Govt. Institutes, Rockville, MD, 1993, 2184-2194.

64. M.D. Rumminger, R.W. Dibble, N.H. Heberle, and D.R. Crosley, Gas temperature above a porous radiant burner: comparison of measurements and model predictions, *Twenty-Sixth Symposium (International) on Combustion*, The Combustion Institute, Pittsburgh, PA, 1996, 1755-1762.

65. L.B. Younis and R. Viskanta, Experimental determination of the volumetric heat transfer coefficient, *Int. J. Heat Mass Transfer*, 36(6), 1425-1434, 1993.

66. R.M. Kendall and J.D. Sullivan, Selective and Enhanced Radiation from Porous Surface Radiant Burners, Gas Research Institute Report GRI-93/0160, Chicago, 1993.

67. R. Dibble, Univ. of CA at Berkeley, private communication, 1997.

68. R.B. Bird and M.D. Graham, General equations of newtonian fluid dynamics, in *The Handbook of Fluid Dynamics*, R.W. Johnson, Ed., CRC Press, Boca Raton, FL, 1998, chap. 3.

69. R.B. Pember, A.S. Almgren, W.Y. Crutchfield, L.H. Howell, J.B. Bell, P. Colella, and V.E. Beckner, An Embedded Boundary Method for the Modeling of Unsteady Combustion in an Industrial Gas-Fired Furnace, U.S. Dept. of Commerce Report No. DE96004155, Springfield, VA, 1995.

70. S.V. Patankar and D.B. Spalding, *Heat and Mass Transfer in Boundary Layers: A General Calculation Procedure*, second edition, Intertext Books, London, 1970.

71. S.V. Patankar, *Numerical Heat Transfer and Fluid Flow*, McGraw-Hill, New York, 1980.

72. G.K. Batchelor, *The Theory of Homogeneous Turbulence*, Cambridge Univ. Press, New York, 1953.

73. J.T. Davies, *Turbulence Phenomena*, Academic Press, New York, 1972.

74. B.E. Launder and D.B. Spalding, *Mathematical Models of Turbulence*, Academic Press, New York, 1972.
75. J.O. Hinze, *Turbulence*, 2nd ed., McGraw-Hill, New York, 1975.
76. P. Bradshaw, Ed., *Turbulence*, Springer-Verlag, New York, 1978.
77. W. Kollmann, *Prediction Methods for Turbulent Flows*, Hemisphere, Washington, D.C., 1980.
78. P. Bradshaw, *Engineering Calculation Methods for Turbulent Flow*, Academic Press, New York, 1981.
79. T. Tatsumi, Ed., *Turbulence and Chaotic Phenomena in Fluids*, North Holland, New York, 1984.
80. M.M. Stanisic, *The Mathematical Theory of Turbulence*, Springer-Verlag, New York, 1985.
81. M. Lesieur, *Turbulence in Fluids*, Dordrecht, Boston, 1987.
82. H.C. Mongia, R.M.C. So, and J.H. Whitelaw, Eds., *Turbulent Reactive Flow Calculations*, Gordon and Breach, New York, 1988.
83. B.S. Petukhov and A.F. Polyakov, *Heat Transfer in Turbulent Mixed Convection*, Hemisphere, New York, 1988.
84. V.R. Kuznetsov, *Turbulence and Combustion*, Hemisphere, New York, 1990.
85. W.D. McComb, *The Physics of Fluid Turbulence*, Oxford Univ. Press, Oxford, U.K., 1990.
86. W. Rodi and E.N. Ganic, Eds., *Engineering Turbulence Modeling and Experiments*, Elsevier Science, Amsterdam, The Netherlands, 1990.
87. M. Landahl, *Turbulence and Random Processes in Fluid Mechanics*, Cambridge Univ. Press, Cambridge, U.K., 1992.
88. S.W. Churchill, *Turbulent Flows*, Butterworth-Heinemann, Boston, 1993.
89. D.C. Wilcox, *Turbulence Modeling for CFD*, DCW Industries, La Cãnada, CA, 1993.
90. A.J. Chorin, *Vorticity and Turbulence*, Springer-Verlag, New York, 1994.
91. R.J. Garde, *Turbulent Flow*, Wiley, New York, 1994.
92. U. Frisch, *Turbulence*, Cambridge Univ. Press, Cambridge, U.K., 1995.
93. K. Hanjalic and J.C.F. Pereira, Eds., *Turbulence, Heat, and Mass Transfer*, Begell House, New York, 1995.
94. C.J. Chen, *Fundamentals of Turbulence Modeling*, Taylor & Francis, Washington, D.C., 1998.
95. C.G. Speziale and R.M.C. So, Turbulence modeling and simulation, in *The Handbook of Fluid Dynamics*, R.W. Johnson, Ed., CRC Press, Boca Raton, FL, 1998, chap. 14.
96. J. Baldyga, *Turbulent Mixing and Chemical Reactions*, Wiley, New York, 1999.
97. D.B. Spalding, Development of the eddy-break-up model of turbulent combustion, *Twentieth Symposium (International) on Combustion*, The Combustion Institute, Pittsburgh, PA, 1976, 1657-1663.
98. A.G. Prudnikov, Flame turbulence, *Seventh Symposium (International) on Combustion*, Butterworths Scientific Publications, London, 1959, 575-582.
99. D.B. Spalding, Mathematical models of turbulent flames: a review, *Combust. Sci. Tech.*, 13(1-6), 3-25, 1976.
100. K.N.C. Bray, The interaction between turbulence and combustion, *Seventeenth Symposium (International) on Combustion*, The Combustion Institute, Pittsburgh, PA, 1978, 223-233.
101. K.N.C. Bray, The challenge of turbulent combustion, *Twenty-Sixth Symposium (International) on Combustion*, The Combustion Institute, Pittsburgh, PA, 1996, 1-26.
102. S.B. Pope, Computations of turbulent combustion: progress and challenges, *Twenty-Third Symposium (International) on Combustion*, The Combustion Institute, Pittsburgh, PA, 1990, 591-612.
103. W.T. Ashurst, Modeling turbulent flame propagation, *Twenty-Fifth Symposium (International) on Combustion*, The Combustion Institute, Pittsburgh, PA, 1994, 1075-1089.
104. W.P. Jones and J.H. Whitelaw, Modelling and measurements in turbulent combustion, *Twentieth Symposium (International) on Combustion*, The Combustion Institute, Pittsburgh, PA, 1984, 233-249.
105. G.M. Faeth, Heat and mass transfer in flames, in *Heat Transfer 1986*, Vol. 1, C.L. Tien, V.P. Carey, and J.K. Ferrell, Eds., Proceedings of the Eighth International Heat Transfer Conf., San Francisco, 1986, 151-160.
106. V.S. Arpaci, Radiative turbulence: radiation affected turbulent forced convection, in *Heat Transfer in Fire and Combustion Systems — 1993*, B. Farouk, M.P. Menguc, R. Viskanta, C. Presser, and S. Chellaiah, Eds., New York, ASME HTD-Vol. 250, 155-160, 1993.
107. T. Yoshimoto, T. Okamoto, and T. Takagi, Numerical simulation of combustion and heat transfer in a furnace and its comparison with experiments, in *Tranport Phenomena in Combustion*, Vol. 2, S.H. Chan, Ed., Taylor & Francis, Washington, D.C., 1996, 1153-1164.

108. R.P. Lindstedt and E.M. Váos, Modeling of premixed turbulent flames with second moment methods, *Comb. Flame*, 116, 461-485, 1999.
109. N. Swaminathan and R.W. Bilger, Assessment of combustion submodels for turbulent nonpremixed hydrocarbon flames, *Comb. Flame*, 116, 519-545, 1999.
110. R.W. Bilger, Turbulent jet diffusion flames, *Prog. Energy Combust. Sci.*, 1, 87-109, 1976.
111. R.D. Cess and S.N. Tiwari, Infrared radiative energy transfer in gases, in *Advances in Heat Transfer*, Vol. 8, J.P. Hartnett and T.F. Irvine, Eds., Academic Press, New York, 1972, 229-283.
112. C.B. Ludwig, W. Malkmus, J.E. Reardon, and J.A.L. Thomson, *Handbook of Infrared Radiation*, National Aeronautics and Space Administration Report SP-3080, Washington, D.C., 1973.
113. T.R. Johnson and J.M. Beer, Radiative heat transfer in furnaces: further development of the zone method of analysis, *Fourteenth Symposium (International) on Combustion*, The Combustion Institute, Pittsburgh, PA, 1972, 639-649.
114. J.M. Beér, Methods for calculating radiative heat transfer from flames in combustors and furnaces, in *Heat Transfer in Flames*, N.H. Afgan and J.M. Beer, Eds., Scripta Book Co., Washington, D.C., 1974, 29-45.
115. T.M. Lowes, H. Bartelds, M.P. Heap, S. Michelfelder, and B.R. Pai, Prediction of radiant heat flux distribution, in *Heat Transfer in Flames*, N.H. Afgan and J.M. Beer, Eds., Scripta Book Company, Washington, D.C., 1974, chap. 10, 179-190.
116. R.O. Buckius and C.L. Tien, Infrared flame radiation, *Int. J. Heat Mass Transfer*, 20, 93-106, 1977.
117. A.L. Crosbie and R.L. Dougherty, Two-dimensional radiative transfer in cylindrical geometry with anisotropic scattering, *J. Quant. Spectrosc. Radiat. Transfer*, 25(6), 551-569, 1981.
118. T.F. Wall, H.T. Duong, I.M. Stewart, and J.S. Truelove, Radiative heat transfer in furnaces: flame and furnace models of the IFRF M1- and M2-trials, *Nineteenth Symposium (International) on Combustion*, The Combustion Institute, Pittsburgh, PA, 1982, 537-547.
119. H. Hayasaka, A direct simulation method for the analysis of radiative heat transfer in furnaces, in *Heat Transfer in Furnaces*, C. Presser and D.G. Lilley, Eds., New YorkASME HTD-Vol. 74, 59-63, 1987.
120. S. Bhattacharjee and W.L. Grosshandler, A simplified model for radiative source term in combusting flows, *Int. J. Heat Mass Transfer*, 33(3), 507-516, 1990.
121. R. Viskanta and M.P. Mengüç, Principles of radiative heat transfer in combustion systems, in *Handbook of Heat and Mass Transfer*, N. Cheremisinoff, Ed., Gulf Publishing, Houston, 1990, Vol. 4, chap. 22, 925-978.
122. W. Komornicki and J. Tomeczek, Modification of the wide-band gas radiation model for flame calculation, *Int. J. Heat Mass Transfer*, 35(7), 1667-1672, 1992.
123. A. Soufiani and E. Djavdan, A comparison between weighted sum of gray gases and statistical narrow-band radiation models for combustion applications, *Comb. Flame*, 97, 240-250, 1994.
124. N. Lallemant, A. Sayre, and R. Weber, Evaluation of emissivity correlations for H_2O-CO_2-N_2/air mixtures and coupling with solution methods of the radiative transfer equation, *Prog. Energy Combust. Sci.*, 22, 543-574, 1996.
125. D.K. Edwards, Molecular gas band radiation, in *Advances in Heat Transfer*, Vol. 12, T.F. Irvine and J.P. Hartnett, Eds., Academic Press, New York, 1976, 115-193.
126. A.T. Modak, Exponential wide band parameters for the pure rotational band of water vapor, *J. Quant. Spectosc. Radiat. Transfer*, 21(2), 131-142, 1979.
127. M.G. Carvalho and T.L. Farias, Modelling of heat transfer in radiating and combusting systems, *Trans. IChemE*, 76A, 175-184, 1998.
128. H.C. Hottel and A.F. Sarofim, *Radiative Transfer*, McGraw-Hill, New York, 1967.
129. J.R. Howell and M. Perlmutter, Monte Carlo solution of thermal transfer through radiant media between gray walls, *J. Heat Transfer*, 86(1), 116–122, 1964.
130. H.C. Hamaker, Philips Research Reports 3, 103, 112, and 142, 1947.
131. A.D. Gosman and F.C. Lockwood, Incorporation of a flux model for radiation into a finite difference procedure for furnace calculations, *Fourteenth Symp. (International) on Combustion*, the Combustion Institute, Pittsburgh, PA, 1973, 661-671.
132. T.M. Lowes, H. Bartelds, M.P. Heap, S. Michelfelder and B.R. Pai, Prediction of radiant heat flux distributions, Int. Flame Research Found. Report GO2/A/26, IJmuiden, The Netherlands, 1973.
133. W. Richter and R. Quack, A mathematical model of a low-volatile pulverised fuel flame, in *Heat Transfer in Flames*, N.H. Afgan and J.M. Beer, Eds., Scripta Book Co., Washington, D.C., 1974, 95-110.

134. R.G. Siddall and N. Selçuk, Two-flux modelling of two-dimensional radiative transfer in axi-symmetrical furnaces, *J. Inst. Fuel*, 49, 10-20, 1976.

135. R.G. Siddall and N. Selçuk, Evaluation of a new six-flux model for radiative transfer in rectangular enclosures, *Trans. IChem*, 57, 163-169, 1979.

136. R. Viskanta, Radiative transfer and interaction of convection with radiation heat transfer, *Advances in Heat Transfer*, T.F. Irvine and J.P. Hartnett, Eds., 1966, 175-252.

137. S. Chandrasekhar, *Radiative Transfer*, Dover Publications, New York, 1960.

138. W.A. Fiveland, Discrete-ordinates solutions of the radiative transport equation for rectangular enclosures, *J. Heat Transfer*, 106, 699-706, 1984.

139. A.S. Jamaluddin and P.J. Smith, Predicting radiative transfer in rectangular enclosures using the discrete ordinates method, *Comb. Sci. Tech.*, 59(4-6), 321-340, 1988.

140. A.S. Jamaluddin and P.J. Smith, Predicting radiative transfer in axisymmetric cylindrical enclosures using the discrete ordinates method, *Comb. Sci. Tech.*, 62(4-6), 173-186, 1988.

141. W.A. Fiveland and A.S. Jamaluddin, Three-dimensional spectral radiative heat transfer solutions by the discrete-ordinates method, in *Heat Transfer Phenomena in Radiation, Combustion, and Fires*, R.K. Shah, Ed., New York, ASME HTD-Vol. 106, 43-48, 1989.

142. G.D. Raithby and E.H. Chui, A finite-volume method for predicting radiant heat transfer in enclosures with participating media, *J. Heat Transfer*, 112, 414-423, 1990.

143. J.C. Chai, H.S. Lee, and S.V. Patankar, Finite volume method for radiation heat transfer, *J. Thermophysics & Heat Transfer*, 8(3), 419-425, 1994.

144. N.G. Shah, New Method of Computation of Radiative Heat Transfer in Combustion Chambers, Ph.D. thesis, Imperial College, London, 1979.

145. F.C. Lockwood and N.G. Shah, A new radiation solution method for incorporation in general combustion prediction procedures, *Eighteenth Symp. (International) on Combustion*, the Combustion Institute, Pittsburgh, PA, 1981, 1405-1414.

146. P. Docherty and M. Fairweather, Predictions of radiative transfer from nonhomogeneous combustion products using the discrete transfer method, *Comb. Flame*, 71, 79-87, 1988.

147. M.H.N. Naraghi and M. Kassemi, Radiative transfer in rectangular enclosures: a discretized exchange factor solution, *ASME Proceedings of the 1988 National Heat Transfer Conf.*, H.R. Jacobs, Ed., New York, 1, 259-267, 1988.

148. M.H.N. Naraghi, Radiative heat transfer in non-rectangular enclosures, in *Heat Transfer Phenomena in Radiation, Combustion, and Fires*, R.K. Shah, Ed., New York, ASME HTD-Vol. 106, 17-25, 1989.

149. M.H.N. Naraghi and B. Litkouhi, Discrete exchange factor solution of radiative heat transfer in three-dimensional enclosures, in *Heat Transfer Phenomena in Radiation, Combustion, and Fires*, R.K. Shah, Ed., New York, ASME HTD-Vol. 106, 221-229, 1989.

150. M.K. Denison and B.W. Webb, A spectral line-based weighted-sum-of-gray-gases model for arbitrary RTE solvers, *J. Heat Transfer*, 115, 1004-1012, 1993.

151. R.G. Siddall and N. Selcuk, The application of flux methods to prediction of the behavior of a process gas heater, in *Heat Transfer in Flames*, N.H. Afgan and J.M. Beer, Eds., Scripta Book Company, Washington, D.C., 1974, chap. 11, 191-200.

152. L.D. Cloutman and L. Brookshaw, Numerical Simulation of Radiative Heat Loss in an Experimental Burner, Lawrence Livermore National Lab, U.S. Dept. of Energy Report UCRL-JC-115048, 1993.

153. A.M. Abdullin and D.V. Vafin, Numerical investigation of the effect of the radiative properties of a tube waterwall and combustion products on heat transfer in tube furnaces, *J. Eng. Physics and Thermophysics*, 65(2), 752-757, 1994.

154. R.K. Ahluwalia and K.H. Im, Radiative Heat Transfer in Coal Furnaces, U.S. Dept. of Energy Report DE92018770, Argonne National Lab, IL, 1992.

155. W.A. Fiveland and A.S. Jamaluddin, Three-dimensional spectral radiative heat transfer solutions by the discrete-ordinate method, *J. Thermophysics*, 5(3), 335-339, 1991.

156. M.F. Modest, Modified differential approximation for radiative transfer in general three-dimensional media, *J. Thermophysics*, 3(3), 283-288, 1989.

157. T.H. Song and R. Viskanta, Interaction of radiation with turbulence: application to a combustion system, *J. Thermophysics*, 1(1), 56-62, 1987.

158. Ü.Ö. Köylü and G.M. Faeth, Radiative properties of flame-generated soot, *J. Heat Transfer*, 115, 409-417, 1993.

159. C.L. Tien and S.C. Lee, Flame radiation, *Prog. Energy Combust. Sci.*, 8, 41-59, 1982.
160. W.H. Dalzell, G.C. Williams, and H.C. Hottel, A light scattering method for soot concentration measurements, *Comb. Flame*, 14, 161-170, 1970.
161. J.E. Martin and A.J. Hurd, Scattering from fractals, *J. Appl. Cryst.*, 20, 61-78, 1987.
162. N.W. Bressloff, J.B. Moss, and P.A. Rubini, CFD prediction of couple radiation heat transfer and soot production in turbulent flames, *Twenty-Sixth Symposium (International) on Combustion*, The Combustion Institute, Pittsburgh, PA, 1996, 2379-2386.
163. X.S. Bai, M. Balthasar, F. Mauss, and L. Fuchs, Detailed soot modeling in turbulent jet diffusion flames, *Twenty-Seventh Symposium (International) on Combustion*, The Combustion Institute, Pittsburgh, PA, 1998, 1623-1630.
164. S.J. Brookes and J.B. Moss, Predictions of soot and thermal radiation properties in confined turbulent jet diffusion flames, *Comb. Flame*, 116, 486-503, 1999.
165. D.K. Edwards, L.K. Glassen, W.S. Hauser, and J.S. Tuchscher, Radiation heat transfer in nonisothermal nongray gases, *J. Heat Transfer*, 86C, 219-229, 1967.
166. B. Leckner, Spectral and total emissivity of water vapor and carbon dioxide, *Comb. Flame*, 19, 33-48, 1972.
167. J. Taine, A line-by-line calculation of low-resolution radiative properties of CO_2-CO transparent nonisothermal gaseous mixtures up to 3000K, *J. Quant. Spectroscopic Radiative Transfer*, 30, 371-379, 1983.
168. J.M. Hartmann, L. Leon, and J. Taine, Line-by-line and narrow-band statistical model calculations for H_2O, *J. Quant. Spectroscopic and Radiative Transfer*, 32(2), 1984, 119-127.
169. R.M. Goody, *Atmospheric Radiation*, Vol. I, Oxford Univ. Press, Oxford, U.K., 1964.
170. G. Song, T. Bjørge, J. Holen, and B.F. Magnussen, Simulation of fluid flow and gaseous radiation heat transfer in a natural gas-fired furnace, *Int. J. Num. Methods for Heat & Fluid Flow*, 7(2/3), 169-180, 1997.
171. R.K. Ahluwalia and K.H. Im, Spectral radiative heat-transfer in coal furnaces using a hybrid technique, *J. Institute of Energy*, 67, 23-29, 1994.
172. R.J. Kee, R.M. Rupley, and J.A. Miller, CHEMKIN-II: A Fortran Chemical Kinetics Package for the Analysis of Gas Phase Chemical Kinetics, Sandia National Laboratory Report SAND89-8009B, Livermore, CA, 1989.
173. C.T. Bowman, R.K. Hanson, D.F. Davidson, W.C. Gardiner, V. Lissianski, G.P. Smith, D.M. Golden, M. Frenklach, and M. Goldberg, http://www.me.berkeley.edu/gri_mech/, 1999.
174. W.G. Mallard, F. Westley, J.T. Herron, R.F. Hampson, and D.H. Frizzell, *NIST Chemical Kinetics Database User's Guide — Windows Version 2Q98*, National Institute of Standards and Technology, Washington, D.C., 1998.
175. D.L. Baulch, C.J. Cobos, R.A. Cox, C. Esser, P. Frank et al., Evaluated kinetic data from combustion modelling, *J. Phys. Chem. Ref. Data*, 21(3), 411-734, 1992.
176. D.L. Baulch, C.J. Cobos, R.A. Cox, P. Frank, G. Hayman, Th. Just, J.A. Kerr, T. Murrells, M.J. Pilling, J. Troe, R.W. Walker, and J. Warnatz, Summary table of evaluated kinetic data for combustion modeling: Supplement 1, *Comb. Flame*, 98, 59-79, 1994.
177. A. Turbiez, P. Desgroux, J.F. Pauwels, L.R. Sochet, S. Poitou, and M. Perrin, GDF.kin®: a new step towards a detailed kinetic mechanism for natural gas combustion modeling, *Proceedings of the 1998 International Gas Research Conf.*, Vol. V: Industrial Utilization, D.A. Dolenc, Ed., Gas Research Institute, Chicago, 1998, 210-221.
178. W.C. Gardiner, Ed., *Combustion Chemistry*, Springer-Verlag, New York, 1984.
179. T.M. Sloane, *The Chemistry of Combustion Processes*, Amer. Chem. Soc., Washington, D.C., 1984.
180. P.A. Libby and F.A. Williams, Eds., *Turbulent Reacting Flows*, Academic Press, London, 1994.
181. D.M. Golden, Evaluation of chemical thermodynamics and rate parameters for use in combustion modeling, in *Fossil Fuel Combustion*, W. Bartok and A.F. Sarofim, Eds., Wiley, New York, 1991, chap. 2.
182. C.K. Westbrook and F.L. Dryer, Chemical kinetics and modeling of combustion processes, *Eighteenth Symposium (International) on Combustion*, The Combustion Institute, Pittsburgh, PA, 1980, 749-767.
183. F. Kaufman, Chemical kinetics and combustion: intricate paths and simple steps, *Nineteenth Symposium (International) on Combustion*, The Combustion Institute, Pittsburgh, PA, 1982, 1-10.
184. J. Wofrum, Chemical kinetics in combustion systems: the specific effect of energy, collisions, and transport processes, *Twentieth Symposium (International) on Combustion*, The Combustion Institute, Pittsburgh, PA, 1984, 559-573.

185. S.W. Benson, Combustion, a chemical and kinetic view, *Twenty-First Symposium (International) on Combustion*, The Combustion Institute, Pittsburgh, PA, 1986, 703-711.

186. V.Y. Basevich, Chemical kinetics in the combustion process, in *Handbook of Heat and Mass Transfer*, N. Cheremisinoff, Ed,, Gulf Publishing, Houston, 1990, Vol. 4, chap. 18.

187. P. Gray, Chemistry and Combustion, *Twenty-Third Symposium (International) on Combustion*, The Combustion Institute, Pittsburgh, PA, 1990, 1-19.

188. F.L. Dryer, The phenomenology of modeling combustion chemistry, in *Fossil Fuel Combustion*, W. Bartok and A.F. Sarofim, Eds., Wiley, New York, 1991, chap. 3.

189. E. Ranzi, A. Sogaro, P. Gaffuri, G. Pennati, C.K. Westbrook, and W.J. Pitz, A new comprehensive reaction mechanism for combustion of hydrocarbon fuels, *Comb. Flame*, 99, 201-211, 1994.

190. H.C. Magel, U. Schnell and K.R.G. Hein, Simulation of detailed chemistry in a turbulent combustor flow, *Twenty-Sixth Symposium (International) on Combustion*, The Combustion Institute, Pittsburgh, PA, 1996, 67-74.

191. J.A. Miller, Theory and modeling in combustion chemistry, *Twenty-Sixth Symposium (International) on Combustion*, The Combustion Institute, Pittsburgh, PA, 1996, 461-480.

192. A.Y. Federov, V.A. Frost, and V.A. Kaminsky, Turbulent transfer modeling in flows with chemical reactions, in *Tranport Phenomena in Combustion*, Vol. 2, S.H. Chan, Ed., Taylor & Francis, Washington, D.C., 1996, 933-944.

193. S.B. Pope, PDF methods for turbulent reactive flows, *Prog. Energy Combust. Sci.*, 11(2), 119-192, 1985.

194. W.P. Jones and M. Kakhi, PDF modeling of finite-rate chemistry effects in turbulent nonpremixed jet flames, *Comb. Flame*, 115, 210-229, 1998.

195. V. Saxena and S.B. Pope, PDF simulations of turbulent combustion incorporating detailed chemistry, *Comb. Flame*, 117, 340-350, 1999.

196. P.J. Coelho and M.G. Carvalho, Modelling of soot formation in turbulent diffusion flames, in *Heat Transfer in Fire and Combustion Systems*, W.W. Yuen and K.S. Ball, Eds., New York, ASME HTD-Vol. 272, 29-39, 1994.

197. I.M. Khan and G. Greeves, A method for calculating the formation and combustion of soot in diesel engines, in *Heat Transfer in Flames*, N.H. Afgan and J.M. Beer, Eds., Scripta Book Co., 1974, chap. 25.

198. C.D. Stewart, K.J. Syed, and J.B. Moss, Modelling soot formation in non-premixed kerosene-air flames, *Comb. Sci. Tech.*, 75, 211-266, 1991

199. B.F. Magnussen and B.H. Hjertager, On mathematical modelling of turbulent combustion with special emphasis on soot formation and combustion, *Sixteenth Symp. (International) on Combustion*, The Combustion Institute, Pittsburgh, PA, 1977, 719-728.

200. K.B. Lee, M.W. Thring, and J.M. Beer, On the rate of combustion of soot in a laminar soot flame, *Comb. Flame*, 6, 137-145, 1962.

201. J. Nagle and R.F. Strickland-Constable, Oxidation of carbon between 1000-2000C, *Proc. 5th Conf. on Carbon,* Pergamon Press, 1, 154, 1961.

202. M.A. Delichatsios and L. Orloff, Effects of turbulence on flame radiation from diffusion flames, *Twenty-Second Symposium (International) on Combustion*, The Combustion Institute, Pittsburgh, PA, 1988, 1271-1279.

203. P. Boerstoel, T.H. van der Meer, and C.J. Hoogendoorn, Numerical simulation of soot-formation and -oxidation in high temperature furnaces, in *Tranport Phenomena in Combustion*, Vol. 2, S.H. Chan, Ed., Taylor & Francis, Washington, D.C., 1996, 1025-1036.

204. R. Said, A. Garo and R. Borghi, Soot formation modeling for turbulent flames, *Combust. Flame*, 108, 71-86, 1997.

205. F. Xu, K.-C. Lin, and G.M. Faeth, Soot formation in laminar premixed methane/oxygen flames at atmospheric pressure, *Combust. Flame*, 115, 195-209, 1998.

206. M. Frenklach and H. Wang, Detailed modeling of soot particle nucleation and growth, *Twenty-Third Symposium (International) on Combustion*, The Combustion Institute, Pittsburgh, PA, 1990, 1559-1566.

207. K.M. Leung and R.P. Lindstedt, Detailed kinetic modeling of C_1C_3 alkane diffusion flames, *Combust. Flame*, 102(1-2), 129-160, 1995.

208. M.N. Özisik, *Finite Difference Methods in Heat Transfer*, CRC Press, Boca Raton, FL, 1994.

209. J.N. Reddy and D.K. Gartling, *The Finite Element Methods in Heat Transfer and Fluid Dynamics*, CRC Press, Boca Raton, FL, 1994.

210. W. Shyy and R. Mittal, Solution Methods for the Incompressible Navier-Stokes Equations, Chapter 31 in *The Handbook of Fluid Dynamics*, R.W. Johnson, Ed., CRC Press, Boca Raton, FL, 1998.

211. P.J. Roache, *Verification and Validation in Computational Science and Engineering*, Hermosa Publishers, Albuquerque, NM, 1998.

212. W.A. Fiveland, K.C. Kaufman, and J.P. Jessee, Validation of an Industrial Flow and Combustion Model, in *Computational Heat Transfer in Combustion Systems*, M.Q. McQuay, W. Schreiber, E. Bigzadeh, K. Annamalai, D. Choudhury, and A. Runchal, Ed., *ASME Proceedings of the 31st Annual National Heat Transfer Conf.*, Vol. 6, New York, ASME HTD-Vol. 328, 147-157, 1996.

213. K.C. Kaufman and W.A. Fiveland, Pilot Scale Data Collection and Burner Model Numerical Code Validation, topical report for Gas Research Institute Contract 5093-260-2729, 1995.

214. A.M. Lankhorst and J.F.M. Velthuis, Ceramic Recuperative Radiant Tube Burners: Simulations and Experiments, in *Tranport Phenomena in Combustion*, Vol. 2, S.H. Chan, Ed., Taylor & Francis, Washington, D.C., 1996, 1330-1341.

215. F. Mei and H. Meunier, Numerical and experimental investigation of a single ended radiant tube, in *ASME Proceedings of the 32nd National Heat Transfer Conf.*, Vol. 3: Fire and Combustion, L. Gritzo and J.-P. Delplanque, Eds., ASME, New York, 1997, 109-118.

216. H. Ramamurthy, S. Ramadhyani, and R. Viskanta, Development of fuel burn-up and wall heat transfer correlations for flows in radiant tubes, *Num. Heat Transfer, Part A*, 31, 563-584, 1997.

217. S. Bortz and A. Hagiwara, Inviscid model for the prediction of the near field region of swirl burners, in *Industrial Combustion Technologies*, M.A. Lukasiewicz, Ed., American Society of Metals, Warren, PA, 1986, 89-97.

218. Y. Tsujimoto and N. Machii, Numerical analysis of pulse combustion burner, *Twenty-First Symposium (International) on Combustion*, The Combustion Institute, Pittsburgh, PA, 1986, 539-546.

219. T.W. Tong, S.B. Sathe, and R.E. Peck, Improving the performance of porous radiant burners through use of sub-micron size fibers, in *Heat Transfer Phenomena in Radiation, Combustion, and Fires*, R.K. Shah, Ed., New York, ASME HTD-Vol. 106, 257-264, 1989.

220. S.B. Sathe, R.E. Peck, and T.W. Tong, A numerical analysis of combustion and heat transfer in porous radiant burners, *Int. J. Heat Mass Transfer*, 33(6), 1331-1338, 1990.

221. S.H. Chan and K. Kumar, Analytical investigation of SER recuperator performance, in *Fossil Fuel Combustion Symposium 1990*, S. Singh, Ed., New York, ASME PD-Vol. 30, 161-168, 1990.

222. P.-F. Hsu, J.R. Howell, and R.D. Matthews, A numerical investigation of premixed combustion within porous inert media, *J. Heat Transfer*, 115(3), 744-750, 1993.

223. R. Mital, J.P. Gore, R. Viskanta, and S. Singh, Radiation efficiency and structure of flames stabilized inside radiant porous ceramic burners, in *Combustion and Fire*, M.Q. McQuay, W. Schreiber, E. Bigzadeh, K. Annamalai, D. Choudhury and A. Runchal, Eds., *ASME Proceedings of the 31st National Heat Transfer Conf.*, Vol. 6, ASME HTD-Vol. 328, 131-137, New York, 1996.

224. C.L. Hackert, J.L. Ellzey, and O.A. Ezekoye, Numerical simulation of a porous honeycomb burner, in *ASME Proceedings of the 32nd National Heat Transfer Conf.*, Vol. 3: Fire and Combustion, L. Gritzo and J.-P. Delplanque, Eds., ASME, New York, 1997, 147-153.

225. M.M. Sidawi, B. Farouk, and U. Parekh, A numerical study of an industrial hydrogen sulfide burner with air- and oxygen-based operations, in *Heat Transfer in Fire and Combustion Systems — 1993*, B. Farouk, M.P. Menguc, R. Viskanta, C. Presser and S. Chellaiah, Eds., New York, ASME HTD-Vol. 250, 227-234, 1993.

226. G.W. Butler, J. Lee, K. Ushimaru, S. Bernstein, and A.D. Gosman, A numerical simulation methodology and its application in natural gas burner design, in *Industrial Combustion Technologies*, M.A. Lukasiewicz, Ed., American Society of Metals, Warren, PA, 109-116, 1986.

227. A. Schmücker and R.E., Leyens, Development of the delta burner using computational fluid dynamics, *Proc. of 1998 International Gas Research Conf.*, Vol. V: Industrial Utilization, D.A. Dolenc, Ed., Gas Research Institute, Chicago, 1998, 516-526.

228. B. Schmidt, B. Spiegelhauer, N.B. Kampp Rasmussen, and F. Giversen, Development of a process adapted gas burner through mathematical modelling and practical experience, *Proc. of 1998 International Gas Research Conf.*, Vol. V: Industrial Utilization, D.A. Dolenc, Ed., Gas Research Institute, Chicago, 1998, 578-584.

229. J.R. Howell, M.J. Hall, and J.L. Ellzey, Combustion within porous media, in *Heat Transfer in Porous Media*, Y. Bayazitoglu and U.B. Sathuvalli, Eds., New York, ASME, HTD-Vol. 302, 1-27, 1995.

230. M. Perrin, P. Lievoux, R. Borghi, and M. Gonzalez, Utilization of a numerical model for the design of a gas immersion tube, in *Industrial Combustion Technologies*, M.A. Lukasiewicz, Ed., Amer. Soc. Metals, Warren, PA, 1986, 127-134.

231. C.L. Hackert, J.L. Ellzey, and O.A. Ezekoye, Combustion and heat transfer in model two-dimensional porous burners, *Comb. Flame*, 116, 177-191, 1999.

232. X. Fu, R. Viskanta, and J.P. Gore, Modeling of thermal performance of a porous radiant burner, in *Combustion and Radiation Heat Transfer*, R.A. Nelson, K.S. Ball, and Z.M. Zhang, Eds., *Proceedings of the ASME Heat Transfer Division — 1998,* Vol. 2, ASME HTD-Vol. 361-2, 11-19, New York, 1998.

233. S.V. Patankar and D.B. Spalding, A computer model for three-dimensional flows in furnaces, *Fourteenth Symposium (International) on Combustion*, The Combustion Institute, Pittsburgh, PA, 1973, 605-614.

234. L.S. Caretto, A.D. Gosman, S.V. Patankar, and D.B. Spalding, Two calculation procedures for steady, three-dimensional flows with recirculation, in proc. *3rd Int. Conf. On Numerical Methods in Fluid Dynamics*, Springer, Berlin, 1972, 60-68.

235. S.V. Patankar and D.B. Spalding, A computer model for three-dimensional flow in furnaces, *Fourteenth Symposium (International) on Combustion*, The Combustion Institute, Pittsburgh, PA, 1972, 605-614.

236. T.M. Lowes, M.P. Heap, S. Michelfelder, and B.R. Pai, Paper 5. Mathematical modelling of combustion chamber performance, *J. Institute Fuel*, 46(38), 343-351, 1973.

237. W. Richter, Prediction of heat and mass transfer in a pulverised fuel furnace, *Letters in Heat & Mass Transfer*, 1, 83-94, 1978.

238. M.M.M. Abou Ellail, A.D. Gosman, F.C. Lockwood, and I.E.A. Megahed, Description and validation of a three-dimensional procedure for combustion chamber flows, *J. Energy*, 2(2), 71-80, 1978.

239. A.D. Gosman, F.C. Lockwood, and A.P. Salooja, The prediction of cylindrical furnaces gaseous fueled with premixed and diffusion burners, *Seventeenth Symposium (International) on Combustion*, The Combustion Institute, Pittsburgh, PA, 1978, 747-760.

240. E.E. Khalil, P. Hutchinson, and J.H. Whitelaw, The calculation of the flow and heat-transfer characteristics of gas-fired furnaces, *Eighteenth Symposium (International) on Combustion*, The Combustion Institute, Pittsburgh, PA, 1980, 1927-1938.

241. K. Görner, Prediction of the turbulent flow, heat release and heat transfer in utility boiler furnaces, in *Coal Combustion*, J. Feng, Ed., Hemisphere Publishing, New York, 1988, 273-282.

242. K. Görner and W. Zinser, Prediction of three-dimensional flows in utility boiler furnaces and comparison with experiments, *Comb. Sci. Tech.*, 58, 43-58, 1988.

243. P.A. Gillis and P.J. Smith, An evaluation of three-dimensional computational combustion and fluid dynamics in industrial furnace geometries, in *Twenty-Third Symp. (International) on Combustion, The Combustion Institute,* Pittsburgh, PA, 1990, 981-991.

244. M.G. Carvalho, J.B. Lopes, and M. Nogueira, A three-dimensional procedure for combustion and heat transfer in industrial furnaces, in *Advanced Computational Methods in Heat Transfer*, Vol. 3: Phase Change and Combustion Simulation, L.C. Wrobel, C.A. Brebbia, and A.J. Nowak, Eds., Springer-Verlag, Berlin, 1990, 171-183.

245. H. Meunier, Modelling of industrial furnaces, proc. of *2nd European Conf. On Industrial Furnaces and Boilers,* Portugal, April, R. Collin, W. Leuckel, A. Reis, and J. Ward, Eds., 1991, 1-21.

246. J.M. Rhine and R.J. Tucker, *Modelling of Gas-fired Furnaces and Boilers and Other Industrial Heating Processes*, British Gas and McGraw-Hill, New York, 1991.

247. V. Sidlauskas and M. Tamonis, Mathematical modeling of the thermal process in industrial combustion chambers, *Heat Transfer — Soviet Research*, 23(7), 897-914, 1991.

248. M. Matsumura, S. Ito, Y. Ichiraku, and T. Saeki, Heat transfer simulation in industrial gas furnaces, *Proc. of 1992 International Gas Research Conf.*, H.A. Thompson, Ed., Govt. Institutes, Rockville, MD, 1993, 2195-2204.

249. C.J. Hoogendoorn, Full modelling of industrial furnaces and boilers, in *Tranport Phenomena in Combustion*, Vol. 2, S.H. Chan, Ed., Taylor & Francis, Washington, D.C., 1996, 1177-1188.

250. P.K. Barr, J.O. Keller, and J.A. Kezerle, SPCDC: a user-friendly computation tool for the design and refinement of practical pulse combustion systems, *Proc. of 1995 International Gas Research Conf.*, D.A. Dolenc, Ed., Government Institutes, Rockville, MD, 1996, 2150-2159.

251. A.D. Gosman, F.C. Lockwood, I.E.A. Megahed, and N.G. Shah, The prediction of the flow, reaction and heat transfer in the combustion chamber of a glass furnace, in *AIAA 18th Aerospace Sciences Meeting*, January, Pasadena, CA, 1980, 14-46.

252. L. Post and C.J. Hoogendoorn, *Heat Transfer in Gas-Fired Glass Furnaces*, VDI, Berichte, No. 645, 457-466, 1987.

253. C.J. Hoogendoorn, L. Post, and J.A. Wieringa, Modelling of combustion and heat transfer in glass furnaces, *Glastech. Ber.*, 63(1), 7-12, 1990.

254. V.B. Kut'in, S.N. Gushchin, and V.G. Lisienko, Heat transfer in the cross-fired glass furnace, *Glass & Ceramics*, 54(5-6), 135-138, 1997.

255. M.G. Carvalho, M. Nogueira and J. Wang, Mathematical modelling of the glass melting industrial process, in *Proc. XVII Int. Congress on Glass*, Beijing, China, International Academic Publishers, 6, 69-74, 1995.

256. V.B. Kut'in, S.N. Gushchin, and V.G. Lisienko, Heat exchange in the cross-fired glass furnace, *Glass & Ceramics*, 54(5-6), 172-174, 1997.

257. V.Y. Gershtein and C.E. Baukal, Model prediction comparison for aluminum reverberatory furnace firing on air-, air-oxy-, and oxy/fuel, presented at *1999 The Minerals, Metals & Materials Society Annual Meeting & Exhibition*, February 28–March 4, 1999, San Diego, CA.

258. Y.K. Lee, H.S. Park, and K.W. Cho, Effect of fuel gas preheating on combustion and heat transfer in reheating furnace, *Proceedings of the 13ᵗʰ Energy Engineering World Congress, Energy & Environmental Strategies for the 1990s,* Atlanta, GA, October 1990, 1991, chap. 78, 461-466.

259. R. Klima, Improved knowledge of gas flow and heat transfer in reheating furnaces, *Scandinavian J. Metallurgy (Suppl.)*, 26, 25-32, 1997.

260. J.M. Blanco and J.M. Sala, Improvement of the efficiency and working conditions for reheating furnaces through computational fluid dynamics, *Industrial Heating*, LXVI(5), 63-67, 1999.

261. H. Ramamurthy, S. Ramadhyani, and R. Viskanta, Thermal system model for a radiant tube batch reheating furance, *Proc. of 1992 International Gas Research Conf.*, H.A. Thompson, Ed., Govt. Institutes, Rockville, MD, 1993, 2205-2216.

262. N.D.H. Munroe, Experimental and numerical modeling of transport phenomena in a particulate reacting system, in *Heat Transfer in Fire and Combustion Systems — 1993*, B. Farouk, M.P. Menguc, R. Viskanta, C. Presser, and S. Chellaiah, Eds., New York, ASME HTD-Vol. 250, 69-78, 1993.

263. T. Ahokainen, A. Jokilaakso, O. Teppo, and Y. Yang, Flow and Heat Transfer Simulation in a Flash Smelting Furnace, NTIS Report DE95779247, U.S. Dept. of Commerce, Springfield, VA, 1994.

264. D.L. Smoot, Pulverized coal diffusion flames: a perspective through modelling, in *Eighteenth Symp. (International) on Combustion*, The Combustion Institute, Pittsburgh, PA, 1981, 1185-1202.

265. W.A. Fiveland and R.A. Wessel, Numerical model for predicting performance of three-dimensional pulverized-fuel fired furnaces, *ASME J. Engng. Gas Turbines and Power*, Vol. 110, 117-126, 1988.

266. S. Li, B. Yu, W. Yao, and W. Song, Numerical modelling for pulverised coal combustion in large furnace, in *Proc. 2nd Int. Symp. On Coal Combustion*, Beijing, China, X. Xu, L. Zhou, and W. Fu, Eds., China Machine Press, Beijing, China, 1991, 167-173.

267. R. Boyd and A. Lowe, Three-dimensional modelling of a pulverised coal fired utility furnace, in *Coal Combustion*, Hemisphere, New York, 1988, 165-172.

268. B.S. Brewster, S.C. Hill, P.T. Radulovic, and L.D. Smoot, Comprehensive modeling, in *Fundamentals of Coal Combustion for Clean and Efficient Use*, L.D. Smoot, Ed., Elsevier, Amsterdam, 1993, chap. 8.

269. K. Kudo, H. Taniguchi, and K. Guo, Heat-transfer simulation in a furnace for steam reformer, *Heat Transfer — Japanese Research*, 20(8), 750-764, 1992.

270. K.N. Theologos and N.C. Markatos, Advanced modeling of fluid catalytic cracking riser-type reactors, *AIChE J.*, 36(6), 1007-1017, 1993.

271. S.L. Chang, C.Q. Zhou, S.A. Lottes, B. Golchert, and M. Petrick, A numerical investigation of the scaled-up effects on flow, heat transfer, and kinetics processes of FCC units, in Combustion and Radiation Heat Transfer, R.A. Nelson, K.S. Ball, and Z.M. Zhang, Eds., *Proceedings of the ASME Heat Transfer Division — 1998*, Vol. 2, ASME HTD-Vol. 361-2, 73-81, 1998.

272. B. Ponizy and S. Wojcicki, On modeling of pulse combustors, *Twentieth Symposium (International) on Combustion*, The Combustion Institute, Pittsburgh, PA, 1984, 2019-2024.

273. P.K. Barr and H.A. Dwyer, Pulse combustor dynamics: a numerical study, in *Numerical Approaches to Combustion Modeling*, E.S. Oran and J.P. Boris, Eds., Vol. 135, Progress in Astronautics and Aeronautics, American Institute of Aeronautics and Astronautics, Washington, D.C., 1991, 673-710.

274. E. Lundgren, U. Marksten, and S.-I. Möller, The enhancement of heat transfer in the tail pipe of a pulse combustor, *Twenty-Seventh Symposium (International) on Combustion*, The Combustion Institute, Pittsburgh, PA, 1998, 3215-3220.

275. H.C. Meisingset, J.G. Balchen, and R. Fernandez, Mathematical modelling of a rotary hearth calciner, *Light Metals 1996*, W. Hale, Ed., The Minerals, Metals & Materials Society, Warren, PA, 1996, 491-497.

276. J.R. Ferron and D.K. Singh, Rotary Kiln Transport Processes, *AIChE J.*, 37(5), 747-758, 1991.

277. A.A. Boateng, On flow induced kinetic diffusion and rotary kiln bed burden heat transport, in *ASME Proceedings of the 32nd National Heat Transfer Conf.*, Vol. 3: Fire and Combustion, L. Gritzo and J.-P. Delplanque, Eds., ASME, New York, 1997, 183-191.

278. P.S. Ghoshdastidar and V.K. Anandan Unni, Heat transfer in the non-reacting zone of a cement rotary kiln, in *Heat Transfer Phenomena in Radiation, Combustion, and Fires*, R.K. Shah, Ed., New York, ASME HTD-Vol. 106, 113-122, 1989.

279. E.E. Khalil, Flow and combustion modeling of vertical lime kiln chambers, in *Industrial Combustion Technologies*, M.A. Lukasiewicz, Ed., American Society of Metals, Warren, PA, 1986, 99-107.

280. B. Farouk and M.M. Sidawi, Effects of nitrogen removal in a natural gas fired industrial furnace: a three dimensional study, in *Heat Transfer in Fire and Combustion Systems — 1993*, B. Farouk, M.P. Menguc, R. Viskanta, C. Presser, and S. Chellaiah, Eds., New York, ASME HTD-Vol. 250, 173-183, 1993.

281. W.D. Owens, G.D. Silcox, J.S. Lighty, X.X. Deng, D.W. Pershing, V.A. Cundy, C.B. Leger, and A.L. Jakway, Thermal analysis of rotary kiln incineration: comparison of theory and experiment, *Comb. Flame*, 86, 101-114, 1991.

282. D. Pal, J.A. Khan, and J.S. Morse, Computational modelling of an industrial rotary kiln incinerator, in *Heat Transfer in Fire and Combustion Systems*, A.M. Kanury and M.Q. Brewster, Eds., New York, ASME HTD-Vol. 199, 167-173, 1992.

283. F.C. Chang and C.A. Rhodes, Computer modeling of radiation and combustion in a rotary solid-waste incinerator, Argonne National Lab, U.S. Dept. of Energy Report ANL/ET/CP-85778, 1995.

284. N. Machii, K. Nishimura, D. Liu, and K. Shibata, Development of CAE for MSWI ash melting furnace, *Proceedings of the 1998 International Gas Research Conf.*, Vol. V: Industrial Utilization, D.A. Dolenc, Ed., Gas Research Institute, Chicago, 1998, 696-704.

285. S.E. Bayley, R.T. Bailey, and D.C. Smith, Heat transfer analysis of hazardous waste containers within a furnace, in Combustion and Fire, M.Q. McQuay, W. Schreiber, E. Bigzadeh, K. Annamalai, D. Choudhury, and A. Runchal, Eds., *ASME Proceedings of the 31st National Heat Transfer Conf.*, Vol. 6, ASME HTD-Vol. 328, 61-69, New York, 1996.

286. M.G. Carvalho and N. Martins, Mathematical modelling of heat and mass transfer phenomena in baking ovens, 5th International conf. on computational methods and experimental measurements, in *Computational Methods and Experimental Measurements V*, A. Sousa, C.A. Brebbia, and G.M. Carlomagno, Eds., Computational Mechanics Publications, 1991, 359-370.

287. M. Carvalho and M. Nogueira, Improvement of energy efficiency in glass-melting furnaces, cement kilns and baking ovens, *Applied Therm. Eng.*, 17(8-10), 921-933, 1997.

288. T.H. Song and R. Viskanta, Parametric study of the thermal performance of a natural gas-fired furnace, in *Fossil Fuel Combustion 1991*, R. Ruiz, Eds., New York, ASME PD-Vol. 33, 135-141, 1991.

289. J.A. Wieringa, J.J. Elich, and C.J. Hoogendoorn, Spectral radiation modelling of gas-fired furnaces, in *Proc. 2nd European Conf. On Industrial Furnaces and Boilers*, Portugal, April, R. Collin, W. Leuckel, A. Reis, and J. Ward, Eds., 1991, 36-53.

290. M.J.S. de Lemos, Computation of heated swirling flows with a fully-coupled numerical scheme, in Combustion and Fire, M.Q. McQuay, W. Schreiber, E. Bigzadeh, K. Annamalai, D. Choudhury, and A. Runchal, Eds., *ASME Proceedings of the 31st National Heat Transfer Conf.*, Vol. 6, New York, ASME HTD-Vol. 328, 139-145, 1996.

291. M.J.S. de Lemos, Simulation of vertical swirling flows in a model furnace with a high performance numerical method, in Combustion and Radiation Heat Transfer, R.A. Nelson, K.S. Ball, and Z.M. Zhang, Eds., *Proceedings of the ASME Heat Transfer Division — 1998*, Vol. 2, New York, ASME HTD-Vol. 361-2, 21-28, 1998.

292. M.G. Carvalho and M. Nogueira, Modelling of glass melting industrial process, *J. de Physique*, 3(7.2), 1357-1366, 1993.

293. K.S. Chapman, S. Ramadhyani, and R. Viskanta, Modeling and analysis of heat transfer in a direct-fired continuous reheating furnace, in *Heat Transfer in Combustion Systems*, N. Ashgriz, J.G. Quintiere, H.G. Semerjian, and S.E. Slezak, Eds., ASME, HTD-Vol. 122, 35-43, 1989.

294. K.S. Chapman, S. Ramadhyani, and R. Viskanta, Modeling and analysis of heat transfer in a direct-fired batch reheating furnace, in *Heat Transfer Phenomena in Radiation, Combustion, and Fires*, R.K. Shah, Ed., New York, ASME HTD-Vol. 106, 265-274, 1989.

295. R. Viskanta, K.S. Chapman, and S. Ramadhyani, Mathematical modeling of heat transfer in high-temperature industrial furnaces, in *Advanced Computational Methods in Heat Transfer*, Vol. 3: Phase Change and Combustion Simulation, L.C. Wrobel, C.A. Brebbia, and A.J. Nowak, Eds., Springer-Verlag, Berlin, 1990, 117-131.

296. K.S. Chapman, S. Ramadhyani, and R. Viskanta, Two-dimensional modeling and parametric studies in a direct-fired furnace with impinging jets, *Comb. Sci. Tech.*, 97, 99-120, 1994.

6 Experimental Techniques

6.1 INTRODUCTION

It is not the purpose here to critically evaluate all of the experimental techniques that may be used in industrial combustion systems. Instead, each technique is briefly discussed, including its advantages and disadvantages. Specific references for each technique are provided that give more details for the interested reader. Several excellent general references have been published concerning combustion diagnostics. An early book by Beér and Chigier (1972) gives detailed descriptions and diagrams of different probes used to make measurements in industrial-scale flames.[1a] The report by Okoh and Brown (1988)[1] reviews and compares many techniques for the measurement of a wide range of variables in combustion systems. It also includes equipment specifications and suppliers. The books edited by Durão et al. (1992)[2] and Taylor (1993)[3] contain chapters on the use of both physical probes and optical techniques. Fristrom (1995) has several chapters devoted to probe and optical measurements in flames.[3a] The books by Eckbreth (1988)[4] and Chigier (1991)[5] focus specifically on optical techniques for diagnosing flames. The papers by Fristrom (1976),[6] Bowman (1977),[7] Gouldin (1980),[8] and Becker (1993)[9] review probe measurements in flames. Newbold et al. (1996) gave a good example of making probe measurements of velocity, species, radiation, and gas temperature in an industrial, gas-fired, flat-glass furnace.[10] Solomon et al. (1986) reviewed Fourier transform infrared (FT-IR) optical techniques for measuring particles, species concentrations, and temperature in combustion processes.[11] Papers are available on the use of laser diagnostics in combustion systems.[12-16]

6.2 HEAT FLUX

For most industrial heating applications, heat flux is one of the most important parameters in the system design. Heat flux to the load is closely related to the material processing rate. There are usually two types of heat flux that are important: radiative and total (i.e., the combination of radiation and convection). In some high-temperature, low gas velocity combustors, the total and radiative fluxes may be nearly the same because of the predominance of radiation over convection. In lower temperature, high gas velocity combustors, the radiative flux may be much lower than the total flux because radiation and convection are comparable in those applications. However, if any heat flux is measured, it is rare that more than one type would be measured. This type of information can be very valuable in understanding the dynamics of a given combustion system and is useful for making design improvements. Total, radiative, and convective heat flux measurements are discussed next.

6.2.1 TOTAL HEAT FLUX

The total heat flux has been measured using both steady-state and transient methods. Three different steady-state methods have been used. The first method calculates the flux from an energy balance on an uncooled solid. The second measures the sensible energy gain of the coolant for a cooled solid. The third directly measures the flux with a calibrated heat flux gage imbedded in a cooled target.

Two different transient methods have been used. In the first method, an uncooled solid is inserted into the flame for either a given amount of time, or until a certain temperature is reached. The flux is calculated from the sensible energy gain of the solid. In the second method, the flux is measured using a gage imbedded in the surface of an uncooled solid. Both of these methods have very short testing times. Both steady-state and transient methods are discussed below.

6.2.1.1 Steady-State Uncooled Solids

In this method, the heat flux is determined from an energy balance on the solid. It was used in two early flame impingement studies.[17,18] Refractory cylinders were coated with different oxides. These were heated until steady-state conditions were reached. Radiation from the cylinders, along with the gas and cylinder surface temperatures, were measured. The total heat flux to the cylinder was then calculated from an energy balance on the cylinder surface. The heat gained by the cylinder was assumed to be by convection and radiation from the flame. The energy lost by the cylinder was assumed to be solely due to radiation. A thermopile was used to measure radiation from the flame, both with and without the cylinder present. The surface radiation from the cylinder was calculated by subtracting the first measurement from the second. This value was assumed to be equal to the total heat flux from the flame to the cylinder.

This technique is very simple and low in cost. However, only the average heat flux over the entire solid is determined. It also relies on an accurate value of the target emissivity. It has only limited practical relevance and it is limited by the permissible maximum material temperature level of the solid. Also, the solid should be uniformly heated to obtain an accurate average measurement. Nonuniform heating complicates the energy balance calculations. In Kilham's studies,[17,18] the cylinders were rotated to minimize surface temperature gradients. However, only a single thermocouple, mounted on the inside diameter of the hollow cylinder to measure the inside cylinder temperature, was used to calculate the surface temperature.

6.2.1.2 Steady-State Cooled Solids

In this method, a cooled solid is used as a calorimeter. It is commonly cooled internally with water or glycol. The flow rate and the temperature rise of the coolant are measured after steady-state conditions have been reached. The heat flux to the solid is calculated from the rate of sensible energy gain of the coolant. The three arrangements that have been used are: a single cooling circuit, multiple cooling circuits, and a surface probe.

6.2.1.2.1 Single cooling circuit

In this technique, the average heat flux over the entire solid surface is calculated from the sensible energy gain of a single cooling circuit. Davies (1965)[19] and Fells and Harker (1968)[20] studied the effects of electrical boost on enhancing heat flux from impinging flames boosted by an AC discharge that produced as much as 3.6 times more heat flux than unboosted flames. The flames impinged on water-cooled stainless steel and copper pipes. Vizioz and Lowes (1971)[21] studied large-scale flames impinging on a large, circular, water-cooled plate with a refractory surface to simulate a hot target. The surface temperature was maintained using an imbedded spiral water-cooled channel. Davies (1979)[22] used water-cooled pipes to study heat flux from impinging flames on a cylinder as a function of equivalence ratio, firing rate, oxidizer/fuel mixing, and oxygen enrichment ratio. Posillico (1986)[23] studied flames impinging on a flat plate with a single internal cooling passage.

6.2.1.2.2 Multiple cooling circuits

This technique has been used for flames impinging normal to the surface of a large metal disk located inside a furnace. The surface temperature of the disk was maintained with concentric cooling channels. The heat flux to each annulus of the target was calculated from the sensible energy gain for each of the individual cooling circuits. Vizioz and Lowes[21] used a steel plate to simulate a cold target. The surface temperature was maintained below 373K, with eight equally spaced concentric cooling channels. The temperature difference between each water inlet and outlet was only about 36°F to 45°F (20K to 25K) to reduce the radial heat transfer between the adjacent channels. Smith and Lowes (1974)[24] used a similar plate design. Air gaps between the channels were provided to further reduce radial conduction. Experiments included both air and water cooling. Rajani et al. (1978)[25] used a smaller version of this plate. Baukal and Gebhart (1997) used a disk segmented into six concentric sections, as described in section 4.2.1.5.[81]

6.2.1.2.3   Surface probe

Anderson and Stresino (1963),[26] Reed (1963),[27] and Fay (1967)[28] studied flames impinging on a so-called surface probe. In those studies, two identical, water-cooled copper plates were connected in series and separated by a thermal barrier. A single water-cooling circuit passed across both sections. The flame was traversed from the first to the second section. The water temperature rise through each plate was measured as a function of distance from the flame center line. The data was inverted to determine the spatial heat flux distribution. See Reference 27 for details of the inversion calculation.

This method is simple and relatively low in cost. It needs no calibration. The heat flux through the surface is measured directly by the sensible energy gain of the internal coolant. It has high accuracy. The spatial resolution depends on the number of separate cooling channels used. However, finer spatial resolution requires more controls. The choices of target material and coolant are limited at higher surface temperatures.

6.2.1.3   Steady-State Cooled Gages

In this method, the local heat flux is determined using a small gage imbedded in a much larger solid. The heat flux gages are commonly made of a high thermal conductivity material, such as copper. The hot end of the gage is exposed to the flame, while the cold end is water-cooled. Three different variations of this method have been used: a gradient through a thin solid rod, a thin disk calorimeter, and a heat flux transducer.

6.2.1.3.1   Gradient through a thin solid rod

The hot end of a thin solid rod is flush with the target surface. Thermocouples, located along the probe axis, are used to determine the axial internal temperature gradient, which is assumed to be linear. The heat flux is calculated using a one-dimensional conduction equation. In most previous measurements, the probe had shields to minimize heat flux from the sides. The probe used by both Cookson and Kilham (1963)[29] and Kilham and Dunham (1967)[30] is described in detail by Cookson et al. (1965).[31] It was a 4.8 mm (0.19 in.) o.d. copper rod, imbedded in a larger diameter, cooled, hemi-nosed brass cylinder. Beér and Chigier (1968)[32] used a 2.5 cm (1.0 in.) o.d. uncoated stainless steel probe. Buhr et al. (1973)[33] and Kremer et al. (1974)[34] used a 3 mm (0.12 in.) o.d. copper rod imbedded in an internally water-cooled flat plate. Hoogendoorn et al. (1978)[35] and Popiel et al. (1980)[36] used a 5.85 mm (0.23 in.) o.d. copper rod imbedded in a water-cooled flat plate. Hargrave et al. (1984, 1987)[37,38] used a 3.2 mm (0.125 in.) o.d. copper rod internally cooled by glycol. The rod was imbedded inside a 22 mm o.d. hemi-nosed cylinder. Kataoka et al. (1984)[39] used a 10.21 mm (0.402 in.) o.d. copper rod imbedded in a water-cooled flat plate.

Cassiano et al. (1994) used a probe of this type to measure the total heat flux in an end-port regenerative glass furnace.[40] The probe consisted of cylindrical block inside a water-cooled housing, which was required because of the high temperatures in the furnace. The front surface was serrated and blackened to maximize the radiant absorptivity. The probe was calibrated using a blackbody furnace. The response time of the probe was ~10 minutes, with an accuracy of approximately 5%.

6.2.1.3.2   Thin disk calorimeter

In this variation, a thin water-cooled disk, usually copper, is imbedded in a much larger target body. The heat flux to the calorimeter is calculated using the flow rate and the temperature rise of the water. Shorin and Pechurkin (1994)[41] used a 20 mm (0.8 in.) o.d. copper disk imbedded in a larger flat plate. The gage and target were both independently water-cooled.

6.2.1.3.3   Heat flux transducer

In this technique, an electrical signal is generated that is proportional to the heat flux. The transducer is often imbedded flush with a solid surface, such as the target in flame impingement heating or a furnace wall. One type of transducer uses an array of thermocouples to measure the heat flux across a thin distance. Schulte and Kohl (1970)[42] described the water-cooled heat flux transducer used in Schulte's (1972)[43] study. It was designed for surface flux rates up to 1000 kW/m² (300,000 Btu/hr-ft²), using a single germanium crystal as the sensing element. Smith and Lowes[24] used a stainless steel disk with 15 spaced heat flux gages, at 8 equidistant radial locations. You (1985)[44] used water-cooled transducers, whose surfaces were coated with either a gold or black foil, to determine either the convective or total heat flux, respectively. The transducers were imbedded flush with the surface of a water-cooled plate. Another type of transducer is known as the Gardon (1960)[45] gage. The basic configuration consists of a thin metal disk. Thermocouples are connected on the back of the disk, at the center and at the perimeter. When heat flux is received at the front of the disk, a linear voltage is produced by the radial conduction from the center of the disk to the perimeter. This voltage is proportional to the heat flux. This gage was originally developed for measuring high radiant fluxes. Its use has been extended to mixed convection and to radiation environments, using proper corrections. Van der Meer (1991)[46] used this type of gage, imbedded flush with the surface of a water-cooled flat plate.

The technique is simple and relatively low in cost. It has both good accuracy and spatial resolution. The response time is fast enough that it can be used for feedback control systems.[47] A heat flux sensor of this type may be superior to a thermocouple, which is most commonly used because of the thermal lag often involved in temperature measurements. There are some potential concerns with heat flux gages. Calibration is required. This may be complicated in a mixed radiation and convection environment, because calibration typically requires a blackbody source. The maximum allowable temperature and heat flux for some of the commercial transducers appear to limit their use in high-intensity flame impingement.

6.2.1.4 Transient Uncooled Targets

In this technique, the transient heat flux to a relatively large uncooled body is determined indirectly. The temperature profiles in the target are measured by an array of imbedded thermocouples, usually at or near the impingement surface. The flux is then calculated from the measured temperature responses, using inverse conduction heat transfer computational techniques (Gebhart, 1971[48]). Giedt et al. (1960)[49] measured the transient heating in an ingot iron plate, whose surface was parallel to the flame. Measurements were made until the plate reached 1170K. This usually took about 21 sec. Woodruff and Giedt (1966)[50] used a blunt-nosed molybdenum plate to simulate an air foil. Test durations were about 50 sec. Milson and Chigier (1973)[51] used a flat, uncooled plate. The back of the plate was coated with lampblack to create a surface boundary condition having a well-defined emissivity. The heat flow through the plate was equated to the natural convection and radiation losses from the back side. Veldman et al. (1975)[52] used a thin steel disk target with an array of thermocouples welded onto the back side. Matsuo et al. (1978)[53] heated rectangular slabs of varying thicknesses, located inside a furnace. Rauenzahn (1986)[54] used a thick copper plate, which remained nearly isothermal during the heating process.

This technique is very simple and low in cost. It does not need calibration. It commonly suffers from poor spatial resolution. It requires very rapid instrument response times; this may cause difficulties if many measurements must be frequently recorded in a very short time period. This technique also relies on accurate thermal conductivity values as a function of temperature.

6.2.1.5 Transient Uncooled Gages

In this technique, a small heat flux gage is imbedded flush with the target. The gage is commonly made of a high thermal conductivity material. It is assumed to be at a uniform temperature level at all times. The sensible heat gained by the gage can be simply calculated. Hornbaker and Rall (1964)[55] have listed the ideal characteristics for this kind of gage, in high-temperature applications as:

1. The measured flux should equal that transferred to the target.
2. The flux that would have been received by the target, with no gage, should not be changed.
3. The gage must have a rapid temporal response.
4. The gage output should be directly proportional to the flux.
5. The gage should be small, simple and ruggedly constructed.

Bachmann et al. (1965)[56] discussed the errors using gages caused by: (1) conduction between the gage and target, and (2) perturbation of the boundary layer resulting from the temperature discontinuity at the gage–target interface. Bachmann used copper and nickel slug calorimeters to measure the heat flux from O_2/C_2H_2 flames flowing parallel to a flat plate. The gage heat flux was 40% to 140% higher than to a calibrated reference calorimeter — as a result of the second error type above. Two different variations of this method have been used: the slug calorimeter and the heat flux transducer.

6.2.1.5.1   Slug calorimeter

In this technique, the slug thermal capacity is assumed to be internally isothermal at any instant in time. This occurs when the thermal conductivity is large compared to the convection heat transfer coefficient. The heat flux to the slug is calculated from the sensible energy gain of the slug. The average flux is usually calculated over a temperature range during which the energy gain is linear. Kilham and Purvis (1971),[57] Nawaz (1973),[58] and Fairweather et al. (1984)[59,60] used a 2.18 mm (0.086 in.) o.d. stainless steel conductivity plug located inside a hemi-nosed cylinder. The target was inserted into the flame using a solenoid coil. The insertion time was estimated by Nawaz to be 0.04 sec. This was compared to the residence time in the flame, which was estimated to be 0.15 to 0.25 sec. Nawaz used a hollow stainless steel cylinder. The conductivity plug was a 0.26-mm thick plate, used to close one end of the cylinder. A thermocouple was welded on the inside surface of the plug to measure the energy gain. The experimental results of Fairweather et al.[60] showed large heat fluxes near the laminar flame reaction zone due to the large and nonequilibrium concentration of reactive species. Lower peaks in flux were measured for turbulent flames. This was due to the more diffuse mixing in the reaction zone. Conolly and Davies (1972)[61] used an assembly of a cylindrical plug (made of a copper/1% beryllium alloy) and a rear-mounted thermocouple. Kremer et al. (1974)[34] used a 3 mm o.d. × 1.5 mm thick copper disk. The disk was insulated from the rest of the target with a ceramic to reduce radial losses. It was mounted flush to the surface of a flat plate. Kilham and Purvis (1978)[62] used 12 mm (0.47 in.) o.d. high thermal conductivity disks, with thicknesses of 1, 2, or 3 mm (0.04, 0.08, or 0.12 in.). These mounted flush with the surface of a large flat plate. They were made of either silicon carbide or platinum to test the importance of hydrogen atom recombination (see section 4.1.2). No difference in heat flux was measured using the two materials. Ivernel and Vernotte (1979)[63] used 15 mm (0.59 in.) o.d. × 2 mm (0.08 in.) thick platinum disks. The temperature rise of the disk was measured with a thermocouple attached to the back. This rise was used to calculate the total heat flux, based on the sensible energy gain of the disk. Horsley et al. (1982)[64] and Hemeson et al. (1984)[65] used a calorimeter consisting of a small oxygen-free, high-conductivity (OFHC) copper disk with a thermocouple attached on the back face. Horsley estimated that the exposure time for the probe varied from 0.32 to 2.0 sec. The mean surface temperature was 360K (190°F). The sensor probe was located in a cylinder that was inserted by a piston normal to the surface of a water-cooled plate. Hemeson used a 4.44 mm (0.175 in.) o.d. × 1.57 mm (0.0618 in.) thick copper disk imbedded flush and contoured to the surface of both cylinders and hemispheres. Hargrave et al. (1987)[38] used a 3.2 mm (0.125 in.) o.d. × 1.0 mm (0.04 in.) thick copper slug, located at the stagnation point of a solid brass cylinder. Hustad et al. (1991, 1992)[66-68] used a probe consisting of a 25 mm (1 in.) o.d. × 3 mm (0.12 in.) thick flat copper disk. It was imbedded flush with the surface of a round steel pipe, at the forward stagnation point of the flow.

6.2.1.5.2   Heat flux transducer

Hornbaker and Rall (1964)[55] measured the heat flux from O_2/C_2H_2 flames, parallel to a flat plate, using both the transient uncooled target and the uncooled gage techniques. The first technique was assumed to indicate the actual heat flux. Comparison with the second technique showed that the gages overpredicted, matched, and underpredicted the true flux, respectively, at surface temperatures of about 300 to 550K, 550 to 600K, and 600 to 900K (80–530, 530–620, and 620–1200°F).

The transient uncooled gage technique is relatively simple. It simulates many industrial heating processes, such as rapid metal heating. However, the instruments must have very fast response times. Some mechanism must be provided for rapidly inserting the target into the flame. For high-temperature flames, the response time must be fractions of a second before melting occurs.

6.2.2   Radiant Heat Flux

Gray and Müller (1974) note that there are four components for most instruments used to measure radiation:[69]

1. A detector that converts the radiation to an electrical signal
2. An optical system that directs the radiation to the detector, and which may include some type of focusing element
3. A filtering system that ensures only the proper wavelength band of radiation required for the detector is received by the detector
4. An amplifier to enhance the signal and some type of recorder/indicator to display the output

In the following chapter sections, only components 1 and 3 are considered; it has been assumed that the optical system and amplifier can be properly designed for a given detector.

6.2.2.1   Heat Flux Gage

There are three general types of radiant emission that can arise in combustion processes. One is nonluminous gaseous radiation. This results from the well-characterized spectral emission characteristics of both CO_2 and H_2O (see section 3.3.2). These are two of the main constituents in the products of combustion of typical hydrocarbons. Even at high flame temperatures, nonluminous radiation is usually small compared to convection, since the emissivity of the gases is very low. The second type of radiation involves luminous emission from particles (see section 3.3.3). Such sooting flames often generate very high radiation fluxes. This effect is usually only important for liquid and solid fuels, but not for gases. The third effect is radiation from solid surfaces, such as from refractory walls inside a furnace (see section 3.3.1). Such radiation depends on the geometry, temperature, and radiant properties of the walls. In some studies, several types of radiation were important. The techniques used to measure such thermal radiation are discussed below.

One method that has been used to measure nonluminous radiation is to directly measure the radiation from free jet flames with a radiometer. An elevation and a plan view of a radiometer aimed at a flame are shown in Figures 6.1 and 6.2, respectively. In one study, a Medtherm Corp. (Huntsville, AL) model 40-5-18T heat flux transducer was used to measure the radiation.[70] The transducer was a Gardon gage[45] with a diameter of 6.35 mm (0.25 in.). The sensor absorptance was 92%, in the spectral range of 0.6 to 15.0 µm. A model 40VRW-7C water-cooled view restrictor was used to restrict the field of view to $\theta_{radm} = 6.45°$. A narrow-angle radiometer was used to ensure that only radiation from the flame, not from the ambient environment — was measured. This was representative of the case for an impinging flame, where the target was engulfed by the hot combustion gases. The radiometer did not have any windows or gas purging, since the ambient environment was clean.

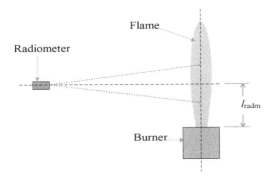

FIGURE 6.1   Elevation view of the radiometer and the flame. (From C.E. Baukal and B. Gebhart, *Int. J. Heat Mass Transfer,* 40(11), 2539-2547, 1997.)

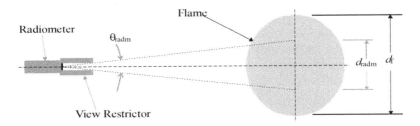

FIGURE 6.2   Plan view of a radiometer and a flame. (From C.E. Baukal and B. Gebhart, *Int. J. Heat Mass Transfer,* 40(11), 2539-2547, 1997.)

Gore et al. (1989) used a Medtherm wide-angle (150°) radiometer to measure luminous radiation from methane/heptane and methane/crude oil flames.[71] The measured radiant fluxes for the methane/heptane flames and for the methane/crude oil flames ranged from 0.34 to 0.64 and 0.53 to 0.68 W/cm^2 (1100–2000 and 1700–2200 Btu/hr-ft^2), respectively. The modeling results significantly underpredicted the experimental data due to lower estimates of temperature, the complexity of predicting soot in turbulent flow, and turbulent radiation interactions, as shown in Figure 6.3.

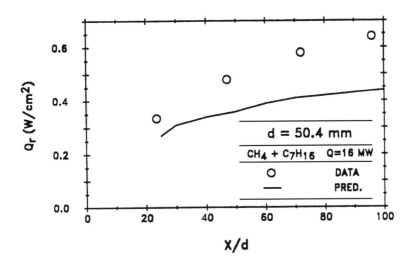

FIGURE 6.3   Radiative heat flux for free jet methane/heptane flames. (Courtesy of ASME, New York.[71] With permission.)

Shorin and Pechurkin (1968)[41] used a steady-state cooled gage to measure the total heat flux from impinging flames. The gage was imbedded flush with the surface of a large plane surface. To determine the importance of radiation, gages with polished, oxidized, and blackened surfaces were tested. The lowest flux was expected to arise for the polished gage, which reflects most of the radiation. Higher fluxes were expected with the oxidized and blackened gages; these are better absorbers. However, no difference in measured heat flux was found. It was concluded that radiation was not important for their experimental conditions. You (1985)[44] measured flame radiation to the surroundings using commercial heat flux gages. The radiant flux to a plane surface was also indirectly measured. For these measurements, commercial heat flux gages were coated with either gold or black foil to measure the convective or total heat flux, respectively. The radiant flux was determined by subtracting the convective flux from the total. Thermal radiation was 13% to 26% of the total flux near the stagnation point; it was negligible at the edge of the plate (R = 7.3).

Heat flux gages are often used as radiometers by putting a glass window in front of them to prevent convective heat transfer to the gage. A variety of materials can be used for the infrared detectors, each having a specific spectral characteristic, as shown in Figure 6.4.[72]

Matthews et al. (1989) described a multihead transpiration radiometer used to make radiative heat flux measurements in a sooty pool fire.[73] The transpiration radiometer[74] uses a transducer to measure the radiation and the transpiration is used to eliminate the effects of convection. The multihead version measured radiation from five different directions in the flame and worked successfully in a sooty environment. The main drawback was insensitivity at low flux rates.

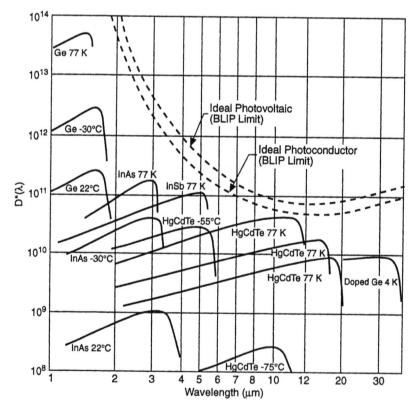

FIGURE 6.4 Spectral characteristics of infrared detectors. (From D.R. Lide, Ed., *CRC Handbook of Chemistry and Physics,* 79th edition, CRC Press, Boca Raton, FL, 1998, 10-216.)

6.2.2.2 Ellipsoidal Radiometer

This probe consists of an ellipsoidal cavity, internally coated with a highly reflective metal such as polished gold. Radiation enters the cavity through a small hole located at one of the foci of the ellipsoid cavity. The radiation is reflected inside the ellipsoid, onto a thermopile, located at the other focal point. The thermopile converts the radiant flux to an electrical signal. The radiometer is commonly calibrated against a blackbody furnace having a known internal surface temperature.

Several flame impingement studies (Beér and Chigier, 1968;[32] Vizioz and Lowes, 1971;[21] Rajani et al., 1978[25]) measured total radiant emission with hollow, gold-plated, ellipsoidal radiometers. These were developed by the International Flame Research Foundation (IFRF) in IJmuiden, The Netherlands (Chedaille and Braud, 1972[75]). Measurements at the IFRF indicated that radiation may account for as much as half of the total heat flux to a target located in a hot furnace. The radiation was primarily from the hot refractory walls. Vizioz attempted to isolate this radiation from the hot combustion gases. The radiation reflected and emitted from a flat plate target has been measured with a narrow-angle radiometer. This consists of a thermopile with a window in front of it to eliminate forced convection. The window and thermopile were arranged so that only radiation from certain angles was intercepted. The radiated flux from the plate was assumed to be a constant fraction of the incident radiation. The absorption and emission of the gas layer, between the plate and the radiometer, were estimated. Smith and Lowes (1974)[24] were unable to obtain consistent results using this ellipsoidal radiometer.

Butler and Webb (1990) reported wall radiant heat flux measurements, ranging from 100 to 500 kW/m^2 (30,000 to 160,000 Btu/ft^2-hr), in an 80 MW$_e$ (2.7 × 10^8 Btu/hr) industrial coal-fired boiler using an ellipsoidal radiometer.[76] Average wall radiant heat fluxes for each of the six elevations in the boiler are shown in Figure 6.5. In experiments located outdoors at ambient conditions, Hustad et al. (1991)[66] measured radiation perpendicular to the flame axis using a wide-angle radiometer. The measurements were made at the middle of the flame height, where soot concentration and, therefore, thermal radiation were at their peaks. It was found that the radiation was higher than the

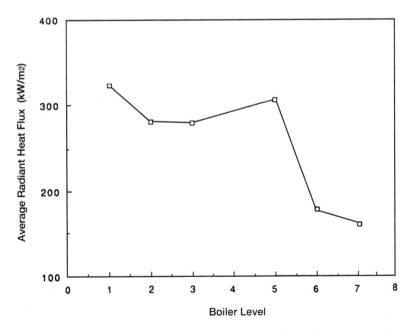

FIGURE 6.5 Total incident radiant heat flux averaged over each of six elevations in an industrial boiler. (Courtesy of ASME, New York.[76] With permission.)

model predictions. It was concluded that the measurement location was not representative of the entire flame. The calculated values were used instead of the actual measurements in subsequent analyses. Hustad (1992)[68] found that radiation perpendicular to the flame axis was 26%, 40%, and 44% of the total heat flux for CH_4, C_4H_{10}, and C_3H_8, respectively at Ma = 1. For C_3H_8, at Ma = 0.6, radiation was 26% of the total flux. These results showed that luminous radiation may be important, even for gaseous fuels. Beltagui et al. (1993) used both a total heat flux probe and calorimetry to measure the total heat flux in a furnace and a radiometer to measure the radiant heat flux in the furnace.[77] They studied the characteristics of swirling flames fired vertically into a water-cooled chamber 3 m (10 ft) long and 1 m (3 ft) i.d., divided into six sections for calorimetry. The firing rate was 400 kW (1.4×10^6 Btu/hr) on natural gas, with equivalence ratios ranging from 0.70 to 1.35 and 5% excess air. The total heat flux probe measurements were in good agreement with the calorimetry calculations for each of the six sections. The radiation measurements showed that the radiant heat flux was at least half of the total flux and in some cases nearly accounted for all of the heat flux. The measurements also showed that the heat flux distribution was much more uniform for the case with no swirl and that the peak flux 70 kW/m^2 (22,000 Btu/ft^2-hr) which occurred in the second section downstream from the burner for the case with high swirl (S = 2.25), was almost double the peak flux 40 kW/m^2 (13,000 Btu/ft^2-hr) for the case without swirl.

Radiometers are generally of low cost and simple to use. They have good spatial resolution. They operate successfully at high temperatures. However, they must be calibrated. Ellipsoidal radiometers are difficult to use in very sooty or in particle-laden environments. That may foul the polished reflectors.

6.2.2.3 Spectral Radiometer

Baukal and Ji (1998) studied the spectral radiation from oxygen-enhanced flames.[78] Spectral radiation was measured with a spectral radiometer from Monolight Instruments (now known as Macam Photometrics in Livingston, Scotland). Three scanning monochromator modules with appropriate sets of order-sorting filters were used to cover the spectral range from 0.25 μm to 5.0 μm. The spectral irradiance calibration was made with a quartz-tungsten-halogen lamp (Oriel Model 6333, Stratford, CT) in the 0.25 to 1.7 μm range, and with an IR emitter (Oriel Model 6580) in the 1.7 to 5.0 μm range. The results of this study were discussed in sections 3.3 and 4.2.1.4.

6.2.2.4 Other Techniques

In the Baukal and Ji (1998) study referred to above, spectral radiometers and the properties of molten glass were used to determine the optimal wavelength range for thermal radiation to penetrate into the glass.[78] However, that method is time-consuming and expensive. A more convenient and reliable way to determine the penetrating radiation is desirable. While the scanning spectral radiometer employed in that work provides quantitative measurements of the penetrating radiation, it requires spectral irradiance calibration, is time-consuming, and may not be as conveniently applied to large-scale flames in industrial furnaces. A more convenient way is to use total radiometers with appropriate spectral filters. Total (heat flux) radiometers have a flat spectral response in the 0.6 to 15.0 μm region. Without any window or filter in front of it, a heat flux meter gives the integrated total radiation intensity over the entire 0.6 to 15.0 μm range. By using an appropriate spectral filter to block out the undesirable, longer wavelength ($\lambda > 2.7$ μm) radiation, a heat flux meter can then be used to conveniently give the integrated radiation intensity that would be most penetrating into molten glass. Ji and Baukal tested several optical filters and compared the results against a spectral radiometer. Among the filters tested, a Schott BK7 optical glass filter exhibited a sharp cutoff transmission at $\lambda = 2.7$ μm. It became opaque at longer wavelengths for filter thickness greater than 6 mm (0.25 in.). Figure 6.6 shows a comparison of the measured flame radiation spectrum after it has passed through a room-temperature BK7 glass window, and the calculated flame radiation

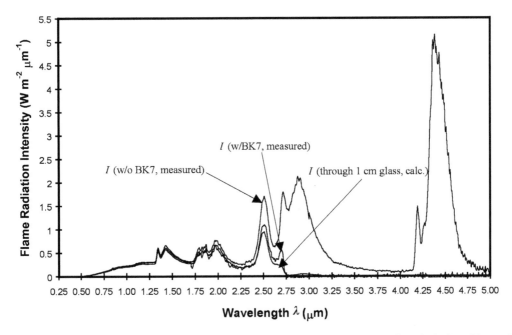

FIGURE 6.6 Transmission of flame radiation through a room-temperature BK7 optical glass filter and through high-temperature molten glass. (From B. Ji and C.E. Baukal, *Proc. of 1998 International Gas Research Conference,* Nov. 8-11, 1998, San Diego, CA, D. Dolenc, Ed., 1998, 422-433.)

spectrum after it has passed through 1 cm thick, high-temperature, low iron content molten glass (with the same spectral absorption coefficient used in Figure 4.6). The BK7 transmitted spectra closely matched that of the molten glass. This technique is a simple, reliable way of determining the penetrating portion of the radiation from a flame. This method was used to measure the penetrating radiation from $q_f = 5.0$ kW flames with different fuel equivalence and oxygen enrichment ratios. Similar to the analysis of spectral radiometer results, the radiation intensities were normalized to that of a flame with $q_f = 5.0$ kW, $\Omega = 1.0$, and $\phi = 1.0$. The agreement between the heat flux meter with a BK7 filter measurement and the numerically integrated spectral radiometer measurement was satisfactory, as shown by the qualitative trends for columns $I_{\text{nrm(pen.)}}$ and $I_{\text{nrm(BK7)}}$ in Table 3.5. This validated the use of a Vatell heat flux meter with a Schott BK7 optical glass filter to determine the penetrating radiation. To estimate the radiation intensity beyond 5.0 μm, a sapphire window was placed in front of the total heat flux meter. The transmission efficiency of a sapphire window is nearly constant at about 84% for $\lambda = 0.25$ to 5.0 μm, but quickly drops beyond 5.0 μm and cuts off at 6.0 μm. The difference between the total radiometer readings with and without the sapphire window can be used to estimate radiation intensity beyond 5.0 μm. The results are listed in Table 6.1. For $\Omega = 1.00$ and $\Omega = 0.60$ flames, radiation beyond 5.0 μm accounted for about 17% of the total flame radiation. For $\Omega = 0.21$ flames, the $I_{(\lambda > 5.0 \, \mu m)}$ was about 25% of the total radiation. Combustion gas radiation at wavelengths longer than 5.0 μm is dominated by an H_2O band at 6.3 μm.[79,80] The decrease in the relative contribution from shorter wavelength vibration band intensity in $\Omega = 0.21$ flames can be attributed to less populations in the highly excited vibrational levels in H_2O molecules. This can be caused by two factors in $\Omega = 0.21$ flames: less violent combustion reactions lead to weaker (if any) nonequilibrium chemiluminescence and a cooler flame temperature leads to fewer equilibrium thermal populations in the upper vibrational levels. Although the estimation given was only approximate, it gave a qualitative indication of the relative contribution from long-wavelength ($\lambda > 5.0$ μm) radiation.

Kilham (1949)[17] and Jackson and Kilham (1956)[18] used a thermopile normal to the surface of heated refractory cylinders. The combustion products from the flame, flowing around the cylinder,

TABLE 6.1
Estimation of the Radiation Intensity Beyond 5.0 μm

Ω	ϕ	$I_{(w/sap.)}$	$I_{(\lambda < 5.0\ \mu m)}$	$I_{(w/o)}$	$I_{(\lambda > 5.0\ \mu m)}$	$(\lambda > 5.0\ \mu m)\%$
1.00	0.67	0.0476	0.0567	0.0672	0.0105	16
1.00	0.80	0.0504	0.0600	0.0723	0.0123	17
1.00	1.00	0.0532	0.0633	0.0751	0.0118	16
1.00	1.33	0.0504	0.0600	0.0727	0.0127	17
1.00	1.50	0.0482	0.0574	0.0684	0.0110	16
1.00	1.60	0.0465	0.0554	0.0674	0.0120	18
1.00	1.79	0.0439	0.0523	0.0636	0.0113	18
1.00	1.89	0.0429	0.0511	0.0609	0.0098	16
1.00	2.00	0.0443	0.0527	0.0620	0.0093	15
0.60	1.00	0.0434	0.0517	0.0631	0.0114	18
0.21	1.00	0.0170	0.0202	0.0269	0.0067	25

Source: From B. Ji and C.E. Baukal, *Proc. of 1998 International Gas Research Conference,* Nov. 8-11, 1998, San Diego, CA, D. Dolenc, Ed., 5, 422-433, 1998.

were also in the field of view. Therefore, the sum of the radiation from the gases — and from the cylinder surface — were measured. Two screens with narrow slit openings were placed between the flame and the detector to limit the field of view to a small region around the heated cylinder. Giedt et al. (1960)[49] measured radiation for a H/H_2/CO mixture flame. From available data and radiation measurements, the gas emissivity was estimated to be 0.001. Nonluminous gas radiation was calculated to be, at most, only 2% of the total heat flux. Woodruff and Giedt (1966)[50] measured gas and surface radiation, with two Leeds and Northrup mirror-type rayotubes. Radiation was found to be a negligible effect. Veldman et al. (1975)[52] concluded that the measured radiation from an air/C_3H_8 flame was negligible. No details of the measurement techniques were given.

Baukal and Gebhart (1997) used surfaces of different emissivities to study flame radiation.[81] As previously discussed in section 4.2.1.5.1, polished metal surfaces were used as the low-emissivity surfaces. Surfaces coated with a high-temperature flat black paint were used as the high-emissivity surfaces. One of the problems with the polished surfaces was very rapid degradation of both the brass and copper surfaces. Within minutes, those surfaces lost much of their reflectivity, due to the contact with the flame gases. The polished stainless steel surfaces remained reflective for over an hour. There was no noticeable degradation of the blackened surfaces for all three target materials after hours of operation.

Sivathanu et al. (1990) measured the soot volume fractions in highly turbulent and strongly radiating flames, which are common industrial-type flames.[82] They used laser extinction measurements made with a three-line optical probe which had laser extinction at 632 nm and two-wavelength emission pyrometry at 900 and 1000 nm.

6.2.3 CONVECTIVE HEAT FLUX

In low-temperature processes, radiation may be insignificant and can be ignored, which means that the convective heat flux can be measured with a standard heat flux gage. However, in high-temperature combustion processes where radiation is very significant, it is difficult to directly measure the convective heat flux. The technique commonly used is to measure the total heat flux and the radiant flux and then to calculate the convective heat flux by subtracting the two. This obviously works only if there are no other modes of heat transfer that contribute to the total heat flux. A related method to determine the convective heat flux is by using other measurements along with a calculation. To do this, the temperatures of the surface where the convection is occurring and of the gas flowing over the surface, along with the gas velocity, are needed. Using the appropriate forced convection heat transfer equation for the given geometry, the convective heat flux can be calculated. This method is useful as a check against other methods that may be used.

Another method of determining the convective heat flux by direct measurement involves adding a radiant shield between the heat source and the heat flux gage to exclude radiation from being received by the gage. The problem is that the shield often disturbs the flow and the measured flux is not representative of the actual convective flow. Another problem is that the surface temperature under the shield will be lower than it would be without the shield, so the measurement would need to be done quickly before the wall temperature under the shield drops too much. In addition, although radiation from the primary source (burners) may be shielded, the rest of the furnace walls may still contribute radiation to the gage.

A third technique that could be used is to make a perfectly reflective heat flux gage so that any radiation received by the gage is reflected away and not measured by the gage. Then, in theory, the gage would only measure convection, assuming no other heat transfer mechanisms were important. Although in theory this sounds plausible, the reality is that it is very difficult to have a perfect reflector in an industrial combustion environment.

A fourth technique that could be used is to use a heat flux gage embedded in the wall where convection is to be measured and then to turn off the radiant source and quickly take measurements with the gage. The radiation from the hot furnace walls can be calculated and subtracted from the total flux. Again, this method suffers from problems similar to the other methods above in that it relies on a transient condition and accurate calculations.

6.3   TEMPERATURE

Many different types of devices are used to measure temperature, depending on the temperature range and specific application. The most common device is the thermocouple.

6.3.1   GAS TEMPERATURE

The gas temperature is not needed if only the heat flux is to be determined. However, it is an important parameter for comparing the measured heat flux to analytical formulations and to determine the gas properties. Also, the temperature profile across the jet gives an indication of the flow uniformity, and of the entrainment rate of ambient air into the flame.

Several techniques have been used for measuring the spatial gas temperature distribution, including thermocouples and the various optical methods, such as Rayleigh and Raman scattering.[2-5] At this time, the light scattering methods are very expensive and difficult to use (Heitor and Moreira, 1992).[83] Their primary use is in turbulent flows, and where probe survivability is a problem. Thermocouples are preferable because of their well-established properties, low cost, and ease of use. Some optical techniques, such as line reversal, may be difficult to use because of the need to seed the flame with particles. Those seeds could plug burners with small passage sizes.

6.3.1.1   Suction Pyrometer

At the desired measurement location, gases are extracted from the flame through a sampling tube. A thermocouple is positioned just inside this tube, typically made of a ceramic or high temperature metal, which acts as a radiation shield. Examples of suction pyrometers are shown in Figure 6.7. This technique is particularly useful when measuring gas temperatures in hot-surfaced enclosures. Surface radiation from the walls to an unshielded thermocouple may introduce large errors in the gas temperature measurement. Chedaille and Braud (1972)[75] and Goldman (1987)[84] specifically discussed the use of a suction pyrometer to measure gas temperatures in combustors. This device has been successfully used to measure gas temperatures in excess of 2200K (3500°F). In flame

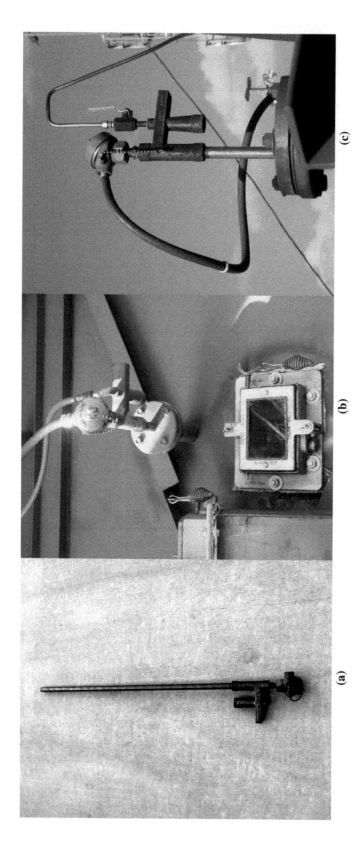

FIGURE 6.7 Examples of suction pyrometers: (a) new, (b) in the sidewall of a furnace, and (c) in the roof of a furnace. (Courtesy of John Zink Co. LLC, Tulsa, OK. With permission.)

impingement studies, Beér and Chigier (1968),[32] Vizioz and Lowes (1971),[21] Smith and Lowes (1974),[24] and Rajani et al. (1978)[25] measured gas temperatures with a suction pyrometer.

Grey (1965)[85] described a technique related to suction pyrometry known as a twin sonic orifice probe. In this technique, the gases were extracted at very high velocity, through two converging-diverging nozzles inside the probe. One limitation of the suction pyrometer is that active chemical reactions may also occur inside the probe passages for highly dissociated gases. With proper corrections, this effect may be overcome by the dual sonic orifice probe. Rajani[25] and Ivernel and Vernotte (1979)[63] used a twin sonic orifice probe to measure gas temperatures in oxygen/fuel flames up to 3060K and 2800K (5050°F and 4600°F), respectively.

6.3.1.2   Optical Techniques

Laurendeau (1987) has given a brief review of temperature measurements using light scattering techniques.[86] The techniques discussed included spontaneous vibrational and rotational Raman scattering, Rayleigh scattering, and laser-induced fluorescence. Fluorescence methods were recommended for temporally resolved planar thermometry.

Chao and Goulard (1974) presented a nonlinear iterative inversion technique for calculating flame temperatures from a single-line-of-sight multifrequency set of radiance measurements.[87] Kondic (1974) developed a method to measure the density and temperature fields in fluids along a light path where the light source and detector were located external to the system.[88] Although a laser is the preferred light source, a conventional light source can also be used. The gas temperature is related to the density field and the index of refraction of the fluid. Hofmann and Leipertz (1996) presented experimental measurements of the temperature field in a sooting, methane/air flame using a new technique referred to as filtered Rayleigh scattering (FRS).[89]

Posillico (1986)[23] made detailed spatial gas temperature measurements using Raman spectroscopy (Eckbreth, 1988[4]), for air/CH_4 flames impinging on a water-cooled flat plate. The highest reported gas temperature was 3200K (5300°F), although the adiabatic flame temperature for a stoichiometric air/CH_4 flame is 2223K (3542°F). No explanation was given for this apparent discrepancy. Hughes et al. (1995) compared a suction pyrometer and CARS in an industrial-scale flame.[90] Rumminger et al. (1996) have used laser-induced fluorescence (LIF) to measure gas temperatures at the exit of porous radiant burners.[91] The measurements were slightly higher than those made with a thermocouple after all corrections had been made to both techniques. The overall uncertainty in the LIF measurements was estimated to be ±110°F (60K). Char and Yeh (1996) described an optical technique of measuring the flame temperature using a combination of measuring the infrared radiation at a selected wavelength (4.3 μm, which has the highest intensity for CO_2 and H_2O) and making some iterative calculations to determine the flame emissivity.[92] The measurements were made on a fuel-lean ($\phi = 0.51$) open propane flame. The results compared favorably with thermocouple measurements. A correlation was developed for the gas emissivity for this flame as a function of the gas temperature position from the burner outlet.

6.3.1.3   Fine Wire Thermocouples

Gas temperature is an important parameter to the extent that it affects the gas properties, which are temperature dependent. Most of the heat transfer equations for heat flux also contain the gas temperature explicitly, as part of the driving potential for heat transfer. A number of thermocouple alloys are available, depending on the temperature range and type of atmosphere (inert, vacuum, oxidizing, or reducing), as shown in Table 6.2.

In one flame impingement study, two types of thermocouples were used to measure the gas temperatures.[93] For the low-temperature flames, uncoated bare wire Type B thermocouples (Pt-60% Rh/Pt-6% Rh) were used. The probes were designed according to the method recommended by Peterson (1981).[94] However, those thermocouples melted at $\Omega > 0.25$. For higher temperature flames,

TABLE 6.2
Characteristics of Common Thermocouple Alloys.

Max T (°F)	Max T (°C)	Allowable Atmos. (Hot)	Material Names	ANSI Type[a]	Color Code	Output (mv/100°F)	Accuracy (%) Sandard[a]	Precision[a]
5072	2800	Inert, H_2, vacuum	Tungsten/tungsten 26% rhenium	—	—	0.86	—	—
5000	2760	Inert, H_2, vacuum	Tungsten 5% rhenium/tungsten 26% rhenium	—	—	0.76	—	—
4000	2210	Inert, H_2	Tungsten 3% rhenium/tungsten 35% rhenium	—	—	0.74	—	—
3720	1800	Oxidizing[b]	Platinum 30% rhodium/platinum 6% rhodium	B	—	0.43	1/2	1/4
2900	1600	Oxidizing[b]	Platinum 13% rhodium/platinum	R	—	0.64	1/4	1/4
2800	1540	Oxidizing[b]	Platinum 10% rhodium/platinum	S	—	0.57	1/4	1/4
2372	1300	Oxidizing[b,c]	Platinel II (5355)/Platinel II (7674)	—	—	2.20	5/8	—
2300	1260	Oxidizing	Chromel/Alumel,[d] Tophel/Nial,[e] Advance T1/T2,[f] Thermo-Kanathal P/N[g]	K	Yellow red	2.20	4°F, or 3/4%	2°F, or 3/8%
1800	980	Reducing[a]	Chromel/constantan	E	Purple red	4.20	1/2	3/8
1600	875	Reducing	Iron/constantan	J	White red	3.00	4°F, or 3/4%	2°F, or 3/8%
750	400	Reducing	Copper/constantan	T	Blue red	2.50	3/4	3/8

[a] Per ANSI C96.1 Standard.
[b] Avoid contact with carbon, hydrogen, metallic vapors, silica, reducing atmosphere.
[c] @ Engelhard Corp.
[d] @ Hoskins Mfg. Co.
[e] Wilber B. Driver Co.
[f] Driver-Harris Co.
[g] The Kanthal Corp.

Source: From F. Kreith, Ed., *CRC Handbook of Mechanical Engineering,* CRC Press, Boca Raton, FL, 1998, 4-183.

an Ir/60% Ir-40% Rh thermocouple was used. Candler (1967)[95] has referred to this thermocouple as a Type L. According to Blackburn and Caldwell (1964),[96] this thermocouple can be used to measure temperatures up to 2400K (3900°F), within ± 20K (± 40°F), using the appropriate reference table. This accuracy only refers to the measured temperature. The actual gas temperature still requires correction for the losses from the thermocouple junction. The Type L thermocouple wires were 0.25 mm (0.01 in.) in diameter.

Heitor and Moreira (1992)[83] provided an extensive review of the use of thermocouples in combustion environments. They stated that flow disturbances and measurement errors can be made negligible in laminar flows. They suggested a coating consisting of 90% Al_2O_3 and 10% MgO, which minimizes catalytic effects and is not toxic. According to Cookson et al. (1964),[97] thermocouple junctions should be coated. The junctions should also be as small as possible to prevent surface recombination effects (see section 4.1.1.2). Kent (1970)[98] has listed the following general requirements for a proper thermocouple coating:

- Noncatalytic to the flame
- Inert to the thermocouple material
- Impermeable to gases to protect the thermocouple
- Poor electrical conductor to prevent electrical shorts
- Stable within the desired temperature range
- Capable of being evenly applied

A coating made from a mixture of approximately 10% BeO/90% Y_2O_3 was recommended.

When using thermocouples to measure high temperatures, the measurements must be corrected for the errors due to radiation, convection, and wire conduction. Moffat (1998) noted the general equations for estimating these types of errors:[99]

$$E_{conv} = (1-\alpha)\frac{v^2}{2g_c Jc_p} \tag{6.1}$$

$$E_{rad} = \frac{\sigma\varepsilon}{h}\left(T_{probe}^4 - T_{surr}^4\right) \tag{6.2}$$

$$E_{cond} = \frac{t_{gas} - t_{probe\ mount}}{\cosh\left[1\sqrt{\dfrac{hA_c}{kA_{cond}}}\right]} \tag{6.3}$$

where E_{conv} = convection error, E_{rad} = radiation error, E_{cond} = conduction error, α = recovery factor, v = gas velocity, g_c = universal gravitational constant, J = Joules constant, c_p = gas specific heat, σ = Stefan-Boltzmann constant, ε = probe emissivity, h = convection coefficient for gas flowing over probe, T_{probe} = absolute temperature of the probe, T_{surr} = absolute temperature of the surroundings, t_{gas} = temperature of the gas flowing past the probe (usually what is being measured), $t_{probe\ mount}$ = temperature of the probe mount, l = length of the exposed junction, A_c = area for convection heat transfer, k = probe thermal conductivity, and A_{cond} = area for conduction heat transfer. Note that these are iterative calculations, as the temperature of the gas is the quantity being measured. Also, some of the quantities may be difficult to know with a high degree of precision. Some of these include the convection coefficient, the gas velocity, the probe emissivity, the mount temperature, and the exposed junction length. Therefore, there will still be some uncertainty even after the measurements have been corrected for these errors. In general, the errors increase as the probe size increases.

Several methods are commonly used to compensate for these errors. One is electrical compensation (Hayhurst and Kittelson, 1977).[100] However, this method cannot be used at very high temperatures, where the thermocouple may already be near its melting point. Any additional heating may cause the thermocouple to fail. Another compensation method is to make measurements using diminishingly smaller diameter thermocouples (Bradley and Matthews, 1968[101]). The thermocouple measurements are plotted against the cross-sectional area of the thermocouple junction. The temperature curves are then extrapolated to a zero-diameter thermocouple. This eliminates the need for correcting the measurements for the losses from the junction. A third method is to compute the correction (Fristrom and Westenberg, 1965[102]). An energy balance at the thermocouple junction may be given as:

$$q_{catalytic} + q_{conduction} + q_{radiation} + q_{convection} = 0 \tag{6.4}$$

Hayhurst and Kittelson (1977)[100] have shown that only surface reactions contribute to catalytic heating. This is commonly neglected when the thermocouple is coated. Bradley and Matthews (1968)[101] have shown that conduction can be neglected for wires over approximately 1 mm (0.04 in.) long. When conduction effects can been neglected, the energy balance then simplifies to:

$$t_j = t_{T/C} + \sigma\varepsilon_{T/C}\, t_{T/C}^4 / h \tag{6.5}$$

where $t_{T/C}$ is the uncorrected temperature as measured with the thermocouple, and t_j is the corrected temperature. The radiation from the environment to the junction has been neglected. An emissivity of 0.14 was recommended for uncoated platinum Type B thermocouple (Sparrow and Cess, 1978[103]). An emissivity of 0.60 was recommended for a coated Type L thermocouple (Peterson and Laurendeau, 1985[104]). The convection coefficient was calculated using the equation recommended by Hinze (1959):[105]

$$Nu = 0.42\, Pr^{0.2} + 0.57\, Pr^{0.33}\, Re^{0.5} \tag{6.6}$$

An iterative procedure is required to calculate t_j because the gas properties used to calculate the convection coefficient are dependent on t_j. The gas temperature t_j may also be required to calculate the gas velocity when, for example, Pitot-static probe measurements are used to calculate the velocity. Son et al. (1986) presented a mathematical method to compensate for the thermal inertia effects of thermocouples.[106] These effects can cause errors if not properly corrected.

Many researchers have used fine wire thermocouples (T/Cs), with and without coatings. Very fine thermocouples minimize the disturbance to the flow. Buhr et al. (1973)[33] and Kremer et al. (1974)[34] used electrical compensation for radiation and conduction heat losses. Horsley et al. (1982),[64] Kataoka et al. (1984),[39] You (1985),[44] Hargrave et al. (1987a),[107] and Hustad et al. (1991, 1992)[66-68] calculated corrections for the heat losses from the thermocouple junction. In some studies, radiation losses were corrected using successively smaller diameter thermocouples. These measurements were extrapolated to a vanishingly small thermocouple junction diameter. This method was suggested by Nichols (1900)[108] and was improved by Bradley and Matthews (1968).[101] Costa et al. (1996) used a bare 300 μm diameter wire uncoated Pt/Pt13%Rh thermocouple to measure gas temperatures inside an industrial glass-melting furnace.[109] Mital et al. (1996) used three different diameter (76, 125, and 200 μm) fine wire, Type R thermocouples to measure the gas temperature inside porous ceramic radiant burners.[110]

Most high-temperature applications require some type of protective coating on the thermocouple unless the atmosphere is inert. Kaskan (1957)[111] described a procedure for applying a silica coating to thermocouples. The junction was immersed in a flame containing a small amount of hexamethyl-disiloxane. This produced a uniform coating of fused quartz. It was found that the coating failed at about 1900K (3000°F), due to softening of the silica. Cookson and Kilham (1963),[29] Kilham and Dunham (1967),[30] and Milson and Chigier (1973)[51] used silica coatings to prevent catalyic hydrogen atom recombination on the platinum in the thermocouple (Fristrom and Westenberg, 1965[102]). Madson and Theby (1984)[112] showed that uncoated platinum alloy thermocouples significantly overpredicted temperature, compared to silica-coated thermocouples in an $O_2/Ar/CH_4$ flame. Pollock (1984)[113] has shown that silica coatings may contaminate platinum-alloy thermocouple wires in chemically reducing environments. Kremer et al. (1974)[34] used Al_2O_3 to prevent radical recombination. Hoogendoorn et al. (1978),[35] Popiel et al. (1980),[36] and Van der Meer (1991)[46] used a yttrium/beryllium oxide (Yt_2O_3/BeO), as recommended by Kent (1970).[98] This eliminates silica contamination. Unfortunately, beryllia is toxic and extremely poisonous.

Rumminger et al. (1996) used fine wire thermocouples to measure the gas temperature above the surface of a porous radiant burner.[91] At 5 mm (0.2 in.) above the surface, they estimated a temperature correction of 270K (490°F), including the effects of re-radiation from the burner surface to the thermocouple bead. Such a large correction resulted from a relatively large bead size (460 μm), high bead emissivity (0.6), and relatively slow gas flow (Re_d 0.9, Nu 0.7).

Fine wire thermocouples are of low cost and easy to use. They have short response times and good spacial resolution. However, they are somewhat intrusive to the flame. Corrections are also required due to heat transfer effects at the junction bead. These include radiation from the bead to the environment and the heat conduction along the thermocouple wires. Often, the junction must be coated to prevent catalytic reactions. There are also material limitations, due to high-temperature oxidizing and reducing conditions.

6.3.1.4   Line Reversal

Many researchers have used this technique, originally suggested by Féry (1903),[114] to measure flame temperatures. The flame is commonly seeded with sodium salt. The salt vaporizes in the flame and dissociates into sodium atoms and other products. These excited atoms emit at a specific wavelength. This radiation is then compared to a calibrated reference source, such as a tungsten filament lamp. The temperature of this source is adjusted until it matches the seed radiation. To the naked eye, the background source appears to disappear when it is at the same temperature as the flame. Strong et al. (1949) have given a general discussion of this technique to measure the temperatures in flames.[115]

Kilham (1949)[17] used a spectrometer to compare the flame with the filament. Fells and Harker (1968)[20] estimated the accuracy of their measurements to be 20K. Nawaz (1973)[58] and Fairweather et al. (1984)[59,60] introduced sodium chloride smoke, by passing N_2 over molten salt, through a tube in the center of the burner. Therefore, only the temperatures along the central axis of the flame were measured. This reduced the influence of ambient air entrainment cooling effects at the flame periphery. Nawaz used a high-pressure xenon arc lamp (maximum temperature of 5000K, or 8500°F) as the comparison radiator, since the temperature levels were out of the range of the conventional tungsten lamps that were available at that time. Measured temperatures were 50K to 100K (90°F to 180°F) below the calculated adiabatic flame temperature. Fairweather (1984)[60] measured a 33K (59°F) and 40K (72°F) reduction, across the equilibrium zone between the reaction zone and the stagnation point, for laminar and turbulent flames, respectively. Atmospheric air entrainment into the flame was thought to be a possible explanation for this effect.

This method is only slightly intrusive, due to the seeding. It is of relatively low cost. It may be effectively used for very high temperatures. However, the seeded field must be uniform. This may require the addition of a carrier gas to transport the seeds into the flame. Only an average gas temperature, across the flame, is measured. A calibrated reference light source is required.

6.3.2   SURFACE TEMPERATURE

There are two common ways to measure the surface temperature of an object, which are discussed next.

6.3.2.1   Embedded Thermocouple

In the flame impingement study by Baukal (1996), a Type K thermocouple was imbedded in the impingement rings of calorimeters.[93] The thermocouples had an 0.5 mm (0.02 in.) o.d. inconel sheath, and the clearance hole in the side of each impingement ring was 0.64 mm (0.025 in.) i.d. A high-conductivity anti-seize sealant was applied to the end of each thermocouple before it was inserted into a ring to ensure good thermal contact. These thermocouples measured the temperatures at the middle of each impingement ring. The hot face temperatures were then calculated using the one-dimensional, steady-state conduction equation. Using the calculated heat flux to each ring, the hot face temperature, t_w, was calculated using:

$$t_w = t_{middle} + \frac{q'' l_1/2}{k_1} \tag{6.7}$$

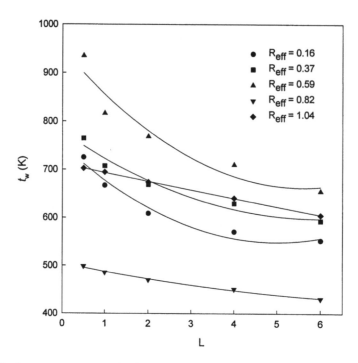

FIGURE 6.8 Surface temperature (t_w) for O_2/natural gas flames ($q_f = 15.0$ kW) impinging on an untreated stainless surface. (From C.E. Baukal, Heat Transfer from Flame Impingement Normal to a Plane Surface, Ph.D. thesis, University of Pennsylvania, 1996.)

where t_{middle} was the measured temperature at the middle of the ring, q'' was the calculated heat flux to the ring, l_1 was the thickness of the impingement ring, and k_1 was the thermal conductivity of the impingement ring. A typical result is shown in Figure 6.8. This graph shows that the wall temperatures decreased as the axial distance between the burner and the target decreased. However, for L > 2, the differences became smaller.

6.3.2.2 Infrared Detectors

Thermal imaging systems have been used for years in a variety of applications, including military and intelligence surveillance activities. Within about the last decade, the cost of these systems has declined enough that they are now being routinely used to measure temperatures in combustion applications. In combustion systems, these infrared imaging systems are typically used to measure the temperature of the furnace or heater walls, the heat load which might be a material like scrap metal or a tube used for process fluid heating, or the burner itself. Imaging systems are primarily used as a diagnostic to detect hot spots that could indicate potential problem areas.

Infrared detectors have a number of potential advantages over thermocouples:[116]

- Faster response times
- Very high accuracies
- Able to measure temperatures on moving objects
- Able to measure higher temperatures than conventional thermocouples
- Able to give two-dimensional temperature profiles
- Do not disturb the surface being measured
- Are not dependent on how they are attached to the surface, as thermocouples are, to get an accurate surface temperature measurement

FIGURE 6.9   Example of an infrared pyrometer measuring the surface temperature of a porous refractory matrix burner.

There are also some potential disadvantages compared to thermocouples:

- The surface emissivity is often required, except for two-color pyrometers.
- Measurements can be impaired by anything, including combustion products, in between the pyrometer and the surface, which could absorb any of the radiation emitted by the surface.
- The front lens of the pyrometer must be kept clean.
- The electronics in the pyrometer must be kept cool.
- Calibration is typically more involved than for a thermocouple.

The theory of radiation pyrometry is available elsewhere.[117]

The original devices used for combustion applications were infrared pyrometers that gave a temperature based on the infrared energy emitted by the object and an input emissivity for the object. An example of an infrared pyrometer is shown in Figure 6.9. Newer devices use imaging technology to produce thermal images of the temperature profile of an object.[118] These devices are capable of producing realtime images of dynamically changing temperature profiles. A typical system consists of a high-resolution infrared camera, a computer, and thermal image analysis software. Besides displaying the temperature profile for an object, the software provides capabilities for analyzing the data. Because of the fast response of the new systems, it is also possible to use them for realtime control of the heating system to produce specific temperature profiles in the load.

Lappe (1997) gave some recommendations for the installation and use of single-color IR thermometers.[119] The most important factor in measuring the correct surface temperature is to know the emissivity of the body being measured. The emissivity of a graybody is constant for all wavelengths and usually does not change significantly with temperature unless there is a change in the material or in its surface characteristics. For example, if a solid material melts and becomes liquid, there may be a significant change in its emissivity. Another example is if a metal oxidizes or becomes roughened from wear, its emissivity also may change dramatically. Lappe recommended using the shortest wavelength instrument possible to measure the surface temperature for two reasons. The first is that the difference in the emissivity of common surfaces such as metals is usually significantly less at shorter wavelengths, with the difference increasing with wavelength. Therefore, there will be less error if the emissivity used on the pyrometer is not properly adjusted for changes in the surface conditions of the body whose temperature is being measured. The second reason is that lower wavelength instruments typically have less error in the temperature reading compared to higher wavelength units. Lappe also gave some recommendations for the installation of these units:

- Ensure that the unit is focused properly on the object of interest and that the spot size is the correct size.
- Ensure that there are no obstructions (e.g., sightports walls, smoke, steam, etc.) between the pyrometer and the surface being measured.
- Fiber optic systems may be appropriate for certain applications.
- Ensure that the interconnecting cable is not overheated or installed with other wires that may carry high voltages that could adversely affect the signal.
- Ensure that the sensing head is kept clean and cool, but not overcooled which could cause condensation in the unit.
- Keep the lens clean with a purge of clean dry air.
- Have the instrument calibrated at least annually.

Two-color pyrometers avoid the problem of determining the emissivity of the surface.

6.4 GAS FLOW

6.4.1 Gas Velocity

The gas velocity of impinging flame jets is not required if only the measured heat flux is of interest. However, the velocity is very important for comparing the measured flux data to analytic models. Also, the velocity profile indicates whether or not the jet is fully developed before it impacts on the target. The two most common techniques for measuring velocity have been Pitot tubes and laser Doppler velocimeters (LDV). Some other techniques are also considered below.

6.4.1.1 Pitot Tubes

Pitot tubes and Prandtl probes are used to measure the total and static pressures, respectively, in a flow. Both sensors are tubes. One end is exposed to the flow and the other end is connected to a pressure transducer. The Pitot tube has a hole at the end of the tube. It is aligned in the main flow direction to measure total head pressure. The Prandtl probe has a hole on the side of the tube. The sensor tube is aligned parallel to the main flow direction to measure the static pressure. The difference between the total and static pressures (i.e., the dynamic pressure) is proportional to the square root of the gas velocity. In flames, these pressures are low but are measurable (Lewis et al., 1956).[120] The two probes may be combined together, as a Pitot-static probe. The Pitot tube and combined Pitot-static probe are shown in Figure 6.10.[121] The combined probe is sometimes referred

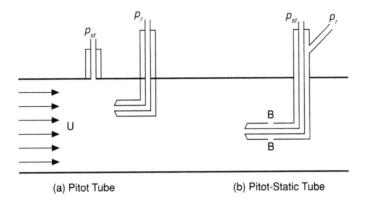

(a) Pitot Tube   (b) Pitot-Static Tube

FIGURE 6.10   Flow measurement using (a) Pitot tube and (b) Pitot-static tube. (Adapted from C. Swan, in *International Encyclopedia of Heat & Mass Transfer*, G.F. Hewitt, G.L. Shires, and Y.V. Polezhaev, Eds., CRC Press, Boca Raton, FL, 1997, 842.)

to as simply a Pitot tube. Some studies have not distinguished between a true Pitot tube and a Pitot-static tube. In some studies, the static pressure at the probe location was assumed to be the same as the ambient. Single-hole Pitot tubes only measure the velocity component parallel to the probe. In addition to velocity, specially designed five-hole Pitot tubes may also indicate the gas flow direction. See Reference 9 for a discussion of both one- and five-hole Pitot tubes, including their use in combustion applications. Becker[9] showed the importance of the probe tip shape, although this feature has rarely been described in the studies considered here.

The equation governing the response of a Pitot probe is commonly given as (Becker and Brown, 1974):[122]

$$v = \sqrt{2(p_t - p_{st})/C_p\rho} \tag{6.8}$$

where p_t is the total or impact pressure, p_{st} is the static pressure, and C_p is the probe calibration constant. Becker and Brown studied the response of Pitot probes in turbulent nonreacting gas streams. The following terminology is used to specify the recommendations of that study:

d_1 = Total pressure hole i.d.
d_2 = Probe o.d.
d_3 = Static pressure hole i.d.
l_1 = Internal length before any change in diameter or direction of total impact tube
l_2 = External length before any change in diameter or direction of probe
l_3 = Length from probe tip to centerline of static pressure hole

In that study, the following conditions were assumed:

1. The probe is long enough ($l_2/d_2 > 6$) so that the downstream geometry has a negligible effect on the response.
2. The internal geometry ($l_1/d_1 > 3$) has a negligible effect on the response.
3. Compressibility effects are small (Ma < 0.3).

A hemispherically nosed, water-cooled P-S probe was used by Baukal (1996).[93] It had the following dimensions: $d_1 = d_3 = 1.40$ mm (0.055 in.); $d_2 = 12.7$ mm (0.5 in.); $l_1 = l_2 = 14.6$ cm (5.75 in.); $l_3 = 22.2$ mm (0.875 in.). These dimensions satisfied conditions 1 and 2 above. The low gas flow rates satisfied condition 3. The probe had a single static pressure inlet, as recommended by Cho and Becker (1985).[123]

For a hemi-nosed probe geometry, Chue (1975)[124] recommended that $Re_p > 1000$, so that $C_p \approx 1$, and no calibration would be required. Becker (1993)[9] recommended $Re_p > 600$ to avoid calibration. Becker and Brown (1974)[122] recommended $d_1/d_2 < 0.3$, with 0.12 as the optimum. Chue (1975)[124] recommended $d_3/d_2 = 0.01$ to 0.50. Cho and Becker (1985)[123] recommended $d_3/d_2 < 0.25$, and preferably < 0.1. Cho and Becker (1985)[123] recommended $l_3/d_2 = 2$.

The above methods are simple and of relatively low cost. However, they are intrusive to the flow. They usually require cooling in combustion environments. This technique has a very limited ability to detect turbulent fluctuations. It also requires careful use in highly recirculating flows.

There are many potential sources of errors when using Pitot tubes, including:[125]

- Errors in the pressure gages
- Misalignment of the Pitot tube
- Wall interference for measurements close to a surface
- Instrument vibrations
- Flaws in the nose of the probe
- Location in recirculation or shear zones
- Viscous effects in low-flow, highly viscous flows or for extremely small tubes

Therefore, care must be used when using P-S probes.

6.4.1.2   Laser Doppler Velocimeter

Laser Doppler velocimetry (LDV) is an optical technique for measuring fluid velocities. For a given velocity direction, two laser beams must cross to create a small scattering volume. As a particle passes through that volume, a Doppler shift is produced. This is detected by receiving optics. With proper processing, the velocity can be calculated from this signal. Baker et al. (1972) presented LDV measurements made on town gas and methane flames from an industrial burner.[126] Chigier (1989),[127] Most (1991),[128] and Heitor et al. (1992)[129] discussed the use of LDVs in combusting flows. Fingerson and Menon (1998) discussed this technique in some detail, although not specifically for combustion measurements.[130]

For air/natural gas flames impinging on a flat plate, Hoogendoorn et al. (1978)[35] and Popiel et al. (1980)[36] used an LDV in turbulent and laminar flames, respectively. The LDV measured turbulent fluctuations. Pitot tubes were used to measure mean velocities. Hoogendoorn (1978),[35] Popiel (1980),[36] and van der Meer (1991)[46] used MgO seeds of unspecified diameter. You (1985)[44] used oil droplets and a small acetylene sooting flame to generate additional particles in the regions of their flames that had inadequate soot particle concentrations for LDV measurements. The mean particle diameter was 0.6 μm. Fairweather et al. (1984)[60] used 1 μm mean diameter alumina particles for seeding. A near-uniform free stream velocity was measured in the equilibrium zone of their laminar air/CH_4 flames. In their turbulent flames, there was a 2% and 40% increase in the free stream velocity and in the turbulent intensity, respectively. Hargrave et al. (1984, 1987)[37,107] used an LDV to measure velocities and turbulent intensities in turbulent flames. The flames were seeded with alumina particles. The mean size was 1 μm, with a maximum size of 2 μm. Posillico (1986)[23] used an LDV, with Al_2O_3 seeds of unspecified size. Van der Meer's (1987)[131] measured velocity profile was not uniform, as stated by others (e.g., see Reference 35) for tunnel burners. Van der Meer (1991)[46] used an LDV to determine the stagnation velocity gradient (discussed in the next section), the velocity field, and the turbulence level. The turbulence level increased with L up to about L = 5, where it flattened out. This behavior was seen even in initially laminar flames.

This technique has excellent temporal resolution; it can be used to measure turbulence, is less intrusive to the flow field than Pitot tubes, and can be used very close to walls without affecting the flow field. An LDV system can also measure multidimensional velocity fields, even in recirculating flows. New fiber-optic techniques make optical alignment easier than before. However, an LDV system is expensive. Flames from gaseous fuels commonly require seeding to ensure sufficient scattering. The choice of seeds is limited in high-temperature oxidizing environments. Optical refraction may occur in regions of large density and temperature gradients in the flame. This may reduce the signal strength and resolution.

6.4.1.3 Other Techniques

Kilham (1949),[17] Jackson and Kilham (1956),[18] and Kilham and Purvis (1971, 1978)[57,62] used shadowgraph photography to estimate the flame cross-sectional area. This data was then used to calculate an average gas velocity, based on the known gas flow rate and the measured temperature. Anderson and Stresino (1963)[26] estimated the gas velocity, based on the measured thrust of a flame impinging on a flat surface. The velocity was calculated in terms of gas dynamic considerations, based on the known flow rates and the estimated gas densities. Reed (1963)[27] calculated the average gas velocity using the gas flow rates, the estimated flame diameter, and the adiabatic flame temperature. Conolly (1971)[132] used a particle tracking photography method. Results agreed well with Pitot tube measurements. Nawaz (1973)[58] calculated the average gas velocity in the same manner as Reed. However, the measured — instead of the adiabatic — flame temperature was used. Fairweather et al. (1984)[59] concluded that the velocity profile for their tunnel burner was very uniform at any location downstream of the reaction zone, based on Pitot tube measurements. This justified using an estimated flame diameter, determined from photographic techniques, to calculate the gas velocity.

6.4.2 Static Pressure Distribution

In many common flow geometries, such as free jets, the static pressure is assumed uniform and constant. However, in impinging jet flows, the static pressure in the stagnation region is often much higher than elsewhere in the flow. This is due to rapid deceleration and deflection of the impinging jet. The actual static pressure distribution has been measured near the stagnation point for two reasons. The first is to determine the velocity gradient adjacent to the surface, β_s. The second is to determine the extent of the stagnation region.

6.4.2.1 Stagnation Velocity Gradient

This is an important parameter in the semianalytical solutions for the heat transfer from high-temperature gases impinging on a blunt-nosed body. These solutions are discussed in Chapter 7. An example is the solution by Sibulkin (1952),[133] given in Equation 7.15. The stagnation velocity gradient has been defined as:

$$\beta_s = \left(\frac{\partial v}{\partial r}\right)_{z=\delta, r=0} \tag{6.9}$$

where the local velocity gradient is evaluated at the edge of the boundary layer ($z = \delta$) along the axis of symmetry. A detailed discussion of this parameter is given by van der Meer (1987).[131]

Two different methods have been used to experimentally determine β_s. The first involves measuring the static pressure distribution near the stagnation point. It can be shown (see Purvis, 1974[134] or Tariq, 1982[135]) that:

$$\beta_s = \left(\frac{d}{dr} \sqrt{\frac{2\{p_{st} - p(r)\}}{\rho}} \right)_{r=0} \quad (6.10)$$

This is estimated by calculating the tangent to the graph of $\sqrt{2(p_{st} - p(r))/\rho}$ as a function of r at $r = 0$. Kilham and Purvis (1978)[62] used 0.88 mm (0.035 in.) i.d. stainless steel tubes. These were mounted within a given radius, surrounding the stagnation point, in a large flat water-cooled copper plate. Using this array of static pressure measurements, they calculated the following relationship using Equation 6.10:

$$\beta_s = v_e/d_j \quad (6.11)$$

Rajani et al. (1978)[25] and Popiel et al. (1980)[36] measured β_s at the stagnation surface, with small holes drilled into a flat plate target. Popiel used a 0.5 mm (0.02 in.) i.d. hole and determined the following relation:

$$\beta_s = 9.7 \, v_e/(x + d_n) \quad (6.12)$$

β_s can also be experimentally determined from LDV measurements. By measuring the velocity distribution near the stagnation point, the gradient in Equation 6.10 can be directly computed. Van der Meer (1991)[46] determined β_s, for air/natural gas flames impinging on a cooled flat plate. The nondimensionalized velocity gradient was given as:

$$B_s = \beta_s \, d_n/v_n \quad (6.13)$$

This increased with L, to a peak at about L = 3. Thereafter, it decreased monotonically. For $1700 \leq Re_n \leq 4250$, B_s showed no dependence on Re_n.

6.4.2.2 Stagnation Zone

The static pressure distribution around the impingement surface is also used to determine the extent of the stagnation zone. This zone is defined as the region where the static pressure, p_{st}, exceeds the distant ambient pressure. The stagnation zone boundary is important in analyzing the mechanisms of the impinging jet process. The flow between the nozzle and the beginning of the stagnation zone, in the direction normal to the target, is a free jet. The boundary of the stagnation zone parallel to the target defines the start of the so-called wall region. This is a boundary layer flow. Both the free jet and the wall regions (see Figure 3.1) can be analyzed by well-established analysis.

Buhr (1969)[136] measured the p_{st} distribution in the radial direction, using a 0.75 mm (0.03 in.) i.d. Pitot tube. It was mounted flush with the surface of a water-cooled plane target. For the fuel-rich turbulent air/CH$_4$ flame, L ranged from 13 to 53 with Re_n = 19,680. Vizioz and Lowes (1971)[21] measured the p_{st} distribution along the radial direction of a plane surface. Holes of unspecified diameter were drilled through a water-cooled steel plate, for flames with and without swirl. The no-swirl flame showed a peak in p_{st}, at R = 0 of nearly 100 Pa (0.0014 psig). The swirl flames showed peaks, at either R = 1.3 or 1.9, for p_{st} between 70 and 80 Pa (0.010 and 0.012 psig), depending on the nozzle design. The stagnation zone appeared to extend to about R $\approx$ 3.8. Vizioz also measured p_{st} in the axial direction. The stagnation zone extended to Z = 2.3 to 4.0, depending on the test configuration. Kilham and Purvis (1978)[62] and Horsley et al. (1982)[64] used 0.88 mm (0.035 in.) i.d., while Hoogendoorn et al. (1978)[35] used 0.5 mm (0.02 in.) i.d. stainless steel tubes, mounted within a given radius surrounding the stagnation point of a large flat water-cooled copper plate. Kilham found that the impingement region extended out to 2.2 < R < 2.7. The measurements were difficult, due to combustion product condensation obstructing the pressure taps.

6.5   GAS SPECIES

Gas composition is useful in determining combustion efficiency. For example, a gas sample at the exit of a tunnel burner measures any fuel that has not been burned. A measurement of the gas composition near the target surface is useful in predicting the amount of thermochemical heat release from the radical species. This composition determines the level of species dissociation in the impinging jet.

Beér and Chigier (1968)[32] measured gas compositions by extracting samples from the flame. Kremer et al. (1974)[34] measured gas concentrations in the stagnation region. In both studies, the gases sampled were not specified and no data were reported. Conolly (1971)[132] estimated the hydrogen atom concentration in flames. This was done by first comparing the spectral intensities of the radiation emitted from flames seeded with sodium and lithium at wavelengths of 589.0 and 670.8 nm, respectively. The composition was then estimated, assuming equilibrium conditions. These results showed super-equilibrium concentrations of H in a small region above the burner. Nawaz (1973)[58] tried unsuccessfully to measure NO with a gas chromatograph (GC). The OH concentration was qualitatively determined — nonintrusively — by a line absorption method. The background source was a water discharge lamp. In general, CO_2, H_2O (calculated), N_2, and O_2 concentrations were higher than adiabatic equilibrium predictions. Also, CO and H_2 concentrations were lower than predicted. Low probe quenching efficiency was suggested as causing CO and H_2 to react with OH in the probe to produce CO_2 and H_2O, respectively. You (1985)[44] extracted gas samples isokinetically through a 1 mm (0.04 in.) i.d. water-cooled stainless steel probe. Posillico (1986)[23] used Raman spectroscopy (see Reference 4 for a discussion of this technique) to spatially measure CO_2, N_2, and O_2 in an impinging air/CH_4 flame. Hustad et al. (1991) measured O_2 concentrations in pure diffusion flames to qualitatively determine air entrainment into the flames.[66] Details of the technique were not given.

There are several potential uncertainties with most of the extractive gas sampling techniques that have been used. The most important one is subsequent chemical effects in the gas sampling system. Thereby, the measurements may not accurately reflect the actual gas composition at the sampling probe inlet. This makes it very difficult to accurately determine the trace radical species. On the other hand, these are of great interest in impingement flame studies. The sampling probe commonly must be cooled in combustion environments. This increases the size of the probe, making it more intrusive in the flame. This also adversely affects the spatial resolution of the measurements. Extractive gas sampling systems are unable to measure large gradients in composition. These gradients occur near the reaction zone. The residence time of the gases in the sampling system is much larger than the time scales characteristic of turbulent flames. Also, the gas chromatographs used in many of the studies commonly have a several minute delay time for the sample analysis. Therefore, only time-averaged measurements can be made.

6.6   OTHER MEASUREMENTS

A variety of other techniques have also been used in combustion studies. This chapter section is merely intended to give a sampling of those techniques and is in no way exhaustive or comprehensive, but is merely intended to be representative. Calcote and King (1955)[137] used a Langmuir[138] probe to study ionization in flames, which is important for calculating thermochemical heat release (see Chapter 4). Anderson and Stresino (1963)[26] determined the flame diameter prior to impingement by an unspecified technique. Dunham (1963)[139] attempted to use Schleiren photography (Reid, 1954)[140] for visualization of a flame produced by a porous disk burner. Beér and Chigier (1968)[32] measured soot concentration by an unspecified technique. Milson and Chigier (1973)[51] photographed and filmed fuel-rich flames. The photographs showed distinct circulation eddy structures. The films showed the pulsating nature of the flow. Nawaz (1973)[58] used photographs, projected on a screen, to estimate the flame diameters. Purvis (1974)[134] used spark Schlieren photography. A capacitance

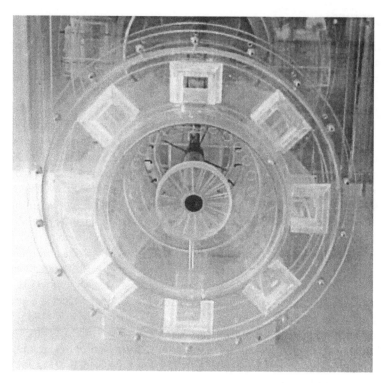

FIGURE 6.11   Plexiglas model of a combustor. (Courtesy of John Zink Co. LLC, Tulsa, OK.)

discharge lamp produced a 0.2 μsec illumination time. Average flame lengths and widths were estimated. Veldman et al. (1975)[52] used shadowgraph visualization (Reid, 1954).[140] Rajani et al. (1978)[25] used a Hubbard probe to determine flame boundaries. This probe is a combination of the two Prandtl probes discussed above. Two static pressure sampling points are located on opposite sides of the probe tip. The tip is aligned perpendicular to the flow direction, with the holes parallel to the burner axis. The stagnation points in the flow occur when there is zero pressure difference between the two holes. This separates the forward and reverse flows. Popiel et al. (1980)[36] used the smoke wire technique to visualize large-scale eddies at about R = 1. These eddies were thought to be the reason why the peak heat flux was not at the stagnation point. You (1985)[44] photographed flames in a darkened room, to determine the flame shape. Horsley et al. (1982)[64] used Schlieren photography to study turbulence in industrial burners. Hargrave et al. (1987)[107] also used Schlieren photography, with an argon spark-jet light source. It gave an illumination period of 0.2 μsec.

Snyder et al. (1989) used laser tomography to measure soot volume fractions in flames that are important in the analysis of luminous radiation from sooting flames.[141] Turns et al. (1989) used a light scattering technique to visualize soot in turbulent diffusion flames.[142] Using a light scattering technique, Hsu et al. (1990) measured the size and fume density of particles containing sodium and potassium that evolve from Kraft recovery furnaces in the combustion of black liquor produced in the production of paper products.[143] Urban and Dryer (1990) used a light-scattering technique to measure the coke particle size distribution produced by the combustion of residual oil in an industrial boiler.[144] Köylü and Faeth (1993) determined the structure of soot produced in buoyant laminar diffusion flames using a combination of thermophoretic sampling and transmission electron microscopy (TEM).[145] Manickavasagam and Mengüç (1993) discussed transmission and scattering techniques using a CO_2-laser nephelometer to measure radiative properties of coal/char particles on premixed ethylene-coal flames.[146] Copin et al. (1998) used helium tracing to determine the residence time in an industrial furnace.[147]

6.7 PHYSICAL MODELING

Before the advent of computers and computer modeling, a common method used to simulate a process was physical modeling. This usually involves the construction of a small-scale model for use in studying some aspect of a combustor. One common technique has been to build a scale replica of a combustor or burner using a see-through material like plexiglas. An example of such a physical model is shown in Figure 6.11. Rather than having an actual flame, a substitute fluid (e.g., ambient-temperature gases [usually air] or water), would be used to study the fluid flow in the system. Neutral-buoyancy tracers would often be added to help in visualizing the flow through the system. Methods were devised to study pseudo heat transfer and chemical reactions using fluids that would either react with each other to produce a distinguishable product or fluids that would react with some type of treatment on the walls of the model to produce a reaction. Because all of these were done at ambient temperatures, they were very difficult to extrapolate to combustion conditions. However, they were useful for studying the fluid flow characteristics of the system.

These small-scale models were often time-consuming and expensive to build. They required the skill of a trained craftsman. They could only be used to reasonably study a few variables of importance in combustion processes, usually fluid flow and sometimes heat transfer and chemical reactions. However, they were not capable of properly simulating, for example, the radiation heat transfer in a high-temperature combustor. Another important limitation is that there may be significant problems scaling-up the results to actual combustor conditions. Significant changes to the model can also be costly, which can make parameter studies expensive. Without some type of scaling correlations based on field data from full-scale systems, it is often difficult to convince end users to accept the results from these small-scale physical models. As a result of these disadvantages and the ready availability of computational fluid dynamic modeling, these small-scale physical models are not often used today.

Another type of physical model sometimes used, is a large-scale mock-up of a combustor and burner. An example is shown in Figure 6.12. These physical models can be full scale, but still do not have actual combustion. They can be used as wind tunnel simulators to study the fluid flow through a burner and a combustor. Measurements in these simulators are useful for calculating things like the pressure drop through a burner and for generating validation data for cold flow CFD models. This type of model is often much less expensive to build than an actual furnace because no insulation, water-cooling, complicated flow controls, gas analysis equipment, or associated safety equipment are required. However, except for the scale-up concerns, these models have similar limitations to the small-scale models discussed above. As a result, they are usually only used in limited circumstances. The latest visualization techniques for CFD modeling now make it possible to create virtual combustors where — like the latest video games — one can travel through the combustion space and view any parameter of interest, including temperature, velocity, heat flux, or species concentration. Although this capability is only in its infancy, as it becomes more affordable for the average combustion design engineer, there will be even less use for the physical models.

REFERENCES

1. C.I. Okoh and R.A. Brown, *Combustion Experimentation Handbook,* Gas Research Institute Report GRI-88/0143, Chicago, IL, 1988.
1a. J.M. Beér and N.A. Chigier, *Combustion Aerodynamics,* Applied Science Publishers, London, 1972.
2. D.F.G. Durão, M.V. Heitor, J.H. Whitelaw, and P.O. Witze, *Combusting Flow Diagnostics,* Kluwer Academic, Dordrecht, The Netherlands, 1992.
3. A.M.K.P. Taylor, *Instrumentation for Flows with Combustion,* Academic Press, London, 1993.
3a. R.M. Fristrom, *Flame Structure and Processes,* Oxford University Press, New York, 1995.
4. A.C. Eckbreth, *Laser Diagnostics for Combustion Temperature and Species,* Abacus Press, Cambridge, MA, 1988.

FIGURE 6.12 Wind tunnel simulator for studying fluid flow in a combustor. (Courtesy of John Zink Co. LLC, Tulsa, OK.)

5. N. Chigier, *Combustion Measurements,* Hemisphere, New York, 1991.
6. R.M. Fristrom, Probe measurements in laminar combustion systems, in *Combustion Measurements — Modern Techniques and Instrumentation*, R. Goulard, Ed., Academic Press, New York, 1976.
7. C.T. Bowman, Probe measurements in flames, *Prog. Astr. Aero.*, 53, 1-24, 1977.
8. F.C. Gouldin, Probe Measurements in multi-dimensional reacting flows, in *Testing and Measurement Techniques in Heat Transfer and Combustion*, AGARD CP-281, Paper #4, 1980.
9. H.A. Becker, Physical probes, in *Instrumentation for Flows with Combustion*, A.M.K.P. Taylor, Ed., Academic Press, London, 1993, 53-112.
10. J. Newbold, M.C. McQuay, B.W. Webb, and A.M. Huber, The experimental characterization of the combustion process in an industrial, gas-fired, flat-glass furnace, *29th Int. ISATA Conf. — Automotive Technology & Automation,* Florence, Italy, June, Automotive Assoc. Ltd., 2, 967-976, 1996.
11. P.R. Solomon, P.E. Best, R.M. Carangelo, J.R. Markham, P.-L. Chien, R.J. Santoro, and H.G. Semerjian, FT-IR emission/transmission spectroscopy for *in situ* combustion diagnostics, *Twenty-First Symposium (International) on Combustion*, The Combustion Institute, Pittsburgh, PA, 1986, 1763-1771.
12. A.C. Eckbreth, Recent advances in laser diagnostics for temperature and species concentrations in combustion, *Eighteenth Symposium (International) on Combustion*, The Combustion Institute, Pittsburgh, PA, 1980, 1471-1488.
13. S.S. Penner, C.P. Wang, and M.Y. Bahadori, Laser diagnostics applied to combustion systems, *Twentieth Symposium (International) on Combustion*, The Combustion Institute, Pittsburgh, PA, 1984, 1149-1176.
14. R.K. Hanson, Combustion diagnostics: planar imaging techniques, *Twenty-First Symposium (International) on Combustion*, The Combustion Institute, Pittsburgh, PA, 1986, 1677-1691.
15. N.R. Fornaciari, R.W. Schefer, P.M. Walsh, and L.E. Claytor, Application of laser-based diagnostics to industrial scale burners, *Proc. of 1995 International Gas Research Conf.*, D.A. Dolenc, Ed., Govt. Institutes, Rockville, MD, 1996, 2398-2405.
16. M. Perrin, J. Imbach, S. Albert, J. Mariasine, and A. Quinquenneau, Application of advanced instantaneous in-flame measurements techniques in an industrial flame with preheated air, *Proc. of 1995 International Gas Research Conf.*, D.A. Dolenc, Ed., Govt. Institutes, Rockville, MD, 1996, 2406-2415.

17. J.K. Kilham, Energy transfer from flame gases to solids, *Third Symposium on Combustion and Flame and Explosion Phenomena*, The Williams and Wilkins Co., Baltimore, MD, 1949, 733-740.

18. E.G. Jackson and J.K. Kilham, Heat transfer from combustion products by forced convection, *Ind. Eng. Chem.*, 48(11), 2077-2079, 1956.

19. R.M. Davies, Heat Transfer measurements on electrically-boosted flames. *Tenth Symposium (International) on Combustion*, The Combustion Institute, Pittsburgh, PA, 1965, 755-766.

20. I. Fells and J.H. Harker, An investigation into heat transfer from unseeded propane-air flames augmented with D.C. electrical power. *Comb. Flame*, 12, 587-596, 1968.

21. J.-P. Vizioz and T.M. Lowes, Convective Heat Transfer from Impinging Flame Jets, Int. Flame Research Found. Report F 35/a/6, IJmuiden, The Netherlands, 1971.

22. D.R. Davies, Heat Transfer from Working Flame Burners, B.S. thesis, Univ. of Salford, Salford, U.K., 1979.

23. C.J. Posillico, Raman Spectroscopic and LDV Measurements of a Methane Jet Impinging Normally on a Flat Water-Cooled Boundary, Ph.D. thesis, Polytechnic Institute of New York, New York, 1986.

24. R.B. Smith and T.M. Lowes, Convective heat transfer from impinging tunnel burner flames — A short report on the NG-4 trials, Int. Flame Research Found. Report F 35/a/9, IJmuiden, The Netherlands, 1974.

25. J.B. Rajani, R. Payne, and S. Michelfelder, Convective heat transfer from impinging oxygen-natural gas flames — Experimental results from the NG5 Trials, Int. Flame Research Found. Report F 35/a/12, IJmuiden, The Netherlands, 1978.

26. J.E. Anderson and E.F. Stresino, Heat transfer from flames impinging on flat and cylindrical surfaces, *J. Heat Trans.*, 85(1), 49-54, 1963.

27. T.B. Reed, Heat-transfer intensity from induction plasma flames and oxy-hydrogen flames, *J. Appl. Phys.*, 34(8), 2266-2269, 1963.

28. R.H. Fay, Heat transfer from fuel gas flames, *Welding J. (Research Supplement)*, 380s-383s, 1967.

29. R.A. Cookson and J.K. Kilham, Energy transfer from hydrogen-air flames, *Ninth Symposium (International) on Combustion*, Academic Press, New York, 1963, 257-263.

30. J.K. Kilham and P.G. Dunham, Energy transfer from carbon monoxide flames, *Eleventh Symposium (International) on Combustion*, The Combustion Institute, Pittsburgh, PA, 1967, 899-905.

31. R.A. Cookson, P.G. Dunham, and J.K. Kilham, Stagnation point heat flow meter, *J. Sci. Instrum.*, 42, 260-262, 1965.

32. J.M. Beér and N.A. Chigier, Impinging jet flames, *Comb. Flame*, 12, 575-586, 1968.

33. E. Buhr, G. Haupt, and H. Kremer, Heat transfer from impinging turbulent jet flames to plane surfaces, in *Combustion Institute European Symposium 1973*, F.J. Weinberg, Ed., Academic Press, New York, 1973, 607-612.

34. H. Kremer, E. Buhr, and R. Haupt, Heat transfer from turbulent free-jet flames to plane surfaces, in *Heat Transfer in Flames*, N.H. Afgan and J.M. Beér, Eds., Scripta Book Company, Washington, D.C., 1974, 463-472.

35. C.J. Hoogendoorn, C.O. Popiel, and T.H. van der Meer, Turbulent heat transfer on a plane surface in impingement round premixed flame jets, *Proc. of 6th Int. Heat Trans. Conf.*, Toronto, 4, 107-112, 1978.

36. C.O. Popiel, T.H. van der Meer, and C.J. Hoogendoorn, Convective heat transfer on a plate in an impinging round hot gas jet of low Reynolds number, *Int. J. Heat Mass Trans.*, 23, 1055-1068, 1980.

37. G.K. Hargrave and J.K. Kilham, The effect of turbulence intensity on convective heat transfer from premixed methane-air flames, *Inst. Chem. Eng. Symp. Ser.*, 2(86), 1025-1034, 1984.

38. G.K. Hargrave, M. Fairweather, and J.K. Kilham, Forced convective heat transfer from premixed flames. 2. Impingement heat transfer, *Int. J. Heat Fluid Flow*, 8(2), 132-138, 1987.

39. K. Kataoka, H. Shundoh, and H. Matsuo, Convective heat transfer between a flat plate and a jet of hot gas impinging on it, in *Drying '84*, A.S. Mujumdar, Ed., Hemisphere/Springer-Verlag, New York, 1984, 218-227.

40. J. Cassiano, M.V. Heitor and T.F. Silva, Combustion tests on an industrial glass-melting furnace, *Fuel*, 73(10), 1638-1642, 1994.

41. S.N. Shorin and V.A. Pechurkin, Effectivnost' teploperenosa na poverkhnost' plity ot vysokotemperaturnoi strui produktov sjoraniya razlichnykh gazov (The effectiveness of heat transfer to the surface of a plate from a high-temperature jet of combustion products of various gases), *Teoriya i Praktika Szhiganiya Gaza*, 4, 134-143, 1968.

42. E.M. Schulte and R.F. Kohl, A transducer for measuring high heat transfer rates, *Rev. Sci. Instrum.*, 41(12), 1732-1740, 1970.

43. E.M. Schulte, Impingement heat transfer rates from torch flames, *J. Heat Trans.*, 94, 231-233, 1972.

44. H.-Z. You, An investigation of fire-plume impingement on a horizontal ceiling: 2. Impingement and ceiling-jet regions, *Fire & Materials*, 9(l), 46-56, 1985.

45. R. Gardon, A transducer for the measurement of heat flow rate, *J. Heat Trans.*, 82, 396-398, 1960.

46. T.H. Van der Meer, Stagnation point heat transfer from turbulent low Reynolds number jets and flame jets, *Exper. Ther. Fluid Sci.*, 4, 115-126, 1991.

47. A. Barnes, Heat Flux Sensors. 2. Applications, *Sensors*, 16(2), 54-57, 1999.

48. B. Gebhart, *Heat Transfer*, McGraw-Hill, New York, 1971.

49. W.H. Giedt, L.L. Cobb, and E.J. Russ, Effect of Hydrogen Recombination on Turbulent Flow Heat Transfer, ASME Paper 60-WA-256, New York, 1960.

50. L.W. Woodruff and W.H. Giedt, Heat transfer measurements from a partially dissociated gas with high Lewis number, *J. Heat Trans.*, 88, 415-420, 1966.

51. A. Milson and N.A. Chigier, Studies of methane-air flames impinging on a cold plate, *Comb. Flame*, 21, 295-305, 1973.

52. C.C. Veldman, T. Kubota and E.E. Zukoski, An Experimental Investigation of the Heat Transfer from a Buoyant Gas Plume to a Horizontal Ceiling. 1. Unobstructed Ceiling, National Bureau of Standards Report NBS-GCR-77-97, Washington, D.C., 1975.

53. M. Matsuo, M. Hattori, T. Ohta, and S. Kishimoto, The Experimental Results of the Heat Transfer by Flame Impingement, Int. Flame Research Found. Report F 29/1a/1, IJmuiden, The Netherlands, 1978.

54. R.M. Rauenzahn, Analysis of Rock Mechanics and Gas Dynamics of Flame-Jet Thermal Spallation Drilling, Ph.D. thesis, MIT, Cambridge, MA, 1986.

55. D.R. Hornbaker and D.L. Rall, Thermal perturbations caused by heat-flux transducers and their effect on the accuracy of heating-rate measurements, *ISA Trans.*, 3, 123-130, 1964.

56. R.C. Bachmann, J.T. Chambers, and W.H. Giedt, Investigation of surface heat-flux measurements with calorimeters, *ISA Trans.*, 4, 143-151, 1965.

57. J.K. Kilham and M.R.I. Purvis, Heat transfer from hydrocarbon-oxygen flames, *Comb. Flame*, 16, 47-54, 1971.

58. S. Nawaz, Heat Transfer from Oxygen Enriched Methane Flames, Ph.D. thesis, The University of Leeds, Leeds, U.K., 1973.

59. M. Fairweather, J.K. Kilham, and S. Nawaz, Stagnation point heat transfer from laminar, high temperature methane flames, *Int. J. Heat Fluid Flow*, 5(1), 21-27, 1984.

60. M. Fairweather, J.K. Kilham, and A. Mohebi-Ashtiani, Stagnation point heat transfer from turbulent methane-air flames. *Comb. Sci. Tech.*, 35, 225-238, 1984.

61. R. Conolly and R.M. Davies, A study of convective heat transfer from flames, *Int. J. Heat Mass Trans.*, 15, 2155-2172, 1972.

62. J.K. Kilham and M.R.I. Purvis, Heat transfer from normally impinging flames, *Comb. Sci. Tech.*, 18, 81-90, 1978.

63. A. Ivernel and P. Vernotte, Etude expérimentale de l'amélioration des transferts convectis dans les fours par suroxygénation du comburant, *Rev. Gén. Therm.*, Fr., 210-211, 375-391, 1979.

64. M.E. Horsley, M.R.I. Purvis, and A.S. Tariq, Convective heat transfer from laminar and turbulent premixed flames, *Heat Transfer 1982*, U. Grigull, E. Hahne, K. Stephan, and J. Straub., Eds., Hemisphere, Washington, D.C., 3, 409-415, 1982.

65. A.O. Hemeson, M.E. Horsley, M.R.I. Purvis, and A.S. Tariq, Heat transfer from flames to convex surfaces, *Inst. of Chem. Eng. Symp. Series*, Publ. by Inst. of Chem. Eng., Rugby, U.K., 2(86), 969-978, 1984.

66. J.E. Hustad, N.A. Røkke, and O.K. Sønju, Heat transfer to pipes submerged in lifted buoyant diffusion flames, in *Experimental Heat Transfer, Fluid Mechanics, and Thermodynamics, 1991*, J.F. Keffer, Ed., Elsevier, New York, 1991, 567-574.

67. J.E. Hustad and O.K. Sønju, Heat transfer to pipes submerged in turbulent jet diffusion flames, in *Heat Transfer in Radiating and Combusting Systems*, Springer-Verlag, Berlin, 1991, 474-490.

68. J.E. Hustad, M. Jacobsen, and O.K. Sønju, Radiation and heat transfer in oil/propane jet diffusion flames, *Inst. Chem. Eng. Symp. Series*, 10(129), 657-663, 1992.

69. W.A. Gray and R. Müller, *Engineering Calculations in Radiative Heat Transfer*, Pergamon, Oxford, U.K., 1974.

70. C.E. Baukal and B. Gebhart, Oxygen-enhanced natural gas flame radiation, *Int. J. Heat Mass Transfer*, 40(11), 2539-2547, 1997.
71. J.P. Gore, S.M. Skinner, D.W. Stroup, D. Madrzykowski, and D.D. Evans, Structure and radiation properties of large two-phase flames, in *Heat Transfer in Combustion Systems*, N. Ashgriz, J.G. Quintiere, H.G. Semerjian, and S.E. Slezak, Eds., Amer. Soc. Mech. Engs., New York, HTD-Vol. 122, 77-86, 1989.
72. D.R. Lide, Ed., *CRC Handbook of Chemistry and Physics*, 79th edition, CRC Press, Boca Raton, FL, 1998.
73. L. Matthews, A. Harris, and G. Garcia, Radiative flux measurements in a sooty pool fire using a multihead transpiration radiometer, in *Heat Transfer Phenomena in Radiation, Combustion, and Fires*, R.K. Shah, Ed., New York, ASME HTD-Vol. 106, 375-380, 1989.
74. R.J. Moffat, B.D. Hunn and J.F. Ayers, Development of a transpiration radiometer, *Instrument Society of America, Conference Proceedings,* Paper No. 613, pp. 613.1-613.7, 1971.
75. J. Chedaille and Y. Braud, *Vol. 1: Measurements in Flames*, Crane, Russak & Co., New York, 1972.
76. B.W. Butler and B.W. Webb, Measurements of local temperature and wall radiant heat flux in an industrial coal-fired boiler, in *Heat Transfer in Combustion Systems — 1990*, B. Farouk, W.L. Grosshandler, D.G. Lilley, and C. Presser, Eds., New York, ASME HTD-Vol. 142, 49-56, 1990.
77. S.A. Beltagui, A.M.A. Kenbar, and N.R.L. Maccallum, Heat transfer and emission studies in a gas fired furnace, in *Boilers & Furnaces, Proc. of Int. Symp. on Combustion & Emissions Control*, Cardiff, U.K., September 1993, Inst. of Energy, 1993, chap. 3, 275-296.
78. B. Ji and C.E. Baukal, Spectral radiation properties of oxygen-enhanced/natural gas flames, *Proc. of 1998 International Gas Research Conference*, 8-11 November 1998, San Diego, CA, Dan Dolenc, Ed., 5, 1998, 422-433.
79. G.M. Faeth, J.P. Gore, S.G. Chuech, and S.-M. Jeng, Radiation from turbulent diffusion flames, *Ann. Rev. of Fluid Mech. & Heat Transfer*, 2, 1-38, 1989.
80. C.B. Ludwig, W. Malkmus, J.E. Reardon, and J.A.L. Thomson, *Handbook of Infrared Radiation from Combustion Gases*, National Aeronautics and Space Administration Report SP-3080, Washington, D.C., 1973.
81. C.E. Baukal and B. Gebhart, Surface Condition Effects on Flame Impingement Heat Transfer, *Therm. Fluid Sci.*, 15, 323-335, 1997.
82. Y.R. Sivathanu, J.P. Gore, and J. Dolinar, Transient scalar properties of strongly radiating jet flames, in *Heat and Mass Transfer in Fires and Combustion Systems*, W.L. Grosshandler and H.G. Semerjian, Eds., ASME, New York, ASME HTD-Vol. 148, 45-56, 1990.
83. M.V. Heitor and A.L.N. Moreira, Probe measurements of scalar properties in reacting flows, in *Combusting Flow Diagnostics*, D.F.G. Durão, M.V. Heitor, J.H. Whitelaw, and P.O. Witze, Eds., Kluwer Academic Publishers, the Netherlands, 1992, 79-136.
84. Y. Goldman, Gas temperature measurement in combustors by use of suction pyrometry, in *Heat Transfer in Furnaces*, C. Presser and D.G. Lilley, Eds., ASME, New York, ASME HTD-Vol. 74, 19-22, 1987.
85. J. Grey, Thermodynamic methods of high temperature measurement, *ISA Trans.*, 4, 102-115, 1965.
86. N.M. Laurendeau, Temperature measurements by light-scattering methods, in *Developments in Experimental Techniques in Heat Transfer and Combustion*, R.O. Warrington, M.M. Chen, J.D. Felske, and W.L. Grosshandler, Eds., ASME, New York, ASME HTD-Vol. 71, 45-65, 1987.
87. C.M. Chao and R. Goulard, Nonlinear inversion techniques in flame temperature measurements, in *Heat Transfer in Flames*, N.H. Afgan and J.M. Beer, Eds., Scripta Book Company, Washington, D.C., 1974, chap. 20, 295-337.
88. N.N. Kondic, Temperature field measurement in flames by external means, in *Heat Transfer in Flames*, N.H. Afgan and J.M. Beer, Eds., Scripta Book Company, Washington, D.C., 1974, chap. 22, 353-363.
89. D. Hofmann and A. Leipertz, Temperature field measurements in a sooting flame by filtered Rayleigh scattering (FRS), *Twenty-Sixth Symposium (International) on Combustion*, The Combustion Institute, Pittsburgh, PA, 1996, 945-950.
90. P.M.J. Hughes, R.J. Lacelle, and T. Parameswaran, A comparison of suction pyrometer and CARS derived temperatures in an industrial scale flame, *Comb. Sci. Tech.*, 105, 131-145, 1995.
91. M.D. Rumminger, R.W. Dibble, N.H. Heberle, and D.R. Crosley, Gas temperature above a porous radiant burner: comparison of measurements and model predictions, *Twenty-Sixth Symposium (International) on Combustion*, The Combustion Institute, Pittsburgh, PA, 1996, 1755-1762.

92. J.-M. Char and J.-H. Yeh, The measurement of open propane flame temperature using infrared technique, *J. Quant. Spectrosc. Radiat. Trans.*, 56(1), 133-144, 1996.
93. C.E. Baukal, Heat Transfer from Flame Impingement Normal to a Plane Surface, Ph.D. thesis, Univ. of Pennsylvania, 1996.
94. R.C. Peterson, Kinetics of Hydrogen-Oxygen-Argon and Hydrogen-Oxygen-Argon-Pyridine Combustion Using a Flat Flame Burner, Ph.D. thesis, Purdue University, 1981.
95. E.M. Candler, Thermocouple Reference Tables, Air Force Flight Dynamics Laboratory Report no. AFFDL-TR-66-178, Wright-Patterson Air Force Base, Ohio, 1967.
96. G.F. Blackburn and F.R. Caldwell, Reference tables for thermocouples of iridium-rhodium alloys versus iridium, *J. Research of NBS — C. Engineering and Instrumentation*, 68C(1), 41-59, 1964.
97. R.A. Cookson, P.G. Dunham, and J.K. Kilham, Non-catalytic coatings for thermocouples, *Comb. Flame*, 8, 168-170, 1964.
98. J.H. Kent, A Noncatalytic Coating for Platinum-Rhodium Thermocouples, *Comb. Flame*, 14, 279-282, 1970.
99. R.J. Moffat, Temperature and heat transfer measurements, in *The CRC Handbook of Mechanical Engineering*, F. Kreith, Ed., CRC Press, Boca Raton, FL, 1998, 4-182–4-205.
100. A.N. Hayhurst and D.B. Kittleson, Heat and mass transfer considerations in the use of electrically heated thermocouples of iridium versus an iridium/rhodium alloy in atmospheric pressure flames, *Comb. Flame*, 28, 301-317, 1977.
101. D. Bradley and K.J. Matthews, Measurements of high gas temperatures with fine wire thermocouples, *J. Mech. Eng. Sci.*, 10(4), 299-305, 1968.
102. R.M. Fristrom and A.A. Westenberg, *Flame Structure*, McGraw-Hill, New York, 1965.
103. E.M. Sparrow and R.D. Cess, *Radiation Heat Transfer*, augmented edition, McGraw-Hill, New York, 1978.
104. R.C. Peterson and N.M. Laurendeau, The emittance of yttrium-beryllium oxide thermocouple coating, *Comb. Flame*, 60, 279-284, 1985.
105. J.O. Hinze, *Turbulence*, McGraw-Hill, New York, 1959.
106. S.F. Son, M. Queiroz, and C.G. Wood, Compensation of Thermocouples for Thermal Inertia Effects Using a Digital Deconvolution, in *Heat Transfer Phenomena in Radiation, Combustion, and Fires*, R.K. Shah, Ed., ASME, New York, ASME HTD-Vol. 106, 515-522, 1989.
107. G.K. Hargrave, M. Fairweather and J.K. Kilham, Forced convective heat transfer from premixed flames. 1. Flame structure, *Int. J. Heat Fluid Flow*, 8(1), 55-63, 1987.
108. E.L. Nichols, On the temperature of the acetylene flame, *Phys. Rev.*, 10, 234-252, 1900.
109. M. Costa, M. Mourão, J. Baltasar, and M.G. Carvalho, Combustion measurements in an industrial glass-melting furnace, *J. Inst. Energy*, 69(479), 80-86, 1996.
110. R. Mital, J.P. Gore, R. Viskanta, and S. Singh, Radiation efficiency and structure of flames stabilized inside radiant porous ceramic burners, in *Combustion and Fire*, M.C. McQuay, W. Schreiber, E. Bigzadeh, K. Annamalai, D. Choudhury, and A. Runchal, Eds., *ASME Proceedings of the 31st National Heat Transfer Conf.*, Vol. 6, ASME, New York, ASME HTD-Vol. 328, 131-137, 1996.
111. W.E. Kaskan, The Dependence of Flame Temperatures on Mass Burning Velocity, *Sixth Symposium (International) on Combustion*, Reinbold, New York, 1957, 134-143.
112. J.M. Madson and E.A. Theby, SiO_2 coated thermocouples, *Comb. Sci. Tech.*, 36, 205-209, 1984.
113. D.D. Pollock, Thermocouples in high-temperature reactive atmospheres, *Comb. Sci. Tech.*, 42, 111-113, 1984.
114. C. Féry, Sur la température des flammes, *Compt. Rend. Acad. Sci.*, No. 22, 909-912, 1903.
115. H.M. Strong, F.P. Bundy, and D.A. Larson, Temperature measurement on complex flames by sodium line reversal and sodium D line intensity contour studies, *Third Symposium (International) on Combustion*, The Williams & Wilkins Co., Baltimore, 1949, 641-647.
116. A.M. Young, Infrared temperature measurement essentials: performance characteristics, calibration and testing, *Industrial Heating*, Vol. LXV, No. 8, 45-60, 1998.
117. D.P. Dewitt and G.D. Nutter, Eds., *Theory & Practice of Radiation Thermometry*, John Wiley & Sons, New York, 1988.
118. D. K. Patrick, Infrared thermal image recording and analysis for thermal processes, *Industrial Heating*, LXV(1), 33-35, 1998.
119. V. Lappe, Installation and maintenance of infrared thermometers, *Industrial Heating*, Vol. LXIV, No. 11, 45-49, 1997.

120. B. Lewis, R.N. Pease, and H.S. Taylor, *Physical Measurements in Gas Dynamics and Combustion*, Princeton Univ. Press, Princeton, NJ, 1956.

121. C. Swan, Pitot tube, in *International Encyclopedia of Heat & Mass Transfer*, G.F. Hewitt, G.L. Shires, and Y.V. Polezhaev, Eds., CRC Press, Boca Raton, FL, 1997, 842.

122. H.A. Becker and A.P.G. Brown, Response of Pitot probes in turbulent streams, *J. Fluid Mech.*, 62, 85-114, 1974.

123. S.H. Cho and H.A. Becker, Response of static pressure probes in turbulent streams, *Exp. Fluids*, 3, 93-102, 1985.

124. S.H. Chue, Pressure probes for fluid measurement, *Prog. Aero. Sci.*, 16(2), 147-223, 1975.

125. M. Gad-el-Hak, Basic Instruments, in *The Handbook of Fluid Dynamics*, R.W. Johnson, Ed., CRC Press, Boca Raton, FL, 1998.

126. R.J. Baker, P.J. Bourke, and J.H. Whitelaw, Application of laser anemometry to the measurement of flow properties in industrial burner flames, *Fourteenth Symposium (International) on Combustion*, The Combustion Institute, Pittsburgh, PA, 1972, 699-706.

127. N. Chigier, Velocity measurements in inhomogeneous combustion systems, *Comb. Flame*, 78, 129-151, 1989.

128. J.-M. Most, Laser Doppler velocimetry in industrial flames, in *Laser Anemometry: Advances and Applications — 1991*, A. Dybbs and B. Ghorashi, Eds., ASME, New York, 1991, 139-146.

129. M.V. Heitor and A.L.N. Moreira, Velocity characteristics of a swirling recirculating flow, *Exper. Therm. Fluid Sci.*, 5, 369-380, 1993.

130. L.M. Fingerson and R.K. Menon, Laser doppler velocimetry, in *The Handbook of Fluid Dynamics*, R.W. Johnson, Ed., CRC Press, Boca Raton, FL, 1998, chap. 35.

131. T.H. van der Meer, Heat Transfer from Impinging Flame Jets, Ph.D. thesis, Technical University of Delft, The Netherlands, 1987.

132. R. Conolly, A Study of Convective Heat Transfer from High Temperature Combustion Products. Ph.D. thesis, University of Aston, Birmingham, U.K., 1971.

133. M. Sibulkin, Heat transfer near the forward stagnation point of a body of revolution, *J. Aero. Sci.*, 19, 570-571, 1952.

134. M.R.I. Purvis, Heat Transfer from Normally Impinging Hydrocarbon Oxygen Flames to Surfaces at Elevated Temperatures, Ph.D. thesis, The University of Leeds, Leeds, U.K., 1974.

135. A.S. Tariq, Impingement Heat Transfer From Turbulent and Laminar Flames, Ph.D. thesis, Portsmouth Polytechnic, Hampshire, U.K., 1982.

136. E. Buhr, Über den Wärmeflub in Staupunkten von turbulenter Freistrahl-Flammen an gekühlten Platten, Ph.D. thesis, University of Trier-Kaiserslautern, Kaiserslautern, Germany, 1969.

137. H.F. Calcote and I.R. King, Studies of ionization in flames by means of Langmuir probes, *Fifth Symposium (International) on Combustion*, Reinhold Publishing, New York, 1955, 423-434.

138. I. Languir and H.M. Mott-Smith, *Gen. Elec. Rev.*, 26, 2726-, 1923; 27, 449, 583, 616, 726, 810, 1924.

139. P.G. Dunham, Convective Heat Transfer from Carbon Monoxide Flames, Ph.D. thesis, The University of Leeds, Leeds, U.K., 1963.

140. W.T. Reid, Flame photography, in *Physical Measurements in Gas Dynamics and Combustion*, R.W. Ladenburg, B. Lewis, R.N. Pease, and H.S. Taylor, Eds., Princeton University Press, Princeton, NJ, 1954.

141. R.E. Snyder, R.G. Joklik, and H.G. Semerjian, Laser tomographic measurements in an unsteady jet-diffusion flame, in *Heat Transfer in Combustion Systems*, N. Ashgriz, J.G. Quintiere, H.G. Semerjian, and S.E. Slezak, Eds., ASME, New York, HTD-Vol. 122, 1-7, 1989.

142. S.R. Turns, J.A. Lovett, and H.J. Sommer, Visualization of soot zones in turbulent diffusion flames, *Combust. Flame*, 77, 405-409, 1989.

143. J.C.L. Hsu, C. Presser and D.T. Clay, *In-situ* fume particle size and density measurements from a synthetic smelt, in *Heat Transfer in Combustion Systems — 1990*, B. Farouk, W.L. Grosshandler, D.G. Lilley, and C. Presser, Eds., ASME, New York, ASME HTD-Vol. 142, 67-75, 1990.

144. D.L. Urban and F.L. Dryer, Formation of coke particulate in the combustion of residual oil: particulate size and mass prediction from initial droplet parameters, in *Heat Transfer in Combustion Systems — 1990*, B. Farouk, W.L. Grosshandler, D.G. Lilley, and C. Presser, Eds., ASME, New York, ASME HTD-Vol. 142, 83-88, 1990.

145. Ü.Ö. Köylü and G.M. Faeth, Optical properties of soot in buoyant laminar diffusion flames, in *Heat Transfer in Fire and Combustion Systems — 1993*, B. Farouk, M.P. Menguc, R. Viskanta, C. Presser, and S. Chellaiah, Eds., ASME, New York, ASME HTD-Vol. 250, 127-136, 1993.
146. S. Manickavasagam and M.P. Mengüç, Optical properties of soot in buoyant laminar diffusion flames, in *Heat Transfer in Fire and Combustion Systems — 1993*, B. Farouk, M.P. Menguc, R. Viskanta, C. Presser, and S. Chellaiah, Eds., ASME, New York, ASME HTD-Vol. 250, 127-136, 1993.
147. C. Copin, S. Ressent, and A. Bahmed, Measuring residence time distribution (RTD) of combustion products in an industrial furnace by helium tracing, *Proc. of the 1998 International Gas Research Conf.*, Vol. V: Industrial Utilization, D.A. Dolenc, Ed., Gas Research Institute, Chicago, 1998, 450-461.

7 Flame Impingement

7.1 INTRODUCTION

Impinging flame jets have been extensively studied because of their importance in a wide range of applications. Figure 7.1 shows a flame impinging normal to a water-cooled plate.[1] Early experiments were used to simulate the extremely high heat fluxes encountered by space vehicles reentering the Earth's atmosphere. These heat levels are caused by the hypersonic flow impact velocities that ionize the highly shocked atmospheric gases.[2] Subsequent studies have investigated the heat fluxes attainable using high-intensity combustion, with pure oxygen instead of air, to increase metal heating and melting rates.[3] High-intensity impinging jet flames have been used in recent years to produce synthetic diamond coatings by chemical vapor deposition.[4] Supersonic-velocity, high-intensity flames have been used in a process known as thermal spallation. In this process, the impinging jet bores through rock by causing it to fragment, due to the large thermal stresses arising from the high heat fluxes on a cold surface.[5] This may be more rapid and economical than traditional mechanical rock drilling, depending on the rock type. High-velocity flames impinging on structural elements have been used to simulate large-scale fires caused by ruptured piping in the chemical process industry.[6] Low-intensity impinging flame jets have been used in safety research to quantify the heating rate caused by buoyant fires impinging on walls and ceilings.[7] Eibeck et al. (1993) studied the impact of pulse impinging flames on the heat transfer to a target.[8]

Most previous research has concerned air/fuel combustion. In these lower intensity flames, the predominant mechanism is forced convection. Much work has also concerned fuels combusted with pure oxygen. These high-intensity flames produce significant amounts of dissociated species (e.g., H, O, OH, etc.) and uncombusted fuel (e.g., CO, H_2, etc.). These reactive gases then impact on a relatively low-temperature target surface. As these species cool, they exothermically combine into products such as CO_2 and H_2O, which are more thermodynamically stable at lower temperatures. This chemical heat release is sometimes referred to as "convection vivre," or live convection[9] and is discussed in more detail in Chapter 4. This mechanism may be comparable in magnitude to the forced convection heat transfer at the surface.[10] Such high-intensity flames require much more complicated analysis. Several experiments have used fuels combusted with an oxidizer having an oxygen content between that of air and pure oxygen. This process is sometimes referred to as oxygen-enriched air combustion (see section 12.2). These medium-intensity flames have received much less attention. They are currently not used in industrial applications.

Six heat transfer mechanisms have been identified in previous flame impingement studies: convection (forced and natural), conduction (steady-state and transient), radiation (surface, luminous, and nonluminous), thermochemical heat release (equilibrium, catalytic, and mixed), water vapor condensation, and boiling (internal and external).[1] These are shown schematically for a water-cooled target in Figure 7.2. All of the mechanisms are not usually present simultaneously and depend on the specific problem. Each of the mechanisms is discussed in more detail in Chapters 3 and 4. Figure 7.3 shows a schematic of forced and natural convection in flame impingement. In natural convection flame impingement, the flame is often far from the surface so that the buoyant hot combustion products blended with cooler ambient air are impinging on the target. Figure 7.4 shows a schematic of steady-state conduction for a water-cooled target and transient conduction for an uncooled target. In the water-cooled case, both the hot and cold sides of the target are approximately constant at steady-state conditions. In the uncooled target, there is a nonlinear temperature gradient through the target during transient heating. Figure 7.5 depicts radiation heat

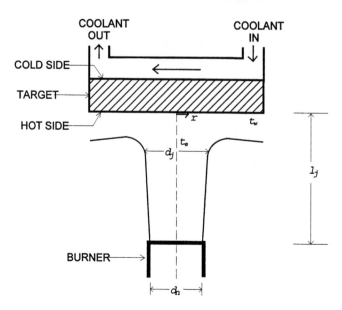

FIGURE 7.1 Flame impingement normal to a cooled target. (From C.E. Baukal, L.K. Farmer, B. Gebhart, and I. Chan, *1995 Int. Gas Res. Conf.,* D.A. Dolenc, Ed., Govt. Institutes, Rockville, MD, II, 2277-2287, 1996.)

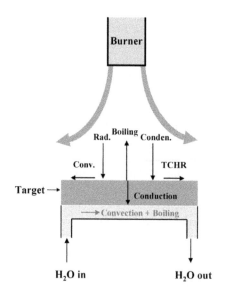

FIGURE 7.2 Heat transfer mechanisms in flame impingement on a water-cooled target.

transfer in flame impingement for luminous flames, nonluminous flames and surface radiation from hot furnace walls. The luminous flame contains particles that glow and radiate heat to the target. The nonluminous flame contains no radiating particles, so that only gaseous radiation from CO_2 and H_2O are commonly present. Figure 7.6 shows TCHR for catalytic, equilibrium, and mixed mechanisms for CO and CO_2 as an example. Similar reactions occur for H_2 and H_2O, as well as other dissociated species, as discussed in Chapter 4. Equilibrium TCHR occurs in the gas phase between the burner and the target, outside the boundary layer, which has been greatly exaggerated. CO combines with radicals to produce CO_2. Catalytic TCHR occurs when the gases contact the surface, which catalytically promotes the reaction of CO to CO_2 in the presence of radical species. Mixed

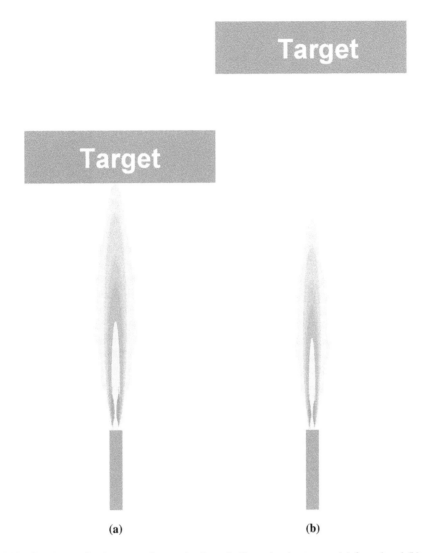

(a)   (b)

FIGURE 7.3 Convective heat transfer mechanisms in flame impingement: (a) forced and (b) natural.

TCHR is a combination of equilibrium and catalytic TCHR. Internal and external boiling are schematically shown in Figure 7.7. Internal boiling occurs on the cold side of a water-cooled target where the coolant is heated above its boiling point by contacting the hot target. External boiling can occur if an initially cold target that has condensed water, from the products of combustion of the flame, on the target surface then heats above the boiling point of water. Water vapor condensation is shown in Figure 7.8. This can occur when the water vapor in the products of combustion from flame condenses onto a target whose temperature is below the dewpoint of the water vapor.

7.2 EXPERIMENTAL CONDITIONS

Many important parameters arise in flame jet impingement processes.[11] The first and most important aspect is the overall geometric configuration. This includes the target shape and its orientation relative to the burner. These operating conditions strongly influence the heat transfer intensity. They also determine which mechanisms will be most important. These conditions include the oxidizer composition, the fuel composition, the equivalence ratio, and the Reynolds number at the nozzle

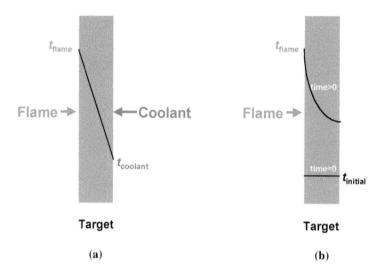

FIGURE 7.4 Conduction heat transfer mechanisms in flame impingement: (a) steady-state and (b) transient.

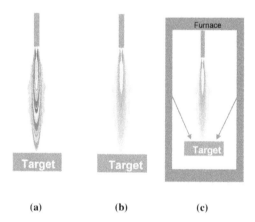

FIGURE 7.5 Radiation heat transfer mechanisms in flame impingement: (a) luminous flame, (b) nonluminous flame, and (c) surface radiation.

exit. Other factors commonly have secondary influences on the heat transfer processes. These include the burner design and the position relative to the target. The characteristics of the target also influence the heat transfer. These include the dimensions, material composition, surface treatments or coatings, and the surface temperature. The above parameters are tabulated and discussed below.

7.2.1 CONFIGURATIONS

The configuration specifies the relative orientation of the target surface and the burner. This is usually the most important consideration for the designer or researcher when reviewing previous experimental results. The four most common geometric configurations in flame jet experiments have been flames impinging (1) normal to a cylinder in crossflow, (2) normal to a hemi-nosed cylinder, (3) normal to a plane surface, and (4) parallel to a plane surface. Other configurations have been tested, including flames at an angle to a plane surface and flames normal to and around the circumference of a large cylindrical furnace.[12] These are not common geometries and little information is available on them, so they are not considered in detail here.

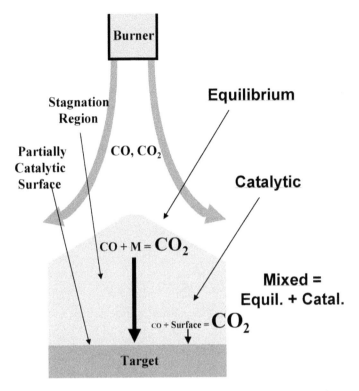

FIGURE 7.6   Thermochemical heat release mechanisms in flame impingement.

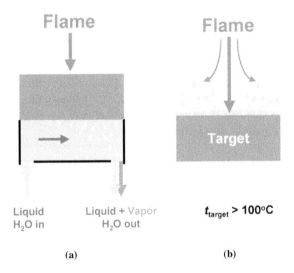

FIGURE 7.7   Boiling heat transfer mechanisms in flame impingement: (a) internal and (b) external.

7.2.1.1   Flame Normal to a Cylinder in Crossflow

In this configuration, shown in Figure 7.9, the cylinder axis is perpendicular to the burner axis. This configuration has been widely studied, as shown in Table 7.1. It applies to industrial processes such as heating round metal billets, and in fires impinging on pipes in chemical plants. Some of the earliest studies investigated impingement on refractory cylinders.[13,14] A pipe was the most

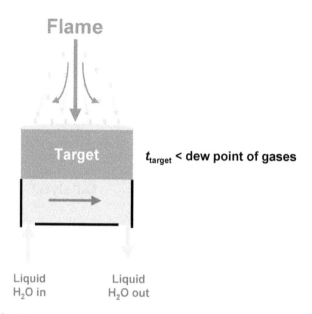

FIGURE 7.8   Water vapor condensation in flame impingement on a water-cooled target.

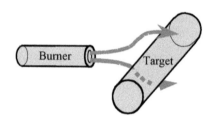

FIGURE 7.9   Flame impinging normal to a cylinder in crossflow.

common target for this geometry (e.g., see Reference 15). In some studies, the local heat flux was measured at the forward stagnation point of the pipe.[6,16,17] In many studies, the pipe was cooled with water circulating through it.[18] The average heat flux over the entire surface was calculated from the sensible energy gain of the cooling water.

7.2.1.2   Flame Normal to a Hemispherically Nosed Cylinder

For this geometry, shown in Figure 7.10, the cylinder axis is parallel to the burner axis. The flame impinges on the end of the cylinder, which is hemispherical. These tests have been very important in aerospace applications. This relatively uncommon geometry (see Table 7.2) for industrial application is important because the heat transfer results can be directly compared to some of the analytical solutions (e.g., see Reference 19) derived for aerospace applications, such as rockets and missiles. In this configuration, the heat flux has been measured only at the forward stagnation point. Most such studies have concerned laminar flames.

7.2.1.3   Flame Normal to a Plane Surface

This configuration, shown in Figure 7.11, has received the most attention (see Table 7.3) since it has been widely used in many industrial processes. It has also been used for the widest range of operating and target surface conditions. Shorin and Pechurkin (1968)[20] and Kremer et al. (1974)[21] also investigated the effect of flame impingement at an angle between parallel and normal, onto

TABLE 7.1

Experimental Conditions for Flame Impingement Normal to Cylindrical Surfaces

Oxidizer	Fuel	φ	q_f (kW)	Burner	d_n (mm)	Re_n	L	R	Ref.
Air	CH_4	1.25–1.67	0.71–1.2	Partial premix	12.7–34.9	Laminar[b]	1.1–4.5	0–0.7	Anderson & Stresino, 1963[15]
		0.8–1.2	1.2–10[c]	Tube	16	2000–8000	3.1–11	0	Hargrave et al., 1987[75]
		∞	37–150	Tube	5–40	200–3600	30–200	0	Hustad et al., 1991[16,17]
		∞	60–1600	Tube	5–40	26,000[b]	80	0	Hustad et al., 1992[6]
	C_3H_8	1.25	3.5–9.4	Tunnel	18	Not given	1.7–6.9	0	Fells & Harker, 1968[80]
		∞	34.3–740	Tube	5–40	200–3600	20–200	0	Hustad et al., 1991[16,17]
		∞	60–1600	Tube	5–40	54,000; 330,000[b]	120, 150	0	Hustad et al., 1992[6]
	C_4H_{10}	∞	60–1600	Tube	5–40	66,000[b]	130	0	Hustad et al., 1992[6]
	CO	1.0, 1.19	0.51–0.75	Slot	4 × 50	Laminar[a]	2.5–7.5	0	Kilham, 1949[13]
	H_2	1.0	2.1, 4.3	Annular port w/precombustor	Not given	Laminar[a]	Not given	0	Jackson & Kilham, 1956[14]
	Natural gas	0.6–1.3	0.56–1.7	Premix, slot-shaped torch	Not given	Not given	[c]	0	Davies, 1979[18]
		1.05	Not given	Flame retention head	22.4, 41.0	12,600	0–4.5	0	Hemeson et al., 1984[45]
	Town gas	Lean–rich[a]	7.1–28	Tunnel	13, 26	Not given	0.5	0	Davies, 1965[34]
Air/O_2	Natural gas	0.5–1.3	0.56–1.7	Premix, slot-shaped torch	not given	Not given	[c]	0	Davies, 1979[18]
O_2	C_2H_2	2.34	1.3–4.1	Partial premix	1.2, 1.8	Laminar[b]	4.3–7.5	0–5.4	Anderson & Stresino, 1963[15]
	C_3H_8	1.0, 1.43	77, 3.3	Tunnel, partial premix	9.5, 1.8	Laminar[b]	0, 8.6	0–0.53; 6.1	Anderson & Stresino, 1963[15]
	CO	1.0, 2.0	2.4	Annular port w/precombustor	Not given	Laminar[a]	Not given	0	Jackson & Kilham, 1956[14]
	H_2	1.0	6.1	Annular port w/precombustor	Not given	Laminar[a]	Not given	0	Jackson & Kilham, 1956[14]
		1.0	670	Tunnel	9.5	Laminar[b]	1.3	0–0.67	Anderson & Stresino, 1963[15]
	Natural gas	0.5–1.16	1.7–2.8	Premix, slot-shaped torch	Not given	Not given	[c]	0	Davies, 1979[18]
	Town gas	1.0	12–16	Tunnel	13	Not given	0.5	0	Davies, 1965[34]

[a] According to author (only qualitative).

[b] Calculated from data given in reference.

[c] Target located at point of maximum heat release.

Source: From C.E. Baukal and B. Gebhart, *Comb. Sci. Tech.*, 104, 339–357, 1995.

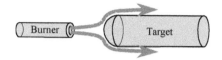

FIGURE 7.10   Flame impinging normal to a hemi-nosed cylinder.

plane surfaces. Only the results for flame impingement normal to a plane surface are included here. The reader is referred to the literature for further information on angled jets.

7.2.1.4   Flame Parallel to a Plane Surface

This configuration, shown in Figure 7.12, has been the least studied (see Table 7.4). However, it is very important for flight applications. It simulates the heat transfer processes on airfoil surfaces. The studies by Giedt and co-workers (1960, 1966)[2,22] investigated flow across the top and bottom of a plane surface. Neither of these studies investigated the heat flux at the leading edge of the target. Beér and Chigier (1968)[23] studied impingement only on one side of a plane surface, which was the hearth of a furnace. The flame was inclined at 20° above the horizontal (see Figure 7.12b). Mohr et al. (1996) studied a special configuration for flames parallel to a plane surface which they called radial jet reattachment flames as shown in Figure 7.13.[24] This type of flame promises to give more uniform heating of the surface, compared to flames impinging normal to a surface. Figure 7.14 shows the effect of the spacing (S) between adjacent nozzles (air inlet pipe inside radius, $R_0 = 12.7$ mm or 0.5 in.) on the maximum heat flux and surface temperatures.

7.2.2   OPERATING CONDITIONS

Operating conditions have been found to strongly influence the heat transfer intensity. The effects include the oxidizer and fuel composition, flame equivalence ratio and firing rate, Reynolds number at the nozzle exit, burner type, nozzle diameter, and location of the target with respect to the burner.

7.2.2.1   Oxidizers

The most important variable, after the physical configuration, is the oxidizer composition. The oxygen mole fraction in the oxidizer, Ω, has a very large influence on heat transfer intensity. Almost all previous studies used either air ($\Omega = 0.21$) or pure oxygen ($\Omega = 1.0$) as the oxidizer. This affects both the flame temperature and the amount of dissociation in the combustion products. As an example, the adiabatic flame temperatures for methane combusted stoichiometrically with air and with pure oxygen are 2220K and 3054K (3537°F and 5038°F), respectively. Figures 2.4 and 2.5 indicate that the products of combustion for a stoichiometric air/CH_4 adiabatic flame contain essentially no unreacted fuel or dissociated species, except at very high flame temperatures approaching adiabatic conditions. However, the combustion products of an O_2/CH_4 flame contain nearly 23 vol.% unreacted fuel (CO and H_2) and over 18 vol.% dissociated species (H, O, and OH). As these products cool down in the boundary layer along the target surface, they exothermically release heat. This heat release rate from TCHR (see section 4.1.2) may be greater in magnitude than that of the forced convection heat transfer mechanism.

A few studies have used oxygen concentration levels between those of air and pure oxygen. Beér and Chigier[23] injected a small, unspecified amount of O_2 into an air/coke oven gas flame to raise the temperature slightly. Vizioz and Lowes (1971)[25] studied oxygen-enriched air for $\Omega = 0.30$. In related studies, Nawaz (1973)[26] and Fairweather et al. (1984)[27] studied oxygen-enriched air flames for Ω ranging from 0.46 to 0.61. Davies (1979)[18] studied flame impingement heat transfer to a water-cooled tube in crossflow for $\Omega = 0.23$ to 0.35. Ivernel and Vernotte (1979)[28] studied oxygen-enriched air/natural gas flames, with Ω ranging from 0.25 to 1.0. Kataoka et al. (1984)[29]

TABLE 7.2
Experimental Conditions for Flame Impingement Normal to Hemi-nosed Cylinders

Oxidizer	Fuel	φ	q_f (kW)	Burner	d_n (mm)	Re_n	L	R	Ref.
Air	CH_4	1.0	1–8[b]	Tube	16	1200–10,000	1.6–3.8	0	Hargrave & Kilham, 1984[41]
		0.943, 1.171	Not given	Converging nozzle	12	Laminar & turbulent[a]	4.6–11	0	Fairweather et al., 1984[27]
		0.8–1.2	1.2–10[b]	Tube	16	2000–12,000	1.9–15	0	Hargrave et al., 1987[5]
	CO	0.7–1.3	Not given	Porous bronze	51	Laminar[b]	0.02–0.3	0	Kilham & Dunham, 1967[49] (cf. Dunham, 1963[76])
	H_2	0.5–0.63	Not given	Porous bronze	51	Laminar[b]	0.02–0.2	0	Cookson & Kilham, 1963[36] (cf. Cookson, 1963[35])
	Natural gas #2	1.0	Not given	Bundle of tubes	Not given	50–500	(1.3 cm)	0	Conolly & Davies, 1972[3]
	Natural gas	1.05	Not given	Flame retention head	22.4, 41.0	12,600	0–4.5	0	Hemeson et al., 1984[45]
Air/O_2	CH_4	0.62–1.14	3.2	Bundle of tubes	10	Laminar[b]	0.3–1.4	0	Nawaz, 1973[26]
		0.84–1.14	Not given	Bundle of tubes	10	Laminar[a]	0.3–1.4	0	Fairweather et al., 1984[27]
	Natural gas	0.95	210, 270[b]	Multi-annular w/swirl	Not given	Not given	(40–140 cm)	0	Ivernel & Vernotte, 1979[28]
O_2	CH_4	0.95–1.31	Not given	Glass torch	10	Laminar[a]	Not given	0	Kilham & Purvis, 1971[39] (cf. Purvis, 1974[77])
		1.0	Not given	Bundle of tubes	Not given	50–500	(1.3 cm)	0	Conolly & Davies, 1972[3]
		0.83–1.70	3.8	Bundle of tubes	10	Laminar[b]	0.3–1.4	0	Nawaz, 1973[26]
		0.83–1.70	Not given	Bundle of tubes	10	Laminar[a]	0.3–1.4	0	Fairweather et al., 1984[74]
	C_2H_4	1.0	Not given	Bundle of tubes	Not given	50–500	(1.3 cm)	0	Conolly & Davies, 1972[3]
	C_3H_8	1.45–1.83	Not given	Glass torch	10	Laminar[a]	Not given	0	Kilham & Purvis, 1971[39] (cf. Purvis, 1974[77])
		1.0	Not given	Bundle of tubes	Not given	50–500	(1.3 cm)	0	Conolly & Davies, 1972[3]
	CO	1.0	Not given	Bundle of tubes	Not given	50–500	(1.3 cm)	0	Conolly & Davies, 1972[3]
	H_2	1.0	Not given	Bundle of tubes	Not given	50–500	(1.3 cm)	0	Conolly & Davies, 1972[3]
	Natural gas	0.95	210[b]	Multi-annular w/swirl	Not given	Not given	(40–140 cm)	0	Ivernel & Vernotte, 1979[28]

[a] According to author (only qualitative).
[b] Calculated from data given in reference.

Source: From C.E. Baukal and B. Gebhart, *Comb. Sci. Tech.*, 104, 339-357, 1995.

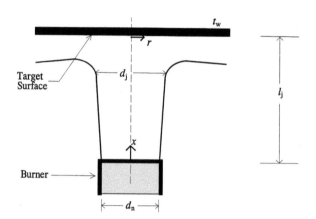

FIGURE 7.11 Flame impinging normal to a plane surface. (From C.E. Baukal and B. Gebhart, *Int. J. Heat Fluid Flow,* 17(4), 386-396, 1996.)

blended O_2 and N_2 to an equivalent Ω of 0.39.The study by Matsuo et al. (1978)[30] was the only study that used a preheated oxidizer. The combustion air was heated to about 500K (440°F) before reaching the burner. Baukal and Gebhart (1998) studied natural gas flames combusted with air, oxygen-enriched air, and pure oxygen.[31] They found that the heat transfer to the target was highly dependent on the oxidizer composition (see Chapter 8).

7.2.2.2 Fuels

Another parameter of interest is the fuel composition. Only experiments using gaseous fuels are considered here. Tables 7.1 to 7.4 indicate that natural gas and methane (the main constituent in natural gas) have been the most widely used. Where specified, the specific natural gas compositions are noted in Table 7.5. An example of another fuel used is acetylene, by Schulte (1972).[32] The study by Fay[33] was the only one in which MAPP® gas was used. This is a stabilized mixture of methylacetylene and propadiene (equivalent to $C_{3.1}H_{5.4}$), which is produced by Dow Chemical Co. Beér and Chigier[23] used another uncommon fuel, coke oven gas, which is a by-product of converting coal to coke. Davies (1965)[34] and Shorin and Pechurkin[20] used town gas, which is also obtained from coal. Both coke oven gas and town gas are roughly half H_2 and a third CH_4, with a mixture of other gaseous hydrocarbons. Both fuels have similar heating values and adiabatic flame temperatures. In one set of tests, Shorin and Pechurkin[20] used a mixture of propane and butane in unspecified proportions.

The combination of fuel type and the equivalence ratio ϕ determines the tendency to produce soot and, therefore, luminous gas radiant emission. This tendency is higher in fuel-rich mixtures ($\phi > 1$). It also increases with higher carbon-to-hydrogen weight ratios in the fuel. For example, C_4H_{10}, which has a C:H weight ratio of 4.8, has a higher propensity to produce soot than CH_4, which has a C:H weight ratio of 3.

7.2.2.3 Equivalence Ratios

This ratio directly affects both the sooting tendency and the level of dissociation in the combustion products. Fuel-rich flames ($\phi > 1$) produce a combination of both luminous and nonluminous thermal radiation. The combustion products of these flames may also contain unreacted fuel components, due to insufficient oxygen. Fuel-lean flames ($\phi < 1$) normally do not produce luminous thermal radiation, due to the absence of soot particles. These flames seldom produce significant quantities of unreacted fuel species unless the flame temperature is high enough to produce dissociation. Flames at or near stoichiometric ($\phi = 1$) produce the highest flame temperatures, due

TABLE 7.3
Experimental Conditions for Flame Impingement Normal to Plane Surfaces

Oxidizer	Fuel	ϕ	q_f (kW)	Burner	d_n (mm)	Re_n	L	R	Ref.
Air	Coke oven gas	Not given	606	Tunnel w/air preheat	100	Turbulent[c]	5–9	0–7	Matsuo et al., 1978[30]
	CH_4	1.33, 1.67	0.71	Partial premix	13, 16	Laminar[c]	2.4	0–1.3	Anderson & Stresino, 1963[15]
		2.58–2.76	1.0–160	Partial premix	7.8	17,680–22,700	13–65	0–4	Buhr et al., 1973[66] (cf. Buhr, 1969[85])
		4.8, ∞	46–220[c]	Tube	12.7	7000–35,300	10, 16	0–60	Milson & Chigier, 1973[9]
		2.75	1.0–160	Partial premix	7.8	17,680–22,700	13–64	0–10	Kremer et al., 1974[21] (cf. Buhr, 1969[85])
		0.51	0.7[c]	Long tube w/precombustor	9.5	3100	7.5	0–3	Posillico, 1986[48]
		0.78	22	Premix	50	6630 9700	2, 3, 4	0–5	Eibeck et al., 1993[8]
	C_2H_2	1.0	Not given	Premix torch	2.0	Laminar[a]	1.3–13	0–2.5	Schulte, 1972[32]
	C_3H_8	1.0	1.2, 1.5	Premix tube w/stainless steel mesh	25	1300[c]	23, 32	0–17	Veldman et al., 1975[42]
	Natural gas #1	0.95	2100	Variable swirl	131.3–152.0	Turbulent[c]	5.7–11	0–4.6	Vizioz & Lowes, 1971[25]
	Natural gas	1.0	1000–1500	Tunnel	172	Turbulent[c]	0–8.4	0–3.5	Smith & Lowes, 1974[37]
	Natural gas #3	1.0	Not given	Tunnel	13.8	1050, 1860	2–20	0–5	Hoogendoorn et al., 1978[52]
	Natural gas #2	1.0	1000, 1500	Tunnel	Not given	Not given	(60–100 cm)	(0–37 cm)	Rajani et al., 1978[38]
	Natural gas #3	1.0	Not given	Tunnel	13.8	1050, 1860	2–20	0–10	Popiel et al., 1980[53]
	Natural gas #5	∞	1.67, 8.51	Tube	55	70, 356	7.3	0–7.3	You, 1985[7]
Air	Natural gas	1.05	0.3, 0.4[c]	Flame retention head	5.6–15.3	7050–16,200	16–27	0	Horsley et al., 1982[46] (cf. Tariq, 1982[47])
	Natural gas #4	1.0	Not given	Tunnel	13.8, 27.6	1700–4250	1–12	0	van der Meer, 1991[54]
	Natural gas	0.65–2.79	39.7	Co-axial radial jet reattachment flame	63.5	3640–24,600	0.93	0–8.11	Mohr et al., 1995[24]
	Natural gas	∞	10.5	Tube	3.81	5000	0–170	0–100	Rigby and Webb, 1995[44]
			5.1, 10.5		6.35	1450, 3000	0–100	0–60	
	Natural gas #6	1.0	5	Flame working torch	38.5	2910	0.5–6.0	0–1.0	Baukal and Gebhart, 1998[31]

TABLE 7.3 (continued)
Experimental Conditions for Flame Impingement Normal to Plane Surfaces

Oxidizer	Fuel	ϕ	q_f (kW)	Burner	d_n (mm)	Re_n	L	R	Ref.
	Town gas +	0.67–0.95	Not given	Water-cooled tube w/precombustor	25, 46	1500–19,000	0–14	0–5	Shorin & Pechurkin, 1968[20]
Air/O₂	CH₄	0.61	Not given	Converging nozzle w/precombustor	10	857–2000	2–20	0–3	Kataoka et al., 1984[29]
	Natural gas #1	0.95	2100, 3000	Variable swirl	115.4	Turbulent[c]	8.7	0–4.8	Vizioz & Lowes, 1971[25]
	Natural gas #6	1.0	5–25	Flame working torch	38.5	1850–9200	0.5–6.0	0–1.0	Baukal and Gebhart, 1998[31]
O₂	CH₄	0.95–1.31	Not given	Premixed multiport	9.5	489–549	0.74, 2.1	0	Kilham & Purvis, 1978[57] (cf. Purvis, 1974[77])
	C₂H₂	2.34	4.1	Partial premix	1.8	Laminar[c]	4.3	0–36	Anderson & Stresino, 1963[15]
		0.9–2.5	2.9	Premix welding torch	Not given	Turbulent[c]	b	(0–1.3 cm)	Fay, 1967[33]
O₂	C₃H₈	1.0	2.7	Tunnel, partial premix	1.8	Laminar[c]	1	0–13	Anderson & Stresino, 1963[15]
		1.43	78		9.5		8.6	0–29	
		1.0–2.0	2.9	Premix torch	Not given	Turbulent[c]	b	(0–1.3 cm)	Fay, 1967[33]
		1.45–1.83	Not given	Premixed multiport	9.5	551–692	0.74, 2.1	0	Kilham & Purvis, 1978[57] (cf. Purvis, 1974[77])
	C₃H₈ + C₄H₁₀	1.0	61, 94	Premix welding torch	25	Turbulent[c]	6, 8	0	Rauenzahn, 1986[5]
		0.83–1.0	34	Water-cooled tube w/precombustor	25	300–1200	0–8	0	Shorin & Pechurkin, 1968[20]
	H₂	1.0	105	Tunnel	9.5	Laminar[c]	1.3	0–12	Anderson & Stresino, 1963[15]
		0.85	5.4	(2) Glass working torches	27	2000[c]	2	0–2.5	Reed, 1963[43]
		1.1	2.8		5.6	4600[c]	9	0–10	
	MAPP gas	0.85–2.2	2.9	Premix welding torch	Not given	Turbulent[c]	b	(0–1.3 cm)	Fay, 1967[33]
	Natural gas	1.0	Not given	Premix torch	0.53, 1.7	Laminar[a]	1.2–29	0–9.5	Schulte, 1972[32]
	Natural gas #2	1.0	1000–1750	(2) Commercial burners	Not given	Not given	(60–100 cm)	(0–37 cm)	Rajani et al., 1978[38]
	Natural gas #6	1.0	5–25	Flame working torch	38.5	810–4050	0.5–6.0	0–1.0	Baukal and Gebhart, 1998[31]

[a] According to author (only qualitative).
[b] Tip of inner flame cone just touching target.
[c] Calculated from data given in reference.

Source: From C.E. Baukal and B. Gebhart, *Comb. Sci. Tech.*, 104, 339–357, 1995.

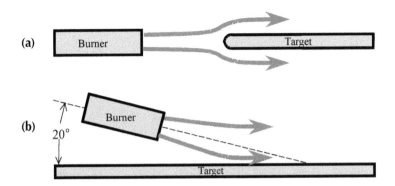

FIGURE 7.12 Flames impinging (a) parallel and (b) oblique to a plane surface.

to complete combustion. They also generally produce only nonluminous radiation since no soot is generated.

Most of the studies used stoichiometric mixtures. Cookson and Kilham (1963)[35,36] and Davies[34] used equivalence ratios as fuel lean as $\phi = 0.5$. Milson and Chigier,[9] You,[7] and Hustad and co-workers[6,16,17] used pure fuel jets ($\phi = \infty$). Many studies have investigated a range of equivalence ratios in order to determine the resulting effects on heat transfer to the target. Davies[34] used lean and rich mixtures. The actual equivalence ratios were not specified.

7.2.2.4   Firing Rates

The firing rate or gross heat release (q_f) of the flames has ranged from 0.3 to 3000 kW (10^3 to 10^7 Btu/hr). Large industrial-scale flames were used by Beér and Chigier,[23] Vizioz and Lowes,[25] Smith and Lowes (1974),[37] and Rajani et al. (1978)[38] at the International Flame Research Foundation (IFRF) in IJmuiden, The Netherlands. Beér's flame impinged downward on the hearth of a furnace. In the other three studies, the flames impinged on water-cooled targets located inside a furnace environment. Many of the studies (e.g., see Reference 39) considered here used torch tips, firing at under 50 kW (170,000 Btu/hr).

7.2.2.5   Reynolds Number

The Reynolds number at the burner nozzle, Re_n, varied from 50 to 330,000. Both laminar (e.g., see Reference 27) and turbulent (e.g., see Reference 40) flow conditions arose. As shown in Tables 7.1 to 7.4, the Reynolds number was not always given. For some studies, the flows were indicated to be either laminar or turbulent. In other studies, the nozzle Reynolds number was estimated from other information, such as nozzle diameter and gas flow rates.

The Reynolds number varies directly with the burner diameter. It is also influenced by burner design. In partially or fully premixed flames, the combustion products leave the burner at an elevated temperature. However, in diffusion flames, the gases leave the burner at essentially ambient conditions. Since the gas viscosity increases with temperature, Re_n is generally lower if the gases have been heated at the burner exit. In some studies, turbulence effects were analyzed (e.g., see Reference 41). However, in most studies, the turbulence effects were not assessed.

7.2.2.6   Burners

Many different types of burners have been used. These range from fully premixed to diffusion mixing, downstream of the burner exit. In fully premixed burners, the fuel and oxidizer mix prior to reaching the nozzle exit (see Figure 1.10). An example was the study by Veldman et al. (1975), where premixed gases were fired in a burner consisting of a tube with a stainless steel mesh.[42] The

TABLE 7.4
Experimental Conditions for Flame Impingement Parallel to Plane Surfaces

Oxidizer	Fuel	ϕ	q_f (kW)	Burner	d_n (mm)	Re_n	L	R	Ref.
Air	Coke oven gas	1.0	1470	Concentric tubes	150	4800–270,000	9	0–10	Beér & Chigier, 1968[23]
O_2	C_2H_2	2.5	Not given	(30) Multihead torch tips	Not given	Turbulent[a]	Not given	(1.3–14 cm)	Giedt et al., 1960[2]
		2.5	Not given	Multihead torch tips	Not given	Laminar[a]	Not given	(1.3–14 cm)	Woodruff & Giedt, 1966[22]

[a] According to author (only qualitative).

Source: From C.E. Baukal and B. Gebhart, *Comb. Sci. Tech.*, 104, 339-357, 1995.

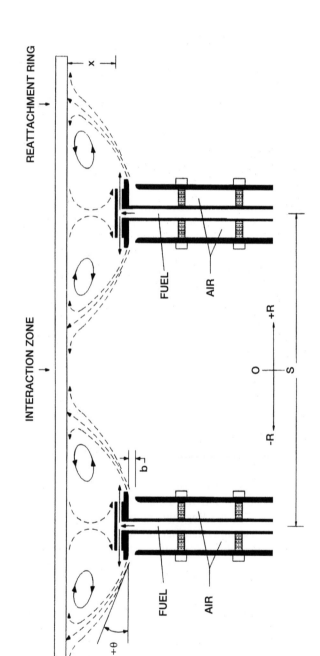

FIGURE 7.13 Pair of radial jet reattachment flames. (Courtesy of ASME, New York.[24] With permission.)

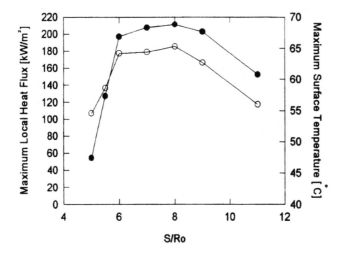

FIGURE 7.14   Effect of nozzle spacing on the maximum surface heat flux and temperature for a radial jet reattachment flame. (Courtesy of ASME, New York.[24] With permission.)

TABLE 7.5
Natural Gas Compositions (% by volume)

No.	CH_4	C_2H_6	C_3H_8	C_4H_{10}	C_mH_n[a]	CO_2	O_2	N_2
1	81.3	2.9	0.4	0.1	0.1	0.8		14.4
2	84.9	5.64			2.0	0.18		7.27
3	81.3	2.85			0.60	0.89	0.01	14.35
4	81.3	2.85						14.35
5	94.9	3.75	0.27	0.9	0.12	0.42		0.41
6	95.3	2.54	0.54	0.2	0.12	0.87	0.002	0.42

[a] C_mH_n = other hydrocarbons.

Source: From C.E. Baukal and B. Gebhart, *Comb. Sci. Tech.,* 104, 339-357, 1995.

resulting flame in premixed burners may have either a uniform or a nonuniform velocity profile, depending on the nozzle design. It also depends on the distance between the ignition point and the exit. In most studies using this burner type, both the temperature and the composition of the combustion products at the burner exit were approximately uniform. This circumstance is often simpler to analyze since the flame conditions are well established at the nozzle exit. There are then few subsequent chemical reactions ahead of the target surface. The tunnel burner is a common fully premixed burner. The gases are mixed and ignited inside the burner. They then travel through a refractory-lined chamber before leaving the burner. The combustion products may equilibrate inside the chamber. The temperature and composition are then uniform at the exit. However, the velocity profile may not be uniform. It may be approximately developed pipe flow, depending on the downstream length of the equilibration chamber.

In partially premixed burners, the fuel and oxidizer mix prior to reaching the nozzle exit (see Figure 1.13). However, only a portion of the stoichiometric amount of oxygen is supplied through the burner. The rest is provided by mixing with the surrounding ambient air, entrained into the flame. At the nozzle exit, the velocity profile is commonly nonuniform. Both uniform and nonuniform outlet temperature profiles and compositions have been reported.

In diffusion-mixing burners, the fuel and oxidizer begin to mix at the nozzle exit (see Figure 1.12), where the velocity is often nonuniform. An example is the glass working torches used by Reed (1963).[43] In diffusion burners, the exit temperature field is commonly homogeneous and equal to ambient conditions. The gas composition at the exit is pure fuel and pure oxidizer, with no combustion products. If the oxidizer is not supplied through the burner, a pure diffusion flame results. The oxygen is provided for combustion by ambient air entrainment into the flame.

7.2.2.7 Nozzle Diameter

In the configurations considered here, the exiting flame shape was round, except for those used by Kilham[13] and Davies,[18] which were slot shaped. The burner nozzle diameter, d_n, ranged from 0.53 to 152 mm (0.02 to 6.0 in.). This dimension was not given in some of the studies listed in Tables 7.1 to 7.4. In many studies (e.g., see Reference 3), the burner outlet consisted of a nested bundle of small tubes or orifices, arranged in a circular pattern. This is common practice for nozzles used on cutting and welding torches.

7.2.2.8 Location

The dimensionless axial distance between the burner exit and the target, L, varied from 0 to 200. In most studies, it was less than about 20. This distance has a strong influence on the resulting heat transfer to the target. Shorter distances result in higher flux rates. One cause is less ambient air entrainment. Another reason is that the flame widens at longer distances. This diffuses the heat flux over a wider cross-sectional area. Fay[33] adjusted the axial position of a premixed torch flame so that the inner flame cone tip touched the target. The specific axial locations were not given. Davies[18] positioned water-cooled tubes at the location in the flame that produced the highest heating rate. This location varied with the oxidizer composition and the equivalence ratio. Again, no dimensions were reported. For flames parallel to plane surfaces, L is taken as the distance from the burner to the leading edge of the target.

The dimensionless radial distance, R, from the nozzle centerline to the location on the target where the heat flux was measured is an important parameter. The heat flux at the stagnation point (R = 0) was measured in every study. In most of the experiments of jet flames impinging normal to plane surfaces, the radial heat flux profile was measured. The radial distance, R, ranged from 0 to as high as 60. In general, the heat flux decreased with R. However, in several studies, the peak heat flux did not occur at the stagnation point of the target. For example, Milson and Chigier[9] measured a peak flux at R ranging from 6 to 29, depending on the nozzle exit Reynolds number. The turbulent mixing of ambient air with the pure jets of CH_4 ($\phi = \infty$) was given as the cause. In the study by Rigby and Webb (1995), the heat flux to a wide range of both axial (L = 0 to 170) and radial (R = 1 to 100) locations was investigated.[44]

Table 7.6 indicates the studies in which the target was located inside a furnace. This influences the heat transfer in two important ways. The radiation from the hot refractory walls to the target is a significant portion of the total heat flux to the target. Vizioz and Lowes[25] and Smith and Lowes[37] showed that this effect was comparable in magnitude to the forced convection effect at the target. Ivernel and Vernotte[28] calculated this effect to be up to 42% of the total heat flux to a hemi-nosed cylinder. Also, any gases entrained into the jet are hot combustion products, instead of ambient air, as in most studies. Therefore, the flame jet is not cooled as much as in tests done at ambient conditions.

7.2.3 STAGNATION TARGETS

Tables 7.7 to 7.10 tabulate the important features of the targets. These include the dimensions, compositions, surface conditions, and temperature level for each of the four configurations. These

TABLE 7.6
Flame Jet Impingement Studies with the Target in a Furnace

Ref.	Furnace Wall Temp. (K)	Furnace Dimensions (m)
Beér and Chigier, 1968[23]	1200–1500	2 W × 2 H × 6.25 L
Vizioz and Lowes, 1971[25]	Not given	2 W × 2 H × 6.25 L
Smith and Lowes, 1974[37]	Not given	2 W × 2 H × 6.25 L
Matsuo et al., 1978[30]	1300–1500	2.16 W × 1.80 H × 1.36 L
Rajani et al., 1978[38]	770–1870	1.0 dia. × 4.5 L
Ivernel and Vernotte, 1979[28]	1650–2000	0.6 dia. × 2.05 L

Source: From C.E. Baukal and B. Gebhart, *Comb. Sci. Tech.,* 104, 339-357, 1995.

TABLE 7.7
Stagnation Targets for Flames Impinging Normal to Cylindrical Surfaces

Ref.	d_b (mm)	Material	Surface Coating(s) or Treatment	t_w (K)
Kilham, 1949[13]	3	Sillimanite	Al_2O_3, FeO, Cr_2O_3, uranium oxide	1220–1630
Jackson & Kilham, 1956[14]	3.2	Sillimanite	Al_2O_3, FeO, Cr_2O_3, uranium oxide	950–1445
Anderson & Stresino, 1963[15]	0.91, 1.5	Copper	None	<373
Davies, 1965[34]	1.5	Stainless steel	None	<373
	3.2	Copper	None	<373
Fells & Harker, 1968[80]	50	Stainless steel	None	<373
Davies, 1979[18]	2.7	Stainless steel	None	300
Hemeson et al., 1984[45]	32.6–59.8	Not given	None	Not given
Hargrave et al., 1987[75]	22	Brass	None	378
Hustad et al., 1991[16,17]	50	Steel	None	520–620
Hustad et al., 1992[6]	50	Steel	None	520–570

Source: From C.E. Baukal and B. Gebhart, *Comb. Sci. Tech.,* 104, 339-357, 1995.

properties varied widely among the studies reviewed here. In some studies, these were varied to examine the effect on the heat transfer mechanisms. These are discussed further below.

7.2.3.1 Size

The targets ranged in size from a 0.56 cm (0.22 in.) o.d. disk to a 200 × 625 cm (6.6 × 20.5 ft) furnace hearth. Most of the cylindrical targets have been hollow pipes. Kilham[13] and Jackson and Kilham[14] used solid refractory rods. Hustad et al.[6,16,17] used solid steel cylinders. All of the heminosed cylinders were under 22 mm (0.87 in.) o.d., except for those used by Hemeson et al. (1984),[45] which were from 50 to 150 mm (2 to 6 in.). Most of the plane surfaces in Table 7.9 were disks of over 16 mm (0.63 in.) in diameter, except for the 0.56 mm (0.022 in.) o.d. disks used by Horsley et al. (1982).[46,47] In addition to disks, many studies used square and rectangular plates as the target surfaces.

TABLE 7.8
Stagnation Targets for Flames Impinging Parallel to Hemi-nosed Cylinders

Ref.	d_b (mm)	Material	Surface Coating(s) or Treatment	t_w (K)
Cookson & Kilham, 1963[35]	14	Brass	None	>373
Kilham & Dunham, 1967[49]	22.2	Brass	None, A418, Pt	>373
Kilham & Purvis, 1971[39]	9.5	Stainless steel	None	290–470
Conolly & Davies, 1972[3] (cf. Conolly, 1971[50])	12.7	Stainless steel	None, SiO$_2$, Pt	400
Nawaz, 1973[26]	9.5	Brass	None	290–420
Ivernel & Vernotte, 1979[28]	46	Inconel	None	600–1400
Fairweather et al., 1984[27]	9.5	Brass	None	340–1600
Fairweather et al., 1984[40]	9.5	Brass	None	~378
Hargrave & Kilham, 1984[41]	22	Brass	None	393
Hemeson et al., 1984[45]	50–150	Not given	None	Not given
Hargrave et al., 1987[75]	22	Brass	None	418

Source: From C.E. Baukal and B. Gebhart, *Comb. Sci. Tech.,* 104, 339-357, 1995.

7.2.3.2   Target Materials

The most commonly used materials were aluminum, copper, brass, and stainless steel. For example, Posillico (1986) studied flames impinging normal to uncoated aluminum disks.[48] Aluminum, copper, and brass have very high internal thermal conductivities. Therefore, they are easier to water-cool since the heat from the impinging flame is conducted away quickly. However, these materials also have relatively low melting points. For that reason, they have not been used at very high surface temperatures. Refractories, including alumina (Al$_2$O$_3$) and sillimanite (Al$_2$SiO$_5$), were also used.

7.2.3.3   Surface Preparation

Most of the target surfaces were untreated. However, in some experiments, the surfaces were treated or coated to study a specific surface effect. Kilham[13] and Jackson and Kilham[14] coated the surface of their refractory cylinders with different oxides. The objective was to estimate the emissivities of the coatings. The surface temperature was approximately determined using a thermocouple imbedded inside the cylinder body. Radiation from the surface was measured with a thermopile. Using the temperature level and the radiation, along with an energy balance on the cylinder, the emissivity of the coatings was calculated as a function of temperature.

Four studies used coatings to determine the effects of surface catalysis on the heat transfer from an impinging flame. Giedt et al.[2] used both uncoated and porcelain-coated iron plates. These surfaces were considered to be catalytic and noncatalytic, respectively. The measured heat flux for these two surface conditions was about the same. It was concluded that surface chemical recombination effects were negligible for those test parameters. Kilham and Dunham (1967)[49] used three different surface conditions to study the effects on air/CO flames impinging on a hemi-nosed cylinder. The results indicated that the lowest heat flux was for the National Bureau of Standards Coating A418, which is a mixture of eight oxides. It was assumed to be noncatalytic. On the other hand, the highest flux was found for a platinum coating, which was considered to be perfectly catalytic. The heat flux to an uncoated copper heat flux gage surface was intermediate between the results for the A418 and platinum coatings. However, the differences were small. It was concluded that the bulk of the chemical reactions occurred in the boundary layer. In addition, as L increased, the heat flux for all three surface conditions converged to the same value. Conolly and Davies[3,50] found no difference in

TABLE 7.9
Targets for Flames Impinging Normal to Plane Surfaces

Ref.	Dimensions (cm)	Material	Surface Coating(s) or Treatment	t_w (K)
Anderson & Stresino, 1963[15]	$20 \times 20 \times 0.64$	Copper	None	<373
Reed, 1963[43]	$13 \times 25 \times 0.16$	Copper	None	<373
Fay, 1967[33]	$13 \times 25 \times 0.32$	Copper	None	<373
Shorin & Pechurkin, 1968[20]	16 diam.			
	50×30	Not given	None	<373
Vizioz & Lowes, 1971[25]	120 diam. $\times$ 1	Steel	None	<373
	120 diam. $\times$ 8	Refractory	None	1280–1420
Schulte, 1972[32]	Not given	Not given	Not given	<370
Buhr et al., 1973[66]				
(cf. Buhr, 1969[85])	32 diam.	Not given	None	<373
Milson & Chigier, 1973[9]	183×183	Steel	None	300–650
Kremer et al., 1974[21]	32 diam.	Not given	None	<373
	Not given	Iron	None	350–1000
Smith & Lowes, 1974[37]	120 diam. $\times$ 0.1	Steel	None	373
		Stainless steel	None	970
Veldman et al., 1975[42]	91 diam. $\times$ 0.16	Steel	None	293–347
Hoogendoorn et al., 1978[52]	32×57	Copper	Polished	303
Kilham & Purvis, 1978[57]	$15.2 \times 15.2 \times 5.6$	Alumina	None	490
Matsuo et al., 1978[30]	$140 \times 80 \times 2.5, 10$	Steel	None	300–1500
Rajani et al., 1978[38]	74 diam. $\times$ 0.8	Steel	None	470–520
Popiel et al., 1980[53]	57.3×32.0	Copper	Polished	<373
Horsley et al., 1982[46]				
(cf. Tariq, 1982[47])	0.56 diam.	Copper	None	360
Kataoka et al., 1984[29]	42 diam.	Copper	None	306
You, 1985[7]	100 diam.	Not given	High emissivity paint	~300
Posillico, 1986[48]	46 diam. $\times$ 0.64	Aluminum	None	<310
Rauenzahn, 1986[5]	38 dia $\times$ 13	Copper	None	290–480
van der Meer, 1991[54]	Not given	Copper	Polished	330
Eibeck et al., 1993[8]	$61 \times 61 \times 3.81$	Alumina	None	300–1205
Mohr et al., 1995[24]	103 diam.	Copper	None	305–338
Rigby & Webb, 1995[44]	76.2 diam.	Copper & aluminum	None	323
Baukal & Gebhart, 1998[31]	10.5 diam.	Brass, copper, stainless	Various	323–1400

Source: From C.E. Baukal and B. Gebhart, *Comb. Sci. Tech.*, 104, 339-357, 1995.

TABLE 7.10
Stagnation Targets for Flames Impinging Parallel to Plane Surfaces

Ref.	Dimensions (cm)	Material	Surface Coating(s) or Treatment	t_w (K)
Giedt et al., 1960[2]	$7.6 \times 7.6 \times 0.16$	Iron	None, porcelain	294–1170
Woodruff & Giedt, 1966[22]	$7.6 \times 15 \times 0.32$	Molybdenum	None	1100–1900
Beér & Chigier, 1968[23]	200×625	Refractory	None	1210–1470

Source: From C.E. Baukal and B. Gebhart, *Comb. Sci. Tech.*, 104, 339-357, 1995.

heat flux between a surface coated with highly catalytic platinum and a surface coated with SiO_2, which was expected to be noncatalytic. This again led to the conclusion that the recombination of radical species occurs in the boundary layer prior to reaching the surface. Baukal and Gebhart (1997) also found no difference in heat flux to surfaces coated with alumina or platinum.[51]

You[7] used a high-emissivity (0.96) paint to maximize the amount of radiant heat absorbed by the target. Three interrelated studies[52-54] used polished copper surfaces to reduce the effects of flame radiation. In some studies, the heat flux gage surface was coated or treated, as discussed in Chapter 6.

7.2.3.4   Surface Temperatures

Surface temperatures ranged from 290K to 1900K (63°F to 2960°F). For many measurements, the temperature, t_w, was maintained below 373K (212°F), using water-cooled targets. In some studies (e.g., Hargrave and Kilham, 1984[41,45]), the surface temperature was slightly above 373K (212°F). This eliminated the possibility of combustion products condensing on the target. Ethylene glycol was used as the target coolant because it has a higher boiling temperature than water. Beér and Chigier[23] and Vizioz and Lowes[25] measured surface temperatures above 1200K (1700°F) for refractory targets. In some studies, the surface temperature level was actually for the heat flux gage, and not the target. For example, Fairweather et al.[27] reported a maximum surface temperature of 1600K (2400°F). However, the target was made of brass, which melts at about 1300K (1900°F). A stainless steel heat flux gage imbedded in the brass target was used to measure the heat flux. Stainless steels have a melting point of about 1700K (2600°F).

7.2.4   MEASUREMENTS

Experimental measurement techniques used in flame impingement studies were discussed in Chapter 6. Table 7.11 shows the measured heat fluxes, gas temperatures, and gas velocities for the air/fuel flame impingement studies for all geometries. Table 7.12 is similar except for air/O_2/fuel studies, and Table 7.13 is for O_2/fuel studies. Table 7.14 is a compilation of the maximum reported ranges for the total heat flux at the given gas temperatures using the five methods of measuring heat flux that were used in the flame impingement studies. Table 7.15 gives the maximum temperature ranges for all studies for the various methods used to measure gas temperature. Table 7.16 lists all of the flame impingement studies that used the line reversal method to measure gas temperature. Table 7.17 gives the details of all experiments that used fine wire thermocouples to measure gas temperatures. Table 7.18 gives details of the experiments that used Pitot probes to measure gas velocity. Table 7.19 shows the studies that measured gas composition, which gases were measured, and what techniques were used to measure them.

7.3   SEMIANALYTICAL HEAT TRANSFER SOLUTIONS

This chapter section is taken from a review paper by Baukal and Gebhart (1996).[10] Heat transfer from high-temperature gases to axisymmetric and blunt-nosed bodies has been studied for many years. These processes are very important in aerospace applications. Aerospace vehicles, such as rockets and missiles, travel at high supersonic velocities. Commonly, the nose of these surfaces is axisymmetric, blunt, and rounded. Gas shock-waves are produced as the vehicles travel through the atmosphere. The resulting temperatures at the stagnation point are generally high enough to cause the atmospheric gases to dissociate into many chemical species. Very high heat fluxes arise in that region. Several semianalytical solutions have been proposed for calculating such fluxes. Those solutions were derived from the laminar, two-dimensional, axisymmetric boundary layer equations applied in the stagnation region. The equations were simplified using similarity flow formulations. In the resulting heat flux solutions, a constant was numerically determined. Therefore, the solutions are referred to as semianalytical.

TABLE 7.11

Experimental Methods and Ranges for Total Heat Flux, Gas Temperature, and Velocity for Air/Fuel Studies

Year	Author(s)	Total Heat Flux, q''		Gas Temperature, t_j		Gas Velocity, v_j	
		Method	(kW/m²)	Method	(K)	Method	(m/sec)
1949	Kilham[13]	Steady-state uncooled target	70–130[a]	Line reversal	1213–1633	Calculated	5–30[a]
1956	Jackson & Kilham[14]	Steady-state uncooled target	20–107[a]	(6) Decreasing diam. uncoated T/Cs	1179–1659	Calculated	8–20[a]
1963	Anderson & Stresino[15]	Steady-state cooled target	32–1100	Adiabatic equilibrium calculation	Not given	Calculated	0.3–38
1965	Cookson & Kilham[35]	Steady-state cooled gage	73–180	(7) Decreasing diam. coated T/Cs	1360–1450	Not measured	—
	Davies[34]	Steady-state cooled target	570–950	Line reversal	2260	Not measured	—
1967	Kilham & Dunham[49]	Steady-state cooled gage	41–95	(7) Decreasing diam. coated T/Cs	1624–1910	Not measured	—
1968	Beér & Chigier[23]	Steady-state cooled gage	15–54	Suction pyrometer	1370–1950	Pitot tube	Not given
	Fells & Harker[80]	Steady-state cooled target	250–840	Line reversal	2500–2750	Not measured	—
1971	Shorin & Pechurkin[20]	Steady-state cooled gage	Nondimen.	Uncoated T/C	Not given	Uncooled Pitot-static probe	Not given
	Vizioz & Lowes[25]	Steady-state cooled target	24–280	Suction pyrometer	1450–1850	5-hole Pitot-static probe	2.8–46
1972	Conolly & Davies[3] (cf. Conolly, 1971[50])	Transient uncooled gage	Nondimen.	Line reversal	2200	Cooled Pitot tube	Not given
	Schulte[32]	Steady-state cooled gage	79–620	Not measured	—	Not measured	—
1973	Buhr et al.[66]	Steady-state cooled gage	110–440	Uncoated T/C	573–1923	Not measured	—
	Milson & Chigier[9]	Transient uncooled target	1.3–13	Coated T/C	430–2200	Not measured	—
1974	Kremer et al.[21]	Steady-state cooled gage	120–450				
		Transient uncooled gage	120–470	Coated T/C	Not given	Not measured	—
	Smith & Lowes[37]	Steady-state cooled target	180–450				
		Steady-state cooled gage	57–150	Suction pyrometer	1170–2020	1- & 5-hole Pitot-static probe	6–195
1975	Veldman et al.[42]	Transient uncooled target	0.086–1.8	Coated T/C	292–419	Not measured	—
1978	Hoogendoorn et al.[52]	Steady-state cooled gage	0–560	Coated T/C	370–2100	Cooled Pitot tube & LDV	2–54
	Matsuo et al.[30]	Transient uncooled target	23–350	Uncoated T/C	1200–1600	Pitot tube	20–82
	Rajani et al.[38]	Steady-state cooled target	262–421	Suction pyrometer	460–1950	1- & 5-hole Pitot-static probe	0.91–87
1979	Davies[18]	Steady-state cooled target	330–650	Adiabatic equilibrium calculation	2225	Not measured	—
1980	Popiel et al.[53]	Steady-state cooled gage	100–630	Coated decreasing diam. T/Cs	370–2100	Cooled Pitot tube & LDV	2–54
1982	Horsley et al.[46] (cf. Tariq, 1982[47])	Transient uncooled gage	220–440	Uncoated T/C	1643–2165	Not measured	—

TABLE 7.11 (continued)
Experimental Methods and Ranges for Total Heat Flux, Gas Temperature, and Velocity for Air/Fuel Studies

Year	Author(s)	Total Heat Flux, q''		Gas Temperature, t_j		Gas Velocity, v_j	
		Method	(kW/m²)	Method	(K)	Method	(m/sec)
1984	Fairweather et al.[40]	Transient uncooled gage	240–580	Line reversal	Not given	LDV	Not given
	Hargrave & Kilham[41]	Steady-state cooled gage	120–400	Line reversal & uncoated T/C	1900–2100	LDV	2–3
	Hemeson et al.[45]	Transient uncooled gage	100–430	Not measured	—	Not measured	—
1985	You[7]	Steady-state cooled gage	0.2–11[a]	Uncoated T/C	293–1500	Pitot-static probe & LDV	0.3–2
1986	Posillico[48]	Steady-state cooled target	21	Raman spectroscopy	600–3200	LDV	6.7–13
1987	Hargrave et al.[74,75]	Steady-state cooled gage Transient uncooled gage	100–460				
1991	Hustad et al.[16,17]	Transient uncooled gage	150–410	Uncoated T/C	400–2350	LDV	3.2–14.3
			0–160	Uncoated T/C	400–1700	Pitot tube	5–30
	van der Meer[54]	Steady-state cooled gage	Nondimen.	Coated decreasing diam. T/Cs	Up to 2130	LDV	Up to 50
1992	Hustad et al.[6]	Transient uncooled gage	35–180	Uncoated T/C	Not given	Not measured	—

[a] Calculated from data given in reference.

Source: From C.E. Baukal and B. Gebhart, *Comb. Sci. Tech.*, 104, 339–357, 1995.

TABLE 7.12
Experimental Methods and Ranges for Total Heat Flux, Gas Temperature, and Velocity for Air/O$_2$/Fuel Studies

Year	Author(s)	Total Heat Flux, q''		Gas Temperature, t_j		Gas Velocity, v_j	
		Method	(kW/m²)	Method	(K)	Method	(m/sec)
1971	Vizioz & Lowes[25]	Steady-state cooled target	26.5	Suction pyrometer	1850–1950	Not measured	—
1973	Nawaz[26]	Transient uncooled gage	1400–2600	Line reversal	2600–2763	Calculated	Not given
1979	Davies[18]	Steady-state cooled target	600–1600	Adiabatic equilibrium calculation	2380–2875	Not measured	—
	Ivernel & Vernotte[28]	Transient uncooled gage	50–1200	Twin sonic orifice probe	1640–2750	Cooled Pitot-static probe	5–51
1984	Fairweather et al.[74]	Transient uncooled gage	1400–2600	Line reversal	2710–2760	Calculated	32–36
	Kataoka et al.[29]	Steady-state cooled gage	Nondimen.	Uncoated T/C	820–1720	Uncooled Pitot tube	7.4–45.1

Source: From C.E. Baukal and B. Gebhart, *Comb. Sci. Tech.,* 104, 339-357, 1995.

The heat transfer from impinging, chemically active flames has also been extensively studied. The experimental conditions and measurements for those studies have been previously reviewed[56] and discussed above. In 12 of those studies,[3,26-28,35,39,41,45,46,49,54,57] the measured heat flux was compared against one or more semianalytical solutions. The stagnation body has commonly been a hemispherically nosed cylinder (see Figure 7.15). This geometry has been used because of its similarity to the shape of aerospace vehicles. Then, the semianalytical heat transfer solutions, derived for aeronautical applications, were used to model the measured heat fluxes in energetic flame impingement. The applicability of those equations to flame heating applications has been determined. In aerospace applications, the vehicle moves through stagnant atmospheric gases. In impinging flames, the combustion products move around a stationary target. Therefore, the relative motion is similar in both applications.

Chen and McGrath (1969)[58] reviewed impingement heat transfer from combustion products containing dissociated species. Sample heat transfer calculations were given for a stoichiometric O$_2$/C$_3$H$_8$ flame impinging on a 2 cm (0.8 in.) o.d. sphere. The equations recommended by McAdams (1954),[59] Altman and Wise (1956),[60] and Rosner (1961)[61] were used. The Lewis number, Le, is the ratio of the mass diffusion rate to the thermal diffusion effect (see Chapter 2). The results were evaluated for both Le = 1 and Le > 1. It was concluded that, for Le = 1, the existing information was sufficient to adequately predict heat transfer in chemically reacting systems. However, for flows in which Le > 1, more analytical and experimental work was recommended. Such systems are important for high-temperature flames, where considerable dissociation occurs.

Two types of heat transfer behavior have been compared with experimental data from the studies considered here (see Table 7.20). The first was forced convection with no chemical dissociation. This is applicable in lower temperature-level flame impingement. The second was forced convection with dissociation. The heat transfer relations for the second type have been variations of those recommended for the first type. These equations are useful in predicting the heat transfer in the absence of experimental data.

TABLE 7.13
Experimental Methods and Ranges for Total Heat Flux, Gas Temperature and Velocity for O_2/Fuel Studies

Year	Author(s)	Total Heat Flux, q''		Gas Temperature, t_j		Gas Velocity, v_j	
		Method	(kW/m²)	Method	(K)	Method	(m/sec)
1956	Jackson & Kilham[14]	Steady-state uncooled target	27–58[a]	(6) Decreasing diam. uncoated T/Cs	1394–1468	Calculated	3–6[a]
1960	Giedt et al.[2]	Transient uncooled target	200–670	Not measured	—	Not measured	—
1963	Anderson & Stresino[15]	Steady-state cooled target	32–41000	Adiabatic equilibrium calculation	Not given	Calculated	43–1400
	Reed[43]	Steady-state cooled target	0–1450	Adiabatic equilibrium calculation	3080	Calculated	7.5, 10
1965	Davies[34]	Steady-state cooled target	3500–5500	Line reversal	2260	Not measured	—
1966	Woodruff & Giedt[22]	Transient uncooled target	Nondimen.	Line reversal	2900–3200	Cooled Pitot tube	40–64
1967	Fay[33]	Steady-state cooled target	0–5600	Not measured	—	Not measured	—
1968	Shorin & Pechurkin[20]	Steady-state cooled gage	Nondimen.	Uncoated T/C	Not given	Uncooled Pitot-static probe	Not given
1971	Kilham & Purvis[39]	Transient uncooled gage	1990–2590	Line reversal	2755–2925	Calculated	21–32
1972	Conolly & Davies[3] (cf. Conolly, 1971[50])	Transient uncooled gage	410–2400	Line reversal	2543–3144	Cooled Pitot tube	9–17
	Schulte[32]	Steady-state cooled gage	140–3500	Not measured	—	Not measured	—
1973	Nawaz[26]	Transient uncooled gage	1700–3600	Line reversal	2795–2996	Calculated	Not given
1978	Kilham & Purvis[57]	Transient uncooled gage	950–1500	Line reversal	2755–2925	Calculated	39.5–48.6
	Rajani et al.[38]	Steady-state cooled target	349–1290	Twin sonic orifice probe	323–3060	Cooled 1 & 5 hole Pitot-static probe	2–135
1979	Davies[18]	Steady-state cooled target	1350–3400	Adiabatic equilibrium calculation	3050	Not measured	—
	Ivernel & Vernotte[28]	Transient uncooled gage	340–1300	Twin sonic orifice probe	2300–2800	Cooled Pitot-static probe	11–66

TABLE 7.13 (continued)
Experimental Methods and Ranges for Total Heat Flux, Gas Temperature and Velocity for O_2/Fuel Studies

Year	Author(s)	Total Heat Flux, q'' Method	(kW/m²)	Gas Temperature, t_j Method	(K)	Gas Velocity, v_j Method	(m/sec)
1984	Fairweather et al.[27]	Transient uncooled gage	1700–3700	Line reversal	2930–2990	Calculated	35–38
1986	Rauenzahn[5]	Transient uncooled target	645–1210	Not measured	—	Not measured	—

[a] Calculated from data given in reference.

Source: From C.E. Baukal and B. Gebhart, *Comb. Sci. Tech.,* 104, 339-357, 1995.

TABLE 7.14
Reported Ranges for Each Heat Flux Technique

Technique	Total Heat Flux (kW/m²)	Gas Temperature (K)
1. Steady-state uncooled target	20–130	1179–1659
2. Steady-state cooled target	0–41000	323–3200
3. Steady-state cooled gage	0–3500	293–2350
4. Transient uncooled target	0.086–1210	292–3200
5. Transient uncooled gage	0–3700	400–3144

Source: From C.E. Baukal and B. Gebhart, *Comb. Sci. Tech.,* 104, 339-357, 1995.

TABLE 7.15
Reported Ranges for Each Temperature Technique

Technique	Gas Temperature (K)
Line reversal	1213–3200
Fine wire thermocouple	292–2350
Suction pyrometer	460–2020
Twin sonic orifice	323–3060
Raman spectroscopy	600–3200

Source: From C.E. Baukal and B. Gebhart, *Comb. Sci. Tech.,* 104, 339-357, 1995.

TABLE 7.16
Line Reversal Measurements

Seeding	Comparison Light Source	Max. t_j (K)	Ref.
Sodium	Not given	3200	Woodruff & Giedt, 1966[22]
Sodium	Tungsten filament	2100	Hargrave & Kilham, 1984[41] (*cf.* Hargrave, 1984[55])
Sodium	Tungsten filament	2260	Davies, 1965[34]
Sodium	Tungsten filament	2925	Kilham & Purvis, 1971[39]
Sodium	Tungsten filament	3144	Conolly & Davies, 1972[3]
Sodium	Xenon arc	2925	Kilham & Purvis, 1978[57]
Sodium carbonate	Carbon arc	2750	Fells & Harker, 1968[80]
Sodium chloride	Tungsten filament	1633	Kilham, 1949[13]
Sodium chloride	Tungsten filament	2990	Fairweather et al., 1984[27,40]
Sodium chloride	Xenon arc	2996	Nawaz, 1973[26]

Source: From C.E. Baukal and B. Gebhart, *Comb. Sci. Tech.*, 104, 339-357, 1995.

TABLE 7.17
Fine Wire Thermocouple Measurements

T/C Alloys	T/C Type	Coating	Max. t_j (K)	O.D. (mm)	Ref.
Chromel/alumel	K	None	Not given	Not given	Hustad et al., 1992[6]
		None	1700	Not given	Hustad et al., 1991[16,17]
Iron/constantan	J	Teflon	419	0.13	Veldman et al., 1975[42]
Pt/Pt-Rh		None	1720	0.1	Kataoka et al., 1984[29]
		None	1923	0.1	Buhr et al., 1973[66]
		Al_2O_3	Not given	0.1	Kremer et al., 1974[21]
Pt/Pt-10%Rh	S	None	1500	0.05	You, 1985[7]
		None	1659	0.061–0.193	Jackson & Kilham, 1956[14]
Pt/Pt-10%Rh & Pt-30%Rh/Pt-6%Rh	S, B	Yt_2O_3/BeO	2100	0.05–0.18	Hoogendoorn et al., 1978[52]
		Yt_2O_3/BeO	2100	0.05–0.18	Popiel et al., 1980[53]
		Yt_2O_3/BeO	2130	0.05–0.18	van der Meer, 1991[54]
Pt/Pt-13%Rh	R	None	Not given	Not given	Hargrave & Kilham, 1984[41] (*cf.* Hargrave, 1984[55])
		None	2165	0.5	Horsley et al., 1982[46]
		None	2350	0.24	Hargrave et al., 1987[74]
		SiO_2	1450	0.127	Cookson & Kilham, 1967[35]
		SiO_2	1910	0.127	Kilham & Dunham, 1967[49]
		SiO_2	2200	0.127	Milson & Chigier, 1973[9]
Pt-30%Rh/Pt-6%Rh	B	None	Not given	0.3	Shorin & Pechurkin, 1968[20]
Not given		None	1600	Not given	Matsuo et al., 1978[30]

Source: From C.E. Baukal and B. Gebhart, *Comb. Sci. Tech.*, 104, 339-357, 1995.

TABLE 7.18
Pitot Measurements

Probe Type	Re_n	max. t_j (K)	Coolant	Material	o.d. (mm)	Ref.
Pitot tube	Laminar	3200	Water	Not given	Not given	Woodruff & Giedt, 1966[22]
	50–500	2200	Water	Copper	3.6	Conolly & Davies, 1972[3]
	200–3600	1700	Not given	Not given	Not given	Hustad et al., 1991[16,17]
	857–2000	1720	None	Quartz	1.4	Kataoka et al., 1984[29]
	1050, 1860	2100	Water	Tantalum	1.2	Hoogendoorn et al., 1978[52]; Popiel et al., 1980[53]
	Turbulent	2020	Not given	Not given	Not given	Smith & Lowes, 1974[37]
	Turbulent	1600	Not given	Not given	Not given	Matsuo et al., 1978[30]
	4800–270,000	1950	Not given	Not given	Not given	Beér & Chigier, 1968[23]
	Not given	3060	Water	Tantalum	12	Rajani et al., 1978[38]
Pitot-static	70, 356	1500	Not given	Not given	Not given	You, 1985[7]
	1500–19,000	Not given	None	Alundum	1.8	Shorin & Pechurkin, 1968[20]
	Not given	2800	Water	Not given	Not given	Ivernel & Vernotte, 1979[28]
5-Hole	Turbulent	1950	Not given	Not given	8	Vizioz & Lowes, 1971[25]
	Turbulent	2020	Not given	Not given	Not given	Smith & Lowes, 1974[37]
	Not given	3060	Water	Tantalum	12	Rajani et al., 1978[38]

Source: From C.E. Baukal and B. Gebhart, *Comb. Sci. Tech.,* 104, 339-357, 1995.

TABLE 7.19
Analyzers Used to Measure Gas Compositions

Ref.	CH_4	CO	CO_2	H	H_2	H_2O	N_2	NO	NO_2	O_2	OH
Buhr, 1969[85]	GC	GC	GC		GC		GC			GC	
Conolly, 1971[50]					Calc.						
Vizioz & Lowes, 1971[25]	GC	GC	GC		GC		GC			GC	
Nawaz, 1973[26]		GC	GC		GC	Calc.	GC	GC		GC	LA
Smith & Lowes, 1974[37]	GC	GC	GC		GC					GC	
Rajani et al., 1978[38]		IR	IR					CL	CL	Para.	
Ivernel & Vernotte, 1979[28]		GC	GC		GC		GC			GC	
You, 1985[7]	GC	GC	GC		GC	GC	GC			GC	
Posillico, 1986[48]			RS				RS			RS	

Note: CL = chemiluminescent; GC = gas chromatograph; IR = nondispersive infrared; LA = line absorption; Para. = paramagnetic; and RS = Raman spectroscopy.

Source: From C.E. Baukal and B. Gebhart, *Comb. Sci. Tech.,* 104, 339-357, 1995.

7.3.1 EQUATION PARAMETERS

The thermophysical properties and the stagnation velocity gradient have been used in all of the semianalytical solutions. Many methods have been used to calculate these parameters. Those methods are discussed here, prior to presenting the semianalytical solutions.

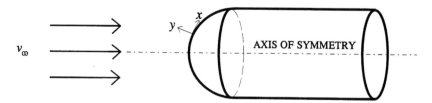

FIGURE 7.15 Stagnation flow around a hemi-nosed, axisymmetric, body of revolution. (From C.E. Baukal and B. Gebhart, *Int. J. Heat Fluid Flow,* 39(14), 2989-3002, 1996.)

7.3.1.1 Thermophysical Properties

These include the viscosity, density, thermal conductivity, Lewis number, and the enthalpy of the gaseous combustion products — all of which are temperature dependent (see Chapter 2). The temperature of the combustion products varies with the oxygen enrichment ratio, Ω. For stoichiometric C_3H_8 flames, Chen (1972)[62] showed how the properties vary as a function of Ω. Various methods have been used to evaluate the thermophysical properties. These methods are discussed below. The nomenclature is shown in Figure 7.16.

In every semianalytical solution, the gas temperature at the edge of the stagnation zone, t_e, has been used to evaluate some of the properties:

$$p_e = p(t_e) \tag{7.1}$$

where p is the property being evaluated. The wall temperature at the stagnation point of the target, t_w, has also been used in every solution:

$$p_w = p(t_w) \tag{7.2}$$

Many studies[3,27,41,45,57] have used a weighted average, over the temperature range between the edge and the wall temperatures, as

$$\bar{p} = \frac{\int_w^e p\,dt}{t_e - t_w} \tag{7.3}$$

The film, that is the mean temperature, between the edge and the wall temperatures has commonly been used:[28,39,46,49]

$$\bar{p} = p\left(\frac{t_e - t_w}{2}\right) \tag{7.4}$$

The reference temperature[63] has been used in several solutions:[52-54]

$$p_{ref} = p\left\{t_e + 0.5(t_w - t_e) + 0.22(t_{rec} - t_e)\right\} \tag{7.5}$$

where

$$t_{rec} = t_e + \frac{v_e^2\,Pr_e^{0.5}}{2c_{p_e}} \tag{7.6}$$

For low-speed flows, $t_{rec} \approx t_e$, so $p_{ref} \approx \bar{p}$. Table 7.21 shows sample calculations, using the various methods, for the thermal conductivity of the combustion products.[64]

TABLE 7.20
Experimental Studies Using Semianalytical Solutions

Target	Oxidizer	Re_n	ϕ	Fuel	β_s	d_b (mm)	v (m/s)	t_j (K)	Ref.
Cylinder	Air	12600	1.05	Natural gas	$2v_e/r_{0.5t}$	32.6–59.8	Not measured	Not measured	Hemeson et al., 1984[45]
Flat plate	Air	1700–4250	1.0	Natural gas	v_e/d_b	Not given	Up to 50	Up to 2130	van der Meer, 1991[54]
	Air	7050–16200	1.05	Natural gas	$v_e/r_{0.5t}$	5.6 diam.	Not measured	1643–2165	Horsley et al., 1982[46]
	O_2	489–549	0.95–1.31	CH_4	v_e/d_j	152 x 152	40–49	2755–2925	Kilham & Purvis, 1978[57]
		551–692	1.45–1.83	C_3H_8					
Hemi-nosed cylinder	Air	Laminar[a]	0.5–0.63	H_2	$3v_e/d_b$	14	Not measured	1360–1450	Cookson & Kilham, 1963[35]
	Air	Laminar[a]	0.7–1.3	CO	$3v_e/d_b$	22.2	Not measured	1624–1910	Kilham & Dunham, 1967[49]
	Air	50–500	1.0	Natural gas	$3v_e/d_b$	12.7	Not given	2200	Conolly & Davies,1972[3]
	O_2			$CH_4,C_2H_4,C_3H_8,CO,H_2$	$(2.67+0.0926Tu)v_e/d_b$		9–17	2543–3144	
	Air	1200–10000	1.0	CH_4	$3v_e/r_{0.5t}$	22	2–3	1900–2100	Hargrave & Kilham, 1984[41]
	Air	12600	1.05	Natural gas	$3v_e/d_b$	50–150	Not measured	Not measured	Hemeson et al.,1984[45]
	Air/O_2	Laminar[a]	0.62–1.14	CH_4	$3v_e/d_b$	9.5	Not given	2600–2763	Nawaz,1973[26]
	O_2		0.83–1.70					2795–2996	
	Air/O_2	Laminar[b]	0.84–1.14	CH_4	$3v_e/d_b$	9.5	32–36	2710–2760	Fairweather et al.,1984[27]
	O_2		0.83–1.70				35–38	2930–2990	
	O_2	Laminar[b]	0.95–1.31	CH_4	$3v_e/d_b$	9.5	21–32	2755–2925	Kilham & Purvis,1971[39]
			1.45–1.83	C_3H_8					
	Air/O_2	Not given	0.95	Natural gas	$2.5v_e/d_b$	46	5–51	1640–2750	Ivernel & Vermotte, 1979[28]
	O_2						11–66	2300–2800	

[a] Calculated from data in the reference.
[b] According to the author.

Source: From C.E. Baukal and B. Gebhart, *Int. J. Heat Mass Transfer*, 39(14), 2989-3002, 1996.

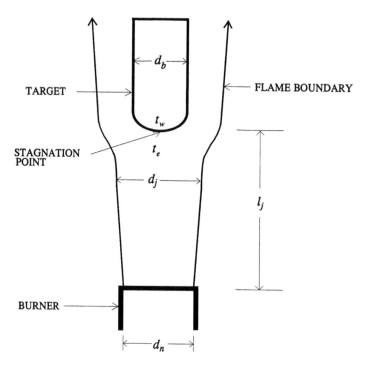

FIGURE 7.16 Flame impingement on a hemi-nosed cylinder. (From C.E. Baukal and B. Gebhart, *Int. J. Heat Mass Transfer*, 39(14), 2989-3002, 1996.)

TABLE 7.21
Thermal Conductivities for the Equilibrium Combustion Products for Stoichiometric CH₄ Flames, Calculated Using Reference 64

		Oxidizer	
Parameter	Units	Air	O_2
t_{AFT}	K	2220	3054
t_w	K	400	400
t_e	K	2000	2800
	K	1200	1600
t_{rec}	K	2000.6	2800.1
t_{ref}	K	1200.1	1600.0
k_w	W/m-K	0.0297	0.1546
k_e	W/m-K	0.1693	1.6818
	W/m-K	0.0860	0.1546
	W/m-K	0.0884	0.3288
k_{ref}	W/m-K	0.0860	0.1546

Note: v_e assumed to be 50 m/sec.

Source: From C.E. Baukal and B. Gebhart, *Int. J. Heat Mass Transfer*, 39(14), 2989-3002, 1996.

7.3.1.2   Stagnation Velocity Gradient

The original semianalytical heat transfer solution was developed for uniform external flows imping-ing normal to a body of revolution.[19] The momentum equation for steady flow in the boundary layer along the surface of an axisymmetric body and near the forward stagnation point[65] is:

$$v_x \frac{\partial v_x}{\partial x} + v_y \frac{\partial v_x}{\partial y} = \beta^2 x + v \frac{\partial^2 v_x}{\partial y^2} \tag{7.7}$$

As shown in Figure 7.15, x is the distance from the stagnation point along the body, and y is the distance normal from the surface. From potential flow theory, near the forward stagnation point, the flow just outside the boundary layer may be given by:[65]

$$v_x = \beta x \tag{7.8}$$

The constant β appears in the semianalytical solutions for the heat flux at the forward stagnation point. At the edge of the stagnation zone, it is defined as:

$$\beta_s = \left(\frac{\partial v_x}{\partial x} \right)_{x=0, y=\delta} \tag{7.9}$$

where δ is the local boundary layer thickness at the axis of symmetry. This constant has been described in a variety of ways:

- Velocity gradient at or near the stagnation point[26,46,66-68]
- Stagnation velocity gradient[26,27,41,69]
- Velocity gradient in the radial direction, outside of the boundary layer, in the vicinity of the stagnation point[52-54]
- Stagnation point radial velocity gradient[3,45]
- Velocity gradient tangential to the potential flow[28]

This gradient has been determined analytically. It has also been determined experimentally, as shown in Table 7.20.

7.3.1.2.1   Analytical solutions

The surface velocity gradient has been calculated for impinging flows normal to several stagnation body shapes. The solutions were developed using potential flow theory. This applies for high-altitude flight, where the flow is approximately inviscid. The following relations have been com-puted:[70,71]

$$\beta_s = 4v_e/d_b \quad \text{for a cylinder in crossflow} \tag{7.10}$$

$$= 3v_e/d_b \quad \text{for a sphere} \tag{7.11}$$

$$= 4v_e/\pi d_b \quad \text{for a disk} \tag{7.12}$$

Most of the flame impingement studies, wherein the target was a hemi-nosed cylinder, assumed the value of β_s for a sphere. No analytical solution was available for the hemi-nosed cylinder. For an axisymmetric jet impinging normal to an infinite plane, β_s has been analytically determined as:[66]

$$\beta_s = \frac{3\pi}{16}\frac{v_e}{d_j} \qquad (7.13)$$

where d_j is the diameter of the jet at the edge of the stagnation zone (see Figure 7.16). As shown in Figure 7.15, the external velocity far from the body of revolution, v_∞, is uniform. In Equations 7.10 to 7.13, the velocity at the edge of the boundary layer, v_e, is equal to v_∞. For flame impingement, the external flow is generally not uniform. This is especially true if the target is large compared to the flame width. Therefore, v_e is commonly a function of the axial distance between the burner and the target, L. Then, for impinging flames, $v_e = v_e(x = 0, y = \delta)$.

7.3.1.2.2   Empirical correlations

In one study,[46] β_s was experimentally determined as a function of $r_{0.5t}$. This is the radius at which the measured gas temperature is halfway between the maximum and the ambient, at a given axial location (see Figure 7.17):

$$\frac{t_{r_{0.5t}} - t_\infty}{t_{max} - t_\infty} = 0.5 \qquad (7.14)$$

In a related study,[45] β_s was determined for flames impinging normal to a cylinder, a hemi-nosed cylinder, and a flat plate, using the same type of formulation. The results are shown in Table 7.22. Two forms of β_s were used. For the hemi-nosed cylinder, the calculated heat transfer, using β_{s_1}, underpredicted the experimental measurements by up to 64%. The calculated heat transfer using β_{s_2} only underpredicted the data by, at most, 29%. It overpredicted the data by up to 54%. Hargrave and Kilham (1984)[41] determined β_s as a function of turbulence intensity, Tu. For laminar flow (Tu = 0), $\beta_s = 2.67v_e/d_b$. This is slightly lower than the value of β_s used in most of the hemi-nosed cylinder studies. In later studies, Hargrave and co-workers measured β_s for heated air impinging normal to a cylinder[72] and parallel to a hemi-nosed cylinder.[73] For the cylinder, β_s was empirically determined as $\beta_s = 3.85 + 4.90$Tu. For the hemi-nosed cylinder, β_s was empirically determined as $\beta_s = 2.67 + 9.62$Tu. These relations were then used in subsequent flame impingement studies.[74,75] In those studies, empirical — not semianalytical — heat transfer equations were determined. Therefore, those equations have not been included here. Van der Meer showed (1991) that $\beta_s = v_e/d_b$ for disks applies within a distance of about five nozzle diameters from the burner outlet.[54]

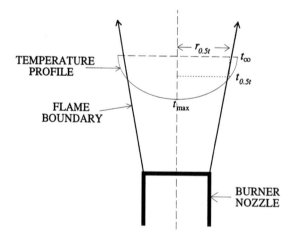

FIGURE 7.17   Gas temperature profiles in a flame jet. (From C.E. Baukal and B. Gebhart, *Int. J. Heat Mass Transfer*, 39(14), 2989-3002, 1996.)

TABLE 7.22
Stagnation Velocity Gradients Used by Hemeson et al.
(1984) in Reference 45

Target	β_{s1}	Predicted q_s'' / Measured q_s''	β_{s2}	Predicted q_s'' / Measured q_s''
Cylinder	$2v_e/d_b$	—	$2v_e/r_{0.5t}$	0.59–1.18
Hemi-nosed cylinder	$3v_e/d_b$	0.36–1.0	$3v_e/r_{0.5t}$	0.71–1.54
Flat plate	—	—	$1.5v_e/r_{0.5t}$	0.56–0.91

Source: From C.E. Baukal and B. Gebhart, *Int. J. Heat Mass Transfer,* 39(14), 2989-3002, 1996.

7.3.2 EQUATIONS

This chapter section discusses the early semianalytical solutions for the heat transfer in stagnation flows. Sibulkin (1952),[19] Fay and Riddell (1958),[69] and Rosner (1961)[61] developed equations to compute the heat flux at the stagnation point of an axisymmetric body in a uniform, external steady flow. Radiation effects were ignored. Sibulkin's equation included only the forced convection effect. Fay and Riddell and Rosner developed solutions for both equilibrium and catalytic TCHR.

7.3.2.1 Sibulkin Results

The heat transfer at the forward stagnation point of a body of revolution was considered. The flow was uniform, except in the boundary layer. The flow around the body in the boundary layer was assumed to be laminar, incompressible, axisymmetric, and of low speed. Using the axisymmetric boundary layer equations, the following relation for the local surface heat transfer was given:[19]

$$q_s'' = 0.763 \left(\beta_s \rho_e \mu_e \right)^{0.5} \mathrm{Pr}_e^{-0.6} c_{p_e} \left(t_e - t_w \right) \tag{7.15}$$

This applies for $0.6 < \mathrm{Pr}_e < 2.0$. This formulation was actually developed for the hypersonic velocities common to space vehicle reentry. It was assumed that the velocity is very low behind the bow shock-wave, near the stagnation point, due to boundary layer flow conditions. The constant 0.763 was determined numerically. All other semianalytical solutions presented here are based on Equation 7.15.

7.3.2.2 Fay and Riddell Results

Fay and Riddell used the same axisymmetric boundary layer equations as Sibulkin. However, chemical dissociation was included. The driving force for heat transfer was enthalpy, instead of temperature. A factor was added that contains the ratio of $\rho\mu$, evaluated at the wall and at the edge of the boundary layer. The heat transfer at the stagnation point for equilibrium TCHR was given as:[69]

$$q_s'' = 0.76 \left(\beta_s \rho_e \mu_e \right)^{0.5} \left(\frac{\rho_w \mu_w}{\rho_e \mu_e} \right)^{0.1} \mathrm{Pr}_e^{-0.6} \left[1 + \left(\mathrm{Le}^{0.52} - 1 \right) \frac{h_e^C - h_w^C}{h_e^T} \right] \left(h_e^T - h_w^T \right) \tag{7.16}$$

The heat transfer at the stagnation point for catalytic TCHR was given as:

$$q_s'' = 0.76 \left(\beta_s \rho_e \mu_e \right)^{0.5} \left(\frac{\rho_w \mu_w}{\rho_e \mu_e} \right)^{0.1} Pr_e^{-0.6} \left[1 + \left(Le_e^{0.63} - 1 \right) \frac{h_e^C - h_w^C}{h_e^T} \right] \left(h_e^T - h_w^T \right) \qquad (7.17)$$

The only difference between the equations is the exponent of the Lewis number, Le_e.

7.3.2.3 Rosner Results

In this formulation, the boundary layer equations were not solved directly, as had been done by Sibulkin and by Fay and Riddell. Instead, Fay and Riddell's equations were modified to include the effects of chemical dissociation. The resulting equation was then the sum of a forced convection term from Sibulkin and a diffusion-chemical reaction term. Two different forms of the solution were given, depending on the nature of the thermochemical heat release.[61] For equilibrium TCHR, the following form was recommended:

$$q_s'' = 0.763 \left(\beta_s \rho_e \mu_e \right)^{0.5} Pr_e^{-0.6} \left[1 + \left(Le_e - 1 \right) \frac{h_e^C - h_w^C}{h_e^T - h_w^T} \right]^{0.6} \left(h_e^T - h_w^T \right) \qquad (7.18)$$

For catalytic TCHR, the recommended form was:

$$q_s'' = 0.763 \left(\beta_s \rho_e \mu_e \right)^{0.5} Pr_e^{-0.6} \left[1 + \left(Le_e^{0.6} - 1 \right) \frac{h_e^C - h_w^C}{h_e^T - h_w^T} \right] \left(h_e^T - h_w^T \right) \qquad (7.19)$$

For equilibrium TCHR, Equation 7.18 differs significantly from that recommended by Fay and Riddell in Equation 7.16. For catalytic TCHR, Equation 7.19 is similar to that recommended in Equation 7.17. It has commonly been assumed that the driving force for energy transport is the total enthalpy difference across the boundary layer. Rosner noted that this is a common misconception. It applies only for $Le_e = 1$, since the terms inside the square brackets above also contain the total and chemical enthalpy differences across the boundary layer. For $Le_e = 1$, Equations 7.18 and 7.19 yield the same result. For the O_2/H_2 system considered by Rosner, the calculated heat flux was very similar using either equation, across a realistic range of values. The application was rocket motors. For that system, it was not important whether the TCHR was equilibrium or catalytic. Rosner showed that the factor $\frac{h_e^C - h_w^C}{h_e^T - h_w^T}$ becomes more important as the target temperature approaches the flame temperature ($t_w/t_e \rightarrow 1$). In that case, assuming $Le_e = 1$ may seriously underestimate the actual heat flux.

7.3.3 COMPARISONS WITH EXPERIMENTS

This chapter section discusses the impingement experiments in which the measured heat flux was compared to some form of a semianalytical solution. There have been two types of experiments. The first was for lower temperature flames. There, forced convection is the dominant mechanism. TCHR is negligible. The second type was for higher temperature flames. There, forced convection and TCHR are both important. Both types are further classified into laminar and turbulent flow regimes.

7.3.3.1 Forced Convection (Negligible TCHR)

This mechanism has been important in air/fuel flames. In some studies,[3,41,45] the driving potential was taken as $(h_e^T - h_w^T)$. For air/fuel flames, this is essentially the same as $(h_e^S - h_w^S)$.

7.3.3.1.1 Laminar flow

In one study,[3] a wide range of heat fluxes was calculated (convection with and without TCHR) for several variations of Equations 7.15 to 7.19. The best agreement with the experimental data was for:

$$q_s'' = 0.763 \left(\beta_s \rho_e \mu_e \right)^{0.5} \frac{\overline{\mu}}{\mu_e} \overline{Pr}^{-0.6} \left(h_e^T - h_w^T \right) \tag{7.20}$$

The maximum deviation between this relation and the experimental data was 4%. Horsley et al. (1982)[46] recommended the following modification of Equation 7.15:

$$q_s'' = 1.67 \left(\beta_s \overline{\rho \mu} \right)^{0.5} \overline{Pr}^{-0.6} \left(h_e^S - h_w^S \right) \tag{7.21}$$

This applies to $7050 \leq Re_n \leq 16200$. These values of Re_n are generally considered to be turbulent,[56] although the flames were described as having a "laminar appearance."

7.3.3.1.2 Turbulent flow

Horsley also inserted metal grids into industrial burners to promote turbulence.[46] A similar modification of Equation 7.15 was determined as,

$$q_s'' = 1.12 \left(\beta_s \overline{\rho \mu} \right)^{0.5} \overline{Pr}^{-0.6} \left(h_e^S - h_w^S \right) \tag{7.22}$$

This also applies to $7050 \leq Re_n \leq 16200$. The burner was located at the axial position, L, that yielded the maximum surface heat flux. The heat flux for the "laminar" flames, given in Equation 7.21, is greater than the flux for the turbulent flames given in Equation 7.22. Hargrave and Kilham (1984)[41] recommended the following relation:

$$q_s'' = 0.763 \left(\beta_s \overline{\rho \mu} \right)^{0.5} \overline{Pr}^{-0.6} \left(h_e^T - h_w^T \right) \tag{7.23}$$

It was shown that TCHR was not important. Therefore, all properties were evaluated at $Le = 1$. Equation 7.23 correlated with the experimental data to within 4%. Hemeson et al. (1984)[45] used a wide variety of burner designs. A range of cylinder and hemi-nosed cylinder diameters were tested. An equation similar to Equation 7.23, was suggested:

$$q_s'' = 0.763 \left(\beta_s \overline{\rho \mu} \right)^{0.5} Pr_e^{-0.6} \left(h_e^T - h_w^T \right) \tag{7.24}$$

It was found that the heat flux did not depend on the radius of curvature for the cylinder or for the hemi-nosed cylinder. The flux was dependent on the burner design. Van der Meer (1991)[54] used linear regression to correlate the experimental heat flux data as:

$$q_s'' = (1 + \gamma) 0.763 \left(\beta_s \rho_{ref} \mu_{ref} \right)^{0.5} Pr_{ref}^{-0.6} \left(h_e^S - h_w^S \right) \tag{7.25}$$

where γ = turbulence enhancement factor. This was an experimentally determined function of (Tu $Re^{0.5}$). It varied from 0.0 for Tu = 0, up to about 0.4 for Tu $Re^{0.5} \cong 30$.

7.3.3.2   Forced Convection with TCHR

The combination of forced convection and TCHR has been most important in O_2/fuel flames. The total enthalpy difference has been used as the driving potential. Some of the equations also included the effect of Le. Figure 2.50 shows the importance of Le for O_2/CH_4 flames, especially at high temperatures.

7.3.3.2.1   Laminar flow

Cookson and Kilham (1963)[36] tested fuel-lean air/H_2 flames. Kilham and Dunham (1967)[49] tested fuel-lean to fuel-rich air/CO flames. A modified form of Equation 7.18 was used in both studies. This included the catalytic TCHR effects for multiple active species:

$$q_s'' = 0.763 \left(\beta_s \rho_e \mu_e \right)^{0.5} \overline{\mathrm{Pr}}^{-0.6} \left[1 + \sum_i^n \Phi_i \left(\overline{\mathrm{Le}_i}^{0.6} - 1 \right) \frac{h_{e,i}^C - h_{w,i}^C}{h_e^T - h_w^T} \right] \left(h_e^T - h_w^T \right) \tag{7.26}$$

The surface catalytic efficiency, Φ_i, is a measure of the ability of the surface to act as a catalyst in a chemical reaction. A value of 0.0 means no radical species will catalytically react as a result of contacting the surface. A value of 1.0 means that all radical species will catalytically react upon contacting the surface. In Equation 7.26, a value of 1.0 was assumed. The actual value was not known. The quasi-equilibrium composition for air/H_2 mixtures was calculated, with a range of assumed concentrations of OH molecules.[35] The equilibrium components included H, H_2, H_2O, N_2, O, O_2, and OH. Using these concentrations and the above equation, the heat flux as a function of OH concentration was calculated. By equating the calculated flux with the measured flux, the theoretical OH mole fraction was estimated to be 1.7% for $\phi = 0.5$ and 3% for $\phi = 0.63$. The experimental and calculated values were in excellent agreement at the location in the flame where the atom concentrations (e.g., O) were negligible.[76] In general, the calculated values were shown to be highly dependent on the H, H_2, O, and OH concentrations. Those concentrations were not measured. Using Equation 7.26 and the measured total heat flux, the O concentration was estimated to be 2% by volume. Kilham and Purvis[39,77] measured the total heat flux from mostly fuel-rich flames. The data were compared to three different equations, using the recommendation by Fay and Riddell[69] for $\rho_e \mu_e$. Only forced convection heat transfer was assumed in the first equation:

$$q_s'' = 0.763 \left(\beta_s \rho_e \mu_e \right)^{0.5} \left(\frac{\rho_w \mu_w}{\rho_e \mu_e} \right)^{0.1} \overline{\mathrm{Pr}}^{-0.6} \left(h_e^S - h_w^S \right) \tag{7.27}$$

Equilibrium TCHR with no Le augmentation (Le = 1) was assumed in the second equation:

$$q_s'' = 0.763 \left(\beta_s \rho_e \mu_e \right)^{0.5} \left(\frac{\rho_w \mu_w}{\rho_e \mu_e} \right)^{0.1} \overline{\mathrm{Pr}}^{-0.6} \left(h_e^T - h_w^T \right) \tag{7.28}$$

Equilibrium TCHR with Le augmentation (Le > 1) was assumed in the third equation:

$$q_s'' = 0.763 \left(\beta_s \rho_e \mu_e \right)^{0.5} \left(\frac{\rho_w \mu_w}{\rho_e \mu_e} \right)^{0.1} \overline{\mathrm{Pr}}^{-0.6} \left[1 + \left(\mathrm{Le}_{e,\mathrm{H}} - 1 \right) \frac{h_{e,\mathrm{H}}^C - h_{w,\mathrm{H}}^C}{h_e^T - h_w^T} \right]^{0.6} \left(h_e^T - h_w^T \right) \tag{7.29}$$

Equation 7.29 is a form of Equation 7.26. Kilham showed that Equation 7.26 could be simplified by calculating Le_e and $(h_e^C - h_w^C)$, based only on H atom recombination. Equation 7.27 underpredicted the experimental data by 24% to 42%. Equation 7.28 underpredicted the data by 3% to 7%. Equation 7.29 overpredicted the data by 2% to 10%. Conolly and Davies (1972)[3] used several variations of Equations 7.15 to 7.20. A wide range of heat fluxes (forced convection with and without TCHR) were calculated. Equation 7.20 gave the best agreement with the experimental data. Nawaz (1973)[26] tested air-O_2/CH_4 ($\Omega = 0.46$–0.61) and O_2/CH_4 flames. Variations of Equations 7.27 to 7.29 were used. For nonreacting flow, the relation was given as:

$$q_s'' = 0.763 \left(\beta_s \rho_e \mu_e\right)^{0.5} \left(\frac{\rho_w \mu_w}{\rho_e \mu_e}\right)^{0.24} Pr_e^{-0.6} \left(h_e^S - h_w^S\right) \tag{7.30}$$

For equilibrium TCHR with Le = 1, the relation was given as:

$$q_s'' = 0.763 \left(\beta_s \rho_e \mu_e\right)^{0.5} \left(\frac{\rho_w \mu_w}{\rho_e \mu_e}\right)^{0.24} Pr_e^{-0.6} \left(h_e^T - h_w^T\right) \tag{7.31}$$

For equilibrium TCHR with Le > 1, the relation was given as:

$$q_s'' = 0.763 \left(\beta_s \rho_e \mu_e\right)^{0.5} \left(\frac{\rho_w \mu_w}{\rho_e \mu_e}\right)^{0.24} Pr_e^{-0.6} \left[1 + \left(Le_{e,H} - 1\right)\frac{h_{e,H}^C - h_{w,H}^C}{h_e^T - h_w^T}\right]\left(h_e^T - h_w^T\right) \tag{7.32}$$

The Le_e was calculated as the weighted average between the flame temperature and 1600K (2400°F). Based on the experimental data, the flame was divided into two regions. The region closest to the burner was assumed to have radicals in excess of that predicted by thermodynamic equilibrium calculations. This is referred to as superequilibrium. All three equations seriously underpredicted the measured data in that region. The far region, at about L > 0.7, appeared to be in chemical equilibrium. In that region, Equation 7.30 underpredicted the data by 17% to 77%; Equation 7.31 underpredicted by 0.6% to 5.5%; and Equation 7.32 overpredicted by 1.0% to 3.5%. It was concluded that the flow was mixed TCHR. As in their earlier study,[39] Kilham and Purvis (1978)[57] used three different equations to simulate various flow conditions. However, the Fay and Riddell[69] recommendation for $\rho_e\mu_e$ was not used. The properties were also evaluated differently. The following relation was given for nonreacting flows:

$$q_s'' = 0.763 \left(\beta_s \overline{\rho\mu}\right)^{0.5} Pr_e^{-0.6} \left(h_e^S - h_w^S\right) \tag{7.33}$$

This underpredicted the experimental data by 17% to 33%. The relation given for equilibrium TCHR, where $Le_e = 1$, was:

$$q_s'' = 0.763 \left(\beta_s \overline{\rho\mu}\right)^{0.5} Pr_e^{-0.6} \left(h_e^T - h_w^T\right) \tag{7.34}$$

For the CH_4 flames and $\phi > 1.12$, this underpredicted the data by at most 2%. For $0.95 \leq \phi \leq 1.12$, it overpredicted the data by as much as 40%. For the C_3H_8 flames and $1.45 \leq \phi \leq 1.83$, it underpredicted the data by 1% to 40%. The relation given for equilibrium TCHR, with $Le_e > 1$, was:

$$q_s'' = 0.763 \left(\beta_s \overline{\rho\mu}\right)^{0.5} Pr_e^{-0.6} \left[1 + \left(Le_{e,H} - 1\right)\frac{h_{e,H}^C - h_{w,H}^C}{h_e^T - h_w^T}\right]\left(h_e^T - h_w^T\right) \tag{7.35}$$

This overpredicted the data for both types of flames by as much as 48%. Fairweather et al. (1984)[27] tested air-O_2/CH_4 ($\Omega = 0.46$–0.61) and O_2/CH_4 flames. A modification of Equation 7.24 was used:

$$q''_s = 0.763\left(\beta_s \overline{\rho\mu}\right)^{0.5} \overline{\text{Pr}}^{-0.6} \left(h_e^T - h_w^T\right) \tag{7.36}$$

This was an upper limit when Le > 1. Equation 7.34 was used as a lower limit when Le = 1. For L > 0.8, Equation 7.34 underpredicted the data by up to 4%. Equation 7.36 overpredicted the data by up to 3%. For L < 0.8, Equations 7.34 and 7.35 underpredicted the data by up to 40%. The experimental data generally fell between these two limits, except near the reaction zone. There, all predictions severely underestimated the heat flux, compared to the measurements. It was determined that the inclusion of only H atom reactions was a good approximation to including all possible radical reactions.

7.3.3.2.2 Turbulent flow

Ivernel and Vernotte (1979)[28] tested air-O_2/natural gas ($\Omega = 0.25$–0.90) and O_2/natural gas flames. Although the flow regime was not specified, it appears to have been turbulent, based on a comparison with other studies.[56] A variation of Equation 7.15 was used:

$$q''_s = 0.763\left(\beta_s \rho_w \mu_w\right)^{0.5} \frac{\overline{\text{Pr}}^{0.4}}{\text{Pr}_w} \left(h_e^T - h_w^T\right) \tag{7.37}$$

This differs from the other semianalytical solutions because most of the properties are evaluated at the wall temperature. Equation 7.37 overpredicted the experimental data by up to 68%.

7.3.4 Sample Calculations

Sample calculations are given here to compare the predicted heat flux, using the semianalytical equations, with experimental measurements. The hemi-nosed cylinder is chosen as the target geometry since it was used in 9 of the 12 studies. Also, CH_4 is chosen as the fuel since it was used in 10 of the 12 studies, either directly as methane or indirectly as natural gas. In all of these studies, the flame was at or near stoichiometric conditions. The first set of calculations is for a stoichiometric air/CH_4 flame. This simulates forced convection without TCHR. Both laminar and turbulent conditions are modeled. The second set of calculations is for a stoichiometric O_2/CH_4 flame. This has been chosen to simulate forced convection with TCHR. Only laminar flames have been considered. This is due to the lack of both correlations and measurements for turbulent flames with TCHR. Tables 7.23 to 7.25 show the comparison between the measured and the computed heat fluxes for the flames that have been modeled. The first section in each table lists the values for the important parameters, such as d_b, that are used in the computations. The measured heat flux is given next. Finally, the predictions are given in order of increasing heat flux.

7.3.4.1 Laminar Flames Without TCHR

Table 7.23 shows the measured and calculated heat fluxes for a laminar flame without TCHR. The measured heat flux by Hargrave and Kilham (1984)[41] was from one set of tests at $\text{Re}_n = 2000$. The rest of those tests were done under turbulent flow conditions. Those results are discussed in the next section. As expected, the correlation by Conolly and Davies (1972)[3] in Equation 7.20, closely matches their own experimental data. The other three correlations significantly overpredict that data. The heat flux measurements by Hargrave and Kilham varied widely with L. The fluxes calculated using Equations 7.15 and 7.25 are within the measured range. Equation 7.21 overpredicts

TABLE 7.23
Heat Flux for Laminar Flames Without TCHR

Parameter	Units	Conolly & Davies (1972)[3]	Hargrave & Kilham (1984)[41]
β_s	sec^{-1}	$3v_e/d_b$	$2.67v_e/d_b$
d_b	mm	12.7	22
v_e	m/sec	11	2.5
t_w	K	400	393
t_e	K	2200	2000
q_s'', measured	kW/m^2	265	110–210
q_s'', Eq. 7.20	kW/m^2	272	81
q_s'', Eq. 7.25, $\gamma = 0$	kW/m^2	590	163
q_s'', Eq. 7.15	kW/m^2	642	157
q_s'', Eq. 7.21	kW/m^2	1225	356

Source: From C.E. Baukal and B. Gebhart, *Int. J. Heat Mass Transfer,* 39(14), 2989-3002, 1996.

TABLE 7.24
Heat Flux for Turbulent Flames Without TCHR

Parameter	Units	Hargrave & Kilham (1984)[41]
β_s, Tu = 0.2	sec^{-1}	$2.7v_e/d_b$
d_b	mm	22
v_e	m/sec	2.5
t_w	K	393
t_e	K	2000
q_s'', measured	kW/m^2	150–410
q_s'', Eq. 7.23	kW/m^2	139
q_s'', Eq. 7.24	kW/m^2	146
q_s'', Eq. 7.25, $\gamma = 0.4$	kW/m^2	229
q_s'', Eq. 7.22	kW/m^2	240

Source: From C.E. Baukal and B. Gebhart, *Int. J. Heat Mass Transfer,* 39(14), 2989-3002, 1996.

the Conolly and Davies experimental data by 360%. It overpredicts the Hargrave and Kilham data by up to 220%. Heat flux measurements from laminar air/fuel flames to hemi-nosed cylinders with a variety of diameters[56] have ranged from 73 to 460 kW/m^2 (23,000 to 146,000 Btu/hr-ft^2). The results using Equations 7.15, 7.21, and 7.25 for the Conolly and Davies[3] conditions exceed that range.

7.3.4.2 Turbulent Flames Without TCHR

The results for this case are shown in Table 7.24. The heat flux measured by Hargrave and Kilham, varied widely with turbulence intensity, Tu, and with axial position, L. The predicted heat flux values, using Equations 7.22 and 7.25, are within the measured range. Hargrave and Kilham recommended the relation given in Equation 7.23. It underpredicts their own experimental data by 7% to 66%. Heat flux measurements from turbulent air/fuel flames to hemi-nosed cylinders of various diameters have ranged from 100 to 580 kW/m^2 (32,000 to 180,000 Btu/hr-ft^2). The calculations in Table 7.24 are all within that range.

TABLE 7.25
Heat Flux for Laminar Flames with TCHR

Parameter	TCHR Type	Units	Kilham & Purvis, 1971[39]	Nawaz, 1973[26]; Fairweather et al., 1984[27]
β_s		sec^{-1}	$3v_e/d_b$	$3v_e/d_b$
d_b		mm	9.5	9.5
v_e		m/sec	27	37
t_w		K	380	360
t_e		K	2800	2900
q_s'', measured		kW/m^2	2530	2700–3640
q_s'', Eq. 7.20		kW/m^2	1163	1520
q_s'', Eq. 7.27		kW/m^2	1240	1520
q_s'', Eq. 7.30		kW/m^2	1630	2030
q_s'', Eq. 7.33		kW/m^2	1910	2440
q_s'', Eq. 7.28		kW/m^2	1950	2580
q_s'', Eq. 7.26	Catalytic	kW/m^2	2020	2690
q_s'', Eq. 7.29	Equil.	kW/m^2	2100	2800
q_s'', Eq. 7.36		kW/m^2	2340	3180
q_s'', Eq. 7.19	Catalytic	kW/m^2	2510	3400
q_s'', Eq. 7.18	Equil.	kW/m^2	2510	3410
q_s'', Eq. 7.31		kW/m^2	2550	3450
q_s'', Eq. 7.16	Equil.	kW/m^2	2560	3490
q_s'', Eq. 7.17	Catalytic	kW/m^2	2600	3540
q_s'', Eq. 7.32	Equil.	kW/m^2	2740	3760
q_s'', Eq. 7.34		kW/m^2	3000	4140
q_s'', Eq. 7.35	Equil.	kW/m^2	3230	4510

Source: From C.E. Baukal and B. Gebhart, *Int. J. Heat Mass Transfer,* 39(14), 2989-3002, 1996.

7.3.4.3 Laminar Flames with TCHR

The results for this case are given in Table 7.25. There is a wide variation in the predicted flux. However, most of the predictions that incorporate TCHR closely approximate the measurements. As expected, the correlations given in Equations 7.28, 7.30, and 7.33, for nonreacting flow significantly underpredict the data. The lowest heat flux prediction is for Equation 7.27. It underpredicts the measurements by 44 to 58%. This is because the sensible, not the total, enthalpy difference was used as the driving force. The highest predictions are for Equation 7.35. It overpredicts the measurements by 24% to 67%. The results using Equations 7.32, 7.34, and 7.35 exceeded the range for the Nawaz[26] and Fairweather[27] measurements.

The heat flux calculated using Fay and Riddell's relation for equilibrium TCHR in Equation 7.16, was lower than that for catalytic TCHR in Equation 7.17. However, the predicted heat flux using Rosner's relation for equilibrium TCHR in Equation 7.18 was higher than that for catalytic TCHR in Equation 7.19. Therefore, no trend in the predictions is apparent when comparing catalytic and equilibrium TCHR. Heat flux measurements from laminar O_2/fuel flames to hemi-nosed cylinders of various diameters have ranged from 410 to 3700 kW/m^2 (130,000 to 1.2×10^6 Btu/hr-ft²).[56]

7.3.5 SUMMARY

Twelve flame impingement experimental studies were reviewed here. In those studies, the measured heat flux was compared against one or more semianalytical solutions. Cylindrical, flat plate, and hemi-nosed cylindrical targets were used in 1, 3, and 9 studies, respectively. Note that Hemeson et al. (1984)[45] used both a cylindrical target and a hemi-nosed cylinder target. Laminar flames have been used more often than turbulent flames.

Catalytic TCHR has been considered in only two studies,[35,49] as given in Equation 7.26. An uncertainty is the lack of information on the surface catalytic efficiency, Φ, for given target materials. None of the studies compared the experimental results to both equilibrium and catalytic TCHR solutions. No semianalytical solutions have been suggested for mixed TCHR.

Sample calculations for laminar and turbulent flows without TCHR, and laminar flows with TCHR, generally showed good agreement with the experimental data. However, in one case, the prediction was nearly 5 times the experimental data. Therefore, caution must be used in the absence of any experimental data.

There are many possible explanations for the discrepancies between the predictions and the experimental data. It has been assumed that the experimental data are reliable. However, none of the studies gave an estimate of the uncertainty in the measurements. In most cases, a complete set of data has not been given for a specific heat flux measurement. Commonly, a range of values or an average value for a particular variable has been given. For example, Hargrave and Kilham (1984)[41] reported gas temperatures ranging from 1900K to 2100K (3000°F to 3300°F). No specific relationship between those temperatures and the reported heat fluxes was given. In the sample calculations, an average value of 2000K (3100°F) was used. Another possible source of error in the calculations is the choice of β_s. In most cases, Equation 7.11 for a sphere has been used, in the absence of an equation for a hemi-nosed cylinder. Also, the gases have been assumed to be at equilibrium conditions. In many cases, this is a reasonable assumption. However, Cookson and Kilham (1963)[36] and Kilham and Dunham (1967)[49] showed that the gases in the region closest to their burner were not in equilibrium. This could dramatically affect the properties used in the calculations.

For high-temperature flames, where dissociation is important, the TCHR effects must be included. Otherwise, the predictions will significantly underpredict the data. Fay and Riddell[69] and Rosner[61] have developed solutions for both equilibrium and catalytic TCHR. However, the sample calculations in Table 7.25 showed that the results are nearly the same for either type of TCHR. Further work should determine if that would be true over a wider range of conditions.

7.4 EMPIRICAL HEAT TRANSFER CORRELATIONS

Empirical heat transfer correlations for impinging flame jets have been generally given in two forms.[78] The first is the Nusselt number, which is a dimensionless heat transfer coefficient. In the studies considered here, it has appeared as:

$$\text{Nu} \approx a \, \text{Pr}^b \, \text{Re}^c$$

where a, b, and c are constants. The second form has been directly in terms of the heat flux, q''. These two forms are related as follows:

$$q'' = (k/d)\{\text{Nu}\} \, \Delta t$$

where k is the thermal conductivity of the fluid and d is a characteristic dimension.

For consistency, most of the correlations given here are written in terms of q''. This makes it easier to directly compare equations. Also, the driving force, or the potential for heat transfer, is explicitly given in the second form. In some correlations, this potential is the temperature difference. In others, it is the enthalpy difference. In the first form, the Nu formulation, the potential is not given explicitly. For the correlations converted from the first form to the second form (i.e., Nu to q''), the Nu relationship is shown inside curly brackets { }.

This chapter section is arranged to aid the designer or researcher in finding the appropriate correlation for a specific set of conditions. The primary consideration is to ensure that the heat transfer correlation applies to the chosen geometric configuration. The next consideration is the location on the target where the desired flux is to be calculated. This has typically been at a specific location, such as the stagnation point. This is referred to as a local heat flux. Correlations have also been determined for the average heat flux over a given portion of the target. Next, the correlations are arranged in terms of the heat transfer mechanism(s). These include convection, TCHR, radiation, and combinations of these. Finally, the correlations are arranged by the flow type (laminar or turbulent). In some studies (e.g., see Reference 27), correlations were given for both laminar and turbulent flows. However, in other studies, the flow type was not specified.

7.4.1 THERMOPHYSICAL PROPERTIES

The transport and thermodynamic properties used in the correlations presented below are discussed in this chapter section. These properties have been evaluated using several rules. These properties include, for example, the specific heat, thermal conductivity, enthalpy, and viscosity of the combustion products. The nomenclature is shown in Figure 7.16. The properties have been evaluated at the gas temperature at the edge of the stagnation zone and the boundary layer, t_e (Equation 7.1), at the wall temperature of the target, t_w (Equation 7.2), at the weighted average over the temperature range between the edge of the stagnation zone and the wall temperatures (Equation 7.3), at the film temperature (Equation 7.4), and at the reference temperature (Equation 7.5). In addition, Beér and Chigier (1968)[23] evaluated the properties at the maximum temperature measured at a given location in the flame jet:

$$p_{max} = p(t_{max}) \tag{7.38}$$

Definitions of dimensionless quantities to be introduced below, such as Re, Nu, and Pr, will require specification of a characteristic length scale and characteristic transport properties. The transport properties will, in turn, require specification of the rule by which they are evaluated (e.g., use of a specific characteristic temperature of the material). When only transport properties (i.e., material temperature) are required, this will be indicated by a single subscript (e.g., Pr_e [i.e., Pr evaluated at t_e]). When both a characteristic length and a property temperature are required, these will be indicated by the first and second of two subscripts, respectively. For example, $Re_{n,e}$ is shorthand notation for:

$$Re_{n,e} = \rho_e v_e d_n / \mu_e$$

For this example, the characteristic dimension is d_n. That dimension is an important parameter in correlating the heat transfer. There are many possible lengths that can be used. The most common one has been the burner nozzle diameter, d_n. Another has been the axial distance from the nozzle edge to the surface stagnation point, l_j. For cylindrical and hemi-nosed cylindrical targets (see Figures 7.9 and 7.10), the diameter of the body, d_b, has been used. For plane targets, the distance along the surface, r, has also been used. Smith and Lowes (1974)[37] used the width of the cooling channel in the target, l_c. Kataoka et al. (1984)[29] used the radius where the velocity was half the axial ($r = 0$) velocity, $r_{1/2v}$.

7.4.2 FLAMES IMPINGING NORMAL TO A CYLINDER

This refers to flames impinging perpendicular to a cylinder, as in Figure 7.9, a cylinder in crossflow. The experimental conditions for previous studies using this geometry are given in Table 7.1. The table also shows which heat transfer mechanisms were correlated in those studies.

7.4.2.1 Local Convection Heat Transfer

7.4.2.1.1 Laminar and turbulent flows

Hustad and co-workers (1991)[17] measured the radiation and the total heat fluxes for both laminar and turbulent flows. The radiation flux was also calculated. The convection heat transfer was calculated by subtracting the calculated radiation from the measured heat flux. Several correlations from the literature were tested for the convective portion of the heat transfer. The best match with the experimental convection at the stagnation point was the correlation from Zukauskas and Ziugzda (1985),[79] assuming Tu = 0.02:

$$q''_{s,conv} = \frac{k_e}{d_b}\left\{0.41\,Re_{b,e}^{0.6}\,Pr_e^{0.35}\,Tu^{0.15}\left(\frac{Pr_e}{Pr_w}\right)^{0.25}\right\}(t_e - t_w)\qquad(7.39)$$

For a pure fuel jet, there may have been significant quantities of uncombusted species in the flame at the impingement point. Therefore, TCHR may have been important, but was not discussed by Hustad and co-workers. If TCHR was present, it would have been included with forced convection in Equation 7.39.

7.4.2.1.2 Turbulent flows

Hargrave et al. (1987)[75] studied the effects of turbulence. The following correlation was determined:

$$q''_{s,conv} = \frac{\bar{k}}{d_b}\left\{\overline{Re_b}^{0.5}\left[1.071+4.669\left(\frac{Tu\,\overline{Re_b}^{0.5}}{100}\right)-7.388\left(\frac{Tu\,\overline{Re_b}^{0.5}}{100}\right)^2\right]\right\}(t_e - t_w)\qquad(7.40)$$

This correlation underpredicted the data by up to 14%.

7.4.2.2 Average Convection Heat Transfer

7.4.2.2.1 Laminar flows

In related studies, Kilham (1949)[13] and Jackson and Kilham (1956)[14] measured the total and radiative heat fluxes. The convective heat transfer was calculated by subtracting the radiation from the total heat flux. Kilham studied air/CO flames. The convection was correlated by:

$$q''_{b,conv} = \frac{\bar{k}}{d_b}\left\{\overline{Pr}^{0.3}\left(0.35+0.47\,\overline{Re_b}^{0.52}\right)\right\}(t_e - t_w)\qquad(7.41)$$

Jackson and Kilham studied air/H$_2$, O$_2$/H$_2$, and O$_2$/CO flames. The convection was correlated by:

$$q''_{b,conv} = \frac{\bar{k}}{d_b}\left\{\overline{Pr}^{0.3}\left(0.35+0.50\,\overline{Re_b}^{0.52}\right)\right\}(t_e - t_w)\qquad(7.42)$$

Both equations were valid for $60 \le Re_b \le 230$. The constant multiplying Re_b varied slightly in the two studies. McAdams' (1954) correlation[59] for this geometry, but for nonreacting jet impingement, had a multiplier of 0.47. In the O$_2$/H$_2$ and O$_2$/CO flames used in Jackson and Kilham's study, there may have been significant quantities of dissociated species. Therefore, TCHR may have been important, but was not discussed. If TCHR was present, it would have been included in Equation 7.42. However, a larger difference in Equations 7.41 and 7.42 would have been expected if TCHR was significant.

7.4.2.2.2   Laminar and turbulent flows

Hustad and co-workers (1991)[17] measured the total heat flux. The radiation heat flux was assumed to be uniformly distributed around the cylinder. The forced convection was calculated by subtracting the calculated radiation from the measured total heat flux. At a turbulence intensity of 2%, the best fit of the data was given by:

$$q''_{b,\text{conv}} = \frac{k_e}{d_b}\left\{0.23\,\text{Re}_{b,e}^{0.6}\,\text{Pr}_e^{0.35}\,\text{Tu}^{0.15}\left(\frac{\text{Pr}_e}{\text{Pr}_w}\right)^{0.25}\right\}(t_e - t_w) \tag{7.43}$$

7.4.2.2.3   Flow type unspecified

Fells and Harker (1968)[80] empirically determined:

$$q''_{b,\text{conv}} = \frac{\bar{k}}{d_b}\left\{\left(0.573 - 0.0179\,l_j\right)\overline{\text{Re}}_b^{0.5}\right\}(t_e - t_w) \tag{7.44}$$

for C_3H_8 flames. This correlated with the experimental data to within 10%. Davies (1979)[18] calculated the forced convection heat transfer using an equation developed by Conolly and Davies (1972)[3] for a water-cooled tube in crossflow:

$$q''_{b,\text{conv}} = \left(\bar{k}/d_b\right)\left\{1.32\,\overline{\text{Pr}}^{0.4}\,\overline{\text{Re}}_b^{0.5}\right\}(t_e - t_w) \tag{7.45}$$

7.4.2.3   Average Convection Heat Transfer with TCHR

7.4.2.3.1   Flow type unspecified

Davies (1965)[34] correlated heat flux data using an equation from McAdams (1954),[59] assuming $Pr = 0.7$:

$$q''_{b,\text{conv+TCHR}} = \left(\bar{k}/d_b\right)\left\{0.615\,\overline{\text{Re}}_b^{0.466}\right\}(t_e - t_w) \tag{7.46}$$

7.4.2.4   Average Radiation Heat Transfer

7.4.2.4.1   Laminar and turbulent flows

Hustad and co-workers (1991)[17] studied pure diffusion flames. The radiation from the CH_4 flames was given as:

$$q''_{b,\text{rad}} = 5 + 3.9\left(l_j - x_{\text{liftoff}}\right)\ \left(\text{kW}/\text{m}^2\right) \tag{7.47}$$

The radiation for the C_3H_8 flames was given as:

$$q''_{b,\text{rad}} = 19 + 7.8\left(l_j - x_{\text{liftoff}}\right)\ \left(\text{kW}/\text{m}^2\right) \tag{7.48}$$

These were based on an analytical model of the flame as a radiating cylinder. The liftoff distance, x_{liftoff}, is the length between the nozzle and the start of the flame. The length $l_j - x_{\text{liftoff}}$ was termed the radiation height. For a conventional burner, x_{liftoff} is zero since the flame is attached to the nozzle exit. The distances used in Equations 7.47 and 7.48 were in meters. In those studies, the fuel gas velocities at the nozzle outlet in terms of the Mach number were Ma ≥ 0.3. This equates to a minimum outlet velocity of 134 and 82 m/sec (440 and 270 ft/sec) for CH_4 and C_3H_8, respectively. The burning velocity in air for CH_4 and C_3H_8 is 0.37 and 0.41 m/sec (1.2 and 1.35 ft/sec), respectively.[81] Therefore, the flames were not attached to the burner because the exit velocities were much higher than the flame speed. This produced a so-called unattached or lifted flame. The flame starts where the local gas velocity is the same as the flame speed.

7.4.2.5 Maximum Convection and Radiation Heat Transfer

7.4.2.5.1 Turbulent flows
Hustad et al. (1992)[6] correlated their experimental data by:

$$q''_{max,conv+rad} = 10q_f^{0.3} \quad (kW/m^2)$$ (7.49)

based on a thermal input of q_f = 60 to 1600 kW (200,000 to 5.5×10^6 Btu/hr). The location of the maximum heat flux was at the center of the flame. For pure fuel jets, there may have been significant quantities of unburned fuel at the target impingement point. Therefore, TCHR may have been important. It would have been included in the above correlation.

7.4.3 FLAMES IMPINGING NORMAL TO A HEMI-NOSED CYLINDER

This pertains to cylinders whose axis is parallel to the flame, as shown in Figure 7.10. One end of the cylinder has a hemispherical nose. The flame impinges on that nose. Experimental conditions for these studies are given in Table 7.2.

7.4.3.1 Local Convection Heat Transfer

7.4.3.1.1 Laminar and turbulent flows
Fairweather et al. (1984)[27] correlated experimental data for laminar and turbulent flames. The correlations were modifications of empirical equations. These were derived for heat transfer from the stagnation point of heated spheres to turbulent air flows. Galloway and Sage (1968)[82] recommended the following equations for nonreacting flows:

$$Nu = 1.255 \, Re_{b,e}^{0.5} \left(Re_{b,e} \, Tu \right)^{0.0214} \quad \text{for } Re_{b,e} \, Tu < 7000$$ (7.50)

$$Nu_{s,e} = 1.128 \, Re_{b,e}^{0.5} \left(Re_{b,e} \, Tu \right)^{0.2838} \quad \text{for } Re_{b,e} \, Tu > 7000$$ (7.51)

Fairweather used modifications of these equations to get:

$$q''_{s,conv} = 1.743 \left(Re_{b,e} \, Tu \right)^{0.0214} \frac{\rho_e v_e c_{p_e}}{Re_{b,e}^{0.5}} \left(t_e - t_w \right)$$ (7.52)

Gostkowski and Costello (1970)[83] recommended the following equation, also for nonreacting flows:

$$Nu_{s,e} = 2 + \left\{ 1.849 \left(\frac{v_e}{v_w} \right)^{0.16} + \left[0.2122 Tu(Tu - 0.1072) + 0.001317 \right] Re_{b,e}^{1/2} \, Pr_e^{1/6} \right\} Re_{b,e}^{1/2} \, Pr_e^{1/3}$$ (7.53)

Fairweather recommended the following modification of Equation 7.53 for laminar and turbulent flames:

$$q''_{s,conv} = \rho_e v_e c_{p_e} \left[\begin{array}{c} \dfrac{2}{\mathrm{Pr}_e \mathrm{Re}_{b,e}} + \dfrac{1.1849}{\mathrm{Pr}_e^{2/3} \mathrm{Re}_{b,e}^{1/2}} \left(\dfrac{v_e}{v_w} \right)^{0.16} \\ + \left[0.2122 \mathrm{Tu}(\mathrm{Tu} - 0.1072) + 0.001317 \right] \mathrm{Pr}_e^{-1/2} \end{array} \right] (t_e - t_w) \qquad (7.54)$$

Both Equations 7.52 and 7.54 tended to overpredict the experimental data.

7.4.3.1.2 Turbulent flows

Hargrave et al. (1987)[75] recommended the following empirical correlation:

$$q''_{s,conv} = \frac{\bar{k}}{d_b} \left\{ \overline{\mathrm{Re}_b}^{0.5} \left[0.993 + 5.465 \left(\frac{\overline{\mathrm{TuRe}_b}^{0.5}}{100} \right) - 2.375 \left(\frac{\overline{\mathrm{TuRe}_b}^{0.5}}{100} \right)^2 \right] \right\} (t_e - t_w) \qquad (7.55)$$

This underpredicted the data by up to 15% near the reaction zone. Agreement was very good further downstream.

7.4.3.2 Local Convection Heat Transfer with TCHR

7.4.3.2.1 Turbulent flows

Ivernel and Vernotte (1979)[28] determined:

$$q''_{s,conv+TCHR} = \frac{k_w}{r_b} \left\{ 0.853 \mathrm{Re}_{r_b,w} \overline{\mathrm{Pr}}^{0.4} \right\} \frac{h_e^T - h_w^T}{c_{p_w}} \qquad (7.56)$$

This was valid for $1000 < \mathrm{Re}_{r_b,w} < 45000$. Equation 7.56 correlated with the experimental data to within 10%. Although the flow conditions were not given, they are believed to have been turbulent. This is based on a comparison of the reported velocities with those of other studies (see Baukal and Gebhart, 1995[56]). Radiation to the target was calculated. It was specifically excluded from the above correlation.

7.4.4 FLAMES IMPINGING NORMAL TO A PLANE SURFACE

This geometry is shown in Figure 7.11. The targets have included both disks and rectangular plates. The experimental conditions are given in Table 7.3.

7.4.4.1 Local Convection Heat Transfer

7.4.4.1.1 Laminar flows

In related studies, Hoogendoorn et al. (1978)[52] and Popiel et al. (1980)[53] determined two empirical correlations for the heat transfer to the stagnation point. These were based on the axial distance between the target and the burner. The correlation for $2 \leq L \leq 5$ was:

$$q''_{s,conv} = \frac{k_{ref}}{d_n} \left\{ (0.65 + 0.084L) \mathrm{Re}_{n,ref}^{0.5} \mathrm{Pr}_{ref}^{0.4} \right\} \frac{h_e^S - h_w^S}{c_{p_{ref}}} \qquad (7.57)$$

The correlation for L > 12 was:

$$q''_{s,conv} = \frac{k_{ref}}{d_n}\left\{(137-1.8L)\cdot 10^{-3}\,\mathrm{Re}_{n,ref}^{0.75}\,\mathrm{Pr}_{ref}^{0.4}\right\}\frac{h_e^S - h_w^S}{c_{p_{ref}}} \qquad (7.58)$$

Both correlations closely matched the data. For $5 < L \le 12$, the data showed a peak for Nu_n. No correlation was given. A modified form of a semianalytical solution was also determined. It was a form of the Equation 7.15 recommended by Sibulkin.[19] The modified correlation was:

$$q''_{s,conv} = \frac{k_{ref}}{d_n}\left\{2.37(L+1)^{-0.5}\,\mathrm{Re}_{n,ref}^{0.5}\,\mathrm{Pr}_{ref}^{0.4}\right\}\frac{h_e^S - h_w^S}{c_{p_{ref}}} \qquad (7.59)$$

valid for $8 \le L \le 20$. This underpredicted the data by as much as 60%. This was believed to have been caused by the failure to include the effects of free jet turbulence. Another correlation was recommended that included those effects:

$$q''_{s,conv} = \frac{k_{ref}}{d_n}\left\{\left[\frac{2.37}{(L+1)^{0.5}} + 2.22\left(\frac{\mathrm{TuRe}_{n,ref}^{0.5}}{100}\right) - 2.76\left(\frac{\mathrm{TuRe}_{n,ref}^{0.5}}{100}\right)^2\right]\mathrm{Re}_{n,ref}^{0.5}\,\mathrm{Pr}_{ref}^{0.4}\right\}\frac{h_e^S - h_w^S}{c_{p_{ref}}} \quad (7.60)$$

valid for $L \ge 8$. This compared favorably with the data. For $L \ge 4$, the maximum heat flux occurred at the stagnation point. For $L = 2$ and 3, the maximum flux occurred at about $R = 0.5$ and 0.2, respectively. Kataoka et al. (1984)[29] studied an air-O_2/CH_4 ($\Omega = 0.39$) flame. The heat transfer to the stagnation point was correlated by:

$$q''_{s,conv} = \frac{k_e}{2r_{1/2v}}\left\{1.44\,\mathrm{Re}_{2r_{1/2v},e}^{0.5}\,\mathrm{Pr}_e^{0.5}\left(L - X_v\right)^{0.12}\right\}(t_e - t_w) \qquad (7.61)$$

$r_{1/2v}$ is the radius from the burner axis near the stagnation point where the velocity is 1/2 of the velocity along the axis of symmetry. This correlation is valid only for $L > X_v$ with X_v correlated by:

$$X_v = 2.82\left(\rho_n/\rho_\infty\right)^{0.29}\,\mathrm{Re}_{n,X=0}^{0.07} \qquad (7.62)$$

In a subsequent study, Kataoka (1985)[84] found that the maximum heat flux for the same flame occurred at $L = X_v$.

7.4.4.1.2 Turbulent flows

Shorin and Pechurkin (1968)[20] empirically determined:

$$q''_{s,conv} = \frac{k_e}{d_n}\left\{4.04\,\mathrm{Re}_{n,e}^{0.2}\,\mathrm{Pr}_e\right\}(t_e - t_w) \qquad (7.63)$$

This was valid for $L \le X_v$. The correlation matched the experimental data within 15%. Another correlation was given for the heat flux for $L > X_v$:

$$\frac{\mathrm{Nu}_{s,e}\left(X_v < L < 14\right)}{\mathrm{Nu}_{s,e}\left(L < X_v\right)} = 0.8e^{-0.36\frac{(L-X_v)^2}{L}} \qquad (7.64)$$

The local heat transfer, as a function of R, was correlated by:

$$q''_{r,\text{conv}} = \frac{k_e}{r}\left\{3.22\,\text{Re}_{n,e}^{0.4}\,\text{Pr}_e\,e^{-0.36\frac{(L-X_v)^2}{L}-3.6\frac{R}{L}}\right\}(t_e - t_w) \tag{7.65}$$

for $0 < R/L < 0.9$. Vizioz and Lowes (1971)[25] investigated industrial-scale, air and oxygen-enriched air ($\Omega = 0.30$) natural gas flames impinging on cooled flat plate targets located inside a furnace. Three different types of flames with various levels of swirl were tested. The data were correlated by:

$$q''_{r,\text{conv}} = (k_e/r)\{\text{Nu}_{r,e}\}(t_e - t_w) \tag{7.66}$$

In one set of tests, the surface temperature of a water-cooled steel flat plate was maintained below 373K (212°F). The convective heat flux as a function of radius from the stagnation point was:

$$0.07\,\text{Re}_{r,e}^{0.8} < \text{Nu}_{r,e} < 0.22\,\text{Re}_{r,e} \tag{7.67}$$

In another set of tests, the surface temperature of an air-cooled refractory plate was maintained between 1280K and 1420K (1850°F and 2100°F). The convection correlation was:

$$0.035\,\text{Re}_{r,e}^{0.8} < \text{Nu}_{r,e} < 0.125\,\text{Re}_{r,e} \tag{7.68}$$

The Prandtl number effect was included in the coefficients.

In related studies, Buhr et al. (1969, 1973)[66,85] and Kremer et al. (1974)[21] determined the following modification of Sibulkin's equation:

$$q''_{s,\text{conv}} = 0.0371\,\text{Pr}_e^{-0.6}\,\rho_e v_e\left(h_e^S - h_s^S\right) \tag{7.69}$$

This was valid for $17680 \leq \text{Re}_n \leq 22700$, $13 \leq L \leq 65$. The experimental data was correlated within 9%. The peak flux was measured near the reaction zone. The heat flux showed a small linear decline with probe surface temperature over the range 320K to 1000K (120°F to 1300°F). Smith and Lowes (1974)[37] determined the following empirical correlation:

$$q''_{s,\text{conv}} = (k_e/l_c)\left\{1.6\,\text{Re}_{l_c,e}^{0.54}\,\text{Pr}_e^{0.52}\right\}(t_e - t_w) \tag{7.70}$$

where l_c was the width of the cooling channels. The correlation was valid for $0 \leq L \leq 8.4$. The fluid properties were taken, as for air, at the gas temperatures measured near the plate. The axial velocity at 10 cm (4 in.) from the plate was used to calculate the velocity in Re_n.

Matsuo et al. (1978)[30] determined the following correlation, for the heat flux to the stagnation point:

$$q''_{s,\text{conv}} = 0.010353\left(q_f^{0.936}/L^{1.032}\right)(t_e - t_w) \tag{7.71}$$

Another correlation was given for the heat flux as a function of the distance r from the stagnation point:

$$q''_{s,\text{conv}} = \left(q_f^{0.939}\Big/K_1 K_2 L^{\left(\frac{6.318}{r^{0.414}}\right)}\right)(t_e - t_w) \tag{7.72}$$

where

$$K_1 = -1.903 \cdot 10^{-7} r^2 + 3.89 \cdot 10^{-3} r + 0.985 \qquad (7.73)$$

$$K_2 = 8.42 \cdot 10^{-7} r^3 - 1.803 \cdot 10^3 r^2 + 1.37r + 5.00 \qquad (7.74)$$

Equations 7.71 and 7.72 were valid for $5 \leq L \leq 9$ and $148 \leq q_f \leq 482$. Equation 7.72 was valid for $0 < r \leq 500$, where r was in millimeters.

7.4.4.2 Local Convection Heat Transfer with TCHR

7.4.4.2.1 Laminar flows

Anderson and Stresino (1963)[15] found that the heat flux decreased exponentially as R increased:

$$q''_{r,\text{conv+TCHR}} = q''_{s,\text{conv+TCHR}} \, e^{-\frac{4r}{d_j} \overline{\text{Re}}_j^{-0.34}} \qquad (7.75)$$

where $q''_{s,\text{conv+TCHR}}$ was measured and d_j was the estimated flame diameter prior to impingement. Shorin and Pechurkin (1968)[20] found for $C_3H_8 + C_4H_{10}$ flames that:

$$q''_{s,\text{conv+TCHR}} = \frac{k_e}{d_n} \left\{ 7.8\,\text{Re}_{n,e}^{0.4} \, \text{Pr}_e \right\} \frac{h_e^T - h_w^T}{\overline{c}_p} \qquad (7.76)$$

for $L \leq X_v$. This correlated with the data to within 10%.

7.4.4.2.2 Turbulent flows

Rauenzahn (1986)[5] calculated the heat flux at the surface for the transient temperature field in a copper block. The assumed form of the surface flux distribution, after Anderson and Stresino,[15] was:

$$q''_{r,\text{conv+TCHR}} = q''_{s,\text{conv+TCHR}} \, e^{-\left(\frac{r}{r_0}\right)^a} \qquad (7.77)$$

r_0 was termed the spreading radius for the heat flux distribution. From the reduction of the data, $a \cong 1$. A transient inverse conduction analysis (see Gebhart, 1993[86]) was used to calculate $q''_{s,\text{conv+TCHR}}$ and $q''_{r,\text{conv+TCHR}}$. For $q_f = 61$ and 94 kW (210,000 and 320,000 Btu/hr) and $L = 6$ and 8, $q''_{s,\text{conv+TCHR}}$ ranged from 645 to 1210 kW/m^2 (205,000 to 384,000 Btu/hr-ft^2), and r_0 ranged from 7.3 to 9.8 cm (2.9 to 3.9 in.). The spreading radius generally decreased as L decreased and as q_f increased. Unfortunately, no correlations were given for $q''_{s,\text{conv+TCHR}}$ and r_0 in terms of L and q_f. Therefore, the constants used in the above equation are expected to be dependent on the specific experimental conditions.

7.4.4.3 Average Convection Heat Transfer

7.4.4.3.1 Laminar flows

You (1985)[7] studied pure diffusion flames impinging on a plate. The fuel flow from the burner nozzle was laminar. However, the buoyant plume impinging on the plate was turbulent. It was found that the convective heat flux in the stagnation zone was essentially constant:

$$q''_{b,\text{conv}} = 31.2 \left(q_f / l_j^2 \right) \text{Ra}_e^{-1/6} \, \text{Pr}_e^{-3/5} \qquad (7.78)$$

for R < 0.16. The flux decreased with R in the wall jet region:

$$q''_{b,conv} = 1.46 \text{R}^{-1.63} \left(q_f / l_j^2 \right) \text{Ra}_e^{-1/6} \text{Pr}_e^{-3/5} \qquad (7.79)$$

for R > 0.16. In both cases, $10^9 < \text{Ra} < 10^{14}$ and $\text{Pr} \cong 0.7$. The Rayleigh number was defined as:

$$\text{Ra} = g \tilde{\beta}_e q_f l_j^2 / \rho_e c_{p_e} v_e^3 \qquad (7.80)$$

No correlations were given for the measured radiation heat flux. This accounted for up to 26% of the total heat flux. The radiant flux was specifically excluded from the above correlations.

7.4.4.3.2   Turbulent flows

Shorin and Pechurkin (1968)[20] determined the average heat transfer as a function of R:

$$q''_{b,conv} = \frac{k_e}{r_b} \left\{ 6.44 \, \text{Re}_{n,e}^{0.4} \, \text{Pr}_e \, e^{0.36 \frac{(L-X_v)^2}{L}} r_b^2 \left[0.08 l_j^2 - 0.08 l_j^2 e^{-3.6 \frac{R}{L}} - 0.28 r_b l_j e^{-3.6 \frac{R}{L}} \right] \right\} (t_e - t_w) \quad (7.81)$$

7.4.5   FLAMES PARALLEL TO A PLANE SURFACE

This configuration, shown in Figure 7.12, is useful for studying the heat transfer to high-speed airfoils where very high temperatures occur at the leading edge. It is also useful for studying the heat transfer from flames to the walls of a furnace. The experimental conditions for these studies are given in Table 7.4.

7.4.5.1   Local Convection Heat Transfer with TCHR

7.4.5.1.1   Laminar flows

Woodruff and Giedt (1966)[22] studied the configuration shown in Figure 7.12a. This simulated an air foil. The best fit of the experimental data was obtained using:

$$q''_{r,conv+TCHR} = \frac{2\mu_e}{\delta} \text{Pr}_e^{-2/3} \left[1 + \left(\text{Le}_{e,H} - 1 \right) \frac{h_{e,H}^C - h_{w,H}^C}{h_e^T - h_w^T} \right]^{2/3} \left(h_e^T - h_w^T \right) \qquad (7.82)$$

The boundary layer thickness, δ, was measured with a Pitot tube. An empirical curve-fit of that data was given as:

$$\delta = 0.0139 e^{-0.312r} + 0.152 \left(\left(e^{0.703r} - 1 \right) / e^{0.625r} \right)^{0.5} \qquad (7.83)$$

r is the distance along the plate, measured in inches, from the leading edge. The best fit of the heat flux data was for Le = 3.0.

7.4.5.1.2   Turbulent flows

Giedt et al. (1960)[2] tested an array of fuel-rich C_2H_2 flames in parallel flow over a flat plate, as shown in Figure 7.12a. The heat flux data was correlated using a form of the equation recommended by Eckert (1956):[63]

$$q''_{r,\text{conv+TCHR}} = \left(\overline{k}/r\right)\left\{0.0296\,\overline{\text{Re}}_r^{0.8}\,\overline{\text{Pr}}^{1/3}\right\}\left(t_e - t_w\right) \tag{7.84}$$

r is the distance from the leading edge. This equation underpredicted the experimental data for by 25%. A derived correction factor was then added for the H atom recombination reaction:

$$q''_{r,\text{conv+TCHR}} = \frac{\overline{k}}{r}\left\{0.0296\,\overline{\text{Re}}_r^{0.8}\,\overline{\text{Pr}}^{1/3}\left[1 + \left\{\frac{Q_\text{H}\left(m_{\text{H},e} - m_{\text{H},w}\right)}{M\overline{\rho}\,\overline{c_p}\left(t_e - t_w\right)}\right\}\right]\right\}\left(t_e - t_w\right) \tag{7.85}$$

$$\text{where}\quad M = \left(m_{\text{H},w} - m_{\text{H},e}\right)\Bigg/\left(\ln\frac{1 - m_{\text{H},w}}{1 - m_{\text{H},e}}\right) \tag{7.86}$$

$$Q_\text{H} = 436 \text{ kJ/kg-mole}$$

Q_H is the H atom heat of recombination and m_H is the H atom mole fraction. This correlated with the data to within 15%.

7.4.5.2   Local Convection and Radiation Heat Transfer

7.4.5.2.1   Turbulent flows
Beér and Chigier (1968)[23] studied flames impinging at a 20° angle from the furnace hearth (see Figure 7.12b). The correlation for the total heat flux was given as:

$$q''_{r,\text{conv+rad}} = \left(k_\text{max}/r\right)\left\{0.13\,\text{Re}_{r,\text{max}}^{0.8}\right\}\left(t_\text{max} - t_w\right) \tag{7.87}$$

The velocity in the Reynolds number was the maximum velocity measured at the axial distance from the point of impingement. The temperature, t_max, was the maximum temperature measured at the axial distance from the point of impingement. A Prandtl number of 0.7 was assumed.

REFERENCES

1. C.E. Baukal, L.K. Farmer, B. Gebhart, and I. Chan, Heat transfer mechanisms in flame impingement heating, in *1995 International Gas Research Conf.*, Vol. II, D.A. Dolenc, Ed., Govt. Institutes, Rockville, MD, 1996, 2277-2287.
2. W.H. Giedt, L.L. Cobb, and E.J. Russ, Effect of Hydrogen Recombination on Turbulent Flow Heat Transfer, ASME Paper 60-WA-256, New York, 1960.
3. R. Conolly and R.M. Davies, A study of convective heat transfer from flames, *Int. J. Heat Mass Trans.*, 15, 2155-2172, 1972.
4. M.A. Cappelli, and P.H. Paul, An investigation of diamond film deposition in a premixed oxyacetylene flame, *J. Appl. Phys.*, 67(5), 2596-2602, 1990.
5. R.M. Rauenzahn, Analysis of Rock Mechanics and Gas Dynamics of Flame-Jet Thermal Spallation Drilling, Ph.D. thesis, Massachusetts Institute of Tech., Cambridge, MA, 1986.
6. J.E., Hustad, M., Jacobsen, and O.K. Sønju, Radiation and heat transfer in oil/propane jet diffusion flames, *Inst. Chem. Eng. Symp. Series*, 10(129), 657-663, 1992.
7. H.-Z. You, Investigation of fire impingement on a horizontal ceiling. 2. Impingement and ceiling-jet regions, *Fire & Materials*, 9(1), 46-56, 1985.
8. P.A. Eibeck, J.O. Keller, T.T. Bramlette, and D.J. Sailor, Pulse combustion: impinging jet heat transfer enhancement, *Comb. Sci. Tech.*, 94, 147-165, 1993.

9. A. Milson and N.A. Chigier, Studies of methane-air flames impinging on a cold plate, *Comb. Flame*, 21, 295-305, 1973.

10. C.E. Baukal and B. Gebhart, A review of semi-analytic solutions for flame impingement heat transfer, *Int. J. Heat Mass Trans.*, 39(14), 2989-3002, 1996.

11. C.E. Baukal and B. Gebhart, A review of flame impingement heat transfer studies. 1. Experimental conditions, *Comb. Sci. Tech.* 104, 339-357, 1995.

12. K.H. Hemsath, A novel gas fired heating system for indirect heating, in *Fossil Fuel Combustion Symposium 1990*, S. Singh, Ed., New York, ASME PD-Vol. 30, 155-159, 1990.

13. J.K. Kilham, Energy transfer from flame gases to solids, *Third Symposium on Combustion and Flame and Explosion Phenomena*, Williams and Wilkins, Baltimore, MD, 1949, 733-740.

14. E.G. Jackson and J.K. Kilham, Heat transfer from combustion products by forced convection, *Ind. Eng. Chem.*, 48(11), 2077-2079, 1956.

15. J.E. Anderson and E.F. Stresino, Heat transfer from flames impinging on flat and cylindrical surfaces, *J. Heat Trans.*, 85(1), 49-54, 1963.

16. J.E. Hustad, and O.K. Sønju, Heat transfer to pipes submerged in turbulent jet diffusion flames, in *Heat Transfer in Radiating and Combusting Systems*, Springer-Verlag, Berlin, 1991, 474-490.

17. J.E. Hustad, N.A. Røkke, and O.K. Sønju, Heat transfer to pipes submerged in lifted buoyant diffusion flames, in *Experimental Heat Transfer, Fluid Mechanics, and Thermodynamics, 1991*, J.F. Keffer et al., Eds., Elsevier, New York, 1991, 567-574.

18. D.R. Davies, Heat Transfer from Working Flame Burners. B.S. thesis, Univ. of Salford, Salford, U.K., 1979.

19. M. Sibulkin, Heat transfer near the forward stagnation point of a body of revolution, *J. Aero. Sci.*, 19, 570-571, 1952.

20. S.N. Shorin, and V.A. Pechurkin, Effectivnost' teploperenosa na poverkhnost' plity ot vysokotemperaturnoi strui produktov sjoraniya razlichnykh gazov, *Teoriya i Praktika Szhiganiya Gaza*, 4, 134-143, 1968.

21. H., Kremer, E., Buhr, and R. Haupt, Heat transfer from turbulent free-jet flames to plane surfaces, in *Heat Transfer in Flames*, N.H. Afgan and J.M. Beér, Eds., Scripta Book Company, Washington, D.C., 1974, 463-472.

22. L.W. Woodruff and W.H. Giedt, Heat transfer measurements from a partially dissociated gas with high Lewis number, *J. Heat Trans.*, 88, 415-420, 1966.

23. J.M. Beér and N.A. Chigier, Impinging jet flames, *Comb. Flame*, 12, 575-586, 1968.

24. J.W. Mohr, J. Seyed-Yagoobi, and R.H. Page, Heat transfer from a pair of radial jet reattachment flames, in *Combustion and Fire, ASME Proceedings of the 31st National Heat Transfer Conf.*, Vol. 6, M. McQuay, K. Annamalai, W. Schreiber, D. Choudhury, E. Bigzadeh, and A. Runchal, Eds., New York, HTD-Vol. 328, 11-17, 1996.

25. J.-P. Vizioz and T.M. Lowes, Convective heat transfer from impinging flame jets, Int'l Flame Research Found. Report F 35/a/6, IJmuiden, The Netherlands, 1971.

26. S. Nawaz, Heat transfer from oxygen enriched methane flames, Ph.D. thesis, The University of Leeds, Leeds, U.K., 1973.

27. M. Fairweather, J.K. Kilham, and S. Nawaz, Stagnation point heat transfer from laminar, high temperature methane flames, *Int. J. Heat Fluid Flow*, 5(1), 21-27, 1984.

28. A. Ivernel and P. Vernotte, Etude expérimentale de l'amélioration des transferts convectis dans les fours par suroxygénation du comburant, *Rev. Gén. Therm.*, Fr., (210-211), 375-391, 1979.

29. K. Kataoka, H. Shundoh, and H. Matsuo, Convective heat transfer between a flat plate and a jet of hot gas impinging on it, in *Drying '84*, A.S., Majumdar, Ed., Hemisphere/Springer-Verlag, New York, 1984, 218-227.

30. M. Matsuo, M. Hattori, T. Ohta, and S. Kishimoto, The Experimental Results of the Heat Transfer by Flame Impingement, Int'l Flame Research Found. Report F 29/1a/1, IJmuiden, The Netherlands, 1978.

31. C.E. Baukal and B. Gebhart, Heat transfer from oxygen-enhanced/natural gas flames impinging normal to a plane surface, *Exp. Therm. & Fluid Sci.*, 16(3), 247-259, 1998.

32. E.M. Schulte, Impingement heat transfer rates from torch flames, *J. Heat Trans.*, 94, 231-233, 1972.

33. R.H. Fay, Heat transfer from fuel gas flames, *Welding J.*, Research Supplement, 380s-383s, 1967.

34. R.M. Davies, Heat transfer measurements on electrically-boosted flames, *Tenth Symposium (International) on Combustion*, The Combustion Institute, Pittsburgh, PA, 1965, 755-766.

35. R.A. Cookson, An Investigation of Heat Transfer from Flames, Ph.D. thesis, The University of Leeds, Leeds, U.K., 1960.

36. R.A. Cookson and J.K. Kilham, Energy transfer from hydrogen-air flames. *Ninth Symposium (International) on Combustion*, Academic Press, New York, 1963, 257-263.

37. R.B. Smith and T.M. Lowes, Convective heat transfer from impinging tunnel burner flames — A short report on the NG-4 trials, Int'l Flame Research Found. Report F 35/a/9, IJmuiden, The Netherlands, 1974.

38. J.B. Rajani, R. Payne, and S. Michelfelder, Convective heat transfer from impinging oxygen-natural gas flames — Experimental results from the NG5 Trials, Int'l Flame Research Found. Report F 35/a/12, IJmuiden, The Netherlands, 1978.

39. J.K. Kilham and M.R.I. Purvis, Heat transfer from hydrocarbon-oxygen flames, *Comb. Flame*, 16, 47-54, 1971.

40. M. Fairweather, J.K. Kilham, and Mohebi-A. Ashtiani, Stagnation point heat transfer from turbulent methane-air flames, *Comb. Sci. Tech.*, 35, 225-238, 1984.

41. G.K. Hargrave and J.K. Kilham, The effect of turbulence intensity on convective heat transfer from premixed methane-air flames, *Inst. Chem. Eng. Symp. Ser.*, 2(86), 1025-1034, 1984.

42. C.C. Veldman, T. Kubota, and E.E. Zukoski, An Experimental Investigation of the Heat Transfer from a Buoyant Gas Plume to a Horizontal Ceiling. 1. Unobstructed Ceiling. National Bureau of Standards Report NBS-GCR-77-97, Washington, D.C., 1975.

43. T.B. Reed, Heat-transfer intensity from induction plasma flames and oxy-hydrogen flames, *J. Appl. Phys.*, 34(8), 2266-2269, 1963.

44. J.R. Rigby and B.W. Webb, An experimental investigation of diffusion flame jet impingement heat transfer, *Proceedings of the ASME/JSME Thermal Engineering Conference*, ASME, New York, 3, 117-126, 1995.

45. A.O. Hemeson, M.E. Horsley, M.R.I. Purvis, and A.S. Tariq, Heat transfer from flames to convex surfaces, *Inst. of Chem. Eng. Symp. Series*, 2(86), 969-978. Publ. by Inst. of Chem. Eng., Rugby, U.K., 1984.

46. M.E. Horsley, M.R.I. Purvis, and A.S. Tariq, Convective heat transfer from laminar and turbulent premixed flames, in *Heat Transfer 1982*, U. Grigull, E. Hahne, K. Stephan, and J. Straub, Eds., Hemisphere, Washington, D.C., 3, 409-415, 1982.

47. A.S. Tariq, Impingement heat transfer from turbulent and laminar flames, Ph.D. thesis, Portsmouth Polytechnic, Hampshire, U.K., 1982.

48. C.J. Posillico, Raman Spectroscopic and LDV Measurements of a Methane Jet Impinging Normally on a Flat Water-Cooled Boundary. Ph.D. thesis, Polytechnic Institute of New York, New York, 1986.

49. J.K. Kilham and P.G. Dunham, Energy transfer from carbon monoxide flames, *Eleventh Symposium (International) on Combustion*, The Combustion Institute, Pittsburgh, PA, 1967, 899-905.

50. R. Conolly, A Study of Convective Heat Transfer from High Temperature Combustion Products, Ph.D. thesis, University of Aston, Birmingham, U.K., 1971.

51. C.E. Baukal and B. Gebhart, Surface condition effects on flame impingement heat transfer, *Therm. & Fluid Sci.*, 15, 323-335, 1997.

52. C.J. Hoogendoorn, C.O. Popiel, and T.H. van der Meer, Turbulent heat transfer on a plane surface in impingement round premixed flame jets, *Proc. of 6th Int. Heat Trans. Conf.*, Toronto, 4, 107-112, 1978.

53. C.O. Popiel, T.H. van der Meer, and C.J. Hoogendoorn, Convective heat transfer on a plate in an impinging round hot gas jet of low Reynolds Number, *Int. J. Heat Mass Trans.*, 23, 1055-1068, 1980.

54. T.H. van der Meer, Stagnation point heat transfer from turbulent low Reynolds number jets and flame jets, *Exper. Ther. Fluid Sci.*, 4, 115-126, 1991.

55. G.K. Hargrave, A Study of Forced Convective Heat Transfer from Turbulent Flames, Ph.D. thesis, University of Leeds, Leeds, U.K., 1984.

56. C.E. Baukal and B. Gebhart, A review of flame impingement heat transfer studies. 2. Measurements, *Comb. Sci. Tech.*, 104, 359-385, 1995.

57. J.K. Kilham and M.R.I. Purvis, Heat transfer from normally impinging flames, *Comb. Sci. Tech.*, 18, 81-90, 1978.

58. D.C.C. Chen and I.A. McGrath, Convective heat transfer in chemically reacting systems, *J. Inst. Fuel*, 42(336), 12-18, 1969.

59. W.H. McAdams, *Heat Transmission,* 3rd edition, McGraw-Hill, New York, 1954, chap. 10.

60. D. Altman and H. Wise, Effect of chemical reactions in the boundary layer on convective heat transfer, *Jet Propulsion*, 26(4), 256-269, 1956.
61. D.E. Rosner, Convective Heat Transfer with Chemical Reaction, Aeron. Res. Lab. Rept. ARL 99, Part 1, AD269816, 1961.
62. D.C.C. Chen, Improvement in convective heat transfer from hydrocarbon flames by oxygen enrichment, *J. Inst. Fuel*, 45(380), 562-567, 1972.
63. E.R.G. Eckert, Engineering relations for heat transfer and friction in high-velocity laminar and turbulent boundary-layer flow over surfaces with constant pressure and temperature, *J. Heat Trans.*, 78, 1273-1283, 1956.
64. S. Gordon, B.J. McBride, and F.J. Zeleznik, Computer Program for Calculation of Complex Chemical Equilibrium Compositions and Applications. Supplement I. Transport Properties, NASA Technical Memorandum 86885, Washington, D.C., 1984.
65. S. Goldstein, *Modern Developments in Fluid Dynamics*, Dover Publications, New York, 1965, 142.
66. E. Buhr, G. Haupt, and H. Kremer, Heat transfer from impinging turbulent jet flames to plane surfaces, in *Combustion Institute European Symposium 1973*, F.J. Weinberg, Ed., Academic Press, New York, 1973, 607-612.
67. P.S. Shadlesky, Stagnation point heat transfer for jet impingement to a plane surface, *AIAA J.*, 21(8), 1214-1215, 1983.
68. R. Viskanta, Heat transfer to impinging isothermal gas and flame jets, *Exper. Therm. Fluid Sci.*, 6, 111-134, 1993.
69. J.A. Fay and F.R. Riddell, Theory of stagnation point heat transfer in dissociated air, *J. Aero. Sci.*, 25, 73-85, 1958.
70. T.H. van der Meer, Heat Transfer from Impinging Flame Jets, Ph.D. thesis, Technical University of Delft, The Netherlands, 1987.
71. W.M. Kays and M.E. Crawford, *Convective Heat and Mass Transfer*, McGraw-Hill, New York, 1993, 141.
72. G.K. Hargrave, M. Fairweather, and J.K. Kilham, Turbulence enhancement of stagnation point heat transfer on a circular cylinder, *Int. J. Heat Fluid Flow*, 7(2), 89-95, 1986.
73. G.K. Hargrave, M. Fairweather, and J.K. Kilham, Turbulence enhancement of stagnation point heat transfer on a body of revolution, *Int. J. Heat Fluid Flow*, 6(2), 91-98, 1985.
74. G.K. Hargrave, M. Fairweather, and J.K. Kilham, Forced convective heat transfer from premixed flames. 1. Flame structure, *Int. J. Heat Fluid Flow*, 8(1), 55-63, 1987.
75. G.K. Hargrave, M. Fairweather, and J.K. Kilham, Forced convective heat transfer from premixed flames. 2. Impingement heat transfer, *Int. J. Heat Fluid Flow*, 8(2), 132-138, 1987.
76. P.G. Dunham, Convective Heat Transfer from Carbon Monoxide Flames, Ph.D. thesis, The University of Leeds, Leeds, U.K., 1963.
77. M.R.I. Purvis, Heat Transfer from Normally Impinging Hydrocarbon Oxygen Flames to Surfaces at Elevated Temperatures, Ph.D. thesis, The University of Leeds, Leeds, U.K., 1974.
78. C.E. Baukal and B. Gebhart, A review of empirical flame impingement heat transfer correlations, *Int. J. Heat Fluid Flow*, 17(4), 386-396, 1996.
79. A. Zukauskas and J. Ziugzda, *Heat Transfer of Crossflow*, Hemisphere, Washington, D.C., 1985.
80. I. Fells and J.H. Harker, An investigation into heat transfer from unseeded propane-air flames augmented with D.C. electrical power, *Comb. Flame*, 12, 587-596, 1968.
81. B. Lewis and G. von Elbe, *Combustion, Flames and Explosions of Gases*, third edition, Academic Press, New York, 1987.
82. T.R. Galloway and B.H. Sage, Thermal and material transfer from spheres, prediction of local transport, *Int. J. Heat Mass Trans.*, 11, 539-549, 1968.
83. V.J. Gostowski and F.A. Costello, The effect of free stream turbulence on the heat transfer from the stagnation point of a sphere, *Int. J. Heat Mass Trans.*, 13, 1382-1386, 1970.
84. K. Kataoka, Optimal nozzle-to-plate spacing for convective heat transfer in nonisothermal, variable-density impinging jets. *Drying Tech.*, 3, 235-254, 1985.
85. E. Buhr, Über den Wärmefluß in Staupunkten von turbulenter Freistrahl-Flammen an gekühlten Platten, Ph.D. thesis, University of Trier-Kaiserslautern, Kaiserslautern, Germany, 1969.
86. B. Gebhart, *Heat Conduction and Mass Diffusion*, McGraw-Hill, New York, 1993.

8 Heat Transfer from Burners

8.1 INTRODUCTION

The purpose of this chapter is to consider the heat transfer from different types of burners, without much consideration for the applications of the burners that are discussed in Chapters 10 and 11. Thus, the emphasis here is to give the reader an idea of the general behavior for different types of burners. This information should be useful when considering new applications or reevaluating existing technologies.

8.2 OPEN-FLAME BURNERS

Here, open-flame burners are ones where the flame is not confined as is the case in radiant tubes, nor are the flames primarily attached to a surface as in porous refractory burners. Open-flame burners are normally visible to the naked eye where the radiant heat from the flame, rather than from a surface heated by the flame, can directly heat the load. An example of a ribbon burner (open flame) is shown in Figure 8.1.

8.2.1 MOMENTUM EFFECTS

There are two aspects to the momentum effects on flames. The first involves the forward momentum normally associated with the average outlet velocity of the combustion products. The second aspect is the lateral momentum caused by swirl. The swirl number is defined as the ratio of the lateral momentum to the forward momentum. Burners with no swirl have a swirl number of zero. Beér and Chigier (1972) defined weak swirl as a swirl number below 0.6 and strong swirl as a swirl number greater than 0.6.[7] Villasenor and Escalera (1998) studied swirl effects on heavy fuel oil combustion in an air-cooled, high-temperature research furnace.[1] The experimental results showed that there was a strong dependence between the incident radiation flux and the swirl number. Burners with intermediate swirl numbers (0.1 and 0.4) produced more uniform heat flux profiles than burners with either no swirl or with higher swirls (0.75 and 1.0). As expected, the burner swirl has an impact on the heat flux distribution from the flame.

8.2.2 FLAME LUMINOSITY

An example of a high-luminosity flame is shown in Figure 8.2. An example of a low-luminosity flame is shown in Figure 8.3. The flame luminosity is a function of many variables, but is especially dependent on the fuel. Solid and liquid fuels tend to make more luminous flames than gaseous fuels because of particles in the flame that radiate like graybodies. A recent trend in the glass industry has been to make more luminous flames, with natural gas as the fuel, to improve the thermal efficiency of the glass-melting process.[2] The burner design also plays a large role in how luminous the flame will be and how heat is transferred from the flame to the load.[3]

The radiant energy from a flame can be approximated by:[4]

$$q_{rad} \propto a_p V_f T_f^4 \tag{8.1}$$

FIGURE 8.1 High-capacity ribbon burner. (Courtesy of Ensign Ribbon Burners LLC, Pelham Manor, NY.)

where a_p is the Planck mean absorption coefficient for an optically thin flame, V_f is the flame volume, and T_f is the absolute temperature of the flame. The total heat release by combustion is:

$$q_{total} = \dot{m}_{fuel}\Delta H_c \qquad (8.2)$$

where $\dot{m}_{fuel}$ is the fuel mass flow rate and ΔH_c is the heat of combustion. The radiant fraction is then defined as the ratio of the radiant heat transfer to the total heat released by combustion:

$$\chi_{rad} = q_{rad}/q_{total} \approx a_p T_f^4 \, d/u \qquad (8.3)$$

where d is the diameter of the nozzle outlet diameter, u is the outlet velocity, and the flame volume was assumed proportional to d^3 and the fuel flow rate to d^2u. Turns and Myhr (1991) studied the interaction between flame radiation and NOx emissions from turbulent jet flames.[14] Radiant fractions were calculated from radiant heat flux measurements made with a transducer having a 150° view angle with a window for free jet flames of C_2H_4, C_3H_8, CH_4, and a blend of 57% CO/43% H_2. A global residence time was defined as:

$$\tau_g = \frac{\rho_f w_f^2 l_f f_s}{2\rho_0 d^2 u} \qquad (8.4)$$

where ρ_f is the flame density, w_f is the flame width, l_f is the flame length, f_s is the fuel mass fraction, and ρ_0 is the cold fuel density. Figure 8.4 shows a plot of the calculated radiant fraction as a function of the global residence time for all four fuels. At small residence times, the radiant fraction is somewhat independent of the fuel composition because the flames are momentum-dominated and

FIGURE 8.2   Example of a high-luminosity oil flame. (Courtesy of John Zink Co. LLC, Tulsa, OK.)

nonluminous. As the residence time increases, the radiant fraction becomes dependent on the sooting tendency of the fuel.

Slavejkov et al. (1993) showed that a new-style oxy/fuel burner can produce significantly more radiant flux than conventional, older-style designs.[5] A comparison of the radiant flux as a function of firing rate is shown in Figure 8.5. In some cases, the measured radiant flux was more than double using the new burner design.

8.2.3   Firing Rate Effects

The main concern in industrial combustors is normally to maintain a certain temperature profile in the material being heated, which often equates to a specific temperature profile inside the combustor for a given burner type. It is often necessary to adjust the firing rate to meet the needs of a given application. For example, for an existing combustion system it may be desirable to increase the material processing rate, which normally means the firing rate must be increased. The design question to be answered is by how much, since this may or may not be a linear relationship. Another example is the modification of a well-known system design for higher or lower throughput rates. Again, the question is how to do the scaling from the known design.

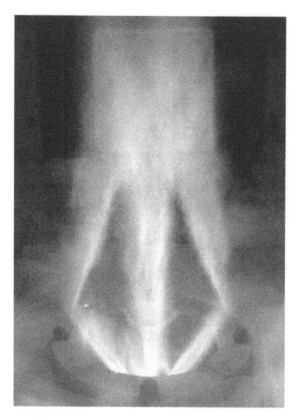

FIGURE 8.3 Example of a low-luminosity gas flame. (Courtesy of John Zink Co. LLC, Tulsa, OK.)

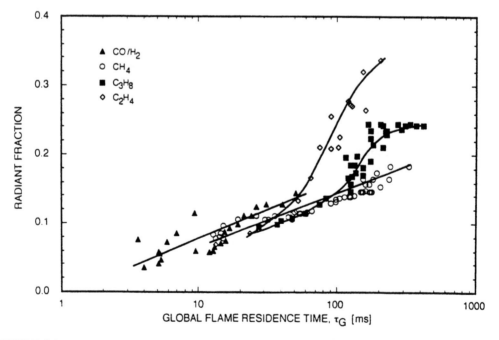

FIGURE 8.4 Flame radiant fractions as functions of global residence time. (Courtesy of The Combustion Institute, Pittsburgh, PA.[4] With permission.)

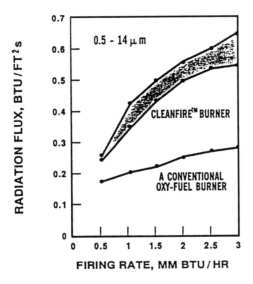

FIGURE 8.5 Comparison of measured flame radiation for a new-style oxy/fuel burner (Cleanfire™ burner and a conventional oxy/fuel burner. (Courtesy of Govt. Institutes, Rockville, MD.[5] With permission.)

There are many possible ways to scale a burner according to changes in the firing rate, which is the primary characteristic of interest in most industrial applications. Spalding (1963),[6] Beér and Chigier (1972),[7] and Damköhler (1936)[8] looked at numerous dimensionless groups based on considerations of the momentum, energy, and mass balances. Some of these groups include the Reynolds, Froude, and Damköhler numbers. However, it is not possible to maintain all of the dimensionless groups constant simultaneously.

The two most common methods used to scale industrial burners are constant velocity and constant residence time. Constant-velocity scaling is by far the most popular. The burner thermal input (Q) can be calculated using:

$$Q_0 = K\rho_0 u_0 d_0^2 \qquad (8.5)$$

where ρ_0, u_0, and d_0 are the inlet average fluid density, characteristic fluid velocity, and characteristic diameter, respectively, and K is a proportionality constant. Assuming that the inlet fluid density is constant, and the characteristic inlet fluid velocity u_0 is held constant, then the constant velocity scaling law can be written as:

$$d_1 \propto \left(Q_1\right)^{0.5} \qquad (8.6)$$

where the new characteristic burner diameter d_1 is proportional to the square root of the new firing rate. This law simply says that the outlet area of the burner is directly proportional to the firing rate for a constant-velocity scaling law. For round outlets, the area is proportional to the square of the diameter.

The principle for constant residence-time scaling is to maintain the ratio of d_0/u_0, which has the units of time (typically seconds). This scaling law, sometimes known as the inertial or convective time scale, can then be written as:

$$d_1 \propto \left(Q_1\right)^{0.33} \qquad (8.7)$$

This approach is not as commonly used because it leads to very low windbox pressures for smaller burners and excessive pressures for larger burners.[9]

The heat transfer from the flame to the furnace and load is often dominated by radiation close to the burner outlet. Further from the burner outlet, radiation and convection may both be important. Another approach to scaling is based on the radiative characteristics of the flame. Markstein (1976) developed a correlation for buoyancy-dominated turbulent flames.[10] Other correlations have been developed for laminar flames.[11,12] Works by Buriko and Kuznetsov (1978),[13] Turns and Mayr (1991),[14] Faeth et al. (1989),[15] and Delichatsios et al. (1988, 1992)[16,17] have quantified radiation from flames for different fuels of different sooting tendencies, using the radiant fraction as a scaling parameter. All of the above were for free jets, without confinement and not inside a hot combustor. The radiation scaling inside a combustor is more complicated because of recirculation effects and re-radiation from the furnace walls to the flame gases.

Weber (1996) reported on burner scaling effects for firing rates ranging from 7 kW to 14 MW (24×10^3 to 48×10^6 Btu/hr), or a 2000:1 turndown range.[18] The goal of the study, known as the Scaling 400 Study for the nominal turndown of 400:1,[19] was to determine the proper method for scaling burner sizes. The burner used in that study was a staged mixing burner for primary, secondary, and tertiary air, with variable swirl capability. The primary conclusions of the study were:

- The fuel-to-air momentum ratio must be maintained.
- The geometrical burner similarity must be maintained (small departures are acceptable).
- Confinement effects of the furnace are secondary if the confinement ratio (furnace:burner diameter ratio) is larger than 3.
- The inlet swirl should be reduced by 20% to 30% for laboratory-scale burners with characteristic diameters less than 2 in. (5 cm).

Baukal and Gebhart (1998) studied oxygen-enhanced natural gas flames impinging normal to a water-cooled flat metal disk.[20] The firing rate of the burner was one of the parameters studied to determine its effect on the heat transfer to the target (see Figures 4.8 and 4.9). The heat flux was determined by calorimetry to concentric water-cooled rings on the target. The firing rate was varied from 5 to 25 kW (17,000 to 85,000 Btu/hr). The upper limit was a function of the flow control equipment. Figure 8.6 shows contours of the heat flux as functions of both the axial and radial distances from the burner, with 35% total O_2 in the oxidizer ($\Omega = 0.35$). This was the lowest oxidizer composition that could be used that produced an acceptable flame through the entire range of firing rates for the burner used (see Figure 3.23). These plots show that for small L, the peak heat flux did not occur at the stagnation point, but at about $R_{eff} = 0.5$ to 0.7. This was caused primarily by a slightly higher heat output from the burner at that radial location.[21] The heat flux increased by 40% to 230%, depending on the axial and radial position, by increasing the firing rate 400% by going from 5 to 25 kW (17,000 to 85,000 Btu/hr). The smallest percentage improvements were for the closest axial (L = 0.5) and radial ($R_{eff} = 0.16$) locations. This was because the heat flux rate was already relatively high there, compared to farther axial and radial locations. The largest percentage improvements occurred at the farthest axial (L = 6) and radial ($R_{eff} = 1.04$) locations because the heat flux rate was initially so low there. At $q_f = 5$ kW and $R_{eff} = 1.04$, the heat flux increased by 91% by decreasing the axial spacing from L = 6 to L = 0.5. However, at $q_f = 25$ kW (85,000 Btu/hr) and $R_{eff} = 0.16$, the heat flux decreased by 14% by decreasing the axial spacing from L = 6 to L = 0.5. In general, at higher firing rates, the heat flux did not have a strong dependence on the axial distance, L, between the flame and the target. At the farthest axial location (L = 6), the heat flux to the inner calorimeter was approximately double the heat flux to the outer calorimeter. However, at the closest axial spacing (L = 0.5), the peak flux occurred at about the middle calorimeter ($R_{eff} = 0.59$). Figure 8.7 shows similar heat flux contours as, but for a pure O_2 oxidizer. The peak fluxes occurred at the stagnation point, except for $q_f = 20$ and 25 kW (68,000 and 85,000 Btu/hr) for L = 0.5 where a slightly higher flux was measured at $R_{eff} = 0.37$. The heat flux increased by 78% to 470%, depending on the axial and radial locations, by increasing the firing rate from 5 to 25 kW (17,000 to 85,000 Btu/hr). Again, the largest percentage improvements occurred at the

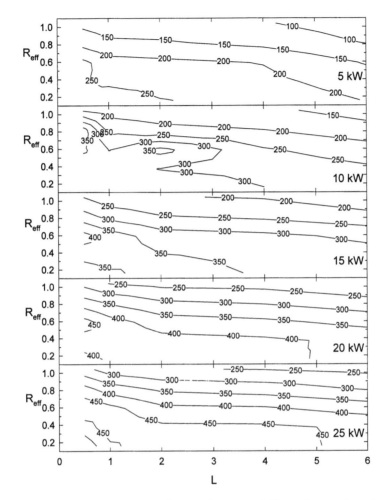

FIGURE 8.6 Contours of the total heat flux, q'' (kW/m²), from stoichiometric natural gas flames ($\Omega = 0.35$) with various firing rates ($q_f = 5$ to 25 kW) impinging on the surface of an untreated stainless target. (From C.E. Baukal and B. Gebhart, *Exp. Therm. Fluid Sci.*, 16(3), 247-259, 1998.)

farthest locations, while the smallest improvements occurred at the closest locations. Unlike the $\Omega = 0.35$ data in, the heat flux strictly increased, by 42% to 230%, as the axial spacing decreased for the $\Omega = 1.00$ data in Figure 8.7. Like the $\Omega = 0.35$ data, at large axial spacings, the heat flux to the innermost calorimeter was approximately double the flux to the outermost calorimeter. Figure 8.8 shows how the heat flux to the innermost calorimeter was affected by the axial location of the target and by the burner firing rate. The heat flux increased with the firing rate and decreased with the axial distance. The shape of the contour lines was significantly different for the $\Omega = 0.35$ flames, compared to the $\Omega = 1.00$ flames. The peak flux occurred at an intermediate axial location ($\approx L = 2$) for the $\Omega = 0.35$ flames. For the $\Omega = 1.00$ flames, the heat flux strictly increased as the axial spacing decreased. Figure 8.9 shows that for $\Omega = 1.00$ and large L, the heat flux increased rapidly with the firing rate. The shape of each curve is similar. In all cases, the heat flux increased with the firing rate and decreased with the axial spacing. The thermal efficiency, η, was defined as:

$$\eta = \frac{\sum_{i=1}^{5} q_i'' A_i}{q_f}$$

(8.8)

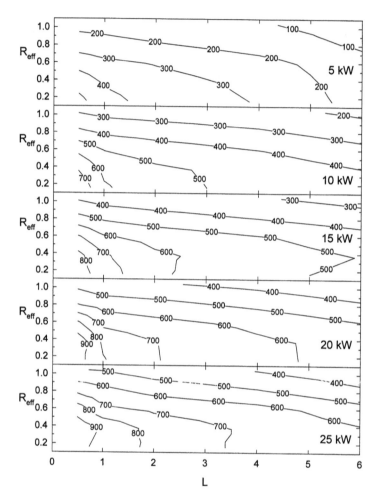

FIGURE 8.7 Contours of the total heat flux, q'' (kW/m²), from stoichiometric natural gas flames ($\Omega = 1.00$) with various firing rates ($q_f = 5$ to 25 kW) impinging on the surface of an untreated stainless target. (From C.E. Baukal and B. Gebhart, *Exp. Therm. Fluid Sci.*, 16(3), 247-259, 1998.)

where q_i'' was the calculated heat flux to ring i, A_i was the impingement surface area of ring i, and q_f was the burner firing rate. The efficiency was then the total energy absorbed by the first five calorimeters, divided by the burner firing rate. Figure 8.10 shows how the thermal efficiency varied with the firing rate, oxidizer composition, and axial location. The efficiency decreased with L, increased with Ω, and decreased with the firing rate. At $\Omega = 1.00$ and $q_f = 5$ kW (17,000 Btu/hr), the efficiency decreased rapidly from $L = 4$ to $L = 6$ because, at $L = 6$, the visible flame length was less than the distance between the burner and the target. Buoyancy effects were visually evident, as the flame shape was highly transient and wavy. Air infiltration may have significantly reduced the heat flux to the target for that set of conditions.

8.2.4 FLAME SHAPE EFFECTS

Hutchinson et al. (1975) experimentally and numerically showed that a quarl (burner block or tile) shortened the flame on an air/natural gas burner with swirl firing into a water-cooled cylindrical furnace.[21a] The recirculation zone in the furnace was pulled closer to the burner with the quarl, compared to the same burner without a quarl. Higher temperatures and heat fluxes were measured

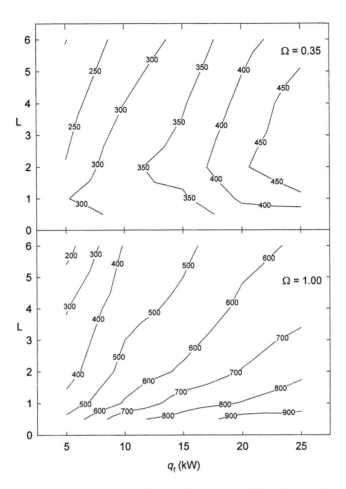

FIGURE 8.8 Contours of the total heat flux, q'' (kW/m^2), from stoichiometric natural gas flames of various firing rates (q_f = 5 to 25 kW) and oxidizer compositions (Ω = 0.35 and 1.00) to the stagnation point (R_{eff} = 0.16) of an untreated brass target. (From C.E. Baukal and B. Gebhart, *Exp. Therm. Fluid Sci.*, 16(3), 247-259, 1998.)

closer to the burner with the quarl, than without the quarl. The quarl shortened the flame and moved the heat release closer to the burner. The calculations showed that the convective heat transfer and the radiative heat transfer to the furnace wall were of the same order with or without the quarl.

One recent burner development trend in about the last decade has been the development of so-called flat-flame burners. This can mean different things, but here it refers to a rectangular flame shape, as opposed to the traditional round flame shape. The goal is to have a higher flame surface area to increase the area that radiates from the flame to the load. This trend in burner design has been particularly evident in the use of oxygen-enriched burners in the glass and aluminum industries. If the trend continues, this type of flame will likely be applied to other industrial heating and melting applications.

There are numerous examples of the use of flat flames in glass melting. Kirilenko et al. (1988) described a flat-flame burner used in a regenerative glass-melting furnace.[22] Ibbotson (1991) described a technique for lancing a fan-shaped jet of oxygen under a fuel port in a glass-melting furnace to improve temperature uniformity and minimize hotspots.[23]

Flat shaped flames have also been used in aluminum melting. Yap and Pourkashanian (1996) reported on a flat-flame oxy/fuel burner, which they described as a large aspect-ratio flame.[24] The burner was claimed to have more uniform heat transfer rates, compared to conventional round flame

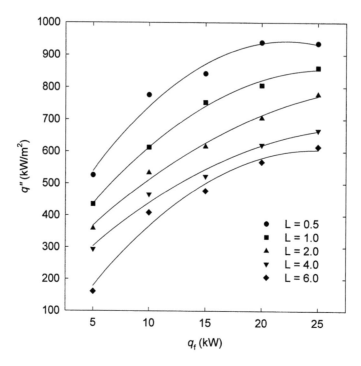

FIGURE 8.9 Total heat flux (q'') from stoichiometric natural gas flames ($\Omega = 1.00$) of various firing rates (q_f) impinging on the stagnation point ($R_{eff} = 0.16$) of an untreated stainless target. (From C.E. Baukal and B. Gebhart, *Exp. Therm. Fluid Sci.*, 16(3), 247-259, 1998.)

burners. The burner also had excellent flame stability, resulting in a wide operating range, low NOx emissions, and high flame luminosity. The combination of low NOx emissions and high luminosity, both of which are important in most heating applications, is a unique aspect of the burner. In addition, the burner minimized oxidation of the molten aluminum, which maximizes the product yield.

The length of the flame influences the heat transfer as longer flames may be more luminous at lower temperature compared to similar flames at the same operating conditions. Blake and McDonald (1993) have correlated visible flame length data for vertical turbulent diffusion flames as follows:[25]

$$\frac{l_f}{d_f} = \alpha_1 \left(\frac{4J}{\pi \rho_\infty g d_f^3} \right)^{\alpha_2} \tag{8.9}$$

where l_f is the flame length, d_f is the flame diameter, J is the momentum of the source jet, ρ_∞ is the density of the ambient, g is the gravitational constant, and α_1 and α_2 are constants as given in Table 8.1. The fraction inside the parentheses in Equation 8.9 is referred to as the density weighted Froude number.

Figure 8.11 shows how the visible flame length for an open-flame diffusion burner varies as a function of the firing rate and oxidizer composition.[26] The flame length was nearly the same for lower firing rates. At higher firing rates, the pure O_2 flames ($\Omega = 1.00$) were longer than the lower purity flames ($\Omega = 0.35$). The flame length affects the flame shape, which influences the radiating area of the flame.

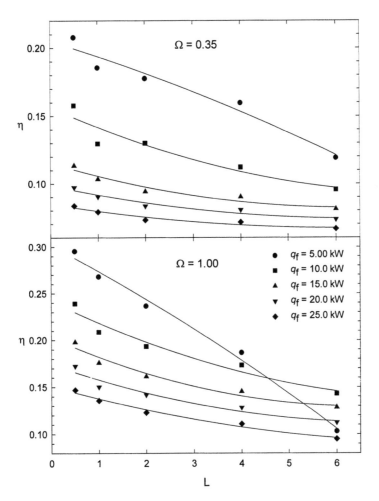

FIGURE 8.10   Thermal efficiency (η) for stoichiometric natural gas flames of various firing rates (q_f = 5 to 25 kW) and oxidizer compositions (Ω = 0.35 and 1.00). (From C.E. Baukal and B. Gebhart, *Exp. Therm. Fluid Sci.*, 16(3), 247-259, 1998.)

TABLE 8.1
Constants for Flame
Length Equation

$4J/\pi\rho_\infty g d_f^3$	α_1	α_2
$10^{-9}-10^{-6}$	6.73	0.209
$10^{-6}-10^{-4}$	7.79	0.218
$10^{-4}-10^{-1}$	7.25	0.184

Source: Adapted from T.R. Blake and M. McDonald, *Comb. Flame,* 94, 426-432, 1993.

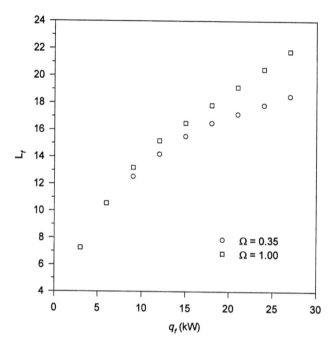

FIGURE 8.11 Flame length as function of firing rate and oxidizer composition for stoichiometric natural gas flames. (From C.E. Baukal and B. Gebhart, *Int. J. Heat Mass Transfer,* 44(11), 2539-2547, 1997.)

8.3 RADIANT BURNERS

Radiant burners operate by combusting a fossil fuel, which heats a solid surface that radiates infrared (IR) energy to a load. These burners are used in a number of lower temperature heating and drying applications including, for example:

- Drying paper and cardboard in a paper mill
- Paper finishing
- Drying wood
- Porcelain frit drying
- Curing ceiling tiles
- Teflon curing
- Drying and curing coatings on paper and metals
- Curing ink on paper and powder coat paints
- Baking in mass food preparation
- Setting dyes in textile and carpet production, sometimes referred to as pre-drying
- Curing lens assemblies in automotive headlamp assembly manufacturing
- Plastics curing

Both gas-fired and electric radiant heaters are commonly used; however, gas-fired radiant burners have lower operating costs due to the difference between the cost of electricity and natural gas in most locations. Pettersson and Stenström (1998) compared gas-fired and electric IR heaters in two paper-coating drying processes.[27] The efficiencies of the gas-fired and electric dryers were 30% and 40%, respectively. However, there was considerable uncertainty in the measurements due to the difficulty in obtaining accurate moisture measurements of the paper being dried. Even if the efficiencies are correct, the gas-fired system may still be more economical, depending on the cost of electricity vs. natural gas.

Radiant burners are designed to produce a uniform surface temperature heat source for heating and melting a variety of materials (see Chapter 10). The uniform surface temperature produces more homogeneous heating of the materials, which normally improves the product quality compared to conventional burners that may produce hotspots. Other advantages of these burners may include:

- High thermal efficiencies
- Low pollutant emissions
- Directional heating
- Very fast response time to load changes
- Very fast heating compared to convective heating
- Burner shape can be tailored to the shape of the heat load to optimize heat transfer
- Ability to segment a burner to produce a nonuniform heat output profile, which may be useful in certain types of heating and drying applications
- Certain types of radiant burners have very rapid heat-up and cool-down times
- No open flames that could ignite certain types of materials (e.g., paper or textiles)
- More control over the heating process because of the known and measurable surface temperature of the radiant surfaces compared to open flames, where the flame temperature is very difficult to measure
- Burners are very modular and can be configured in a wide variety of geometries to accommodate the process heating requirements

The primary parameters of interest for radiant burners are the power density (firing rate per unit area), radiant efficiency (fraction of fuel heating value converted to thermal radiation), heat-up and cool-down times, and pollutant emissions. Other factors of importance include cost, durability, and longevity.

There are also some important limitations of porous refractory burners, compared to more conventional open-flame burners, which may include:

- Relatively low temperature limit for the radiant surface due to the limits of the refractory material
- Fuel and oxidizer must be clean to avoid plugging the porous radiant surface, which essentially precludes the use of fuels like coal or heavy fuel oil
- Some of the radiant surfaces can be damaged by water or by contact with solid materials, which may be prevalent in certain applications
- Holes in the radiant surface can cause flashback since these burners use premix
- Some designs may have high pressure drops, which means more energy is needed for the blower to flow the combustion air through the ceramic burner material
- Due to the limits in radiant surface temperatures, the firing rate density is usually limited
- Some types of radiant burners using hard ceramic surfaces may have high heat capacitances that could ignite certain load materials upon a sudden line stoppage

In these burners, fuel and air are premixed and combusted either just inside a radiating surface or just above the surface, depending on the operating conditions and specific radiant burner design. If the mixture velocity is too low, flashback or flame extinguishment may occur, depending on the design of the burner. Besides the operational considerations, flashback is an obvious safety concern. If the mixture velocity is too high, the flame may blow off or the radiant performance may be severely reduced because the burner surface is not being directly heated by the hot exhaust products. Depending on the specific design of the burner, optimum performance is achieved when the flame is stabilized just inside or just above the outer burner outlet.

Howell et al. (1995) referred to the burner material as "porous inert media" and presented a review of these types of burners, which have been made from a wide variety of ceramics that often

TABLE 8.2
Comparison of Thermal Efficiencies for Radiant Burners

Radiant Burner Type	Combustion Intensity (kW/m²)	Thermal Efficiency Range (%)
Metal fiber		
No perforations	150–540	18–26
With perforations	150–540	16–27
With perforations & front metal screen	100–440	22–38
Reticulated ceramic	150–630	27–39
Ported ceramic	140–520	37–54
Flame impingement	150–430	46–52

Source: Adapted from R.F. Speyer, W.-Y. Lin, and G. Agarwal, *Exp. Heat Transfer,* 9, 213-245, 1996.

include alumina, zirconia, or silicon carbide.[28] Hsu and Howell (1993) developed the following equation for the effective thermal conductivity of partially stabilized zirconia (PSZ) (zirconium oxide plus <3% magnesium oxide) for temperatures in the range of 290K to 890K (63°F to 1140°F) and for sample pore sizes from 10 to 65 pores per inch (ppi):[29]

$$k \text{ (W/m-K)} = 0.188 - 0.0175\, d \tag{8.10}$$

where d is the actual pore size of the material in (in millimeters) and $0.3 < d < 1.5$ mm.

Speyer et al. (1997) studied the performance of four different types of commercial gas-fired radiant burners: a metal (Fe-Cr-Al) fiber (~40 μm diameter) burner, a reticulated ceramic burner made of a porous cordierite ($Mg_2Al_4Si_5O_{18}$), a porous single unit mullite ($Al_6Si_2O_{13}$) tile burner, and a flame impingement burner where the products of combustion are forced around a metal screen that radiates to the load.[30] The study was sponsored by the Gas Research Institute (Chicago, IL). The thermal efficiencies as a function of the combustion intensity are shown in Table 8.2. The efficiency was determined by measuring the radiant output or radiosity from the burner surface and was calculated using:

$$e = \frac{R_T}{\dot{V}_g \Delta H_c / A} \tag{8.11}$$

where e is the calculated efficiency, R_T is the measured radiosity, $\dot{V}_g$ is the volume flow rate of the fuel, H_c is the heating value of the fuel, and A is the surface area of the burner. The experimental results showed that the peak radiosity measurements as a function of the air-to-fuel mixture ratio occurred for percent excess fuels ranging from 6% to 7.5%, depending on the specific burner design. Similar results were calculated for the radiant efficiencies. Preheating the air/fuel mixture increased both the radiosity and thermal efficiencies, essentially in a linear manner for mixture temperatures ranging from 40°C to 155°C (100°F to 311°F). In most cases, the efficiency declined with combustion intensity, except for the reticulated ceramic and metal fiber with perforations and a front metal screen, both of which had peak efficiencies at intermediate combustion intensities. The flame impingement and ported ceramic burners were the most efficient, while the metal fiber with and without perforations was the least efficient. The study also showed that the thermal efficiency generally increased as the fraction of closed area increased on the burner surface.

The heat transfer coefficient between the hot exhaust products and the radiant burner material is difficult to predict and measure due to the uncertainty in the surface area of the ceramic structure. These coefficients have traditionally been presented in terms of a volumetric coefficient (Btu/hr-°F-ft³ or kW/°C-m³). Chen et al. (1987) assumed the coefficient was large enough that the temperature of the solid was essentially the same as the temperature of the hot flowing gas.[31] Sathe et al. (1990) used a value of 2×10^9 W/m³-K (1×10^8 Btu/hr-°F-ft³) computed for cylinders in crossflow.[32] Hsu et al. (1993) used 10^7 W/m³-K (5×10^5 Btu/hr-°F-ft³).[33] Younis and Viskanta (1993) empirically determined the following correlation for alumina and cordierite ceramic foams with pore diameters ranging from 0.29 to 1.52 mm (0.011 to 0.0598 in.):[34]

$$\mathrm{Nu} = 0.819 \left[1 - 7.33 (d/L)\right] \mathrm{Re}^{0.36[1+15(d/L)]} \tag{8.12}$$

where d is the actual pore diameter and L is the thickness of the specimen in the flow direction. Fu et al. (1998) noted the relationship between the volumetric heat transfer coefficient h_v and the convection coefficient h:[35]

$$h_v = a_v h \tag{8.13}$$

where a_v can be computed using the following empirical equation:

$$a_v = 169.4 \text{ PPC (m}^2/\text{m}^3) \tag{8.14}$$

where PPC is the number of pores per centimeter for cellular ceramics. Four different choices for the characteristic dimension were identified:

$$\text{the reciprocal of the surface area: } d_a = 1/a_v \tag{8.15}$$

$$\text{the hydraulic diameter: } d_h = 4\phi/a_v \tag{8.16}$$

$$\text{the mean pore diameter: } d_m = \frac{\sqrt{4\phi/\pi}}{\text{PPC}} \text{(cm)} \tag{8.17}$$

and

$$\text{the ratio of the inertial to the viscous friction coefficients: } d_r = \beta/\alpha \tag{8.18}$$

where ϕ is the porosity, β is the inertia coefficient in the Reynolds-Forchheimer equation, and α is the viscous coefficient in the Reynolds-Forchheimer equation. The following empirical correlations, using each of these characteristic dimensions, were developed for specimens having a PPC ranging from 4 to 26:

$$\mathrm{Nu}_v = \left[0.0426 + \frac{1.236}{L/d_m}\right] \mathrm{Re}_{d_m} \text{ for } 2 \leq \mathrm{Re}_{d_m} \leq 836 \tag{8.19}$$

$$\mathrm{Nu}_v = \left[0.0730 + \frac{1.302}{L/d_h}\right] \mathrm{Re}_{d_h} \text{ for } 3 \leq \mathrm{Re}_{d_h} \leq 1594 \tag{8.20}$$

$$\mathrm{Nu}_v = \left[0.0252 + \frac{1.280}{L/d_a}\right] \mathrm{Re}_{d_a} \text{ for } 1 \leq \mathrm{Re}_{d_a} \leq 480 \tag{8.21}$$

$$\mathrm{Nu}_v = \left[0.000267 + \frac{1.447}{L/d_r}\right] \mathrm{Re}_{d_r} \text{ for } 0.02 \leq \mathrm{Re}_{d_r} \leq 2.4 \tag{8.22}$$

where Nu_v is the volumetric Nusselt number, defined as:

$$Nu_v = \frac{h_v l_c^2}{k} \qquad (8.23)$$

where l_c is the characteristic pore length and k is the thermal conductivity of the gas. The uncertainty in the correlations was largest for the smaller Reynolds numbers.

One of the challenges of gas-fired infrared burners is determining the radiant efficiencies. Mital et al. (1998) noted a wide discrepancy in the reported radiant efficiencies (varying by more than 200%) for radiant burners.[36] Part of the discrepancy was attributed to a lack of a standard measurement technique to determining the efficiency. Other problems include nondiffuse radiation from the burners and nonuniform burner surfaces. They presented a technique that is not sensitive to burner surface nonuniformities. Typical results for a reticulated ceramic foam radiant burner are shown in Figure 8.12. A calorimetric method was used to check the consistency of the data. It was shown that single-point radiation measurements can deviate considerably from more rigorous multi-point measurement techniques. Yetman (1993) described a simple technique for measuring the total radiant output of an infrared burner using a narrow angle pyrometer.[37] Johansson (1993) presented a method for measuring the spectral output of radiant burners using an infrared spectrometer.[38]

Madsen et al. (1996) measured the spectral radiation from several types of radiant burners as shown in Figure 8.13.[39] As can be seen, the spectra are fairly similar that have distinctive peaks around 3 and 4.5 μm. This shows the selective emittance of certain types of radiant burners and the possibility of matching the burner to the load spectral absorptivity to optimize heat transfer efficiency.

8.3.1 PERFORATED CERAMIC OR WIRE MESH RADIANT BURNERS

Examples of these types of burners are shown in Figure 8.14. Perforated or ported ceramic burners may consist of a pressed ceramic plate, which may include prepunched holes, where the flames heat the surface directly.[40] The surface can be textured to further enhance the radiant efficiency of the burner. New developments in ceramic foams are being applied to this type of burner. These foams are often less expensive to make than perforated ceramics. They provide a higher surface area for radiation and a more uniform heating surface, compared to perforated ceramics. Many shapes are possible with the ceramic foams and the pore size is adjustable. Flanagan et al. (1992) described the use of a ported ceramic burner, shown in Figure 8.15, to achieve low NOx emissions.[41] However, the burner was actually fired at up to 15 times its normal maximum design firing rate so that the flames were highly lifted and therefore did not radiate from the surface as in normal operation. Mital and Gore (1994) discussed the use of a reticulated ceramic insert to enhance radiation heat transfer in direct-fired furnaces.[42] The insert was placed downstream of the outlet of a laminar diffusion flame, as shown in Figure 8.16. Experimental results showed up to a 60% improvement in radiative heat flux compared to the case with no insert. The radiative heat flux with and without the insert is shown in Figure 8.17. Kataoka (1998) described a new type of high-temperature porous ceramic burner made of aluminum titanate (Al_2TiO_5), capable of surface temperatures up to 1100°C (2000°F).[43]

Wire mesh burners are made from high-temperature resistant metals, such as stainless steels or inconel. The open area in the mesh serves as the port area for the burner. However, due to the high thermal conductivity of metals, several layers of mesh are often required to prevent flashback. The thermal conductivity between the layers is much less than through the mesh itself because of the contact resistance between the layers. An important problem with wire mesh radiant burners are the lower temperature limits compared to ceramic burners, due to the temperature limits of the metals.

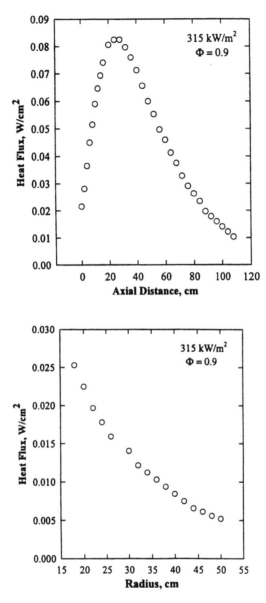

FIGURE 8.12 Heat flux measurements from a ceramic foam radiant burner as a function of (A) the axial distance from the burner surface, and (B) the radial distance from the center of the burner. (Courtesy of Gas Research Institute, Chicago, IL.[36] With permission.)

8.3.2 FLAME IMPINGEMENT RADIANT BURNERS

This burner is sometimes referred to a direct-fired refractory burner. An example of this type of burner is shown schematically in Figure 8.18. An actual burner is shown in Figure 8.19. In this type of radiant burner, the flame impinges on a hard ceramic surface, which then radiates to the load. The heat transfer from the flame to the tile is mostly by convection due to the direct flame impingement (see Chapter 7). The heat transfer from the burner tile to the load is purely by radiation. The hot exhaust gases from the burner heat up the surrounding wall by convection, but at a much lower rate than to the burner tile. One advantage of this type of burner is that there is no metal

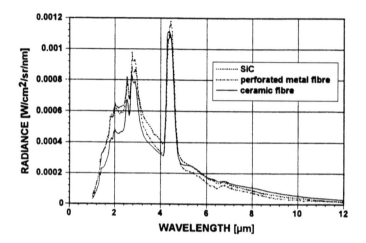

FIGURE 8.13 Measured radiance from several types of radiant burners with air/methane combustion at 400 kW/m² (127,000 Btu/hr-ft²) and 10% excess air. (Courtesy of Government Institutes, Rockville, MD.[39] With permission.)

matrix with lower temperature limits, compared to ceramic that has high-temperature limits. Therefore, this type of burner can often be used in higher temperature applications. There is also no matrix of porous ceramic fiber refractory that could get plugged up as in the next type of burner. This type of burner can also fire a liquid fuel, as opposed to many other radiant burners that use only gas. A disadvantage is that the heat flux is not as uniform as other types of radiant burners. Another problem is that the burner tile has the typical problem of thermal cycling, which can cause the tile to disintegrate.

8.3.3 PorOus REFRACTORY RADIANT BURNERS

In this type of burner, the surface is made of a porous ceramic fiber that is often made in a vacuum-forming process. A relatively new type is now available, made from a woven ceramic fiber mesh similar to the wire mesh radiant burners except that ceramic fiber is used instead of metal. The predominant shape used in porous refractory burners is a flat panel. An example is shown in Figure 8.20.

Besides flat panels, other shapes are also available. Alzeta Corp. (Santa Clara, CA) makes many cylindrically shaped porous refractory burners.[44,45] For their Duratherm™ burners, sizes range from as small as 2 in. (5 cm) in diameter by 4.5 in. (11 cm) long, to as large as 30 in. (76 cm) in diameter by 180 in. (460 cm) long. Firing rates range from 23,000 Btu/hr (6.7 kW) to 16.5 × 10⁶ Btu/hr (4.83 MW). Bartz et al. (1992) described the use of an Alzeta porous refractory radiant burner to incinerate volatile organic compounds (VOCs).[46] The unique aspect of the combustion system was that the burner was formed into a cylinder and fired inwardly, while the VOCs flowed into that inner core with the combustion products where the VOCs were then destroyed, as shown in Figure 8.21. Very high VOC destruction efficiencies were measured.

The American Gas Association (Cleveland, OH) made an extensive study of radiant burners, both gas-fired and electric.[47] For the gas-fired burners, they determined what was termed a "gas infrared radiation" factor or GIR, which was determined as follows:

$$\text{GIR} = \frac{W}{\sigma T_b^4} \qquad (8.24)$$

FIGURE 8.14 Examples of porous ceramic and wire mesh radiant burners. (From F. Ahmady, *Process Heating*, 1(2), 38-43, 1994. Courtesy of Solaronics, Rochester, MI. With permission.)

where $W = Q/A$ is the total normal infrared radiation from the burner measured by a spectrophotometer, σ is the Stefan-Boltzman constant, and T_b is the absolute brightness temperature of the burner surface measured with an optical pyrometer. The GIR factor was similar to a burner emissivity. However, for some of the burners, the GIR factor exceeded 1.0. The GIR factor was approximately 15% higher than the true burner surface emissivity. This is due to the gaseous nonluminous radiation in addition to the surface radiation from the burner. For the burners tested, the GIR factor ranged from 0.36 to 1.17, depending on the burner type.

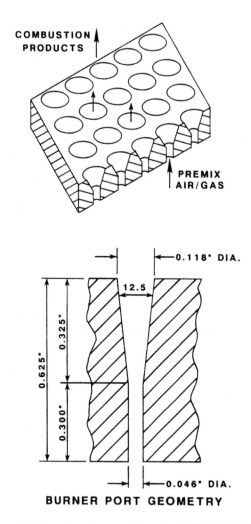

FIGURE 8.15   Ported ceramic tile radiant burner. (Courtesy of ASME, New York.[41] With permission.)

Sathe et al. (1990) made experimental measurements and numerical predictions on a porous radiant burner.[48] They used the following correlation for flow over circular cylinders for the convection heat transfer coefficient from the flame to the porous refractory:

$$Nu = 0.989 \ Re^{0.33} \tag{8.25}$$

where the characteristic length d is the equivalent particle diameter. Figure 8.22 shows a comparison of the measured and predicted result as a function of the axial location of the flame (x_f) normalized by the porous layer length (L). The figure shows that both the radiant output and the flame speed are fairly independent of the flame location, but are dependent on the equivalence ratio. Zabielski et al. (1991) used an optical technique to measure the radiant characteristics of a porous ceramic fiber radiant burner.[49] The macroscopic emittance of the burner was estimated to be 0.70. Xiong and Viskanta (1992) measured the heat flux and calculated the thermal efficiency of a porous matrix ceramic combustor, with typical results shown in Figure 8.23.[50]

Jugjai and Sanitjai (1996) presented a concept for a new type of porous radiant burner, incorporating internal heat recirculation to increase thermal efficiency.[51] They studied the effects of the optical thickness of the porous medium on the radiant output of porous radiant burners. The

FIGURE 8.16   Laminar flame with and without a ceramic insert. (Courtesy of ASME, New York.[42] With permission.)

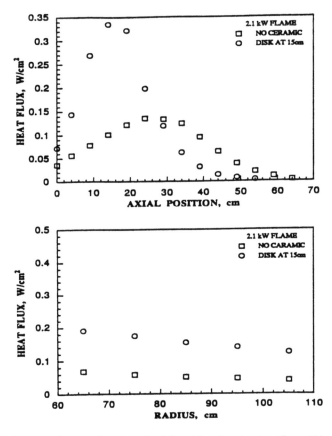

FIGURE 8.17   Radiant heat flux as a function of axial position for a laminar flame (A) with and (B) without a ceramic insert. (Courtesy of ASME, New York.[42] With permission.)

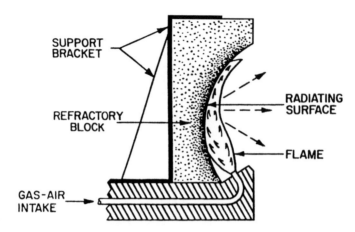

FIGURE 8.18 Schematic of a flame impingement radiant burner. (Courtesy of American Gas Association, Cleveland, OH.[47] With permission.)

FIGURE 8.19 InfraRad ceramic radiant infrared burner. (Courtesy of Eclipse, Rockford, IL.)

optical thickness was increased by added layers of stainless steel wire mesh that formed the porous medium for the burners tested. The added layers improved the heat transfer from the hot gases to the porous medium.

Mital et al. (1996) claimed to have made the first measurements of temperature and species distributions inside submerged flames stabilized inside porous ceramic burners.[52] They noted the wide discrepancy in the performance of these burners in the literature and in manufacturers' litererature.

Rumminger et al. (1996) developed a one-dimensional model of a bilayered reticulated ceramics radiant burner.[53] The model results showed that for a "submerged flame" where the flame is anchored inside the porous refractory, nearly all of the important reactions occur inside the porous medium. This complicates the study for this type of burner because of the difficulty in making measurements inside the porous medium. They speculate that it is possible the large internal surface area of the porous medium could affect the chemistry, but there is no data at this time to confirm that possibility.

Van der Drift et al. (1997) studied various coatings on porous foam ceramic infrared burners to determine the effects on heat transfer to wet paper and to three colors (white, blue, and black).[54] They showed that there can be up to a 10% improvement using coated burners compared to a base-case metal fiber burner.

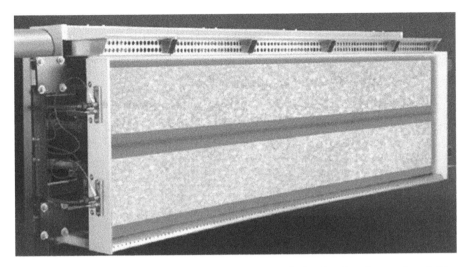

FIGURE 8.20 Flat-panel infrared burner. (Courtesy of Marsden, Inc., Pennsauken, NJ.)

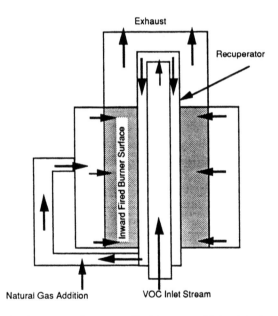

FIGURE 8.21 Inwardly fired porous refractory radiant burner, with heat recovery, for VOC destruction. (Courtesy of ASME, New York.[46] With permission.)

Mital et al. (1998) made experimental measurements on a bilayered reticulated ceramic foam made of Cordierite.[55] Measurements included radiation efficiency using a heat flux gage and burner surface measurements with a type R thermocouple. Radiation efficiency results are shown in Figure 8.24, which include both single and multi-point measurements and model predictions.

8.3.4 Advanced Ceramic Radiant Burners

Tong et al. (1989) showed through computer modeling that the performance of porous radiant burners can be improved by as much as 109% using submicron diameter ceramic fibers.[56] Bell et al. (1992) described a staged porous ceramic burner that demonstrated low NOx emissions.[57] Kendall and Sullivan (1993) studied enhancing the radiant performance of porous surface radiant burners using improved ceramic, high-temperature, high-emissivity fibers.[58] Selective emissivity

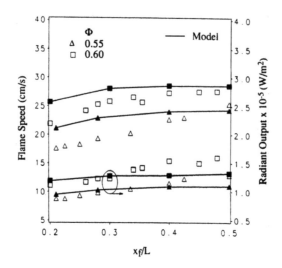

FIGURE 8.22   Measured and predicted flame speed and radiant output for a porous radiant burner. (Courtesy of The Combustion Institute, Pittsburgh, PA.[48] With permission.)

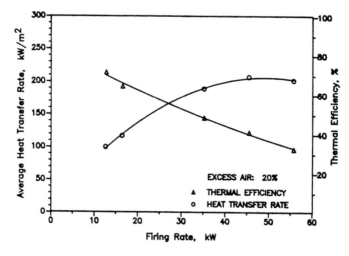

FIGURE 8.23   Average heat transfer rate and thermal efficiency as a function of firing rate for a porous ceramic matrix burners. (Courtesy of ASME, New York.[50] With permission.)

was achieved by coating a standard ceramic fiber burner with an outer layer of ytterbia. The output of the uncoated vs. the coated burner is shown in Figure 8.25. As can be seen, there is a large spike in the output of the coated burner at around 1 μm. However, there was not a significant improvement in performance and the durability was in question. In certain applications, there may be a need to have a burner with selective radiant emissions corresponding to selective absorption in the load. Potential applications include glass melting, glass bending and lamination, thin-film drying, and indirect heating. Xiong et al. (1993) described a porous ceramic surface combustor-heater with a built-in heat exchanger for improving the thermal efficiency, as shown in Figure 8.26.[59] Ruiz and Singh (1993) described an advanced infrared burner shown in Figure 8.27.[60] Figure 8.28 shows that the new burner design has considerably higher radiant outputs than the previous style design. The new burner increased a powder paint drying process by 40% and a paper drying process by 200%. Severens et al. (1995) modeled porous radiant burners.[61] The 1-D model results compared favorably with experimental data for the radiant fraction as a function of the gas velocity through the burner.

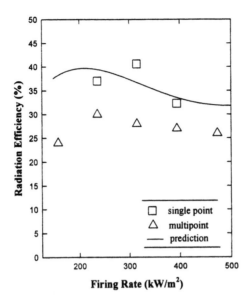

FIGURE 8.24 Predicted and measured radiation efficiency for a porous radiant burner at an equivalence ratio of 0.9. (Courtesy of The Combustion Institute. Pittsburgh, PA.[55] With permission.)

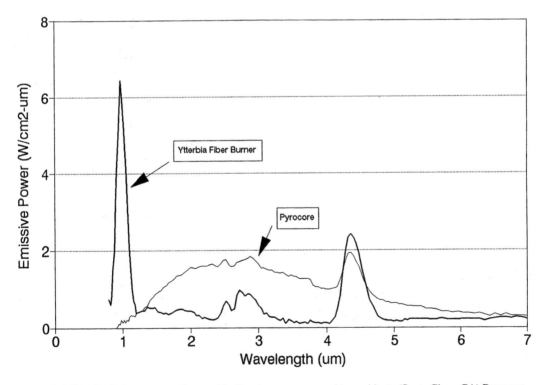

FIGURE 8.25 Emission spectra of a ytterbia fiber burner compared to an Alzeta (Santa Clara, CA) Pyrocore burner, both operating at 127,000 Btu/hr-ft[2] (400 kW/m[2]). (Courtesy of GRI Gas Research Institute, Chicago, IL.[58] With permission.)

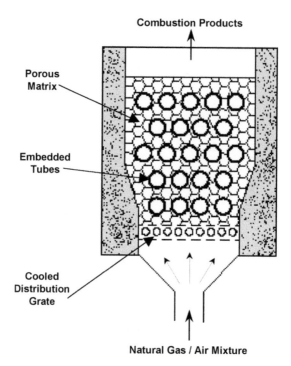

FIGURE 8.26 Surface combustor-heater concept. (Courtesy of Government Institutes, Rockville, MD.[59] With permission.)

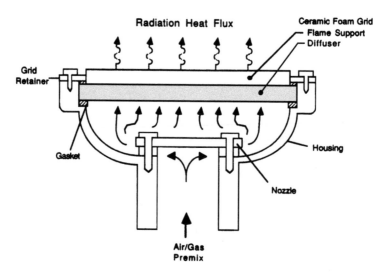

FIGURE 8.27 Advanced infrared burner. (Courtesy of Government Institutes, Rockville, MD.[60] With permission.)

Bogstra (1998) described a new type of infrared radiant burner, referred to as CHERUB, with a closed surface to separate the combustion exhaust gases from the product being heated.[62] The burner had an enclosed, flat, multi-burner system that heated a ceramic radiant plate. The burner was reported to have a radiant efficiency of 80%, compared to an efficiency of 40% for conventional high-temperature radiant burners.

As previously discussed, one of the limitations of porous refractory burners is the burner surface temperature. If the firing rate density is to be increased, new refractory materials are needed that

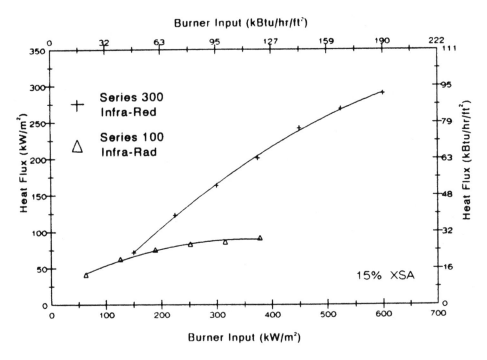

FIGURE 8.28 Radiant flux vs. burner output for the new design series 300 compared to the old model series 100 infrared burner. (Courtesy of GRI, Gas Research Institute, Chicago, IL.[60] With permission.)

can withstand continuous operation at higher temperatures, while maintaining their integrity during thermal cycling. An example of research to develop improved radiant burner materials is by Superkinetic, Inc. (Albuquerque, NM) with funding from the U.S. Dept. of Energy.[63] Three single-crystal ceramic fibers were produced and two fiber materials were successfully made into felt for testing as radiant burner screen surfaces. The materials were alpha-alumina and alpha-silicon carbide, which were successfully bonded with a high-temperature ceramic to form burner screens that were 95% porous. The purpose of this project was to develop materials capable of radiant burner service near 3000°F (1900K), compared to conventional radiant burner surface temperatures of about 1800°F (1300K). The new materials performed well in actual burner operation, but more research was recommended.

8.3.5 RADIANT WALL BURNERS

A typical radiant wall burner is shown schematically in Figure 8.29.[64] This type of burner is commonly used in process heaters where they heat a refractory wall that radiates heat to tubes parallel to the wall. The tubes contain a fluid, typically a hydrocarbon, that is being heated. These burners are similar to flame impingement radiant burners except that the flame in a radiant wall burner is directed along the wall and not at the wall or burner tile as in the case of impingement burners. The object of a radiant wall burner is to distribute heat as evenly as possible over a fairly wide area. The impingement burner primarily heats its burner tile, which then radiates to the load.

8.3.6 RADIANT TUBE BURNERS

In some heating processes, it is not desirable to have the products of combustion come in contact with the load. One example is in certain types of heat treating applications where the exhaust gases from a combustion process could contaminate the surface of the parts being heated. In those cases, an indirect method of heating is needed. Electric heaters are sometimes used; however, the energy

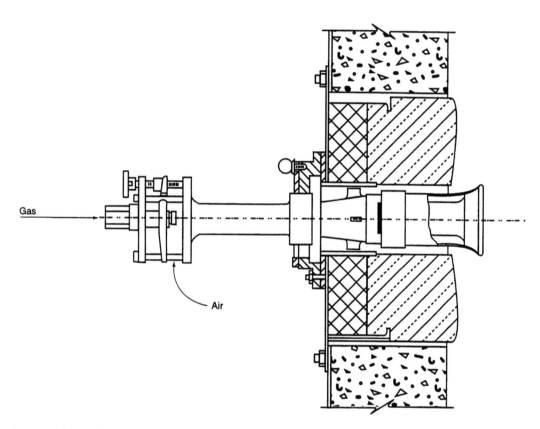

FIGURE 8.29 Typical radiant wall burner. (Courtesy of American Petroleum Institute, Washington, D.C.[64] With permission.)

costs are considerably higher than the cost of fossil-fuel-fired heaters. The burner of choice for indirect heating is typically a radiant tube.

The objective of the radiant tube burner is to efficiently transfer heat from the combustion gases to the radiant tube, and then to efficiently radiate that energy to the load. There are several challenges that need to be considered when using radiant tubes. The biggest challenge is the material of construction for the tube itself. Typical tubes are constructed of high-temperature metal alloys or of ceramics. Metal alloys can be expensive and typically do not have as high a continuous operating temperature as ceramics. Metal tube burners typically operate at temperatures around 2000°F (1400K). These tubes can fail, due to a variety of metallurgical problems related to the high-temperature operation and thermal cycling of the burners. Metal tubes are more commonly used than ceramic tubes, and research continues into higher temperature metals for use in metallic radiant tube burners.[65,66] Ceramic tubes tend to have higher temperature operating limits compared to the metal tubes, but are even more susceptible to thermal shock. Research also continues on new ceramic materials for radiant tube burners.[66a,67,68] There may also be problems joining the ceramic tubes to the metal burner body due to the differences in thermal expansion that can cause the ceramic tubes to crack. The Institute of Gas Technology (Chicago, IL) is working on a new composite material for radiant tubes that is silicon carbide based and has a working temperature exceeding 2450°F (1620K).[69] These new tubes have excellent shock resistance and are capable of heat flux rates up to 150 Btu/hr-in.^{2} (68 kW/m^2), compared to rates of 55 Btu/hr-in.^{2} (25 kW/m^2) for metallic radiant tubes. Tube lives may be increased by as much as 3 times with the new tubes. The Gas Research Institute (1998) has funded many projects related to the use of advanced ceramic composite materials for radiant tube burners.[70] GRI estimated that about 40% of the more than

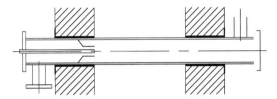

FIGURE 8.30 Straight-through radiant tube burner (From S.S. Singh and L.M. Gorski, Ceramic Single Ended Recuperative Radiant Tube (Phase 1), Final Report, NTIS PB91222554, 1990.)

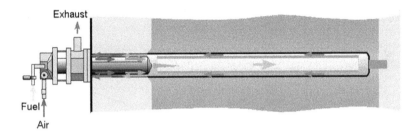

FIGURE 8.31 Example of a single-ended recuperative radiant tube burner. (Courtesy of WS Thermal Process Technology, Inc., Elyria, OH.)

50,000 heat treating furnaces are indirectly heated using an estimated 250,000 radiant tube burners. The new ceramic composites promise higher operating temperatures and longer lives.

There are three common configurations for radiant tube heaters. The first and simplest is known as a straight-through tube where there is a burner at one end of a tube with the exhaust gases from the burner traveling through the tube and exiting at the other end (see Figure 8.30). The challenge when using this geometry is to get efficient heat transfer from the flame gases to the tube, which then radiates to the load. Another potential disadvantage of this type, compared to the other two types, is that connections for the supply gases are on one side while the connections for the exhaust gases are on the other side. The second type of radiant tube heater is known as a single-ended recuperative burner. Here, the burner and exhaust are located on the same side of the tube. A commercial burner is shown in Figure 8.31. The burner fires down an inner tube and then returns back to the starting end through an outer annulus. The exhaust gases are used to preheat the incoming fuel and oxidizer, which increases the overall efficiency. A potential difficulty of this design is that the tube is cantilevered from the wall, which puts some additional stress on the tube. Louis et al. (1998) described a single-ended recuperative radiant tube burner for use in a steam-cracking ethylene production plant.[71] The third radiant tube design is known as a U-tube because of its shape (see Figure 8.32). Again, the supply inlets and exhaust gas outlet are on the same side. One of the difficulties of this design is making a single monolithic U-tube that avoids the need for the 180° elbow where leaks and failures may occur. Abbasi et al. (1998) described an advanced, high-efficiency, low emissions U-tube radiant tube burner made of either a metal alloy or a ceramic composite.[72] The new burner promises to have better temperature uniformity and lower pollutant emissions than conventional burners by using internal exhaust gas recirculation.

The heat transfer from radiant tube burners is important because it not only affects the energy transfer to the load, but it also affects the life and performance of the radiant tube. If the heat flux profile along the radiant tube is highly nonuniform, with high- and low-temperature regions along the length of the tube, then the life of the tube will be significantly reduced due to the high thermal stresses and possible overheating. A more uniform heat flux from the tube will give a higher tube

U-SHAPED RADIANT TUBE

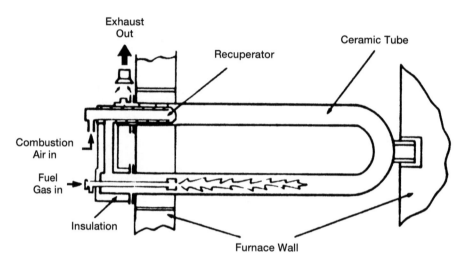

FIGURE 8.32 U-tube radiant tube burner. (Courtesy of American Society of Metals, Warren, PA.[65] With permission.)

life and is normally also more desirable for the load as well to produce uniform heating. The fuel burn-up in the tube is needed to calculate the heat flux from the tube. Ramamurthy et al. (1997) developed fuel burn-up and wall heat transfer correlations for gas-fired radiant tube burners with turbulent flows inside the tube.[73] They studied several parameters, including the burner area ratio (ratio of the outer annular air inlet port area to the inner circular fuel port inlet area), excess combustion air ratio, combustion air preheat temperature, and the fuel firing rate. The fuel burn-up ratio increased for both high and low burner area ratios where there was a higher difference between the fuel and combustion air velocities, as compared to the base-case burner area ratio where the velocities were more similar. The fuel burn-up ratio increased with the excess air ratio and with the air preheat temperature, but decreased as the fuel firing rate increased. The following correlation was developed for the fuel burn-up:

$$\kappa(z) = 1 - \exp\left[-8.24 \times 10^{-4} \frac{A_b^{0.54}}{\phi^{1.53}} \left(\frac{\dot{m}_{air}\bar{v}_{air,z=0}}{\dot{m}_{fuel}\bar{v}_{fuel,z=0}}\right)^{0.41} \left(\frac{z}{d}\right)^{1.6} \mathrm{Re}_{AFT}^{-0.08}\right]$$ (8.26)

where κ is the fuel burn-up coefficient (fraction of the fuel which has been consumed), z is the distance from the burner outlet, A_b is the burner area ratio (defined above), ϕ is the fuel equivalence ratio, $\dot{m}_{air}$ is the combustion air mass flow, $\dot{m}_{fuel}$ is the fuel mass flow, $\bar{v}_{air,z=0}$ is the average combustion air velocity at the burner outlet, $\bar{v}_{air,z=0}$ is the average fuel velocity at the burner outlet, d is the tube inner diameter, and Re_{AFT} is the Reynolds number of the combustion products at the stoichiometric adiabatic flame temperature conditions. They also developed a correlation for the one-dimensional heat flux from the tube to an isothermal bounding wall:

$$q_{rad,1D} = \frac{\varepsilon_w \sigma\left(\varepsilon_{CP}T_{CP}^4 - \alpha_{CP}T_w^4\right)}{1 - \left(1 - \varepsilon_w\right)\left(1 - \alpha_{CP}\right)}$$ (8.27)

where ε_w is the emissivity of the wall, σ is the Stefan-Boltzmann constant, ε_{CP} is the emissivity of the combustion products, T_{CP} is the absolute temperature of the combustion products, α_{CP} is the absorptivity of the combustion products, and T_w is the absolute temperature of the wall. This correlation was then modified for two dimensions by calculating two factors, depending on the temperature of the combustion products compared to the temperature of the isothermal bounding wall:

$$F_1 = \frac{\int_{z=0}^{z=l_1} q_{rad,2D}\,dz}{\int_{z=0}^{z=l_1} q_{rad,1D}\,dz} \quad \text{for } T_{CP} < T_w \tag{8.28}$$

$$F_2 = \frac{\int_{z=l_1}^{z=l_2} q_{rad,2D}\,dz}{\int_{z=l_1}^{z=l_2} q_{rad,1D}\,dz} \quad \text{for } T_{CP} > T_w \tag{8.29}$$

where $0 \le l < l_1$ is the length along the tube where $T_{CP} < T_w$, and $l_1 \le l < l_2$ is the length along the tube where $T_{CP} > T_w$. The average values for F_1 and F_2 were found to be 1.39 and 1.26, respectively. There was excellent agreement between the measured and predicted heat fluxes using the two-dimensional correction factors.

Schultz et al. (1992) reported the use of a single-ended radiant tube burner in a vacuum furnace used for heat treating metals.[74] The radiant flux from the burner was 150 Btu/hr-in.[2] (68 kW/m[2]). Mei and Meunier (1997) tested and modeled the single-ended radiant tube shown in Figure 8.33.[75] They used the Reynolds Stress Model to simulate the burner. The results showed that detailed flame chemistry modeling in the near-flame region is important to get an accurate representation of the heat release pattern. The model was in good agreement with the measurements for the total heat flux from the outer tube to the environment, but that the choice of a turbulence model was important for getting agreement on the tube wall temperatures.

There are a number of new developments being made to improve radiant tube burners. One concept involves coating the inside of the tube with a catalyst so that the premixed preheated air and fuel partially burn on the catalyst surface and thermally conduct that energy to the outside of the tube to radiate to the load.[76] Advantages include more uniform tube heating and lower NOx emissions. Further research was recommended to find a platinum catalyst with a higher melting point. Huebner et al. (1986) described the results of tests using oxygen-enriched air to enhance the performance of radiant tube burners.[77] The oxygen content in the oxidizer ranged from 21% (air) to 80%, by volume. The furnace temperatures ranged from 1300°F to 2400°F (700°C to 1315°C), and radiant tube diameters of 4, 7⅛, and 10 in. (10.2, 18.1, and 25.4 cm) were tested. Large

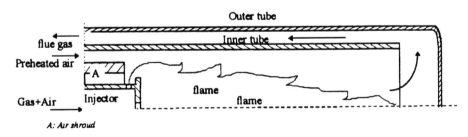

FIGURE 8.33 Single-ended radiant tube burner. (Courtesy of ASME, New York.[75] With permission.)

improvements in thermal efficiency and relatively little change in the tube temperature were found. Larger efficiency improvements were found for the larger diameter tubes.

8.4   EFFECTS ON HEAT TRANSFER

There are many variables that affect the heat transfer from flames. In this chapter section, the following effects are briefly discussed: fuel and oxidizer composition and temperature, fuel and oxidizer staging, burner orientation, heat recuperation, and using pulse combustion.

8.4.1   Fuel Effects

The fuel can have a significant impact on the heat transfer from flames. The three major types of fuels (solids, liquids, and gases) are considered in this section.

8.4.1.1   Solid Fuels

Solid fuels like coal and coke are well-known for producing luminous flames because of the particles in the flame that radiate (see Chapter 3). Solid fuels are not often used in industrial combustion applications and are most used in large power generation plants. The solid particles in solid-fuel combustion radiate as graybodies and give a more uniform spectral radiation output than gaseous flames, which tend to have discrete wavelength bands for water and carbon dioxide radiation. The luminous radiation from solid-fuel combustion tends to make these flames more efficient, although there are challenges to completing combusting the fuel and handling any residues generated from the flame.

Another aspect of solid fuels that indirectly impacts the heat transfer in a combustor is the deposition of ash, produced by the combustion of the solid fuel, onto the inside of the combustor.[78] These deposits may improve the radiant absorptivity of the materials they adhere to, since the emissivity of the deposit is often high. However, the deposits may have a lower thermal conductivity than the base material, particularly when the substrate is a metal tube containing a fluid to be heated, such as water. Then, the deposits can significantly impede the heat transfer to the load and must be periodically removed. Wall et al. (1994) note that there can actually be a slight increase in the thermal conductivity of the deposit with time as its physical and chemical nature change, but the conductivity is still considerably less than that of the metal tube.[78]

Books are available on solid-fuel combustion, particularly coal combustion, although most of them have very little on heat transfer and tend to focus more on the physics of the combustion process itself.[79-86]

8.4.1.2   Liquid Fuels

Boersma (1973) showed that the heat transfer from liquid oil flames can be considerably higher than for comparable gas flames (see Figure 8.34).[87] This is caused by particle generation in the flames where the particles radiate like graybodies, similar to solid-fuel flames. An example of an oil flame is shown in Figure 8.2.

Zung (1978) has edited a book about the evaporation and combustion of liquid fuels.[88] However, the book has almost nothing on heat transfer from flames or in furnaces. Williams (1990) has written a book that has a chapter specifically devoted to liquid fuel combustion in furnaces, although there is very little on heat transfer.[89]

8.4.1.3   Gaseous Fuels

One of the benefits of the gaseous fuels used in industrial combustion applications is that they are very clean-burning and normally generate very few particulates. However, this is a detriment when

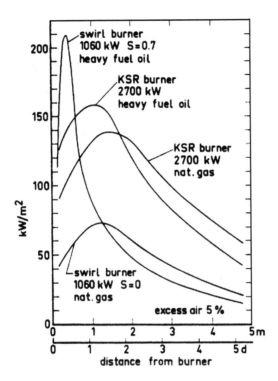

FIGURE 8.34 Measured heat flux rates from different fuels and burners in a cylindrical furnace. (Courtesy of Academic Press, London.[87] With permission.)

it comes to heat transfer from flames because gaseous flames are often very nonluminous and may only radiate in a few narrow wavelength bands, as discussed in Chapter 3. New burner designs have been introduced in recent years to increase the flame luminosity from natural gas flames. This has improved thermal efficiencies and increased product throughputs. One major way this has been achieved is through staged combustion by making the inner part of the flame closest to the burner be fuel rich, and then burn out the fuel downstream with staged oxidizer injection. This technique is well-known for NOx emission reduction as well. Other techniques for increasing the luminosity of gaseous flames include particle injection into the flames, oil injection into the flames, and preheating and cracking the fuel prior to combustion. As discussed in Chapter 1, this continues to be an important area of combustion research.

8.4.1.4 Fuel Temperature

Preheating the fuel influences the flame temperature by increasing the adiabatic flame temperature (see Figure 2.17). This influences both the radiation and convection from the combustion products to the load. The radiation is often increased because of the higher gas temperature. However, different gas species are often produced at higher temperatures (see Figures 2.9 and 2.10), which also influences the radiant heat flux, especially if more or less soot is produced. Amin et al. (1995) experimentally determined that the radiant fraction from an oxygen-enhanced methane flame decreased as the fuel preheat temperature was increased from 300°C to 500°C (570°F to 930°F).[90] The measured radiant fraction actually increased slightly by increasing the fuel preheat temperature from 500°C to 600°C (930°F to 1100°F), due to increased soot production. The convective heat transfer from the combustion products to the load is also influenced by fuel preheating. By preheating the fuel, the transport properties change with both temperature and composition, which directly influence the convective heat transfer coefficient. The gas velocity increases with temperature because of the gas expansion at higher temperatures, which can increase the convective heat transfer to the load.

8.4.2   OXIDIZER EFFECTS

The oxidizer composition and temperature both play important roles in the heat transfer from flames to the load and are briefly discussed next.

8.4.2.1   Oxidizer Composition

This chapter section briefly discusses the effects of the oxidizer composition on the heat transfer from flames. Related discussions are also given in Chapters 3, 6, and 12. Chedaille (1965) showed that an air/oil flame can be enhanced by injecting pure oxygen between the flame and the load.[91] The experiments were conducted in a tunnel furnace where the load consisted of water-cooled tubes located 7 cm (0.3 in.) below the top of the hearth. The flame was angled, at angles ranging from 16° to 30°, from an end wall down toward the hearth. The total combined oxygen content of the combustion air and lanced O_2 was 30% by volume. The experiments showed that the total heat transfer was highest by injecting the oxygen under the flame, but the heat transfer was more uniform by premixing the oxygen in with the combustion air. Arnold (1967) calculated the heat transfer rates for burners fired on town gas with either air or pure oxygen, as shown in Figure 8.35.[92] The oxy/fuel flames produced heating rates as much as 6 times higher than the air/fuel flames when the surface temperature of the load was the highest. Kobayashi et al. (1986) showed theoretically how the heat flux to the furnace heat load can be significantly improved using higher levels of O_2 in the oxidizer, as shown in Figure 8.36.[93] De Lucia (1991) showed that using oxygen to enhance the performance of industrial furnaces can result in some dramatic fuel reductions per unit of output:[94]

- 30% to 50% in glass melting
- More than 50% in pig iron melting
- More than 30% in copper alloy production
- As much as 39% in ceramic production

In the study by Baukal and Gebhart discussed in Chapter 8.2.3, the oxidizer composition was also a parameter of interest.[20] Figure 8.37 shows how the heat flux to the target varied as a function of the oxidizer composition and as a function of the geometry. The heat flux intensity increased by 54% to 230% as Ω increased from 0.30 to 1.00. The average improvement was approximately 80%. The largest improvement occurred at L = 0.5 and R_{eff} = 0.16. At higher values of Ω, the peak heat flux was measured at the closest axial and radial locations. At lower values of Ω, the peak flux occurred at intermediate axial and radial positions. Therefore, at lower Ω, there was clearly an optimum position to maximize the heat transfer and the thermal efficiency. Figure 8.38 depicts the increase in heat flux intensity as the O_2 content in the oxidizer (Ω) increased. For L = 0.5, the heat flux increased by 192% as the O_2 in the oxidizer increased from Ω = 0.30 to Ω = 1.00. As L increased, the oxidizer composition had less influence on the heat flux intensity. The slope of each curve increased as L decreased. Therefore, the oxidizer composition was more important for closer axial spacings. Figure 8.39 shows that the thermal efficiency increased with Ω. This was a consequence of removing the diluent N_2 from the oxidizer. The efficiency decreased with L. The shapes of the curves are similar.

8.4.2.2   Oxidizer Temperature

Preheating the oxidizer is commonly done to recover energy from the exhaust products and to increase the adiabatic flame temperature of the flame, as shown in Figure 2.16. Preheating the oxidizer also improves the thermal efficiency of a process, as shown in Figure 2.21. Higher flame temperatures can dramatically increase the radiant heat flux from the flame because of its dependence on the absolute temperature raised to the fourth power. Guénebaut and Gaydon (1957) studied

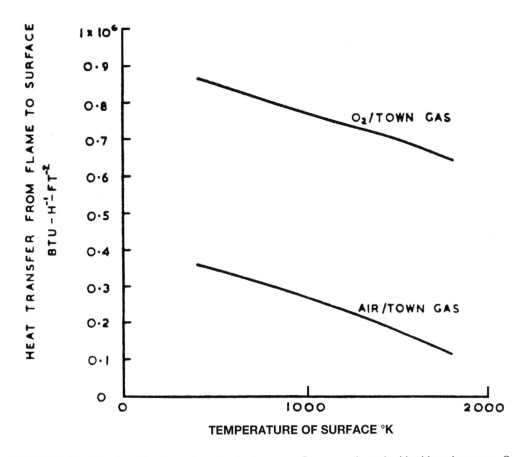

FIGURE 8.35 Calculated heat transfer rates for town gas flames combusted with either air or pure O_2. (From G.D. Arnold, J. Inst. Fuel, 40, 117-121, 1967. With permission.)

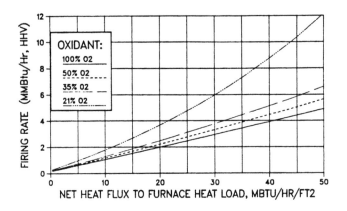

FIGURE 8.36 Calculated heat flux vs. firing rate for different oxidant compositions. (Courtesy of American Society of Metals, Warren, PA.[93] With permission.)

the effect of air preheating on the shape and radiant output from methane, bunsen-burner-type flames.[95] As shown in Figure 8.40, preheating the air to 550°C (1020°F) caused the flame to become twice as long and the radiant heat flux profile to become more uniform compared to the burner with no air preheat.

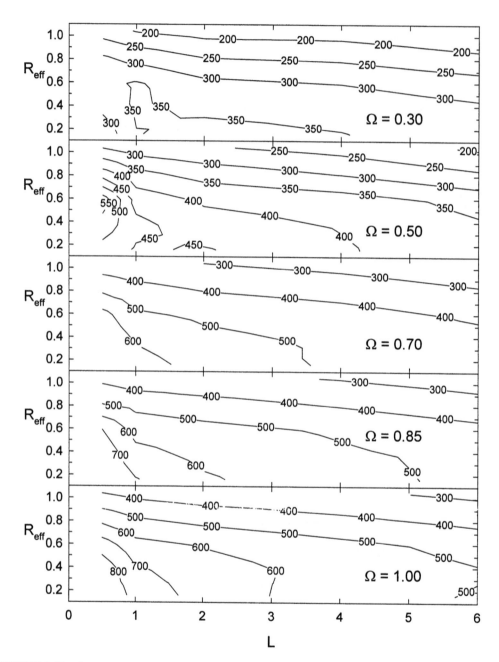

FIGURE 8.37   Contours of the total heat flux, q'' (kW/m²), for stoichiometric natural gas flames ($q_f = 15.0$ kW), with a variable oxidizer composition impinging on an untreated stainless target. (From C.E. Baukal and B. Gebhart, *Exp. Therm. Fluid Sci.*, 16(3), 247-259, 1998.)

8.4.3   Staging Effects

Staging the fuel[96] or the oxidizer[97,98] is a common technique for reducing NOx emissions. This staging also impacts the heat transfer of the system, which is briefly considered next. In either type of staging, the flame is usually longer than unstaged flames. It is assumed in the discussions for both types of staging that both the furnace geometry and the heating process permit longer flames without either flame impingement on the walls of the combustor or without damaging the product quality.

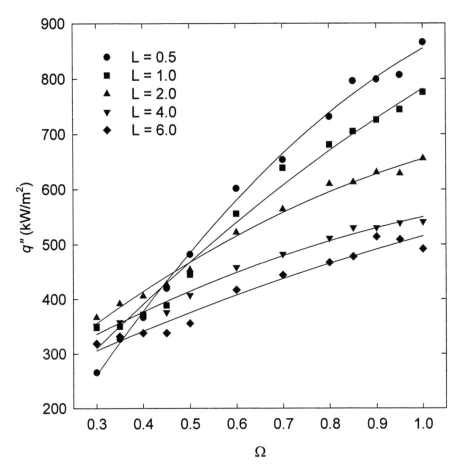

FIGURE 8.38 Total heat flux (q'') to the stagnation point (R_{eff} = 0.16) of an untreated stainless target, for stoichiometric natural gas flames (q_f = 15.0 kW) with a variable oxidizer composition (Ω). (From C.E. Baukal and B. Gebhart, *Exp. Therm. Fluid Sci.*, 16(3), 247–259, 1998).

8.4.3.1 Fuel Staging

Examples of fuel staging are shown in Figures 1.16 and 1.17. Staging the fuel normally means that the inner flame region has excess oxidizer or is fuel-lean. Fuel-lean flames tend to be very nonluminous, depending on the mixing, and therefore only generate gaseous radiation, with little or no soot formation. The balance of the oxidizer is added downstream of the main flame region and normally brings the overall combustion process from a very fuel-lean condition to slightly lean conditions. Again, this does not favor luminous flame radiation even in the secondary flame region. Therefore, fuel staged flames tend to be nonluminous and the heat transfer from these flames is more dominated by forced convection compared to luminous flames. Depending on the application, this may not only be acceptable, but desirable. If putting too much heat near the beginning of the flame can cause overheating, then fuel staging may be an option for stretching out the heat flux over a longer length, while simultaneously reducing the heat flux near the base of the flame. An example would be a counterflow rotary kiln. The material enters the kiln at one end and the burner is located at the other end. This means that the processed material exits at the burner end. Too much heat applied at the material exit could cause slagging, which is partial melting and agglomeration. Pushing the flame more toward the material feed end puts more of the heat where there is much less chance of overheating.

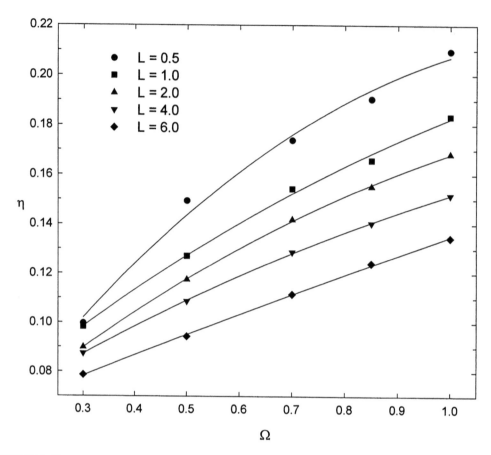

FIGURE 8.39 Thermal efficiency (η) for stoichiometric natural gas flames ($q_f = 15.0$ kW) with a variable oxidizer composition (Ω) impinging on the stagnation point ($R_{eff} = 0.16$) of an untreated stainless target. (From C.E. Baukal and B. Gebhart, *Exp. Therm. Fluid Sci.,* 16(3), 247-259, 1998).

8.4.3.2 Oxidizer Staging

Examples of air staging are shown in Figures 1.14 and 1.15. In this type of flame, the main flame region closest to the burner does not have enough oxygen and is fuel-rich. This normally produces soot and therefore a luminous flame. This type of flame has a higher heat flux from the flame, although the overall efficiency of the process is still dependent on the system geometry and any heat recovery devices that have been incorporated into the system. A staged oxygen flame then commonly has higher heat flux near the burner outlet than the staged fuel flame. Again, depending on the application, this may be desirable or undesirable. In the counterflow rotary kiln process discussed above, this would be undesirable and could lead to slagging. However, for a co-flow rotary kiln, an initially fuel-rich and luminous flame is often desirable because the incoming cold/wet material can absorb higher heat fluxes without affecting the product quality. In the staged oxidizer flame, the secondary flame zone combusts the unburned fuel coming out the primary, fuel-rich flame region to avoid emitting particulates or CO from the exhaust products. This secondary flame zone helps lengthen the flame and the resulting heat flux profile, compared to an unstaged flame.

8.4.4 BURNER ORIENTATION

The choice of burner orientation can vary widely, depending on the industry. Some examples will suffice to illustrate. In the metals and minerals industries, the burners are usually mounted in the

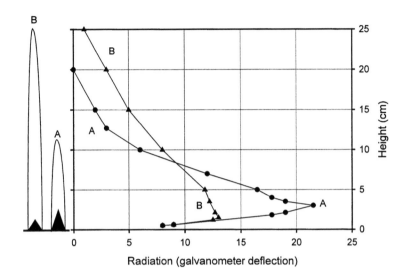

FIGURE 8.40 Flame shapes and radiant heat flux from a methane flame produced by a bunsen burner: (A) no air preheating, and (B) air preheated to 550°C (1020°F). (From H. Guénebaut and A.G. Gaydon, *Sixth Symposium (International) on Combustion,* Reinhold, New York, 1957, 292-295. With permission.)

side wall and either fire parallel to or angled toward the materials being heated. In the heat-treating industry, the burners may fire over or at a muffle that separates the combustion products from the materials being processed. In the petrochemical industry, a more common configuration is for the burners to be mounted in the floor of the furnaces and fire vertically upward. Several common configurations are briefly discussed in this chapter section to show their impact on the heat transfer in the process.

8.4.4.1 Hearth-Fired Burners

A cartoon of a hearth-fired burner is shown in Figure 8.41. Pictures of hearth-fired burners are shown in Figures 8.2 and 8.3. These burners are commonly used in the petrochemical industry and are often natural draft. Depending on the heater design, the burners can be arranged in a variety of ways. A smaller cylindrical heater may have a single burner in the middle of the floor, while a larger cylindrical heater usually has multiple burners located at some radius from the centerline,

FIGURE 8.41 Elevation view of a hearth-fired burner configuration.

arranged at equal angular distances apart (see Figure 9.10). Rectangular heaters generally have one or more rows of burners in the floor. These arrangements are discussed further in Chapter 9.

Normally, the flames from hearth-fired burners are directly vertical, in line with the natural buoyancy force. These burners are designed to release the majority of their heating value in the lower part of the furnace or heater, in what is usually referred to as the radiant section. Many of these heaters have a convection section near the top, where further heat is released from the combustion products to the tubes in the convection section.

Often, the vertical heat flux profile from hearth-fired burners to the tubes in the radiant section is an important design consideration (see Chapter 10). An important heat transfer consideration for this type of burner is the distance from the burners to the tubes. If the spacing is too close, then the tubes may fail prematurely due to overheating. If the spacing is too far, then not enough heat will be transferred in the radiant section and too much may be transferred in the convection section. The former reduces productivity, while the latter can cause damage to the tubes in the convection section. The spacing between burners is also important in order to maximize the power density without causing adverse interactions between flames.

8.4.4.2 Wall-Fired Burners

There are a variety of wall-fired burner configurations. A sketch of a perpendicular, wall-fired burner is shown in Figure 8.42. A schematic is shown in Figure 8.29. The flame comes out in all angular directions, flowing radially outward from the burner. A variation of this type of burner that is sometimes considered to be a wall burner is an infrared burner (see Chapter 8.3). In that case, the burner tile (sometimes referred to as a block or quarl) is heated by flames and radiates toward the load (see Figure 8.19). A sketch of a parallel, wall-fired burner is shown in Figure 8.43. That is a variation of the hearth-fired burner where the burner is located next to and fires along a vertical wall. A sketch of a terrace-fired burner is shown in Figure 8.44. The schematic shows a single terrace, but furnaces using this technology usually have multiple terraces along the vertical wall. There the combustion products from lower terrace burners also flow against upper terraces.

The basic principle of wall-fired burners is to heat a refractory wall, which then radiates to the load. It is often important to have fairly uniform heat flux to the walls to prevent overheating the refractory, which could cause the walls to fail prematurely. Uniform heating of the wall is often desired in order to uniformly heat the load. In vertical wall-fired burners used in the petrochemical industry, there may be a desired heat flux profile along the wall to optimize the heat transfer to the tubes in the radiant section.

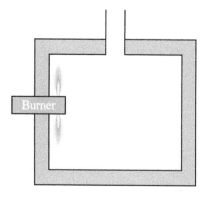

FIGURE 8.42 Elevation view of a perpendicular, wall-fired burner configuration.

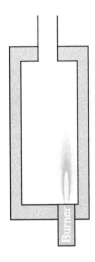

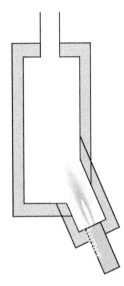

FIGURE 8.43 Elevation view of a parallel, wall-fired burner configuration.

FIGURE 8.44 Elevation view of a terrace-fired burner configuration.

8.4.4.3 Roof-Fired Burners

A photograph of a roof- or down-fired burner is shown in Figure 8.45. These burners are used in a variety of applications. One application is to supplement electric arc furnaces (EAF) (see Chapter 11). There, burners are often mounted in the roof or upper side wall of an EAF and fire down onto the scrap metal. The flames may impinge directly on the scrap. Another application is in reforming furnaces where the burners are often natural draft. In this type of configuration, the

FIGURE 8.45 Example of a down-fired, forced draft, preheated air burner. (Courtesy of John Zink Company, Tulsa, OK.)

flow of the combustion gases is against the buoyancy force. This can affect the flame shape if there is not adequate draft in the furnace. This configuration is not used in many applications because of the difficulties in mounting burners in the roof of a furnace.

An important heat transfer consideration with this design is to ensure that the roof is not overheated, which could become a safety issue if the hot combustion products leak through the roof where they could damage the burner or the fuel gas piping. Another challenge is to get the proper heat flux distribution with the desired flame length, especially in a furnace like a down-fired reformer.

8.4.4.4   Side-Fired Burners

This is a common configuration in a wide range of applications. Examples are shown in Figures 9.2, 9.4, 9.5, 9.7 to 9.9, 9.11, and 9.16. The burners are mounted on the side walls and fired perpendicular to those walls, in contrast to wall-fired burners that fire along the wall. Side-fired burners often fire parallel to and over the top of the load. The location of the burners on opposing side walls may be either directly opposed or staggered, depending on furnace design. Low-momentum flames are often more luminous and have lower NOx emissions compared to high-momentum flames. However, there are potential problems with lower momentum flames. Buoyancy effects can lift the flames toward the roof and away from the load. This may reduce the roof refractory life and reduce the heating efficiency to the load. Another problem is that lower momentum flames are also more easily disturbed by neighboring flames. Therefore, the burner design is a balance of many factors affected by flame momentum.

Traditional side-fired burners were and in many cases still are round flame designs. A relatively newer side-fired burner is a rectangular design with a wide, fan shape, often referred to as a flat-flame burner. The wide part of the flame is parallel to the load below it to maximize the flame radiation surface area, which maximizes the heat transfer to the load. It is usually desirable to have a fairly uniform heat flux output from the flame and to minimize hotspots in the flame. The flame length and surrounding burners also play important roles in determining the heat flux to the load.

8.4.5   Heat Recuperation

There are two strategies commonly used to recuperate heat from combustion exhaust products. The first strategy is to use those hot gases to preheat an incoming feed stream to the combustor, such as the incoming fuel, oxidizer, both the fuel and the oxidizer, or the raw materials being processed in the combustor. In this strategy, a heat exchanger is used so that the hot exhaust gases do not contact the material they are preheating. This is commonly done using either recuperative or regenerative techniques, which are discussed below. Freeman (1986) compared these two techniques for an innovative air preheating system for use in glass-melting furnaces, where in that case the regenerative system performed slightly better than the recuperative systems, as shown in Figure 8.46.[99] The second strategy for recuperating heat from the hot exhaust gases is to actually blend them into a feed stream coming into the furnace. This is commonly done by either furnace or flue gas recirculation.

There are two primary reasons for recuperating the sensible energy in those exhaust products. The first is simply to improve the thermal efficiency of the combustion system. The increase in the available heat for the stoichiometric combustion of methane by preheating the incoming combustion air is shown in Figure 2.21. The second reason that heat may be recuperated is to increase the flame temperature for processes that need higher temperatures, such as in the melting of raw materials like sand to produce glass. Figure 2.15 shows how the adiabatic flame temperature for a stoichiometric air/methane flame increases as the combustion air temperature increases.

Heat recuperation is frequently done using heat exchangers and does not involve the burners at all. However, there are also systems that incorporate heat recuperation in the burner itself. This

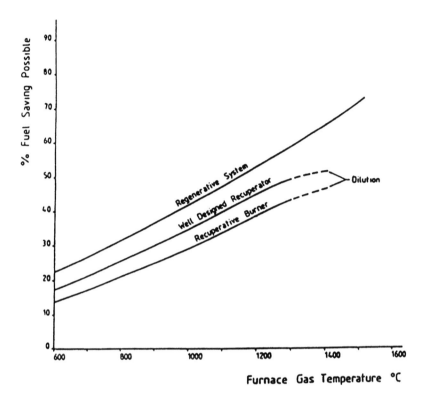

FIGURE 8.46 Calculated fuel savings from regenerative and recuperative systems in a glass-melting furnace. (Courtesy of the American Society of Metals, Warren, PA.[99] With permission.)

heat recuperation can have a significant impact on the heat transfer from the flame to the load, which is briefly discussed next for three types of recuperative burners. Brooks and Winter (1990) discussed the use of cellular ceramic materials (CCMs) to recuperate heat in a furnace.[100] Combustion products from burners flow through the CCM, which is separate from the burner and which then radiates energy to the furnace. This is not discussed further here as it has apparently not become commercially popular at this time, but it may have merit for future applications.

8.4.5.1 Regenerative Burners

Regenerative burners are designed to fire intermittently where half of the burners are on at any given time while half are off. The exhaust products from the burners that are firing flow through the burners that are off. The burners contain some type of heat storage material, usually some type of porous ceramic or ceramic beads, that removes much of the heat from the hot exhaust products flowing through them. An example is shown in Figure 8.47. After a certain amount of time, usually on the order of 0.5 to 15 minutes, the firing pattern is reversed. The combustion air flows through the heat storage material and is preheated prior to combustion. As previously shown, the preheated air can significantly raise the flame temperature. Although the level of dissociation may increase in the flame because of the higher temperatures, any dissociated species are normally in trace quantities so that the bulk composition of the flame gases is essentially the same with or without preheating (see Table 8.3). The main difference is the flame temperature, which affects the radiation from the flame.

Davies (1986) described the use of a regenerative burner for radiant tubes for heating galvanized steel coils.[101] The cycle frequency for switching the firing between burners was 20 sec. Fuel savings of more than 50% were reported. Newby (1986) reported on an all-ceramic, high-temperature regenerative burner for use in applications requiring temperatures up to 3000°F (1650°C).[102] The

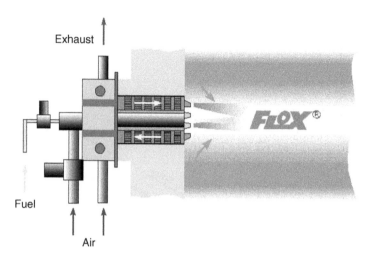

FIGURE 8.47   Example of a regenerative burner. (Courtesy of WS Thermal Process Technology, Inc., Elyria, OH.)

TABLE 8.3
Adiabatic Equilibrium Combustion
Products (mole fraction) for the
Stoichiometric Combustion of Methane
With and Without Air Preheating

	Combustion Air Temperature	
Species	77°F	2000°F
CO	0.012	0.041
CO_2	0.103	0.071
H	0.000	0.005
H_2	0.003	0.011
H_2O	0.148	0.129
NO	0.002	0.008
N_2	0.721	0.699
O	0.000	0.004
O_2	0.006	0.018
OH	0.003	0.014
Adiabatic Flame Temp. (°F)	3620	4311

burner has been proven to operate in dirty environments without plugging the regenerators or causing corrosion damage in the burner. Examples were given of its use in an open pot glass-melting furnace with a reported fuel savings of 20%, an argon-oxygen decarburization vessel preheater, and in (5) small aluminum-melting furnaces. Morita (1996) described Japanese research into high-temperature regenerative air preheaters capable of air temperatures exceeding 1000°C (2100°F).[103] Katsuki and Hasegawa (1998) discussed a regenerative preheater for industrial burners capable of air preheats above 1300K (1900°F).[104] Takamichi (1998) described the use of regenerative burners for improving the performance of forging furnaces.[105] Energy savings of up to 40% were reported.

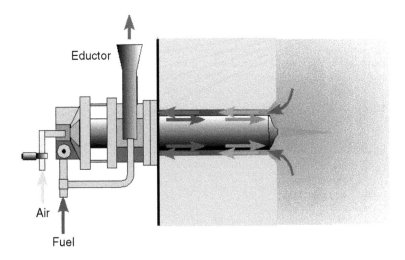

FIGURE 8.48   Example of a recuperative burner. (Courtesy of WS Thermal Process Technology, Inc., Elyria, OH.)

8.4.5.2   Recuperative Burners

An example of a recuperative burner is shown in Figure 8.48. In a recuperative burner, a heat exchanger is built into the burner so that the hot combustion products are exhausted through the burner where they preheat either the fuel, the oxidizer, or both. This is a challenging design due to the large temperature and pressure differences in the various streams. Using heat recuperation in the burner has some advantages over using an external heat exchanger for recovering energy from the furnace, which requires a significant amount of large, insulated ductwork. However, the heat exchanger built into the burner may be less efficient than an external exchanger and therefore tends to have lower preheat temperatures. As discussed above, preheating the incoming fuel or oxidizer raises the flame temperature, which enhances flame radiation.

Singh et al. (1986) reported on the use of a recuperator for single-ended radiant tube burners with a mullite outer tube and silicon carbide inner recuperator.[106] The heat flux rates from the tube ranged from 55 to 69 Btu/hr-in.^{2} (25 to 31 kW/m^2) and had exhaust gas temperatures up to 2000°F (1100°C). Flamme et al. (1998) compared several different types of ceramic heat exchangers for use in self-recuperative burners.[107] They defined the combustion efficiency (η) as:

$$\eta = \frac{h_{\text{fuel(LHV)}} + h_{\text{air}} - h_{\text{flue gas (inlet)}}}{h_{\text{fuel (LHV)}}} \tag{8.30}$$

where $h_{\text{fuel (LHV)}}$ is the lower heating value of the fuel, h_{air} is the enthalpy of the preheated combustion air, and $h_{\text{flue gas (inlet)}}$ is the enthalpy of the flue gas at the inlet of the recuperator. The relative air preheat rate (ε) was expressed as:

$$\varepsilon = \frac{T_{\text{air (outlet)}} - T_{\text{air (inlet)}}}{T_{\text{flue gas (inlet)}} - T_{\text{air (inlet)}}} \tag{8.31}$$

where $T_{\text{air (outlet)}}$ is the combustion air temperature at the outlet of the recuperator, $T_{\text{air (inlet)}}$ is the combustion air at the inlet of the recuperator, and $T_{\text{flue gas (inlet)}}$ is the temperature of the flue gas at the inlet of the recuperator. Figure 8.49 shows the relationship between the combustion efficiency

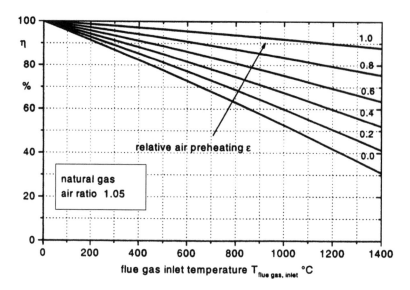

FIGURE 8.49 Combustion efficiency as a function of flue gas temperature at the inlet of the recuperator. (Courtesy of Gas Research Institute, Chicago, IL.[107] With permission.)

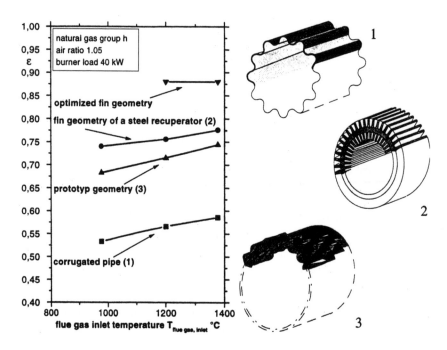

FIGURE 8.50 Calculated relative air preheating rates for different recuperator geometries. (Courtesy of Gas Research Institute, Chicago, IL.[107] With permission.)

and the relative air preheating rate for various flue gas inlet temperatures. Figure 8.50 shows the calculated relative air preheating rates for three different ceramic recuperators as a function of the flue gas inlet temperature. Louis et al. (1998) described a new self-recuperative, single-ended radiant tube burner named CERAJET, which is made of silicon carbide and capable of combustion air preheat temperatures up to 900°C (1650°F) and a thermal efficiency up to 75%.[71]

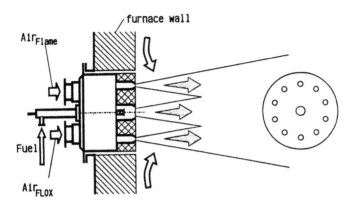

FIGURE 8.51 Example of furnace gas recirculation into an air/fuel flame. (Courtesy of Government Institutes, Rockville, MD.[108] With permission.)

8.4.5.3 Furnace or Flue Gas Recirculation

In flue gas recirculation, some of the hot combustion products from the exhaust stack are recirculated back through the burner. This requires an external fan and insulated ductwork, which are often fairly large because of the hot expanded gases and low gas pressures. Although there are added costs for the fan and ductwork, this method has better control because the amount of flue gas recirculation can be controlled independently of the burner.

In furnace gas recirculation, the hot combustion products inside the furnace are recirculated back into either the flame or inside the burner to mix with either the fuel or the oxidizer. An example is shown in Figure 8.51.[108] No external ductwork or fans are required, which makes it much less expensive than furnace gas recirculation. The fluid pumping is accomplished by the design of the burner, where the fuel or the oxidizer flows through a venturi to create a vacuum to induce the furnace gases to flow back toward the burner. This method is less controllable than flue gas recirculation because the amount of recirculated gas is dependent on the burner conditions. Also, at least one of the incoming gases must have a high enough supply pressure to create suction in the venturi. An example of an oxygen/fuel burner[109] utilizing furnace gas recirculation is shown in Figure 8.52.[110] The measured heat flux rates from that burner using a total heat flux probe inserted into a pilot-scale furnace are shown in Figure 8.53. Plessing et al. (1998) described a combustion system that uses furnace gas recirculation and is known as flameless oxidation (FLOX), where the combustor temperature must be a minimum of 800°C (1500°F) for safe operation.[111] The primary advantage of FLOX is low NOx emissions. Although no specific measurements are given, the radiation from the flame was said to remain the same although the flame is completely nonluminous. Abbasi et al. (1998) described an advanced U-tube radiant tube burner that uses internal furnace gas recirculation to improve temperature uniformity and lower pollutant emissions.[72]

8.4.6 PULSE COMBUSTION

Pulse combustion refers to the periodic change in the outlet flow of the exhaust products from a combustion process to produce a pulsating flow. Putnam (1971) has written an entire book on the subject.[121] The main purpose of that book was how to suppress unwanted combustion-driven oscillations that cause noise and vibration, which are detrimental to processes not designed to pulse. Putnam discussed how oscillations can be generated in a wide range of industrial combustors. Some of the early papers focused more on the acoustics and fluid dynamics than on the heat transfer from pulse combustion.[113]

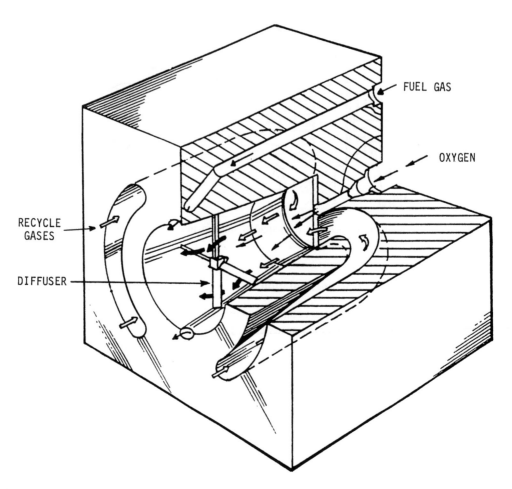

FIGURE 8.52 Oxygen/fuel burner incorporating furnace gas recirculation. (Courtesy of ASME, New York.[110] With permission.)

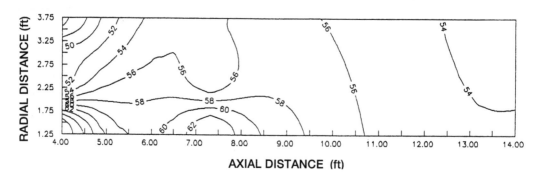

FIGURE 8.53 Measured heat flux rates (MBtu/ft³-hr) from an oxygen/fuel burner incorporating furnace gas recirculation. (Courtesy of ASME, New York.[110] With permission.)

Some of the benefits of combustion specifically designed to pulse include:[114]

- Ability to burn various fuels
- High combustion intensities
- Low NOx formation

- Low excess air requirements
- Self aspiration which eliminates the need for combustion air fans
- Improved heat transfer

Note that for some pulse combustors, the burner design is integrally linked to the combustion chamber. Therefore, both the burner and, where appropriate, the combustor are considered together here for the sake of consistency and to avoid redundancy. Zinn et al. (1993) described a tunable pulse combustor capable of incinerating waste materials with high efficiencies.[115]

There are two common types of "pulse" combustion: natural and mechanical. In natural pulse combustion, the combustion system is designed in such as way as to deliberately create a resonance that causes the flow to pulse. The heat released in the combustion process excites the fundamental acoustic mode of the combustor. The resulting oscillations interact with the combustion process in such a way as to cause periodic reaction and heat release rates. When the periodicity of the heat release is in phase with the acoustic pressure oscillations, the pulsing process is self-sustaining. There are three common types of natural resonant combustor designs: the quarter-wave or Schmidt combustor, the Helmholtz combustor, and the Rijke combustor.[116] The resonance in the chamber causes the exhaust gases to pulse in a regular and periodic fashion. The pressure initially rises during the first part of the cycle, forcing the exhaust gases out the tail pipe. After the initial pressure wave, a partial vacuum is created that pulls in fresh combustion air. The fuel valve (sometimes referred to as a flapper valve) is simultaneously opened by the vacuum pressure and fuel is injected into the incoming combustion air, creating a flammable mixture. The cycle frequency rates may be over 100 Hz, often appearing as steady flow to the naked eye and to the ear. Neumeier et al. (1991) discussed modeling of the flapper valve in pulse combustors.[117]

The second type of pulse combustor is where a mechanical system causes the flow pulsations. This is often accomplished by opening and closing a valve at fairly high frequencies. One common way is to open and close the fuel valve while keeping a constant flow of oxidizer to a burner. This causes successive regions of fuel-rich and fuel-lean combustion, which can lead to significant reductions in NOx emissions. This type of pulse combustion has a much lower cycle frequency and is often detectable by the human eye and ear. Two concerns with this type of pulse combustion system are the life of the mechanical valve because of the number of times it opens and closes, as well as the increase in noise that is often associated with lower frequency pulsing. George and Putnam (1991) studied the development of a rotary valve for pulse combustors and looked at pulse frequencies from 25 to 90 Hz.[118] They noted that significant improvements in heat transfer can be expected using a rotary valve pulse combustor. The Institute of Gas Technology (Chicago, IL) has developed an oscillating combustion process that uses a fast cycling valve to pulse the incoming fuel to a burner, while having a steady flow of oxidant.[119] By pulsing the fuel, alternating zones of fuel-rich and fuel-lean combustion are produced (see Figure 8.54). The preferred pulse rate is about 5 to 30 cycles/sec. The primary reason to do this is to reduce NOx emissions because NOx formation

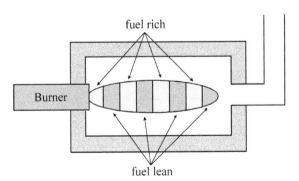

FIGURE 8.54   Alternating fuel-rich and fuel-lean zones in a pulse combustor.

is less favorable at either fuel-rich or very fuel-lean conditions. However, a secondary benefit of the process is often an increase in heat transfer from the flame to the load. This may occur for two reasons. The first is that the fuel-rich regions often produce a luminous flame that radiates heat more efficiently to the load compared to less luminous or nonluminous flames. A second reason is that the oscillating causes the flame gases to pulse, which tends to minimize boundary layer formation and increase convective heat transfer.

In either type of pulse combustion, an important benefit of the process is increased convective heat transfer rates. This is explained by analyzing the flow of the combustion products over a surface (the load). In continuous forced convective flow, a boundary layer builds up as the gases flow over a surface. This boundary layer causes thermal resistance to heat transfer from the gases to the solid surface. In pulsed flow, the boundary layer is continually broken up by the constant changes in flow over the surface. In the higher pressure part of the cycle, the flow goes in one direction, while in the low pressure (vacuum) portion of the cycle, the flow may actually reverse. Breaking up or reducing the thickness of the boundary layer lowers the resistance to heat transfer between the gases and the solid, which enhances the heat transfer rate for a given system compared to non-pulse flow. Hanby (1969) reported heat transfer improvements of up to 100% for an air/propane flame in a straight tube for pulse combustion with a frequency of 100 Hz.[120] The convective heat transfer coefficient was 15 Btu/hr-ft^2-°F (85 W/m^2-°C) for steady flow and 35 Btu/hr-ft^2-°F (200 W/m^2-°C) for pulse flow. Blomquist et al. (1982) measured an improvement in heat transfer ranging from 25% to 40% for a Helmholtz-type pulse combustor operating at about 70 Hz and firing on natural gas.[121] Corliss and Putnam (1986) noted an increase in the convective heat transfer coefficient by as much as 100% using pulse combustion.[122] The results also showed that when considering both heat and mass transfer, a process can be improved by up to a factor of 5. Brinckman and Miller (1989) reported experimental measurements on methane combustion in a Rijke pulse combustor.[123] Results showed that pulse combustion could be maintained over a range of fuel flow rates, equivalence ratios, nozzle exit positions, and combustor lengths. Xu et al. (1991) discussed pulse combustion of heavy liquid oil fuels in a Rijke type pulse combustor.[124] Arpaci et al. (1993) developed an empirical equation for the convective heat transfer in a pulse combustor tailpipe[124a] to correlate the experimental data of Dec and Keller (1989).[124b] The Nusselt number increased with combustion chamber pressure (0.5 to 9 kPa) and oscillation frequency (54 to 101 Hz). Barr et al. (1996) described a computational tool for designing pulse combustors.[125] Figure 8.55 shows how the predicted heat transfer efficiency of a pulse combustion system can be improved

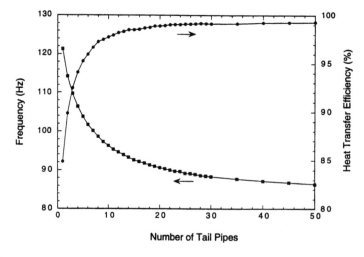

FIGURE 8.55 Effect of number of tail pipes on pulse combustor efficiency. (Courtesy of Government Institutes, Rockville, MD.[125] With permission.)

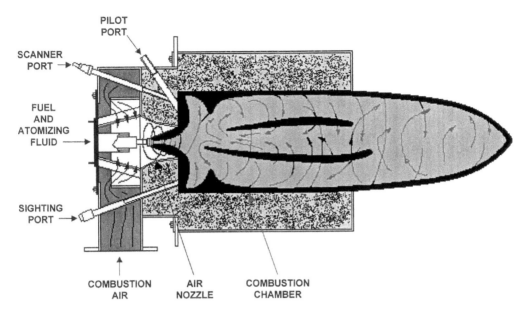

FIGURE 8.56   Swirl burner with waste liquid injection. (Courtesy of the American Society of Metals, Warren, PA.[130] With permission.)

by increasing the number of tail pipes in the system. However, there is a diminishing return as the number of tail pipes increases above about 10. Marsano et al. (1998) reported experimental measurements on a 250-kW Helmholtz-type pulse combustor.[126] Lundgren et al. (1998) analytically showed that a Helmholtz-type combustor can increase the heat transfer by a factor of approximately 2 to 6.[127] Grosman et al. (1998) successfully demonstrated the use of an oscillating combustion system using oxygen-enhanced combustion (see Chapter 12) in a rotary iron melter, ladle preheater, and batch annealing furnace in the steel industry.[128] In laboratory tests, increases in heat transfer up to 59% were demonstrated.

8.5   IN-FLAME TREATMENT

In some applications, particularly waste incineration, it is possible to inject the material to be heated directly into the flame. The material may be a solid, a liquid, a gas, or a combination of the three. For example, Steward and Guruz (1974) described the injection of solid particles (alumina or magnesia) into the fuel stream of a burner to study the effects on radiation.[129] The injected particles had relatively little effect on the measured radiation in the furnace. In many cases, the objective is to fully combust the injected material. For example, Santoleri (1986) described the incineration of waste liquid fuels injected through a swirl burner, as shown in Figure 8.56.[130] Although the waste materials in these systems have some heating value, it may be fairly low with a high water content. This can make the combustion process more difficult due to the high heat extraction from the water. The reason for injecting the waste materials into the flame is to improve the heat and mass transfer processes and more efficiently destruct the waste.

   Another type of in-flame treatment involves injecting a solid material into the flame for some type of heat treatment. For example, an inorganic solid can be injected through the flame to be heated and possibly even melted. That type of material is strictly a heat load and does not contribute any heating through combustion reactions. Although the heat transfer to the load can be significantly increased using in-flame treatment — because of the intimate contact with the hot exhaust gases — the main challenge is transporting and injecting the solids into the flame. Wagner et al. (1996) described a process for injecting spent aluminum potliner (SPL) into a cyclonic combustor (see

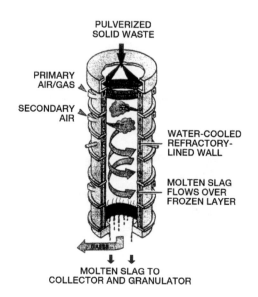

FIGURE 8.57 Slagging cyclonic combustor. (Courtesy of Government Institutes, Rockville, MD.[131] With permission.)

Figure 8.57) for destruction of the SPL, which is a hazardous waste.[131] SPL is a refractory-like substance containing roughly 30% carbon, 15% highly leachable fluoride, and 0.2% cyanide, and is employed as the liner in furnaces used to reduce alumina into aluminum in the Hall reduction cell process. The SPL was fed in the top of the cyclonic combustor and traveled through the flame zone, which destroyed the hazardous waste with high efficiency.

Zhang et al. (1996) presented a novel method for incinerating solid wastes by feeding them through a counter-rotating air-oxy/fuel (see Chapter 12) burner.[132] A sketch of the system is shown in Figure 8.58. High temperatures are important in destructing waste materials. The process was tested in a 50-kW (170,000-Btu/hr) pilot-scale furnace using phenolic-coated foundry sand as the waste material to simulate sand reclamation. The furnace had an internal diameter of 0.2 m (0.7 ft) and a height of 2.5 m (8.2 ft). Complete destruction of the resin was demonstrated and clean sand was produced.

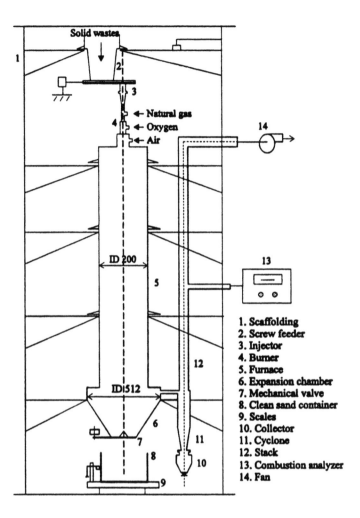

FIGURE 8.58 Schematic of an air-oxy/gas waste incinerator. (Courtesy of Government Institutes, Rockville, MD.[132] With permission.)

REFERENCES

1. R. Villasenor and R. Escalera, A highly radiative combustion chamber for heavy fuel oil combustion, *Int. J. Heat Mass Transfer*, 41, 3087-3097, 1998.
2. A.G. Slavejkov, C.E. Baukal, M.L. Joshi, and J.K. Nabors, Oxy-fuel glass melting with a high-performance burner, *Ceramic Bulletin*, 71(3), 340-343, 1992.
3. D. Neff, P. Mohr, J. Gaddonc, A. Fridman, S. Nester, R. Viskanta, R. Jain, D. Rue, and O. Loo, High-luminosity, low-NOx oxy-natural gas burner for high-temperature furnaces, *IGRC98*, Vol. V: Industrial Utilization, D.A. Dolenc, Ed., Gas Research Institute, Chicago, 1998, 34-44.
4. S.R. Turns, *An Introduction to Combustion,* McGraw-Hill, New York, 1996, chap. 13.
5. A.G. Slavejkov, C.E. Baukal, M.L. Joshi, and J.K. Nabors, Advanced oxygen-natural gas burner for glass melting, *Proc. of 1992 International Gas Research Conf.*, H.A. Thompson, Ed., Govt. Institutes, Rockville, MD, 1993, 2269-2278.
6. D.B. Spalding, The art of partial modeling, *Ninth Symposium (International) on Combustion*, Academic Press, New York, 1963, 833-843.
7. J.M. Beér and N.A. Chigier, *Combustion Aerodynamics*, Applied Science Publishers Ltd., London, 1972.

8. G. Damköhler, Elektrochem, 42, 846, 1936.

9. C.J. Lawn, T.S. Cunningham, P.J. Street, K.J. Matthews, M. Sarjeant, and A.M. Godridge, in *Principles of Combustion Engineering in Boilers*, C.J. Lawn, Ed., Academic Press, New York, 1987.

10. G.H. Markstein, Scaling of radiative characteristics of turbulent diffusion flames, *Sixteenth Symposium (International) on Combustion*, The Combustion Institute, Pittsburgh, PA, 1976, 1407-1419.

11. G.H. Markstein and J. De Ris, Radiant emission and absorption by laminar ethylene and propylene diffusion flames, *Twentieth Symposium (International) on Combustion*, The Combustion Institute, Pittsburgh, PA, 1984, 1637-1646.

12. J.H. Kent and H.G. Wagner, Temperature and fuel effects in sooting diffusion flames, *Twentieth Symposium (International) on Combustion*, The Combustion Institute, Pittsburgh, PA, 1984, 1007-1015.

13. Y.Y. Buriko and V.R. Kuznetsov, Possible mechanisms of the formation of nitrogen oxides in turbulent diffusional combustion, *Combust. Explos. Shockwaves*, 14(3), 296-303, 1978.

14. S.R. Turns and F.H. Myhr, Oxides of nitrogen emissions from turbulent jet flames. I. Fuel effects and flame radiation, *Comb. Flame*, 87, 319-335, 1991.

15. G.M. Faeth, J.P. Gore, S.G. Chuech, and S.M. Jeng, Radiation from turbulent diffusion flames, *Annu. Rev. Numerical Fluid Mech. Heat Transfer*, C.L. Tien and T.C. Chawla, Eds., Hemisphere, New York, Vol. 2, 1-38, 1989.

16. M.A. Delichatsios and L. Orloff, Effects of turbulence on flame radiation from diffusion flames, *Twenty-Second Symposium (International) on Combustion*, The Combustion Institute, Pittsburgh, PA, 1988, 1271-1279.

17. M.A. Delichatsios, J. De Ris, and L. Orloff, An enhanced flame radiation burner, *Twenty-Fourth Symposium (International) on Combustion*, The Combustion Institute, Pittsburgh, PA, 1992, 1075-1082.

18. R. Weber, Scaling characteristics of aerodynamic, heat transfer, and pollution emissions in industrial flames, *Twenty-Sixth Symposium (International) on Combustion*, The Combustion Institute, Pittsburgh, PA, 1996, 3343-3354.

19. R. Weber, J.F. Driscoll, W.J.A. Dahm, and R.T. Waibel, Scaling Characteristics of the Aerodynamics and Low NOx Properties of Industrial Natural Gas Burners. The Scaling 400 Study. I. Test Plan, Gas Research Institute, Chicago, IL, Report GRI-93/0227, 1993.

20. C.E. Baukal, and B. Gebhart, Heat transfer from oxygen-enhanced/natural gas flames impinging normal to a plane surface, *Exp. Therm. & Fluid Sci.*, 16(3), 247-259, 1998.

21. C.E. Baukal, Heat Transfer from Flames Impinging Normal to a Plane Surface, Ph.D. thesis, Univ. of Pennsylvania, Philadelphia, PA, 1996.

21a. P. Hutchinson, E.E. Khalil, J.H. Whitelaw, and G. Wigley, Influence of burner geometry on the performance of small furnaces, *2nd European Symp. on Combustion*, 659-665, 1975.

22. V.I. Kirilenko, I.S. Il'yashenko, A.I. Es'kov, I.B. Smulyanskii, and V.I. Basov, Flat-flame combustion of natural gas in a recuperative glass-melting furnace, *Glass & Ceramics*, Nov., 121-123, 1988.

23. A. Ibbotson, Flat-jet oxygen lancing for homogeneous melt, *Glass*, 68(6), 217, 1991.

24. L.T. Yap and M. Pourkashanian, Low-NOx oxy-fuel flames for uniform heat transfer, *Light Metals 1996*, W. Hale, Ed., Minerals, Metals & Materials Society, 1996, 655-660.

25. T.R. Blake and M. McDonald, An examination of flame length data from vertical turbulent diffusion flames, *Comb. Flame*, 94, 426-432, 1993.

26. C.E. Baukal and B. Gebhart, Oxygen-enhanced/natural gas flame radiation, *Int. J. Heat Mass Transfer*, 44(11), 2539-2547, 1997.

27. M. Pettersson and S. Stenström, Evaluation of gas-fired and electrically heated industrial infrared paper dryers, *proceedings of the 1998 International Gas Research Conf.*, Vol. V: Industrial Utilization, D.A. Dolenc, Ed., Gas Research Institute, Chicago, 1998, 210-221.

28. J.R. Howell, M.J. Hall, and J.L. Ellzey, Combustion within porous media, in *Heat Transfer in Porous Media*, Y. Bayazitoglu and U.B. Sathuvalli, Eds., New York, ASME, HTD-Vol. 302, 1-27, 1995.

29. P.F. Hsu and J.R. Howell, Measurement of thermal conductivity and optical properties of porous partially stabilized zirconia, *Exper. Heat Transfer*, 5, 293-313, 1993.

30. R.F. Speyer, W.-Y. Lin, and G. Agarwal, Radiant efficiencies and performance considerations of commercially manufactured gas radiant burners, *Exp. Heat Transfer*, 9, 213-245, 1996.

31. Y.-K. Chen, R.D. Matthews, and J.R. Howell, The effect of radiation on the structure of a premixed flame within a highly porous inert medium, in *Radiation, Phase Change Heat Transfer and Thermal Systems*, Y. Jaluria, V.P. Carey, W.A. Fiveland, and W. Yuen, Eds., ASME HTD-Vol. 81, 35-42, 1987.

32. S.B. Sathe, R.E. Peck, and T.W. Tong, Flame stabilization and multimode heat transfer in inert porous media, a numerical study, *Comb. Sci. Tech.*, 70, 93-109, 1990.

33. P.F. Hsu, J.R. Howell, and R.D. Matthews, A numerical investigation of pre-mixed combustion within porous media, *J. Heat Transfer*, 115(3), 744-750, 1993.

34. L.B. Younis and R. Viskanta, Experimental determination of the volumetric heat transfer coefficient between stream of air and ceramic foam, *Int. J. Heat Mass Transfer*, 36(6), 1425-1434, 1993.

35. X. Fu, R. Viskanta, and J.P. Gore, Measurement and correlation of volumetric heat transfer coefficients of cellular ceramics, *Exper. Thermal Fluid Sci.*, 17, 285-293, 1998.

36. R. Mital, J.P. Gore, and R. Viskanta, A radiation measurement procedure for gas-fired radiant burners, *Proceedings of the 1998 International Gas Research Conf.*, Vol. V: Industrial Utilization, D.A. Dolenc, Ed., Gas Research Institute, Chicago, 1998, 197-209.

37. M.E. Yetman, Evaluation of infrared generators, *Proc. of 1992 International Gas Research Conf.*, H.A. Thompson, Ed., Govt. Institutes, Rockville, MD, 1993, 2401-2409.

38. M. Johansson, Spectral characteristics and efficiency of gasfired infrared heaters, *Proc. of 1992 International Gas Research Conf.*, H.A. Thompson, Ed., Govt. Institutes, Rockville, MD, 1993, 2420-2426.

39. O.H. Madsen, K. Jørgensen, N.B.K. Rasmussen, and A. van der Drift, Improved efficiency heat transfer using selective emittance radiant burners, *Proc. of 1995 International Gas Research Conf.*, D.A. Dolenc, Ed., Govt. Institutes, Rockville, MD, 1996, 2208-2217.

40. F. Ahmady, Gas-fired IR burners are worth considering, *Process Heating*, 1(2), 38-43, 1994.

41. P. Flanagan, K. Gretsinger, H.A. Abbasi, and D. Cygan, Factors influencing low emissions combustion, in *Fossil Fuels Combustion — 1992*, R. Ruiz, Ed., New York, ASME PD-Vol. 39, 13-22, 1992.

42. R. Mital and J.P. Gore, An experimental study of laminar diffusion flames with enhanced heat transfer by reticulated ceramic inserts, in *Heat Transfer in Fire and Combustion Systems*, W.W. Yuen and K.S. Ball, Eds., New York, ASME HTD-Vol. 272, 7-12, 1994.

43. A. Kataoka, New type of high temperature surface combustion burner, *Proceedings of the 1998 International Gas Research Conf.*, Vol. V: Industrial Utilization, D.A. Dolenc, Ed., Gas Research Institute, Chicago, 1998, 115-124.

44. W.V. Krill and R. Pam, Industrial applications of the pyrocore burner, in *Industrial Combustion Technologies*, M.A. Lukasiewicz, Ed., American Society of Metals, Warren, PA, 1986, 267-271.

45. Alzeta Corp., brochure, Santa Clara, CA, 1998.

46. D.F. Bartz, F.E. Moreno, and P.A. Duggan, Ultra-Low, NOx ultra-high VOC destruction with adiabatic radiant combustors, in *Fossil Fuels Combustion — 1992*, R. Ruiz, Ed., New York, ASME PD-Vol. 39, 7-12, 1992.

47. D.W. DeWerth, A Study of Infra-Red Energy Generated by Radiant Gas Burners, American Gas Association Research Bulletin 92, Catalog No. 41/IR, Cleveland, OH, 1962.

48. S.B. Sathe, M.R. Kulkarni, R.E. Peck, and T.W. Tong, An experimental and theoretical study of porous radiant burner performance, *Twenty-Third Symposium (International) on Combustion*, The Combustion Institute, Pittsburgh, PA, 1990, 1011-1018.

49. M.F. Zabielski, J.D. Freihaut, and C.J. Egolf, Fuel/air control of industrial fiber matrix burners using optical emission, in *Fossil Fuel Combustion 1991*, R. Ruiz, Ed., New York, ASME PD-Vol. 33, 41-48, 1991.

50. T.-Y. Xiong and R. Viskanta, A basic study of a porous-matrix combustor-heater, in *Fossil Fuels Combustion — 1992*, R. Ruiz, Ed., New York, ASME PD-Vol. 39, 31-39, 1992.

51. S. Jugjai and S. Sanitjai, Parametric studies of thermal efficiency in a proposed porous radiant recirculated burner (PRRB): a design concept for the future burner, *RERIC Int. Energy J.*, 18(2), 97-111, 1996.

52. R. Mital, J.P. Gore, R. Viskanta, and S. Singh, Radiation efficiency and structure of flames stabilized inside radiant porous ceramic burners, in Combustion and Fire, M.Q. McQuay, W. Schreiber, E. Bigzadeh, K. Annamalai, D. Choudhury, and A. Runchal, Eds., *ASME Proceedings of the 31st National Heat Transfer Conf.*, Vol. 6, ASME HTD-Vol. 328, 131-137, 1996.

53. M.D. Rumminger, R.W. Dibble, N.H. Heberle, and D.R. Crosley, Gas temperature above a porous radiant burner: comparison of measurements and model predictions, *Twenty-Sixth Symposium (International) on Combustion*, The Combustion Institute, Pittsburgh, PA, 1996, 1755-1762.

54. A. van der Drift, N.B.K. Rasmussen, and K. Jorgensen, Improved efficiency drying using selective emittance radiant burners, *Applied Therm. Engrg.*, 17(8-10), 911-920, 1997.

55. R. Mital, J.P. Gore, R. Viskanta, and A.C. McIntosh, An experimental evaluation of asymptotic analysis of radiant burners, *Twenty-Seventh Symposium (International) on Combustion*, The Combustion Institute, Pittsburgh, PA, 1998, 3163-3171.

56. T.W. Tong, S.B. Sathe, and R.E. Peck, Improving the performance of porous radiant burners through use of sub-micron size fibers, in *Heat Transfer Phenomena in Radiation, Combustion, and Fires*, R.K. Shah, Ed., New York, ASME HTD-Vol. 106, 257-264, 1989.

57. R.D. Bell, C. Chaffin, and M. Koeroghlian, Experimental investigation of a staged porous ceramic burner, in *Fossil Fuels Combustion — 1992*, R. Ruiz, Ed., New York, ASME PD-Vol. 39, 41-46, 1992.

58. R.M. Kendall and J.D. Sullivan, Selective and Enhanced Radiation from Porous Surface Radiant Burners, Gas Research Institute Report GRI-93/0160, Chicago, 1993.

59. T.-Y. Xiong, M.J. Khinkis, B.M. Kramnik, R. Viskanta, and F.F. Fish, An experimental and theoretical study of a porous-matrix combustor-heater, *Proc. of 1992 International Gas Research Conf.*, H.A. Thompson, Ed., Govt. Institutes, Rockville, MD, 1993, 2013-2024.

60. R. Ruiz and S.N. Singh, Enhanced Infrared Burner System, *Proc. of 1992 International Gas Research Conf.*, H.A. Thompson, Ed., Govt. Institutes, Rockville, MD, 1993, 2410-2419.

61. P.F.J. Severens, P.H. Bouma, C.J.H. van de Ven, L.P.H. de Goey, and A. van der Drift, Modeling of twofold flame behavior of ceramic foam surface burners, *J. Energy Resources Tech.*, 117(1), 29-36, 1995.

62. A.N. Bogstra, Development of a new prototype of a flat closed high efficient infrared radiant burner, *Proc. of 1998 International Gas Research Conf.*, Vol. V: Industrial Utilization, D.A. Dolenc, Ed., Gas Research Institute, Chicago, 1998, 24-33.

63. J.V. Milewski, R.A. Shoultz, M.M.B. McConnell, and E.B. Milewski, Improved Radiant Burner Material, Superkinetic, Inc., Albuquerque, NM, Final Report, U.S. Dept. of Energy Report DOE/EE/15643-T2, 1998.

64. Amer. Petroleum Institute Publication 535: Burners for Fired Heaters in General Refinery Services, first edition, API, Washington, D.C., July 1995.

65. W.W. Liang and M.E. Schreiner, Advanced materials development for radiant tube applications, in *Industrial Combustion Technologies*, M.A. Lukasiewicz, Ed., American Society of Metals, Warren, PA, 1986, 305-311.

66. I. Alliat, R. Rezakhanlou, M. Gutmann, and M. Boussuge, Predictive model of delayed failure of ceramic materials at high temperature, *Proc. of 1998 International Gas Research Conf.*, Vol. V: Industrial Utilization, D.A. Dolenc, Ed., Gas Research Institute, Chicago, 1998, 169-177.

66a. J.E. Peters, M.Q. Brewster, and R.O. Buckius, Radiative Heat Transfer Augmentation in High Temperature Combustion Systems with Application to Radiant Tube Burners, Gas Research Institute report GRI-91/0101, Chicago, IL, June 1990.

67. S.S. Singh and L.M. Gorski, Final Report — Ceramic Single Ended Recuperative Radiant Tube (Phase 1), NTIS Document PB91222554, 1990.

68. J.C. Mocsari, H.A. Abbasi, S.M. Nelson, and S.J. Sikirica, Performance evaluation of an advanced hybrid U-tube for industrial heating applications, *Proc. of 1995 International Gas Research Conf.*, D.A. Dolenc, Ed., Govt. Institutes, Rockville, MD, 1996, 2225-2231.

69. Institute of Gas Technology, Composite Radiant Tubes, brochure, Chicago, IL, 1998.

70. S. Sikirica, Benefits of furnace conversions using composite radiant tubes, *Proc. of the 1998 International Gas Research Conf.,* Vol. V: Industrial Utilization, D.A. Dolenc, Ed., Gas Research Institute, Chicago, 1998, 178-188.

71. G. Louis, J. Peureux, T. Landais, J.P. Burzynski, and C. Busson, A new high temperature furnace technology for application in the steam-cracking area, *Proc. 1998 International Gas Research Conf.*, Vol. V: Industrial Utilization, D.A. Dolenc, Ed., Gas Research Institute, Chicago, 1998, 136-146.

72. H. Abbasi, H. Kurek, M. Khinkis, A.E. Yerynov, and O.M. Semerin, Advanced, high-efficiency, low-emissions burner for radiant tube applications, *Proc. 1998 International Gas Research Conf.*, Vol. V: Industrial Utilization, D.A. Dolenc, Ed., Gas Research Institute, Chicago, 1998, 147-157.

73. H. Ramamurthy, S. Ramadhyani, and R. Viskanta, Development of fuel burn-up and wall heat transfer correlations for flows in radiant tubes, *Num Heat Transfer, Part A*, 31, 563-584, 1997.

74. T.J. Schultz, R.A. Schmall, and I. Chan, Selection of a heating system for a high temperature gas fired soft vacuum furnace, in *Fossil Fuels Combustion — 1992*, R. Ruiz, Ed., New York, ASME PD-Vol. 39, 13-22, 1992.

75. F. Mei and H. Meunier, Numerical and experimental investigation of a single ended radiant tube, in *ASME Proc. 32nd National Heat Transfer Conf.*, Vol. 3: Fire and Combustion, L. Gritzo and J.-P. Delplanque, Eds., ASME, New York, 1997, 109-118.

76. J.L. Lannutti, R.J. Schreiber, and M.A. Lukasiewicz, Catalytic radiant tube for industrial process heating applications, in *Industrial Combustion Technologies*, M.A. Lukasiewicz, Ed., American Society of Metals, Warren, PA, 1986, 29-37.

77. S.R. Huebner, C.A. Hersch, and M.A. Lukasiewicz, Experimental evaluation of a radiant tube combustion system fired with oxygen enriched combustion air, in *Industrial Combustion Technologies*, M.A. Lukasiewicz, Ed., American Society of Metals, Warren, PA, 1986, 49-53.

78. T.F. Wall, L.L. Baxter, G. Richards, and J.N. Harb, Ash Deposits, Coal blends and the thermal performance of furnaces, *Eng. Foundation Conf. on Coal-Blending and Switching of Low-Sulfur Western Coals*, Snowbird, UT, Sept./Oct. 1993, 1994, 453-463.

79. D. Merrick, *Coal Combustion and Conversion Technology*, Elsevier, New York, 1984.

80. S. Singer, *Pulverized Coal Combustion — Recent Developments*, Noyes Publications, Park Ridge, NJ, 1984.

81. L.D. Smoot and P.J. Smith, *Coal Combustion and Gasification*, Plenum Press, New York, 1985.

82. J. Feng, Ed., *Coal Combustion*, Hemisphere, New York, 1988.

83. D.A. Tillman, *The Combustion of Solid Fuels and Wastes*, Academic Press, San Diego, CA, 1991.

84. L.D. Smoot, *Fundamentals of Coal Combustion*, Elsevier, New York, 1993.

85. J. Tomeczek, *Coal Combustion*, Krieger Pub., Malabar, FL, 1994.

86. A. Williams, M. Pourkashanian, J.M. Jones, and N. Skorupska, *Combustion and Gasification of Coal*, Taylor & Francis, New York, 1999.

87. D. Boersma, Flame stabilization and heat transfer in a cylindrical furnace, in *Combustion Institute European Symposium 1973*, F.J. Weinberg, Ed., Academic Press, London, 1973, 615-620.

88. J.T. Zung, Ed., *Evaporation — Combustion of Fuels*, American Chemical Society, Washington, D.C., 1978.

89. A. Williams, *Combustion of Liquid Fuel Sprays,* Butterworths, London, 1990.

90. E.M. Amin, M. Pourkashanian, A.P. Richardson, A. Williams, L.T. Yap, R.A. Yetter, and N.A. Moussa, Fuel preheat as NOx abatement strategy for oxygen enriched turbulent diffusion flames, EC-Vol.3/FACT-Vol. 20, *1995 Joint Power Generation Conf.*, S.M. Smouse and W.F. Frazier, Eds., ASME, 1995, 259-269.

91. J. Chedaille, Experimental Study of the Influence on Heat Transfer of the Injection of Pure Oxygen under Industrial Oil Flames, Inter. Flame Research Found. (IJmuiden, The Netherlands) Report K 20/a/28, 1965.

92. G.D. Arnold, Developments in the use of oxy/gas burners, *J. Inst. Fuel*, 40, 117-121, 1967.

93. H. Kobayashi, J.G. Boyle, J.G. Keller, J.B. Patton, and R.C. Jain, Technical and economic evaluation of oxygen enriched combustion systems for industrial furnace applications, in *Industrial Combustion Technologies*, M.A. Lukasiewicz, Ed., American Society of Metals, Warren, PA, 1986, 153-163.

94. M. De Lucia, Oxygen enrichment in combustion processes: comparative experimental results from several application fields, *J. Energy Resources Technology*, 113, 122-126, 1991.

95. H. Guénebaut and A.G. Gaydon, The effect of preheating on flame radiation and flame shape, *Sixth Symposium (International) on Combustion*, Reinhold, New York, 1957, 292-295.

96. B.E. Cain, T.F. Robertson, and J.N. Newby, Reducing NOx emissions in high-temperature furnaces, *Proc. of 1998 International Gas Research Conf.*, Vol. V: Industrial Utilization, D.A. Dolenc, Ed., Gas Research Institute, Chicago, 1998, 237-253.

97. A. Quinqueneau, P.F. Miquel, L.M. Dearden, M. Pourkashanian, G.T. Spence, A. Williams, and B.J. Wills, Experimental and theoretical investigation of a low-NOx high temperature industrial burner, *Proc. of 1998 International Gas Research Conf.*, Vol. V: Industrial Utilization, D.A. Dolenc, Ed., Gas Research Institute, Chicago, 1998, 225-236.

98. D.S. Neff, P.J. Mohr, D. Rue, H. Abbasi, and L. Donaldson, Oxygen-enriched air staging for NOx reduction in regenerative glass melters, *Proc. of 1998 International Gas Research Conf.*, Vol. V: Industrial Utilization, D.A. Dolenc, Ed., Gas Research Institute, Chicago, 1998, 254-261.

99. R.A. Freeman, Some innovative techniques in the glass industry, in *Industrial Combustion Technologies*, M.A. Lukasiewicz, Ed., American Society of Metals, Warren, PA, 1986, 273-277.

100. D.L. Brooks and E.M. Winter, Material selection of cellular ceramics for a high temperature furnace, in *Fossil Fuel Combustion Symposium 1990*, S. Singh, Ed., New York, ASME PD-Vol. 30, 117-125, 1990.

101. T. Davies, Regenerative burners for radiant tubes — field test experience, in *Industrial Combustion Technologies*, M.A. Lukasiewicz, Ed., American Society of Metals, Warren, PA, 1986, 65-70.

102. J.N. Newby, Regen Regenerative Burner for High Performance, Heat Recovery in Aggressive Environments, in *Industrial Combustion Technologies*, M.A. Lukasiewicz, Ed., American Society of Metals, Warren, PA, 1986, 77-85.

103. M. Morita, Present state of high-temperature air combustion and regenerative combustion technology, *Science & Technology in Japan*, No. 58, 11-26, 1996.

104. M. Katsuki and T. Hasegawa, The science and technology of combustion in highly preheated air, *Twenty-Seventh Symposium (International) on Combustion*, The Combustion Institute, Pittsburgh, PA, 1998, 3135-3146.

105. S. Takamichi, Development of High Performance Forging Furnaces, *Proc. 1998 International Gas Research Conf.*, Vol. V: Industrial Utilization, D.A. Dolenc, Ed., Gas Research Institute, Chicago, 1998, 100-112.

106. S. Singh, S. Yokosh, T. Briselden, and S.S. Singh, Improved combustion/thermal efficiency with compact recuperator design, in *Industrial Combustion Technologies*, M.A. Lukasiewicz, Ed., American Society of Metals, Warren, PA, 1986, 71-75.

107. M. Flamme, M. Boß, M. Brune, A. Lynen, J. Heym, J.A. Wünning, J.G. Wünning, and H.J. Dittman, Improvement of energy saving with new ceramic self-recuperative burners, *Proc. 1998 International Gas Research Conf.*, Vol. V: Industrial Utilization, D.A. Dolenc, Ed., Gas Research Institute, Chicago, 1998, 88-99.

108. J.A. Wünning and J.G. Wünning, Regenerative burner using flameless oxidation, *Proc. 1995 International Gas Research Conf.*, D.A. Dolenc, Ed., Govt. Institutes, Rockville, MD, 1996, 2487-2495.

109. K. J. Fioravanti, L. S. Zelson, and C. E. Baukal, Flame Stabilized Oxy-Fuel Recirculating Burner, U.S. Patent 4,954,076 issued 04 September 1990.

110. C.E. Baukal, K.J. Fioravanti, and L. Vazquez del Mercado, The Reflex® Burner, in *Fossil Fuel Combustion — 1991*, R. Ruiz, Ed., ASME PD-Vol. 33, 61-67, 1991.

111. T. Plessing, N. Peters, and J.G. Wünning, Laseroptical investigation of highly preheated combustion with strong exhaust gas recirculation, *Twenty-Seventh Symposium (International) on Combustion*, The Combustion Institute, Pittsburgh, PA, 1998, 3197-3204.

112. A.A. Putnam, *Combustion-Driven Oscillations in Industry*, Elsevier, New York, 1971.

113. W.W. Sipowicz, N.W. Ryan, and A.D. Baer, Combustion-driven acoustic oscillations in a gas-fired burner, *Thirteenth Symposium (International) on Combustion*, The Combustion Institute, Pittsburgh, PA, 1970, 559-564.

114. B.T. Zinn, Applications of pulse combustion in industry, in *Industrial Combustion Technologies*, M.A. Lukasiewicz, Ed., American Society of Metals, Warren, PA, 1986, 55-61.

115. B.T. Zinn, B.R. Daniel, A.B. Rabhan, M.A. Lukasiewicz, P.M. Lemieux, and R.E. Hall, Applications of pulse combustion in industrial and incineration processes, *Proc. 1992 International Gas Research Conf.*, H.A. Thompson, Ed., Govt. Institutes, Rockville, MD, 1993, 2383-2391.

116. B.T. Zinn, Pulse combustion: recent applications and research issues, *Twenty-Fourth Symposium (International) on Combustion*, The Combustion Institute, Pittsburgh, PA, 1992, 1297-1305.

117. Y. Neumeier, J.I. Jagoda, and B.T. Zinn, Modelling of pulse combustor flapper valves, in *Fossil Fuel Combustion 1991*, R. Ruiz, Ed., New York, ASME PD-Vol. 33, 117-125, 1991.

118. P.E. George and A.A. Putnam, Development of a rotary valve for industrial pulse combustors, in *Fossil Fuel Combustion 1991*, R. Ruiz, Ed., New York, ASME PD-Vol. 33, 27-33, 1991.

119. H.A. Abbasi, Fuel Combustion, U.S. Patent 4,846,665, 11 July 1989.

120. V.I. Hanby, Convective heat transfer in a gas-fired pulsating combustor, Trans. ASME, *J. Eng. Power*, 1, 48-51, 1969.

121. C.A. Blomquist and J.M. Clinch, Operational and heat-transfer results from an experimental pulse-combustion burner, *Proc. Pulse-Combustion Applications*, Gas Research Institute Report GRI-82/0009.2, Vol. 1, Chicago, 1982.

122. J.M. Corliss and A.A. Putnam, Heat-transfer enhancement by pulse combustion in industrial processes, in *Industrial Combustion Technologies*, M.A. Lukasiewicz, Ed., American Society of Metals, Warren, PA, 1986, 39-48.

123. G.A. Brinckman and D.L. Miller, Combustion of methane in a Rijke pulsating combustor, in *Heat Transfer Phenomena in Radiation, Combustion, and Fires*, R.K. Shah, Ed., New York, ASME HTD-Vol. 106, 487-491, 1989.

124. Z.X. Xu, D. Reiner, A. Su, T. Bai, B.R. Daniel and B.T. Zinn, Flame stabilization and combustion of heavy liquid fuels in a Rijke type pulse combustor, in *Fossil Fuel Combustion 1991*, R. Ruiz, Ed., New York, ASME PD-Vol. 33, 17-26, 1991.

124a. V.S. Arpaci, J.E. Dec, and J.O. Keller, Heat transfer in pulse combustor tailpipes, *Comb. Sci. Tech.*, 94(1-6), 131-146, 1993.

124b. J.E. Dec and J.O. Keller, Pulse combustor tail-pipe heat-transfer dependence on frequency, amplitude, and mean flow rate, *Comb. Flame*, 77, 359-374, 1989.

125. P.K. Barr, J.O. Keller, and J.A. Kezerle, SPCDC: a user-friendly computation tool for the design and refinement of practical pulse combustion systems, *Proc. 1995 International Gas Research Conf.*, D.A. Dolenc, Ed., Government Institutes, Rockville, MD, 1996, 2150-2159.

126. S. Marsano, P.J. Bowen, and T. O'Doherty, Cyclic modulation characteristics of pulse combustors, *Twenty-Seventh Symposium (International) on Combustion*, The Combustion Institute, Pittsburgh, PA, 1998, 3155-3162.

127. E. Lundgren, U. Marksten, and S.-I. Möller, The enhancement of heat transfer in the tail pipe of a pulse combustor, *Twenty-Seventh Symposium (International) on Combustion*, The Combustion Institute, Pittsburgh, PA, 1998, 3215-3220.

128. R.E. Grosman, M.L. Joshi, J.C. Wagner, H.A. Abbasi, L.W. Donaldson, C.F. Youssef, and G.Varga, Oscillating combustion to increase heat transfer and reduce NOx emissions from conventional burners, *Proc. 1998 International Gas Research Conf.*, Vol. V: Industrial Utilization, D.A. Dolenc, Ed., Gas Research Institute, Chicago, 1998, 1-14.

129. F.R. Steward and K.H. Guruz, The effect of solid particles on radiative transfer in a cylindrical test furnace, *Fifteenth Symposium (International) on Combustion*, The Combustion Institute, Pittsburgh, PA, 1974, 1271-1283.

130. J.J. Santoleri, Burner/atomizer requirements for combustion of waste fuels, in *Industrial Combustion Technologies*, M.A. Lukasiewicz, Ed., American Society of Metals, Warren, PA, 1986, 335-343.

131. J.C. Wagner, H.A. Abbasi, and D. Sager, Thermal treatment of spent aluminum potliner (SPL) in a natural gas-fired slagging cyclonic combustor, *Proc. 1995 International Gas Research Conf.*, D.A. Dolenc, Ed., Govt. Institutes, Rockville, MD, 1996, 2677-2686.

132. B.L. Zhang, C. Guy, J. Chaouki, and L. Mauillon, Heat treatment of divided solid wastes in an oxy-gas reactor, *Proc. 1995 International Gas Research Conf.*, Vol. II, D.A. Dolenc, Ed., Govt. Institutes, Rockville, MD, 1996, 2667-2676.

9 Heat Transfer in Furnaces

9.1 INTRODUCTION

There are many different names used for the combustion chamber containing the heating system and the material being heated, often referred to as the load. Dryers are combustors that typically operate at lower temperatures and are used to remove moisture from materials. Furnaces are refractory-lined chambers usually considered to be higher temperature combustors (see Chapter 11) used for heating and melting solids. Kilns (pronounced "kills") are combustors used to bring about physical and chemical changes in materials and are a type of furnace usually associated with thermal processing of nonmetallic solids such as the ceramic, cement, and lime.[1] The chemical changes are often through calcination that may involve water evaporation, volatile evolution, and even partial combustion of the load material. Heaters are typically refractory-lined chambers usually considered to be lower temperature combustors (see Chapter 10) used to heat solids and liquids.

Hottel (1961) developed a relatively simple equation to calculate the heat transfer in a furnace.[2] The model assumes that the gas in the combustion space is isothermal and is referred to as a one-gas-zone furnace model. The wall and load temperatures are both assumed to be constant, and the furnace walls are assumed to be radiatively adiabatic. A sketch of the combustion system is shown in Figure 9.1. The heat transfer to the load can be calculated as follows:

$$q_l = \frac{\sigma\left(T_g^4 - T_l^4\right)}{\dfrac{1-\varepsilon_l}{A_l\varepsilon_l} + \dfrac{K_s}{4K_aV} + \dfrac{1}{A_l\varepsilon_g + \dfrac{1}{\dfrac{1}{A_lF_{lw}\left(1-\varepsilon_g\right)} + \dfrac{1}{A_w\varepsilon_g}}}} + h_lA_l\left(T_g - T_l\right) \tag{9.1}$$

where σ is the Stefan-Boltzmann constant; T_g and T_l are the absolute temperatures of the gas and load, respectively; ε_l and ε_g are the emissivities of the load and the gas, respectively; A_l and A_w are the surface areas of the load and the wall, respectively; K_s and K_a are the scattering and absorption coefficients of the gas, respectively; V is the volume of the gas; F_{lw} is the view factor to the wall from the load; and h_l is the convection coefficient to the load. The equation is only valid when multiple scatter is negligible ($K_sL \ll 1$). An enthalpy balance on the gas can be written as:

$$q = H_{in} - \dot{m}c_p\left(T_g - T_\infty\right) \tag{9.2}$$

where H_{in} is the energy contained in the fuel, $\dot{m}$ is the gas flow rate, c_p is the average specific heat of the gas between T_g and T_0, and T_g and T_∞ are the absolute temperatures of the gas in the furnace and at ambient, respectively. Note that in this chapter, the heat losses through openings and through the furnace walls are not considered, as they are discussed in Chapters 3 and 4.

Bigzadeh (1993) experimentally studied the heat transfer coefficient in the recirculation zone near the burner outlet in an industrial furnace.[3] The analogy between heat and mass transfer was used by studying the sublimation of thymol in a plexiglass square duct 200 mm (8 in.) wide by 200 mm (8 in.) high by 1000 mm (40 in.) long. The following correlation was determined:

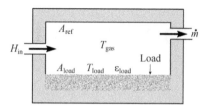

FIGURE 9.1   Elevation view of one-gas-zone furnace model.

$$\overline{\mathrm{Nu}} = 0.637\,\mathrm{Re}^{0.78}\!\left(1.8\,\mathrm{Pr}^{0.3} - 0.8\right)\!\left(\frac{d_n}{B}\right) \tag{9.3}$$

where d_n is the burner nozzle diameter and B is the width of furnace with a square cross-sectional area.

9.2   FURNACES

The choice of the proper furnace type depends on the specific application, as there are many available possibilities. One important consideration is the processing method and whether materials will be processed in batch or continuous mode. Another important consideration is the kind of fuel that will be used.[4] Other considerations include the furnace geometry and required temperature range. The processing method and furnace geometry are briefly considered next, while the temperature ranges are considered in Chapters 10 and 11. The two-volume set of books by Trinks and Mawhinney is considered by many to be the bible of furnaces and is a good reference that deals with all aspects of furnaces, including heat transfer and heat recovery.[5,6] However, Trinks and Mawhinney (1961, 1967) have specifically limited their definition of industrial furnaces to only those processes that raise the temperature of the material (mainly metals) but do not cause any phase changes (melting or vaporization) or any chemical changes (e.g., calcining). Even with that narrow definition of furnace, they list 61 applicable processes, ranging in material temperatures from 300°F to 3250°F (150°F to 1790°C). Here, industrial furnace is more broadly defined to include processes that not only raise the material temperature, but that may also cause phase changes and chemical reactions. Trinks and Mawhinney also include electric heating, which is not considered here.

DeBellis (1993) studied the heat transfer in furnaces having high-emittance coatings applied to the refractory walls.[7] The emittance of the uncoated refractory was approximately 0.5 and the emittance with the coating was approximately 0.8. However, the coating was not stable and the emittance decreased with time. Experiments were conducted both in a small-scale lab furnace and a large-scale metal reheating furnace. In both cases, furnace performance was measured with and without coatings. The results showed that there was no significant change in the rotary hearth reheat furnace efficiency under normal production conditions after the application of the coating.

Khavkin (1996) devoted a significant portion of his book on combustion system design to the use of unlined combustors.[8] These unlined chambers are commonly used, for example, in gas turbine engines and internal combustion engines, but they are not commonly used in industrial combustion because of the high temperatures. Khavkin presented a number of designs where no refractory lining is needed, which can save weight, space, and cost. These designs were developed in Russia, but at the present time have not become popular elsewhere because of the technical challenges and potential problems associated with the high heat flux rates from flames to unprotected metal walls.

Lehrman et al. (1998) noted the importance of controlling air infiltration into furnaces.[9] This "tramp" air not only reduces thermal efficiency, but in metals production it can also aggravate the problem of surface oxidation (scaling) that lowers product quality and yields. Air infiltration was discussed in Chapter 4 and therefore will not be discussed further here.

9.2.1 Firing Method

There are two main methods used to heat load materials that are usually referred to as direct and indirect firing. The method chosen is normally dictated by the process requirements. If the material being processed must not come into contact with the combustion products, then indirect firing is used. If there are no such requirements, direct firing is preferred as it is normally more efficient and straightforward.

9.2.1.1 Direct Firing

In a direct-fired furnace, there is nothing between the combustion products generated by the burners and the load, as shown in Figure 9.2. This is the predominant type used in most industrial heating applications and essentially all of the high-temperature processes (see Chapter 11).

In one respect, the heat transfer analysis of direct-fired furnaces is simpler compared to indirect-fired furnaces as there is no intermediate heat-exchanging surface between the flames and the load. However, in another respect, the analysis of the heat transfer to the load is more complicated as convective heat transfer between the flame gases and the load, and therefore fluid flow, must be included. Figure 9.3 shows a cartoon of the heat transfer in a direct-fired furnace using nonluminous flames. As can be seen, the nonluminous radiation from the flame is larger than the luminous radiation. The radiation from the walls to the load is surface radiation. Figure 9.4 shows a cartoon of the heat transfer in a direct-fired furnace from a luminous flame. In that case, the luminous radiation is comparable to, and in many cases more than, the nonluminous flame radiation.

9.2.1.2 Indirect Firing

In an indirect-fired furnace, there is some intermediate heat transfer surface between the combustion products and the load, as shown in Figure 9.5. That surface is commonly some type of ceramic due to the high temperatures, although metals are used in some cases. The surface is designed to

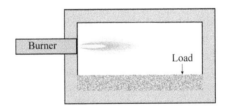

FIGURE 9.2 Elevation view of a direct-fired furnace.

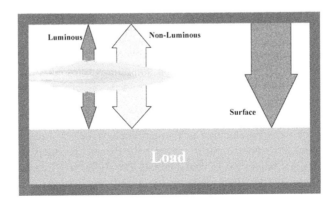

FIGURE 9.3 Heat transfer in a direct-fired furnace from a nonluminous flame (elevation view).

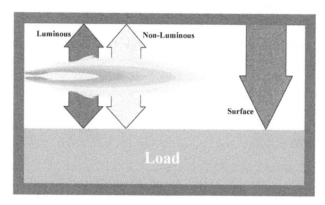

FIGURE 9.4   Heat transfer in a direct-fired furnace from a luminous flame (elevation view).

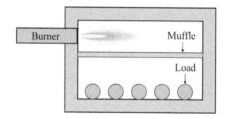

FIGURE 9.5   Elevation view of an indirect-fired furnace.

prevent the combustion products from contacting the load and reducing the quality of the finished product. That quality can be reduced in two ways: (1) by chemically altering the product, and (2) by physically changing the surface. An example of the first is in many metallurgical processes, especially heat treating, where the metal product must be heated in a protective atmosphere containing H_2, N_2, and CO, with only negligible quantities of O_2 and H_2O, which are detrimental to the quality of the metal during heat treating (see Chapter 10).[10] An example of the second type of product quality reduction is in drying a coating or an ink on a web surface in a dryer where the combustion gases could disturb the surface before the coating is dry or the ink is set. Another type of indirect-fired furnace is where the flame gases are separated from the load for either transport or safety reasons. For example, in a process fluid heater, the fluid is transported through metal tubes located inside a furnace. The metal tubes separate the combustion products from the fluid. One reason for the tubes is to transport the fluids at an elevated pressure through the heaters that are fired at approximately atmospheric pressure. Another reason is related to safety, where many of the fluids being heated are hydrocarbons that are potentially explosive if overheated and exposed to enough oxygen.

Two methods are commonly used to separate the combustion products from the load. One is to use open-flame burners but to have a separator, sometimes referred to as a muffle, across the entire combustion space between the flames and the load. Some of the challenges of this method include supporting the separator because of the high temperatures, maximizing the heat transfer from the flames to the separator to optimize the thermal efficiency, and getting a good gas seal around the perimeter of the separator due to thermal expansion. The second method commonly used in industrial combustion applications to separate the exhaust products from the load is to use radiant tube burners (see Chapter 8). In that method, the flame from each individual burner is contained in a ceramic tube. This often improves the overall heat transfer to the separator because of the improved forced convection inside the tube. As with the muffle, supporting long radiant tubes can be a problem as well as the seal between the typically metal burner and ceramic tube.

A different challenge compared to the muffle is getting uniform heat flux from the radiant tubes as the hottest gases are produced at the burner end with the coldest gases exiting from the tube. This can be mitigated through proper design of the radiant tube burner.

The analysis of an indirect-fired furnace may be more complicated than for a direct-fired furnace due to the presence of the heat exchange surface between the flames and the load. The heat transfer from the flames to that surface is normally by both radiation and forced convection. Then, the heat must be conducted through that surface, where it subsequently radiates to the load. This simplifies the analysis of the heat transfer to the load, which is essentially all by radiation.

9.2.1.3   Heat Distribution

The distribution of burners within a furnace has a significant impact on how heat will be transferred. As discussed later in this chapter, the furnace geometry is a large determining factor on where burners will be placed. For example, in a rotary furnace (see Figure 11.15), which is often long and relatively narrow in diameter, a single burner centered on one end of the furnace and firing toward the other end is often the preferred geometry to get proper heat distribution. The desired flame length is often a specific fraction (e.g., half) of the length of the cylinder. In a rectangularly shaped furnace, multiple burners are often used and distributed on one or more walls. In a side-port glass-melting furnace (see Figure 11.20), burners are typically located on two opposite side walls and directly opposed to each other.

The furnace designer, in conjunction with the burner manufacturer, must decide how many burners will be used and where they will be located in the furnace. There are several factors that must be considered. As with most things, economics is an important consideration. It is generally more expensive to use two smaller burners than it is to use one larger burner — not only because of the cost of the burners which is incrementally higher, but also because of the associated extra piping and controls equipment. Another consideration is the physical layout of the furnace structural steel, which limits where burners can be placed in order to avoid steel support members. Still another consideration is the heat transfer distribution within the combustion chamber. Multiple smaller burners often have a more uniform heat distribution pattern than single larger burners, as shown in Figure 9.6 where the size of the flame is proportional to the firing rate. In both cases, the average amount of energy released per unit volume of combustion space is the same; only the distribution of the energy release is different. As can be seen in this example, more uniform heating is usually achieved with more smaller burners, as compared to fewer larger burners.

The actual burner layout is also a function of the burner design and the firing rate of each burner. In certain applications, it may be preferable to have different burners firing at different rates. As an example, Figure 9.7 shows the burner closest to the incoming materials firing the

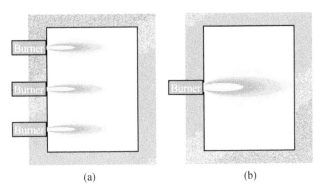

(a)   (b)

FIGURE 9.6   Burner distribution effects shown in a plan view of two furnaces with equal total firing rates: (a) three smaller burners vs. (b) one larger burner.

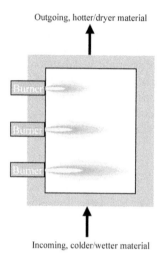

Outgoing, hotter/dryer material

Incoming, colder/wetter material

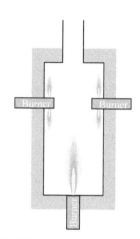

FIGURE 9.7 Plan view of burner profiling where burners are fired at different rates to get a certain heat flux profile.

FIGURE 9.8 Elevation view of a hearth burner and wall burners in a process heater.

hardest, and the burner closest to the outgoing materials firing at the lowest rate. In some cases, the incoming material can absorb as much heat as is available, without overheating. It is also sometimes the case that too much heat applied to the outgoing materials can reduce the product quality by overheating. In some applications, different burner types may also be used in a single furnace. In fired heaters, it is fairly common to have one type of burner in the hearth and a different type of burner in the side walls, as shown in Figure 9.8. Another example is shown in Figure 9.9, which is referred to as oxy/fuel (see Chapter 12) boosting. In that process, oxy/fuel burners are located at the feed end of the furnace to provide more heat than with conventional air/fuel burners. This is used in glass-melting furnaces to increase productivity. In that application, the heat transfer distribution in the furnace is even more complicated because the air/fuel burners fire on only one side at a time and cycle to the other side about every 15 to 30 minutes. Another example of a variation of the heat distribution in the furnace is shown in Figure 9.10. The scrap aluminum is loaded through the top of the furnace. The burners do not fire perpendicular to the wall, but fire at

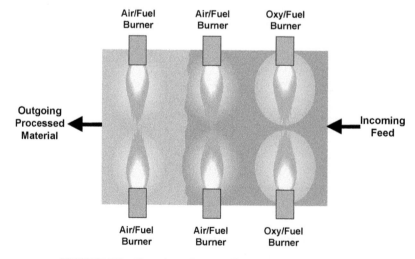

FIGURE 9.9 Plan view of an oxy/fuel boosting in a furnace.

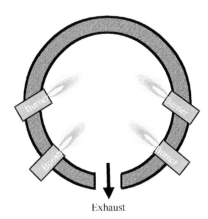

FIGURE 9.10   Cross-section of a round-top, aluminum-melting furnace.

an angle to create a specific flow pattern in the pattern, instead of having all four burners fire directly toward the center of the furnace. This furnace relies heavily on both convection and radiation, depending on the portion of the melting cycle. Right after a new charge of scrap is added to the furnace (which may be piled above the height of the burners), forced convection plays an important role by flowing through the gaps in the scrap pile. As the pile melts down to a flat bath, radiation becomes more dominant.

9.2.2   Load Processing Method

Industrial furnaces are commonly either batch or continuous processes. The specific method has a strong influence on the heat transfer mechanisms that will be important for each type of system. There are also some hybrid processes that are a combination of batch and continuous methods.

9.2.2.1   Batch Processing

In a batch process, materials are loaded into a combustor, heat is then applied, and the materials are removed from the combustor at the completion of the heating cycle. An example is an electric arc furnace where scrap metal is loaded into the furnace, melted, and then tapped out for casting. Batch processes are typically more complicated systems to analyze because of their time-dependent nature. Since the initially loaded material is usually much colder than the combustor, convection and radiation are important modes of heat transfer at the beginning of the heating cycle. The temperature inside the combustor is also transient, where colder temperatures are present near the opened charging door, and higher temperatures are more prevalent farther from the door because they have been less affected by the heat lost through the open door. During the heating cycle, the temperature distribution inside the combustor and in the load are both dynamically changing. Later in the heating cycle, as the load temperature comes closer to that of the combustor, radiation often becomes the predominant heat transfer mechanism.

9.2.2.2   Continuous Processing

In a continuous heating process, the material is constantly fed through the combustor, heated, and withdrawn from the combustor. During normal operations, the combustor is at steady-state conditions with little variation, for example, in the wall temperatures, except at start-up, shutdown, during product changes, or when there are process problems. An example of a continuous process is the production of glass, where raw batch material is continuously fed into the furnace inlet and molten glass is removed at the furnace outlet. A continuous heating process is often more straightforward to analyze and model because of the steady-state nature of the system. However, this does not necessarily mean that analysis is simple, merely that time dependence can be excluded. Continuous heating processes may have a more complicated temperature distribution in the combustor to accommodate the heating of the load. Where a batch combustion process may have fairly uniform furnace wall temperatures inside the combustor, continuous furnaces may have a wider range of wall temperatures corresponding to the wide spatial temperature differences in the load, which is typically much colder at the feed end than at the discharge end.

Swithenbank et al. (1972) presented a zero-dimensional model for designing continuous combustors.[11] The model predicted the following primary performance variables: blow-off stability limits, the overall thermal efficiency, the combustion intensity, and the overall pressure loss. Secondary performance variables that were predicted included ignition capability, pressure and temperature effects, pollutant generation, noise, acoustic oscillations, and heat transfer to the wall. The model consisted of interconnected reactors. This is an example of a simplified model used to analyze continuous processes. Other examples are given in Chapter 5.

9.2.2.3 Hybrid Processing

Some processes, like reverberatory furnaces (see Figure 9.12), are hybrids because they have some features of both batch and continuous furnaces. Side-well reverbs often have a provision for continuously feeding new scrap into a well on the side of the furnace, outside of the combustion chamber, so that the door to that chamber does not need to be opened to charge new materials. However, heats are done in discrete cycles, which take on average from 4 to 10 hours, depending on the alloy being made, the furnace conditions, and the operating procedures at the given plant. At the end of the heat, the molten aluminum is tapped out of the furnace through a tap hole and then poured into molds to make ingots for later use, or directly into a specific part like an aluminum wheel. During the tapping portion of the cycle, the burners are normally on a low-fire mode to maintain the heat in the furnace without overheating the aluminum. The heating portion of the cycle is more continuous in nature. The heat transfer inside the furnace varies more during certain parts of the cycle than in others.

Hybrid processes may be more complicated to model and analyze than batch and continuous processes as the hybrid has features of both. One strategy that is often used is to divide the cycle into several subcycles that can be approximated by a quasi steady-state analysis.

9.2.3 HEAT TRANSFER MEDIUM

Another factor that has a significant influence on the heat transfer in the furnace is the medium used to transport the energy from the burners to the load. In industrial combustion, this medium is most commonly a gas, but liquids and solids are also used. In some heat-treating processes, no medium at all is used (vacuum furnaces). The effect of the heat transfer medium on the overall heat transfer in the furnace is briefly considered here.

9.2.3.1 Gaseous Medium

Gases are by far the most common medium used to transport heat from the burners to the load and to the furnace walls. In most direct-fired processes, the combustion products themselves transport heat to the load by forced convection. The actual rate of heat transfer depends on the flow geometry (gas velocity, how the gas flows over the load, and the shape of the load), the temperatures of both the gases and the load, and the gas composition. Processes are often specifically designed to maximize this convective heat transfer. For example, in a counter-flow rotary kiln, the combustion products flow in one direction while the load flows in the other direction to maximize the temperature difference between the combustion products and the load over the entire length of the kiln. Another example is in heat-treating furnaces where the furnace atmosphere is recycled and blown onto the parts being treated to improve temperature uniformity and to reduce the heating cycle time. Certain gases (CO_2, H_2O, SO_2, CH_4, etc.) also transfer heat to the load and to the furnace walls by gaseous radiation (see Chapter 3). This radiation is dependent on the wavelength, gas temperature, pressure and composition, and on the path length from the gas to the surface being heated.

However, a gaseous medium can also be a hindrance to heat transfer. In processes where gases evolve during the heating, these gases can absorb some of the radiation that does not need a medium to transport energy. For example, it is well known that water selectively absorbs radiation in specific wavelength bands (see Chapter 3). In some industrial heating processes, large quantities of water are vaporized out of the heat load. This water vapor can then absorb some of the radiation going from the flame to the load and to the furnace walls. Carbon dioxide is another gas whose selective radiant absorptivity is well known. Some processes like calcining and glass-making may generate significant quantities of CO_2, which can also absorb radiation going toward the load and walls.

The effect on heat transfer of water vapor evolving from the load is dependent on the specific conditions. As previously discussed, the water vapor may absorb some of the radiation targeted at the load, but it may also re-radiate that energy to the load. The water vapor is at a lower temperature than the gaseous combustion products, so the evolving water vapor may lower the temperature of the convective gases heating the load, which would reduce the convective heat transfer. However, the water vapor may actually improve the convective heating by increasing the mass of the convective gases. Another effect that needs to be considered is the effect of the evolving water vapor on mass transfer. As the partial pressure of water over the load increases, the mass transfer rate from water vaporization decreases. In many drying processes, the evolving water vapor is convected away from the load to minimize its effect on mass transfer.

9.2.3.2 Vacuum

Some heat treating furnaces have a vacuum atmosphere to protect the material being heated from reacting with any gaseous materials. Vacuum heat-treating furnaces rely solely on thermal radiation to heat the parts being treated. The walls of the furnace are heated and those walls radiate to the load. Then, the radiant heat transfer is determined by the temperature and emissivity of both the furnace walls and the load, and on the geometrical view factor between the walls and the load. There will be thermal conduction and radiation between the parts, but not convection as there is no fluid to transport the energy. Therefore, in vacuum furnaces, it is important to properly load the furnace to distribute the heat as uniformly as possible.

9.2.3.3 Liquid Medium

There are several ways that liquid mediums can transport heat in an industrial furnace. One way is by convection from molten materials. For example, in a glass-making furnace, raw cold batch materials are distributed over the surface of the molten glass at the feed end of the furnace. The batch materials are typically melted into the molten glass before they reach halfway across the furnace. Another example of heat transfer from liquids is in a scrap aluminum-melting furnace. As the aluminum melts and flows down to the bottom of the furnace, it convectively heats the solid scrap that it flows over. If molten metal is poured into a cold ladle transfer vessel (see Figure 11.5), there is rapid heat transfer from the liquid metal to the ladle refractory walls which usually have a high thermal conductivity. In some metal production processes, there is a slag layer on the top of the molten metal. This layer is deliberately created by the steel-maker to minimize oxidizing the desired metal. Then, heat must be transferred to the liquid slag layer, which in turn transfers heat to the molten metal. Heat may also be conducted through a liquid medium by conduction. For example, in a reverberatory furnace used for producing aluminum, the energy is received at the surface of the molten liquid aluminum. Because of the high thermal conductivity of aluminum, the heat is quickly conducted to lower levels through the liquid aluminum.

9.2.3.4 Solid Medium

There are several ways that heat can be transferred through a solid medium in an industrial furnace. One method is by luminous radiation produced by solid soot particles generated in the flame (see Chapter 3). Another method is by thermal conduction between contacting solid parts. For example, in a vacuum furnace, energy is received at the outer boundaries of the pile of parts. Then heat is conducted into the parts by thermal conduction between the parts in contact with each other. Solid parts can also transfer heat by radiating to each other. Although not considered here, in a fluidized bed combustor, heat is transferred by circulating suspended solid particles.

9.2.4 GEOMETRY

The geometry refers to both the shape and the size of the combustion chamber. An important parameter that determines the heat loss from a combustor is the external surface area. Heat is lost by both convection and radiation from the outside surface walls. Combustor geometries with higher surface areas will normally have higher losses than those with lower surface areas, assuming everything else is equivalent. One way that combustors are designed is based on the heat release per unit of combustor volume (e.g., kW/m^3 or $Btu/hr\text{-}ft^3$). It is then possible to compare the ratio of the surface area to the volume for different geometries:

$$SV = \frac{\text{surface area}}{\text{volume}} \tag{9.4}$$

The higher the SV ratio, the higher the proportion of heat loss for that combustor design. A spherical combustor would have the following surface area to volume ratio:

$$SV_{\text{sphere}} = \frac{4\pi r^2}{\frac{4}{3}\pi r^3} = \frac{3}{r} \tag{9.5}$$

where r is the radius of the sphere. This shows that the larger the sphere, the lower the SV_{sphere}, which normally means proportionally lower heat losses. Similarly, a cubical combustor would have the following surface area to volume ratio:

$$SV_{\text{cube}} = \frac{6l^2}{l^3} = \frac{6}{l} \tag{9.6}$$

where l is the length of each side of the cube. As the cube size increases, SV_{cube} decreases, which again should result in proportionately lower heat losses (see Figure 9.11). This implies that larger furnaces should have proportionally lower heat losses and therefore be more thermally efficient than smaller furnaces, all other factors being equivalent. This is true in most cases, although real industrial processes are often more complicated than strictly considering the size. For example, in a batch process, loading and unloading the furnace may take proportionately more or less time for a larger furnace, depending on the material handling methods and downstream processes.

It is also instructive to compare the relative shapes of combustors and their impact on heat losses. For the sake of comparison, the volume of the furnace will be held constant as the heat release per unit furnace volume will be assumed to be fixed. This would not be strictly true if the heat losses varied for different furnace geometries as more heat would be required for furnaces with higher heat losses. However, the heat losses for a well-insulated furnace are commonly relatively small, so fixing the furnace volume is reasonable to compare surface area effects. The base case will be a cubical furnace with the length of each side equal to 1 unit ($l = 1$). Then, the

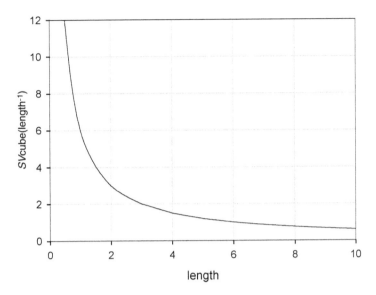

FIGURE 9.11 Surface area-volume ratio (SV) for a cubical furnace as a function of the length of each side of the cube.

volume of that cube would be 1 unit^3 and the surface area would be 6 unit^2, with an $SV_{cube} = 6$ unit^{-1}. For a spherical combustor with an equal volume of 1 unit^3, the radius would be $r = 0.62$ units. Then the SV_{sphere} would be 4.8 unit^{-1}. The ratio of surface to volume ratios for the sphere to the cube (SV_{sphere}/SV_{cube}) equals 0.81. This means that a cubical furnace would normally have proportionally higher heat losses than a spherical furnace, for equal volumes. However, it is not as practical to build a spherical furnace, compared to a cubical furnace.

One can also compare a rectangular furnace to a cubical furnace to see how the heat loss would be affected. Again assume a base case of a cubical furnace with a unit volume. Choose a rectangular furnace that has two sides of equal length l_1 with the remaining side twice as long or $l_2 = 2l_1$, also with a volume of 1 unit^3. Then l_1 would be 0.79 units to get an equivalent volume to the cube and the $SV_{rectangular} = 6.3$. The ratio of the surface to volume ratios for this rectangular shape and the cube of equal volume $(SV_{rectangular}/SV_{cube})$ equals 1.05. This indicates that the heat loss should be higher for a rectangular furnace compared to a cubical furnace of equal volume and shows a comparison of the cubical, spherical, and rectangular furnaces that all have a unit volume of 1 unit^3. Figure 9.12 shows a scale drawing with a comparison of the three furnace geometries of equal volume.

There are numerous combustor geometries in industrial combustion processes, many of which are considered in other sections of this book (especially Chapters 10 and 11). Only a few representative geometries are considered next to study their effects on heat transfer.

9.2.4.1 Rotary Geometry

Rotary kilns are typically cylindrical, refractory-lined vessels that are used to heat solid materials like cement, lime, and scrap aluminum. They typically have high length-to-diameter ratios, are slightly inclined from the horizontal, and rotate. The material is fed in at the higher end and exits at the lower end of the kiln. This geometry is normally characterized by a single burner located in or near the center of one end of the kiln and firing parallel to the axis of the kiln. In counter-current rotary kilns, the material is fed in on the opposite end of the burner. In co-current rotary kilns, the material inlet and the burner are at the same end. The flames are usually round — to match the cylindrical furnace geometry and to transfer heat uniformly around the furnace. Longer flame lengths are often desired to get better heat transfer coverage over a longer length of the furnace.

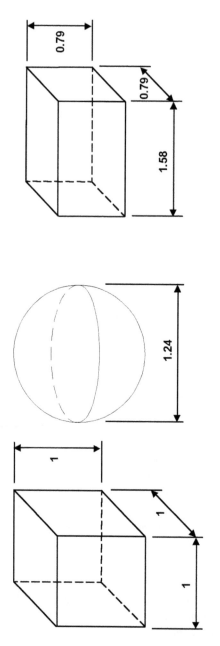

FIGURE 9.12 Cubical, spherical, and rectangular furnaces with volumes of 1 unit[3].

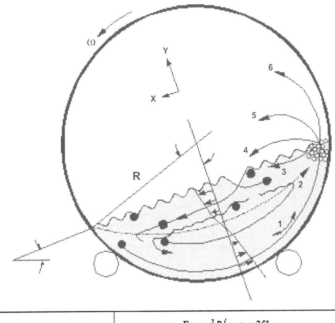

Mode	$Fr = \omega^2 R/g, \varphi = 35°$
1. Slipping	$Fr < 1 \times 10^{-5}$
2. Slumping	$1 \times 10^{-5} < Fr < 0.3 \times 10^{-3}$
3. Rolling	$0.5 \times 10^{-3} < Fr < 0.2 \times 10^{-1}$
4. Cascading	$0.4 \times 10^{-1} < Fr < 0.8 \times 10^{-1}$
5. Cataracting	$0.9 \times 10^{-1} < Fr < 1$
6. Centrifuging	$Fr \geq 1$

FIGURE 9.13 Schematic of a cross-sectional view of granular flow in a rotary kiln, including the bed behavior at 12% loading. (Courtesy of ASME, New York.[13] With permission.)

The heat transfer in a rotary kiln is unique compared to other geometries because it transfers heat under the load, as well as on top. The flame transfers heat to the load and to the exposed furnace walls. Because the furnace rotates, the heated walls rotate under the load and transfer heat to the bottom of the load by thermal conduction. In addition, the solids typically tumble inside the kiln, which constantly exposes new material to the heat source. Tumbling the load and heating from below help increase the efficiency of this process.

Jenkins and Moles (1981) developed an analytical model to predict the heat transfer from a large flame in a rotary kiln.[12] Hottel's zone method was used to simulate radiation heat transfer in the kiln. The model results compared reasonably well with experimental data. Boateng (1997) numerically studied the heat transfer in rotary kilns processing a granular bed.[13] Figure 9.13 shows a sketch of a rotary kiln with 12% loading of a granular material. Two distinct regions are present in the cross-section of the kiln, which are referred to as the freeboard and the bed. The gases flow in the freeboard space and the solid material occupies the bed. The bed material is heated from above by radiation from the flame and the upper kiln walls, and by convection from the combustion gases. The bed is also heated from below by thermal conduction from the lower refractory kiln

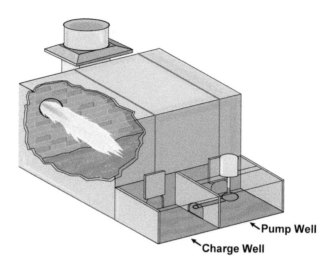

FIGURE 9.14   Sketch of a reverb furnace.

wall in contact with the bed. In the figure, six different bed behaviors are given, depending on the bed rotational Froude number, which is defined as:

$$Fr = \frac{\omega^2 r}{g} \tag{9.4}$$

where ω is the angular velocity of the kiln (sec^{-1}), r is the cylinder radius (m), and g is the acceleration due to gravity (m/sec²).

9.2.4.2   Rectangular Geometry

The rectangular shape is one of the workhorses in industrial combustion. One example of a horizontally fired rectangular furnace is the reverb furnace, commonly used in aluminum melting, which is shown in Figure 9.14. An example of a vertically fired rectangular furnace is the cabin heater. Figure 9.15 shows several different burner arrangements that are used in cabin heaters, depending on both the width of the cabin and the size of the burners. If more than two rows are used in a cabin-type heater, the inner rows of burners may be fired at full rate, while the outer burners may be fired at a reduced rate to prevent overheating the outer walls.

As with other furnace geometries, much of the energy received by the load comes from surface radiation from the refractory-lined furnace walls. In most cases, there is very little heat transfer to or from the furnace below the level of the material loading. The objective is normally to make the furnace as adiabatic as possible to minimize heat losses and to maintain the material temperature.

As discussed above, burners may be positioned in a variety of locations, depending on the process. In some process heaters, there are burners in the floor that fire upward. In glass furnaces, burners are located in the side walls that fire parallel to the floor and roof. In reverberatory furnaces used to melt aluminum, burners are located in the side wall, but are typically angled down toward the load. In reformers, burners are located in the roof and fire downward.

9.2.4.3   Ladle Geometry

Ladles are commonly used to transport molten metal from melting furnaces to casting stations in metal production facilities (see Chapter 11). They are refractory-lined, vertical cylindrical vessels that are closed at one end and usually open at the top. Many metal-melting operations are batch

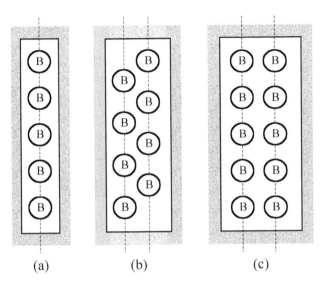

FIGURE 9.15 Burner B arrangement in the floor of cabin heaters (plan view): (a) single row of burners in a narrower heater; (b) two rows of staggered burners in a slightly wider heater; and (c) two rows of parallel burners in an even wider heater.

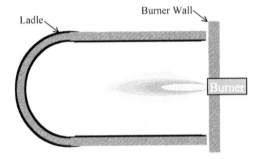

FIGURE 9.16 Ladle preheater stand with empty ladle on its side and temporary burner wall against ladle.

processes. When a heat is ready to be tapped from the furnace, the molten metal is poured into a ladle and transported, usually via an overhead crane, to the casting station. A major challenge with ladles is to minimize the amount of metal that solidifies on the relatively cold (prior to receiving the molten metal) ladle wall. One way to minimize the thermal gradient between the ladle wall and the molten metal is to preheat the ladle with a burner just prior to pouring the metal into the ladle. This preheating is done in a preheat stand as shown in Figure 9.16. The burner wall may have an opening above the burner that acts as the exhaust flue. In some cases, the gap between the ladle and the burner wall acts as the vent. The top edge of a ladle is often far from flat due to what are called "skulls," which are frozen pieces of molten metal around the rim. This makes it difficult to get a tight seal between the burner wall and the ladle.

From a heat transfer perspective, ladles are somewhat unique in that when they are used to transport molten metal, the heat transfer is essentially all by convection and conduction. There is very little radiation heating because the vessel is typically filled close to the top. If the vessel is not covered, then heat is lost by radiation from the top layer of molten metal. When the ladle is being preheated with a burner, the heat transfer is primarily by radiation with some convection.

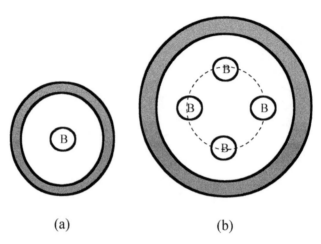

(a)   (b)

FIGURE 9.17   Burner B arrangement in the floor of vertical cylindrical furnaces (plan view): (a) small-diameter furnace with a single centered burner, and (b) larger diameter furnace with four burners symmetrically arranged at a radius from the centerline.

9.2.4.4   Vertical Cylindrical Geometry

Vertical cylindrical (VC) furnaces are commonly used in the chemical and petrochemical industries for heating hydrocarbon fluids. These furnaces usually have burners in the floor of the furnace, either centered or arranged in a circle pattern (see Figure 9.17), and fire straight upward. The tubes carrying the fluids are vertical and against the cylinder wall. The heat transfer to the tubes is primarily by radiation from both the VC walls and from the flames.

Maesawa et al. (1972) measured the radiation heat flux from industrial residual oil flames in a VC furnace.[14] The peak emissivity and radiant heat flux were found at approximately one-third the length of the furnace away from the burner. Steward et al. (1972) made heat transfer measurements in a small VC test furnace where the walls were cooled by an oil.[15] Radiative and total heat fluxes were measured with a custom-built radiometer. A single vertical burner firing propane with 20% excess air was used. Sample radiative heat flux results are shown in Figure 9.18.

9.2.5   Furnace Types

A few common types of furnaces are briefly discussed in this chapter section. Although there are many types of furnaces, the ones chosen for discussion here have significant differences in the heat transfer of the process. These are representative of a variety of furnaces that are often variations of those discussed here.

9.2.5.1   Reverberatory Furnace

A reverberatory or "reverb" furnace (see Figure 9.14) is a rectangular refractory-lined box that is typically used to process non-ferrous metals. Burners are located in one wall or in the roof. The exhaust may be in the roof or on one of the walls, often in between the burners. Reverberatory furnaces are operated as semicontinuous processes, as discussed above. The burners are used primarily to heat the refractory walls of the reverb. The hot walls then re-radiate or reverberate the energy to the load inside the furnace. In the case of aluminum, the surface emissivity of molten aluminum ranges from 0.004 to 0.55.[16] However, molten aluminum readily forms a slag layer on the surface consisting of aluminum oxide, which has an emissivity of 0.11 to 0.19. This results in poor heat transfer qualities as the primary mode of heat transfer is radiation to the surface. Therefore, the dross layer thickness should be minimized. The furnace is designed to compensate for this by having a high bath surface area-to-melt ratio, compared to some other furnace designs.

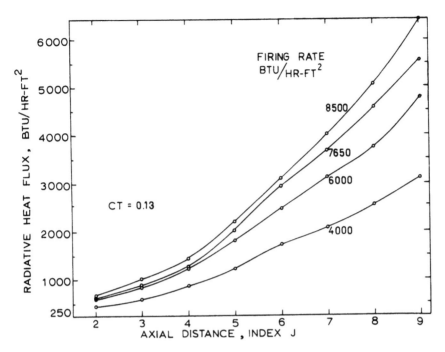

FIGURE 9.18 Radiative flux as a function of axial distance and firing rate for propane firing in a vertical cylindrical furnace. (Courtesy of the Combustion Institute, Pittsburgh, PA.[15] With permission.)

There are two common methods for loading reverbs: direct charging and indirect charging. In direct charging, a charging door to the combustion chamber is opened and the load is directly placed into the combustion chamber. The burners in that chamber normally go to a low-fire mode during the charging and then go back to a high-fire mode after the door has been closed. This method of operation is less efficient as the heat losses from the opened door are substantial. This is also much more of a batch-type operation because of the discrete load charges.

Indirect charge reverbs are often more efficient and are closer to a continuous process because of the charging method, although they are still batch processes because there are discrete times when the load is tapped from the furnace. The load is commonly charged into a chamber attached to the side of the combustion chamber, often referred to as a side well. This type of reverb is called a side-well reverb. These are more thermally efficient as no doors need to be opened on the combustion chamber. The solid load is placed into the side well and then carried into the combustion chamber by circulating molten metal.

Forced convection also plays an important role in reverbs, especially during the initial part of the heating cycle after a new charge has been added directly into the combustion chamber. High-velocity burners direct hot combustion products toward the load to enhance convective heating. As the solids melt, radiation becomes more and more important. Thermal conduction also plays an important role in reverbs since the heat transferred to the surface of the load must be conducted down into the load. In the case of aluminum, which has a very high thermal conductivity, the heat is conducted rapidly through the load, resulting in a more uniform temperature distribution.

Reverb furnaces also have large wall surface areas to maximize the heat transfer to the bath. This also increases the wall losses. For furnaces that do not have special submerged pumps to circulate the molten metal, the bath depth is often shallow to ensure good heat transfer through the bath. For furnaces with pumps, the bath depth is usually thicker because the increased mixing enhances the heat transfer through the bath. It is not uncommon for reverbs to have some type of heat recuperation for preheating the incoming combustion air, to improve the thermal efficiency.

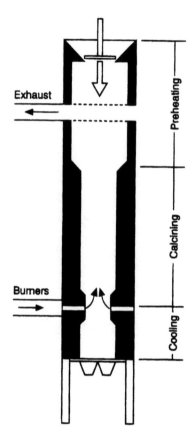

FIGURE 9.19   Elevation view of a vertical shaft kiln. (Courtesy of CRC Press, Boca Raton, FL.[1])

9.2.5.2 Shaft Kiln

A vertical shaft kiln is shown in Figure 9.19. It acts like a counterflow heat exchanger, where the hot exhaust products from the burners located at the bottom flow upward while the incoming cold materials flow downward. These vertical shaft kilns have been used for years for calcining limestone and for manufacturing cement. The heat transfer is fairly unique compared to other industrial heating processes because of the vertical counterflow of materials going down and exhaust products going up. Convection plays a more important role and wall radiation is much less important, compared to many other processes. Thermal conduction is also more important because of the direct contact between moving layers of materials. Feeding solid materials from the top of the furnace is a unique aspect of this furnace configuration.

9.2.5.3 Rotary Furnace

A rotary kiln is shown in Figure 9.20. The raw materials are fed in at the top of the inclined, slowly rotating cylinder and exit at the opposite end. These are often continuous processes where raw material is continuously fed in one end and treated materials exit at the other end. The raw materials are solid materials that are often granular in composition so that they can easily flow through the kiln. These kilns also normally operate as counterflow heat exchangers with the burner at the discharge end and the incoming feed at the opposite end. However, there are also co-flow versions of this type of kiln. Rotary kilns are used for processing bulk materials like cement, lime, and other minerals. Large cement kilns can be up to 20 ft (6 m) in diameter and 500 ft (160 m) in length.

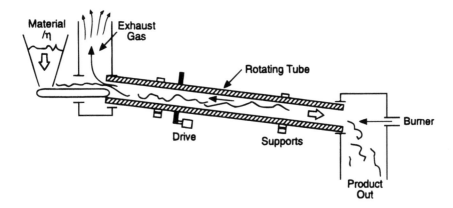

FIGURE 9.20 Elevation view of a rotary kiln. (Courtesy of CRC Press, Boca Raton, FL.[1])

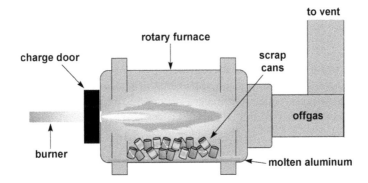

FIGURE 9.21 Elevation view of a rotary scrap aluminum-melting furnace.

An important design feature of the rotary furnace is that the granular load is constantly agitated and new material is continuously being exposed to the heat. The material in contact with the furnace shell also gets heating from the bottom by thermal conduction. In some industrial processes, the load may melt and change from a solid to a liquid. An example is a rotary kiln (see Figure 9.21) furnace used to melt scrap aluminum, especially used beverage containers (UBCs). In those kilns, the scrap aluminum is fed in one end and molten aluminum is tapped off the other end. That particular process is batch in nature as raw materials are fed into the kiln at discrete intervals. When the kiln is fully loaded, it will continue heating the aluminum until it is fully melted, at which time the furnace is tapped.

There are some unique challenges associated with the heat transfer in rotary kilns. The kiln is at a slight angle to the horizontal and the load is at an angle to the kiln because of the furnace rotation. The load moves, but not in as well-defined a manner as a liquid or a gas. The load is constantly mixed by the furnace rotation and is heated from above by the walls and the burner and from below by the wall. Volatiles may evolve from the load, which can be an additional source of energy in the heating process. Some materials processes in rotary kilns may have some endothermic chemical reactions, such as calcining, pyrolysis, and sintering.[17] Some processes can be three-phase if the solid feed material melts and becomes liquid in the furnace, in addition to the combustion gases and possible gases evolving from the process.

9.3 HEAT RECOVERY

Heat recovery devices are often used to improve the efficiency of combustion systems. Some of these devices are incorporated into the burners, but more commonly they are another component

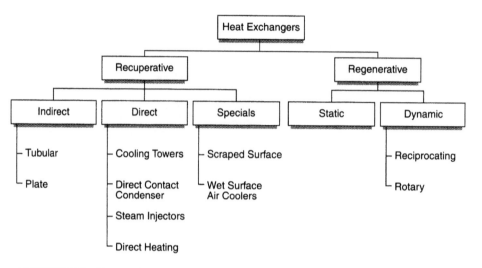

FIGURE 9.22   Heat exchanger classifications. (Courtesy of CRC Press, Boca Raton, FL.[18])

in the combustion system, separate from the burners. These heat recovery devices incorporate some type of heat exchanger, depending on the application. The two most common types have been recuperators and regenerators, which are discussed next. The classifications of these heat exchangers are shown in Figure 9.22. Gas recirculation effects on heat transfer are also considered in this chapter section.

9.3.1   RECUPERATORS

A recuperator is a low- to medium-temperature (up to about 1300°F, or 700°C), continuous heat exchanger that uses the sensible energy from hot combustion products to preheat the incoming combustion air. These heat exchangers are commonly counterflow, where the highest temperatures for both the combustion products and the combustion air are at one end of the exchanger and the coldest temperatures are at the other end. Lower temperature recuperators are normally made of metal, while higher temperature recuperators may be made of ceramics. Recuperators are typically used in lower temperature applications because of the limitations of the metals used to construct these heat exchangers.

9.3.2   REGENERATORS

A regenerator is a higher temperature, transient heat exchanger used to improve the energy efficiency of higher temperature heating and melting processes, particularly in the high-temperature processing industries like glass production. Regenerators are sometimes referred to as "capacitive heat exchangers" and are mainly used in gas/gas heat recovery.[18] In a regenerator, energy from the hot combustion products is temporarily stored in a unit constructed of some type of packing, such as firebricks. This energy is then used to heat the incoming combustion air during a given part of the firing cycle up to temperatures in excess of 2400°F (1300°C).[19]

   Regenerators constructed of a honeycomb of firebricks are often called a chequer work or simply a chequer. Other packing materials can also be used, depending on the application. The packing material must be able to tolerate the thermal shock of constant thermal cycling and may need to withstand a potentially corrosive environment, such as in the manufacturing of glass. Many different patterns, such as the "square chimney" or "closed basket weave" are used in arranging the packing material, depending on the material and the pressure drop requirements. An additional

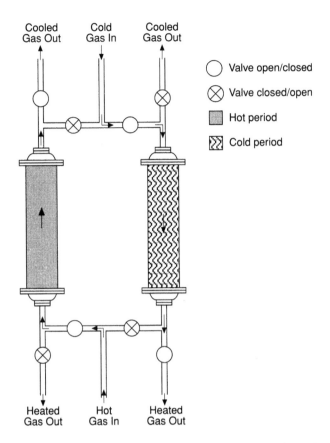

Cooled   Cold   Cooled
Gas Out   Gas In   Gas Out

○ Valve open/closed

⊗ Valve closed/open

▦ Hot period

▨ Cold period

Heated   Hot   Heated
Gas Out   Gas In   Gas Out

FIGURE 9.23   Schematic of a fixed-bed regenerator. (Courtesy of CRC Press, Boca Raton, FL.[20])

requirement is that the regenerators should either not be easily plugged or should be easy to clean if they do get plugged — in order to minimize downtime.

Regenerators are normally operated in pairs because of the normal requirement for a continuous stream of preheated air. During one part of the cycle, the hot combustion gases are flowing through one of the regenerators and heating up the refractory bricks (known as the "hot blow" or "hot period"), while the combustion air is flowing through and cooling down the refractory bricks in the second regenerator (known as the "cold blow" or "cold period"). Both the exhaust gases and the combustion air directly contact the bricks in the regenerators, although not both at the same time since each is in a different regenerator at any given time. After a sufficient amount of time (usually from 5 to 30 min), the cycle is reversed so that the cooler bricks in the second regenerator are then reheated, while the hotter bricks in the first regenerator exchange their heat with the incoming combustion air. A reversing valve is used to change the flow from one gas to another in each regenerator. This is shown schematically in Figure 9.23.

In the more efficient counter-flow regenerator operation, the hot gases pass through the regenerator in the opposite direction to the cold gases that go through the regenerator; while in the less efficient parallel flow operation, both the hot and cold gases pass through the regenerator in the same direction.[20] There are also rotary regenerators that are single units that use the hot combustion products to preheat a rotating ring which is then used to preheat the incoming combustion air. The ring rotates so that it is continually heating and cooling. Then, only a single regenerator is needed to provide a continuous stream of preheated air. The basic problem with rotary regenerators is that they are not completely gas-tight so that exhaust products leak into the combustion air stream and vice versa.

Miyami et al. (1998) described the heat transfer characteristics of a rotary regenerative combustion system where the heat from the exhaust products was used to preheat the incoming combustion air in a single burner.[21] Their model, which included both convection and radiation, was able to predict the performance of a rotary regenerative air preheating system within 10%. Radiation was much less than forced convection in this system. The emissivity of the heat accumulator was assumed to be 1.0. The new RRX system was designed to eliminate some potential disadvantages of high-frequency (every 20 to 30 seconds) regenerative systems, including:

- Furnace pressure fluctuations due to repeated ignition and extinguishment of the burner
- Reliability concerns for the changeover valve, which might actuate as many as a million times in a year
- Higher than desired NOx emissions
- Large space requirements because of the size of the regenerators and the fact that traditional regenerative burners must be installed in pairs

9.3.3 Gas Recirculation

There are two common types of gas recirculation in industrial combustors: furnace and flue gas. These are briefly considered next.

9.3.3.1 Flue Gas Recirculation

Flue gas recirculation is schematically shown in Figure 1.9. In this process, the exhaust products are taken out of the flue and recirculated back into the furnace, often through the burners. The recirculation requires an external, high-temperature fan to move the gases and the associated high-temperature ductwork to transport the gases. There are two common reasons for employing this type of recirculation. In the past, the major reason was for heat recovery. The hot combustion exhaust products were given a second chance to transfer their energy to a lower temperature heat load. A more recent reason to use this technology is to reduce NOx emissions. Although the recirculated gases may be hot, they are not nearly as hot as the flame gases, so the recirculated gases actually cool the flame. The reduced flame temperature reduces NOx emissions, which are highly temperature dependent.

This type of recirculation can have a significant impact on the heat transfer in the furnace. One primary effect is that the forced convection heat transfer is often increased due to the added mass of gases flowing through the furnace and the higher overall gas velocities. However, the forced convection may be tempered by the lower overall gas temperature by diluting the hot flame products with colder flue gases. Since the thermal efficiency increases for nearly all processes employing flue gas recirculation, the overall effect on convection is to increase it.

9.3.3.2 Furnace Gas Recirculation

This is a related process where combustion products are recirculated inside the combustor. An example of a burner designed to produce this type of recirculation is shown in Figure 8.48. There, the high-velocity core jet of pure oxygen induces furnace gases to flow into the burner to dilute the flame temperature. Other burner designs use the flow inside the furnace to recirculate the exhaust products back toward the burners. This technique also tends to improve thermal efficiencies and reduce NOx emissions, but to a lesser extent compared to flue gas recirculation. However, the capital and operating costs are also less since no external fan and ductwork are required. Therefore, the technique chosen is a balance of economics and desired benefits. The heat transfer in the furnace is similarly affected as with flue gas recirculation, but the magnitude of the change is usually smaller.

REFERENCES

1. G.L. Shires, Kilns, in *International Encyclopedia of Heat & Mass Transfer*, G.F. Hewitt, G.L. Shires, and Y.V. Polezhaev, Eds., CRC Press, Boca Raton, FL, 1997, 651-653.
2. H.C. Hottel, The Melchett Lecture for 1960: Radiative Transfer in Combustion Chambers, *J. Inst. Fuel*, 34, 220-234, 1961.
3. E. Bigzadeh, An experimental determination of the heat transfer coefficient in the recirculation zone of an industrial furnace with a square cross-section, in *Heat Transfer in Fire and Combustion Systems — 1993*, B. Farouk, M. Pinar Menguc, R. Viskanta, C. Presser, and S. Chellaiah, Eds., New York, ASME HTD-Vol. 250, 185-189, 1993.
4. A.J. Johnson and G.H. Auth, *Fuels and Combustion Handbook*, first edition, McGraw-Hill, New York, 1951.
5. W. Trinks and M.H. Mawhinney, *Industrial Furnaces*, Vol. 1, 5th edition, Wiley, New York, 1961.
6. W. Trinks and M.H. Mawhinney, *Industrial Furnaces*, Vol. 2, 4th edition, Wiley, New York, 1967.
7. C.L. DeBellis, Evaluation of high-emittance coatings in a large industrial furnace, in *Heat Transfer in Fire and Combustion Systems — 1993*, B. Farouk, M. Pinar Menguc, R. Viskanta, C. Presser, and S. Chellaiah, Eds., ASME HTD-Vol. 250, New York, 1993, 190-198.
8. Y.I. Khavkin, *Combustion System Design: A New Approach*, PennWell Books, Tulsa, OK, 1996.
9. A. Lehrman, C.D. Blumenschein, D.J. Doran, and S.E. Stewart, Steel plant fuels and water requirements, in *The Making, Shaping and Treating of Steel*, 11th edition, Steelmaking and Refining Volume, R.J. Fruehan, Ed., AISE Steel Foundation, Pittsburgh, PA, 1998, 311-412.
10. H.S. Kurek, M. Khinkis, W. Kunc, A. Touzet, A. de La Faire, T. Landais, A. Yerinov, and O. Semernin, Flat radiant panels for improving temperature uniformity and product quality in indirect-fired furnaces, *Proc. 1998 International Gas Research Conf.*, Vol. V: Industrial Utilization, D.A. Dolenc, Ed., Gas Research Institute, Chicago, 1998, 15-23.
11. J. Swithenbank, I. Poll, M.W. Vincent, and D.D. Wright, Combustion design fundamentals, *Fourteenth Symposium (International) on Combustion*, The Combustion Institute, Pittsburgh, PA, 1972, 627-638.
12. B.G. Jenkins and F.D. Moles, Modelling of heat transfer from a large enclosed flame in a rotary kiln, *Trans. Inst. Chem. Eng.*, 59(1), 17-25, 1981.
13. A.A. Boateng, On flow induced kinetic diffusion and rotary kiln bed burden heat transfer, in *Proc. 32nd National Heat Transfer Conf.*, Vol. 3: Fire and Combustion Systems, L. Gritzo and J.-P. Delplanque, Eds., New York, ASME HTD-Vol. 341, 183-191, 1997.
14. M. Maesawa, Y. Tanaka, Y. Ogisu, and Y. Tsukamoto, Radiation from the luminous flames of liquid fuel jets in a combustion chamber, *Twelfth Symposium (International) on Combustion*, The Combustion Institute, Pittsburgh, PA, 1968, 1229-1237.
15. F.R. Steward, S. Osuwan, and J.J.C. Picot, Heat-transfer measurements in a cylindrical test furnace, *Fourteenth Symposium (International) on Combustion*, The Combustion Institute, Pittsburgh, PA, 1972, 651-660.
16. E. Lange, Stack melter or reverb … choosing a new central melter, *Die Casting Engineer*, 38(4), 12-16, 1994.
17. J.R. Ferron and D.K. Singh, Rotary kiln transport processes, *AIChE J.*, 37(5), 747-758, 1991.
18. R.J. Brogan, Heat exchangers, in *International Encyclopedia of Heat & Mass Transfer*, G.F. Hewitt, G.L. Shires, and Y.V. Polezhaev, Eds., CRC Press, Boca Raton, FL, 1997, 546-550.
19. Y. Suzukawa, S. Sugiyama, and I. Mori, Heat transfer improvement and NOx reduction in an industrial furnace by regenerative combustion system, *Proc. 31st Intersociety Energy Conversion Eng. Conf.*, Institute of Electrical & Electronics Eng., Washington, D.C., August, 1996, 804-809.
20. A.J. Willmot, Regenerative heat exchangers, in *International Encyclopedia of Heat & Mass Transfer*, G.F. Hewitt, G.L. Shires, and Y.V. Polezhaev, Eds., CRC Press, Boca Raton, FL, 1997, 944-953.
21. H. Miyami, H. Kaji, Y. Hirose, and N. Arai, Heat transfer characteristics of a rotary regenerative combustion system (RRX), *Heat Transfer — Japanese Research*, 27(8), pp. 584-596, 1998.

10 Lower Temperature Applications

10.1 INTRODUCTION

The designation "lower temperature application" is somewhat relative, but here it refers to processes with furnace temperatures below about 2200°F (1200°C). The major applications considered in this chapter include ovens and dryers, fired heaters, and heat-treating furnaces. In general, the dominant heat transfer mechanisms are radiation and forced convection, depending on the specific application. This is in contrast to higher temperature applications that are dominated almost exclusively by thermal radiation.

10.2 OVENS AND DRYERS

An important consideration for the choice of the type of oven or dryer for a specific application depends on whether direct or indirect heating is needed.[1] Direct heating means the product is exposed directly to the flame (see Figure 10.1), while indirect heating means the products of combustion do not come in contact with the material being heated (see Figure 10.2). In indirect heating, an intermediate carrier or type of heat exchanger is used to transfer the heat. Normally, an intermediate medium between the flame and the load reduces the thermal efficiency compared to direct heating. The most common reason for keeping the exhaust products separate from the heat load is for product quality, wherein the product could be contaminated by the exhaust gases. Another reason is that the product may be emitting combustible volatile vapors that could be ignited if they come into contact with open flames. Still another reason might be that exhaust gases could carry fine particles from the load out of the combustor. These particles would then have to be scrubbed out of the exhaust before the exhaust enters the atmosphere. That means that not only are the raw material costs higher, but there is also the added cost of removing the unused raw materials in the exhaust. As expected, the heat transfer mechanisms will be different, depending on the type of heating used in a given process.

10.2.1 PREDRYER

Predryers are used in some applications prior to the final drying of the product. An example of this type of application is the use of infrared (IR) burners to set the dyes in the dyeing of fabrics in textile manufacturing.[2] After the dyes are applied to the fabric, they must be set prior to contact with the dryer; otherwise, the dyes will migrate to drier areas of the fabric, which reduces the quality of the textile. The IR burners in the predryer are used to rapidly set the dyes without the need to contact the material, as would be the case with, for example, a drum dryer.

Another example of predrying is in the paper industry where IR burners are installed after the coating machine and are used to set the coatings on the paper prior to the paper contacting a steam cylinder drum dryer used to complete the drying (see Chapter 10.2.2).[3] The IR predryer is primarily used to increase productivity and improve the paper coating quality. The productivity is increased because of the added heat. The quality is improved because the IR energy does not disturb the coating as convection or conduction heat transfer methods would, which lets the coating set on the paper prior to contact with the steam cylinder, which relies on thermal conduction heat transfer. Figure 10.3 shows the typical spectral radiant transmissivity for uncoated paper. The paper is less absorbent at longer wavelengths and more absorbent at shorter wavelengths. Since IR burners

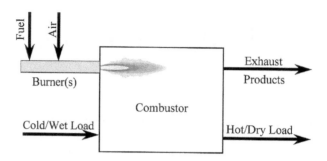

FIGURE 10.1   Schematic of a direct heating system.

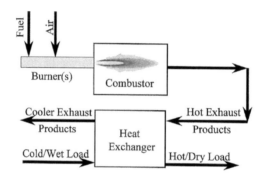

FIGURE 10.2   Schematic of an indirect heating system.

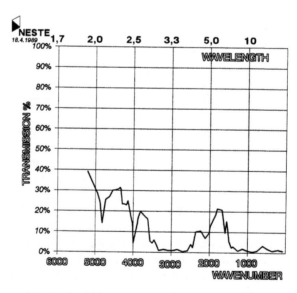

FIGURE 10.3   Spectral transmissivity for typical uncoated paper. (Courtesy of Government Institutes, Rockville, MD.[3] With permission.)

typically have their peak radiant output at wavelengths less than 3 μm, this leads to good thermal efficiencies because the paper is highly absorbent to those wavelengths. Mottle is a measure of the quality of the finished paper product and refers to the finish. Higher mottle means the paper has more blemishes on the surface than lower mottle. Figure 10.4 shows how mottle, measured by a

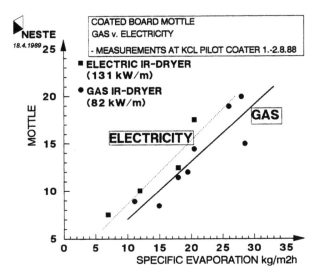

FIGURE 10.4 Mottle as a function of evaporation rate and heat source type. (Courtesy of Government Institutes, Rockville, MD.[3] With permission.)

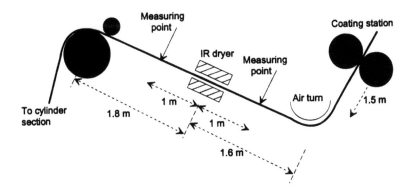

FIGURE 10.5 Schematic of IR heaters used to set coating in a paper machine where the paper is traveling from right to left. (Courtesy of Gas Research Institutes, Chicago, IL.[4] With permission.)

surface image analyzer, was affected by the evaporation rate (related to the firing rate of the burners) and the type of heat source. This shows that gas-fired IR burners produced less mottle, and therefore higher quality, than electric IR burners.

Pettersson and Stenström (1998) compared the use of gas-fired and electric IR burners used to set coating before the paper reached the next cylinder in a paper line.[4] Figure 10.5 shows the IR heaters between the coating station and the next steel cylinder in the paper machine. IR burners are preferred in this application, instead of convective or conductive dryers, because they are non-contact and have high power densities (10 to 40 kW/m², or 3000 to 13,000 Btu/hr-ft²). The thermal efficiencies were calculated as 30% and 40%, respectively, for the gas-fired (propane) and electric IR burners. However, the burners were tested on two different machines and there was some uncertainty in the measurements. That being said, the gas-fired IR burners may still be more economical than the electric IR burners because electricity is often much more expensive than natural gas on an equal energy basis.

Riikonen et al. (1987) developed a computer model to simulate both the convection and radiation heat transfer in IR heating systems used in paper drying.[5] Experiments were conducted to test the model. The model significantly underpredicted the time for the sample to dry. The error in the calculations was identified as the moisture isotherm predictions. The explanation was that IR heating quickly removes the water between fibers but more slowly removes water inside the fibers. A suggested correction was to use an apparent diffusivity to correct for the difference in moisture evaporation rates in and around the pulp fibers.

10.2.2 DRYER

Dryers are commonly used to remove moisture from products during the manufacturing process. Common industries using dryers include the color, pigment, and dyestuff industries, the pharmaceutical and fine chemicals industry, natural ores, minerals, and heavy chemicals, paper and allied products, foodstuffs and agricultural products, and the ceramics and textile industries.[6] If the heat source is some type of burner, dryers may be directly or indirectly fired, depending on the application. Dryers may also be batch or continuous.[7] Figure 10.6 shows some examples of continuous dryers. Sloan (1967) reviewed the following classes of dryers: air-suspended systems that rely primarily on using air for mass transfer, direct-heat continuous systems, direct-heat batch systems, indirect-heat continuous systems, and indirect-heat batch systems.[8] Infrared burners can be used to boost the performance of conventional hot-air convective dryers.[8a] Direct flame impingement is even used in limited cases, such as drying refractory-lined vessels and moulds used in metals and minerals production.[8b]

There are commonly three periods in the drying process, as shown in Figure 10.7. The period from A to B is the warming-up period, where the material is heated up to the evaporation temperature. The period from B to C is referred to as the constant rate period, where the drying rate is constant. The rate of drying is essentially that for the liquid and behaves as if there is no solid present. The liquid wicks rapidly during this period. The period from C to D is referred to as the falling rate period, where the free liquid has been essentially removed and the remaining liquid must move to the surface by diffusion and capillarity. During this period, the solid is an impediment to the liquid removal.

Dryers are sometimes classified by the predominant heat transfer mechanism: convection, conduction, or radiation. They are then subclassified into batch and continuous dryers. An example of a batch convection dryer is a tray dryer. Continuous convection dryers include truck-and-tray tunnels, continuous through-circulation, rotary dryers, fluid bed dryers, spray dryers, and pneumatic or flash dryers. Batch conduction dryers include vacuum tray dryers, vacuum double-cone dryers, trough dryers, pan dryers, and rotary dryers. Continuous conduction dryers include film drum or roller dryers, cylinder or can dryers, rotary dryers, and trough dryers. There are both batch and continuous radiation dryers that use IR heating. While not all of these dryers use combustion, many of the convection dryers preheat the air with burners and the radiation dryers often use gas-fired IR burners.

Dryers are usually low-temperature applications as the material is usually no more than about 300°F (400K). One example is in the drying of paper and board in a paper mill. The paper products are made from pulp in a slurry that has a very high moisture content. The paper is then dried in one of several ways. First, as much water as possible is removed mechanically, typically with some type of press. The remaining moisture is removed by heating, using some type of dryer.

The most common way of drying paper traveling at high velocities is by contact with steam-heated drums or cans, usually referred to as steam cylinders. The paper wraps around the drums in a serpentine fashion to maximize the contact area with the drum. In this type of dryer, the primary method of drying is by thermal conduction.[9] One problem with this technique is that as the paper dries, the thermal conductivity goes down, which makes it more difficult to conduct the heat into the paper. This is known as the "falling rate period," where the downstream steam cylinders are much less effective at removing moisture than the upstream cylinders.

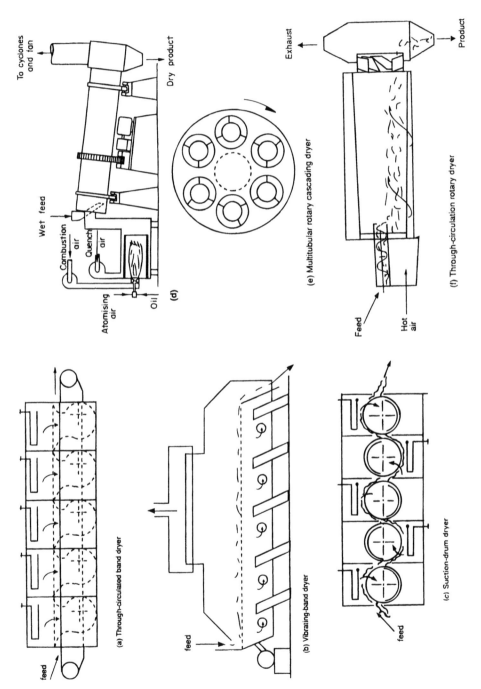

FIGURE 10.6 Examples of continuous dryers. (Courtesy of CRC Press, Boca Raton, FL.[7])

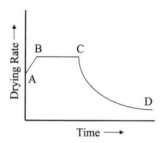

FIGURE 10.7   Periods of the drying process.

IR burners are often used to supplement these dryers because the IR radiation can penetrate into the paper better when it is dry since there is less water to absorb the radiation. In a survey of paper makers by the Gas Research Institute (Chicago, IL), respondents believed the best place to install IR burners on a paper drying line is in the pre-heat zone.[10] Other locations identified in the survey included in the forming section, above and below the steam cylinders in the constant rate zones, and above and below the steam cylinders in the falling rate zone.

One reason for the popularity of steam cylinder dryers is that there is usually plenty of steam available in paper mills because much of the waste bark and liquor from the trees used to make the paper is burned in hog fuel boilers. Another reason is that the cylinders help to guide and transport the paper. One disadvantage includes the large thermal inertia of the stainless steel cylinders, which causes longer start-up times and a reduced ability to quickly change drying rate. An important disadvantage is the reduced drying effectiveness as the moisture content of the paper decreases because of the reduction in the thermal conductivity. There is a potential reduction in paper quality due to contact with the steel cylinders. These dryers do not typically have the capability to vary the drying capacity across the width of the paper. Also, for thicker materials, the drying rate is reduced because the evolving water vapor is trapped between the cylinder and the paper because it is unable to come out the side of the paper in contact with the cylinder.

Another type of dryer uses very high-velocity hot air impingement on both sides of a moving web. The web "floats" through the nozzles, which is where this type of dryer gets its name — *floater dryer*. The primary mode of heat transfer for this dryer is convection. This technique combines heat and mass transfer in the same apparatus as the hot air both heats the web and carries away the moisture that evolves from it. This type of dryer has several potential advantages over other types of dryers. No additional systems are needed to remove the volatiles vaporizing from the material being dried, and there is no direct contact with the product that could reduce the quality. It is possible to segment this type of dryer to vary the moisture removal rate across the width, although the reaction time is slow compared to IR dryers. There are also potential disadvantages. The air nozzles can become plugged because, typically, they are fairly small to achieve the high gas velocities. In drying materials like papers and textiles, this method also relies ultimately on conduction for the energy to get to the core of the product whose thermal conductivity decreases as the moisture content decreases.

Hawlader et al. (1997) presented design charts for determining food drying rates in tunnel dryers.[11] They used a combination of mathematical modeling and experimental results from two heat pump dryers of different drying capacities to develop the charts. The charts relate air temperature and humidity to the required energy for drying and to the dryer length. A chart was developed that relates the energy required for drying and the dryer length to the air temperature (40°C to 47.5°C, 104 to 117.5°F) and relative humidity (30% to 45%). The importance of both heat and mass transfer in the drying process was noted.

10.3 FIRED HEATERS

Fired or tubestill heaters are used in the petrochemical and hydrocarbon industries to heat fluids in tubes for further processing. A fired heater consists of three major components: the heating coil, the furnace enclosure, and the combustion equipment. The objective is to transfer heat to the fluids in the heating coil. The heat is produced by the combustion equipment (burners), which is transferred directly to the tubes and also to the furnace enclosure, which in turn also radiates heat to the tubes. The design of all three components is optimized for efficiently and uniformly transferring heat to the fluids in the tubes.

Garg (1997) has given some typical energy consumptions for fired heaters in several common processes: 0.32×10^6 Btu/bbl (94 kW/bbl) crude oil in the refining industry, 22×10^6 Btu/ton (7.1 MW/m-ton) ethylene, and 28.5×10^6 Btu/ton (9.21 MW/m-ton) ammonia.[12] In these types of processes, fluids flow through an array of tubes located inside a furnace or heater. The tubes are heated by direct-fired burners that often use fuels that are by-products from processes in the plant and that vary widely in composition. Using tubes to contain the load is somewhat unique compared to the other types of industrial combustion applications considered in this book. It was found that heating the fluids in tubes had many advantages over heating them in the shell of a furnace.[13] These include better suitability for continuous operation, better controllability, higher heating rates, more flexibility, less chance of fire, and more compact equipment.

One of the problems encountered in refinery-fired heaters is an imbalance in the heat flux in the individual heater passes.[14] This imbalance can cause high coke formation rates and high tube metal temperatures, which reduce a unit's capacity and can cause premature failures. Coke formation on the inside of the heater tubes reduces the heat transfer through the tubes that leads to the reduced capacity. One cause of coking is flame impingement directly on a tube, which causes localized heating and increases coke formation there. This flame impingement may be caused by operating without all the burners in service, insufficient primary or secondary air to the burner, operating the heater at higher firing rates, fouled burner tips, eroded burner tip orifices, or insufficient draft. This problem shows the importance of proper design[15] for even heat flux distribution inside the fired heater.

Fewer new fired heaters are being built, so the major emphasis has been on increasing the capacity of existing heaters. The limitations of overfiring a heater are:[16]

- High tube metal temperatures
- Flame impingement, causing high coke formation rates
- Positive pressure at the arch of the heater
- Exceeding the capacity of induced-draft and forced-draft fans
- Exceeding the capacity of the feed pump

Hottel (1974) presented an analytical analysis for the heat transfer in tubular heaters.[17] In the analysis, the heat sink is a row or rows of tubes mounted parallel to and near a refractory wall. The effective emissivity of the tube-wall combination was given by:

$$\varepsilon_{\mathrm{eff}} = \cfrac{1}{\cfrac{1}{F+(1-F)F} + \cfrac{B}{\pi}\left(\cfrac{1}{\varepsilon'}-1\right)} \tag{10.1}$$

where ε' is the emissivity of the metal tube, B is the ratio of the tube center-to-center distance divided by the tube diameter, and F is the fraction of radiation incident on the tube plane intercepted by the tubes. Hottel discussed several tube geometries in some detail.

Garg (1998) reported on increasing the process heat duty of an existing fired heater from 26.4 to 52.2×10^6 Btu/hr (7.74 to 15.3 MW).[16] The average radiant heat flux increased from 8000 to 12,000 Btu/hr-ft^2 (25 to 38 kW/m^2). The thermal efficiency increased from 70% to 90%. One method used to increase the capacity was to replace the original bare tubes with extended-surface or finned tubes to increase the heat transfer rates. The number of tubes and the number of burners in the heater were increased. Natural draft was replaced by forced draft.

10.3.1 REFORMER

As the name indicates, reformers are used to reformulate a material into another product. For example, a hydrogen reformer takes natural gas and reformulates it into hydrogen in a catalytic chemical process that involves a significant amount of heat. A sample set of reactions is given below for converting propane to hydrogen:[18]

$$C_3H_8 \rightarrow C_2H_4 + CH_4$$

$$C_2H_4 + 2H_2O \rightarrow 2CO + 4H_2$$

$$CH_4 + H_2O \rightarrow CO + 3H_2$$

$$CO + H_2O \rightarrow CO_2 + H_2$$

The reformer is an indirect-fired combustor (because the tubes are heated, which then heat the fluid) containing numerous tubes, filled with catalyst, inside the combustor.[19] The reformer is heated with burners, firing either vertically downward or upward, with the exhaust on the opposite end, depending on the specific design of the unit. The raw feed material flows through the catalyst in the tubes which, under the proper conditions, converts that material to the desired end-product. The heat needed for the highly endothermic chemical reactions is provided by the burners. The fluid being reformulated typically flows through a reformer combustor containing many tubes (see Figures 8.45 and 10.8). The side-fired reformer (Figure 10.8A) has multiple burners on the side of the furnace, with a single row of tubes centrally located. The heat is transferred primarily by radiation from the hot refractory walls to the tubes. Top-fired reformers (Figure 10.8B) have multiple rows of tubes in the firebox. In that design, the heat is transferred primarily from radiation from

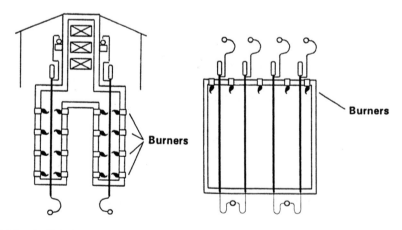

FIGURE 10.8 A: Side-fired and B: top-fired reformers. (Courtesy of Marcel Dekker, New York.[19] With permission.)

the flame to the tubes. Burners may be located in the side wall but be firing up the wall at a slight angle, which is sometimes referred to as terrace firing (see Figure 8.44). Foster Wheeler uses terrace wall reformers in the production of hydrogen by steam reformation of natural gas or light refinery gas.[20] Heat flux rates over 20,000 Btu/hr-ft^2 (63 kW/m^2) are typical, which means that heat flux uniformity is needed to avoid hotspots.

The reformer tubes are a critical element in the overall design of the reformer. They are typically made of a high-temperature and -pressure (reformers operate at pressures up to 350 psig or 24 barg) nickel alloy like inconel to ensure that they can withstand the operating conditions inside the reformer. Failure of the tubes can be very expensive because of the downtime of the unit, lost product, damaged catalyst, and possibly damaged reformer.

There are several important aspects of heat transfer involved in this process. One is the heat transfer from the flame to the tubes that occurs mainly by radiation and also by convection. It is critical, however, that the heating be uniform to prevent damage to the tubes that could occur if the flames directly impinged on them. Another aspect of heat transfer is that from the tube to the fluid; this could be affected if the tubes become coked with soot or other deposits. Although the radiant absorptivity might be significantly enhanced by the deposits, the conduction through them is normally impeded because the thermal conductivity of these deposits is usually lower than the tube metal. The catalyst in the tubes further complicates the heat transfer to the fluid. The catalyst is heated by thermal conduction from the metal tubes and the fluids are heated by forced convection while flowing through the catalyst.

Kudo et al. (1992) modeled the reforming of propane to hydrogen in a furnace 10-ft (3 m) in diameter and 26-ft (8 m) long, containing 14 tubes equally spaced 4.160 ft (1.268 m) from the center of the furnace.[21] Some of the key heat transfer parameters used in the model were:

- Combustion gas absorption coefficient of 0.067 ft^{-1} (0.22 m^{-1})
- Flame absorption coefficient of 0.24 ft^{-1} (0.80 m^{-1})
- Adiabatic wall
- Furnace wall emissivity of 0.3
- Tube surface emissivity of 0.8
- Heat generation in the flame 0.5853 × 10^6 Btu/hr (171.5 kW)
- Convection coefficient for heat transfer to the tubes of 3.5 Btu/hr-ft^2-°F (20 W/m^2-K)

The convection coefficient was calculated using:

$$Nu = 0.023 \, Re_{e,d}^{0.8} \, Pr^{1/3} \left\{ 1 + \left(\frac{d_e}{l_w} \right)^{0.7} \right\} \tag{10.2}$$

where subscript e refers to the equivalent or characteristic for that value. The results showed the heat flux to the tubes ranged from about 6300 to 38,000 Btu/ft^2-hr (20 to 120 kW/m^2), with the vast majority coming from radiation and very little from convection. The peak flux occurred just downstream of the burner and declined fairly quickly after that. The calculated exhaust temperature was 1790°F (980°C), which corresponded reasonably well to the measured exhaust temperature of 1700°F (930°C).

Louis et al. (1998) presented the application of a new burner called the Cerajet and of a ceramic furnace, to high temperature, ethane, steam-cracking furnaces used in the production of ethylene.[22] The self-recuperative, ceramic radiant tube, jet burner is used to boost the ethane temperature from 870°C (1600°F) to 1000°C (1800°F), which improves the conversion rate of ethane from about 70% to 95%. The power density of the last radiant tube was measured as 38 kW/m^2 (12,000 Btu/hr-ft^2). The heat transfer coefficient was calculated to be 160 W/m^2-°C (28 Btu/hr-ft^2-°F).

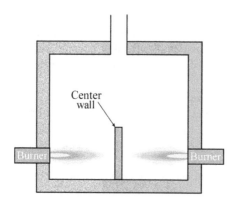

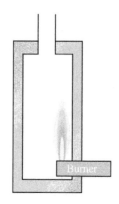

FIGURE 10.9 Elevation view of center or target wall-firing configuration.

FIGURE 10.10 Elevation view of a horizontally mounted, vertically fired burner configuration.

10.3.2 PROCESS HEATER

Process heaters are designed to heat petroleum products, chemicals, and other liquids and gases. Typical petroleum fluids include gasoline, naphtha, kerosene, distillate oil, lube oil, gas oil, and light ends.[23] Kern noted that refinery heaters may carry liquids at temperatures as high as 1500°F (1100K) and pressures up to 1600 psig (110 barg). The primary modes of heat transfer in process heaters are radiation and convection. Figure 10.9 shows one common firing configuration where burners fire against a wall (sometimes referred to as a center or target wall) in the middle of the heater and the wall re-radiates to the process tubes. Figure 10.10 shows another configuration where a horizontally mounted burner fires up along a wall which then radiates to the process tubes. Other configurations, previously discussed, are shown in Figures 8.41, 8.42, 8.43, and 8.44.

The initial part of the fluid heating is done in the convection section of the furnace, while the latter heating is done in the radiant section. Each section has a bank of tubes in it through which the fluids flow, as shown in Figure 10.11.[24] Early heater designs had only a single bank of tubes that failed prematurely because designers did not understand the importance of radiation on the

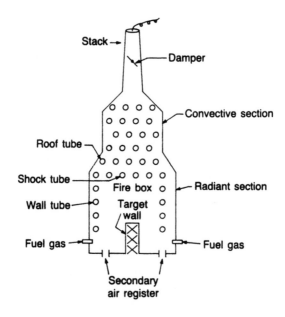

FIGURE 10.11 Typical process heater. (Courtesy of PennWell Books, Tulsa, OK.[24] With permission.)

process.[13] The tubes closest to the burners failed due to overheating that caused the hydrocarbons to coke inside the tube, which further aggravated the problem by reducing the cooling ability of the fluids due to the reduction in thermal conductivity through the coke layer inside the tube. One of the key challenges for the heater designer is to get even heat distribution inside the combustor to prevent coking inside the tubes. Bell and Lowy (1967) estimated that approximately 70% of the energy is transferred to the fluids in the radiant section of a typical heater and the balance in the convection section.[25] The tubes in the convection section often have fins to improve convective heat transfer efficiency. These fins are designed to withstand temperatures up to about 1200°F (650°C). If delayed combustion occurs in the convection section, the fins can be exposed to temperatures up to 2000°F (1100°C), which can damage the fins.[24]

Vertical heaters can be broadly categorized into two types: cylindrical and box heaters.[25a] Both have tubes along the walls in the radiant section. In cylindrical heaters, the tubes are installed vertically. In box heaters, the tubes are generally installed horizontally. In both designs the burners are usually installed in the floor and fire vertically upwards. Both designs normally have a convection section near or in the entrance of the stack. A rule-of-thumb principal for the heat distribution is that greater than 60 to 70% of the total heat duty should be in the radiant section of the heater.[25b] The calculated maximum radiant flux ranges from 1.70 to 1.85 times the average radiant flux. A tall, narrow heater is more economical to build, but the radiant flux is often poorly distributed with much higher fluxes in the lower part of the heater. The recommended height-to-diameter ratio for a cylindrical heater is between 1.5 and 2.75 to get more uniform heat flux loading. Garg (1989) noted that fired heater performance can be enhanced by installing a convection section to an all-radiant heater, enlarging the heat transfer area of an existing convection section, converting a natural-draft heater to forced-draft, and adding air preheating or steam-generation equipment.[25c]

Garg (1989) noted a number of factors that need to be considered when specifying burners for fired heaters.[25d] Choosing the right burners can increase the heater capacity by 5 to 10% and increase the efficiency by 2 to 3%. These factors include burner type, heat release and turndown, air supply, excess air, fuel specifications, firing position, flame dimensions, ignition mode, atomization media for liquid fuel burners, noise, NOx emission, and waste gas firing. Most of those factors directly influence the heat transfer in the heater.

Hoogendoorn et al. (1970) made heat flux measurements in a rectangular, vertical tube furnace with two round burners firing vertically upward.[26] Both oil and gas flames were tested. The objective of the study was to determine the validity of the assumption of a constant furnace temperature often used to calculate the heat flux to process tubes. A 25-mm (1 in.) diameter, water-cooled heat flux probe with and without air screens was used to measure radiation and total heat flux, respectively. Forced convection was calculated from the difference between the total and radiant flux measurements. The heat flux was found to be significantly nonuniform in the furnace. Gas flames were found to have a more uniform heat flux distribution than oil flames.

Selçuk et al. (1975) studied the effect of flame length on the radiative heat flux distribution in a process fluid heater.[27] This information is important to the design of the heater to prevent premature damage to the process tubes, due to improper flame heights, and to optimize the heat transfer rate to the tubes to maximize thermal efficiency. A two-flux radiation model was used to predict the radiant heat transfer in the heater. The predictions were in good agreement with a set of experimental data. The results showed that the radiant flux at the tube surfaces was a strong function of the flame height.

Kern noted that process heaters are typically designed around the burners.[23] Berman (1979) discussed the different burner designs used in fired heaters.[28] Burners may be located in the floor, firing vertically upward. In vertical cylindrical (VC) furnaces, those burners are located in a circle in the floor of the furnace (see Figure 9.15). The VC furnace itself serves as a part of the exhaust stack to help create draft to increase the chimney effect.[29] In cabin heaters, which are rectangular, there are one or more rows of burners located in the floor (see Figure 9.13). Burners may be at a

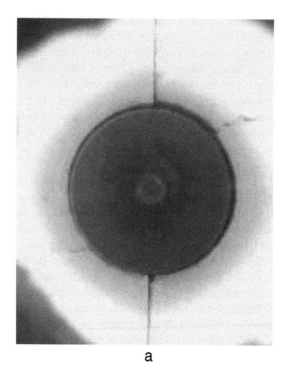

a

b

FIGURE 10.12 Wall-fired burner: (a) side view, (b) front view. (Courtesy of John Zink Co. LLC, Tulsa, OK. With permission.)

low level, firing parallel to the floor. In that configuration, they may be firing from two opposite sides toward a partial wall in the middle of the furnace that acts as a radiator to distribute the heat (see Figure 10.9). Burners may be located on the wall firing radially along the wall (see Figure 10.12), which are referred to as radiant wall burners. There are also combinations of the above in certain heater designs. For example, in ethylene production heaters, both floor-mounted, vertically fired burners and radiant wall burners are used in the same heater.

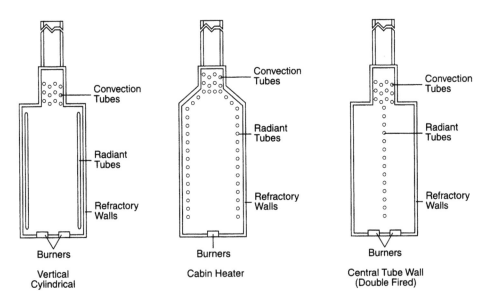

FIGURE 10.13 Examples of process heaters. (Courtesy of CRC Press, Boca Raton, FL.[32])

Typical examples of process heaters are shown in Figure 10.13. Berman (1979) noted the following categories of process heaters: column reboilers, fractionating-column feed preheaters, reacter-feed preheaters including reformers, heat supplied to heat transfer media (e.g., a circulating fluid or molten salt), heat supplied to viscous fluids, and fired reactors including steam reformers and pyrolysis heaters.[30] Six types of vertical-cylindrical fired heaters were given: all radiant, helical coil, crossflow with convection section, integral convection section, arbor or wicket type, and single-row/double-fired. Six basic designs were also given for horizontal-tube fired heaters: cabin, two-cell box, cabin with dividing bridgewall, end-fired box, end-fired box with side-mounted convection section, and horizontal-tube/single-row/double-fired. Many commonly used process heaters typi-cally have a radiant section and a convection section. Burners are fired in the radiant section to heat up tubes. Fluids flow through the tubes and are heated to the desired temperature for further processing. The fluids are preheated in the convection section and heated to the desired process temperature in the radiant section. Radiant heat transfer from the flames to the tubes is the most critical aspect of this heater because overheating of the tubes leads to tube failure and shutdown of the heater.[31] The tubes may be horizontally or vertically oriented, depending on the particular heater design. Permissible average radiant flux rates range from 5000 to 18,000 Btu/hr-ft^2 (16 to 57 kW/m^2) of circumferential tube area, depending on the type of furnace.[23]

Feintuch and Negin (1996) discussed process heater design for delayed cokers.[50] The traditional allowable average radiant flux rates range from 10,000 to 12,000 Btu/hr-ft^2 (32 to 38 kW/m^2). However, there has been a recent trend to reduce that to about 9000 Btu/hr-ft^2 (28 kW/m^2) to provide for longer run lengths, future capacity expansion, and a more conservative heater design. In delayed cokers, there is a gradual build-up of coke on the inside of the heater tubes that affects both the heat transfer to the oil in the tubes and the life of the tube. The coke build-up becomes a thermal resistance to heat flow. This reduces the heat transfer to the fluid in the tube, while simultaneously causing the tube wall to overheat because the heat cannot be dissipated as quickly as when there is no coke build-up.

A unique aspect of process heaters is that they are often natural draft. This means that no combustion air blower is used. The air is inspirated into the furnace by the suction created by the hot gases rising through the combustion chamber and exhausting to the atmosphere. Another unique aspect of these heaters is the wide range of fuels used, which are often by-products of the petroleum

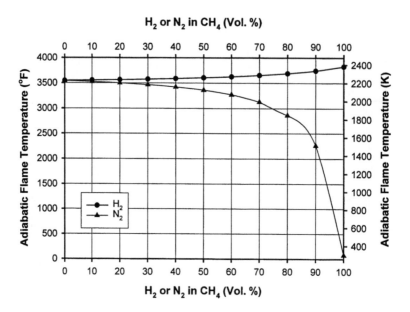

FIGURE 10.14 Adiabatic equilibrium flame temperature for the stoichiometric combustion of ambient air with blends of H_2/CH_4 and N_2/CH_4.

refining process. These fuels may contain significant amounts of hydrogen, which has a large impact on the burner design. It is also fairly common for multiple fuel compositions to be used, depending on the operating conditions of the plant at any given time. In addition to hydrocarbons ranging up to C_5, the gaseous fuels may also contain hydrogen and inerts (like CO_2 or N_2). The compositions can range from gases containing high levels of inerts to fuels containing high levels of H_2. The flame characteristics for fuels with high levels of inerts are very different from fuels with high levels of H_2. Figure 10.14 shows that the adiabatic flame temperature increases rapidly and non-linearly as the hydrogen content in CH_4/H_2 flames increases. The figure also shows that the adiabatic flame temperature decreases rapidly and nonlinearly as the nitrogen content in CH_4/N_2 flames increases. Add to that the requirement for turndown conditions, and the burner design becomes very challenging to maintain stability, low emissions, and the desired heat flux distribution over the range of conditions that are possible. Some plants use liquid fuels, like no. 2 to no. 6 fuel oil, sometimes by themselves and sometimes in combination with gaseous fuels. So-called combination burners use both a liquid and a gaseous fuel, which are normally injected separately through each burner.

Shires gave a general heat balance for a process heater:[32]

$$\dot{Q}_f = \dot{Q}_g + \dot{Q}_l + \dot{Q}_p \tag{10.3}$$

where $\dot{Q}_f$ is the heat generated by combusting the fuel, $\dot{Q}_g$ is the heat going to the load, $\dot{Q}_l$ is the heat lost through the walls, and $\dot{Q}_p$ is the heat carried out by the exhaust products. This is shown schematically in Figure 10.15.

The heat transfer from the hot combustion products to the tubes in the convection section is forced convection flow over a bank of tubes. Colburn's equation for this type of convection is:

$$Nu = a\,Re_{f,d}^{0.6}\,Pr_f^{0.33} \tag{10.4}$$

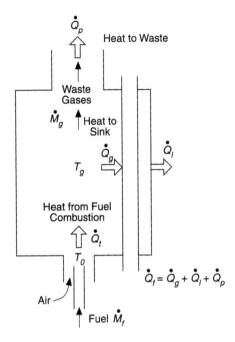

FIGURE 10.15   Process heater heat balance. (Courtesy of CRC Press, Boca Raton, FL.[32])

TABLE 10.1
Values of B and N

Range of Re	In-line Bank of Tubes		Staggered Bank of Tubes	
	B	N	B	N
10–300	0.742	0.431	1.309	0.360
$300–2 \times 10^5$	0.211	0.651	0.273	0.635
$2 \times 10^5–2 \times 10^6$	0.116	0.700	0.124	0.700

Source: From J.W. Rose and J.R. Cooper, *Technical Data on Fuel,* 7th
edition, British National Committee, London, 1977.

where $a = 0.33$ for staggered tubes and 0.26 for in-line tubes, with the properties evaluated at the
film temperature. A similar but more detailed correlation, is given by:[33]

$$\mathrm{Nu} = BF_1F_2\,\mathrm{Re}_{f,d}^{N}\,\mathrm{Pr}_f^{0.34} \tag{10.5}$$

where B and N are given in Table 10.1, $F_2 = 1, 0.97, 0.95$, or 0.90 for 10, 8, 6, or 4 rows of tubes,
respectively, and F_1 is given by:

$$F_1 = \left(\frac{\mathrm{Pr}_f}{\mathrm{Pr}_w}\right)^{0.26} \tag{10.6}$$

where Pr_w is the Prandtl number evaluated at the wall or tube surface temperature.

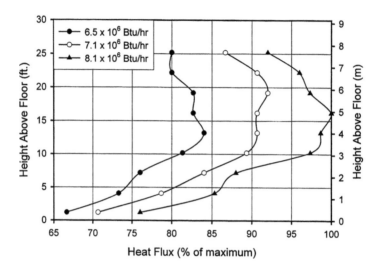

FIGURE 10.16 Heat flux profiles as a function of firing rate for a natural draft burner mounted in the floor of a furnace and firing vertically upward, using a fuel blend with 78% natural gas (Tulsa) and 22% H_2 with 2% O_2 in the exhaust gases.

Berman (1979) gave equations and sample calculations for determining the heat transfer split between the radiant and convection sections.[34] Graphs were provided for determining the average radiant heat transfer rate as functions of both the tube metal temperature and target wall (wall fired on in middle of heater to radiate to tubes) temperature for both vertical-cylindrical and horizontal-tube heaters. A graph was also provided for determining the maximum heat flux on a tube as a function of the tube geometry. Wimpress (1979) also gave a simplified method for calculating the performance of a fired heater.[35]

Talmor (1982) has written a book dealing with how to predict, control, and troubleshoot hot-spots in process heaters.[36] The book gave a method for estimating the magnitude and location of the maximum combustion zone heat flux. It takes into account the firing rate of each burner, the number of burners, the flame length, the flame emissivity, the spacing between the burner and the tubes, the spacing between the burners, and the geometry of the firebox. The book included much empirical data specific to a variety of different process heaters and also gave many detailed examples that have been worked.

Figure 10.16 shows how the measured heat flux to the wall (normalized to the peak flux) of a furnace varied as a function of the vertical elevation and the firing rate for a fuel containing 78% natural gas and 22% H_2. As expected, the heat flux increased with firing rate. The curves show the peak fluxes for a given firing rate were between about 10 and 20 ft (3 and 6 m). Figure 10.17 shows a similar graph where the fuel was 86% natural gas and 14% H_2. The results are similar to the blend containing 22% H_2. Figure 10.18 is a comparison of the measured heat fluxes for the two fuel blends at a fixed firing rate. The profiles are similar but the peaks occur at different vertical elevations.

10.4   HEAT TREATING

Heat treating involves the thermal treatment of metal to produce some type of enhanced performance characteristic. Typical material improvements include surface hardening, strengthening the part, relieving stresses, and improving ductility. Typical thermal treatments include annealing, brazing, carburizing, normalizing, sintering, and tempering. *Annealing* is a heat treatment used to remove internal stresses and to make a material less brittle by heating a part and then cooling it. This may be done under a protective atmosphere (commonly nitrogen) to prevent surface reactions if the

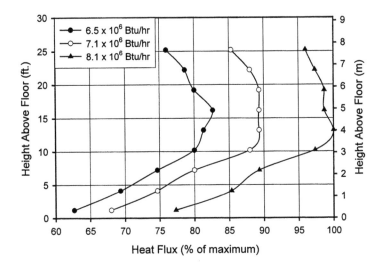

FIGURE 10.17 Heat flux profiles as a function of firing rate for a natural draft burner mounted in the floor of a furnace and firing vertically upward, using a fuel blend with 86% natural gas (Tulsa) and 14% H_2 with 2% O_2 in the exhaust gases.

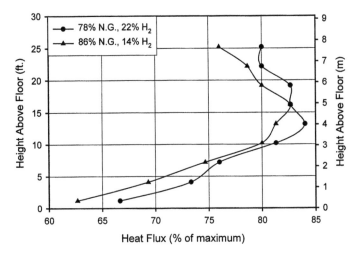

FIGURE 10.18 Heat flux profiles as a function of fuel blend for a natural draft burner mounted in the floor of a furnace and firing vertically upward at 6.5×10^6 Btu/hr with 2% O_2 in the exhaust gases.

application is particularly sensitive or critical. Sometimes, a reducing agent like hydrogen may be added to the protective atmosphere to produce clean, bright surfaces. *Brazing* is a method of joining metals together using a non-ferrous filler material such as brass or a brazing alloy. The melting point of the braze material is lower than that of the metals being joined, which dictates the furnace temperature. Again, a protective atmosphere of nitrogen, possibly with small amounts of hydrogen or a hydrocarbon, may be used to produce high-quality parts. Dewpoint control in the atmosphere is often important in certain applications to minimize surface reactions with water at higher temperatures that would reduce the surface quality. *Carburizing* and *neutral hardening* refer to surface hardening of steels by adding carbon just to the surface of a part at a temperature high enough for the carbon to convert the surface from a lower carbon steel to a higher carbon steel. This must be done under an atmosphere, sometimes referred to as an endothermic gas, that contains carbon. The endothermic gas may have a composition of approximately 20% CO, 40% H_2, and

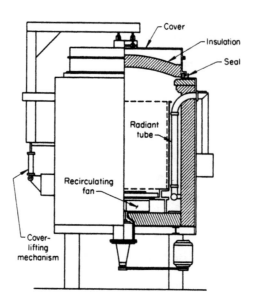

FIGURE 10.19   Sketch of a pit furnace. (Courtesy of American Society of Metals, Warren, PA.[37] With permission.)

40% N_2. *Normalizing* involves taking a ferrous alloy above its upper transformation temperature and cooling it well below that temperature. This refines the grain structure of the metal and is usually done after a part has been hot-worked to soften the material. *Sintering* is used to produce solid metal parts from metal powders by heating them to a high enough temperature, often under a protective atmosphere depending upon the metals being used, that the powder fuses together without melting. *Tempering* is the heat treatment of ferrous alloys by heating them below the lower transformation temperature to obtain certain mechanical properties. Although it is similar to annealing and stress relieving, the resulting material properties are distinct in all three processes.

One method that has been used to classify heat-treating furnaces is based on the heat transfer medium: gaseous (inert, oxidizing, or reducing), vacuum (no medium), liquid (molten metal or molten salt), and solid (fluidized bed).[37] The dominant heat transfer mechanism varies with the medium (see Chapter 9). In gaseous-medium furnaces, the dominant mode is often convection, but radiation may also be important. In vacuum furnaces, radiation is by far the dominating mode of heat transfer. In liquid-medium furnaces, convection is the dominant mode. In fluidized beds, conduction is significant and convection is also usually important.

Heat-treating furnaces may be either continuous or batch.[38] The major types of continuous furnaces include rotary retort, rotary hearth, pusher, roller hearth, conveyor, and continuous strand. A rotary retort furnace has a gas-tight cylindrical retort that revolves inside a firebrick and steel enclosed shell. The rotary hearth furnace has a cylindrical gas-tight shell that rotates. In a pusher furnace, the load is usually loaded on trays and pushed into the furnace. The major types of batch furnaces include box, rotary retort, rotary hearth, car bottom, bell, pit, liquid bath (molten lead and salt), fluidized bed, and vacuum. A pit furnace is located below the floor and extends up to or slightly beyond the floor. An example of a pit furnace is shown in Figure 10.19. The challenge in either batch or continuous processes requiring a specific atmosphere composition is to maintain that atmosphere during the heat treating cycle. For example, in vacuum furnaces (batch operation), fairly elaborate door seals are needed to maintain high levels of vacuum inside the furnace. In belt furnaces, gas curtains are usually employed at both ends of the furnace to prevent ambient air infiltration into the process that would reduce product quality. Heat-treating furnaces are either direct- or indirect-fired. If a special atmosphere is required, the furnace is indirect-fired. If no special atmosphere is required, the furnace is often direct-fired.

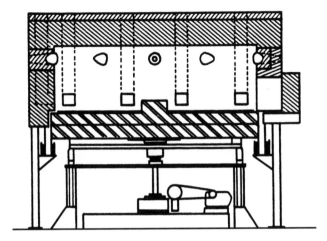

FIGURE 10.20   Sketch of a rotary hearth furnace. (Courtesy of American Society of Metals, Warren, PA.[37] With permission.)

Orzechowski and Hughes (1999) showed how the importance of forced convection and thermal radiation vary as a function of the furnace temperature in heat treating applications.[39] Forced convection heating is recommended for furnace temperatures below 1000°F (540°C). Radiation heating is recommended for furnace temperatures above 1000°F (540°C).

10.4.1   STANDARD ATMOSPHERE

There are some heat-treating processes that do not require a special atmosphere, where the gas around the parts being treated is air and the products of combustion. One example is a car bottom furnace, which is often used to relieve stresses in large welded parts. Furnaces used for annealing also do not usually require any specific gas composition around the parts being treated. Another example is a rotary hearth, continuous heat treating furnace, shown in Figure 10.20, which operates like a revolving door with parts entering on one part of the circle and exiting at another part after rotating some angular distance. Rotary hearth furnaces are often used for heating larger workpieces.

Ferguson (1979) compared the energy requirements for indirect-fired (radiant tube) and direct-fired (with recuperation) furnaces for batch annealing furnaces.[40] The direct-fired furnace used about a third less energy than the indirect-fired furnace. In the radiant tube furnace, over half of the energy is lost in the flue gases, compared to only 30% for the direct-fired recuperative furnace. This shows the importance of the heating method on energy consumption. The analysis assumed that no special atmosphere was required for the heating application.

10.4.2   SPECIAL ATMOSPHERE

In this type of process, the combustion products cannot come in contact with the metal load, as they would reduce the product quality. There are two basic types of special atmospheres that are commonly used, depending on the type of heat treating process. In the first type of special atmosphere, the objective is to prevent certain chemicals from reaching the part that could contaminate the surface. One such type is sometimes referred to as a neutral atmosphere, where for example, an inert gas like helium is used to protect metal products from oxidation, which causes scale formation on the metal surface that reduces the metal yield and product quality. Another type of protective atmosphere is the absence of any gases, created by a vacuum, contacting the part surface. Some specific contaminants that can detrimentally affect a part's surface quality include oxygen (air), water, carbon-containing materials, hydrogen, and nitrogen. Oxygen and water can cause surface oxidation, carbon-containing materials can cause surface hardening, hydrogen can

TABLE 10.2
Special Atmosphere Compositions and Applications

Class	Description	N_2	CO	CO_2	H_2	CH_4	Application
		Nominal Composition (% by volume)					
101	Lean exothermic	86.8	1.5	10.5	1.2	0	Oxide coating of steel
102	Rich exothermic	71.5	10.5	5.0	12.5	0.5	Bright annealing; copper brazing; sintering
201	Lean prepared nitrogen	97.1	1.7	0	1.2	0	Neutral heating
202	Rich prepared nitrogen	75.3	11.0	0	13.2	0.5	Annealing and brazing of stainless steels
301	Lean endothermic	45.1	19.6	0.4	34.6	0.3	Clean hardening
302	Rich endothermic	39.8	20.7	0	38.7	0.8	Carburizing (enriched 301)
402	Charcoal	64.1	34.7	0	1.2	0	Carburizing
501	Lean exothermic-endothermic	63.0	17.0	0	20.0	0	Clean hardening
502	Rich exothermic-endothermic	60.0	19.0	0	21.0	0	Carburizing
601	Dissociated ammonia	25.0	0	0	75.0	0	Brazing; sintering
621	Lean combusted ammonia	99.0	0	0	1.0	0	Neutral heating
622	Rich combusted ammonia	80.0	0	0	20.0	0	Sintering of stainless powders

Source: From ASM Committee on Furnace Atmospheres, *Metals Handbook,* 8th edition, Vol. 2, American Society for Metals, Metals Park, OH, 1964, 67-84.

cause embrittlement, and nitrogen can cause nitriding. The latter two are sometimes desirable, but they can also be detrimental, depending on the desired part characteristics.

The second type of special atmosphere is where a specific reaction is desired by the atmosphere contacting the part surface at an elevated temperature. One example is a carburizing atmosphere that contains a specific amount of gases containing carbon, which diffuses into the metal surface for hardening. This is used, for example, in tool steels that require hard outer surfaces. An atmosphere containing nitrogen can be used to nitride a metal. Table 10.2 shows common special furnace atmosphere compositions and their applications.[41] Many of these atmospheres are directly or indirectly produced by combusting hydrocarbon fuels. One example of a furnace that has a protective atmosphere is a bell furnace. The inner part of the furnace has a heated muffle that contains a protective atmosphere — often hydrogen or hydrogen blends. These furnaces are commonly used for annealing steel coils. Another example of a furnace using a protective atmosphere is a roller hearth, as shown in Figure 10.21. The photograph shows an entrance vestibule that helps contain the protective atmosphere. Furnace pressure control is especially important in furnaces containing special atmospheres, not only to ensure that ambient air does not leak into the furnace but also that not much special atmosphere leaks out of the furnace, which would be wasted cost.[42]

Thekdi et al. (1985) described an indirect, natural gas-fired heating system used where controlled atmospheres are needed for heat treating.[43] The heat source was a high-intensity radiant tube burner capable of temperatures greater than 2000°F (1100°C). Heat flux improvements of up to three times conventional heating systems were measured. Peak fluxes up to 21,900 Btu/hr-ft^2 (69 kW/m^2) were measured which compared to fluxes from conventional systems of 7930 Btu/hr-ft^2 (25 kW/m^2). As shown in Figure 10.22, the heat flux varied along the length of the tube.

Liang et al. (1987) discussed the application of ceramic tube burners in indirect-fired heat-treating furnaces.[44] They described developments of advanced materials to increase the maximum operating temperatures of metallic radiant tubes that were limited to 1800°F (1000°C). Furnaces operating at higher temperatures typically used electric heating, which is often more expensive than using natural gas. Advanced ceramic radiant tube burners need to be capable of radiant outputs of at least 95 kW/m^2 (30,000 Btu/hr-ft^2).

FIGURE 10.21 Photograph of a roller hearth furnace. (Courtesy of Electric Furnace, Salem, OH.)

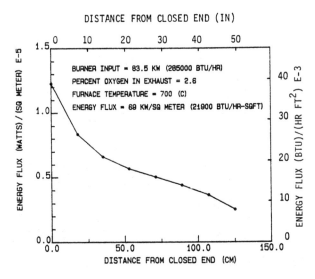

FIGURE 10.22 Heat flux as a function of distance from the end of a ceramic radiant tube. (Courtesy of Gas Research Institute, Chicago, IL.[43] With permission.)

Hemsath (1990) discussed a new gas-fired heating system for indirect heating under a protective or reactive atmosphere, or under a vacuum.[45] Important advantages of the system include lower operating costs compared to electrically heated indirect furnaces and uniform heating rates. Con-

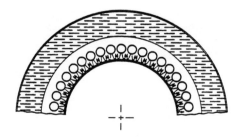

FIGURE 10.23 Convection heating of an indirect-fired furnace. (Courtesy of ASME, New York.[45] With permission.)

PART	DESCRIPTION		
1	DOOR ASSEMBLY	6	VACUUM VESSEL
2	DOOR MANIPULATOR	7	FLUE GAS PLENUM
3	FURNACE SHELL	8	FIBER INSULATION
4	CONVECTION TUBES	9	FAN ASSEMBLY
5	COMBUSTION CHAMBER	10	GAS BURNER
		11	INTERNAL CONVECTION FAN

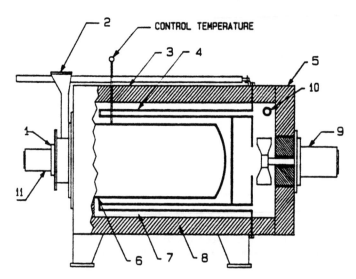

FIGURE 10.24 Schematic of an indirect-fired furnace. (Courtesy of ASME, New York.[45] With permission.)

ventional gas-fired indirect furnaces use radiant tubes. The system described by Hemsath used forced convection to heat the vacuum vessel, as shown in Figure 10.23. The impinging jets consist of the exhaust products from a single burner as noted in the schematic of the furnace shown in Figure 10.24. The maximum reported temperature for a pilot-scale furnace was 1250°F (677°C), which was expected to be raised to 1750°F (954°C) by subsequent development. The application of this novel furnace design was discussed for batch coil annealing and as a thermal cleaner for aluminum borings and scrap.

Copes and Brooks (1991) tested a ceramic composite material called Siconex™ for use in a controlled-atmosphere furnace.[46] The material, which is a fiber-reinforced composite, was tested in endothermic, exothermic, and nitrogen-based atmospheres. Tubular shapes were tested with and without coatings (23 different coatings were tested) to determine their properties under the different atmospheres. The performance tests for the coatings included burst strength, permeability, adhesion, stability, cracking, and maintenance of emissivity. Only two of the coatings performed well over a wide range of conditions. The inert atmosphere proved to be the most detrimental to the strength of the ceramic samples.

Schultz et al. (1992) studied heat transfer in a gas-fired, high-temperature "soft" vacuum (1 to 100 torr) furnace with a design operating temperature of 2350°F (1290°C).[47] The furnace was designed for hardening tool steels, annealing stainless steel, and for sintering and brazing applications. The furnace was heated with single-ended radiant tube burners. The object of the study was to model the heat transfer in the furnace with various silicon carbide radiant tube sizes ranging from 3.25 to 6 in. (8.26 to 15 mm) o.d., quantities, and locations. The following parameters were used in the evaluation: empty furnace temperature uniformity, surface temperature uniformity during heating, process or cycle time, and temperature uniformity within the load. A combination of modeling and experiments showed that for the given furnace configuration, four 6-in. (15 mm) o.d. single-ended radiant tubes provided the most optimized results.

Erinov et al. (1996) described the development of a low-inertia, high-efficiency heat-treating furnace using indirect heating with contained, flat, gas-fired IR burners.[48] The claimed benefits included fast heat-up and cool-down, better heating uniformity, higher thermal efficiency, increased productivity, and low NOx emissions. The burner design incorporated flue gas heat recuperation and had high surface temperature uniformity.

Kurek et al. (1998) presented the development of high-temperature uniformity, natural gas-fired, flat radiant panels for use in indirect-fired heat treating furnaces.[49] These panels were designed as a potential replacement for radiant tube burners. They had increased radiating surface area, which means they could be operated at lower temperatures to achieve the same heat flux output. The lower radiating temperature and high surface area improved the temperature uniformity to the load, which improved product quality. Because of the higher surface area, less refractory was required in the furnace, which meant faster heat-up and cool-down. Self-heat recuperation increased the thermal efficiency to over 70%. The radiant output was approximately 120 kW/m² (38,000 Btu/hr-ft²).

REFERENCES

1. D. Traub, Indirect vs. direct: the heat transfer method does affect the process, *Process Heating*, 6(2), 26-29, 1999.
2. Anonymous, Ceramic tile burner improves performance of infrared predryer, *Process Heating*, 4(7), 43-45, 1997.
3. P. Mattsson, J. Perkonen, and A. Riikonen, Infrared drying of coated paper, *Proc. 1989 International Gas Research Conf.*, T.L. Cramer, Ed., Govt. Institutes, Rockville, MD, 1989, 1308-1316.
4. M. Pettersson and S. Stenström, Evaluation of gas-fired and electrically heated industrial infrared paper dryers, *Proc. 1998 International Gas Research Conf.*, Vol. V: Industrial Utilization and Power Generation, D. Dolenc, Ed., Govt. Institutes, Rockville, MD, 1998, 100-112.
5. J. Riikonen, E. Härkönen, and S. Palosaari, Modeling of infrared drying of pulp, in *Drying '87*, A.S. Majumdar, Ed., Hemisphere, Washington, D.C., 1987, 18-23.
6. A. Williams-Gardner, *Industrial Drying*, Leonard Hill, London, 1971.
7. R.B. Keey, Dryers, in *International Encyclopedia of Heat & Mass Transfer*, G.F. Hewitt, G.L. Shires, and Y.V. Polezhaev, Eds., CRC Press, Boca Raton, FL, 1997, 337-342.
8. C.E. Sloan, Drying systems and equipment, *Chem. Eng.*, 74(14), 169-200, 1967.
8a. B.J. Ezerski and G.P. Megan, Boosting the performance of your convection oven, *Process Heating*, 7(1), 22–26, 2000.
8b. J.R. Cornforth, Ed., *Combustion Engineering and Gas Utilisation*, E&FN Spon, London, 1992.
9. T. Berntsson, P.-A. Franck, and A. Åsblad, *Learning from Experiences with Process Heating in the Low and Medium Temperature Ranges*, CADDET Energy Efficiency, Sittard, The Netherlands, 1997.
10. C.E. Bean and J.M. Cocagne, Assessment of Gas-Fired Infrared Heaters in the Paper Industry, Gas Research Institute Report GRI-96/0087, Chicago, IL, 1996.
11. M.N.A. Hawlader, S.K. Chou, and K.J. Chua, Development of design charts for tunnel dryers, *Int. J. Energy Research*, 21(11), 1023-1037, 1997.

12. A. Garg, Optimize fired heater operations to save money, *Hydrocarbon Processing*, 76(6), 97-104, 1997.
13. W.L. Nelson, *Petroleum Refinery Engineering*, second edition, McGraw-Hill, New York, 1941.
14. G.R. Martin, Heat-flux imbalances in fired heaters cause operating problems, *Hydrocarbon Processing*, 77(5), 103-109, 1998.
15. R. Nogay and A. Prasad, Better design method for fired heaters, *Hydrocarbon Processing*, 64(11), 91-95, 1985.
16. A. Garg, Revamp fired heaters to increase capacity, *Hydrocarbon Processing*, 77(6), 67-80, 1998.
17. H.C. Hottel, First estimates of industrial furnace performance — The one-gas-zone model reexamined, in *Heat Transfer in Flames*, N.H. Afgan and J.M. Beer, Eds., Scripta Book Co., Washington, D.C., 1974, 5-28.
18. H. Futami, R. Hashimoto, and H. Uchida, Development of new catalyst and heat-transfer design method for steam reformer, *J. Fuel Soc.* Japan, 68(743), 236-243, 1989.
19. H. Gunardson, *Industrial Gases in Petrochemical Processing*, Marcel Dekker, New York, 1998.
20. J.D. Fleshman, FW hydrogen production, in *Handbook of Petroleum Refining Processes*, 2nd edition, R.A. Myers, Ed., McGraw-Hill, New York, 1996, chap. 6.2.
21. K. Kudo, H. Taniguchi, and K. Guo, Heat-transfer simulation in a furnace for steam reformer, *Heat Transfer — Japanese Research*, 20(8), 750-764, 1992.
22. H.S. Kurek, M. Khinkis, W. Kunc, A. Touzet, A. de La Faire, T. Landais, A. Yerinov, and O. Semernin, Flat radiant panels for improving temperature uniformity and product quality in indirect-fired furnaces, *Proc. 1998 International Gas Research Conf.*, Vol. V: Industrial Utilization and Power Generation, D. Dolenc, Ed., Gas Research Institute, Chicago, 1998, 100-112.
23. D.Q. Kern, *Process Heat Transfer*, McGraw-Hill, New York, 1950.
24. N.P. Lieberman, *Troubleshooting Process Operations*, PennWell Books, Tulsa, OK, 1991.
25. H.S. Bell and L. Lowy, Equipment, in *Petroleum Processing Handbook*, W.F. Bland and R.L. Davidson, Eds., McGraw-Hill, New York, 1967, chap. 4.
25a. A. Garg, Trimming NOx from furnaces, *Chem. Eng.*, 99(11), 122–124, 1992.
25b. A. Garg and H. Ghosh, Good heater specifications pay off, *Chem. Eng.*, 95(10), 77–80, 1988.
25c. A. Garg, How to boost the performance of fired heaters, *Chem. Eng.*, 96(11), 239–244, 1989.
25d. A. Garg, Better burner specifications, *Hydrocarbon Processing*, 68(8), 71–72, 1989.
26. C.J. Hoogendoorn, C.M. Ballintijn, and W.R. Dorresteijn, Heat-flux studies in vertical tube furnaces, *J. Inst. Fuel*, 43, 511-516, 1970.
27. N. Selçuk, R.G. Siddall, and J.M. Beér, Prediction of the effect of flame length on temperature and radiative heat flux distributions in a process fluid heater, *J. Inst. Fuel*, 48, 89-96, 1975.
28. H.L. Berman, Fired heaters. II. Construction materials, mechanical features, performance monitoring, in *Process Heat Exchange*, V. Cavaseno, Ed., McGraw-Hill, New York, 1979, 293-302.
29. A.J. Johnson and G.H. Auth, *Fuels and Combustion Handbook*, first edition, McGraw-Hill, New York, 1951.
30. H.L. Berman, Fired heaters. I. Finding the basic design for your application, in *Process Heat Exchange*, V. Cavaseno, Ed., McGraw-Hill, New York, 1979, 287-292.
31. V. Ganapathy, *Applied Heat Transfer*, PennWell Books, Tulsa, OK, 1982.
32. G.L. Shires, Furnaces, in *The International Encyclopedia of Heat & Mass Transfer*, G.F. Hewitt, G.L. Shires, and Y.V. Polezhaev, Eds., CRC Press, Boca Raton, FL, 1997, 493-497.
33. J.W. Rose and J.R. Cooper, *Technical Data on Fuel*, 7th edition, British National Committee, London, 1977.
34. H.L. Berman, Fired heaters. III. How combustion conditions influence design and operation, in *Process Heat Exchange*, V. Cavaseno, Ed., McGraw-Hill, New York, 1979, 303-314.
35. N. Wimpress, Generalized method predicts fired-heater performance, in *Process Heat Exchange*, V. Cavaseno, Ed., McGraw-Hill, New York, 1979, 320-327.
36. E. Talmor, *Combustion Hot Spot Analysis for Fired Process Heaters*, Gulf Publishing, Houston, 1982.
37. H.E. Boyer, *Practical Heat Treating*, American Society for Metals, Metals Park, OH, 1984.
38. ASM Committee on Gas Carburizing, Case Hardening of Steel, in *Metals Handbook*, 8th edition, Vol. 2: Heat Treating, Cleaning and Finishing, Amer. Soc. For Metals, Metals Park, Ohio, 1964, 93-114.
39. N.J. Orzechowski and D.L. Hughes, Selecting the proper furnace for vacuum tempering applications, *Industrial Heating*, LXVI(9), 45-49, 1999.

40. N.T. Ferguson, Energy and furnace design for batch annealing and heat treating furnaces, *Iron & Steel Engineer*, 56(6), 31-33, 1979.

41. ASM Committee on Furnace Atmospheres, Furnace atmospheres and carbon control, in *Metals Handbook*, 8th edition, Vol. 2: Heat Treating, Cleaning and Finishing, American Society for Metals, Metals Park, OH, 1964, 67-84.

42. W. Trinks and M.H. Mawhinney, *Industrial Combustion*, Vol. II, 4th edition, Wiley, New York, 1967.

43. A.C. Thekdi, S.R. Huebner, and M.A. Lukasiewicz, Development of an indirect gas-fired high temperature heating system, *Proc. 1984 International Gas Research Conf.*, Govt. Institutes, Rockville, MD, 1985, 709-718.

44. W.W. Liang, M.E. Schreiner, S.J. Sikirica and E.S. Tabb, Application of ceramic tubes in high temperature furnaces, *Proc. 1986 International Gas Research Conf.*, T.L. Cramer, Ed., Govt. Institutes, Rockville, MD, 1987, 875-888.

45. K.H. Hemsath, A novel gas fired heating system for indirect heating, in *Fossil Fuel Combustion Symposium 1990*, S.N. Singh, Ed., New York, ASME PD-Vol. 30, 155-159, 1990.

46. J.S. Copes and D.L. Brooks, Siconex™ coupon testing in controlled-atmosphere furnaces, in *Fossil Fuel Combustion — 1991*, R. Ruiz, Ed., New York, ASME PD-Vol. 33, 155-160, 1991.

47. T.J. Schultz, R.A. Schmall, and I. Chan, Selection of a heating system for a high temperature gas fired soft vaccum furnace, in *Fossil Fuels Combustion — 1992*, R. Ruiz, Ed., New York, ASME PD-Vol. 39, 23-30, 1992.

48. A.E. Erinov, A.M. Semernin, V.A. Povarenkov, M.J. Khinkis, and H.A. Abbasi, Development of a gas-fired, low-inertia, high-efficiency heat treating furnace, *Proc. of 1995 International Gas Research Conf.*, D.A. Dolenc, Ed., Govt. Institutes, Rockville, MD, 1986, 2774-2773.

49. H.S. Kurek, M. Khinkis, W. Kunc, A. Touzet, A. de La Faire, T. Landais, A. Yerinov, and O. Semernin, Flat radiant panels for improving temperature uniformity and product quality in indirect-fired furnaces, *Proc. 1998 International Gas Research Conf.*, Vol. V: Industrial Utilization and Power Generation, D. Dolenc, Ed., Govt. Institutes, Rockville, MD, 1998, 100-112.

50. H.M. Feintuch and K.M. Negin, FW Delayed-Coking Process, in *Handbook of Petroleum Refining Processes,* 2nd edition, R.A. Meyers, Ed., McGraw-Hill, New York, 1996, chap. 12.2.

11 Higher Temperature Applications

11.1 INTRODUCTION

The designation "higher temperature application" is somewhat relative, but here it refers to processes with furnace temperatures above about 2200°F (1200°C). The major applications considered in this chapter include furnaces and combustors used in the metals, minerals, and waste incineration industries. In general, the dominant heat transfer mechanism in those industries is thermal radiation. This is in contrast to lower temperature applications where both radiation and forced convection are often important.

11.1.1 FURNACES

A furnace is a device where heat is transferred to a material to effect physical or chemical changes in the material.[1] For the purposes of this book, the energy is generated by the combustion of a fossil fuel inside the furnace. Industrial furnaces are refractory-lined vessels designed both to keep energy inside to minimize heat losses *and* to radiate energy to the load. The heat transfer mechanisms that are important in furnaces are discussed in Chapter 9.

11.1.2 INDUSTRIES

The three primary higher temperature industries considered in this chapter include the metals, minerals, and waste incineration industries. Each of these requires higher temperatures to either melt high-melting-point materials or destroy waste materials processed in their furnaces. Each of these industries has slightly different requirements and operating conditions compared to the others that makes the heat transfer analysis slightly different in each case.

In the metals industries, there are generally two types of high temperature processes: metal production and parts production from that metal. Metal heat treatment was considered in Chapter 10 and is not included here as it generally is done at lower temperatures. Final finished parts are not usually made directly in a single process starting with raw materials. In the case of metals production, many of these processes are batch in nature. Solid raw ore materials or scrap metals are heated and melted to the liquid state. The molten metal is then blended with the appropriate ingredients to get the desired grade of metal. Chemical reactions may occur in the molten metal. In the second type of high temperature metal heating process, a specific grade of metal is melted and then formed into a specific shape or part. For example, a given grade of steel may be cast into sheet, plate, I-beams, bar stock, or other shapes. Another example would be taking a specific grade of aluminum and casting it to make a part like an engine component. The batch nature of many high temperature metals processes complicates the heat transfer analysis because of the transient nature and the high temperature gradients and nonuniformity caused by the discontinuous processing.

In the minerals industries, solids are often blended together in specific ratios and then processed in a high-temperature furnace in a continuous process and then made directly into the final finished part. Unlike most metals processes, the mineral products go from raw materials to the final finished part in a single process. For example, in glass manufacturing, scrap glass, sand, and many other components are preblended and fed into a continuous furnace where they are made into molten glass, which is then fed directly into a casting process to make finished glass parts like windows or car windshields. There are often chemical reactions occurring during the heating and melting portion of the cycle. The continuous nature of high temperature minerals processing simplifies the heat transfer analysis because of the steady-state operating conditions.

The waste incineration industry is somewhat unique in that the desired result is usually not an end product but the destruction of contaminants rather than the production of a new material. In some cases, the end product may be a cleaned solid material such as in soil decontamination, for example, resulting from the destruction of the contaminants. Waste incineration processes are usually continuous, which makes heat transfer analysis a little simpler, but mass transfer is often important as liquid and solid waste materials are vaporized and combusted. If the contaminants contain significant quantities of hydrocarbons, these may also be used as heat sources in addition to the fuel supplied through the burners. Also, enough air must be supplied to burn both the fuel and the hydrocarbons to ensure complete destruction of the contaminants (see section 4.1.1.2).

11.2 METALS INDUSTRY

The heating and melting of metal has been around for centuries. Many of the techniques still being used today were developed years ago and are often little-changed. As is customary, the metals industry has been broken into two categories: ferrous and non-ferrous. The ferrous industry includes iron- and steelmaking. The non-ferrous industry includes essentially all other metals, but the more common ones include aluminum, brass, copper, and lead.

11.2.1 FERROUS METAL PRODUCTION

There are many processes in the iron and steel industries that use large quantities of fossil fuel energy. Examples include the blast furnace, the open hearth furnace, and the basic oxygen furnace (BOF). However, these processes do not use burners. The combustion occurs inside the vessels in conjunction with processing the materials inside. Therefore, these processes are not considered here as they are not industrial combustion processes in the sense of the definition used in this book. Although a furnace like the electric arc furnace, which is discussed next, uses electricity as the primary energy source, it also often uses supplemental burners and is therefore considered here. Lehrman et al. (1998) have listed the following applications in steel mills that use gaseous fuels: coke-oven heating, blast-furnace stoves, gas turbines for power generation, boilers, soaking pits, reheating furnaces, forge and blacksmith furnaces, normalizing and annealing furnaces, controlled-cooling pits, foundry core ovens, blast furnace and steel ladle drying, drying of blast-furnace runners, hot-top drying, ladle preheating, and oxy/fuel burners.[2] Several of those that are "conventional" combustion applications as defined here are considered next. A relatively recent batch steelmaking process, known as the energy optimizing furnace (EOF), uses oxy/fuel burners to assist in melting a blend of approximately 50% scrap and 50% hot metal.[3] However, since the EOF is primarily used only in Brazil, it is not included here.

11.2.1.1 Electric Arc Furnace

Even before the arrival of the BOF, a major alternative to conventional ore-based steelmaking had begun to gain acceptance.[4] The electric arc furnace (EAF) was conceived as a unit to melt scrap steel and recycle the iron units back to useful service. It did not rely on molten iron from a blast furnace, but rather solid scrap was melted by energy input from electrical arcs passed between three carbon electrodes. Originally, this furnace was conceived as an all-electric melter. Freedom from the blast furnace and comparatively cheap electric power fueled the growth of the EAF, particularly in what came to be known as mini-mills.

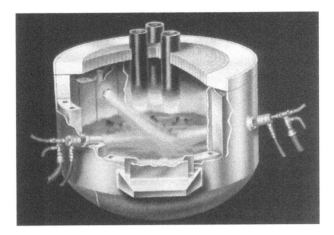

FIGURE 11.1 Electric arc furnace (EAF) with three supplemental burners mounted in the sidewalls and firing between the three electrodes. (Courtesy of Air Products, Allentown, PA. With permission.)

Due to the very high temperature requirements in most ferrous metal processes, oxy/fuel burners (see Chapter 12) have become standard equipment on most EAFs.[5] Initially, firing through the slag door, oxy/fuel burners mounted on a boom or a carriage were introduced to increase scrap melting. Later, to efficiently target the coldspots in the EAF, the burner location was moved to either the roof or sidewalls, to aim at the coldspots associated with the spaces between the electrodes, with one to four burners in a furnace providing supplemental energy. Figure 11.1 shows a sketch of a three-burner, wall-fired supplemental system firing between the electrodes. In addition to productivity improvements of 5% to 20%, the burners provide economical energy for melting scrap at lower cost.

Heat transfer from the oxy/fuel burners to the scrap metal occurs by forced convection from the combustion products to the scrap and by radiation from the flame to the scrap.[6] Since the heat transfer is dependent on the temperature difference between the scrap and the heat source (combustion gases and flame), the oxy/fuel burners are most effective immediately after new cold scrap has been added to the EAF. The burners are often discontinued after half of the meltdown period is complete in order to get reasonable efficiencies.[7] There is also thermal conduction between contacting pieces of scrap and from the heated iron oxide (scale) layer on the outside of the scrap to the scrap metal under the scale layer. This mode of heat transfer is obviously undesirable because the oxidation layer on the scrap reduces the product quality and the yield and is therefore itself undesirable. The reported overall heat transfer efficiencies for the oxy/fuel scrap heating range from 50% to 75%.[8] Figure 11.2 shows how the efficiency decreases with the time into the meltdown.[9] By 50% into the meltdown, the instantaneous efficiency is below 40%. This is a result of the reduction in heat transfer as the scrap temperature increases.

Oxy/fuel burners are most efficient during cold scrap heating and meltdown. Typically, burner effectiveness decreases during the latter half of the meltdown period from an initial 60%–80% to less than 20% when firing on a flat molten bath. Excessive use of burners with low efficiency can result in yield loss and potential furnace damage due to higher offgas temperatures.

With regard to design considerations, it is important to maximize flame velocity for maximum heat transfer efficiency (i.e., deeper scrap penetration and higher convective heat flux). By using the smaller diameter burners for the available oxygen and fuel supply pressures and desired firing rate, the highest flame velocity is achieved. Converging/diverging nozzle designs have also been successfully used in a number of burner designs.

An efficient oxy/fuel practice typically supplies 25% of the total energy required to melt the steel. This may be higher or lower, depending on the desired results and characteristics of the furnace and operation in question. The optimum burner firing rate is limited by how quickly the

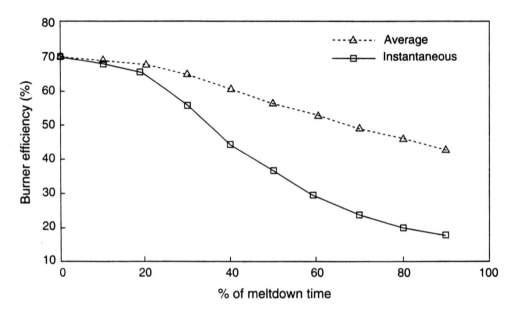

FIGURE 11.2 Oxy/fuel burner efficiency vs. time into meltdown for an electric arc furnace. (Courtesy of AISE Steel Foundation, Pittsburgh, PA.[5] With permission.)

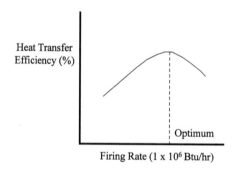

FIGURE 11.3 Optimum firing rate is at maximum heat transfer efficiency. (Courtesy of CRC Press, Boca Raton, FL.[4])

scrap can absorb the heat supplied by the burners. In other words, the optimum firing rate is at the point where the heat transfer efficiency to the scrap is at a maximum (see Figure 11.3). Experience and energy balance-based computer models are often used as aids in predicting the optimum firing rate. In practice, the firing rate is further optimized during burner start-up by experimenting with different firing rates while monitoring melting rate, yield, and offgas temperatures.

Burners must have the flexibility of operating in either the fuel-rich or fuel-lean mode. They can be fired fuel-rich with a bushy, luminous flame early in the heat to preheat scrap without excessive bridging or welding of the scrap, and then switched to an oxygen-rich mode to cut away scrap and assist with uniform melt-in. Operating in an oxygen-rich mode is beneficial for slag door installations to assist with cutting away the scrap to allow earlier use of an oxygen/carbon lance manipulator to produce a foamy slag.

Battles and Knowles (1985) reported the use of oxy/gas burners to improve the productivity in electric arc furnaces by supplemental heating.[10] A schematic of the wall-mounted burner system is shown in Figure 11.4. A water-cooled, oxy/gas burner was chosen to accelerate the melting rate in

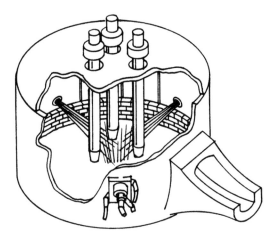

FIGURE 11.4 Gas-fired burners for supplemental heating in an electric arc furnace. (Courtesy of Government Institutes, Rockville, MD.[10] With permission.)

the colder areas of the furnace. On a 100-ton EAF, melting times were reduced by 20% to 35% and overall fuel consumption (electricity and natural gas) decreased by 10%. Another important benefit included reduced consumption of refractory and electrodes.

Sikirica (1987) described field test results for the use of oxy/gas burners in smaller EAFs (<30 tons) often used in foundries.[11] Conventional oxy/gas burners can potentially damage smaller EAFs because the flame lengths would impinge on the furnace walls. Normally, the oxy/gas burners would be fired simultaneously with the arcing electrodes. For the smaller EAFs, the process was modified so that the oxy/gas burners fired before using the electrodes, to minimize the possibility of flame impingement. The burners preheated the scrap metal in the colder parts of the furnace. Results showed a reduction in melting time ranging from 16% to 20% compared to all-electric melting.

11.2.1.2 Smelting

Smelting is a bulk production process where solid metal ore mixtures are heated which causes a chemical change that results in liquid metal. This process is done in a furnace often referred to as a smelter. The liquid phase may be either a liquid metal sulphide (matte) common in the production of copper and nickel, or a metallic phase common for the production of iron, steel, lead, zinc, and aluminum. The two most common smelting processes are flash smelting and bath smelting. Depending on the specific production method, large quantities of heat are normally required to melt the metal ore, which is where heat transfer from a combustion process is important.

Davila et al. (1995) described an oxy/fuel rotary furnace process for smelting white cast-iron and low alloy cast-carbon steels.[12] The main parameters studied were the furnace lining, raw material composition, load protection (slag), flame types and temperature, and furnace rotation speed. The main measurements included the chemical analysis of each heat, analysis of the gas content, heat performance, metal yield, and slag analysis. The flames were only slightly oxidizing to minimize metal oxidation, which would reduce yields. This process reduced energy consumption by more than 35% compared to electric induction melting.

11.2.1.3 Ladle Preheating

Another major use of burners in the steel industry's melt shops is for ladle preheating, as shown in Figure 9.16. Figure 11.5[13] shows a ladle that is used to transport molten metal in steel mills from the melting furnace to the casting station. Ladles are refractory-lined cylindrical vessels closed

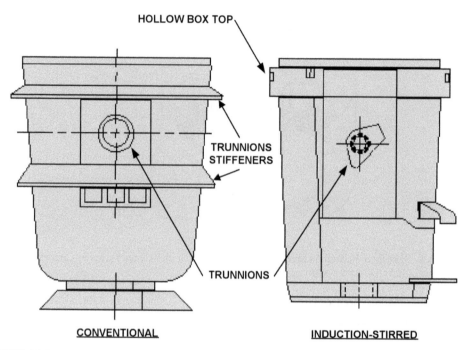

FIGURE 11.5   Conventional ladle used in steel melting processes. (Courtesy of AISE Steel Foundation, Pittsburgh, PA.[13] With permission.)

at one end. The molten metal is poured into the top and then the ladle is usually transported via an overhead crane for pouring. There are four primary reasons for preheating ladles prior to pouring molten metal into them:

1. To minimize the cooling effect on the molten metal. A cold ladle could cool the liquid metal enough to cause the metal to solidify inside the ladle. This reduces the metal yield and increases maintenance costs to clean out the frozen metal. The relatively high thermal conductivity of the refractories used in ladles means that excessive heat losses would result if the ladles were not preheated.[14]
2. To minimize the thermal shock on the refractory, which could go from ambient conditions to molten metal temperatures (>2000°F or 1100°C) in a matter of seconds. This rapid temperature rise often damages the hard ceramic refractory used to line ladles.
3. To remove any moisture that may have accumulated in the vessel. As any metal producer knows, water is anathema to molten metal. Any liquid water in the ladle could quickly turn to steam with the addition of molten metal. The rapid expansion from liquid water to steam often produces violent explosions that can damage equipment and threaten the safety of personnel.
4. To make the process more consistent. Without preheating, the temperature of a given ladle would vary significantly from one heat to the next, depending on how long between uses, the ambient conditions in the plant, how long the last heat was held in the vessel, how much metal was poured into the vessel, etc.

The benefits of ladle preheating include lower tap temperatures, increased ladle lining life, lower refractory maintenance costs, and increased productivity and quality in casting due to more consistent steel temperatures.[13]

Today's refractories and steelmaking processes often require ladle preheat temperatures above 2000°F (1100°C), which can be difficult to obtain with the relatively low temperatures generated by conventional air/fuel flames. Because of their higher flame temperature, oxy/fuel burners provide the energy required for ladle preheating to reach the desired temperature in a shorter period of time with minimum flue losses and maximum fuel efficiency (see Chapter 12). Even with high-temperature flames, Tomazin et al. (1986) showed that although the inside ladle refractory temperature (hot face) of an initially cold dry ladle heats up to 2000°F (1100°C) in about 8 hours, the outside ladle refractory temperature next to the steel shell (cold face) only heats up to about 300°F (150°C).[15] Steady-state may not be reached in the ladle refractory until 18 hours after the start of the heating cycle. Heating the refractory too quickly should be avoided because of the nonequilibrium temperature profile in the refractory, excessive stresses on the shell, and excessive shock on the refractory that may cause damage.[14]

A steel ladle is a relatively small, refractory-lined vessel that, during preheating behaves very much like a small furnace. Heat transfer by the preheater to the ladle is primarily by convection and radiation. The convection can be described using:

$$q_c = h_c A \Delta T \tag{11.1}$$

where q_c is the convective heat flow, h_c is the convective heat flow coefficient, A is the hot-face surface area of the ladle, and ΔT is the temperature difference between the ladle's hot face and the combustion gases. To maximize heat flow (q_c), the preheater must maximize h_c and ΔT. The convective heat flow coefficient (h_c) is a function of the combustion gas velocity and properties, and of the flow geometry. The gas velocity is a function of the burner firing rate and design. The gas properties are functions of the gas composition and temperature (the pressure is essentially atmospheric). To maximize h_c, a ladle preheater should incorporate a "high-fire/low-fire" logic. This logic allows the burner to fire at high-fire whenever the ladle requires energy so that maximum combustion gas velocity is maintained. This approach enables the preheater to achieve maximum heat flow at the most efficient times.

Since the hot-face temperature cycle is the same whether the ladle is heated with air/fuel or oxy/fuel, ΔT is a function only of the combustion gas temperature, which is a direct function of the flame temperature. Therefore, to maximize ΔT, the flame temperature must be maximized. This can be done with preheated air or with oxy/fuel combustion.

The radiative heat transfer from the flame to the ladle walls can be calculated using:

$$q_r = 0.1713 F_e F_a 10^{-8} \left(T_f^4 - T_l^4 \right) \tag{11.2}$$

where q_r is the radiative heat flux, F_e is the emissivity factor, F_a is the radiation coefficient due to geometrical arrangement, T_f is the absolute temperature of the preheater flame, and T_l is the absolute temperature of the ladle's hot face. To optimize radiant heat flow to the ladle (q_r), thus making the most efficient use of the energy input, F_e and T_f must be maximized. T_l cannot be manipulated because the hot-face temperature cycle is the same, regardless of the heating method.

Oxy/fuel ladle preheating installations provide the following benefits: faster heating times, hotter ladle bottoms, fuel savings of over 70%, decreased offgas of 90% compared to air/fuel combustion, reduced maintenance, and no water-cooling requirements. Although fuel savings are a major economic benefit, other advantages associated with the oxy/fuel ladle preheating system are just as important. For example, ladle bottom preheat improves. The firewall life also improves dramatically. In one reported example, the firewall in an oxy/fuel system lasted more than 70% longer than a typical air-fuel firewall in this particular shop. This result is due to the lower amount of offgas volume flowing across the firewall.

11.2.1.4 Reheating Furnace

The main objective of a reheating furnace is to deliver properly reheated slabs, blooms, or tubes to a rolling mill.[16] The desired characteristics of a reheating furnace include:

- Correct stock discharge temperature
- Proper temperature distribution along the length of the stock
- Low temperature difference between the surface and the core of the stock
- Thermally efficient
- Minimum scale formation on the stock surface
- Low maintenance and high availability
- Minimized operating and capital costs

A standard steel reheat furnace consists of the following components: heating chamber or furnace, hearth or support for carrying the charge, controls to maintain a specified temperature, distribution system for heating and waste gas removal, and a materials handling system for moving the charge into and out of the furnace.[17] The required material outlet temperature depends on its composition and geometry and can be as high as 2350°F (1290°C).

Handa and Tomita (1985) discussed the development of a highly luminous, natural gas-fired burner for use in large reheat furnaces.[18] One application for the technology is replacing oil with natural gas as the fuel for reheat furnaces. A major problem with switching these fuels is the large disparity that often exists in the flame luminosity since oil flames are normally very luminous and natural gas flames are normally fairly nonluminous (see Chapter 3). Handa and Tomita showed that while the thermal efficiency increases with flame emissivity, there is very little increase in efficiency for emissivities above about 0.5. Therefore, the goal of their technology development was to make a natural gas flame with an emissivity of at least 0.5. The resulting burner design had staged air combustion with the natural gas injected through the center of the burner as shown in Figure 11.6. Figure 11.7 shows that the high-luminosity natural gas burner had a higher emissivity than both an oil and a conventional lower luminosity, natural gas burner. The high-luminosity natural gas burner had a flame shape about 10% larger than the oil flame. The calculated heat flux to 140 mm (5.5 in.) wide × 140 mm (5.5 in.) high × 900 mm (35 in.) long steel billets was

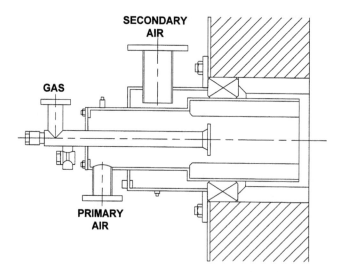

FIGURE 11.6 High-luminosity natural gas burner for use in steel reheat furnaces. (Courtesy of Government Institutes, Rockville, MD.[18] With permission.)

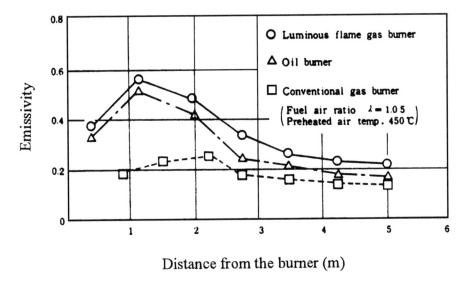

Distance from the burner (m)

FIGURE 11.7   Flame luminosity as a function of axial distance from the burner for several burner types used in steel reheat furnaces. (Courtesy of Government Institutes, Rockville, MD.[18] With permission.)

76.4 kW/m^2 (Btu/ft^2-hr). The thermal efficiency increased by 8% by converting a furnace from oil firing to the high-luminosity natural gas burner technology.

Walsh et al. (1986) discussed the use of oxygen/fuel burners in a continuous steel reheat furnace.[19] The furnace, which was about 100 ft (33 m) long and 34 ft (11 m) wide, had four top-fired zones and one bottom-fired zone. The design production rate for 8 in. × 6 in. × 30 ft (3.1 cm × 2.4 cm × 9.8 m) blooms was 200 ton/hr (180 mton/hr). An overall fuel savings of 28% was demonstrated, as compared to the base case using ambient air for combustion.

Chapman et al. (1989) modeled a direct-fired continuous reheating furnace.[20] To determine their effects on the furnace efficiency, a variety of parameters were studied, including load velocity, load emissivity, furnace combustion space height, and refractory emissivity. Figure 11.8 shows that the load velocity has relatively little effect on the heat flux to the load — except at low velocities, where the shape of the curve is unlike the other four faster velocities given. Figure 11.9 shows that the height of the combustion space has a fairly minimal impact on the heat flux to the load for the range of heights computed. Lee et al. (1991) showed experimentally and computationally that preheating a variety of fuel gases used in reheat furnaces can have a significant impact on the heat transfer distribution inside the furnace.[21]

Chapman et al. (1989) also modeled the heat transfer in a direct-fired, batch reheating furnace.[22] The furnace was modeled as a well-stirred, single-gas-zone enclosure with a uniform gas temperature and included a simplified single-step combustion reaction. Several parameters were varied, including the load heat capacity, load emissivity, furnace space height, and refractory emissivity. Figure 11.10 shows that the load heat capacity has a large impact on the heat transfer to the load. As expected, Figure 5.10 shows that increasing the load emissivity increases the heat transfer to the load. Figure 11.11 shows that decreasing the height of the combustion space can significantly improve the heat transfer to the load. Figure 11.12 shows that the lower the refractory surface emissivity, the higher the heat transfer rate to the load because the energy from the flame is reflected off the walls onto the load.

Blanco and Sala (1999) used computer modeling to optimize the performance of steel reheat furnaces used for heating round billets for making chains.[23] One of the problems they were trying to solve was excessive energy consumption. It was estimated that the furnace efficiency was reduced by as much as 10% by air infiltration alone. As a result of the modeling, the cross-section of the

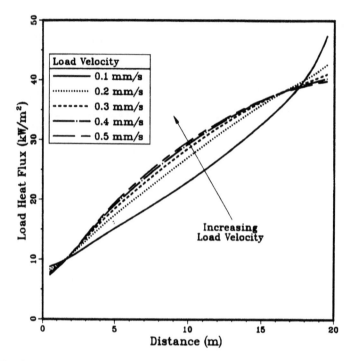

FIGURE 11.8 Predicted effect of load velocity on load heat flux in a direct-fired, continuous reheat furnace. (Courtesy of ASME, New York.[20] With permission.)

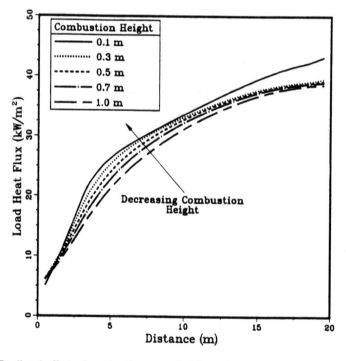

FIGURE 11.9 Predicted effect of combustion space height on load heat flux in a direct-fired continuous reheat furnaces. (Courtesy of ASME, New York.[20] With permission.)

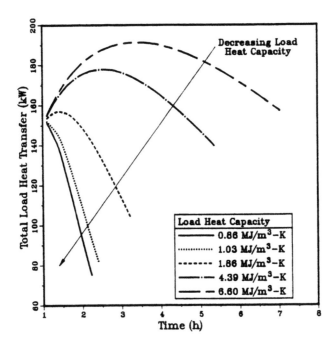

FIGURE 11.10 Predicted effect of load heat capacity on the heat transfer to the load in a batch reheating furnace. (Courtesy of ASME, New York.[22] With permission.)

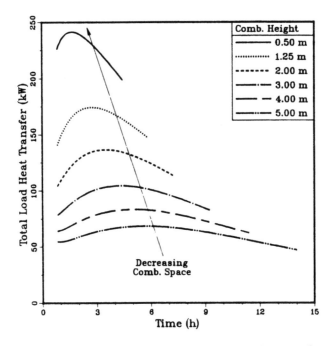

FIGURE 11.11 Predicted effect of the combustion space height on the heat transfer to the load in a batch reheating furnace. (Courtesy of ASME, New York.[22] With permission.)

furnace was narrowed and the burners above the load were tilted down 20° from horizontal toward the load. A heat recovery system was added to preheat the combustion air to 400°C (750°F). These and other modifications resulted in improving the overall thermal efficiency from 40% to 64%.

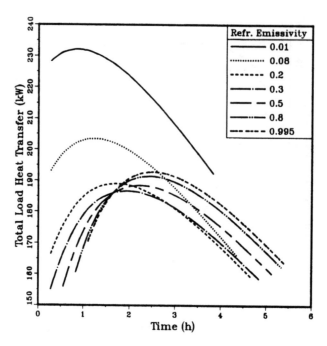

FIGURE 11.12 Predicted effect of refractory surface emissivity on the heat transfer to the load in a batch reheating furnace. (Courtesy of ASME, New York.[22] With permission.)

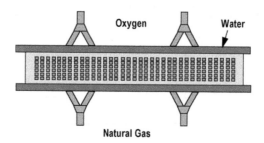

FIGURE 11.13   Air Products Rapidfire™ burner. (Courtesy of CRC Press, Boca Raton, FL.[29])

Lisin and Marino (1999) have also developed a simplified model to predict the heat transfer performance of direct-fired, batch steel reheat furnaces.[24]

A novel concept using a matrix burner in preheating copper strip at the continuous annealing furnace entrance has been successfully tested.[25] The Rapidfire™ heating system has been introduced to the steel and copper industries. The burner contains a multitude of ports which produces a uniform, high-temperature oxy/natural gas flame (see Figure 11.13). Burners are positioned so that the flame impinges directly onto the solid surface of the part to be heated. This approach of applying high-intensity heat using a compact, variable-geometry burner permits the application of heat precisely where needed. This is of particular importance to rolling, forging, and heat treatment lines to increase line speed, maintain edge, corner, and surface temperatures of slabs and transfer bars between forming operations, reheat components after line stoppages, and to heat bars and billets prior to hot-working. The Rapidfire™ system uses water-cooled matrix burners of width similar to the sheet width and could be about 2 in. (50 mm) high. The matrix burner approach gives great flexibility to the shaping of the flame front, which can be designed to fit the component to be heated. This leads to a high degree of control over the heat input geometry — important in preventing localized overheating or surface oxidation.

11.2.1.5 Forging

Forging is a metal working process used to change the shape of a metal part by using compressive forces, often with the use of a die, to get the desired shape. This can be done in either a batch or continuous process, depending upon the original and final part shapes, the material temperature and the material composition. Forging may be done when the part is hot or cold. Hot parts may require energy to heat them up if they started cold, but require lower compressive forces to do the shaping. Cold forging does not require any energy to heat the part, but does require higher compressive forces for shaping. Hot forging is of interest here because it usually requires fossil fuel combustion to do the heating, because electric induction heating is often expensive.

Ward et al. (1985) showed how the thermal efficiency and productivity of a forge furnace can be increased using a high-temperature metallic recuperator for heat recuperation.[26] A schematic of the system is shown in Figure 11.14. Combustion air preheat temperatures up to 700°F (370°C) were achieved. The fuel consumption was reduced by 28% compared to the base case without heat

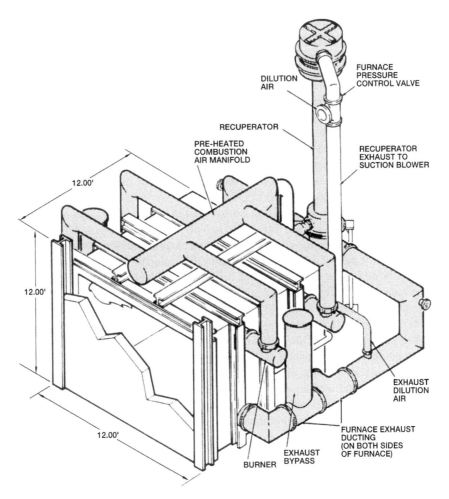

FIGURE 11.14 Heat recuperation system for a forge furnace. (Courtesy of Government Institutes, Rockville, MD.[26] With permission.)

recuperation. The heating time was reduced by 8% and scale formation on the outside of the billets was reduced by one third.

Salama and Desai (1985) studied electric induction heating and gas metal heating in forging applications.[27] The following market factors were considered: workpiece size (temperature uniformity of large workpieces and the need to reheat partially forged products from large workpieces), product-run (frequency of changes in shape and size of the workpieces for short-run products), quality tolerance (surface defects), and energy price. The first two factors were considered to be the most important. Two extremes of gas furnace sophistication were chosen to compare against electric induction: the slot furnace with low efficiency (4% to 25%) and low capital cost, and the rotary hearth furnace with high energy efficiency (20% to 45%) but high capital cost (almost 5 times higher than the slot furnace). Electric induction furnaces have high efficiency (45% to 55%) but even higher capital costs (nearly 6.5 times the cost of a slot furnace). The gas-fired rotary hearth furnace is comparable to the electric induction furnace. A big advantage of the electric induction furnace is better product quality because of reduced surface defects. Therefore, the biggest recommended area for research is to reduce scale formation in gas-fired systems. One method would be to use indirect heating with radiant tube burners.

Takamichi (1998) described the development of a high-performance forging furnace utilizing twin regenerative burners.[28] Energy savings of 40% were reported for heating tips of round steel in a forge furnace and 25% for heating aluminum in a roller hearth furnace. The combustion air was preheated to 1000°C (1800°F), compared to 150°C to 350°C (300°F to 630°F) for conventional preheat systems on forge furnaces. Better temperature uniformity was also measured in the furnace using the regenerative burners.

11.2.2 ALUMINUM METAL PRODUCTION

Non-ferrous metal production includes everything but iron and steel. Common non-ferrous processes incorporating industrial combustion equipment include aluminum, copper, brass, and lead production. Only aluminum production is considered here as it is representative of the other processes and because more information has been published on it. Some of this chapter section has been adapted from Saha and Baukal (1998).[29]

A typical rotary aluminum-melting furnace is shown in Figure 11.15. Radiation and convection from the hot combustion products are transferred to both the furnace walls and the charge materials. The heat in the furnace wall is then transferred to the charge material as the furnace rotates on its horizontal axis. On the burner side is the charge door for adding the feed materials into the furnace. The exhaust is located at the opposite end of the furnace. A typical charge contains mixed scrap such as aluminum foils, castings, turnings, and dross. The aluminum producer would like to recover as much of that aluminum as possible to maximize the yield.

Conventional practice is to use air/fuel burners in these rotary furnaces. The concept of using an oxygen/fuel burner within an air/fuel burner provides the aluminum melter with the flexibility of using either burner or a combination of both burners (see Figure 11.16). This innovative approach has proved very successful in the aluminum industry.* This differs considerably from both conventional oxygen-enriched techniques and the oxy/fuel burners typically used for melting in the metals industries. During meltdown, the oxygen/fuel burner is used along with the air/fuel burner to provide maximum melt rates and minimum flue gas volumes; while during holding and casting, only the air/fuel burner is used. For the melting of aluminum, a combination of air-oxy/fuel combustion provides the best results. The outer air/fuel flame envelope provides a shielding effect for the hot oxygen/fuel inner flame. By firing the oxygen/fuel inner burner on a fuel-rich mixture,

---

* Bazarian, E. R., Heffron, J. F. and Baukal, C. E., Method for Reducing NOx Production During Air-Fuel Combustion Processes, U.S. Patent 5,308,239 issued 03 May 1994.

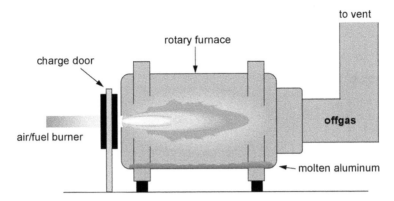

FIGURE 11.15 Conventional single-pass, rotary aluminum-melting furnace.

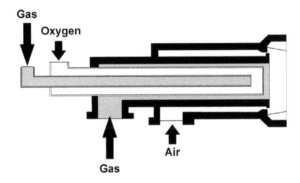

FIGURE 11.16 Air Products EZ-Fire™ burner. (Courtesy of CRC Press, Boca Raton, FL.[29])

flame luminosity is increased as combustion radicals and gas particles are heated to incandescence by the hot oxy/fuel flame. This in turn improves the radiative heat transfer of the flame to the charge material and furnace wall while preventing excess oxygen from contacting the load and causing metal oxidation. This air-oxy/fuel burner concept was first tried in 1989 at a Canadian aluminum dross reclamation plant where drosses and skimmings are smelted.[31] The dross, with 10% salt, was charged through the burner end of a 25,000-lb (11,000-kg) rotary furnace, fired by a dual-fuel burner of 20×10^6 Btu/hr (6 MW) capacity for melting. By firing the oxy/fuel burner at 9×10^6 Btu/hr (2.6 MW) and the air/fuel burner at 4×10^6 Btu/hr (1.2 MW), the combined air-oxygen/fuel input of 13×10^6 Btu/hr (3.8 MW) provided a meltdown reduction of 50% to 65% from 3 to 4 hours per heat to 1 to 2 hours per heat, depending on the charge material. Phillips et al. (1993) also reported the benefits of using an air-oxy/fuel burner for aluminum dross processing.[32] Compared to conventional air/fuel operation, the melting time was reduced by 50%, productivity was increased by 38%, and natural gas consumption was reduced by 45%.

An aluminum smelter, with three open-well reverberatory furnaces melting scrap and needed increased production, implemented oxygen-assisted combustion for increased furnace productivity, reduced cost per pound of material melted, and improved metal yield.[33] Two furnaces of 150,000-lb (68,000-kg) capacity each and one of 70,000-lb (32,000 kg) capacity were sidewell fed reverberatories (see Figure 9.12). Four dual-fuel burners fired along one furnace wall in the larger furnaces, while the smaller furnace was equipped with two burners. For the first furnace, the EZ-Fire™ burners (see Figure 11.17) were installed within the existing air/fuel burners without interrupting production. Adding molten metal recirculating pumps can significantly improve melt rates by up to 15% and reduce fuel consumption by more than 20%, with melt yields up to 96%.[34] This is due to the increased heat transfer rates by mixing the metal which has a more uniform temperature instead of much higher surface temperatures as is often the case in naturally circulating systems.

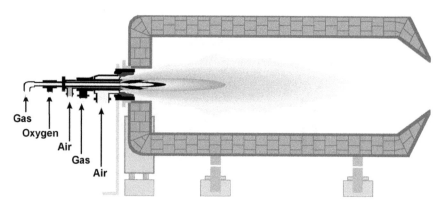

FIGURE 11.17   Air Products EZ-Fire™ burner in a rotary furnace. (Courtesy of CRC Press, Boca Raton, FL.[29])

Stala and Hindman (1990) described a gas-fired immersion tube for melting aluminum.[35] One of the problems with conventional aluminum melting processes is that they rely primarily on radiation heat transfer. This mode of heat transfer is often not optimal for a material like aluminum, which can be highly reflective when it is polished or when it is molten, both of which may exist at certain times in the aluminum melting cycle. Another possible problem with radiant heating is overheating of the surface, which can produce dross that lowers yields. A heating method that utilizes the high thermal conductivity of aluminum could increase the thermal efficiency of the aluminum melting process. An immersion-type burner not only heats primarily by thermal conduction, but it also separates the combustion products from the gases evolving from the aluminum melting process. This makes heat recuperation easier, especially in melting processes where chlorine is injected into the aluminum for contaminant removal because the excess chlorine gases coming out of the molten aluminum are highly corrosive and can damage recuperative burners. The key to the development of an immersion burner is to find the proper ceramic material that has high thermal conductivity, good thermal shock resistance, good corrosion resistance, and is easy to fabricate. Stala and Hindman found that alpha-silicon carbide had the best characteristics during testing in an aluminum holding furnace that showed the feasibility of the technology despite some problems encountered during testing. Kimura and Taniwaki (1990) described a similar immersion-type burner for melting smaller quantities of aluminum used in some segments of the aluminum casting industry. The burner tubes only survived up to 6 months in the aggressive environment of aluminum melting processes.[91] Both super-alumina and hi-alumina materials showed similar performance characteristics in the single-ended radiant tube burners. Melting efficiencies up to 60% were measured. Other benefits included low metal loss, more uniform aluminum temperature, higher product quality, reduced exhaust volume, and smaller equipment size.

Ward et al. (1987) described the development of a new, modular ceramic recuperator for use in an aluminum remelt furnace.[36] With thermal efficiencies ranging from 10% to 30% for nonrecuperated furnaces, this development was designed to increase efficiencies in reverberatory furnaces by 40% with heat recuperation. The difficulty with conventional metal recuperators is that the atmosphere in aluminum furnaces is moderately corrosive, so the life of the metal recuperators is shortened. The ceramic recuperator was installed on an aluminum-melting furnace and operated for 4000 hr. A payback of less than 3 years was estimated for a production system.

11.3   MINERALS INDUSTRY

Minerals production encompasses a wide range of materials, including glass, cement, lime, bricks, ceramics, and refractories.

11.3.1 GLASS

The glass industry consists primarily of four segments: container, flat, pressed and blown, and fiberglass. Glass is formed from raw materials that are fed into a high-temperature melting furnace to produce glass that may be clear or colored. The specific furnace and heat transfer issues vary slightly by segment. There are four common furnace designs used to make glass: direct-fired, regenerative side-port, regenerative end-port, and all-electric furnaces.[37]

One of the unique aspects of glass-melting is that the load not only absorbs radiation, but it also transmits radiation. Gardon has referred to such materials as diathermanous and argued that the absorption and emission in glass can be treated as bulk phenomena.[38] For glass-melting, the heat input should be prompt, intense, and uniformly distributed, which favors luminous flames.[39]

It has been shown by Russian scientists that the heat transfer in a glass-melting furnace with firing ports can be improved by injecting the fuel under the port instead of on the side of the port, as is the common practice.[40-42] The specific efficiency depends on the design of the fuel nozzle (e.g., the angle of attack). Another Russian paper determined the optimum flame length for both luminous and nonluminous flames to be one quarter of the length of an end-fired recuperative glass-melting furnace.[43] Computer modeling results in another paper by Russian scientists suggested an optimum flame length of between half and three quarters of the length of the furnace width on side-port regenerative glass melters.[44]

Cassiano et al. (1994) made extensive measurements on an end-port regenerative melting furnace used for making soda-lime glass.[45] Two sets of tests were performed: one for clear glass at a pull rate of 90 tons/day and one for amber glass at 73 tons/day. The furnace was air/fuel-fired. Crown temperatures ranged from 2786°F to 2912°F (1530 to 1600°C). Bottom temperatures ranged from 2206°F to 2343°F (1208 to 1284°C). In addition to the wall temperatures, detailed measurements were also made of gas temperatures in the furnace, including in the flame; concentrations of CO, CO_2, O_2, and NOx; and heat flux at the wall. A total heat flux meter was used to measure heat fluxes ranging from 291 to 308 kW/m^2 (92,300 to 97,700 $Btu/hr\text{-}ft^2$) for the clear glass, and 235 to 286 kW/m^2 (74,500 to 90,700 $Btu/hr\text{-}ft^2$) for the amber glass. Assuming that the dominant mode of heat transfer was radiation, the overall gas emissivity was estimated at ~0.6, which is considerably higher than the estimated 0.2 for clean gases composed of the major measured species.

Because of the high temperatures required to melt glass, two common approaches have been used in designing the combustion equipment. The conventional approach has been to use high air preheat temperatures to significantly increase the adiabatic flame temperature (see Figure 2.15). The second and more recent approach has been to use oxygen-enhanced combustion. Industrial oxygen has been used for several decades to enhance combustion in the glass industry.[46] Most of these installations utilized supplemental oxy/fuel burners, premixed oxygen enrichment of the combustion air, or under-shot lancing of oxygen to the port or burner (see Chapter 12). Supplemental oxy/fuel is the practice of installing one or more oxy/fuel burners into an air/fuel furnace. Premixed oxygen enrichment is the practice of introducing oxygen into the combustion air to a level up to 27% total contained oxygen (i.e., 6% oxygen enrichment). The amount of oxygen enrichment is limited by materials compatibility issues in highly oxidizing environments. Under-shot lancing is the practice of strategically injecting oxygen through a lance into the combustion region.

Frit furnaces generally produce less than 30 tons per day per furnace and are not continuous, with frequent start-ups and shutdowns of the furnace. Therefore, heat recovery/air preheat is uncommon. Without heat recovery, fuel savings alone have justified the use of full oxy/fuel. In parallel with the conversions in frit, Corning converted a large number of their smaller specialty glass furnaces — mainly ones that had no heat recovery — to 100% oxy/fuel firing.[47]

Pincus (1980) has compiled a series of articles, appearing in the magazine *The Glass Industry*, related to combustion in the glass industry.[48] The compilation contains major sections on fuels, combustion, heat balances, and emissions control. It is noted that glass-melting requires the heat input to a batch composed of raw materials plus cullet to bring about the desired chemical and physical changes to yield a chemically homogenous liquid that can be formed into the desired shape.

11.3.1.1 Types of Traditional Glass-Melting Furnaces

The type of furnace for melting glass typically depends on the type and quantity of glass being produced, and the local fuel and utility costs. While there are exceptions, the following discussion describes the primary furnace types and the glass segments that most commonly use each style.

11.3.1.2 Unit Melter

The term "unit melter" is generally given to any fuel-fired, glass-melting furnace that has no heat recovery device. Typically the air/fuel-fired unit melters are relatively small in size and fired with 2 to 16 burners. Furnaces range in production from as large as 40 tons (36 metric tons) of glass per day to as small as 500 lb (230 kg) of glass per day. Larger air/fuel unit melters are found in areas where fuel is extremely cheap. Frit, tableware, opthamalic glass, fiberglass, and specialty glasses with highly volatile and corrosive components are produced in unit melters.

Due to the very low energy efficiency and the use of individual burners, the air/fuel unit melters are very amenable to oxygen-enhanced combustion techniques, including supplemental oxy/fuel boosting, premix oxygen enrichment, and full oxy/fuel combustion. Oxy/fuel unit melters have been built large enough to produce as much as 350 tons (320 metric tons) of glass per day and small enough to produce as little as 500 pounds (230 kg) of glass per day.

11.3.1.3 Recuperative Melter

A recuperative melter is a unit melter equipped with a recuperator. Typically, the recuperator is a metallic shell and tube style heat exchanger that preheats the combustion air to 1000°F to 1400°F (540°C to 760°C). The furnace is fired with 4 to 20 individual burners. These furnaces range in size from as large as 280 tons (250 metric tons) of glass per day to as small as 20 tons (18 metric tons) of glass per day. These furnaces are common in fiberglass production but can also be used to produce frit. Some recuperative furnaces are used in the container industry, although this is not common. Furnace life is a function of glass type being produced. For example, a 6-year furnace life is typical for wool fiberglass. A typical recuperative melter is shown in Figure 11.18. Booker (1982) noted that production rates were increased by 25% using oxy/fuel instead of air/fuel combustion in an end-fired recuperative furnace employed to make mineral fiber.[49]

11.3.1.4 Regenerative or Siemens Furnace

In a regenerative (or Siemens) furnace, air for combustion is preheated by being passed over hot regenerator bricks, typically called checkers (sometimes spelled chequers). This heated air then enters an inlet port to the furnace. Using one or more burners, fuel is injected at the port opening, mixes with the preheated air, and burns over the surface of the glass. Products of combustion exhaust out the furnace through non-firing ports and pass through a second set of checkers, thereby heating them (see Figure 1.5). After a period of 15 to 30 minutes, a reversing valve changes the flow and the combustion air is passed over the hot checkers that were previously on the exhaust side of the process. The fuel injection system also reverses. After reversing, the exhaust gases pass through and heat the checkers that had previously heated the combustion air.

The Siemens furnace is the workhorse of the glass industry. Most flat and container glasses are produced in this furnace type. Regenerative furnaces are also used in the production of TV products, tableware, lighting products, and sodium silicates. There are two common variants of the Siemens furnace: the side-port regenerative melter and the end-port regenerative melter.

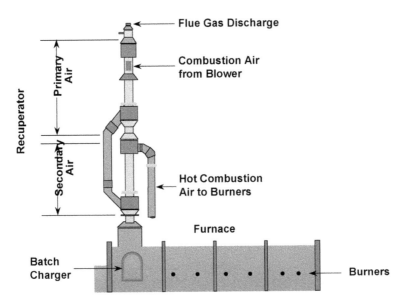

FIGURE 11.18   Typical recuperative glass melter. (Courtesy of CRC Press, Boca Raton, FL.[46])

11.3.1.4.1   End-port regenerative furnace

End-port regenerative furnaces are typically used for producing less than 250 tons (230 metric tons) of glass per day. In an end-port furnace, the ports are located on the furnace back wall. Batch is charged into the furnace near the back wall on one or both of the side walls. Figure 11.19 shows the layout of a typical end-port furnace. These furnaces are commonly used for producing container glass, but are also used for producing tableware and sodium silicates. For container production, a furnace campaign typically lasts 8 years. Undershot of oxygen through lances and supplemental oxy/fuel have been used successfully on this type of furnace.[50]

11.3.1.4.2   Side-port regenerative furnace

Side-port regenerative furnaces have ports located on the furnace side walls. Batch is charged into the furnace from the back wall. Figure 11.20 shows the layout of a typical side-port furnace. Side-port

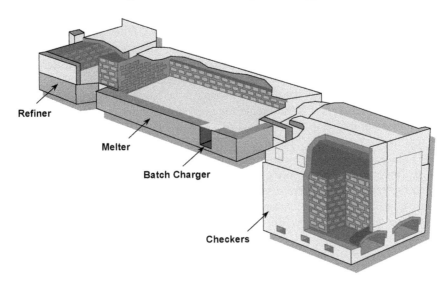

FIGURE 11.19   Typical end-port glass-melting furnace. (Courtesy of CRC Press, Boca Raton, FL.[46])

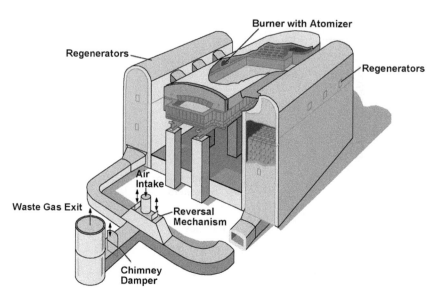

FIGURE 11.20   Typical side-port glass-melting furnace. (Courtesy of CRC Press, Boca Raton, FL.[46])

regenerative furnaces are typically used for producing greater than 250 tons (230 metric tons) of glass per day. A side-port furnace for float glass commonly produces 500 to 700 tons (460 to 630 metric tons) of glass per day. For container glass, side-port furnaces ordinarily produce between 250 and 350 tons (230 to 320 metric tons) of glass per day. These furnaces are commonly used in container and float glass production, but are also used for the production of tableware and sodium silicates. Undershot and supplemental oxy/fuel oxygen enrichment have been successfully used in this type of furnace.[51] These furnaces have also been converted to full oxy/fuel.

11.3.1.5   Oxygen-Enhanced Combustion for Glass Production

Production increase is the most common motivation for adding oxygen-based combustion technologies to an air/fuel furnace. These technologies are a low capital cost method for achieving production increases. Any of the methods described in the introduction — including full oxy/fuel — could be used, depending on the goals and furnace design. It should be noted that although oxygen-enhanced combustion allows for increased energy to the glass, other limitations in the furnace design may prevent production increase. Computational fluid dynamics simulations of the combustion space coupled to the glass melt can be used to determine whether the flow patterns in the glass would change in a way that would prevent increased production. An example is the paper by Carvalho et al. (1987), who predicted the heat transfer in an oxy/fuel glass furnace using CFD.[52]

Premixed oxygen enrichment is the simplest method to apply. To use this method, one requires only a basic flow control skid to control pressure and flow, and a diffuser for proper distribution of oxygen into the air main. This method is not foolproof, however. Problems such as shortening of the burner flame and increased temperature of the burner nozzle can occur if not properly implemented. There is also a limit on the level of oxygen enrichment that can be used because of the increased oxidizing nature of the combustion air. Typically, oxygen enrichment is limited to 6% on nozzle-mix burners and 3% on regenerative-style burners because of operational concerns. Experience has shown that higher levels of enrichment cause burner nozzles to wear at an accelerated rate.

Undershot oxygen enrichment and supplemental oxy/fuel are more common methods of achieving a production increase in an existing furnace. From an equipment standpoint, undershot enrichment is less expensive since no burners or additional flow controls are required. It is not trivial, however, to adjust the excess air and balance fuel. Typically, undershot enrichment is focused on

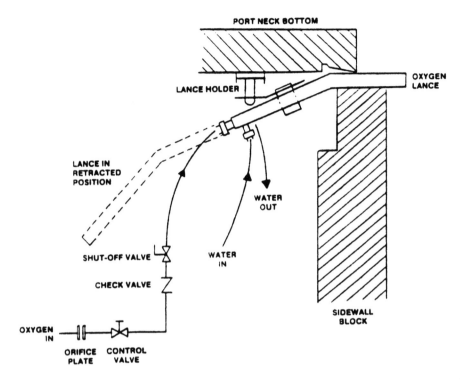

FIGURE 11.21 Oxygen injection system on a float glass furnace. (Courtesy of Government Institutes, Rockville, MD.[53] With permission.)

the burners near the furnace hotspot. Henry et al. (1987) described the use of undershot enrichment on a float glass tank.[53] A sketch of the oxygen injection system is shown in Figure 11.21.

Supplemental oxy/fuel has also been used to increase production with great success, especially in flat-glass furnaces. The advantage of this technique is that the additional energy can be focused directly where it is desired, without concern for balancing other burners.

Conversion to full oxy/fuel also provides an opportunity for production increase. The change in pull rate achieved with an oxy/fuel furnace — in comparison to an air/fuel furnace — varies, depending on furnace type. Pull rate increases of up to 60% have been observed for unit melters. Cross-fired regenerative furnaces have seen increases as little as 10%. End-fired regenerative furnaces converted to oxy/fuel increase pull capacity by 20%. Recuperative melters typically achieve a 30% pull rate increase.

These improvements in pull capacity are due to the improved heat flux density to the glass, which results in a faster melt rate. In addition, the ability of an oxy/fuel furnace to distribute heat input to the most ideal locations within the furnace helps to improve pull capacity. Some claim that the water content in the glass increases because of the different combustion atmosphere for oxy/fuel.[47] More water or hydroxyl ions in the glass decrease the viscosity of the melt allowing improved circulation of the melt.

If high-momentum oxy/fuel burners are used, the wall opposing the burner may overheat and damage may occur to the refractory.[54] To minimize this problem, low-momentum burners are recommended. Overheating can still occur with low-momentum burners if the flame impinges on the opposite wall. In this case, the firing rate must be reduced. If the wall on which the burner is mounted is overheating, the flame is simply too hot and/or too short. This can occur when applying premix, supplemental oxy/fuel, or full oxy/fuel. If the enrichment technique is premix, a lower level of enrichment is required. If an oxy/fuel burner is being used, a lower momentum, slower mixing burner is recommended.

Crown overheating is commonly caused by either burner lofting, or by two opposing burner flames colliding and deflecting toward the roof. If the flame appears to be lifting toward the crown due to buoyancy, then the burner is likely to be operating below the lower firing rate of the burner. A different burner size or higher firing rate is required. If the problem is due to opposed burners, the burner firing rate must be decreased. Generally, opposed burners are not recommended for furnaces less than 24 feet (7.3 m) wide.[55]

11.3.1.6 Advanced Techniques for Glass Production

The Gas Research Institute (Chicago, IL) funded Avco Research Lab[56] and Vortec Corp. (Collegeville, PA)[57] to work on an Advanced Glass Melter (AGM) design that promised to reduce pollution emissions, increase thermal efficiency, reduce capital costs, and improve operational flexibility compared to conventional furnace designs. Figure 11.22 shows a schematic of the AGM process. The major steps include:

1. Rapid suspension of the glass-forming ingredients (batch) in a high-intensity combustor,
2. Acceleration of the gas-solids suspension in a converging nozzle, which directs the gas-solids flow at a vertically oriented center (bluff) body,
3. Impact and separation of the batch on the center body, which serves as a site for the glass-forming reactions
4. Homogenization and fining of the thin glass layer, which moves down the center body.

Both radiation and convection heat the glass-forming ingredients as they are in intimate proximity of the exhaust products generated by the combustion of the fuel (natural gas) and the preheated

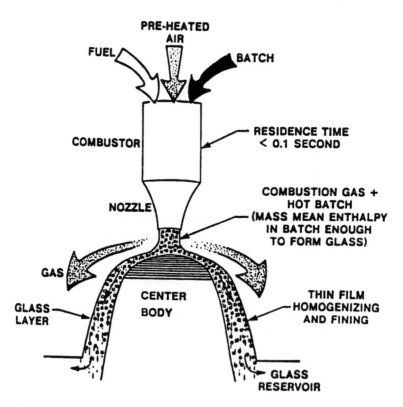

FIGURE 11.22 Advanced glass-melting process. (Courtesy of Government Institutes, Rockville, MD.[57] With permission.)

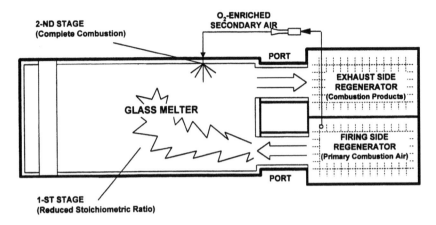

FIGURE 11.23 Plan view of oxygen-enriched air staging process on an end-port glass-melting furnace. (Courtesy of Government Institutes, Rockville, MD.[58] With permission.)

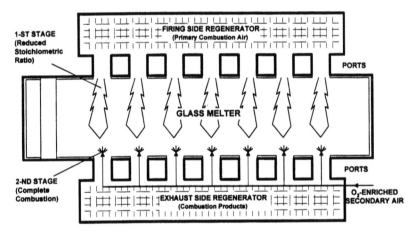

FIGURE 11.24 Plan view of oxygen-enriched air-staging process on a side-port, glass-melting furnace. (Courtesy of Government Institutes, Rockville, MD.[58] With permission.)

air. The AGM process takes on the order of minutes, compared to conventional glass melting processes which take on the order of hours up to days. The AGM claims 20% to 25% higher thermal efficiencies than conventional melting processes due to the improved heat transfer.

Abbasi et al. (1996) described a newly developed process known as oxygen-enriched air staging (OEAS).[58] A schematic of the process for an end-port furnace is shown in Figure 11.23. A schematic for the application of this process on a side-port furnace is shown in Figure 11.24. While the primary objective of the technology is to reduce NOx emissions (50% to 70% reduction was measured), there are also some effects on heat transfer. Since the preheated air flames are now initially fuel-rich, they are more luminous and therefore more radiative. The secondary oxygen added to the exhaust products completes the combustion before the gases exit the furnace. This secondary combustion zone helps to spread out the heat transfer over a wider region in the furnace, which helps to make the heat flux more uniform.

11.3.2 CEMENT AND LIME

Cement and lime are commonly produced in rotary kilns. Pearce (1973) studied heat transfer in rotary kiln flames used in limestone production.[59] Coal, gas, and oil flames were studied. Swirl

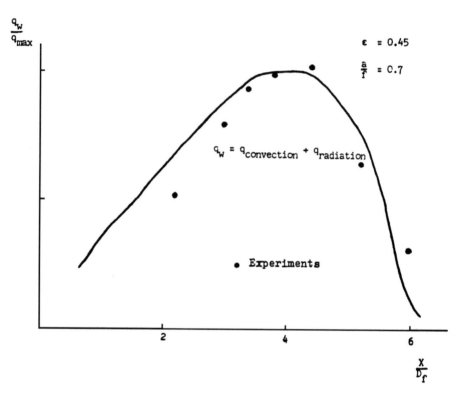

FIGURE 11.25 Predicted and measured ratio of heat flux to peak heat flux. (Courtesy of American Society of Metals, Warrendale, PA.[61] With permission.)

was needed for the coal flames. The measured gas temperatures along the length of the kiln were relatively uniform, with only about a 100°C (180°F) variation over 20 m (66 ft). The estimated convective heat transfer coefficient to the solids was 1.10 kW/m²-°C (190 Btu/ft²-hr-°F).

Richter and Quack (1974) presented modeling and experimental results for a pulverized fuel flame commonly used in cement kilns.[60] The experimental results were obtained in the large research furnace at the International Flame Research Foundation in IJmuiden, The Netherlands. The burner was described as a double concentric jet where secondary air is introduced at low momentum around a primary air jet. The modeling results showed good agreement with the experimental measurements. The temperature field was fairly symmetrical, with the highest temperatures (1500°C [2700°F]) in the center of the furnace and the temperatures declining radially from there. The heat flux to the furnace walls was higher toward the exit of the furnace than near the burner.

Khalil (1986) modeled the flow and combustion in a vertical lime kiln.[61] The furnace diameter was 5 m (16 ft) and the length was 30 m (98 ft). Two sets of burners at the lower level of the kiln, located 3 m (10 ft) apart, fired toward the centerline. Additional air for combustion was provided by a fan blowing air from the bottom. The limestone was fed into the top of the kiln. Figure 11.25 shows that the peak wall heat flux was at approximately 4 kiln diameters from the bottom of the kiln and dropped off rapidly in either direction. There was good agreement between the measurements and predictions — except near the inlet. Since the actual heat flux rates were not given, it is not known how well the model predicted the magnitude of the fluxes.

Ghoshdastidar and Unni (1989) modeled the heat transfer in the nonreacting zone of a cement rotary kiln used for heating and drying wet solids.[62] The model included the radiant exchange among the hot gases, the refractory wall, and the solids in the kiln; the transient conduction in the wall; and the mass and energy balances of the hot gas and solids. The kiln diameter was 1.95 m (6.40 ft), with a refractory thickness of 0.3 m (1.0 ft) and a refractory emissivity of 0.9. The solids

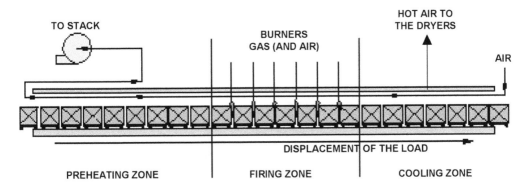

FIGURE 11.26   Tunnel kiln. (From *Appl. Therm. Eng.*, 17(8-10), 921-933, 1997. With permission.)

had an emissivity of 0.95 and initially contained 18% water by weight. The convection coefficient between the inner refractory surface and the solids was chosen as 64 W/m²-K (11 Btu/ft²-hr-°F). The convection coefficient between the outer refractory surface and the surrounding environment was chosen as 10 W/m²-K (1.8 Btu/ft²-hr-°F). The predicted length to completely dry the solids in the kiln was shorter than the measured length. There was good agreement between the actual and predicted gas and solids temperatures as a function of axial position.

11.3.3 BRICKS, REFRACTORIES, AND CERAMICS

Bricks, refractories, and ceramics are often heated in a tunnel kiln (see Figure 11.26). The materials to be heated — sometimes referred to as ware — are loaded onto cars and pushed through the kiln at regular intervals. The ware passes through three zones: preheating, firing, and cooling. In the preheating zone, the ware is preheated by the exhaust products from the firing zone. The primary function of the preheating zone is to remove moisture and burn out any organic material that might be present. In the firing zone, roof-mounted burners are fired in the free space around the ware. The function of this zone is to promote high-temperature chemical transformations (e.g., the decomposition of carbonates), and to heat the ware to the sintering temperature. In the third zone — the cooling zone — the ware is cooled down by the incoming combustion air, which is simultaneously preheated by the hot ware to improve the thermal efficiency of the system. Some of the preheated air is removed from this zone and used in other parts of the plant.

An important aspect of the process is how the ware is stacked on the cars traveling through the tunnel kiln. The ware is arranged to maximize the convective heat and mass transfer of the combustion gases flowing around it. The ware must be arranged so that all pieces receive adequate convection and are properly treated. This precludes dense packing, which would not give adequate exposure to the innermost pieces of ware. Carvalho and Nogueira (1997) have modeled a tunnel kiln and shown how the stacking arrangement of the ware affects the load temperature.[63]

11.4 WASTE INCINERATION

Some of the material in this chapter section has been partially adapted from previous publications.[64,65] Incinerators are designed to burn — and in many cases destroy — waste materials that may sometimes be contaminated with hazardous substances. The waste materials usually have some heating value. However, nearly all incineration processes require a substantial amount of auxiliary heat that is commonly generated by the combustion of hydrocarbon fuels such as natural gas or oil. An example of the important processes in a rotary kiln incinerator is shown in Figure 11.27.[66]

Incineration is a common method for treating waste materials. An ASME Committee on Industrial and Municipal Wastes published a general guide to combustion in waste incineration.[67]

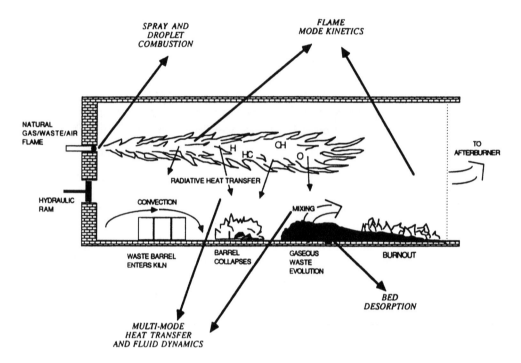

SPRAY AND
DROPLET
COMBUSTION

FLAME
MODE KINETICS

NATURAL
GAS/WASTE/AIR
FLAME

TO
AFTERBURNER

RADIATIVE HEAT TRANSFER

H   HC → CH   O

HYDRAULIC
RAM

CONVECTION

MIXING

WASTE BARREL
ENTERS KILN

BARREL
COLLAPSES

GASEOUS
WASTE
EVOLUTION

BURNOUT

BED
DESORPTION

MULTI-MODE
HEAT TRANSFER
AND FLUID DYNAMICS

FIGURE 11.27   Fundamental processes in rotary kiln incinerators. (Courtesy of CRC Press, Boca Raton, FL.[66])

Seeker (1990)[68] reviewed the use of combustion to dispose of waste materials and offered the following advantages compared to other methods of disposal:

- Destruction of hazardous constituents such as organics and pathogens in the waste
- Reduction of the volume and mass of the waste
- Potential for energy recovery by burning the wastes having heating value
- Making the waste unrecognizable from its original form (often a requirement for medical waste)

The objective of most waste incineration processes is to destroy or burn any of the organic materials in the waste, leaving only an inert residue. The waste combustion system consists of the following components:

1. Waste preparation and feeding,
2. Combustion equipment,
3. Heat recovery equipment,
4. Air pollution control equipment,
5. Ash/solid residue stabilization and disposal.

Wendt (1996) has also given a review of combustion in waste incineration.[69] Although the paper does not specifically discuss heat transfer in waste incineration, it gives a good overview of the technology, with a specific emphasis on the pollutant emissions. Vrable (1987) has noted the importance of recovering heat from waste incineration processes to improve thermal efficiency.[70]

Historically, air/fuel combustion has been used in waste processing to provide heat for thermal destruction of solid, liquid, and gaseous waste streams. Examples include medical and municipal waste (solid), spent solvents (liquid), and offgas or vent streams (gaseous). Oxygen-based combustion systems are becoming more common in waste processing applications.[71] When traditional air/fuel combustion systems have been modified for oxygen-enhanced combustion (OEC), many benefits can be demonstrated. Typical improvements include higher destruction and removal efficiencies of the waste, increased thermal efficiency, increased processing rates, lower NOx and particulates emissions, and less downtime for maintenance.

There are many benefits of using oxygen-enriched combustion in waste incineration. OEC has been used in several different types of incineration applications to overcome thermal limitations in the process.[72] One example is a fixed rotary incinerator where kiln instabilities were produced by variations in the incoming waste, which was sometimes cold and wet. The existing combustion system in the primary combustion chamber (PCC) was unable to handle these transient variations. O_2 was injected into the kiln through a lance. The O_2 flow was automatically controlled, based on feedback from the temperature at the exit. This improved the kiln stability. It also improved the refractory life in the afterburner due to the more uniform temperature profile in the overall system. In general, OEC can be used to increase incinerator throughput of low-heating-value waste materials.

Another application of OEC to overcome thermal limitations is in thermal pyrolysis of municipal solid waste or refuse-derived fuel (RDF).[73] The N_2 in air impedes the pyrolysis process, which is commonly used to recover chemicals and energy. Using oxygen enrichment, high-quality char or gas can be produced from the high ash and moisture content fuels. OEC can increase the heating value of the gas produced in gasification of municipal solid waste (MSW) by enhancing the devolatilization and evolution of the gaseous products.[73]

In many cases, the volatilization characteristics of waste materials are unpredictable so that incinerators are generally run with large amounts of excess air to ensure complete destruction of the volatiles.[74] This excess air puts a large heat load on the incinerator. When some or all of the air is replaced with oxygen, some or all of the ballast nitrogen is eliminated, which reduces the heat load on the furnace and increases the thermal efficiency of the process. Eliminating ballast nitrogen also improves the destruction efficiency of the volatiles so that less excess oxygen is usually required, which further reduces the unproductive heat load on the incinerator. For an existing incinerator, this reduction in heat load may either mean more waste can be processed or the auxiliary fuel consumption can be reduced. For a new incinerator, the equipment size and cost can be reduced.

The destruction and removal efficiency (DRE) is typically regulated for hazardous pollutants as defined by the Resource Conservation and Recovery Act (RCRA). These pollutants include, for example, polychlorinated biphenyls (PCBs) and principal organic hazardous constituents (POHCs) that may be found, for example, in contaminated soil. OEC can increase the DRE in an incineration process. This is accomplished by a combination of increased residence time within the incinerator and higher gas temperatures. These are two of the "three T's of incineration." The higher residence time is a result of the dramatically reduced flue gas volume that occurs when some or all of the combustion air in an existing system is replaced with oxygen. For a given PCC, the gas velocities through the system will be lower with OEC compared to air/fuel combustion. Therefore, the residence time will increase, which improves the mass transfer within the system. This increases the destruction efficiency of the hazardous pollutants in the process. The higher flame temperatures associated with oxygen-enhanced combustion ensure that all hydrocarbons in the incinerator will be well above their ignition temperatures. Therefore, if there is sufficient oxygen for combustion, the hazardous pollutants should combust. The turbulence in the flame may be higher for OEC compared to conventional air/fuel combustion. However, the overall turbulence level in the combustion chamber will be reduced because of lower gas velocities. This must be offset by the increased residence time and higher gas temperatures in order to ensure adequate DRE. As will be shown, the DRE generally increases using OEC.

There are several potential problems that may result from OEC. Depending on the burner design, oxygen-enhanced flames may be much hotter than air/fuel flames. This could lead to potential equipment damage due to overheating. Refractory damage may occur as a result of localized heating. A related problem is slagging. This may occur if the waste material is overheated and begins to vitrify. Both of these overheating problems can be avoided by the choice of a properly designed burner, adequate waste flow through the incinerator to absorb the heat, and the proper choice of auxiliary heat input into the incinerator.

Another issue to consider when using OEC in an incinerator is the potentially detrimental effects of reducing the amount of convection in the combustion chambers resulting from the reduction in the nitrogen in the combustion products. Convection is important to both the heat and mass transfer within the incinerator. Gas circulation helps to homogenize the temperature in the combustor, which minimizes hotspots that can damage refractories and cause nonuniform heating. A reduction in the gas volume could also reduce the mass transfer from the waste to the gas stream as volatiles are carried away from the surface of the waste. The gas flow over the waste material in the incinerator carries away the volatiles so they can be incinerated and so that more volatiles can diffuse to the surface of the waste. Therefore, a reduction in the gas circulation within the gas space of the combustor could potentially reduce the heat and mass transfer in the incinerator. This problem is partially offset by the greater volume expansion of the gases because of the higher gas temperatures associated with OEC. Higher gas expansion increases the gas velocity in the chamber, which increases convection. Proper selection of equipment can also overcome this potential problem so that gas circulation in the incinerator is maintained using OEC.[75] New incinerator designs for OEC could have a smaller combustion chamber compared to air/fuel systems for a given waste throughput in order to achieve the desired average gas velocity.

11.4.1 TYPES OF INCINERATORS

This chapter section discusses some typical types of incinerators, including municipal solid waste incinerators, mobile incinerators, transportable incinerators, and fixed incinerators.

11.4.1.1 Municipal Waste Incinerators

An example of a municipal solid waste (MSW) incineration process is shown in Figure 11.28. The performance of these incinerators can be improved using OEC.[76] The economic incentives include increased waste-processing capacity, greater thermal efficiency, increased production in a waste-to-energy facility, reducing the demand on the exhaust system, and a smaller air pollution control

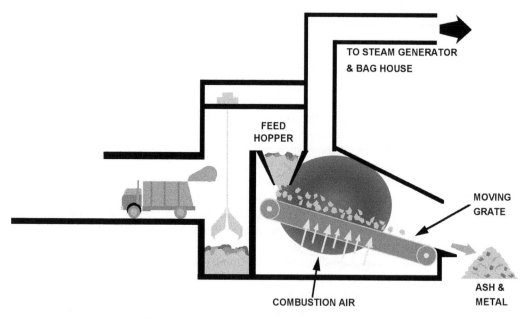

FIGURE 11.28   Municipal solid waste incinerator.

system. Increased capacity can be particularly important for many waste processors that are at their maximum capacity, since it is usually difficult to obtain permits to build new facilities. The environmental incentives include improved ash burnout, lower hydrocarbon emissions, lower CO, greater flexibility and control, and the ability to burn low-heating-value wastes such as dewatered sludge.

One of the earliest tests of oxygen enrichment in an MSW incinerator occurred at the Harrisburg, Pennsylvania, Waste-to-Energy Facility in 1987.[77] The combustion air was enriched with 2% O_2 ($\Omega = 0.23$). The waste throughput, steam production, boiler efficiency, and sludge throughput all increased. The flame stability improved with OEC. OEC has been used in an MSW incinerator to overcome thermal limitations.[72] O_2 was injected, through a diffuser, into the air plenums beneath the waste bed. It was also injected, through a lance, directly onto the bed. This resulted in a 10% increase in the waste processing capacity, an increase in steam production, better overall boiler efficiency, and more complete burnout of the ash. Significant cost savings were realized in ash disposal due to its lower volume and increased density.

The U.S. Environmental Protection Agency (EPA) sponsored a demonstration program to investigate OEC in a pilot-scale incinerator.[78] With only 3% O_2 enrichment ($\Omega = 0.24$), the waste processing rate increased by 24%. OEC did not seem to have any effect on the metal content of the ash. There were some concerns about higher hydrocarbon emissions at the higher throughputs. Further research was recommended. One important commercial consideration was the impact this technology might have on permitting. New or amended permits normally require a lengthy and usually costly review process.

11.4.1.2 Sludge Incinerators

A schematic of a sludge incineration process is shown in Figure 11.29. The biggest challenge in sludge incineration is the large amount of energy required to evaporate the large quantity of water contained in the sludge. The heating value of the sludge is minimal. Therefore, large quantities of auxiliary fuel are required. Oxygen has been used to increase the capacity of a multiple-hearth sludge incinerator by 35% to 55% at a sewage treatment plant in Rochester, New York.[79] Oxygen was injected into the sludge-drying zone at a rate of 1 ton/hr through a series of lances. The amount of auxiliary natural gas fuel used to dry and burn the sludge was reduced by 57%. Emissions (per mass of dry sludge) of total hydrocarbons, NOx, and CO were reduced by 58%, 62%, and 39%, respectively.

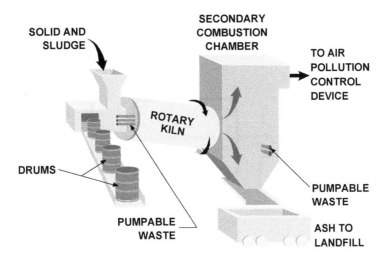

FIGURE 11.29 Sludge processing in a rotary kiln incinerator.

11.4.1.3 Mobile Incinerators

Mobile incinerators are commonly used to clean up contaminated soil and water at Superfund sites. The entire incineration system, including the PCC, the SCC, and the pollution control equipment, is small enough to be transported over-the-road. It can be quickly set up and is usually preferred for smaller sized cleanups. One of the first applications to use OEC in incineration was at the Superfund cleanup site at the Denney Farm in McDowell, Missouri, starting in 1987.[80] The EPA mobile incinerator was used, along with an OEC incineration technology that was later awarded the prestigious Kirkpatrick Chemical Engineering Achievement Award for the results obtained at this cleanup.[81] Dioxin-contaminated liquids and solids were successfully treated. The OEC system showed impressive performance compared to the original air/fuel system. The throughput was increased by 171%, the specific fuel consumption decreased 61%, the residence time in the SCC increased by 21%, CO spikes were reduced, while NOx levels were unaffected.

A more recent example is a trial burn to destroy PCB-containing electrical transformers and related contaminated materials.[82] The waste material was fed into one end of a rotary kiln. A single oxy/fuel burner, located at the kiln entrance, fired co-current with the feed material. The ash was collected at the kiln exit. The combustion and process offgases from the kiln were fed into the secondary chamber, which operated at a higher temperature than the primary chamber, to maximize destruction of any remaining combustible gases. A block diagram of the process is shown in Figure 11.30. Before the contaminated soil was processed, the system was tested using surrogate wastes to ensure the emission requirements could be met. Three different series of tests were conducted using various combinations of fuels and wastes. An oxy/propane burner in the PCC was used to incinerate PCB-contaminated soil for the series A and B tests. The only difference between A and B was that propane and oil with 1% PCB, respectively, were used as fuels in the SCC. In series C, the PCC was not operated, while the fuel for the SCC was oil with 42% PCB.

Vesta Technology, Ltd. (Fort Lauderdale, FL) is a hazardous waste incineration company that provides on-site services throughout North America. Vesta's mobile systems use high-temperature rotary kiln incinerator technologies, coupled with innovative, proprietary designs for flue gas scrubbing. This proven technology destroys PCBs, dioxins, oil sludges, and other hazardous wastes with DREs that exceed Resource Conservation and Recovery Act (RCRA) and Toxic Substances Control Act (TSCA) standards. Vesta has used this technology with both air/fuel and oxy/fuel systems. The decision to try oxygen was mainly influenced by the requirement to reduce the flue gas volume within the system. The result was twofold. First, the lower flue gas velocity in the kiln resulted in less particulate carryover to the secondary combustion chamber (SCC) and the air pollution control system. Actual site operations showed that SCC clean-outs were reduced with the

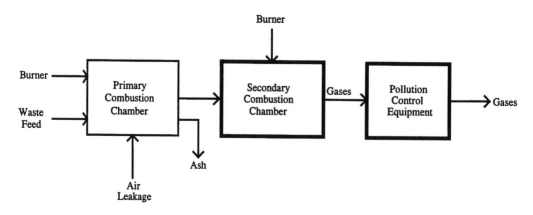

FIGURE 11.30 Mobile incineration system block diagram.

lower flue gas flow, while actually increasing soil processing rates. Second, the flue gas residence time in the SCC increased by 50%, which was expected to give higher DREs. Representative data from two similar remediation projects using the same equipment showed that the instantaneous soil throughput rates increased by 50% using oxygen, compared to the air/fuel base case. Of greater importance, however was the overall average hourly throughput, which increased by 150% due to the elimination of downtime for SCC particulate clean-out when using oxy/fuel. This rate enabled Vesta to complete the remediation project 28 days ahead of schedule.[83] The hourly NOx emissions were reduced by 52%, while the pounds of NOx per ton of soil processed were reduced by 66%.

11.4.1.4   Transportable Incinerators

Transportable incinerators are also commonly used at Superfund cleanup sites. They are larger in size than mobile incinerators and take longer to set up. Therefore, they are used at larger sites because of their increased processing capacity. The trend is to use these instead of mobile incinerators.[84] OEC has been used to maximize the transportable incinerator throughput by reducing the gas volume and improving the heat transfer performance. At the Bayou Bonfouca Superfund site in Saint Tammany Parish, Louisiana, cost savings were estimated to be nearly $3 million using OEC instead of air/fuel.

The example given here shows how OEC reduced particulates in a transportable incinerator. Williams Environmental Services (WES), located in Auburn, Alabama, used an incineration system consisting of a co-current rotary kiln, a hot cyclone, a secondary combustion chamber, a quench tower, baghouses, an induced draft fan, an acid gas absorber, and an exhaust stack. Air/fuel and oxy/fuel were used in this incineration system at two Superfund sites: a bankrupt wood-treating operation in Prentiss, Missouri, and at the Bog Creek Farm site in Howell, New Jersey. At the Prentiss site, the kiln was equipped with two air/fuel burners, while at the Bog Creek Farm site, a single oxy/fuel burner was installed. The Prentiss site contained 9200 tons (8300 metric tons) of creosote-containing soil. The Bog Creek Farm site contained 25,000 tons (23,000 metric tons) of soil that contained VOCs, including benzene derivatives, chlorinated hydrocarbons, and semivolatile organics such as naphthalene and phthalates.

OEC was selected for the Bog Creek site, due to the particulate emissions criteria set by the New Jersey Department of Environmental Protection (NJDEP). The primary problem encountered during start-up at the Prentiss site was higher than expected fines carryover. About 50% of the ash output was from the air pollution control system. The Bog Creek site was located near the New Jersey coastline where the soil is naturally sandy. The existing particulate emission limit of 0.03 grains per dry standard cubic foot (gr/dscf) or 0.07 grams per dry standard cubic meter (g/dscm) was reduced to a more rigorous standard of 0.015 gr/dscf (0.034 g/dscm) for this site, to prevent contaminated sand from entering the atmosphere. At a soil feed rate of 20 tons/hr (18 metric tons/hr), the equivalent Prentiss data indicated the estimated combined emissions (soil and metals) would have to be reduced by 65% to fall below the newly prescribed limit. OEC technology was selected to meet the tougher particulate standard without causing a delay in the schedule.[85] The Bog Creek site was the first Superfund site to use OEC incineration in the northeastern region of the U.S. and the first North American site to use OEC in a commercial, transportable incinerator for the entire project. The processing rate increased using OEC. All emission and ash requirements were satisfied. The site was cleaned up 60 days ahead of schedule. In this case, there was no reduction in NOx by using oxy/fuel, which was probably due to high air leakage into the kiln, as noted by the increased oxygen in the flue gas.

11.4.1.5   Fixed Hazardous Waste Incinerators

Oxygen-enhanced combustion has been used to reduce NOx in a fixed based resource recovery process.[86] Giant Resource Recovery (GRR) is a subsidiary of the Giant Group, Ltd., which is involved in cement manufacture and the use of waste materials as fuel and raw materials supplements. GRR processes creosote-contaminated soil through counter-current rotary kilns. By a patented process, the decontaminated soil is then used as a raw material for cement production, thus replacing a certain portion of the traditional feed material stream. The combustion products are ducted into the cement kilns. The kilns' processing rates were limited by two factors. Since the contaminated soil was high in moisture content and low in heating value, more heat transfer was required to increase the throughput while ensuring the soil's creosote concentration did not exceed permitted levels at the kiln discharge. If the creosote concentration is too high, the soil has to be reprocessed before being sent to the cement kiln. Second, the flue gas volume needed to be minimized to prevent upsetting the cement kiln operation. These criteria were met by using an oxy/fuel burner. The kiln back-end temperature increased over 100°F (38°C) using an oxy/fuel burner. Over a range of several tests, NOx emissions per ton of material processed were from 5% to 35% less using oxy/fuel. In this case, large amounts of air infiltration, evidenced by high O_2 concentrations in the flue gas, limited the NOx reduction using oxygen.

Oxygen has been injected into rotary kiln incinerators and the secondary combustion chamber to reduce CO emissions by more than 60%, while increasing waste throughput by more than 15% at a German merchant incinerator.[87] In the process, 530 ft³ (15 m³) of oxygen is injected for each 84-lb (38-kg) drum of waste material.

11.4.2 HEAT TRANSFER IN WASTE INCINERATION

Owens et al. (1991) studied a pilot-scale rotary kiln incinerator, both computationally and experimentally.[88] They studied the following parameters: kiln rotation rate (0.1 to 0.9 rpm), percent fill fraction (3 to 8), feed moisture content (0% to 20%), and inner wall temperature (190°C to 790°C [370 to 1450°F]). The experimental results showed that the moisture content of the solids significantly affects the temperature profile, the time for the solids to reach a given temperature is linearly dependent on the fill fraction, and the solids' temperature is most sensitive at lower temperatures where convection heat transfer is dominant. The model results showed that geometrical scaling is not sufficient, assuming a well-mixed bed is a reasonable approximation, and existing correlations for the wall-to-solids convection coefficient give useful bed temperature profiles. They found the characteristic time for convection-dominated heat transfer was:

$$\tau_{\text{conv}} = \frac{\rho c_p V}{h_{ws} A_{\text{conv}}}$$

(11.3)

where ρ is the density of the solids, c_p is the specific heat of the dry solids, V is the solids volume per unit length, h_{ws} is the convection heat transfer coefficient between the wall and the solids, and A_{conv} is the area for convection heat transfer between the wall and the solids. The characteristic time for radiation-dominated heat transfer was:

$$\tau_{\text{rad}} = \frac{\rho c_p V}{\sigma T_g^4} \left\{ \frac{1-\varepsilon_s}{\varepsilon_s A_s} + \frac{A_s\left(1-\varepsilon_g\right)+\varepsilon_g A_w}{\varepsilon_g A_s\left[A_s\left(1-\varepsilon_g\right)+A_w\right]} \right\}$$

(11.4)

where ε_s is the emissivity of the solids, T_g is the absolute temperature of the kiln gases, ε_g is the emissivity of the kiln gases, and A_w is the inner wall area excluding the area covered by the solids.

Kuo (1996) developed a simple lumped-parameter model to simulate a batch-fed solid waste incinerator.[89] The model could be used to study the response of an incinerator under a variety of conditions and could also be used to control batch-fed incinerators, which are often controlled by the skill and experience of the operators. One of the problems with batch-fed incinerators that are not steady-state processes is that there is a possibility of emitting puffs of unburned hydrocarbons

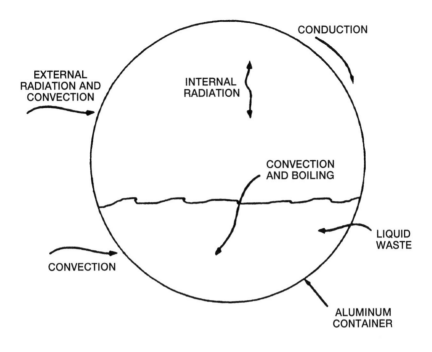

CONDUCTION

EXTERNAL
RADIATION AND
CONVECTION

INTERNAL
RADIATION

CONVECTION
AND BOILING

LIQUID
WASTE

CONVECTION

ALUMINUM
CONTAINER

(2) PHYSICAL SITUATION

FIGURE 11.31 Schematic of waste liquid incineration process. (Courtesy of ASME, New York.[90] With permission.)

due to changes in the waste feed. Therefore, control of the combustion air feed to the incinerator is important to get complete combustion. However, too much air reduces the combustion efficiency, which is important if the energy produced by the incinerator is recovered and used elsewhere, as is commonly the case. The challenge in batch-fed incineration processes is that the incoming waste composition is often in bags and unknown, so the amount of air required for combustion cannot be determined a priori. The model assumes uniform temperature, pressure, and composition inside the incinerator. Although the model is a fairly crude overall heat and material balance on the system, it is accurate and fast enough to be used in a control algorithm.

Bayley et al. (1996) developed a computer model to simulate the radiative, conductive, and convective heat transfer effects in a waste incineration furnace designed to destruct hazardous liquids.[90] The reason for the study was to determine if there would be a problem decontaminating the aluminum containers used to store the hazardous liquids. A schematic of the liquid-filled container and heating process is shown in Figure 11.31. It is known that a potential "vapor explosion" hazard exists when processing molten metals (especially aluminum) in the presence of volatile liquids. The convective heat transfer to the upper surface of the container was modeled using:

$$\text{Nu} = 0.029 \, \text{Re}^{4/5} \, \text{Pr}^{1/3} \tag{11.5}$$

The emissivity of the waste containers that had been in storage for some time was estimated to be 0.1 to 0.5. The analysis showed that the aluminum would probably melt before the liquid would be vaporized, which could lead to a vapor explosion, although the chances were not likely. The most favorable condition to minimizing the chance of an explosion was by using an external container emissivity of 0.1 and an internal emissivity of 0.5. Since these conditions cannot be guaranteed, further analysis was recommended before proceeding with the original plan to decontaminate the aluminum containers.

REFERENCES

1. G.L. Shires, Furnaces, in *International Encyclopedia of Heat and Mass Transfer*, G.F. Hewitt, G.L. Shires, and Y.V. Polezhaev, Eds., CRC Press, Boca Raton, FL, 1997, 493-497.

2. A. Lehrman, C.D. Blumenschein, D.J. Doran, and S.E. Stewart, Steel plant fuels and water requirements, in *The Making, Shaping and Treating of Steel*, 11th edition, Steelmaking and Refining Volume, R.J. Fruehan, Ed., AISE Steel Foundation, Pittsburgh, PA, 1998, 311-412.

3. R.J. Fruehan and C.L. Nassaralla, Alternative Oxygen Steelmaking Processes, in *The Making, Shaping and Treating of Steel*, 11th edition, Steelmaking and Refining Volume, R.J. Fruehan, Ed., AISE Steel Foundation, Pittsburgh, PA, 1998, 743-759.

4. M.D. Kistler and J.S. Becker, Ferrous metals, in *Oxygen-Enhanced Combustion*, C.E. Baukal, Ed., CRC Press, Boca Raton, FL, 1998, 165-180.

5. J.A.T. Jones, B. Bowman, and P.A. Lefrank, Electric furnace steelmaking, in *The Making, Shaping and Treating of Steel*, 11th edition, Steelmaking and Refining Volume, R.J. Fruehan, Ed., AISE Steel Foundation, Pittsburgh, PA, 1998, 525-660.

6. M.B. Wells, and F.A. Vonesh, Oxy-fuel burner technology for electric arc furnaces, *Iron & Steelmaker*, 13(11), 13-22, 1986.

7. H. Gripenberg, M. Brunner, and M. Petersson, Optimal distribution of oxygen in high efficiency arc furnaces, *Iron & Steel Engineer*, 67(7), 33-37, 1990.

8. D.L. Schroeder, Use of energies in electric steelmaking shops, *Electric Furnace Conf. Proceedings 49*, Iron & Steel Society, Warrendale, PA, 1991, 417-428.

9. K. Bergman and R. Gottardi, Design criteria for the modern UHP electric arc furnace with auxiliaries, *Ironmaking and Steelmaking*, 17(4), 282-287, 1990.

10. D.D. Battles and D.F. Knowles, New oxy-gas burner shows significant improvement in electric arc furnace productivity, *Proc. 1984 International Gas Research Conf.*, Govt. Institutes, Rockville, MD, 1985, 796-802.

11. S.J. Sikirica, Field testing of oxygen-gas burner system on foundry and specialty steel electric arc furnaces, *Proc. 1986 International Gas Research Conf.*, T.L. Cramer, Ed., Govt. Institutes, Rockville, MD, 1987, 852-855.

12. J.R. Davila, A. Garrote, A.M. Gutierrez, P. Carnicer, L. Cobos, and I. Erauskin, New applications of the oxygas rotary furnace as a means of production in melting, *Proc. of 1995 International Gas Research Conf.*, D.A. Dolenc, Ed., Govt. Institutes, Rockville, MD, 1996, 2764-2773.

13. D.H. Hubble, R.O. Russell, H.L. Vernon, and R.J. Marr, Steelmaking refractories, in *The Making, Shaping and Treating of Steel*, 11th edition, Steelmaking and Refining Volume, R.J. Fruehan, Ed., AISE Steel Foundation, Pittsburgh, PA, 1998, 227-290.

14. G.J.W. Kor and P.C. Glaws, Ladle refining and vacuum degassing, in *The Making, Shaping and Treating of Steel*, 11th edition, Steelmaking and Refining Volume, R.J. Fruehan, Ed., AISE Steel Foundation, Pittsburgh, PA, 1998, 661-713.

15. C.E. Tomazin, E.A. Upton, and R.A. Willis, *Steelmaking Conf. Proc.*, 69, Iron and Steel Society, Warrendale, PA, 1986, 223.

16. R. Klima, Improved knowledge of gas flow and heat transfer in reheating furnaces, *Scandinavian J. Metallurgy (Suppl.)*, 26, 25-32, 1997.

17. G.S. Koch and J.H. Williams, Characterization of natural gas-fired and alternative steel reheat technologies, in *Industrial Combustion Technologies*, M.A. Lukasiewicz, Ed., American Society of Metals, Warren, PA, 1986, 295-301.

18. K. Handa and Y. Tomita, Development of highly luminous gas burner and its application to save energy in the steel industry, *Proc. 1984 International Gas Research Conf.*, Govt. Institutes, Rockville, MD, 1985, 729-739.

19. L.T. Walsh, M. Ho, M.G. Ding, Demonstrated fuel savings and uniform heating with 100% oxygen burners in a continuous steel reheat furnace, in *Industrial Combustion Technologies*, M.A. Lukasiewicz, Ed., American Society of Metals, Warren, PA, 1986, 259-265.

20. K.S. Chapman, S. Ramadhyani, and R. Viskanta, Modeling and analysis of heat transfer in a direct-fired continuous reheating furnace, in *Heat Transfer in Combustion Systems*, N. Ashgriz, J.G. Quintiere, H.G. Semerjian, and S.E. Slezak, Eds., New York, ASME HTD-Vol. 122, 35-43, 1989.

21. Y.K. Lee, H.S. Park, and K.W. Cho, Effect of fuel gas preheating on combustion and heat transfer in reheating furnace, *Proceedings of the 13th Energy Engineering World Congress, Energy & Environmental Strategies for the 1990s,* Atlanta, GA, October 1990, 1991, chap. 78, 461-466.

22. K.S. Chapman, S. Ramadhyani, and R. Viskanta, Modeling and analysis of heat transfer in a direct-fired batch reheating furnace, in *Heat Transfer Phenomena in Radiation, Combustion and Fires*, R.K. Shah, Ed., ASME HTD-Vol. 106, 1989, 265-274.

23. J.M. Blanco and J.M. Sala, Improvement of the efficiency and working conditions for reheating furnaces through computational fluid dynamics, *Industrial Heating*, LXVI(5), 63-67, 1999.

24. F. Lisin and J.A. Marino, Heat transfer analysis in a batch steel reheat furnace, *Industrial Heating*, LXVI(10), 99-102, 1999.

25. C.D. Smith and D. Saha, *Advances in Oxy/fuel Combustion: International Wrought Copper Council Conference,* Singapore, October 1997.

26. M.E. Ward, D. Knowles, S.R. Davis, and J. Bohn, Effect of combustion air preheat on a forge furnace productivity, *Proc. 1984 International Gas Research Conf.*, Govt. Institutes, Rockville, MD, 1985, 763-776.

27. S.Y. Salama and T.M. Desai, Electric induction versus gas metal heating: a case study, *Proc. 1984 International Gas Research Conf.*, Govt. Institutes, Rockville, MD, 1985, 777-786.

28. S. Takamichi, Development of high performance forging furnaces, *Proc. 1998 International Gas Research Conf.*, Vol. V. Industrial Utilization and Power Generation, D. Dolenc, Ed., Govt. Institutes, Rockville, MD, 1998, 100-112.

29. D. Saha and C.E. Baukal, Nonferrous metals, in *Oxygen-Enhanced Combustion*, C.E. Baukal, Ed., CRC Press, Boca Raton, FL, 1998, 181-214.

30. E.R. Bazarian, J.F. Heffron, and C.E. Baukal, Method for Reducing NOx Production During Air-Fuel Combustion Processes, U.S. Patent 5,308,239 issued 03 May 1994.

31. M.W. Paget, J.F. Heffron, M. Lefebre, and C. Bazinet, A novel burner retrofit used to increase productivity in an aluminum rotary furnace and reduce baghouse loading, *TMS Conference,* 1994.

32. B. Phillips, D. Pakulski, M. Mazzei, E. Lepoutre, and H. Spoon, Use of air/oxy/fuel burners for aluminum dross processing, *Industrial Heating*, LX(3), 65-69, 1993.

33. D.J. Krichten, W.J. Baxter, and C.E. Baukal, Oxygen enhancement of burners for improved productivity, *EPD Congress 1997, Proceedings of the 1997 TMS Annual Meeting,* B. Mishra, Ed., February 9-13, Orlando, FL, 1997, 665-672.

34. L.D. Areaux and P. Corio, Aluminum scrap melting system provides high metal quality and melt yield, *Industrial Heating*, LXVI(8), 39-41, 1999.

35. C. Stala and D.L. Hindman, Gas-fired immersed ceramic tube aluminum melter, *Proc. 1989 International Gas Research Conf.*, T.L. Cramer, Ed., Govt. Institutes, Rockville, MD, 1990, 1453-1461.

36. M.E. Ward, R.E. Gildersleeve, S.J. Sikirica, and W.W. Liang, Application of a new ceramic recuperator technology for use on an aluminum remelt furnace, *Proc. 1986 International Gas Research Conf.*, T.L. Cramer, Ed., Govt. Institutes, Rockville, MD, 1987, 864-874.

37. R.A. Drake, Combustion Progress, Problems, Needs in the Glass Industry, in *Industrial Combustion Technologies*, M.A. Lukasiewicz, Ed., American Society of Metals, Warren, PA, 1986, 23-25.

38. R. Gardon, A review of radiant heat transfer in glass, *J. Amer. Ceramic Soc.*, 44(7), 305-312, 1961.

39. R.J. Reid, What you should know about combustion, *Glass Industry*, 70(7), 24-35, 1989.

40. V.K. Gegelashvili, V.V. Zhukovskii, and G.A. Svidzinskii, Separate admission of gas and air to a glass furnace, *Steklo Keram.*, No. 11, 38-39, 1968.

41. A.I. Kukarkin, V.E. Dunduchenko, V.S. Marinskii, et al., A rational way to burn fuel in glass tank furnaces, *Steklo Keram.*, No. 6, 4-5, 1972.

42. V.B. Kut'in, S.N. Gushchin, and V.G. Lisienko, Heat transfer in the cross-fired glass furnace, *Glass & Ceramics*, 54(5-6), 135-138, 1997.

43. V.G. Lisienko, V.B. Kut'in, and V.Ya. Dzyuzer, Influence of flame length and luminosity on heat transfer in glass tank furnaces, *Steklo Keram.*, No. 3, 6-8, 1981.

44. V.B. Kut'in, S.N. Gushchin, and V.G. Lisienko, Heat exchange in the cross-fired glass furnace, *Glass & Ceramics*, 54(5-6), 172-174, 1997.

45. J. Cassiano, M.V. Heitor, and T.F. Silva, Combustion tests on an industrial glass-melting furnace, *Fuel*, 73(10), 1638-1642, 1994.

46. P.B. Eleazer and B.C. Hoke, Glass, in *Oxygen-Enhanced Combustion*, C.E. Baukal, Ed., CRC Press, Boca Raton, FL, 1998, 215-236.

47. J.T. Brown, 100% Oxygen-Fuel Combustion for Glass Furnaces, *Collected Papers from the 51st Conference on Glass Problems*, 1990, 202-217.

48. A.G. Pincus, Ed., *Combustion Melting in the Glass Industry*, Magazines for Industry, New York, 1980.

49. P.I. Booker, Developments in the use of oxygen in glass furnace combustion systems, *Glass*, 59(5), 172-178, 1982.

50. D. Ertl and A. McMahon, Conversion of a fiberglass furnace from 100% electric to oxy/fuel combustion, *Proceedings from 54th Conference on Glass Problems*, Oct. 1993, 186-190.

51. S. Hope and S. Schemberg, Oxygen-fuel boosting on float furnaces, *International Glass Review*, Spring/Summer, 1997, 63-66.

52. M.G. Carvalho, D.F.G. Durão, and J.C.F. Pereira, Prediction of the flow, reaction and heat transfer in an oxy-fuel glass furnace, *Eng. Comput.*, 4(1), 23-34, 1987.

53. V.I. Henry, W.J. Horbatch, C.V. Ramachandran, and L.W. Donaldson, Oxygen enrichment of a float glass furnace. Phase I: System design and implementation, *Proc. 1986 International Gas Research Conf.*, T.L. Cramer, Ed., Govt. Institutes, Rockville, MD, 1987, 845-851.

54. D.E. Shamp and D.H. Davis, Application of 100% oxygen firing at Parkersburg, West Virginia, *Collected Papers from the 51st Conference on Glass Problems*, 1990, 218-239.

55. Slavejkov, A. G., Baukal, C. E., Joshi, M. L. and Nabors, J. K., Oxy/fuel glass melting with a high performance burner, *Ceram. Soc. Bull.*, 71(3), 340-343, 1992.

56. L.F. Westra and L.W. Donaldson, Development of an advanced glass melter system, *Proc. 1986 International Gas Research Conf.*, T.L. Cramer, Ed., Govt. Institutes, Rockville, MD, 1987, 889-897.

57. J. Hnat, D. Bender, L. Donaldson, A. Bendre, D. Tessari, J. Sacks, and A. Litka, Development of an advanced gas-fired glass melting furnace, *Proc. 1989 International Gas Research Conf.*, T.L. Cramer, Ed., Govt. Institutes, Rockville, MD, 1990, 1381-1390.

58. H.A. Abbasi, R.E. Grosman, L.W. Donaldson, C.F. Youssef, M.L. Joshi, and S.R. Hope, A low-NOx retrofit technology for regenerative glass melters, *Proc. 1995 International Gas Research Conf.*, Vol. II, D.A. Dolenc, Ed., Govt. Institutes, Rockville, MD, 1996, 2322-2331.

59. K.W. Pearce, Heat transfer from rotary kiln flames, in *Combustion Institute European Symposium 1973*, F.J. Weinberg, Ed., Academic Press, London, 1973, 663-668.

60. W. Richter and R. Quack, A mathematical model of a low-volatile pulverised fuel flame, in *Heat Transfer in Flames*, N.H. Afgan and J.M. Beer, Eds., Scripta Book Co., Washington, D.C., 1974, 95-109.

61. E.E. Khalil, Flow and combustion modeling of vertical lime kiln chambers, in *Industrial Combustion Technologies*, M.A. Lukasiewicz, Ed., American Society of Metals, Warrendale, PA, 1986, 99-107.

62. P.S. Ghoshdastidar and V.K. Anandan Unni, Heat transfer in the non-reacting zone of a cement rotary kiln, in *Heat Transfer Phenomena in Radiation, Combustion and Fires*, R.K. Shah, Ed., New York, ASME HTD-Vol. 106, 1989, 113-122.

63. M. Carvalho and M. Nogueira, Improvement of energy efficiency in glass-melting furnaces, cement kilns and baking ovens, *Appl. Therm. Eng.*, 17(8-10), 921-933, 1997.

64. C.E. Baukal, Oxygen-enhanced waste incineration, in *Encyclopedia of Environmental Analysis and Remediation*, R.A. Myers, Ed., Wiley, New York, 1998, 3283-3305.

65. C.E. Baukal, Waste incineration, in *Oxygen-Enhanced Combustion*, C.E. Baukal, Ed., CRC Press, Boca Raton, FL, 1998, 237-259.

66. A.M. Sterling, V.A. Cundy, T.W. Lester, A.N. Montestruc, J.S. Morse, C. Leger, and S. Acharya, *In situ* sampling from an industrial-scale rotary kiln incinerator, in *Emissions from Combustion Processes: Origin, Measurement, Control*, R. Clement and R. Kagel, Eds., CRC Press, Boca Raton, FL, 1990, 319-335.

67. ASME Research Committee on Industrial and Municipal Wastes, *Combustion Fundamentals for Waste Incineration*, ASME, New York, 1974.

68. W.R. Seeker, Waste combustion, *Twenty-Third Symposium (International) on Combustion*, The Combustion Institute, Pittsburgh, PA, 1990, 867-885.

69. J.O.L. Wendt, Combustion science for incineration technology, *Twenty-Fifth Symposium (International) on Combustion*, The Combustion Institute, Pittsburgh, PA, 1994, 277-289.

70. D.L. Vrable, Applications of high-temperature heat transfer — hazardous waste incineration, in *Heat Transfer in High Technology and Power Engineering*, W.-J. Yang and Y. Mori, Eds., Hemisphere, Washington, D.C., 1987.

71. D. Fusaro, Incineration technology: still hot, getting hotter, *Chemical Processing*, 54(6), 26-32, 1991.
72. S.D. Reese, Diverse experience using oxygen systems in waste incineration, presented at the *Fourth Annual National Symposium on Incineration of Industrial Wastes,* February 28–March 2, Houston, Texas, 1990.
73. A.K. Gupta, Thermal destruction of solid wastes, *J. Energy Resources Technology*, 118, 187-192, 1996.
74. G. Gitman, M. Zwecker, F. Kontz, and T. Wechsler, Oxygen enhancement of hazardous waste incineration with the PYRETRON thermal destruction system, *Thermal Processes*, Vol. 1, H. M. Freeman, Ed., Technomic Publishing, Lancaster, PA, 1990, 207-225.
75. M.G. Ding, The use of oxygen for hazardous waste incineration, *Thermal Processes*, Vol. 1, H. M. Freeman, Ed., Technomic Publishing, Lancaster, PA, 1990, 181-190.
76. G.H. Shahani, D. Bucci, D. DeVincentis, S. Goff, and M.B. Mucher, Intensify waste combustion with oxygen enrichment, *Chem. Eng.*, special supplement to 101(2), 18-24, 1994.
77. W.S. Strauss, J.A. Lukens, F.K. Young, and F.B. Bingham, Oxygen enrichment of combustion air in a 360 TPD mass burn refuse-fired waterwall furnace, *Proc. of 1988 National Waste Processing Conference*, 13th Bi-Annual Conference, Philadelphia, PA, May 1-4, 1988, 315-320, 1988.
78. CSI Resource Systems and Solid Waste Assoc. of North America, Evaluation of Oxygen-Enriched MSW/Sewage Sludge Co-Incineration Demonstration Program, U.S. Environmental Protection Agency Report EPA/600/R-94/145, Office of Research and Development, Cincinnati, OH, Sept. 1994.
79. G. Parkinson, Oxygen enrichment enhances sludge incineration, *Chem. Eng.*, 103(12), 25, 1996.
80. M.-D. Ho and M.G. Ding, Field testing and computer modeling of an oxygen combustion system, *J. Air Pollution & Waste Mgmt.*, 38(9), 1185-1191, 1988.
81. N.P. Chopey, The tops in chemical engineering achievement, *Chem. Eng.*, 96(12), 79-83, 1989.
82. C.E. Baukal, L.L. Schafer, and E.P. Papadelis, PCB cleanup using an oxygen/fuel-fired mobile incinerator, *Envir. Progress*, 13(3), 188-191, 1994.
83. C.R. Griffith, PCB and PCP destruction using oxygen in mobile incinerators, *Proceedings of the 1990 Incineration Conference,* San Diego, May 4-18, 1990.
84. P. Acharya, D. Fogo, and C. McBride, Process challenges in rotary kiln-based incinerators in soil remediation projects, *Envir. Progress*, 15(4), 267-276, 1996.
85. F.J. Romano and B.M. McLeod, The use of oxygen to reduce particulate emissions without reducing throughput, *Proceedings of 1990 Incineration Conference,* paper 3.3, San Diego, May 14-18, 1990, 589-596.
86. C.E. Baukal and F.J. Romano, Reducing NOx and particulates, *Pollution Eng.*, 24(15), 76-79, 1992.
87. K. Fouhy and G. Ondrey, Incineration: turning up the heat on hazardous waste, *Chem. Eng.*, 101(5), 39-43, 1994.
88. W.D. Owens, G.D. Silcox, J.S. Lighty, X.X. Deng, D.W. Pershing, V.A. Cundy, C.B. Leger, and A.L. Jakway, Thermal analysis of rotary kiln incineration: comparison of theory and experiment, *Comb. Flame*, 86, 101-114, 1991.
89. J.T. Kuo, System simulation and control of batch-fed solid waste incinerators, in *Heat Transfer in Fire and Combustion Systems — 1994*, W.W. Yuen and K.S. Ball, Eds., New York, ASME HTD-Vol. 272, 55-62, 1994.
90. S.E. Bayley, R.T. Bailey, and D.C. Smith, Heat transfer analysis of hazardous waste containers within a furnace, in Combustion and Fire, *ASME Proceedings of the 31st National Heat Transfer Conf.*, Vol. 6, M. McQuay, K. Annamalai, W. Schreiber, D. Choudhury, E. Bigzadek, and A. Runchal, Eds., ASME, New York, HTD-Vol. 328, 61-69, 1996.
91. Y. Kimura and K. Taniwaki, Development of immersion type aluminum melting furnace, *Proc. 1989 International Gas Research Conf.,* T.L. Cramer, Ed., Govt. Institutes, Rockville, MD, 1990, 1470–1480.

12 AdvancedCombustionSystems

12.1 INTRODUCTION

There are many types of so-called "advanced" combustion techniques. The term "advanced" is somewhat arbitrary, but here refers to technologies that are in use in industry (i.e., they are beyond the research and development stage), but are relatively new. The ones chosen for discussion here have a significant impact on the heat transfer from the flame to the load. This chapter describes technologies that are either in their infancy or are only used in a relatively narrow set of applications. In the future, the technologies described here may have broader appeal as advancements are made and as the technologies are accepted and implemented in industry.

12.2 OXYGEN-ENHANCED COMBUSTION

Most industrial heating processes require substantial amounts of energy, which are commonly generated by combusting hydrocarbon fuels such as natural gas or oil. Most combustion processes use air as the oxidant. In many cases, these processes can be enhanced using an oxidant that contains a higher proportion of O_2 than that in air. This is known as *oxygen-enhanced combustion*, or OEC.[1] Air consists of approximately 21% O_2 and 79% N_2 (by volume). One example of OEC is using an oxidant consisting of air blended with pure O_2. Another example is using high-purity O_2 as the oxidant — instead of air. This is usually referred to as *oxy/fuel* combustion.

New developments have made oxy/fuel combustion technology more amenable to a wide range of applications. In the past, the benefits of using oxygen did not always offset the added costs. New oxygen-generation technologies, such as pressure and vacuum swing adsorption, have substantially reduced the cost of separating O_2 from air. This has increased the number of applications where using oxygen to enhance performance is cost justified. Another important development is the increased emphasis on the environment. In many cases, OEC can substantially reduce pollutant emissions. This has also increased the number of cost-effective applications. The Gas Research Institute in Chicago, IL,[2] and the U.S. Dept. of Energy[3] have sponsored independent studies that predict that OEC will be a critical combustion technology in the very near future.

Historically, air/fuel combustion has been the conventional technology used in nearly all industrial heating processes. OEC systems are becoming more common in a variety of industries. When traditional air/fuel combustion systems have been modified for OEC, many benefits have been demonstrated. Typical improvements include increased thermal efficiency, increased processing rates, reduced flue gas volumes, and reduced pollutant emissions.

The use of oxygen in combustion has received relatively little attention from the academic combustion community. This may be for several reasons. Probably the most basic reason is the lack of research interest and funding to study OEC. The industrial gas companies that produce oxygen have been conducting research into OEC for many years; this has been mostly applied R&D. Very little basic research has been done, compared to air/fuel combustion, to study the fundamental processes in atmospheric flames utilizing OEC. The aerospace industry has done a considerable amount of work, for example, to study the high-pressure combustion of liquid oxygen and liquid hydrogen used to propel space vehicles. However, that work has little relevance to the low-pressure combustion of fuels other than hydrogen in industrial furnace applications. Another reason why minimal research has been done may be due to concerns about the safety issues of using oxygen, as well as the very high-temperature flames that may be encountered using OEC.

Another reason may be a cost issue since the small quantities of oxygen that might be used can be relatively expensive. Handling oxygen cylinders takes more effort than using either a houseline source of air or a small blower for the air used in small-scale flames.

Many industrial heating processes may be enhanced by replacing some or all of the air with high-purity oxygen.[1,4] Typical applications include metal heating and melting, glass melting, and calcining. In a report[5] done for the Gas Research Institute (1996), the following applications were identified as possible candidates for OEC:

- Processes that have high flue gas temperatures, typically in excess of 2000°F (1100°C)
- Processes that have low thermal efficiencies, typically due to heat transfer limitations
- Processes that have throughput limitations which could benefit from additional heat transfer without adversely affecting product quality
- Processes that have dirty flue gases, high NOx emissions, or flue gas volume limitations

When air is used as the oxidizer, only the O_2 is needed in the combustion process. By eliminating N_2 from the oxidizer, many benefits may be realized.

12.2.1 TYPICAL USE METHODS

Oxygen has been commonly used to enhance combustion processes in four primary ways: (1) adding O_2 into the incoming combustion air stream, (2) injecting O_2 into an air/fuel flame, (3) replacing the combustion air with high-purity O_2, and (4) separately providing combustion air and O_2 to the burner. These methods are discussed next.

12.2.1.1 Air Enrichment

Figure 12.1 shows an air/fuel process where the air is enriched with O_2. This may be referred to as low-level O_2 enrichment or premix enrichment. Many conventional air/fuel burners can be adapted for this technology.[6] The O_2 is injected into the incoming combustion air supply, usually through a diffuser to ensure adequate mixing. This is usually an inexpensive retrofit that can provide substantial benefits. Typically, the added O_2 will shorten and intensify the flame. However, there may be some concerns if too much O_2 is added to a burner designed for air/fuel. The flame shape may become unacceptably short. The higher flame temperature may damage the burner or burner block. The air piping may need to be modified for safety reasons to handle higher levels of O_2. One of the earliest experimental studies reported on oxygen enrichment of combustion air was done at the International Flame Research Foundation (IJmuiden, The Netherlands) on oil flames.[7] They also studied O_2 lancing under the flame, which is discussed next.

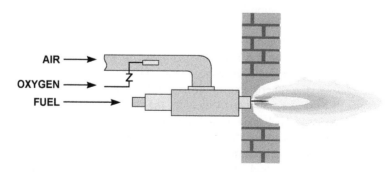

FIGURE 12.1   Schematic of premixing O_2 with air. (Courtesy of CRC Press, Boca Raton, FL.[1])

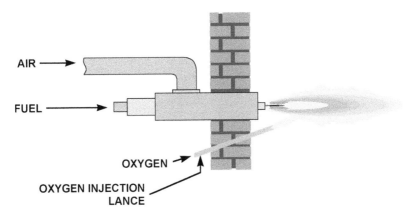

FIGURE 12.2   Schematic of O_2 lancing. (Courtesy of CRC Press, Boca Raton, FL.[1])

12.2.1.2 O_2 Lancing

Figure 12.2 shows another method for enriching an air/fuel process with O_2. As in the first method, this is also generally used for lower levels of O_2 enrichment. However, oxygen lancing may have several advantages over air enrichment. No modifications to the existing air/fuel burner need to be made. Typically, the NOx emissions are lower using O_2 lancing compared to premixing since this is a form of staging, which is a well-accepted technique for reducing NOx.[8] Depending on the injection location, the flame shape may be lengthened by staging the combustion reactions. The flame heat release is generally more evenly distributed than with premix O_2 enrichment. Under certain conditions, O_2 lancing between the flame and the load causes the flame to be pulled toward the material. This improves the heat transfer efficiency. Therefore, there is less likelihood of overheating the air/fuel burner, the burner block, and the refractory in the combustion chamber. Another variant of this staging method involves lancing O_2 not into the flame but somewhere else in the combustion chamber. One example of this technique is known as oxygen-enriched air staging, or OEAS. This O_2 lancing technology is an inexpensive retrofit for existing processes. One potential disadvantage is the cost to add another hole in the combustion chamber for the lance. This includes both the installation cost and the lost productivity. However, the hole is typically very small.

One specific embodiment of O_2 lancing is known as undershot enrichment, where O_2 is lanced into the flame from below. The lance is located between the burner and the material being heated. While air enrichment increases the flame temperature uniformly, the undershot technique selectively enriches the underside of the conventional flame, thereby concentrating extra heat downward toward the material being heated. While the mixing of oxygen and combustion air is not as complete with undershot oxygen as with premixing, this disadvantage is often outweighed by the more effective placement of the extra heat. Another benefit is that the refractory in the roof of the furnace generally receives less heat compared to air enrichment. This usually increases the life of the roof.

12.2.1.3   Oxy/Fuel

Figure 12.3 shows a third method using OEC, commonly referred to as oxy/fuel combustion. In nearly all cases, the fuel and the oxygen remain separated inside the burner. They do not mix until reaching the outlet of the burner. This is commonly referred to as a nozzle-mix burner (see Figure 1.12), which produces a diffusion flame. There is no premixing of the gases — for safety reasons. Because of the extremely high reactivity of pure O_2, there is the potential for an explosion if the gases are premixed. In this method, high-purity oxygen (>90% O_2 by volume) is used to combust the fuel. As discussed later, there are several ways of generating the O_2. In an oxy/fuel

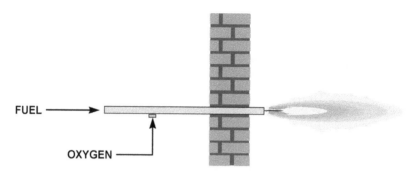

FIGURE 12.3   Schematic of oxy/fuel combustion. (Courtesy of CRC Press, Boca Raton, FL.[1])

system, the actual purity of the oxidizer will depend on which method has been chosen to generate the O_2. As shown later, oxy/fuel combustion has the greatest potential for improving a process, but it also may have the highest operating cost.

One specific variation of oxy/fuel combustion, known as "dilute oxygen combustion," is where fuel and oxygen are separately injected into the combustion chamber.[9] In order to ensure ignition, the chamber temperature must be above the auto-ignition temperature of the fuel. Depending on the exact geometry, this can produce an almost invisible flame, sometimes referred to as flameless oxidation. The advantage of this technique is very low NOx emissions because hotspots in the "flame" are minimized, which generally reduces NOx. A potential disadvantage, besides the safety concern, is a reduction in heat transfer as both the temperature and effective emissivity of the flame may be reduced.

12.2.1.4   Air-Oxy/Fuel

The fourth common method using OEC involves separately injecting air and O_2 through a burner, as shown in Figure 12.4. It is sometimes referred to as an air-oxy/fuel burner. This is a variation of the first three methods. In some cases, an existing air/fuel burner can be easily retrofitted by inserting an oxy/fuel burner through it (see Figure 11.16).[10] In other cases, a specially designed burner can be used.[11] This method of OEC can have several advantages. It can typically use higher levels of O_2 than the air enrichment and O_2 lancing methods, which yields higher benefits. However, the operating costs are less than for oxy/fuel, which uses very high levels of O_2. The flame shape and heat release pattern may be adjusted by controlling the amount of O_2 used in the process. It is also a generally inexpensive retrofit. Many air/fuel burners are designed for dual fuels, usually a liquid fuel like oil and a gaseous fuel like natural gas. The oil gun in the center of the dual-fuel burner can usually be easily removed and replaced by either an O_2 lance or an oxy/fuel burner.

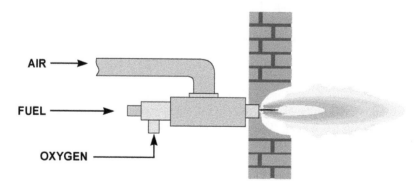

FIGURE 12.4   Schematic of air-oxy/fuel burner. (Courtesy of CRC Press, Boca Raton, FL.[1])

With this method of using OEC, the oxidizer composition can be specified in an alternate way. Instead of giving the overall O_2 concentration in the oxidizer, the oxidizer can be given as the fraction of the total oxidizer, which is air and the fraction of the total oxidizer that is pure O_2. The equivalent overall O_2 in the oxidizer can be calculated as follows:

$$\Omega = \frac{20.9}{0.209\left(\text{vol.}\% \ O_2\right)+\left(\text{vol.}\% \ \text{Air}\right)} \tag{12.1}$$

For example, the oxidizer may be specified as a blend of 60% O_2 and 40% air. The ratio of O_2 to air produces an equivalent of 39.8% overall O_2 in the oxidizer.

12.2.2   Operating Regimes

There are two common operating regimes for oxygen-enhanced combustion The first, or lower, regime is usually referred to as *low-level enrichment* ($\Omega < 0.30$). This is commonly used in retrofit applications where only a few modifications need to be made to the existing combustion equipment. It is used when only incremental benefits are required. For example, in many cases, the production rate in a heating process can be significantly increased even with only relatively small amounts of oxygen enrichment. In most cases, air/fuel burners can successfully operate up to about $\Omega = 0.28$ with no modifications.[12] For $\Omega > 0.28$, the flame may become unstable or the flame temperature may become too high for a burner designed to operate under air/fuel conditions. In some cases, it may be possible to make minor burner modifications to permit operation at slightly higher O_2 concentrations.

The other common operating regime is usually referred to as *high-level enrichment*, where high-purity oxygen ($\Omega > 0.90$) is used. This is used in higher temperature applications where the benefits of higher purity oxygen justify the added costs. The heating process is greatly intensified by the high-purity oxygen. In a retrofit situation, existing air/fuel burners are replaced by burners specifically designed to use the higher levels of O_2.

It has only been in the last decade that a significant number of combustion systems have been operated in the intermediate oxygen regime, or *medium-level enrichment* ($0.30 < \Omega < 0.90$). Again, these usually require specially designed burners or retrofits of existing burners.

12.2.3   Heat Transfer Benefits

12.2.3.1   Increased Productivity

In most high-temperature heating processes, flame radiation is the dominant mode of heat transfer. Radiation is dependent on the fourth power of the absolute temperature (see Chapter 3). The higher temperatures associated with OEC increase the radiation from the flame to the load.[13] This then increases the heat transfer to the load, which leads to increased material processing rates through the combustion chamber. Therefore, more material can be processed in an existing system, or new systems can be made smaller for a given processing rate. This is particularly important where space in a plant is limited. The environment benefits because of the reduction in the size of the equipment since less material and energy are needed in fabricating the combustion system. In order to take advantage of increased processing rates, the rest of the production system must be capable of handling the increased throughput. For example, the material handling system may need to be modified to handle the increased material flow rates.

The incremental costs of adding OEC to an existing combustion system are usually small compared to the capital costs of either expanding an existing system or adding new equipment in order to increase production. This has historically been one of the most popular reasons for using OEC. The advantage of OEC is that it can also be used intermittently to meet periodic demands for increased production. For example, if the demand for aluminum cans increases during the summer because of increased beverage consumption, OEC can be used at that time to meet the increased demand for aluminum. If the demand decreases in the winter, then the OEC system could be throttled back or turned off until it is needed again. In most cases, using OEC would be much more economical than adding new equipment for increased capacity if the increased production demands are only temporary.

12.2.3.2   Higher Thermal Efficiencies

By using oxygen — instead of air — more energy goes into the load instead of being wasted in heating up N_2. The energy needed to make O_2 from air is only a small fraction of the energy used in the combustion process. Therefore, the overall process uses less energy for a given amount of production due to the higher available heat. In some instances, the cost of the oxygen can be offset by the reduction in fuel costs because of the increase in energy efficiency. This is often the case when OEC is used to supplement a process that uses electricity which is generally more expensive than fossil fuel combustion. In some cases, it is possible to substitute a less-expensive source of energy for an existing source of energy. For example, in the glass industry, furnaces are fueled primarily by oil or natural gas and commonly "boosted" with electrical energy. It has been shown that some or all of the more-expensive electrical energy can be eliminated by the proper use of OEC.

One common reason for using OEC is a reduction in the specific fuel consumption where less fuel is required for a given unit of production because of the improvement in available heat. A recent study projects that, for example, the cost of natural gas is expected to rise by about 10% from 1993 to the year 2000.[4] At the same time, the cost of oxygen is expected to decrease by 10% due to lower electricity costs and improvements in oxygen production technology. The rising cost of fuel and the lowering cost of oxygen both make OEC a more attractive technology based solely on fuel savings.

Oxy/fuel combustion has been used in many applications in the steel industry, including both continuous and batch reheat furnaces, soaking pits, and ladle preheaters. Fuel savings of up to 60% have been reported.[14,15] Typical fuel savings achieved by converting from air/fuel to oxy/fuel combustion are given in Table 12.1.[16] Another example, from the glass industry, is that of Spectrum Glass, which reported fuel savings of 50% when it converted from air/fuel to oxy/fuel in its glass-melting furnaces.[17]

12.2.3.3   Higher Heat Transfer Efficiency

It has been argued that the efficiency of transferring heat from the flame to the load may be increased using OEC.[18] In a nonluminous flame, the flame emissivity is higher for an oxygen-enhanced flame, compared to an air/fuel flame. This is due to the higher concentrations of CO_2 and H_2O, which are the gases that radiate in a flame.[19] There is no radiation from the N_2 in the flame. These effects are discussed in Chapter 3. Javorka (1990) gave an example of an oxygen-enrichment combustion system used to improve the heat transfer in a brass-smelting furnace.[20] Farouk and Sidawi (1993) numerically showed that the radiative heat flux from oxy/natural gas flames is about an order of magnitude higher than for air/natural gas flames at the same firing rate.[21]

12.2.3.4   Increased Flexibility

There are many other benefits that can be achieved and are specific to a given process. OEC can increase the flexibility of a heating system.[22] In some cases, a wider range of materials can be processed with OEC compared to air/fuel combustion. In other cases, OEC may be required if very high melting temperatures are required. For example, some ceramic and refractory products require firing temperatures of 2900°F (1600°C) and higher.[23] Such temperatures are difficult if not impossible to achieve with standard air/fuel combustion with no air preheating. A heating system may

TABLE 12.1
Industrial Applications of Oxygen-Enhanced Combustion

Industry	Furnaces/Kilns	Primary Benefits[a]
Aluminum	Remelting	1, 2
	Coke calcining	1
Cement	Calcining	1
Chemical	Incineration	1, 2, 3, 4
Clay	Brick firing	1, 2, 3
Copper	Smelting	1, 2, 3
	Anode	2
Glass	Regenerative melters	1, 2, 4
	Unit melters	1, 2, 4
	Day tanks	1, 2, 4
Iron and steel	Soaking pits	2, 1
	Reheat furnaces	2, 1
	Ladle preheat	1
	Electric arc melters	1, 2
	Forging furnaces	1, 2
Petroleum	FCC regenerator	1
	Claus sulfur	1
Pulp and paper	Lime kilns	1, 2, 3
	Black liquor	1, 2

[a] Benefits of oxygen: 1 = productivity improvement, 2 = energy savings, 3 = quality improvement, 4 = emissions reduction.

Source: Courtesy of CRC Press, Boca Raton, FL.[1]

also be brought up to operating conditions more quickly with OEC compared to air/fuel systems because of the higher heating intensity. For example, it has been shown that using OEC in metal reheat furnaces can substantially improve the ability to start up and shut down quickly.[24]

A combustion process can react more quickly to changes because of the higher heating intensity. This reduces the time, for example, that it takes to change processing rates or to change the product mix. OEC can give tighter control of the heating profile because of the higher intensity.

12.2.4 POTENTIAL HEAT TRANSFER PROBLEMS

There are potential problems associated with the use of oxygen-enhanced combustion if the system is not properly designed. Many of the potential problems can be generally attributed to the increased combustion intensity.

12.2.4.1 Refractory Damage

As previously shown, oxygen-enhanced flames generally have significantly higher flame temperatures compared to conventional air/fuel flames. If the heat is not properly distributed, the intensified radiant output from the flame can cause refractory damage. Today's OEC burners are designed for uniform heat distribution to avoid overheating the refractory surrounding the burner. The burners normally are mounted in a refractory burner block, which is then mounted into the combustion chamber. The burner blocks are made of advanced refractory materials, such as zirconia or alumina, and are designed for long, maintenance-free operation.

If the burner position and firing rate in the furnace are improperly chosen, refractory damage can result. For example, if the flame from an OEC burner is allowed to impinge directly on the wall of a furnace, most typical refractory materials would be damaged. This can be prevented by the proper choice of the burner design and positioning. The flame length should not be so long that it impinges on the opposite wall. The burner mounting position in the furnace should be chosen to avoid aiming the flame directly at furnace refractories.

12.2.4.2   Nonuniform Heating

This is an important concern when retrofitting existing systems that were originally designed for air/fuel combustion. By intensifying the combustion process with OEC, there is the possibility of adversely affecting the heat and mass transfer characteristics within the combustion chamber. These issues are only briefly discussed here.

12.2.4.2.1   Hotspots

OEC normally increases the flame temperature, which also increases the radiant heat flux from the flame to the load. If the increased radiant output is very localized, then there is the possibility of producing hotspots on the load. This could lead to overheating, which may damage or degrade the product quality. Today, burners for OEC have been specifically designed to avoid this problem. More information on burner design issues is given in Chapter 8.

12.2.4.2.2   Reduction in convection

As shown in Figure 2.23, the total volume flow rate of exhaust products is significantly reduced using OEC. However, the average gas temperature is usually higher, but not by enough to offset the reduced gas flow rate. The convective heat transfer from the exhaust gases to the load may be reduced as a result. Another important aspect of convection is mass transfer. In some heating processes, especially those related to drying or removing volatiles, the reduced gas flow rate in the combustion chamber could adversely affect the mass transfer process. This can be offset using a burner that incorporates furnace gas recirculation, which increases the bulk volume flow inside the combustion chamber to help in removing volatiles that evolve from the load during the heating process.

12.2.5   Industrial Heating Applications

Oxygen-enhanced combustion is used in a wide range of industrial heating applications. In general, OEC has been used in high-temperature heating and melting processes that are either very inefficient or not possible with air/fuel combustion. Table 12.1 shows some of the common reasons why OEC has been used in a variety of industrial applications.[25] This chapter section is only intended to give a brief introduction to those applications that are broadly categorized here as metals, minerals, incineration, and other. These applications are discussed in detail elsewhere in the book.

12.2.5.1   Metals

Heating and melting metal was one of the first industrial uses of oxygen-enhanced combustion and continues to be an important application today. OEC has been widely used in both large integrated steel mills as well as in smaller mini-mills. It has also been used in the production of non-ferrous metals such as aluminum, brass, copper, and lead. Some of these applications are discussed in Chapter 11.

12.2.5.2   Minerals

Here, minerals refer to glass, cement, lime, bricks, and other related materials that require high-temperature heating and melting during their manufacture. OEC has been used in all of those applications. The use of OEC in glass is discussed in Chapter 11.

12.2.5.3   Incineration

This is a relatively new area for OEC. Initially, OEC was used to enhance the performance of portable incinerators used to clean up contaminated soil. It has also been used in municipal solid waste incinerators and in boilers burning waste fuels. The use of OEC in incineration is discussed in Chapter 11.

12.2.5.4   Other

OEC has been used in a wide variety of specialty applications that are not discussed in detail in this book. Some of these include gasifying organic materials and vitrifying residual ash,[26] removing unburned carbon from fly ash,[27] oxygen enrichment of fluid catalytic crackers,[28] and oxygen enrichment of combustion air to combust H_2S in sulfury recovery units (SRUs) to increase throughput.[29]

12.3   SUBMERGED COMBUSTION

Many types of submerged combustion processes have been tried, although it has been difficult to commercialize many of these due to the inherent problems with this type of process. Submerged combustion involves locating a burner in the bottom of a vessel filled with solids or liquids to be heated. The combustion products travel through the material and liberate their heat in a very efficient manner, which is usually the primary objective of this technique. Typical problems include finding the proper balance of flame intensity and gas velocity high enough to keep the material from flowing down into the burner, yet low enough that the gases do not bore a hole through the materials without effectively transferring their heat. A more mundane yet important problem is igniting the burner located underneath a bath or pile of material.

12.3.1   Metals Production

One example of a submerged combustion process was designed and tested by Air Products and Chemicals, Inc. (Allentown, PA) with partial funding from the Gas Research Institute (Chicago, IL).[30] The process was designed to use natural gas combusted with high-purity oxygen in a burner located at the bottom of a metal heating vessel. The objective of the process was to more economically melt scrap metal using a fossil fuel by partially or fully replacing the electrical energy used in electric arc furnaces (EAF). The biggest challenge of the project was to design a submerged burner to operate over the entire regime from a packed bed of scrap metal to a full bath of molten metal. This was successfully achieved with a high-velocity burner design concept.[31,32] As part of the research program, extensive testing was done using an oxy/fuel burner, firing vertically upward, mounted at the bottom of a 7.5-ton transfer ladle filled with scrap metal.[33] A schematic of the system is shown in Figure 12.5. The scrap was hand-loaded into the vessel and included punchings, miscellaneous heavy scrap, and scrap containing instrumentation. The overall height of the load was 50 in. (130 cm) and the total weight was 2910 lb (1320 kg). Gas and scrap metal temperatures were measured at 15 different locations in the load at five different heights and three different radii. The firing rates ranged from 0.5 to 4×10^6 Btu/hr (0.15 to 1.2 MW). Using the measured gas and scrap metal temperatures, the convective heat transfer coefficient ranged from 5 to 20 Btu/hr-ft²-°F (30 to 110 W/m²-K). The testing demonstrated the feasibility of using a submerged oxy/fuel burner to heat and melt scrap metal.

Another example of submerged combustion heating is in a pickling operation used for the surface treatment of metallic products. Chouinard and Bocherel (1990) described a submerged natural gas burner used to heat a tank containing a 15% sulfuric acid solution for pickling steel wire coils.[34] A sketch of the system is shown in Figure 12.6. The exhaust products are distributed in a perforated tube array so they can bubble through the acid. Energy savings of up to 73%

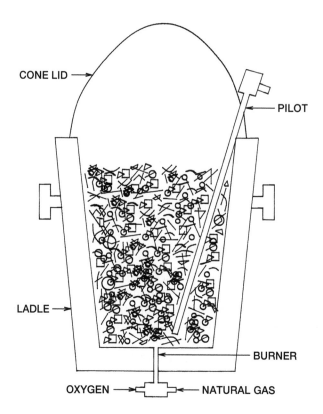

FIGURE 12.5   Submerged combustion steelmaking system experimental setup.

compared to conventional immersed steam coil heating techniques were demonstrated. This is due to the increased heat transfer effectiveness of bubbling the hot gases through the liquid pickling solution.

One type of vessel used to heat and melt in the metals industry is known as a shaft furnace. In that furnace, raw materials to be heated and melted are fed into the top of a tall refractory-lined vessel. Burners are located in the bottom of the vessel so that, similar to the process described above, the combustion products travel up through the packed bed of solids. However, one key distinction is that this process is continuous, with the melted raw materials being tapped off at the bottom of the shaft furnace. The process described previously is a batch process where the molten materials must be periodically tapped off by tipping the vessel on its side and pouring them out and then refilling the vessel with a new charge of raw scrap materials. In the shaft furnace, raw materials are added to the top on at least a semicontinuous, if not continuous, basis. One such shaft furnace is under development at the Institute of Gas Technology.[35] Their process also incorporates heat recovery technology, where the combustion products leaving the top of the shaft are recycled down toward the burners where the heat is recovered and used to preheat the incoming combustion air. Advantages of this technology include uniform heat rates and high thermal efficiencies due to the counterflow heating of the packed bed and the heat recovery used to preheat the incoming combustion air.

Tatsumi (1996) described an immersion burner used to melt non-ferrous metals.[36] The burner shown in Figure 12.7 is essentially a single-ended ceramic radiant tube. This can be used in applications like dip galvanizing, wire-rope galvanizing, and aluminum die casting. This new technology claims a 60% improvement over conventional nonsubmerged heating methods.

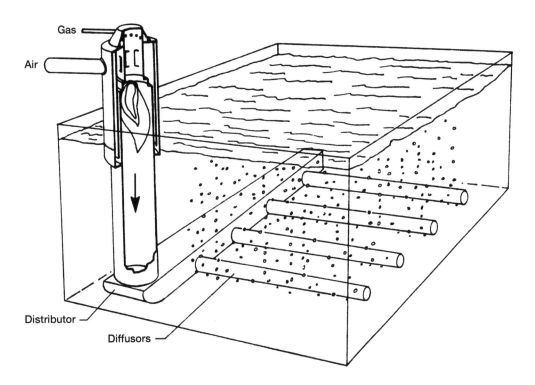

FIGURE 12.6   Submerged combustion acid heating system. (Courtesy of Government Institutes, Rockville, MD.[34] With permissiuon.)

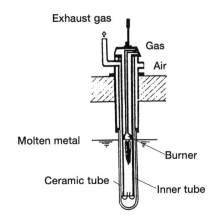

FIGURE 12.7   Ceramic gas-fired immersion burner for heating non-ferrous metal. (Courtesy of Government Institutes, Rockville, MD.[36] With permissiuon.)

12.3.2   MINERALS PRODUCTION

Olabin et al. (1998) described the use of a submerged combustion melting (SCM) process for use in producing mineral products.[37] SCM was developed by the National Academy of Sciences of Ukraine and is now being extended to other applications, including the melting of cement kiln dust mixtures and sodium silicate using air/fuel and oxy/fuel firing. The SCM furnace is a bubbling type that is claimed to have better efficiency and lower capital costs compared to conventional melting furnaces. Burners are fired under a molten bath so that the gaseous combustion products bubble through the melt, which helps homogenize the mix. The high effective surface area promotes high heat transfer rates. The process can be run continuously or in a batch mode. The main technical challenge was establishing stable combustion, which was overcome using special stabilizers, multiple smaller burner jets, and preheating the fuel/oxidant mixture. The intense bubbling fluidizes about a quarter of the melt above the normal melt height. This would normally erode the refractory very quickly. That problem was overcome by using water-cooled walls which causes a layer of melt material to freeze on the walls, to act as a protective layer. For silica melts, heat flux rates of 20,000 to 60,000 kcal/m^2-hr (7400 to 22,000 Btu/hr-ft^2) were reported. System thermal efficiencies for mineral wool SCM furnaces ranged from 19% to 65%.

12.3.3 LIQUID HEATING

Guy et al. (1987) presented a computer model that included the heat and mass transfer processes in a submerged combustion system for heating and evaporating liquids.[38] Important advantages of submerged combustion in liquids include very high heating efficiencies due to high transfer coefficients and transfer area, efficient mixing of the liquid caused by the bubbling of the gas through the liquid, and the absence of metallic surfaces in the bath, which is important where the liquid is corrosive. Potential drawbacks include undesired evaporation of the liquid in some applications and chemical reactions from the combustion process that can contaminate the liquid. Experimental measurements for the heating time and evaporation rate were 20% to 30% lower than the computed results.

12.4 MISCELLANEOUS

At the 1998 IGRC Conference, an entire session of the program was devoted to "Novel Combustion Techniques," all of which use natural gas as the fuel. Some of those related to improved heat transfer are briefly discussed next. Adelt (1998) reported on the use of a gas-fired ceramic immersion tube (comparable to a single-ended radiant tube burner) for heating molten aluminum in a holding furnace that is normally heated electrically.[39] A 53% reduction in operating expenses was reported. Neff et al. (1998) described a high-luminosity oxy/gas burner system designed to produce higher efficiencies and lower pollutant emissions than comparable technologies.[40] Figure 12.8 depicts a novel high-luminosity burner that transfers more heat to a load than a conventional burner. Flamme et al. (1998) showed that improved ceramic heat exchangers for use in recuperative burners can significantly improve the thermal efficiencies in high-temperature industrial heating processes.[41]

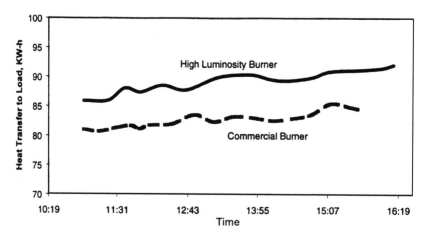

FIGURE 12.8   Heat transfer to the furnace load for a high-luminosity burner and a commercial burner. (Courtesy of Gas Research Institute, Chicago, IL.[40] With permissiuon.)

12.4.1 SURFACE COMBUSTOR-HEATER

In this technology, tubes are embedded inside a packed bed of ceramic materials.[42] Flame gases flow through the packed bed and transfer heat to the ceramic packing and to the embedded tubes. Some type of fluid, such as water or petroleum products, flows through the tubes and is heated. This process can be very thermally efficient because the products of combustion transfer a high proportion of their energy to either the packed bed or the embedded tubes and exit the combustor at lower temperatures than other combustion processes. Although the goal is to transfer heat to the fluid flowing through the tubes and not to the ceramic packing, the ceramic packing transfers its heat to the tubes by conduction so that most of the energy does eventually go into the fluid. The combustion intensity is claimed to be 10 times higher than conventional gas burners. The Institute of Gas Technology has demonstrated high heat transfer rates (100,000 Btu/hr-ft^2 or 315 kW/m^2) for this technology.[43] Important benefits of the technology are more uniform heating due to the porous bed and lower pollutant emissions.

12.4.2 DIRECT-FIRED CYLINDER DRYER

As discussed in Chapter 10, steam-heated cylinders are commonly used to dry paper and paper products. Important limitations of steam cylinders include fairly low thermal efficiencies and low drying temperatures, which are limited to below about 360°F (180°C). Gas-fired infrared burners have been used to supplement these cylinders to increase line speeds and improve moisture profiles across the paper webs. Somervell (1996) described a new technique for firing a linear ribbon burner inside a steel cylinder for use in paper drying.[44] Figure 12.9 shows how the amount of moisture removed from paper during the drying process increases dramatically as the cylinder temperature increases. Lab-scale testing showed cylinder temperatures up to 200°C (390°F). Efficiencies in the range of 70% to 75% were claimed, in comparison to efficiencies rarely exceeding 65% for conventional steam-heated cylinders.

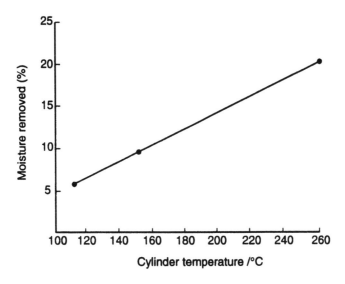

FIGURE 12.9   Moisture removed as a function of the cylinder temperature in paper drying. (Courtesy of Government Institutes, Rockville, MD.[44] With permissiuon.)

REFERENCES

1. C.E. Baukal Ed., *Oxygen-Enhanced Combustion*, CRC Press, Boca Raton, FL, 1998.
2. S.J. Williams, L.A. Cuervo, and M.A. Chapman, High-Temperature Industrial Process Heating: Oxygen-Gas Combustion and Plasma Heating Systems, Gas Research Institute Report GRI-89/0256, Chicago, IL, July 1989.
3. A.S. Chace, H.R. Hazard, A. Levy, A.C. Thekdi, and E.W. Ungar, Combustion Research Opportunities for Industrial Applications — Phase II, U.S. Dept. of Energy Report DOE/ID-10204-2, Washington, D.C., 1989.
4. C.E. Baukal, P.B. Eleazer, and L.K. Farmer, Basis for enhancing combustion by oxygen enrichment, *Industrial Heating*, LIX(2), 22-24, 1992.
5. K.R. Benedek and R.P. Wilson, The Competitive Position of Natural-Gas in Oxy-Fuel Burner Applications, Gas Research Institute Report no. GRI-96-0350, Chicago, IL, September 1996.
6. S.V. Joshi, J.S. Becker, and G.C. Lytle, Effects of oxygen enrichment on the performance of air-fuel burners, in *Industrial Combustion Technologies*, M.A. Lukasiewicz, Ed., Amer. Soc. Metals, 165-170, 1986.
7. R.R. Kissel and M. Michaud, International Flame Research Foundation: first experiments at IJmuiden on the combustion of oil using oxygen, *J. Inst. Fuel*, March, 1962, 109-120.
8. U.S. Environmental Protection Agency, Alternative Control Techniques Document — NOx Emissions from Utility Boilers, EPA Report EPA-453/R-94-023, Research Triangle Park, NC, 1994.
9. H. Kobayashi, Segregated Zoning Combustion, U.S. Patent 5,076,779, issued December 31, 1991.
10. E.R. Bazarian, J.F. Heffron, and C.E. Baukal, Method for Reducing NOx Production During Air-Fuel Combustion Processes, U.S. Patent 5,308,239, 1994.
11. G.M. Gitman, Method and Apparatus for Generating Highly Luminous Flame, U.S. Patent 4,797,087, 1989.
12. A.I. Dalton and D.W. Tyndall, Oxygen Enriched Air/Natural Gas Burner System Development, NTIS Report PB91-167510, Springfield, VA, 1989.
13. B.N. Gibbs and A. Williams, Fundamental aspects on the use of oxygen in combustion process — a review, *J. Inst. Energy*, 56(427), 74-83, 1983.
14. L.M. Farrell, T.T. Pavlack, and L. Rich, Operational and environmental benefits of oxy-fuel combustion in the steel industry, *Iron & Steel Engineer*, 72(3), 35-42, 1995.
15. L.K. Klingensmith, Direct-fired melter performance improved by gas/oxygen firing, *Glass Industry*, 67(4), 14–18, 1986.
16. M.G. Ding and Z. Du, Energy and Environmental Benefits of Oxy-Fuel Combustion, *Proc. of 1995 International Conference on Energy & Environment*, Shanghai, China, May 1995, Begell House Inc., New York, 1995, 674-684.
17. S. Grisham, Oxy-conversion for a smaller furnace yields big results, *American Glass Review*, 117(6), 12, 1997.
18. H. Kobayashi, Oxygen Enriched Combustion System Performance Study, Phase I Interim/Final Report, Vol. I. Technical and Economic Analysis, U.S. Dept. of Energy Report DOE/ID/12597, Washington, D.C., March 1987.
19. H.C. Hottel and A.F. Sarofim, *Radiative Transfer*, McGraw-Hill, New York, 1967.
20. D.J. Javorka, Brass smelter efficiency improved with new air-oxygen-fuel combustion system, *Industrial Heating*, 57(5), 36-38, 1990.
21. B. Farouk and M.M. Sidawi, Effects of nitrogen removal in a natural gas fired industrial furnace: a three dimensional study, in *Heat Transfer in Fire and Combustion Systems — 1993*, B. Farouk, M. Pinar Menguc, R. Viskanta, C. Presser, and S. Chellaiah, Eds., New York, ASME HTD-Vol. 250, 173-183, 1993.
22. M. DeLucia, Oxygen enrichment in combustion processes: comparative experimental results from several application fields, *J. Energy Resources Tech.*, 113, 122-126, 1991.
23. Industrial Heating Equipment Association, *Combustion Technology Manual*, fifth edition, Arlington, VA, 1994.
24. J. S. Becker and L.K. Farmer, Rapid fire heating system uses oxy-gas burners for efficient metal heating, *Industrial Heating*, 62(3), 74-78, 1995.
25. R.J. Reed, *North American Combustion Handbook*, third edition, 1997, Vol. II, Part 13.

26. N.G. Bishop and G. Taylor, Method and Apparatus for Gasifying Organic Materials and Vitrifying Residual Ash, U.S. Patent 5,584,255, issued Dec. 17, 1996.
27. M.P. Martinez, Apparatus and Process for Removing Unburned Carbon in Fly Ash, U.S. Patent 5,555,821, issued Sept. 17, 1996.
28. S. Tamhankar, R. Menon, T. Chou, R. Ramachandran, R. Hull, and R. Watson, Enrichment can decrease NOx, SOx formation, *Oil & Gas J.*, 94(10), 60-68, 1996.
29. G. Parkinson, R. D'Aquino, and G. Ondrey, O_2 Breathes New Life into Processes, *Chem. Eng.*, 106(10), 28-31, 1999.
30. M.B. DeGregorio and M.D. Lanyi, Development of a Submerged Combustion Steelmaking System: Phases I and II, Gas Research Institute Report GRI-91/0067, Chicago, 1991.
31. D.C. Winchester, L.K. Farmer, M.D. Lanyi, and C.E. Baukal, Accretion Controlling Tuyere, U.S. Patent 5,431,709 issued 11 July 1995.
32. D.C. Winchester, L.K. Farmer, M.D. Lanyi, and C.E. Baukal, Accretion Controlling Tuyere, U.S. Patent 5,458,320 issued 17 October 1995.
33. C.E. Baukal, M.D. Lanyi, and D.C. Winchester, Scrap preheating with a submerged oxy-fuel burner, *Iron & Steelmaking*, 17(10), 51-56, 1990.
34. J.-G. Chouinard and P. Bocherel, Submerged combustion system for acid tank heating, *Proc. 1989 International Gas Research Conf.*, T.L. Cramer, Ed., Govt. Institutes, Rockville, MD, 1990, 1481-1488.
35. M. J. Khinkis, Vertical Shaft Melting Furnace and Method of Melting, U.S. patent 4,877,449, issued 31 October 1989.
36. N. Tatsumi, Development of non-ferrous metal melting technique by gas-fired ceramic immersion heater, *Proc. of 1995 International Gas Research Conf.*, D.A. Dolenc, Ed., Govt. Institutes, Rockville, MD, 1996, 2351-2358.
37. V. Olabin, A. Maximuk, D. Rue, and W. Kunc, Development of submerged combustion technology for producing mineral melts, *Proc. of 1998 International Gas Research Conf.*, Vol. V. Industrial Utilization, D.A. Dolenc, Ed., Gas Research Institute, Chicago, 1998, 685-695.
38. C. Guy, P.J. Carreau, and J. Paris, Heating and evaporation of liquids by submerged combustion, *Proc. 1986 International Gas Research Conf.*, T.L. Cramer, Ed., Govt. Institutes, Rockville, MD, 1987, 903-912.
39. M. Adelt, Aluminum holding with ceramic immersion tubes, *Proc. of 1998 International Gas Research Conf.*, Vol. V: Industrial Utilization, D.A. Dolenc, Ed., Gas Research Institute, Chicago, 1998, 15-23.
40. D. Neff, P. Mohr, J. Gaddone, A. Fridman, S. Nestor, R. Viskanta, R. Jain, D. Rue, and O. Loo, High-luminosity, low-NOx oxy-natural gas burner for high-temperature furnaces, *Proc. of 1998 International Gas Research Conf.*, Vol. V: Industrial Utilization, D.A. Dolenc, Ed., Gas Research Institute, Chicago, 1998, 34-44.
41. M. Flamme, M. Boß, M. Brune, A. Lynen, J. Heym, J.A. Wünning, J.G. Wünning, and H.J. Dittman, Improvement of energy saving with new ceramic self-recuperative burners, *Proc. of 1998 International Gas Research Conf.*, Vol. V: Industrial Utilization, D.A. Dolenc, Ed., Gas Research Institute, Chicago, 1998, 88-99.
42. M. J. Khinkis, Gas-Fired, Porous Matrix, Surface Combustor-Fluid Heater, U.S. patent 5,375,563, issued 27 December 1994.
43. Institute of Gas Technology, Compact Ultra-Low-Emission Process Heater, brochure, Chicago, IL, 1998.
44. T.R. Somervell, Experimental evaluation of direct gas fired drying cylinder performance, *Proc. of 1995 International Gas Research Conf.*, D.A. Dolenc, Ed., Govt. Institutes, Rockville, MD, 1996, 2359-2368.

Appendix A

Reference Sources for Further Information

The list below is not intended to be exhaustive as there are other journals that may occasionally have papers on industrial combustion, but the ones listed here are more likely to have such articles.

1. Heat transfer journals

    Applied Thermal Engineering
    Calore é Tecnologica (Heat and Technology)
    Experimental Heat Transfer
    Experimental Thermal and Fluid Science
    Heat Transfer — Japanese Research
    Heat Transfer — Russian Research
    International Communications in Heat and Mass Transfer
    International Journal of Heat and Fluid Flow
    International Journal of Heat and Mass Transfer
    Journal of Heat Transfer
    Journal of Thermophysics and Heat Transfer
    Wärme und Stoffübertragung (Heat and Mass Transfer)

2. Combustion journals

    Combustion and Flame
    Combustion Science and Technology
    Fuel (formerly *Journal of the Institute of Fuel*)
    Journal of the Institute of Energy
    Progress in Energy and Combustion Science

3. Industrial trade journals

    Ceramic Industry
    Glass
    Glass Industry
    Glass Technology
    Hydrocarbon Processing
    Industrial Heating
    Light Metals
    Oil & Gas Journal
    Process Heating

4. Miscellaneous sources

American Flame Research Committee conference proceedings
American Society of Mechanical Engineers (ASME) combustion conference proceedings (New York, NY)
Gas Research Institute reports (Chicago, IL)
International Flame Research Foundation reports (IJmuiden, the Netherlands)
International (Symposia) on Combustion proceedings (The Combustion Institute, Pittsburgh, PA)
U.S. Dept. of Energy reports (Washington, D.C.)
U.S. Patent & Trademark Office (Washington, D.C.)

Appendix B

Common Conversions

1 Btu =	252.0 cal		1 in. =	2.540 cm
	1055 J			25.40 mm
1 Btu/ft^3 =	0.00890 cal/cm^3		1 J =	0.000948 Btu
	0.0373 MJ/m^3			0.239 cal
1 Btu/hr =	0.0003931 hp			1 W/sec
	0.2520 kcal/hr		1 kcal =	3.968 Btu
	0.2931 W			1000 cal
1,000,000 Btu/hr =	0.293 MW			4187 J
1 Btu/hr-ft^2 =	0.003153 kW/m^2		1 kcal/hr =	3.968 Btu/hr
1 Btu/hr-ft-°F =	1.730 W/m-K			1.162 J/sec
1 Btu/hr-ft^2-°F =	5.67 W/m^2-K		1 kcal/m^3 =	0.1124 Btu/ft^3
1 Btu/lb =	0.5556 cal/g			4187 J/m^3
	2326 J/kg		1 kg =	2.205 lb
1 Btu/lb-°F =	1 cal/g-°C		1 kg/hr-m =	0.00278 g/sec-cm
	4187 J/kg-K			0.672 lb/hr-ft
1 cal =	0.003968 Btu		1 kg/m^3 =	0.06243 lb/ft^3
	4.187 J		1 kW =	3413 Btu/hr
1 cal/cm^2-sec =	3.687 Btu/ft^2-sec			1.341 hp
	41.87 kW/m^2			660.6 kcal/hr
1 cal/cm-sec-°C =	241.9 Btu/ft-hr-°F		1 kW/m^2 =	317.2 Btu/hr-ft^2
	418.7 W/m-K		1 kW/m^2-°C =	176.2 Btu/hr-ft^2-°F
1 cal/g =	1.80 Btu/lb		1 lb =	0.4536 kg
	4187 J/kg		1 lb/ft^3 =	0.0160 g/cm^3
1 cal/g-°C =	1 Btu/lb-°F			16.02 kg/m^3
	4187 J/kg-K		1 lbm/hr-ft =	0.413 centipoise
1 centipoise =	2.421 lbm/hr-ft		1 m =	3.281 ft
1 cm^2/sec =	100 centistokes		1 mm =	0.03937 in.
	3.874 ft^2/hr		1 m^2/sec =	10.76 ft^2/sec
1 ft =	0.3048 m		1 mton =	1000 kg
1 ft^2/sec =	0.0929 m^2/sec			2205 lb
1 g/cm^3 =	1000 kg/m^3		1 MW =	3,413,000 Btu/hr
	62.43 lb/ft^3			1000 kW
	0.03613 lb/in.^{3}		1 therm =	100,000 Btu
1 hp =	33,000 ft-lb/min		1 W =	1 J/sec
	550 ft-lb/sec		1 W/m-K =	0.5778 Btu/ft-hr-°F
	641.4 kcal/hr			
	745.7 W			

TEMPERATURE CONVERSIONS

$^{\circ}C = 5/9 \ (^{\circ}F - 32)$ $^{\circ}F = 9/5 \ (^{\circ}C + 32)$

$K = {}^{\circ}C + 273.15$ $^{\circ}R = {}^{\circ}F + 459.67$

Appendix C

Methods of Expressing Mixture Ratios for CH$_4$, C$_3$H$_8$, and H$_2$

TABLE C1a
Methods of Expressing Fuel-Rich Combustion Mixtures for CH$_4$

Oxidizer		S	S$_1$	S$_2$	ϕ	λ		X$_A$	X$_{O_2}$
Vol.% O$_2$ in Oxid.	Vol.% N$_2$ in Oxid.	Stoich. Vol. O$_2$/ Fuel	Actual Vol. O$_2$/ Fuel	Actual Vol. Oxid./ Fuel	Theor. Stoich./ Act. Stoich.	Act. Stoich./ Theor. Stoich.	Vol. % Fuel in Mix	Vol.% Air above Stoich.	Vol.% O$_2$ above Stoich.
0.21	0.79	2.00	0.500	2.381	4.000	0.250	0.2958	−75%	−15.75%
0.21	0.79	2.00	1.000	4.762	2.000	0.500	0.1736	−50%	−10.50%
0.21	0.79	2.00	1.053	5.012	1.900	0.526	0.1663	−47%	−9.95%
0.21	0.79	2.00	1.100	5.238	1.818	0.550	0.1603	−45%	−9.45%
0.21	0.79	2.00	1.111	5.290	1.800	0.556	0.1590	−44%	−9.33%
0.21	0.79	2.00	1.177	5.602	1.700	0.588	0.1515	−41%	−8.65%
0.21	0.79	2.00	1.200	5.714	1.667	0.600	0.1489	−40%	−8.40%
0.21	0.79	2.00	1.250	5.952	1.600	0.625	0.1438	−38%	−7.88%
0.21	0.79	2.00	1.300	6.190	1.538	0.650	0.1391	−35%	−7.35%
0.21	0.79	2.00	1.333	6.349	1.500	0.667	0.1361	−33%	−7.00%
0.21	0.79	2.00	1.400	6.667	1.429	0.700	0.1304	−30%	−6.30%
0.21	0.79	2.00	1.429	6.805	1.400	0.715	0.1281	−29%	−6.00%
0.21	0.79	2.00	1.500	7.143	1.333	0.750	0.1228	−25%	−5.25%
0.21	0.79	2.00	1.538	7.324	1.300	0.769	0.1201	−23%	−4.85%
0.21	0.79	2.00	1.600	7.619	1.250	0.800	0.1160	−20%	−4.20%
0.21	0.79	2.00	1.667	7.937	1.200	0.833	0.1119	−17%	−3.50%
0.21	0.79	2.00	1.700	8.095	1.176	0.850	0.1099	−15%	−3.15%
0.21	0.79	2.00	1.800	8.571	1.111	0.900	0.1045	−10%	−2.10%
0.21	0.79	2.00	1.818	8.657	1.100	0.909	0.1036	−9%	−1.91%
0.21	0.79	2.00	1.889	8.995	1.059	0.945	0.1000	−6%	−1.17%
0.21	0.79	2.00	1.900	9.048	1.053	0.950	0.0995	−5%	−1.05%

TABLE C1b
Methods of Expressing Stoichiometric and Fuel-Lean Combustion Mixtures for CH_4

Oxidizer		S	S_1	S_2	ϕ	λ		X_A	X_{O_2}
Vol.% O_2 in Oxid.	Vol.% N_2 in Oxid.	Stoich. Vol. O_2/ Fuel	Actual Vol. O_2/ Fuel	Actual Vol. Oxid./ Fuel	Theor. Stoich./ Act. Stoich.	Act. Stoich./ Theor. Stoich.	Vol. % Fuel in Mix	Vol.% Air above Stoich.	Vol.% O_2 above Stoich.
0.21	0.79	2.00	2.000	9.524	1.000	1.000	0.0950	0%	0.00%
0.21	0.79	2.00	2.095	9.976	0.955	1.048	0.0911	5%	1.00%
0.21	0.79	2.00	2.100	10.000	0.952	1.050	0.0909	5%	1.05%
0.21	0.79	2.00	2.124	10.114	0.942	1.062	0.0900	6%	1.30%
0.21	0.79	2.00	2.190	10.429	0.913	1.095	0.0875	10%	2.00%
0.21	0.79	2.00	2.200	10.476	0.909	1.100	0.0871	10%	2.10%
0.21	0.79	2.00	2.221	10.576	0.900	1.111	0.0864	11%	2.32%
0.21	0.79	2.00	2.286	10.886	0.875	1.143	0.0841	14%	3.00%
0.21	0.79	2.00	2.300	10.952	0.870	1.150	0.0837	15%	3.15%
0.21	0.79	2.00	2.400	11.429	0.833	1.200	0.0805	20%	4.20%
0.21	0.79	2.00	2.500	11.905	0.800	1.250	0.0775	25%	5.25%
0.21	0.79	2.00	2.600	12.381	0.769	1.300	0.0747	30%	6.30%
0.21	0.79	2.00	2.667	12.698	0.750	1.333	0.0730	33%	7.00%
0.21	0.79	2.00	2.700	12.857	0.741	1.350	0.0722	35%	7.35%
0.21	0.79	2.00	2.800	13.333	0.714	1.400	0.0698	40%	8.40%
0.21	0.79	2.00	2.859	13.614	0.700	1.430	0.0684	43%	9.02%
0.21	0.79	2.00	2.900	13.810	0.690	1.450	0.0675	45%	9.45%
0.21	0.79	2.00	3.000	14.286	0.667	1.500	0.0654	50%	10.50%
0.21	0.79	2.00	3.500	16.667	0.571	1.750	0.0566	75%	15.75%
0.21	0.79	2.00	4.000	19.048	0.500	2.000	0.0499	100%	21.00%
0.30	0.70	2.00	2.000	6.667	1.000	1.000	0.1304	0%	0.00%
0.40	0.60	2.00	2.000	5.000	1.000	1.000	0.1667	0%	0.00%
0.50	0.50	2.00	2.000	4.000	1.000	1.000	0.2000	0%	0.00%
0.60	0.40	2.00	2.000	3.333	1.000	1.000	0.2308	0%	0.00%
0.70	0.30	2.00	2.000	2.857	1.000	1.000	0.2593	0%	0.00%
0.80	0.20	2.00	2.000	2.500	1.000	1.000	0.2857	0%	0.00%
0.90	0.10	2.00	2.000	2.222	1.000	1.000	0.3103	0%	0.00%
1.00	0.00	2.00	2.000	2.000	1.000	1.000	0.3333	0%	0.00%

TABLE C2a
Methods of Expressing Fuel-Rich Combustion Mixtures for C_3H_8

Oxidizer		S	S_1	S_2	ϕ	λ	Vol. %	X_A	X_{O_2}
Vol.% O_2 in Oxid.	Vol.% N_2 in Oxid.	Stoich. Vol. O_2/ Fuel	Actual Vol. O_2/ Fuel	Actual Vol. Oxid./ Fuel	Theor. Stoich./ Act. Stoich.	Act. Stoich./ Theor. Stoich.	Fuel in Mix	Vol.% Air above Stoich.	Vol.% O_2 above Stoich.
0.21	0.79	5.00	1.250	5.952	4.000	0.250	0.1438	−75%	−15.75%
0.21	0.79	5.00	2.500	11.905	2.000	0.500	0.0775	−50%	−10.50%
0.21	0.79	5.00	2.631	12.530	1.900	0.526	0.0739	−47%	−9.95%
0.21	0.79	5.00	2.750	13.095	1.818	0.550	0.0709	−45%	−9.45%
0.21	0.79	5.00	2.778	13.226	1.800	0.556	0.0703	−44%	−9.33%
0.21	0.79	5.00	2.941	14.006	1.700	0.588	0.0666	−41%	−8.65%
0.21	0.79	5.00	3.000	14.286	1.667	0.600	0.0654	−40%	−8.40%
0.21	0.79	5.00	3.125	14.881	1.600	0.625	0.0630	−38%	−7.88%
0.21	0.79	5.00	3.250	15.476	1.538	0.650	0.0607	−35%	−7.35%
0.21	0.79	5.00	3.333	15.873	1.500	0.667	0.0593	−33%	−7.00%
0.21	0.79	5.00	3.500	16.667	1.429	0.700	0.0566	−30%	−6.30%
0.21	0.79	5.00	3.573	17.012	1.400	0.715	0.0555	−29%	−6.00%
0.21	0.79	5.00	3.750	17.857	1.333	0.750	0.0530	−25%	−5.25%
0.21	0.79	5.00	3.845	18.310	1.300	0.769	0.0518	−23%	−4.85%
0.21	0.79	5.00	4.000	19.048	1.250	0.800	0.0499	−20%	−4.20%
0.21	0.79	5.00	4.167	19.842	1.200	0.833	0.0480	−17%	−3.50%
0.21	0.79	5.00	4.250	20.238	1.176	0.850	0.0471	−15%	−3.15%
0.21	0.79	5.00	4.500	21.429	1.111	0.900	0.0446	−10%	−2.10%
0.21	0.79	5.00	4.545	21.643	1.100	0.909	0.0442	−9%	−1.91%
0.21	0.79	5.00	4.723	22.488	1.059	0.945	0.0426	−6%	−1.17%
0.21	0.79	5.00	4.750	22.619	1.053	0.950	0.0423	−5%	−1.05%

TABLE C2b
Methods of Expressing Stoichiometric and Fuel-Lean Combustion Mixtures for C_3H_8

Oxidizer		S	S_1	S_2	ϕ	λ		X_A	X_{O_2}
Vol.% O_2 in Oxid.	Vol.% N_2 in Oxid.	Stoich. Vol. O_2/ Fuel	Actual Vol. O_2/ Fuel	Actual Vol. Oxid./ Fuel	Theor. Stoich./ Act. Stoich.	Act. Stoich./ Theor. Stoich.	Vol. % Fuel in Mix	Vol.% Air above Stoich.	Vol.% O_2 above Stoich.
0.21	0.79	5.00	5.000	23.810	1.000	1.000	0.0403	0%	0.00%
0.21	0.79	5.00	5.238	24.940	0.955	1.048	0.0385	5%	1.00%
0.21	0.79	5.00	5.250	25.000	0.952	1.050	0.0385	5%	1.05%
0.21	0.79	5.00	5.310	25.286	0.942	1.062	0.0380	6%	1.30%
0.21	0.79	5.00	5.475	26.071	0.913	1.095	0.0369	10%	2.00%
0.21	0.79	5.00	5.500	26.190	0.909	1.100	0.0368	10%	2.10%
0.21	0.79	5.00	5.553	26.440	0.900	1.111	0.0364	11%	2.32%
0.21	0.79	5.00	5.715	27.214	0.875	1.143	0.0354	14%	3.00%
0.21	0.79	5.00	5.750	27.381	0.870	1.150	0.0352	15%	3.15%
0.21	0.79	5.00	6.000	28.571	0.833	1.200	0.0338	20%	4.20%
0.21	0.79	5.00	6.250	29.762	0.800	1.250	0.0325	25%	5.25%
0.21	0.79	5.00	6.500	30.952	0.769	1.300	0.0313	30%	6.30%
0.21	0.79	5.00	6.667	31.746	0.750	1.333	0.0305	33%	7.00%
0.21	0.79	5.00	6.750	32.143	0.741	1.350	0.0302	35%	7.35%
0.21	0.79	5.00	7.000	33.333	0.714	1.400	0.0291	40%	8.40%
0.21	0.79	5.00	7.148	34.036	0.700	1.430	0.0285	43%	9.02%
0.21	0.79	5.00	7.250	34.524	0.690	1.450	0.0282	45%	9.45%
0.21	0.79	5.00	7.500	35.714	0.667	1.500	0.0272	50%	10.50%
0.21	0.79	5.00	8.750	41.667	0.571	1.750	0.0234	75%	15.75%
0.21	0.79	5.00	10.000	47.619	0.500	2.000	0.0206	100%	21.00%
0.30	0.70	5.00	5.000	16.667	1.000	1.000	0.0566	0%	0.00%
0.40	0.60	5.00	5.000	12.500	1.000	1.000	0.0741	0%	0.00%
0.50	0.50	5.00	5.000	10.000	1.000	1.000	0.0909	0%	0.00%
0.60	0.40	5.00	5.000	8.333	1.000	1.000	0.1071	0%	0.00%
0.70	0.30	5.00	5.000	7.143	1.000	1.000	0.1228	0%	0.00%
0.80	0.20	5.00	5.000	6.250	1.000	1.000	0.1379	0%	0.00%
0.90	0.10	5.00	5.000	5.556	1.000	1.000	0.1525	0%	0.00%
1.00	0.00	5.00	5.000	5.000	1.000	1.000	0.1667	0%	0.00%

TABLE C3a
Methods of Expressing Fuel-Rich Combustion Mixtures for H_2

Oxidizer		S	S_1	S_2	ϕ	λ	Vol. %	X_A	X_{O_2}
Vol.% O_2 in Oxid.	Vol.% N_2 in Oxid.	Stoich. Vol. O_2/ Fuel	Actual Vol. O_2/ Fuel	Actual Vol. Oxid./ Fuel	Theor. Stoich./ Act. Stoich.	Act. Stoich./ Theor. Stoich.	Fuel in Mix	Vol.% Air above Stoich.	Vol.% O_2 above Stoich.
0.21	0.79	0.50	0.125	0.595	4.000	0.250	0.6269	−75%	−15.75%
0.21	0.79	0.50	0.250	1.190	2.000	0.500	0.4565	−50%	−10.50%
0.21	0.79	0.50	0.263	1.253	1.900	0.526	0.4439	−47%	−9.95%
0.21	0.79	0.50	0.275	1.310	1.818	0.550	0.4330	−45%	−9.45%
0.21	0.79	0.50	0.278	1.323	1.800	0.556	0.4305	−44%	−9.33%
0.21	0.79	0.50	0.294	1.401	1.700	0.588	0.4166	−41%	−8.65%
0.21	0.79	0.50	0.300	1.429	1.667	0.600	0.4118	−40%	−8.40%
0.21	0.79	0.50	0.313	1.488	1.600	0.625	0.4019	−38%	−7.88%
0.21	0.79	0.50	0.325	1.548	1.538	0.650	0.3925	−35%	−7.35%
0.21	0.79	0.50	0.333	1.587	1.500	0.667	0.3865	−33%	−7.00%
0.21	0.79	0.50	0.350	1.667	1.429	0.700	0.3750	−30%	−6.30%
0.21	0.79	0.50	0.357	1.701	1.400	0.715	0.3702	−29%	−6.00%
0.21	0.79	0.50	0.375	1.786	1.333	0.750	0.3590	−25%	−5.25%
0.21	0.79	0.50	0.385	1.831	1.300	0.769	0.3532	−23%	−4.85%
0.21	0.79	0.50	0.400	1.905	1.250	0.800	0.3443	−20%	−4.20%
0.21	0.79	0.50	0.417	1.984	1.200	0.833	0.3351	−17%	−3.50%
0.21	0.79	0.50	0.425	2.024	1.176	0.850	0.3307	−15%	−3.15%
0.21	0.79	0.50	0.450	2.143	1.111	0.900	0.3182	−10%	−2.10%
0.21	0.79	0.50	0.455	2.164	1.100	0.909	0.3160	−9%	−1.91%
0.21	0.79	0.50	0.472	2.249	1.059	0.945	0.3078	−6%	−1.17%
0.21	0.79	0.50	0.475	2.262	1.053	0.950	0.3066	−5%	−1.05%

TABLE C3b
Methods of Expressing Stoichiometric and Fuel-Lean Combustion Mixtures for H_2

Oxidizer		S	S_1	S_2	ϕ	λ		X_A	X_{O_2}
Vol.% O_2 in Oxid.	Vol.% N_2 in Oxid.	Stoich. Vol. O_2/ Fuel	Actual Vol. O_2/ Fuel	Actual Vol. Oxid./ Fuel	Theor. Stoich./ Act. Stoich.	Act. Stoich./ Theor. Stoich.	Vol. % Fuel in Mix	Vol.% Air above Stoich.	Vol.% O_2 above Stoich.
0.21	0.79	0.50	0.500	2.381	1.000	1.000	0.2958	0%	0.00%
0.21	0.79	0.50	0.524	2.494	0.955	1.048	0.2862	5%	1.00%
0.21	0.79	0.50	0.525	2.500	0.952	1.050	0.2857	5%	1.05%
0.21	0.79	0.50	0.531	2.529	0.942	1.062	0.2834	6%	1.30%
0.21	0.79	0.50	0.548	2.607	0.913	1.095	0.2772	10%	2.00%
0.21	0.79	0.50	0.550	2.619	0.909	1.100	0.2763	10%	2.10%
0.21	0.79	0.50	0.555	2.644	0.900	1.111	0.2744	11%	2.32%
0.21	0.79	0.50	0.572	2.721	0.875	1.143	0.2687	14%	3.00%
0.21	0.79	0.50	0.575	2.738	0.870	1.150	0.2675	15%	3.15%
0.21	0.79	0.50	0.600	2.857	0.833	1.200	0.2593	20%	4.20%
0.21	0.79	0.50	0.625	2.976	0.800	1.250	0.2515	25%	5.25%
0.21	0.79	0.50	0.650	3.095	0.769	1.300	0.2442	30%	6.30%
0.21	0.79	0.50	0.667	3.175	0.750	1.333	0.2395	33%	7.00%
0.21	0.79	0.50	0.675	3.214	0.741	1.350	0.2373	35%	7.35%
0.21	0.79	0.50	0.700	3.333	0.714	1.400	0.2308	40%	8.40%
0.21	0.79	0.50	0.715	3.404	0.700	1.430	0.2271	43%	9.02%
0.21	0.79	0.50	0.725	3.452	0.690	1.450	0.2246	45%	9.45%
0.21	0.79	0.50	0.750	3.571	0.667	1.500	0.2188	50%	10.50%
0.21	0.79	0.50	0.875	4.167	0.571	1.750	0.1935	75%	15.75%
0.21	0.79	0.50	1.000	4.762	0.500	2.000	0.1736	100%	21.00%
0.30	0.70	0.50	0.500	1.667	1.000	1.000	0.3750	0%	0.00%
0.40	0.60	0.50	0.500	1.250	1.000	1.000	0.4444	0%	0.00%
0.50	0.50	0.50	0.500	1.000	1.000	1.000	0.5000	0%	0.00%
0.60	0.40	0.50	0.500	0.833	1.000	1.000	0.5455	0%	0.00%
0.70	0.30	0.50	0.500	0.714	1.000	1.000	0.5833	0%	0.00%
0.80	0.20	0.50	0.500	0.625	1.000	1.000	0.6154	0%	0.00%
0.90	0.10	0.50	0.500	0.556	1.000	1.000	0.6429	0%	0.00%
1.00	0.00	0.50	0.500	0.500	1.000	1.000	0.6667	0%	0.00%

Appendix D

Properties for CH_4, C_3H_8, and H_2 Flames

TABLE D1a

H₂ (298K) Adiabatically and Stoichiometrically (φ = 1.00) Combusted with a Variable-Composition Oxidizer (298K) (Metric Units)

O_2	N_2	Flame Temp. (K)	H	HO_2	H_2	$H_2O(g)$	NO	N_2	O	OH	O_2	Enthalpy, H (kJ/kg)	Density (kg/m^3)	Viscosity (kg/m-s)	Equil. Spec. Heat (kJ/kg-K)	Equil. Therm. Cond. (W/m-K)	Equil. Pr	Le
0.21	0.79	2382	0.00179	0.00000	0.01526	0.32366	0.00264	0.64440	0.00054	0.00696	0.00475	2.09E-04	0.124	7.53E-05	2.72	0.383	0.536	1.28
0.30	0.70	2648	0.00933	0.00001	0.04074	0.38912	0.00607	0.51684	0.00321	0.02217	0.01251	0.00E+00	0.104	8.19E-05	4.72	0.851	0.454	1.45
0.40	0.60	2799	0.02092	0.00002	0.06662	0.43867	0.00851	0.39742	0.00770	0.03962	0.02052	4.19E-05	0.091	8.60E-05	7.11	1.420	0.431	1.49
0.50	0.50	2889	0.03269	0.00003	0.08793	0.47585	0.00960	0.29903	0.01253	0.05505	0.02729	4.19E-05	0.082	8.88E-05	9.47	1.962	0.429	1.49
0.60	0.40	2950	0.04357	0.00003	0.10540	0.50534	0.00973	0.21737	0.01717	0.06834	0.03303	8.37E-05	0.075	9.08E-05	11.75	2.459	0.434	1.47
0.70	0.30	2994	0.05333	0.00004	0.11988	0.52948	0.00911	0.14886	0.02148	0.07983	0.03798	1.26E-04	0.070	9.25E-05	13.94	2.913	0.442	1.45
0.80	0.20	3028	0.06203	0.00005	0.13199	0.54964	0.00782	0.09078	0.02545	0.08987	0.04236	1.67E-04	0.066	9.38E-05	16.04	3.328	0.452	1.43
0.90	0.10	3055	0.06980	0.00005	0.14217	0.56673	0.00569	0.04122	0.02915	0.09880	0.04637	0.00E+00	0.062	9.50E-05	18.06	3.709	0.462	1.41
1.00	0.00	3079	0.07695	0.00006	0.15027	0.58120	0.00000	0.00000	0.03302	0.10758	0.05092	0.00E+00	0.059	9.60E-05	20.01	4.0670	0.472	1.40

TABLE D1b

H₂ (77°F) Adiabatically and Stoichiometrically (φ = 1.00) Combusted with a Variable-Composition Oxidizer (77°F) (English Units)

O_2	N_2	Flame Temp. (°F)	H	HO_2	H_2	$H_2O(g)$	NO	N_2	O	OH	O_2	Enthalpy, H (Btu/lb)	Density (lb/ft^3)	Viscosity (lbm/ft-hr)	Equil. Spec. Heat (Btu/lb-F)	Equil. Therm. Cond. (Btu/ft-hr-F)	Equil. Pr	Le
0.21	0.79	3827	0.00179	0.00000	0.01526	0.32366	0.00264	0.64440	0.00054	0.00696	0.00475	0	0.00775	5.06E-05	0.650	0.221	0.536	1.28
0.30	0.70	4307	0.00933	0.00001	0.04074	0.38912	0.00607	0.51684	0.00321	0.02217	0.01251	0	0.00649	5.50E-05	1.127	0.492	0.454	1.45
0.40	0.60	4579	0.02092	0.00002	0.06662	0.43867	0.00851	0.39742	0.00770	0.03962	0.02052	0	0.00568	5.78E-05	1.699	0.821	0.431	1.49
0.50	0.50	4741	0.03269	0.00003	0.08793	0.47585	0.00960	0.29903	0.01253	0.05505	0.02729	0	0.00512	5.97E-05	2.262	1.134	0.429	1.49
0.60	0.40	4850	0.04357	0.00003	0.10540	0.50534	0.00973	0.21737	0.01717	0.06834	0.03303	0	0.00470	6.11E-05	2.808	1.421	0.434	1.47
0.70	0.30	4929	0.05333	0.00004	0.11988	0.52948	0.00911	0.14886	0.02148	0.07983	0.03798	0	0.00437	6.22E-05	3.328	1.683	0.442	1.45
0.80	0.20	4990	0.06203	0.00005	0.13199	0.54964	0.00782	0.09078	0.02545	0.08987	0.04236	0	0.00410	6.31E-05	3.830	1.923	0.452	1.43
0.90	0.10	5039	0.06980	0.00005	0.14217	0.56673	0.00569	0.04122	0.02915	0.09880	0.04637	0	0.00387	6.39E-05	4.312	2.143	0.462	1.41
1.00	0.00	5082	0.07695	0.00006	0.15027	0.58120	0.00000	0.00000	0.03302	0.10758	0.05092	0	0.00367	6.45E-05	4.780	2.350	0.472	1.40

TABLE D2a

H$_2$ (298K) Adiabatically and Stoichiometrically (ϕ = 1.00) Combusted with Air (Variable Temp.) (Metric Units)

Air Temp. (K)	Flame Temp. (K)	H	HO$_2$	H$_2$	H$_2$O(g)	NO	N$_2$	O	OH	O$_2$	Enthalpy, H (kJ/kg)	Density (kg/m³)	Viscosity (kg/m-s)	Equil. Spec. Heat (kJ/kg-K)	Equil. Therm. Cond. (W/m-K)	Equil. Pr	Le
298	2382	0.00179	0.00000	0.01526	0.32366	0.00264	0.64440	0.00054	0.00696	0.00475	2.09E-04	0.124	7.53E-05	2.72	0.383	0.536	1.28
366	2406	0.00209	0.00000	0.01657	0.32145	0.00287	0.64356	0.00064	0.00768	0.00513	6.69E+01	0.123	7.58E-05	2.81	0.404	0.527	1.30
478	2444	0.00267	0.00000	0.01882	0.31760	0.00327	0.64207	0.00083	0.00895	0.00579	1.78E+02	0.121	7.66E-05	2.96	0.442	0.513	1.32
589	2482	0.00335	0.00000	0.02116	0.31350	0.00369	0.64047	0.00106	0.01030	0.00647	2.91E+02	0.119	7.74E-05	3.11	0.482	0.500	1.35
700	2518	0.00415	0.00000	0.02361	0.30913	0.00413	0.63874	0.00133	0.01174	0.00717	4.06E+02	0.117	7.82E-05	3.28	0.526	0.488	1.38
811	2553	0.00507	0.00000	0.02615	0.30448	0.00460	0.63687	0.00165	0.01328	0.00790	5.24E+02	0.115	7.89E-05	3.46	0.573	0.476	1.41
922	2587	0.00613	0.00000	0.02877	0.29958	0.00508	0.63487	0.00202	0.01490	0.00864	6.46E+02	0.113	7.96E-05	3.64	0.623	0.465	1.43
1033	2620	0.00733	0.00001	0.03145	0.29446	0.00557	0.63274	0.00245	0.01659	0.00939	7.69E+02	0.111	8.03E-05	3.83	0.677	0.455	1.46
1144	2652	0.00867	0.00001	0.03416	0.28915	0.00608	0.63050	0.00294	0.01833	0.01015	8.95E+02	0.109	8.09E-05	4.03	0.733	0.445	1.48
1255	2683	0.01015	0.00001	0.03688	0.28369	0.00659	0.62816	0.00348	0.02011	0.01091	1.02E+03	0.108	8.15E-05	4.24	0.791	0.437	1.50
1366	2712	0.01177	0.00001	0.03960	0.27811	0.00711	0.62573	0.00408	0.02193	0.01166	1.15E+03	0.106	8.21E-05	4.45	0.852	0.429	1.52

TABLE D2b

H_2 (77°F) Adiabatically and Stoichiometrically (φ = 1.00) Combusted with Air (Variable Temp.) (English Units)

Air Temp. (°F)	Flame Temp. (°F)	H	HO_2	H_2	$H_2O(g)$	NO	N_2	O	OH	O_2	Enthalpy, H (Btu/lb)	Density (lb/ft³)	Viscosity (lbm/ft-hr)	Equil. Spec. Heat (Btu/lb-F)	Equil. Therm. Cond. (Btu/ft-hr-F)	Equil. Pr	Le
77	3827	0.00179	0.00000	0.01526	0.32366	0.00264	0.64440	0.00054	0.00696	0.00475	0	0.00775	5.06E-05	0.650	0.221	0.536	1.28
200	3871	0.00209	0.00000	0.01657	0.32145	0.00287	0.64356	0.00064	0.00768	0.00513	29	0.00767	5.10E-05	0.671	0.233	0.527	1.30
400	3940	0.00267	0.00000	0.01882	0.31760	0.00327	0.64207	0.00083	0.00895	0.00579	77	0.00753	5.15E-05	0.706	0.255	0.513	1.32
600	4007	0.00335	0.00000	0.02116	0.31350	0.00369	0.64047	0.00106	0.01030	0.00647	125	0.00740	5.21E-05	0.744	0.279	0.500	1.35
800	4072	0.00415	0.00000	0.02361	0.30913	0.00413	0.63874	0.00133	0.01174	0.00717	175	0.00728	5.26E-05	0.783	0.304	0.488	1.38
1000	4135	0.00507	0.00000	0.02615	0.30448	0.00460	0.63687	0.00165	0.01328	0.00790	225	0.00716	5.31E-05	0.825	0.331	0.476	1.41
1200	4197	0.00613	0.00000	0.02877	0.29958	0.00508	0.63487	0.00202	0.01490	0.00864	278	0.00705	5.35E-05	0.869	0.360	0.465	1.43
1400	4256	0.00733	0.00001	0.03145	0.29446	0.00557	0.63274	0.00245	0.01659	0.00939	331	0.00694	5.40E-05	0.915	0.391	0.455	1.46
1600	4314	0.00867	0.00001	0.03416	0.28915	0.00608	0.63050	0.00294	0.01833	0.01015	385	0.00683	5.44E-05	0.963	0.423	0.445	1.48
1800	4369	0.01015	0.00001	0.03688	0.28369	0.00659	0.62816	0.00348	0.02011	0.01091	439	0.00673	5.48E-05	1.012	0.457	0.437	1.50
2000	4422	0.01177	0.00001	0.03960	0.27811	0.00711	0.62573	0.00408	0.02193	0.01166	495	0.00664	5.52E-05	1.062	0.492	0.429	1.52

TABLE D3a
H_2 (Variable Temp.) Adiabatically and Stoichiometrically ($\phi = 1.00$) Combusted with Air (298K) (Metric Units)

Fuel Temp. (K)	Flame Temp. (K)	H	H_2	$H_2O(g)$	NO	N_2	O	OH	O_2	Enthalpy, H (kJ/kg)	Density (kg/m³)	Viscosity (kg/m-s)	Equil. Spec. Heat (kJ/kg-K)	Equil. Therm. Cond. (W/m-K)	Equil. Pr	Le
298	2382	0.00179	0.01526	0.32366	0.00264	0.64440	0.00054	0.00696	0.00475	2.09E-04	0.124	7.53E-05	2.72	0.383	0.536	1.28
311	2384	0.00181	0.01537	0.32348	0.00266	0.64434	0.00055	0.00702	0.00478	5.29E+00	0.124	7.53E-05	2.73	0.384	0.535	1.28
366	2392	0.00191	0.01580	0.32275	0.00274	0.64406	0.00058	0.00726	0.00491	2.79E+01	0.124	7.55E-05	2.76	0.391	0.532	1.28
422	2400	0.00202	0.01626	0.32198	0.00282	0.64376	0.00061	0.00751	0.00504	5.10E+01	0.123	7.57E-05	2.79	0.399	0.529	1.29
478	2408	0.00213	0.01672	0.32121	0.00290	0.64347	0.00065	0.00776	0.00517	7.41E+01	0.123	7.59E-05	2.82	0.406	0.526	1.30
533	2416	0.00224	0.01717	0.32043	0.00298	0.64317	0.00069	0.00802	0.00531	9.69E+01	0.122	7.60E-05	2.85	0.414	0.523	1.30
589	2425	0.00236	0.01764	0.31963	0.00306	0.64286	0.00072	0.00828	0.00544	1.20E+02	0.122	7.62E-05	2.88	0.422	0.520	1.31
644	2432	0.00248	0.01810	0.31884	0.00314	0.64256	0.00076	0.00854	0.00558	1.43E+02	0.121	7.64E-05	2.91	0.429	0.517	1.31
700	2440	0.00260	0.01857	0.31802	0.00323	0.64224	0.00081	0.00881	0.00572	1.66E+02	0.121	7.65E-05	2.94	0.437	0.514	1.32
755	2448	0.00273	0.01904	0.31721	0.00331	0.64192	0.00085	0.00907	0.00585	1.89E+02	0.120	7.67E-05	2.97	0.445	0.512	1.33
811	2456	0.00287	0.01953	0.31637	0.00340	0.64160	0.00089	0.00935	0.00599	2.13E+02	0.120	7.69E-05	3.00	0.454	0.509	1.33

TABLE D3b
H_2 (Variable Temp.) Adiabatically and Stoichiometrically ($\phi = 1.00$) Combusted with Air (77°F) (English Units)

Fuel Temp. (K)	Flame Temp. (K)	H	H_2	$H_2O(g)$	NO	N_2	O	OH	O_2	Enthalpy, H (Btu/lb)	Density (lb/ft³)	Viscosity (lbm/ft-hr)	Equil. Spec. Heat (Btu/lb-F)	Equil. Therm. Cond. (Btu/ft-hr-F)	Equil. Pr	Le
77	3827	0.00179	0.01526	0.32366	0.00264	0.64440	0.00054	0.00696	0.00475	0	0.00775	5.06E-05	0.650	0.221	0.536	1.28
100	3831	0.00181	0.01537	0.32348	0.00266	0.64434	0.00055	0.00702	0.00478	2	0.00775	5.07E-05	0.652	0.222	0.535	1.28
200	3846	0.00191	0.01580	0.32275	0.00274	0.64406	0.00058	0.00726	0.00491	12	0.00772	5.08E-05	0.659	0.226	0.532	1.28
300	3861	0.00202	0.01626	0.32198	0.00282	0.64376	0.00061	0.00751	0.00504	22	0.00769	5.09E-05	0.666	0.230	0.529	1.29
400	3875	0.00213	0.01672	0.32121	0.00290	0.64347	0.00065	0.00776	0.00517	32	0.00766	5.10E-05	0.673	0.235	0.526	1.30
500	3890	0.00224	0.01717	0.32043	0.00298	0.64317	0.00069	0.00802	0.00531	42	0.00763	5.11E-05	0.680	0.239	0.523	1.30
600	3904	0.00236	0.01764	0.31963	0.00306	0.64286	0.00072	0.00828	0.00544	52	0.00760	5.12E-05	0.687	0.244	0.520	1.31
700	3919	0.00248	0.01810	0.31884	0.00314	0.64256	0.00076	0.00854	0.00558	61	0.00757	5.13E-05	0.695	0.248	0.517	1.31
800	3933	0.00260	0.01857	0.31802	0.00323	0.64224	0.00081	0.00881	0.00572	71	0.00755	5.14E-05	0.702	0.253	0.514	1.32
900	3947	0.00273	0.01904	0.31721	0.00331	0.64192	0.00085	0.00907	0.00585	81	0.00752	5.16E-05	0.710	0.257	0.512	1.33
1000	3961	0.00287	0.01953	0.31637	0.00340	0.64160	0.00089	0.00935	0.00599	91	0.00749	5.17E-05	0.717	0.262	0.509	1.33

TABLE D4a

H$_2$ (298K) Stoichiometrically (ϕ = 1.00) Combusted with Air (298K) (Metric Units)

Flame Temp. (K)	H	H$_2$	H$_2$O(g)	NO	N$_2$	O	OH	O$_2$	Enthalpy, H (kJ/kg)	Density (kg/m³)	Viscosity (kg/m-s)	Equil. Spec. Heat (kJ/kg-K)	Equil. Therm. Cond. (W/m-K)	Equil. Pr	Le	Avail. Heat
366	0.00000	0.00000	0.34711	0.00000	0.65289	0.00000	0.00000	0.00000	-3.34E+03	0.817	1.78E-05	1.26	0.027	0.825	1.00	82.2%
478	0.00000	0.00000	0.34711	0.00000	0.65289	0.00000	0.00000	0.00000	-3.20E+03	0.626	2.23E-05	1.28	0.035	0.808	1.00	78.7%
589	0.00000	0.00000	0.34711	0.00000	0.65289	0.00000	0.00000	0.00000	-3.05E+03	0.508	2.65E-05	1.31	0.044	0.795	1.00	75.2%
700	0.00000	0.00000	0.34711	0.00000	0.65289	0.00000	0.00000	0.00000	-2.90E+03	0.427	3.06E-05	1.35	0.053	0.785	1.00	71.5%
811	0.00000	0.00000	0.34711	0.00000	0.65289	0.00000	0.00000	0.00000	-2.75E+03	0.369	3.44E-05	1.39	0.061	0.777	1.00	67.8%
922	0.00000	0.00000	0.34711	0.00000	0.65289	0.00000	0.00000	0.00000	-2.60E+03	0.324	3.80E-05	1.43	0.071	0.769	1.00	63.9%
1033	0.00000	0.00000	0.34711	0.00000	0.65289	0.00000	0.00000	0.00000	-2.43E+03	0.290	4.15E-05	1.46	0.080	0.760	1.00	60.0%
1144	0.00000	0.00000	0.34711	0.00000	0.65289	0.00000	0.00000	0.00000	-2.27E+03	0.261	4.48E-05	1.50	0.089	0.755	1.00	55.9%
1255	0.00000	0.00001	0.34710	0.00000	0.65289	0.00000	0.00000	0.00000	-2.10E+03	0.238	4.80E-05	1.53	0.098	0.750	1.00	51.8%
1366	0.00000	0.00003	0.34707	0.00000	0.65288	0.00000	0.00000	0.00001	-1.93E+03	0.219	5.10E-05	1.56	0.107	0.744	1.00	47.5%
1478	0.00000	0.00008	0.34700	0.00001	0.65286	0.00000	0.00001	0.00003	-1.75E+03	0.202	5.40E-05	1.59	0.117	0.737	1.00	43.2%
1589	0.00000	0.00022	0.34682	0.00004	0.65279	0.00000	0.00004	0.00008	-1.57E+03	0.188	5.69E-05	1.63	0.128	0.728	1.01	38.7%
1700	0.00000	0.00051	0.34644	0.00008	0.65266	0.00000	0.00012	0.00018	-1.39E+03	0.176	5.97E-05	1.68	0.140	0.715	1.02	34.2%
1811	0.00001	0.00106	0.34568	0.00018	0.65240	0.00000	0.00029	0.00037	-1.20E+03	0.165	6.24E-05	1.74	0.155	0.701	1.03	29.5%
1922	0.00004	0.00203	0.34433	0.00034	0.65193	0.00001	0.00062	0.00069	-1.00E+03	0.155	6.50E-05	1.82	0.175	0.678	1.05	24.7%
2033	0.00012	0.00361	0.34204	0.00061	0.65114	0.00003	0.00125	0.00120	-7.93E+02	0.147	6.76E-05	1.94	0.201	0.652	1.09	19.5%
2144	0.00031	0.00605	0.33841	0.00103	0.64985	0.00008	0.00230	0.00197	-5.68E+02	0.139	7.01E-05	2.12	0.240	0.619	1.13	14.0%
2255	0.00074	0.00958	0.33292	0.00164	0.64786	0.00021	0.00399	0.00305	-3.20E+02	0.132	7.26E-05	2.35	0.293	0.581	1.19	7.9%
2382	0.00179	0.01526	0.32366	0.00264	0.64440	0.00054	0.00696	0.00475	2.09E-04	0.124	7.53E-05	2.72	0.383	0.536	1.28	0.0%

TABLE D4b
H$_2$ (Variable Temp.) Adiabatically and Stoichiometrically (ϕ = 1.00) Combusted with Air (77°F) (English Units)

Flame Temp. (°F)	H	H$_2$	H$_2$O(g)	NO	N$_2$	O	OH	O$_2$	Enthalpy, H (Btu/lb)	Density (lb/ft³)	Viscosity (lbm/ft-hr)	Equil. Spec. Heat (Btu/lb-F)	Equil. Therm. Cond. (Btu/ft-hr-F)	Equil. Pr	Le	Avail. Heat
200	0.00000	0.00000	0.34711	0.00000	0.65289	0.00000	0.00000	0.00000	-1435	0.05102	1.19E-05	0.300	0.016	0.825	1.00	82.2%
400	0.00000	0.00000	0.34711	0.00000	0.65289	0.00000	0.00000	0.00000	-1374	0.03906	1.50E-05	0.306	0.020	0.808	1.00	78.7%
600	0.00000	0.00000	0.34711	0.00000	0.65289	0.00000	0.00000	0.00000	-1312	0.03170	1.79E-05	0.313	0.025	0.795	1.00	75.2%
800	0.00000	0.00000	0.34711	0.00000	0.65289	0.00000	0.00000	0.00000	-1248	0.02668	2.06E-05	0.322	0.030	0.785	1.00	71.5%
1000	0.00000	0.00000	0.34711	0.00000	0.65289	0.00000	0.00000	0.00000	-1183	0.02302	2.31E-05	0.331	0.036	0.777	1.00	67.8%
1200	0.00000	0.00000	0.34711	0.00000	0.65289	0.00000	0.00000	0.00000	-1116	0.02025	2.56E-05	0.341	0.041	0.769	1.00	63.9%
1400	0.00000	0.00000	0.34711	0.00000	0.65289	0.00000	0.00000	0.00000	-1047	0.01808	2.79E-05	0.350	0.046	0.760	1.00	60.0%
1600	0.00000	0.00000	0.34711	0.00000	0.65289	0.00000	0.00000	0.00000	-976	0.01632	3.01E-05	0.358	0.051	0.755	1.00	55.9%
1800	0.00000	0.00001	0.34710	0.00000	0.65289	0.00000	0.00000	0.00000	-904	0.01488	3.23E-05	0.366	0.057	0.750	1.00	51.8%
2000	0.00000	0.00003	0.34707	0.00000	0.65288	0.00000	0.00000	0.00001	-830	0.01367	3.43E-05	0.373	0.062	0.744	1.00	47.5%
2200	0.00000	0.00008	0.34700	0.00001	0.65286	0.00000	0.00001	0.00003	-753	0.01263	3.63E-05	0.381	0.068	0.737	1.00	43.2%
2400	0.00000	0.00022	0.34682	0.00004	0.65279	0.00000	0.00004	0.00008	-676	0.01175	3.83E-05	0.390	0.074	0.728	1.01	38.7%
2600	0.00000	0.00051	0.34644	0.00008	0.65266	0.00000	0.00012	0.00018	-597	0.01098	4.01E-05	0.401	0.081	0.715	1.02	34.2%
2800	0.00001	0.00106	0.34568	0.00018	0.65240	0.00000	0.00029	0.00037	-516	0.01030	4.20E-05	0.415	0.089	0.701	1.03	29.5%
3000	0.00004	0.00203	0.34433	0.00034	0.65193	0.00001	0.00062	0.00069	-431	0.00970	4.37E-05	0.436	0.101	0.678	1.05	24.7%
3200	0.00012	0.00361	0.34204	0.00061	0.65114	0.00003	0.00125	0.00120	-341	0.00916	4.55E-05	0.464	0.116	0.652	1.09	19.5%
3400	0.00031	0.00605	0.33841	0.00103	0.64985	0.00008	0.00230	0.00197	-244	0.00868	4.72E-05	0.506	0.139	0.619	1.13	14.0%
3600	0.00074	0.00958	0.33292	0.00164	0.64786	0.00021	0.00399	0.00305	-138	0.00823	4.88E-05	0.561	0.170	0.581	1.19	7.9%
3827	0.00179	0.01526	0.32366	0.00264	0.64440	0.00054	0.00696	0.00475	0	0.00775	5.06E-05	0.650	0.221	0.536	1.28	0.0%

TABLE D5a

H₂ (298K) Adiabatically Combusted with Air (298K) at Various Equivalence Ratios (Metric Units)

Equiv. Ratio	Flame Temp. (K)	H	H₂	H₂O(g)	NH₃	NO	N₂	O	OH	O₂	Enthalpy, H (kJ/kg)	Density (kg/m³)	Viscosity (kg/m-s)	Equil. Spec. Heat (kJ/kg-K)	Equil. Therm. Cond. (W/m-K)	Equil. Pr	Le
4.00	1559	0.00002	0.51009	0.17004	0.00001	0.00000	0.31983	0.00000	0.00000	0.00000	4.19E-05	0.120	5.27E-05	2.75	0.285	0.509	1.00
2.00	2060	0.00122	0.25696	0.25735	0.00006	0.00001	0.48433	0.00000	0.00014	0.00000	8.37E-05	0.111	6.71E-05	2.23	0.283	0.529	1.09
1.33	2314	0.00332	0.10285	0.30829	0.00000	0.00021	0.58357	0.00004	0.00167	0.00005	2.51E-04	0.117	7.35E-05	2.22	0.312	0.523	1.22
1.00	2382	0.00179	0.01526	0.32366	0.00000	0.00264	0.64440	0.00054	0.00696	0.00475	2.09E-04	0.124	7.53E-05	2.72	0.383	0.536	1.28
0.80	2168	0.00018	0.00144	0.28314	0.00000	0.00454	0.67259	0.00040	0.00471	0.03300	1.67E-04	0.142	7.08E-05	1.92	0.223	0.610	1.15
0.67	1957	0.00002	0.00021	0.24439	0.00000	0.00358	0.69070	0.00012	0.00192	0.05907	0.00E+00	0.161	6.61E-05	1.66	0.158	0.694	1.03
0.57	1782	0.00000	0.00003	0.21370	0.00000	0.00242	0.70405	0.00003	0.00072	0.07903	4.19E-05	0.179	6.22E-05	1.53	0.133	0.718	1.00
0.50	1641	0.00000	0.00001	0.18989	0.00000	0.00158	0.71409	0.00001	0.00028	0.09415	0.00E+00	0.197	5.89E-05	1.46	0.119	0.723	1.00

TABLE D5b

H₂ (77°F) Combusted with Air (77°F) at Various Equivalence Ratios (English Units)

Equiv. Ratio	Flame Temp. (°F)	H	H₂	H₂O(g)	NH₃	NO	N₂	O	OH	O₂	Enthalpy, H (Btu/lb)	Density (lb/ft³)	Viscosity (lbm/ft-hr)	Equil. Spec. Heat (Btu/lb-F)	Equil. Therm. Cond. (Btu/ft-hr-F)	Equil. Pr	Le
4.00	2347	0.00002	0.51009	0.17004	0.00001	0.00000	0.31983	0.00000	0.00000	0.00000	0	0.00750	3.54E-05	0.657	0.165	0.509	1.00
2.00	3249	0.00122	0.25696	0.25735	0.00000	0.00001	0.48433	0.00000	0.00014	0.00000	0	0.00692	4.51E-05	0.532	0.163	0.529	1.09
1.33	3706	0.00332	0.10285	0.30829	0.00000	0.00021	0.58357	0.00004	0.00167	0.00005	0	0.00728	4.95E-05	0.530	0.180	0.523	1.22
1.00	3827	0.00179	0.01526	0.32366	0.00000	0.00264	0.64440	0.00054	0.00696	0.00475	0	0.00775	5.06E-05	0.650	0.221	0.536	1.28
0.80	3442	0.00018	0.00144	0.28314	0.00000	0.00454	0.67259	0.00040	0.00471	0.03300	0	0.00885	4.76E-05	0.459	0.129	0.610	1.15
0.67	3063	0.00002	0.00021	0.24439	0.00000	0.00358	0.69070	0.00012	0.00192	0.05907	0	0.01002	4.45E-05	0.396	0.091	0.694	1.03
0.57	2748	0.00000	0.00003	0.21370	0.00000	0.00242	0.70405	0.00003	0.00072	0.07903	0	0.01118	4.18E-05	0.366	0.077	0.718	1.00
0.50	2494	0.00000	0.00001	0.18989	0.00000	0.00158	0.71409	0.00001	0.00028	0.09415	0	0.01228	3.96E-05	0.350	0.069	0.723	1.00

TABLE D6a

H_2 (298K) Stoichiometrically ($\phi = 1.00$) Combusted with Air (Variable Temp.) with an Exhaust Temp. of 1367K (Metric Units)

Oxid. Temp. (K)	H_2	$H_2O(g)$	N_2	O_2	Enthalpy, H (kJ/kg)	Density (kg/m³)	Viscosity (kg/m-s)	Equil. Spec. Heat (kJ/kg-K)	Equil. Therm. Cond. (W/m-K)	Equil. Pr	Le	Avail. Heat
298	0.00003	0.34707	0.65288	0.00001	-1.93E+03	0.219	5.11E-05	1.56	0.107	0.744	1.00	47.5%
366	0.00003	0.34707	0.65288	0.00001	-1.93E+03	0.219	5.11E-05	1.56	0.107	0.744	1.00	49.2%
478	0.00003	0.34707	0.65288	0.00001	-1.93E+03	0.219	5.11E-05	1.56	0.107	0.744	1.00	51.9%
589	0.00003	0.34707	0.65288	0.00001	-1.93E+03	0.219	5.11E-05	1.56	0.107	0.744	1.00	54.7%
700	0.00003	0.34707	0.65288	0.00001	-1.93E+03	0.219	5.11E-05	1.56	0.107	0.744	1.00	57.5%
811	0.00003	0.34707	0.65288	0.00001	-1.93E+03	0.219	5.11E-05	1.56	0.107	0.744	1.00	60.4%
922	0.00003	0.34707	0.65288	0.00001	-1.93E+03	0.219	5.11E-05	1.56	0.107	0.744	1.00	63.4%
1033	0.00003	0.34707	0.65288	0.00001	-1.93E+03	0.219	5.11E-05	1.56	0.107	0.744	1.00	66.4%
1144	0.00003	0.34707	0.65288	0.00001	-1.93E+03	0.219	5.11E-05	1.56	0.107	0.744	1.00	69.5%
1255	0.00003	0.34707	0.65288	0.00001	-1.93E+03	0.219	5.11E-05	1.56	0.107	0.744	1.00	72.7%
1366	0.00003	0.34707	0.65288	0.00001	-1.93E+03	0.219	5.11E-05	1.56	0.107	0.744	1.00	75.9%

TABLE D6b

H₂ (77°F) Stoichiometrically (φ = 1.00) Combusted with Air (Variable Temp.) with an Exhaust Temp. of 2000°F (English Units)

Oxid. Temp. (°F)	H2	H2O(g)	N2	O2	Enthalpy, H (Btu/lb)	Density (lb/ft³)	Viscosity (lbm/ft-hr)	Equil. Spec. Heat (Btu/lb-°F)	Equil. Therm. Cond. (Btu/ft-hr-F)	Equil. Pr	Le	Avail. Heat
77	0.00003	0.34707	0.65288	0.00001	-829	0.01366	3.43E-05	0.373	0.062	0.744	1.00	47.5%
200	0.00003	0.34707	0.65288	0.00001	-829	0.01366	3.43E-05	0.373	0.062	0.744	1.00	49.2%
400	0.00003	0.34707	0.65288	0.00001	-829	0.01366	3.43E-05	0.373	0.062	0.744	1.00	51.9%
600	0.00003	0.34707	0.65288	0.00001	-829	0.01366	3.43E-05	0.373	0.062	0.744	1.00	54.7%
800	0.00003	0.34707	0.65288	0.00001	-829	0.01366	3.43E-05	0.373	0.062	0.744	1.00	57.5%
1000	0.00003	0.34707	0.65288	0.00001	-829	0.01366	3.43E-05	0.373	0.062	0.744	1.00	60.4%
1200	0.00003	0.34707	0.65288	0.00001	-829	0.01366	3.43E-05	0.373	0.062	0.744	1.00	63.4%
1400	0.00003	0.34707	0.65288	0.00001	-829	0.01366	3.43E-05	0.373	0.062	0.744	1.00	66.4%
1600	0.00003	0.34707	0.65288	0.00001	-829	0.01366	3.43E-05	0.373	0.062	0.744	1.00	69.5%
1800	0.00003	0.34707	0.65288	0.00001	-829	0.01366	3.43E-05	0.373	0.062	0.744	1.00	72.7%
2000	0.00003	0.34707	0.65288	0.00001	-829	0.01366	3.43E-05	0.373	0.062	0.744	1.00	75.9%

TABLE D7a

CH$_4$ (298K) Adiabatically and Stoichiometrically (ϕ = 1.00) Combusted with a Variable Composition Oxidizer (298K) (Metric Units)

O$_2$	N$_2$	Flame Temp. (K)	CO	CO$_2$	H	HO$_2$	H$_2$	H$_2$O(g)	NO	N$_2$	O	OH	O$_2$	Enthalpy, H (kJ/kg)	Density (kg/m^3)	Viscosity (kg/m-s)	Equil. Spec. Heat (kJ/kg-K)	Equil. Therm. Cond. (W/m-K)	Equil. Pr	Le
0.21	0.79	2225	0.00893	0.08539	0.00039	0.00000	0.00361	0.18338	0.00197	0.70866	0.00021	0.00291	0.00455	-2.58E+02	0.150	7.12E-05	2.19	0.248	0.630	1.12
0.30	0.70	2525	0.03158	0.09513	0.00312	0.00001	0.01249	0.23302	0.00594	0.58834	0.00204	0.01269	0.01565	-3.56E+02	0.129	7.82E-05	3.63	0.533	0.533	1.29
0.40	0.60	2703	0.05780	0.09934	0.00884	0.00001	0.02389	0.27268	0.00949	0.46666	0.00616	0.02653	0.02859	-4.57E+02	0.116	8.27E-05	5.27	0.909	0.479	1.39
0.50	0.50	2813	0.08072	0.10220	0.01587	0.00002	0.03474	0.30299	0.01155	0.36006	0.01138	0.04032	0.04013	-5.51E+02	0.108	8.57E-05	6.82	1.286	0.454	1.44
0.60	0.40	2889	0.10031	0.10472	0.02321	0.00003	0.04449	0.32738	0.01231	0.26721	0.01697	0.05312	0.05025	-6.38E+02	0.102	8.80E-05	8.28	1.646	0.443	1.46
0.70	0.30	2944	0.11712	0.10704	0.03039	0.00004	0.05311	0.34761	0.01197	0.18615	0.02255	0.06479	0.05922	-7.20E+02	0.097	8.99E-05	9.65	1.983	0.437	1.46
0.80	0.20	2988	0.13168	0.10920	0.03721	0.00005	0.06070	0.36473	0.01059	0.11514	0.02798	0.07540	0.06732	-7.97E+02	0.092	9.14E-05	10.95	2.297	0.436	1.45
0.90	0.10	3023	0.14438	0.11120	0.04363	0.00006	0.06736	0.37937	0.00789	0.05285	0.03325	0.08515	0.07487	-8.69E+02	0.089	9.27E-05	12.18	2.589	0.436	1.44
1.00	0.00	3054	0.15534	0.11310	0.04983	0.00007	0.07296	0.39154	0.00000	0.00000	0.03893	0.09486	0.08338	-9.36E+02	0.086	9.39E-05	13.35	2.868	0.437	1.44

TABLE D7b

CH$_4$ (77°F) Adiabatically and Stoichiometrically (ϕ = 1.00) Combusted with a Variable Composition Oxidizer (77°F) (English Units)

O$_2$	N$_2$	Flame Temp. (°F)	CO	CO$_2$	H	HO$_2$	H$_2$	H$_2$O(g)	NO	N$_2$	O	OH	O$_2$	Enthalpy H (Btu/lb)	Density (lb/ft^3)	Viscosity (lbm/ft-hr)	Equil. Spec. Heat (Btu/lb-F)	Equil. Therm. Cond. (Btu/lb-hr-F)	Equil. Pr	Le
0.21	0.79	3545	0.00893	0.08539	0.00039	0.00000	0.00361	0.18338	0.00197	0.70866	0.00021	0.00291	0.00455	-111	0.00938	4.79E-05	0.524	0.143	0.536	1.28
0.30	0.70	4086	0.03158	0.09513	0.00312	0.00001	0.01249	0.23302	0.00594	0.58834	0.00204	0.01269	0.01565	-153	0.00805	5.26E-05	0.867	0.308	0.454	1.45
0.40	0.60	4406	0.05780	0.09934	0.00884	0.00001	0.02389	0.27268	0.00949	0.46666	0.00616	0.02653	0.02859	-196	0.00726	5.56E-05	1.258	0.525	0.431	1.49
0.50	0.50	4604	0.08072	0.10220	0.01587	0.00002	0.03474	0.30299	0.01155	0.36006	0.01138	0.04032	0.04013	-237	0.00673	5.76E-05	1.628	0.743	0.429	1.49
0.60	0.40	4740	0.10031	0.10472	0.02321	0.00003	0.04449	0.32738	0.01231	0.26721	0.01697	0.05312	0.05025	-274	0.00634	5.92E-05	1.977	0.951	0.434	1.47
0.70	0.30	4840	0.11712	0.10704	0.03039	0.00004	0.05311	0.34761	0.01197	0.18615	0.02255	0.06479	0.05922	-310	0.00603	6.04E-05	2.305	1.146	0.442	1.45
0.80	0.20	4918	0.13168	0.10920	0.03721	0.00005	0.06070	0.36473	0.01059	0.11514	0.02798	0.07540	0.06732	-343	0.00577	6.15E-05	2.615	1.327	0.452	1.43
0.90	0.10	4981	0.14438	0.11120	0.04363	0.00006	0.06736	0.37937	0.00789	0.05285	0.03325	0.08515	0.07487	-373	0.00555	6.24E-05	2.908	1.496	0.462	1.41
1.00	0.00	5037	0.15534	0.11310	0.04983	0.00007	0.07296	0.39154	0.00000	0.00000	0.03893	0.09486	0.08338	-403	0.00535	6.32E-05	3.189	1.657	0.472	1.40

TABLE D8a
CH₄ (298K) Adiabatically and Stoichiometrically (φ = 1.00) Combusted with Air (Variable Temp.) (Metric Units)

Air Temp. (K)	Flame Temp. (K)	CO	CO₂	H	HO₂	H₂	H₂O(g)	NO	N₂	O	OH	O₂	Enthalpy, H (kJ/kg)	Density (kg/m³)	Viscosity (kg/m-s)	Equil. Spec. Heat (kJ/kg-K)	Equil. Therm. Cond. (W/m-K)	Equil. Pr	Le
298	2225	0.00893	0.08539	0.00039	0.00000	0.00361	0.18338	0.00197	0.70866	0.00021	0.00291	0.00455	-2.58E+02	0.150	7.12E-05	2.19	0.248	0.630	1.12
366	2254	0.01008	0.08415	0.00048	0.00000	0.00404	0.18249	0.00222	0.70782	0.00027	0.00335	0.00510	-1.93E+02	0.148	7.19E-05	2.27	0.262	0.622	1.14
478	2301	0.01212	0.08193	0.00067	0.00000	0.00482	0.18088	0.00266	0.70631	0.00039	0.00415	0.00607	-8.43E+01	0.145	7.28E-05	2.39	0.286	0.608	1.16
589	2345	0.01431	0.07955	0.00091	0.00000	0.00566	0.17908	0.00315	0.70464	0.00054	0.00505	0.00711	2.52E+01	0.142	7.38E-05	2.52	0.313	0.595	1.18
700	2389	0.01665	0.07700	0.00121	0.00000	0.00659	0.17709	0.00367	0.70279	0.00073	0.00605	0.00821	1.37E+02	0.139	7.47E-05	2.66	0.342	0.580	1.21
811	2431	0.01913	0.07429	0.00159	0.00000	0.00759	0.17488	0.00424	0.70078	0.00098	0.00715	0.00936	2.52E+02	0.136	7.56E-05	2.80	0.374	0.566	1.23
922	2471	0.02172	0.07146	0.00205	0.00000	0.00867	0.17248	0.00484	0.69860	0.00128	0.00836	0.01055	3.70E+02	0.134	7.64E-05	2.95	0.408	0.552	1.26
1033	2511	0.02437	0.06853	0.00260	0.00000	0.00981	0.16987	0.00547	0.69627	0.00165	0.00966	0.01177	4.90E+02	0.131	7.72E-05	3.10	0.445	0.538	1.29
1144	2550	0.02706	0.06556	0.00325	0.00000	0.01102	0.16708	0.00612	0.69380	0.00208	0.01103	0.01299	6.12E+02	0.129	7.80E-05	3.26	0.484	0.524	1.32
1255	2587	0.02975	0.06257	0.00400	0.00000	0.01228	0.16411	0.00679	0.69120	0.00259	0.01248	0.01420	7.36E+02	0.126	7.87E-05	3.41	0.526	0.511	1.34
1366	2623	0.03242	0.05958	0.00487	0.00001	0.01360	0.16098	0.00747	0.68849	0.00318	0.01399	0.01540	8.62E+02	0.124	7.95E-05	3.57	0.569	0.498	1.37

TABLE D8b

CH$_4$ (77°F) Adiabatically and Stoichiometrically (φ = 1.00) Combusted with Air (Variable Temp.) (English Units)

Air Temp. (°F)	Flame Temp. (°F)	CO	CO$_2$	H	HO$_2$	H$_2$	H$_2$O(g)	NO	N$_2$	O	OH	O$_2$	Enthalpy, H (Btu/lb)	Density (lb/ft³)	Viscosity (lbm/ft-hr)	Equil. Spec. Heat (Btu/lb-F)	Equil. Therm. Cond. (Btu/ft-hr-F)	Equil. Pr	Le
77	3545	0.00893	0.08539	0.00039	0.00000	0.00361	0.18338	0.00197	0.70866	0.00021	0.00291	0.00455	-111	0.00938	4.79E-05	0.524	0.143	0.630	1.12
200	3598	0.01008	0.08415	0.00048	0.00000	0.00404	0.18249	0.00222	0.70782	0.00027	0.00335	0.00510	-83	0.00925	4.83E-05	0.541	0.151	0.622	1.14
400	3682	0.01212	0.08193	0.00067	0.00000	0.00482	0.18088	0.00266	0.70631	0.00039	0.00415	0.00607	-36	0.00904	4.90E-05	0.571	0.165	0.608	1.16
600	3762	0.01431	0.07955	0.00091	0.00000	0.00566	0.17908	0.00315	0.70464	0.00054	0.00505	0.00711	11	0.00886	4.96E-05	0.603	0.181	0.595	1.18
800	3840	0.01665	0.07700	0.00121	0.00000	0.00659	0.17709	0.00367	0.70279	0.00073	0.00605	0.00821	59	0.00868	5.02E-05	0.636	0.198	0.580	1.21
1000	3915	0.01913	0.07429	0.00159	0.00000	0.00759	0.17488	0.00424	0.70078	0.00098	0.00715	0.00936	108	0.00850	5.08E-05	0.670	0.216	0.566	1.23
1200	3989	0.02172	0.07146	0.00205	0.00000	0.00867	0.17248	0.00484	0.69860	0.00128	0.00836	0.01055	159	0.00834	5.14E-05	0.705	0.236	0.552	1.26
1400	4060	0.02437	0.06853	0.00260	0.00000	0.00981	0.16987	0.00547	0.69627	0.00165	0.00966	0.01177	211	0.00819	5.19E-05	0.741	0.257	0.538	1.29
1600	4129	0.02706	0.06556	0.00325	0.00000	0.01102	0.16708	0.00612	0.69380	0.00208	0.01103	0.01299	263	0.00804	5.25E-05	0.778	0.280	0.524	1.32
1800	4196	0.02975	0.06257	0.00400	0.00000	0.01228	0.16411	0.00679	0.69120	0.00259	0.01248	0.01420	317	0.00790	5.30E-05	0.815	0.304	0.511	1.34
2000	4261	0.03242	0.05958	0.00487	0.00001	0.01360	0.16098	0.00747	0.68849	0.00318	0.01399	0.01540	371	0.00776	5.34E-05	0.852	0.329	0.498	1.37

TABLE D9a

CH_4 (Variable Temp.) Adiabatically and Stoichiometrically ($\phi = 1.00$) Combusted with Air (298K) (Metric Units)

Fuel Temp. (K)	Flame Temp. (K)	CO	CO_2	H	H_2	H_2O(g)	NO	N_2	O	OH	O_2	Enthalpy, H (kJ/kg)	Density (kg/m³)	Viscosity (kg/m-s)	Equil. Spec. Heat (kJ/kg-K)	Equil. Therm. Cond. (W/m-K)	Equil. Pr	Le
298	2225	0.00893	0.08539	0.00039	0.00361	0.18338	0.00197	0.70866	0.00021	0.00291	0.00455	-2.58E+02	0.150	7.12E-05	2.19	0.248	0.630	1.12
311	2226	0.00896	0.08536	0.00039	0.00362	0.18336	0.00197	0.70864	0.00021	0.00292	0.00456	-2.56E+02	0.150	7.13E-05	2.19	0.248	0.630	1.12
366	2229	0.00908	0.08523	0.00040	0.00366	0.18327	0.00200	0.70856	0.00022	0.00297	0.00462	-2.49E+02	0.150	7.13E-05	2.20	0.250	0.629	1.13
422	2233	0.00922	0.08508	0.00041	0.00371	0.18316	0.00203	0.70846	0.00023	0.00302	0.00468	-2.41E+02	0.150	7.14E-05	2.21	0.251	0.628	1.13
478	2236	0.00936	0.08492	0.00042	0.00377	0.18305	0.00206	0.70835	0.00023	0.00308	0.00476	-2.33E+02	0.149	7.15E-05	2.22	0.253	0.627	1.13
533	2240	0.00952	0.08475	0.00043	0.00383	0.18293	0.00209	0.70823	0.00024	0.00313	0.00483	-2.24E+02	0.149	7.16E-05	2.23	0.255	0.626	1.13
589	2245	0.00969	0.08457	0.00045	0.00389	0.18280	0.00213	0.70811	0.00025	0.00320	0.00491	-2.14E+02	0.149	7.17E-05	2.24	0.257	0.625	1.13
644	2249	0.00987	0.08437	0.00046	0.00396	0.18266	0.00217	0.70798	0.00026	0.00327	0.00500	-2.04E+02	0.149	7.18E-05	2.25	0.259	0.624	1.13
700	2254	0.01007	0.08416	0.00048	0.00404	0.18250	0.00221	0.70783	0.00027	0.00334	0.00509	-1.93E+02	0.148	7.19E-05	2.26	0.262	0.622	1.14
755	2259	0.01027	0.08394	0.00050	0.00412	0.18234	0.00226	0.70768	0.00028	0.00342	0.00519	-1.82E+02	0.148	7.20E-05	2.28	0.264	0.621	1.14
811	2264	0.01050	0.08369	0.00052	0.00420	0.18217	0.00231	0.70752	0.00029	0.00351	0.00530	-1.70E+02	0.147	7.21E-05	2.29	0.267	0.619	1.14

TABLE D9b

CH_4 (Variable Temp.) Adiabatically and Stoichiometrically ($\phi = 1.00$) Combusted with Air (77°F) (English Units)

Fuel Temp. (°F)	Flame Temp. (°F)	CO	CO_2	H	H_2	H_2O(g)	NO	N_2	O	OH	O_2	Enthalpy, H (Btu/lb)	Density (lb/ft²)	Viscosity (lbm/ft-hr)	Equil. Spec. Heat (Btu/lb-F)	Equil. Therm. Cond. (Btu/ft-hr-F)	Equil. Pr	Le
77	2225	0.00893	0.08539	0.00039	0.00361	0.18338	0.00197	0.70866	0.00021	0.00291	0.00455	-111	0.00938	4.79E-05	0.524	0.143	0.630	1.12
100	2226	0.00896	0.08536	0.00039	0.00362	0.18336	0.00197	0.70864	0.00021	0.00292	0.00456	-110	0.00938	4.79E-05	0.524	0.143	0.630	1.12
200	2229	0.00908	0.08523	0.00040	0.00366	0.18327	0.00200	0.70856	0.00022	0.00297	0.00462	-107	0.00936	4.80E-05	0.526	0.144	0.629	1.13
300	2233	0.00922	0.08508	0.00041	0.00371	0.18316	0.00203	0.70846	0.00023	0.00302	0.00468	-104	0.00935	4.80E-05	0.528	0.145	0.628	1.13
400	2236	0.00936	0.08492	0.00042	0.00377	0.18305	0.00206	0.70835	0.00023	0.00308	0.00476	-100	0.00933	4.81E-05	0.530	0.146	0.627	1.13
500	2240	0.00952	0.08475	0.00043	0.00383	0.18293	0.00209	0.70823	0.00024	0.00313	0.00483	-96	0.00931	4.81E-05	0.533	0.147	0.626	1.13
600	2245	0.00969	0.08457	0.00045	0.00389	0.18280	0.00213	0.70811	0.00025	0.00320	0.00491	-92	0.00929	4.82E-05	0.535	0.149	0.625	1.13
700	2249	0.00987	0.08437	0.00046	0.00396	0.18266	0.00217	0.70798	0.00026	0.00327	0.00500	-88	0.00927	4.83E-05	0.538	0.150	0.624	1.13
800	2254	0.01007	0.08416	0.00048	0.00404	0.18250	0.00221	0.70783	0.00027	0.00334	0.00509	-83	0.00925	4.83E-05	0.541	0.151	0.622	1.14
900	2259	0.01027	0.08394	0.00050	0.00412	0.18234	0.00226	0.70768	0.00028	0.00342	0.00519	-78	0.00923	4.84E-05	0.544	0.153	0.621	1.14
1000	2264	0.01050	0.08369	0.00052	0.00420	0.18217	0.00231	0.70752	0.00029	0.00351	0.00530	-73	0.00920	4.85E-05	0.547	0.154	0.619	1.14

TABLE D10a

CH$_4$ (298K) Stoichiometrically (ϕ = 1.00) Combusted with Air (298K) (Metric Units)

Gas Temp. (K)	CO	CO$_2$	H	H$_2$	H$_2$O(g)	NO	N$_2$	O	OH	O$_2$	Enthalpy, H (kJ/kg)	Density (kg/m^3)	Viscosity (kg/m-s)	Equil. Spec. Heat (kJ/kg-K)	Equil. Therm. Cond. (W/m-K)	Equil. Pr	Le	Avail. Heat
366	0.00000	0.09502	0.00000	0.00000	0.19004	0.00000	0.71493	0.00000	0.00000	0.00000	-2.94E+03	0.920	1.89E-05	1.13	0.027	0.782	1.00	87.6%
478	0.00000	0.09502	0.00000	0.00000	0.19004	0.00000	0.71493	0.00000	0.00000	0.00000	-2.81E+03	0.705	2.35E-05	1.15	0.035	0.772	1.00	83.5%
589	0.00000	0.09502	0.00000	0.00000	0.19005	0.00000	0.71493	0.00000	0.00000	0.00000	-2.68E+03	0.572	2.77E-05	1.19	0.043	0.766	1.00	79.2%
700	0.00000	0.09502	0.00000	0.00000	0.19005	0.00000	0.71493	0.00000	0.00000	0.00000	-2.55E+03	0.481	3.16E-05	1.22	0.051	0.761	1.00	74.8%
811	0.00000	0.09502	0.00000	0.00000	0.19005	0.00000	0.71493	0.00000	0.00000	0.00000	-2.41E+03	0.415	3.53E-05	1.26	0.059	0.756	1.00	70.3%
922	0.00000	0.09502	0.00000	0.00000	0.19005	0.00000	0.71493	0.00000	0.00000	0.00000	-2.27E+03	0.365	3.87E-05	1.29	0.067	0.751	1.00	65.7%
1033	0.00000	0.09502	0.00000	0.00000	0.19005	0.00000	0.71493	0.00000	0.00000	0.00000	-2.13E+03	0.326	4.21E-05	1.33	0.075	0.747	1.00	61.0%
1144	0.00000	0.09502	0.00000	0.00000	0.19005	0.00000	0.71493	0.00000	0.00000	0.00000	-1.98E+03	0.294	4.52E-05	1.35	0.082	0.744	1.00	56.1%
1255	0.00000	0.09502	0.00000	0.00000	0.19005	0.00000	0.71493	0.00000	0.00000	0.00000	-1.82E+03	0.268	4.83E-05	1.38	0.090	0.741	1.00	51.1%
1366	0.00001	0.09501	0.00000	0.00001	0.19003	0.00000	0.71492	0.00000	0.00000	0.00001	-1.67E+03	0.247	5.12E-05	1.40	0.097	0.738	1.00	46.1%
1478	0.00005	0.09496	0.00000	0.00004	0.18999	0.00002	0.71489	0.00000	0.00001	0.00004	-1.51E+03	0.228	5.40E-05	1.43	0.106	0.732	1.00	40.9%
1589	0.00016	0.09485	0.00000	0.00011	0.18989	0.00004	0.71481	0.00000	0.00003	0.00010	-1.35E+03	0.212	5.68E-05	1.46	0.114	0.728	1.00	35.7%
1700	0.00041	0.09458	0.00000	0.00024	0.18969	0.00010	0.71463	0.00000	0.00010	0.00025	-1.19E+03	0.198	5.94E-05	1.51	0.125	0.719	1.01	30.3%
1811	0.00092	0.09403	0.00001	0.00048	0.18929	0.00022	0.71427	0.00000	0.00023	0.00054	-1.01E+03	0.186	6.20E-05	1.58	0.138	0.710	1.02	24.7%
1922	0.00190	0.09298	0.00003	0.00090	0.18858	0.00044	0.71360	0.00001	0.00051	0.00105	-8.34E+02	0.175	6.46E-05	1.67	0.155	0.698	1.03	18.8%
2033	0.00357	0.09118	0.00008	0.00157	0.18736	0.00081	0.71245	0.00004	0.00103	0.00191	-6.40E+02	0.165	6.71E-05	1.82	0.181	0.675	1.06	12.5%
2144	0.00622	0.08832	0.00021	0.00259	0.18542	0.00138	0.71060	0.00011	0.00193	0.00323	-4.28E+02	0.156	6.95E-05	2.01	0.215	0.651	1.09	5.6%
2255	0.00893	0.08539	0.00039	0.00361	0.18338	0.00197	0.70866	0.00021	0.00291	0.00455	-2.58E+02	0.150	7.12E-05	2.19	0.248	0.630	1.12	0.0%

TABLE D10b

CH₄ (Variable Temp.) Adiabatically and Stoichiometrically (φ = 1.00) Combusted with Air (77°F) (English Units)

Flame Temp. (°F)	CO	CO₂	H	H₂	H₂O(g)	NO	N₂	O	OH	O₂	Enthalpy, H (Btu/lb)	Density (lb/ft³)	Viscosity (lbm/ft-hr)	Equil. Spec. Heat (Btu/lb-F)	Equil. Therm. Cond. (Btu/ft-hr-F)	Equil. Pr	Le	Avail. Heat
200	0.00000	0.09502	0.00000	0.00000	0.19004	0.00000	0.71493	0.00000	0.00000	0.00000	-1265	0.05744	1.27E-05	0.269	0.016	0.782	1.00	87.6%
400	0.00000	0.09502	0.00000	0.00000	0.19004	0.00000	0.71493	0.00000	0.00000	0.00000	-1210	0.04398	1.58E-05	0.276	0.020	0.772	1.00	83.5%
600	0.00000	0.09502	0.00000	0.00000	0.19005	0.00000	0.71493	0.00000	0.00000	0.00000	-1154	0.03569	1.86E-05	0.284	0.025	0.766	1.00	79.2%
800	0.00000	0.09502	0.00000	0.00000	0.19005	0.00000	0.71493	0.00000	0.00000	0.00000	-1096	0.03003	2.12E-05	0.292	0.029	0.761	1.00	74.8%
1000	0.00000	0.09502	0.00000	0.00000	0.19005	0.00000	0.71493	0.00000	0.00000	0.00000	-1037	0.02592	2.37E-05	0.301	0.034	0.756	1.00	70.3%
1200	0.00000	0.09502	0.00000	0.00000	0.19005	0.00000	0.71493	0.00000	0.00000	0.00000	-976	0.02280	2.61E-05	0.309	0.039	0.751	1.00	65.7%
1400	0.00000	0.09502	0.00000	0.00000	0.19005	0.00000	0.71493	0.00000	0.00000	0.00000	-914	0.02035	2.83E-05	0.317	0.043	0.747	1.00	61.0%
1600	0.00000	0.09502	0.00000	0.00000	0.19005	0.00000	0.71493	0.00000	0.00000	0.00000	-850	0.01838	3.04E-05	0.323	0.048	0.744	1.00	56.1%
1800	0.00000	0.09502	0.00000	0.00000	0.19005	0.00000	0.71493	0.00000	0.00000	0.00000	-784	0.01675	3.25E-05	0.329	0.052	0.741	1.00	51.1%
2000	0.00001	0.09501	0.00000	0.00001	0.19003	0.00002	0.71492	0.00000	0.00000	0.00001	-718	0.01539	3.44E-05	0.335	0.056	0.738	1.00	46.1%
2200	0.00005	0.09496	0.00000	0.00004	0.18999	0.00004	0.71489	0.00000	0.00001	0.00004	-650	0.01422	3.63E-05	0.342	0.061	0.732	1.00	40.9%
2400	0.00016	0.09485	0.00000	0.00011	0.18989	0.00010	0.71481	0.00000	0.00003	0.00010	-580	0.01323	3.82E-05	0.349	0.066	0.728	1.00	35.7%
2600	0.00041	0.09458	0.00000	0.00024	0.18969	0.00010	0.71463	0.00000	0.00010	0.00025	-510	0.01236	4.00E-05	0.361	0.072	0.719	1.01	30.3%
2800	0.00092	0.09403	0.00001	0.00048	0.18929	0.00023	0.71427	0.00000	0.00023	0.00054	-436	0.01160	4.17E-05	0.376	0.080	0.710	1.02	24.7%
3000	0.00190	0.09298	0.00003	0.00090	0.18858	0.00044	0.71360	0.00001	0.00051	0.00105	-358	0.01092	4.34E-05	0.399	0.089	0.698	1.03	18.8%
3200	0.00357	0.09118	0.00008	0.00157	0.18736	0.00081	0.71245	0.00004	0.00103	0.00191	-275	0.01031	4.51E-05	0.434	0.104	0.675	1.06	12.5%
3400	0.00622	0.08832	0.00021	0.00259	0.18542	0.00138	0.71060	0.00011	0.00193	0.00323	-184	0.00976	4.67E-05	0.481	0.124	0.651	1.09	5.6%
3600	0.00893	0.08539	0.00039	0.00361	0.18338	0.00197	0.70866	0.00021	0.00291	0.00455	-111	0.00938	4.79E-05	0.524	0.143	0.630	1.12	0.0%

TABLE D11a

CH₄ (298K) Adiabatically Combusted with Air (298K) at Various Equivalence Ratios (Metric Units)

Equiv. Ratio	Flame Temp. (K)	C(gr)	CH₄	CO	CO₂	H	H₂	H₂O(g)	NH₃	NO	N₂	O	OH	O₂	Enthalpy, H (kJ/kg)	Density (kg/m³)	Viscosity (kg/m-s)	Equil. Spec. Heat (kJ/kg-K)	Equil. Therm. Cond. (W/m-K)	Equil. Pr	Le
4.00	944	0.0483	0.02281	0.11559	0.02752	0.00000	0.33907	0.04363	0.00012	0.00000	0.40293	0.00000	0.00000	0.00000	-884	0.246	3.85E-05	6.71	0.401	0.644	0.80
2.00	1563	0.0000	0.00000	0.11939	0.02850	0.00002	0.17637	0.11939	0.00000	0.00000	0.55633	0.00000	0.00017	0.00000	-488	0.177	5.55E-05	1.66	0.152	0.603	0.99
1.33	2030	0.0000	0.00000	0.06524	0.05041	0.00044	0.05021	0.18077	0.00000	0.00002	0.65273	0.00000	0.00017	0.00000	-337	0.154	6.68E-05	1.61	0.168	0.640	1.06
1.00	2225	0.0000	0.00000	0.00893	0.08539	0.00039	0.00361	0.18338	0.00000	0.00197	0.70866	0.00021	0.00291	0.00455	-258	0.150	7.12E-05	2.19	0.248	0.630	1.12
0.80	1995	0.0000	0.00000	0.00050	0.07692	0.00002	0.00022	0.15380	0.00000	0.00323	0.72654	0.00012	0.00164	0.03699	-208	0.170	6.65E-05	1.59	0.151	0.698	1.03
0.67	1781	0.0000	0.00000	0.00004	0.06540	0.00000	0.00002	0.13059	0.00000	0.00223	0.73705	0.00003	0.00053	0.06411	-175	0.192	6.18E-05	1.44	0.123	0.723	1.00
0.57	1612	0.0000	0.00000	0.00000	0.05656	0.00000	0.00000	0.11304	0.00000	0.00135	0.74460	0.00000	0.00017	0.08427	-151	0.213	5.79E-05	1.38	0.110	0.726	1.00
0.50	1478	0.0000	0.00000	0.00000	0.04988	0.00000	0.00000	0.09973	0.00000	0.00080	0.75018	0.00000	0.00005	0.09934	-133	0.233	5.48E-05	1.33	0.100	0.729	1.00

TABLE D11b

CH₄ (77°F) Combusted with Air (77°F) at Various Equivalence Ratios (English Units)

Equiv. Ratio	Flame Temp. (°F)	C(gr)	CH₄	CO	CO₂	H	H₂	H₂O(g)	NH₃	NO	N₂	O	OH	O₂	Enth-alpy, H(Btu/lb)	Density (lb/ft³)	Viscosity (lbm/ft-hr)	Equil. Spec. Heat (Btu/lb-°F)	Equil. Therm. Cond. (Btu/ft-hr-F)	Equil. Pr	Le
4.00	1240	0.0483	0.02281	0.11559	0.02752	0.00000	0.33907	0.04363	0.00012	0.00000	0.40293	0.00000	0.00000	0.00000	-380	0.01537	2.59E-05	1.602	0.232	0.644	0.80
2.00	2353	0.0000	0.00000	0.11939	0.02850	0.00002	0.17637	0.11939	0.00000	0.00000	0.55633	0.00000	0.00000	0.00000	-210	0.01105	3.73E-05	0.395	0.088	0.603	0.99
1.33	3195	0.0000	0.00000	0.06524	0.05041	0.00044	0.05021	0.18077	0.00000	0.00002	0.65273	0.00000	0.00017	0.00000	-145	0.00963	4.49E-05	0.385	0.097	0.640	1.06
1.00	3545	0.0000	0.00000	0.00893	0.08539	0.00039	0.00361	0.18338	0.00000	0.00197	0.70866	0.00021	0.00291	0.00455	-111	0.00938	4.79E-05	0.524	0.143	0.630	1.12
0.80	3132	0.0000	0.00000	0.00050	0.07692	0.00002	0.00022	0.15380	0.00000	0.00323	0.72654	0.00012	0.00164	0.03699	-90	0.01061	4.47E-05	0.379	0.087	0.698	1.03
0.67	2747	0.0000	0.00000	0.00004	0.06540	0.00000	0.00002	0.13059	0.00000	0.00223	0.73705	0.00003	0.00053	0.06411	-75	0.01196	4.16E-05	0.344	0.071	0.723	1.00
0.57	2441	0.0000	0.00000	0.00000	0.05656	0.00000	0.00000	0.11304	0.00000	0.00135	0.74460	0.00000	0.00017	0.08427	-65	0.01328	3.90E-05	0.330	0.064	0.726	1.00
0.50	2201	0.0000	0.00000	0.00000	0.04988	0.00000	0.00000	0.09973	0.00000	0.00080	0.75018	0.00000	0.00005	0.09934	-57	0.01452	3.68E-05	0.319	0.058	0.729	1.00

TABLE D12a

CH$_4$ (298K) Stoichiometrically (ϕ = 1.00) Combusted with Air (Variable Temp.) with an Exhaust Temp. of 1367K (Metric Units)

Oxid. Temp. (K)	CO	CO$_2$	H$_2$	H$_2$O(g)	N$_2$	O$_2$	Enthalpy, H (kJ/kg)	Density (kg/m^3)	Viscosity (kg/m-s)	Equil. Spec. Heat (kJ/kg-K)	Equil. Therm. Cond. (W/m-K)	Equil. Pr	Le	Avail. Heat
298	0.00001	0.09501	0.00001	0.19003	0.71492	0.00001	-1.67E+03	0.246	5.12E-05	1.40	0.097	0.737	1.00	46.1%
366	0.00001	0.09501	0.00001	0.19003	0.71492	0.00001	-1.67E+03	0.246	5.12E-05	1.40	0.097	0.737	1.00	48.2%
478	0.00001	0.09501	0.00001	0.19003	0.71492	0.00001	-1.67E+03	0.246	5.12E-05	1.40	0.097	0.737	1.00	51.7%
589	0.00001	0.09501	0.00001	0.19003	0.71492	0.00001	-1.67E+03	0.246	5.12E-05	1.40	0.097	0.737	1.00	55.3%
700	0.00001	0.09501	0.00001	0.19003	0.71492	0.00001	-1.67E+03	0.246	5.12E-05	1.40	0.097	0.737	1.00	59.0%
811	0.00001	0.09501	0.00001	0.19003	0.71492	0.00001	-1.67E+03	0.246	5.12E-05	1.40	0.097	0.737	1.00	62.7%
922	0.00001	0.09501	0.00001	0.19003	0.71492	0.00001	-1.67E+03	0.246	5.12E-05	1.40	0.097	0.737	1.00	66.6%
1033	0.00001	0.09501	0.00001	0.19003	0.71492	0.00001	-1.67E+03	0.246	5.12E-05	1.40	0.097	0.737	1.00	70.5%
1144	0.00001	0.09501	0.00001	0.19003	0.71492	0.00001	-1.67E+03	0.246	5.12E-05	1.40	0.097	0.737	1.00	74.5%
1255	0.00001	0.09501	0.00001	0.19003	0.71492	0.00001	-1.67E+03	0.246	5.12E-05	1.40	0.097	0.737	1.00	78.5%
1366	0.00001	0.09501	0.00001	0.19003	0.71492	0.00001	-1.67E+03	0.246	5.12E-05	1.40	0.097	0.737	1.00	82.6%

TABLE D12b
CH$_4$ (77°F) Stoichiometrically (ϕ = 1.00) Combusted with Air (Variable Temp.) with an Exhaust Temp. of 2000°F (English Units)

Oxid. Temp. (°F)	CO	CO$_2$	H$_2$	H$_2$O(g)	N$_2$	O$_2$	Enthalpy, H (Btu/lb)	Density (lb/ft³)	Viscosity (lbm/ft-hr)	Equil. Spec. Heat (Btu/lb-°F)	Equil. Therm. Cond. (Btu/ft-hr-F)	Equil. Pr	Le	Avail. Heat
77	0.00001	0.09501	0.00001	0.19003	0.71492	0.00001	-718	0.01538	3.44E-05	0.335	0.056	0.737	1.00	46.1%
200	0.00001	0.09501	0.00001	0.19003	0.71492	0.00001	-718	0.01538	3.44E-05	0.335	0.056	0.737	1.00	48.2%
400	0.00001	0.09501	0.00001	0.19003	0.71492	0.00001	-718	0.01538	3.44E-05	0.335	0.056	0.737	1.00	51.7%
600	0.00001	0.09501	0.00001	0.19003	0.71492	0.00001	-718	0.01538	3.44E-05	0.335	0.056	0.737	1.00	55.3%
800	0.00001	0.09501	0.00001	0.19003	0.71492	0.00001	-718	0.01538	3.44E-05	0.335	0.056	0.737	1.00	59.0%
1000	0.00001	0.09501	0.00001	0.19003	0.71492	0.00001	-718	0.01538	3.44E-05	0.335	0.056	0.737	1.00	62.7%
1200	0.00001	0.09501	0.00001	0.19003	0.71492	0.00001	-718	0.01538	3.44E-05	0.335	0.056	0.737	1.00	66.6%
1400	0.00001	0.09501	0.00001	0.19003	0.71492	0.00001	-718	0.01538	3.44E-05	0.335	0.056	0.737	1.00	70.5%
1600	0.00001	0.09501	0.00001	0.19003	0.71492	0.00001	-718	0.01538	3.44E-05	0.335	0.056	0.737	1.00	74.5%
1800	0.00001	0.09501	0.00001	0.19003	0.71492	0.00001	-718	0.01538	3.44E-05	0.335	0.056	0.737	1.00	78.5%
2000	0.00001	0.09501	0.00001	0.19003	0.71492	0.00001	-718	0.01538	3.44E-05	0.335	0.056	0.737	1.00	82.6%

TABLE D13a

C_3H_8 (298K) Adiabatically and Stoichiometrically (ϕ = 1.00) Combusted with a Variable-Composition Oxidizer (298K) (Metric Units)

O_2	N_2	Flame Temp. (K)	CO	CO_2	H	HO_2	H_2	H_2O(g)	NO	N_2	O	OH	O_2	Enthalpy H (kJ/kg)	Density (kg/m³)	Viscosity (kg/m-s)	Equil. Spec. Heat (kJ/kg-K)	Equil. Therm. Cond. (W/m-K)	Equil. Pr	Le
0.21	0.79	2266	0.01248	0.10269	0.00046	0.00000	0.00331	0.14838	0.00245	0.72084	0.00031	0.00327	0.00581	-1.42E+02	0.151	7.20E-05	2.27	0.259	0.631	1.13
0.30	0.70	2562	0.04149	0.11386	0.00340	0.00001	0.01089	0.18788	0.00700	0.60063	0.00266	0.01332	0.01887	-1.96E+02	0.131	7.87E-05	3.76	0.550	0.538	1.29
0.40	0.60	2739	0.07453	0.11889	0.00932	0.00002	0.02044	0.21919	0.01097	0.47806	0.00773	0.02718	0.03366	-2.51E+02	0.119	8.31E-05	5.37	0.920	0.484	1.39
0.50	0.50	2849	0.10359	0.12235	0.01654	0.00003	0.02954	0.24298	0.01326	0.36994	0.01409	0.04090	0.04677	-3.02E+02	0.111	8.61E-05	6.86	1.288	0.458	1.44
0.60	0.40	2925	0.12862	0.12538	0.02410	0.00004	0.03774	0.26203	0.01410	0.27517	0.02089	0.05366	0.05827	-3.49E+02	0.105	8.83E-05	8.25	1.637	0.445	1.45
0.70	0.30	2982	0.15027	0.12816	0.03152	0.00005	0.04501	0.27779	0.01370	0.19203	0.02770	0.06531	0.06846	-3.93E+02	0.100	9.01E-05	9.56	1.962	0.439	1.46
0.80	0.20	3027	0.16913	0.13074	0.03860	0.00006	0.05144	0.29108	0.01211	0.11888	0.03435	0.07594	0.07768	-4.35E+02	0.096	9.17E-05	10.78	2.265	0.436	1.45
0.90	0.10	3063	0.18565	0.13312	0.04529	0.00006	0.05709	0.30240	0.00901	0.05452	0.04083	0.08571	0.08629	-4.73E+02	0.093	9.30E-05	11.94	2.548	0.436	1.44
1.00	0.00	3095	0.19999	0.13537	0.05178	0.00007	0.06184	0.31164	0.00000	0.00000	0.04783	0.09546	0.09602	-5.09E+02	0.090	9.42E-05	13.05	2.818	0.436	1.43

TABLE D13b

C_3H_8 (77°F) Adiabatically and Stoichiometrically (ϕ = 1.00) Combusted with a Variable-Composition Oxidizer (77°F) (English Units)

O_2	N_2	Flame Temp. (°F)	CO	CO_2	H	HO_2	H_2	H_2O(g)	NO	N_2	O	OH	O_2	Enthalpy H (Btu/lb)	Density (lb/ft³)	Viscosity (lbm/ft-hr)	Equil. Spec. Heat (Btu/lb-F)	Equil. Therm. Cond. (Btu/ft-hr-F)	Equil. Pr	Le
0.21	0.79	3620	0.01248	0.10269	0.00046	0.00000	0.00331	0.14838	0.00245	0.72084	0.00031	0.00327	0.00581	-61	0.00942	4.84E-05	0.543	0.150	0.631	1.13
0.30	0.70	4151	0.04149	0.11386	0.00340	0.00001	0.01089	0.18788	0.00700	0.60063	0.00266	0.01332	0.01887	-84	0.00817	5.29E-05	0.897	0.318	0.538	1.29
0.40	0.60	4470	0.07453	0.11889	0.00932	0.00002	0.02044	0.21919	0.01097	0.47806	0.00773	0.02718	0.03366	-108	0.00742	5.59E-05	1.281	0.532	0.484	1.39
0.50	0.50	4668	0.10359	0.12235	0.01654	0.00003	0.02954	0.24298	0.01326	0.36994	0.01409	0.04090	0.04677	-130	0.00692	5.79E-05	1.639	0.744	0.458	1.44
0.60	0.40	4806	0.12862	0.12538	0.02410	0.00004	0.03774	0.26203	0.01410	0.27517	0.02089	0.05366	0.05827	-150	0.00655	5.94E-05	1.971	0.946	0.445	1.45
0.70	0.30	4908	0.15027	0.12816	0.03152	0.00005	0.04501	0.27779	0.01370	0.19203	0.02770	0.06531	0.06846	-169	0.00625	6.06E-05	2.283	1.134	0.439	1.46
0.80	0.20	4988	0.16913	0.13074	0.03860	0.00006	0.05144	0.29108	0.01211	0.11888	0.03435	0.07594	0.07768	-187	0.00601	6.17E-05	2.575	1.309	0.436	1.45
0.90	0.10	5053	0.18565	0.13312	0.04529	0.00006	0.05709	0.30240	0.00901	0.05452	0.04083	0.08571	0.08629	-203	0.00580	6.26E-05	2.852	1.472	0.436	1.44
1.00	0.00	5111	0.19999	0.13537	0.05178	0.00007	0.06184	0.31164	0.00000	0.00000	0.04783	0.09546	0.09602	-219	0.00561	6.34E-05	3.116	1.628	0.436	1.43

TABLE D14a
C$_3$H$_8$ (298K) Adiabatically and Stoichiometrically (ϕ = 1.00) Combusted with Air (Variable Temp.) (Metric Units)

Air Temp. (K)	Flame Temp. (K)	CO	CO$_2$	H	HO$_2$	H$_2$	H$_2$O(g)	NO	N$_2$	O	OH	O$_2$	Enthalpy, H (kJ/kg)	Density (kg/m^3)	Viscosity (kg/m-s)	Equil. Spec. Heat (kJ/kg-K)	Equil. Therm. Cond. (W/m-K)	Equil. Pr	Le
298	2266	0.01248	0.10269	0.00046	0.00000	0.00331	0.14838	0.00245	0.72084	0.00031	0.00327	0.00581	-1.42E-02	0.151	7.20E-05	2.27	0.259	0.631	1.13
366	2294	0.01394	0.10110	0.00057	0.00000	0.00368	0.14756	0.00274	0.71987	0.00038	0.00372	0.00646	-7.75E+01	0.149	7.25E-05	2.35	0.273	0.624	1.14
478	2339	0.01649	0.09831	0.00077	0.00000	0.00432	0.14609	0.00324	0.71814	0.00054	0.00453	0.00758	3.03E+01	0.146	7.35E-05	2.48	0.298	0.611	1.16
589	2382	0.01920	0.09534	0.00103	0.00000	0.00502	0.14447	0.00378	0.71625	0.00073	0.00542	0.00875	1.39E+02	0.143	7.44E-05	2.61	0.325	0.598	1.18
700	2423	0.02207	0.09219	0.00134	0.00000	0.00578	0.14269	0.00436	0.71420	0.00097	0.00641	0.00998	2.51E+02	0.140	7.52E-05	2.75	0.354	0.585	1.20
811	2464	0.02508	0.08888	0.00173	0.00000	0.00659	0.14074	0.00498	0.71198	0.00128	0.00749	0.01126	3.65E+02	0.137	7.61E-05	2.89	0.385	0.572	1.23
922	2503	0.02820	0.08543	0.00219	0.00000	0.00746	0.13862	0.00563	0.70960	0.00164	0.00865	0.01257	4.82E+02	0.135	7.69E-05	3.04	0.418	0.558	1.25
1033	2542	0.03139	0.08189	0.00274	0.00000	0.00838	0.13634	0.00631	0.70708	0.00208	0.00990	0.01390	6.02E+02	0.132	7.76E-05	3.18	0.453	0.545	1.28
1144	2579	0.03462	0.07829	0.00338	0.00000	0.00935	0.13390	0.00701	0.70442	0.00259	0.01121	0.01522	7.23E+02	0.130	7.84E-05	3.33	0.491	0.532	1.31
1255	2615	0.03784	0.07468	0.00413	0.00001	0.01036	0.13132	0.00772	0.70165	0.00318	0.01259	0.01652	8.46E+02	0.128	7.91E-05	3.48	0.530	0.519	1.33
1366	2650	0.04104	0.07109	0.00498	0.00001	0.01140	0.12860	0.00845	0.69877	0.00386	0.01401	0.01780	9.71E+02	0.126	7.98E-05	3.62	0.571	0.507	1.36

TABLE D14b

C_3H_8 (77°F) Adiabatically and Stoichiometrically (ϕ = 1.00) Combusted with Air (Variable Temp.) (English Units)

Air Temp. (°F)	Flame Temp. (°F)	CO	CO_2	H	HO_2	H_2	$H_2O(g)$	NO	N_2	O	OH	O_2	Enthalpy, H (Btu/lb)	Density (lb/ft³)	Viscosity (lbm/ft-hr)	Equil. Spec. Heat (Btu/lb-F)	Equil. Therm. Cond. (Btu/ft-hr-F)	Equil. Pr	Le
77	3620	0.01248	0.10269	0.00046	0.00000	0.00331	0.14838	0.00245	0.72084	0.00031	0.00327	0.00581	-61	0.00942	4.84E-05	0.543	0.150	0.631	1.13
200	3670	0.01394	0.10110	0.00057	0.00000	0.00368	0.14756	0.00274	0.71987	0.00038	0.00372	0.00646	-33	0.00930	4.88E-05	0.561	0.158	0.624	1.14
400	3750	0.01649	0.09831	0.00077	0.00000	0.00432	0.14609	0.00324	0.71814	0.00054	0.00453	0.00758	13	0.00910	4.94E-05	0.592	0.172	0.611	1.16
600	3827	0.01920	0.09534	0.00103	0.00000	0.00502	0.14447	0.00378	0.71625	0.00073	0.00542	0.00875	60	0.00892	5.00E-05	0.624	0.188	0.598	1.18
800	3902	0.02207	0.09219	0.00134	0.00000	0.00578	0.14269	0.00436	0.71420	0.00097	0.00641	0.00998	108	0.00874	5.06E-05	0.657	0.204	0.585	1.20
1000	3975	0.02508	0.08888	0.00173	0.00000	0.00659	0.14074	0.00498	0.71198	0.00128	0.00749	0.01126	157	0.00857	5.12E-05	0.691	0.222	0.572	1.23
1200	4046	0.02820	0.08543	0.00219	0.00000	0.00746	0.13862	0.00563	0.70960	0.00164	0.00865	0.01257	207	0.00841	5.17E-05	0.725	0.242	0.558	1.25
1400	4115	0.03139	0.08189	0.00274	0.00000	0.00838	0.13634	0.00631	0.70708	0.00208	0.00990	0.01390	259	0.00826	5.22E-05	0.760	0.262	0.545	1.28
1600	4182	0.03462	0.07829	0.00338	0.00000	0.00935	0.13390	0.00701	0.70442	0.00259	0.01121	0.01522	311	0.00812	5.27E-05	0.795	0.284	0.532	1.31
1800	4248	0.03784	0.07468	0.00413	0.00001	0.01036	0.13132	0.00772	0.70165	0.00318	0.01259	0.01652	364	0.00798	5.32E-05	0.830	0.306	0.519	1.33
2000	4311	0.04104	0.07109	0.00498	0.00001	0.01140	0.12860	0.00845	0.69877	0.00386	0.01401	0.01780	418	0.00784	5.37E-05	0.865	0.330	0.507	1.36

TABLE D15a

C_3H_8 (Variable Temp.) Adiabatically and Stoichiometrically (ϕ = 1.00) Combusted with Air (298K) (Metric Units)

Fuel Temp. (K)	Flame Temp. (K)	CO	CO_2	H	H_2	$H_2O(g)$	NO	N_2	O	OH	O_2	Enthalpy, H (kJ/kg)	Density (kg/m^3)	Viscosity (kg/m-s)	Equil. Spec. Heat (kJ/kg-K)	Equil. Therm. Cond. (W/m-K)	Equil. Pr	Le
298	2266	0.01248	0.10269	0.00046	0.00331	0.14838	0.00245	0.72084	0.00031	0.00327	0.00581	-1.42E+02	0.151	7.20E-05	2.27	0.259	0.631	1.13
311	2267	0.01251	0.10266	0.00047	0.00332	0.14836	0.00246	0.72082	0.00031	0.00328	0.00582	-1.41E+02	0.151	7.20E-05	2.27	0.259	0.631	1.13
366	2270	0.01264	0.10251	0.00047	0.00335	0.14829	0.00248	0.72073	0.00032	0.00332	0.00588	-1.35E+02	0.151	7.20E-05	2.28	0.261	0.631	1.13
422	2273	0.01280	0.10233	0.00049	0.00339	0.14820	0.00251	0.72063	0.00032	0.00337	0.00596	-1.28E+02	0.150	7.21E-05	2.29	0.262	0.630	1.13
478	2276	0.01298	0.10214	0.00050	0.00344	0.14810	0.00255	0.72051	0.00033	0.00342	0.00603	-1.20E+02	0.150	7.22E-05	2.30	0.264	0.629	1.13
533	2280	0.01317	0.10193	0.00051	0.00348	0.14799	0.00259	0.72038	0.00034	0.00348	0.00612	-1.11E+02	0.150	7.22E-05	2.31	0.266	0.628	1.13
589	2284	0.01338	0.10170	0.00053	0.00354	0.14787	0.00263	0.72024	0.00035	0.00355	0.00621	-1.02E+02	0.150	7.23E-05	2.32	0.268	0.627	1.13
644	2288	0.01361	0.10146	0.00054	0.00359	0.14774	0.00267	0.72009	0.00037	0.00362	0.00631	-9.18E+01	0.149	7.24E-05	2.33	0.270	0.626	1.13
700	2293	0.01385	0.10119	0.00056	0.00365	0.14761	0.00272	0.71993	0.00038	0.00369	0.00642	-8.12E+01	0.149	7.25E-05	2.35	0.272	0.624	1.14
755	2297	0.01410	0.10091	0.00058	0.00372	0.14746	0.00277	0.71976	0.00039	0.00377	0.00653	-7.02E+01	0.149	7.26E-05	2.36	0.275	0.623	1.14
811	2302	0.01438	0.10062	0.00060	0.00379	0.14731	0.00282	0.71957	0.00041	0.00385	0.00665	-5.84E+01	0.148	7.27E-05	2.37	0.278	0.622	1.14

TABLE D15b

C_3H_8 (Variable Temp.) Adiabatically and Stoichiometrically (ϕ = 1.00) Combusted with Air (77°F) (English Units)

Fuel Temp. (°F)	Flame Temp. (°F)	CO	CO₂	H	H₂	H₂O(g)	NO	N₂	O	OH	O₂	Enthalpy, H (Btu/lb)	Density (lb/ft³)	Viscosity (lbm/ft-hr)	Equil. Spec. Heat (Btu/lb-F)	Equil. Therm. Cond. (Btu/ft-hr-F)	Equil. Pr	Le
77	3620	0.01248	0.10269	0.00046	0.00331	0.14838	0.00245	0.72084	0.00031	0.00327	0.00581	-61	0.00942	4.84E-05	0.543	0.150	0.631	1.13
100	3621	0.01251	0.10266	0.00047	0.00332	0.14836	0.00246	0.72082	0.00031	0.00328	0.00582	-61	0.00942	4.84E-05	0.543	0.150	0.631	1.13
200	3626	0.01264	0.10251	0.00047	0.00335	0.14829	0.00248	0.72073	0.00032	0.00332	0.00588	-58	0.00941	4.84E-05	0.545	0.151	0.631	1.13
300	3631	0.01280	0.10233	0.00049	0.00339	0.14820	0.00251	0.72063	0.00032	0.00337	0.00596	-55	0.00939	4.85E-05	0.547	0.152	0.630	1.13
400	3637	0.01298	0.10214	0.00050	0.00344	0.14810	0.00255	0.72051	0.00033	0.00342	0.00603	-51	0.00938	4.85E-05	0.549	0.152	0.629	1.13
500	3644	0.01317	0.10193	0.00051	0.00348	0.14799	0.00259	0.72038	0.00034	0.00348	0.00612	-48	0.00936	4.86E-05	0.552	0.154	0.628	1.13
600	3652	0.01338	0.10170	0.00053	0.00354	0.14787	0.00263	0.72024	0.00035	0.00355	0.00621	-44	0.00934	4.86E-05	0.554	0.155	0.627	1.13
700	3659	0.01361	0.10146	0.00054	0.00359	0.14774	0.00267	0.72009	0.00037	0.00362	0.00631	-39	0.00932	4.87E-05	0.557	0.156	0.626	1.13
800	3667	0.01385	0.10119	0.00056	0.00365	0.14761	0.00272	0.71993	0.00038	0.00369	0.00642	-35	0.00930	4.88E-05	0.560	0.157	0.624	1.14
900	3676	0.01410	0.10091	0.00058	0.00372	0.14746	0.00277	0.71976	0.00039	0.00377	0.00653	-30	0.00928	4.88E-05	0.563	0.159	0.623	1.14
1000	3685	0.01438	0.10062	0.00060	0.00379	0.14731	0.00282	0.71957	0.00041	0.00385	0.00665	-25	0.00926	4.89E-05	0.567	0.160	0.622	1.14

TABLE D16a
C_3H_8 (298K) Stoichiometrically (ϕ = 1.00) Combusted with Air (298K) (Metric Units)

Flame Temp. (K)	CO	CO_2	H	H_2	$H_2O(g)$	NO	N_2	O	OH	O_2	Enthalpy, H (kJ/kg)	Density (kg/m^3)	Viscosity (kg/m-s)	Equil. Spec. Heat (kJ/kg-K)	Equil. Therm. Cond. (W/m-K)	Equil. Pr	Le	Avail. Heat
366	0.00000	0.11624	0.00000	0.00000	0.15498	0.00000	0.72878	0.00000	0.00000	0.00000	-2.87E+03	0.943	1.92E-05	1.10	0.027	0.772	1.00	89.6%
478	0.00000	0.11624	0.00000	0.00000	0.15498	0.00000	0.72878	0.00000	0.00000	0.00000	-2.74E+03	0.722	2.38E-05	1.13	0.035	0.764	1.00	85.5%
589	0.00000	0.11624	0.00000	0.00000	0.15498	0.00000	0.72878	0.00000	0.00000	0.00000	-2.61E+03	0.586	2.79E-05	1.16	0.043	0.760	1.00	81.3%
700	0.00000	0.11624	0.00000	0.00000	0.15498	0.00000	0.72878	0.00000	0.00000	0.00000	-2.48E+03	0.493	3.18E-05	1.20	0.051	0.756	1.00	76.9%
811	0.00000	0.11624	0.00000	0.00000	0.15498	0.00000	0.72878	0.00000	0.00000	0.00000	-2.35E+03	0.426	3.54E-05	1.24	0.058	0.752	1.00	72.5%
922	0.00000	0.11624	0.00000	0.00000	0.15498	0.00000	0.72878	0.00000	0.00000	0.00000	-2.21E+03	0.374	3.89E-05	1.27	0.066	0.748	1.00	67.9%
1033	0.00000	0.11624	0.00000	0.00000	0.15498	0.00000	0.72878	0.00000	0.00000	0.00000	-2.07E+03	0.334	4.22E-05	1.30	0.074	0.744	1.00	63.2%
1144	0.00000	0.11624	0.00000	0.00000	0.15498	0.00000	0.72878	0.00000	0.00000	0.00000	-1.92E+03	0.302	4.53E-05	1.32	0.081	0.742	1.00	58.4%
1255	0.00000	0.11624	0.00000	0.00000	0.15498	0.00000	0.72878	0.00000	0.00000	0.00000	-1.77E+03	0.275	4.83E-05	1.35	0.088	0.739	1.00	53.6%
1366	0.00002	0.11622	0.00000	0.00001	0.15497	0.00000	0.72877	0.00000	0.00000	0.00001	-1.62E+03	0.253	5.12E-05	1.37	0.095	0.737	1.00	48.6%
1478	0.00006	0.11617	0.00000	0.00003	0.15493	0.00002	0.72874	0.00000	0.00001	0.00004	-1.46E+03	0.234	5.40E-05	1.39	0.102	0.734	1.00	43.5%
1589	0.00019	0.11603	0.00000	0.00009	0.15486	0.00004	0.72865	0.00000	0.00003	0.00011	-1.31E+03	0.217	5.67E-05	1.43	0.111	0.729	1.00	38.3%
1700	0.00048	0.11571	0.00000	0.00019	0.15469	0.00011	0.72847	0.00000	0.00009	0.00026	-1.15E+03	0.203	5.94E-05	1.48	0.122	0.722	1.01	33.0%
1811	0.00110	0.11505	0.00001	0.00038	0.15436	0.00023	0.72808	0.00000	0.00021	0.00057	-9.78E+02	0.190	6.20E-05	1.54	0.134	0.713	1.02	27.5%
1922	0.00224	0.11381	0.00002	0.00071	0.15377	0.00046	0.72737	0.00001	0.00047	0.00113	-8.01E+02	0.179	6.45E-05	1.64	0.150	0.704	1.03	21.7%
2033	0.00421	0.11167	0.00007	0.00123	0.15277	0.00084	0.72615	0.00004	0.00095	0.00206	-6.11E+02	0.169	6.69E-05	1.79	0.175	0.682	1.05	15.4%
2144	0.00733	0.10829	0.00018	0.00203	0.15115	0.00145	0.72418	0.00011	0.00178	0.00350	-4.02E+02	0.160	6.93E-05	1.99	0.208	0.661	1.08	8.5%
2266	0.01248	0.10269	0.00046	0.00331	0.14838	0.00245	0.72084	0.00031	0.00327	0.00581	-1.42E+02	0.151	7.20E-05	2.27	0.259	0.631	1.13	0.0%

TABLE D16b

C_3H_8 (Variable Temp.) Adiabatically and Stoichiometrically ($\phi = 1.00$) Combusted with Air (77°F) (English Units)

Flame Temp. (°F)	CO	CO_2	H	H_2	$H_2O(g)$	NO	N_2	O	OH	O_2	Enthalpy, H (Btu/lb)	Density (lb/ft³)	Viscosity (lbm/ft-hr)	Equil. Spec. Heat (Btu/lb-F)	Equil. Therm. Cond. (Btu/ft-hr-F)	Equil. Pr	Le	Avail. Heat
200	0.00000	0.11624	0.00000	0.00000	0.15498	0.00000	0.72878	0.00000	0.00000	0.00000	-1232	0.05888	1.29E-05	0.263	0.016	0.772	1.00	89.6%
400	0.00000	0.11624	0.00000	0.00000	0.15498	0.00000	0.72878	0.00000	0.00000	0.00000	-1178	0.04508	1.60E-05	0.270	0.020	0.764	1.00	85.5%
600	0.00000	0.11624	0.00000	0.00000	0.15498	0.00000	0.72878	0.00000	0.00000	0.00000	-1124	0.03659	1.88E-05	0.278	0.025	0.760	1.00	81.3%
800	0.00000	0.11624	0.00000	0.00000	0.15498	0.00000	0.72878	0.00000	0.00000	0.00000	-1067	0.03078	2.14E-05	0.287	0.029	0.756	1.00	76.9%
1000	0.00000	0.11624	0.00000	0.00000	0.15498	0.00000	0.72878	0.00000	0.00000	0.00000	-1009	0.02657	2.38E-05	0.295	0.034	0.752	1.00	72.5%
1200	0.00000	0.11624	0.00000	0.00000	0.15498	0.00000	0.72878	0.00000	0.00000	0.00000	-949	0.02337	2.61E-05	0.303	0.038	0.748	1.00	67.9%
1400	0.00000	0.11624	0.00000	0.00000	0.15498	0.00000	0.72878	0.00000	0.00000	0.00000	-888	0.02086	2.83E-05	0.310	0.042	0.744	1.00	63.2%
1600	0.00000	0.11624	0.00000	0.00000	0.15498	0.00000	0.72878	0.00000	0.00000	0.00000	-825	0.01884	3.05E-05	0.316	0.047	0.742	1.00	58.4%
1800	0.00000	0.11624	0.00000	0.00000	0.15498	0.00000	0.72878	0.00000	0.00000	0.00000	-761	0.01717	3.25E-05	0.322	0.051	0.739	1.00	53.6%
2000	0.00002	0.11622	0.00000	0.00001	0.15497	0.00000	0.72877	0.00000	0.00000	0.00001	-696	0.01577	3.44E-05	0.327	0.055	0.737	1.00	48.6%
2200	0.00006	0.11617	0.00000	0.00003	0.15493	0.00002	0.72874	0.00000	0.00001	0.00004	-630	0.01458	3.63E-05	0.332	0.059	0.734	1.00	43.5%
2400	0.00019	0.11603	0.00000	0.00009	0.15486	0.00004	0.72865	0.00000	0.00003	0.00011	-562	0.01356	3.82E-05	0.341	0.064	0.729	1.00	38.3%
2600	0.00048	0.11571	0.00000	0.00019	0.15469	0.00011	0.72847	0.00000	0.00009	0.00026	-493	0.01267	3.99E-05	0.353	0.070	0.722	1.01	33.0%
2800	0.00110	0.11505	0.00001	0.00038	0.15436	0.00023	0.72808	0.00000	0.00021	0.00057	-420	0.01189	4.17E-05	0.369	0.077	0.713	1.02	27.5%
3000	0.00224	0.11381	0.00002	0.00071	0.15377	0.00046	0.72737	0.00001	0.00047	0.00113	-345	0.01119	4.34E-05	0.392	0.087	0.704	1.03	21.7%
3200	0.00421	0.11167	0.00007	0.00123	0.15277	0.00084	0.72615	0.00004	0.00095	0.00206	-263	0.01057	4.50E-05	0.427	0.101	0.682	1.05	15.4%
3400	0.00733	0.10829	0.00018	0.00203	0.15115	0.00145	0.72418	0.00011	0.00178	0.00350	-173	0.01000	4.66E-05	0.474	0.120	0.661	1.08	8.5%
3620	0.01248	0.10269	0.00046	0.00331	0.14838	0.00245	0.72084	0.00031	0.00327	0.00581	-61	0.00942	4.84E-05	0.543	0.150	0.631	1.13	0.0%

TABLE D17a
C_3H_8 (298K) Adiabatically Combusted with Air (298K) at Various Equivalence Ratios (Metric Units)

Equiv. Ratio	Flame Temp. (K)	C(gr)	CH4	CO	CO2	H	H2	H2O(g)	NH3	NO	N2	O	OH	O2	Enthalpy, H (kJ/kg)	Density (kg/m³)	Viscosity (kg/m-s)	Equil. Spec. Heat (kJ/kg-K)	Equil. Therm. Cond. (W/m-K)	Equil. Pr	Le
4.00	1004	0.0701	0.00975	0.16583	0.01569	0.00000	0.30829	0.02061	0.00007	0.00000	0.40966	0.00000	0.00000	0.00000	-4.82E+02	0.245	4.06E-05	4.18	0.269	0.631	0.82
2.00	1633	0.0000	0.00000	0.15310	0.02977	0.00003	0.15167	0.09215	0.00000	0.00000	0.57329	0.00000	0.00000	0.00000	-2.68E+02	0.176	5.72E-05	1.59	0.148	0.613	1.00
1.33	2098	0.0000	0.00000	0.07936	0.06268	0.00060	0.03881	0.15013	0.00000	0.00005	0.66807	0.00000	0.00028	0.00001	-1.86E+02	0.154	6.81E-05	1.58	0.169	0.635	1.08
1.00	2266	0.0000	0.00000	0.01248	0.10269	0.00046	0.00331	0.14838	0.00000	0.00245	0.72084	0.00031	0.00327	0.00581	-1.42E+02	0.151	7.20E-05	2.27	0.259	0.631	1.13
0.80	2042	0.0000	0.00000	0.00089	0.09346	0.00003	0.00025	0.12460	0.00000	0.00371	0.73755	0.00018	0.00185	0.03748	-1.15E+02	0.169	6.74E-05	1.60	0.154	0.697	1.03
0.67	1821	0.0000	0.00000	0.00008	0.07949	0.00000	0.00003	0.10575	0.00000	0.00258	0.74663	0.00004	0.00061	0.06479	-9.68E+01	0.191	6.26E-05	1.44	0.125	0.720	1.01
0.57	1646	0.0000	0.00000	0.00001	0.06864	0.00000	0.00000	0.09143	0.00000	0.00157	0.75302	0.00001	0.00019	0.08513	-8.33E+01	0.211	5.87E-05	1.37	0.110	0.727	1.00
0.50	1508	0.0000	0.00000	0.00000	0.06046	0.00000	0.00000	0.08058	0.00000	0.00093	0.75768	0.00000	0.00006	0.10028	-7.33E+01	0.231	5.55E-05	1.32	0.101	0.729	1.00

TABLE D17b
C_3H_8 (77°F) Combusted with Air (77°F) at Various Equivalence Ratios (English Units)

Equiv. Ratio	Flame Temp. (°F)	C(gr)	CH4	CO	CO2	H	H2	H2O(g)	NH3	NO	N2	O	OH	O2	Enthalpy, H (Btu/lb)	Density (lb/ft3)	Viscosity (lbm/ft-hr)	Equil. Spec. Heat (Btu/lb-°F)	Equil. Therm. Cond. (Btu/ft-hr-F)	Equil. Pr	Le
4.00	1348	0.0701	0.00975	0.16583	0.01569	0.00000	0.30829	0.02061	0.00007	0.00000	0.40966	0.00000	0.00000	0.00000	-207	0.01532	2.73E-05	0.997	0.155	0.631	0.82
2.00	2479	0.0000	0.00000	0.15310	0.02977	0.00003	0.15167	0.09215	0.00000	0.00000	0.57329	0.00000	0.00000	0.00000	-115	0.01101	3.84E-05	0.379	0.086	0.613	1.00
1.33	3317	0.0000	0.00000	0.07936	0.06268	0.00060	0.03881	0.15013	0.00000	0.00005	0.66807	0.00000	0.00028	0.00001	-80	0.00960	4.58E-05	0.377	0.098	0.635	1.08
1.00	3620	0.0000	0.00000	0.01248	0.10269	0.00046	0.00331	0.14838	0.00000	0.00245	0.72084	0.00031	0.00327	0.00581	-61	0.00942	4.84E-05	0.543	0.150	0.631	1.13
0.80	3215	0.0000	0.00000	0.00089	0.09346	0.00003	0.00025	0.12460	0.00000	0.00371	0.73755	0.00018	0.00185	0.03748	-49	0.01058	4.53E-05	0.382	0.089	0.697	1.03
0.67	2818	0.0000	0.00000	0.00008	0.07949	0.00000	0.00003	0.10575	0.00000	0.00258	0.74663	0.00004	0.00061	0.06479	-42	0.01190	4.21E-05	0.343	0.072	0.720	1.01
0.57	2502	0.0000	0.00000	0.00001	0.06864	0.00000	0.00000	0.09143	0.00000	0.00157	0.75302	0.00001	0.00019	0.08513	-36	0.01319	3.95E-05	0.327	0.064	0.727	1.00
0.50	2254	0.0000	0.00000	0.00000	0.06046	0.00000	0.00000	0.08058	0.00000	0.00093	0.75768	0.00000	0.00006	0.10028	-32	0.01442	3.73E-05	0.316	0.058	0.729	1.00

TABLE D18a
C_3H_8 (298K) Stoichiometrically (ϕ = 1.00) Combusted with Air (Variable Temp.) with an Exhaust Temp. of 1367K (Metric Units)

Oxid. Temp. (K)	CO	CO_2	H_2	$H_2O(g)$	N_2	O_2	Enthalpy, H (kJ/kg)	Density (kg/m³)	Viscosity (kg/m-s)	Equil. Spec. Heat (kJ/kg-K)	Equil. Therm. Cond. (W/m-K)	Equil. Pr	Le	Avail. Heat
298	0.00002	0.11622	0.00001	0.15497	0.72877	0.00001	-1.62E+03	0.253	5.12E-05	1.37	0.095	0.737	1.00	48.6%
366	0.00002	0.11622	0.00001	0.15497	0.72877	0.00001	-1.62E+03	0.253	5.12E-05	1.37	0.095	0.737	1.00	50.7%
478	0.00002	0.11622	0.00001	0.15497	0.72877	0.00001	-1.62E+03	0.253	5.12E-05	1.37	0.095	0.737	1.00	54.2%
589	0.00002	0.11622	0.00001	0.15497	0.72877	0.00001	-1.62E+03	0.253	5.12E-05	1.37	0.095	0.737	1.00	57.8%
700	0.00002	0.11622	0.00001	0.15497	0.72877	0.00001	-1.62E+03	0.253	5.12E-05	1.37	0.095	0.737	1.00	61.5%
811	0.00002	0.11622	0.00001	0.15497	0.72877	0.00001	-1.62E+03	0.253	5.12E-05	1.37	0.095	0.737	1.00	65.2%
922	0.00002	0.11622	0.00001	0.15497	0.72877	0.00001	-1.62E+03	0.253	5.12E-05	1.37	0.095	0.737	1.00	69.1%
1033	0.00002	0.11622	0.00001	0.15497	0.72877	0.00001	-1.62E+03	0.253	5.12E-05	1.37	0.095	0.737	1.00	73.0%
1144	0.00002	0.11622	0.00001	0.15497	0.72877	0.00001	-1.62E+03	0.253	5.12E-05	1.37	0.095	0.737	1.00	77.0%
1255	0.00002	0.11622	0.00001	0.15497	0.72877	0.00001	-1.62E+03	0.253	5.12E-05	1.37	0.095	0.737	1.00	81.1%
1366	0.00002	0.11622	0.00001	0.15497	0.72877	0.00001	-1.62E+03	0.253	5.12E-05	1.37	0.095	0.737	1.00	85.2%

TABLE D18b

C_3H_8 (77°F) Stoichiometrically ($\phi = 1.00$) Combusted with Air (Variable Temp.) with an Exhaust Temp. of 2000°F (English Units)

Oxid. Temp. (°F)	CO	CO_2	H_2	$H_2O(g)$	N_2	O_2	Enthalpy, H (Btu/lb)	Density (lb/ft³)	Viscosity (lbm/ft-hr)	Equil. Spec. Heat (Btu/lb-°F)	Equil. Therm. Cond. (Btu/ft-hr-F)	Equil. Pr	Le	Avail. Heat
77	0.00002	0.11622	0.00001	0.15497	0.72877	0.00001	-696	0.01577	3.44E-05	0.327	0.055	0.737	1.00	48.6%
200	0.00002	0.11622	0.00001	0.15497	0.72877	0.00001	-696	0.01577	3.44E-05	0.327	0.055	0.737	1.00	50.7%
400	0.00002	0.11622	0.00001	0.15497	0.72877	0.00001	-696	0.01577	3.44E-05	0.327	0.055	0.737	1.00	54.2%
600	0.00002	0.11622	0.00001	0.15497	0.72877	0.00001	-696	0.01577	3.44E-05	0.327	0.055	0.737	1.00	57.8%
800	0.00002	0.11622	0.00001	0.15497	0.72877	0.00001	-696	0.01577	3.44E-05	0.327	0.055	0.737	1.00	61.5%
1000	0.00002	0.11622	0.00001	0.15497	0.72877	0.00001	-696	0.01577	3.44E-05	0.327	0.055	0.737	1.00	65.2%
1200	0.00002	0.11622	0.00001	0.15497	0.72877	0.00001	-696	0.01577	3.44E-05	0.327	0.055	0.737	1.00	69.1%
1400	0.00002	0.11622	0.00001	0.15497	0.72877	0.00001	-696	0.01577	3.44E-05	0.327	0.055	0.737	1.00	73.0%
1600	0.00002	0.11622	0.00001	0.15497	0.72877	0.00001	-696	0.01577	3.44E-05	0.327	0.055	0.737	1.00	77.0%
1800	0.00002	0.11622	0.00001	0.15497	0.72877	0.00001	-696	0.01577	3.44E-05	0.327	0.055	0.737	1.00	81.1%
2000	0.00002	0.11622	0.00001	0.15497	0.72877	0.00001	-696	0.01577	3.44E-05	0.327	0.055	0.737	1.00	85.2%

TABLE D19a

H$_2$ + CH$_4$ (298K) Stoichiometrically (ϕ = 1.00) Combusted with Air (298K) (Metric Units)

H$_2$	CH$_4$	Exhaust Temp. (K)	CO	CO$_2$	H	H$_2$	H$_2$O(g)	NO	N$_2$	O	OH	O$_2$	Enthalpy, H (kJ/kg)	Density (kg/m^3)	Viscosity (kg/m-s)	Equil. Spec. Heat (kJ/kg-K)	Equil. Therm. Cond. (W/m-K)	Equil. Pr	Le
0.0	1.0	2225	0.00893	0.08539	0.00039	0.00361	0.18338	0.00197	0.70866	0.00021	0.00291	0.00455	-2.58E+02	0.150	7.12E-05	2.19	0.248	0.630	1.12
0.1	0.9	2230	0.00888	0.08265	0.00041	0.00378	0.18773	0.00199	0.70676	0.00022	0.00300	0.00458	-2.51E+02	0.149	7.14E-05	2.21	0.251	0.627	1.13
0.2	0.8	2235	0.00881	0.07945	0.00043	0.00400	0.19281	0.00202	0.70453	0.00023	0.00311	0.00461	-2.43E+02	0.148	7.15E-05	2.22	0.255	0.624	1.13
0.3	0.7	2241	0.00872	0.07567	0.00046	0.00427	0.19881	0.00205	0.70189	0.00024	0.00325	0.00465	-2.33E+02	0.147	7.17E-05	2.24	0.259	0.620	1.14
0.4	0.6	2249	0.00858	0.07114	0.00050	0.00461	0.20602	0.00209	0.69871	0.00025	0.00341	0.00469	-2.22E+02	0.146	7.19E-05	2.27	0.265	0.616	1.14
0.5	0.5	2258	0.00837	0.06562	0.00055	0.00505	0.21483	0.00214	0.69480	0.00027	0.00362	0.00474	-2.07E+02	0.144	7.21E-05	2.30	0.272	0.610	1.15
0.6	0.4	2270	0.00805	0.05875	0.00062	0.00567	0.22585	0.00219	0.68990	0.00029	0.00390	0.00479	-1.89E+02	0.142	7.24E-05	2.34	0.281	0.603	1.16
0.7	0.3	2285	0.00751	0.04997	0.00072	0.00655	0.24001	0.00227	0.68354	0.00032	0.00427	0.00485	-1.64E+02	0.140	7.28E-05	2.39	0.293	0.593	1.18
0.8	0.2	2306	0.00654	0.03839	0.00088	0.00792	0.25886	0.00236	0.67500	0.00036	0.00479	0.00489	-1.31E+02	0.136	7.33E-05	2.45	0.310	0.580	1.20
0.9	0.1	2336	0.00461	0.02255	0.00117	0.01030	0.28507	0.00249	0.66289	0.00042	0.00560	0.00489	-8.08E+01	0.131	7.41E-05	2.55	0.337	0.562	1.23
1.0	0.0	2382	0.00000	0.00000	0.00179	0.01526	0.32366	0.00264	0.64440	0.00054	0.00696	0.00475	2.09E-04	0.124	7.53E-05	2.72	0.383	0.536	1.28

TABLE D19b
H₂ + CH₄ (298K) Stoichiometrically (φ = 1.00) Combusted with Air (77°F) (English Units)

H₂	CH₄	Product Temp. (°F)	CO	CO₂	H	H₂	H₂O(g)	NO	N₂	O	OH	O₂	Enthalpy, H (Btu/lb)	Density (lb/ft³)	Viscosity (lbm/ft-hr)	Equil. Spec. Heat (Btu/lb-°F)	Equil. Therm. Cond. (Btu/ft-hr-°F)	Equil. Pr	Le
0.0	1.0	3545	0.00893	0.08539	0.00039	0.00361	0.18338	0.00197	0.70866	0.00021	0.00291	0.00455	-111	0.00938	4.79E-05	0.524	0.143	0.630	1.12
0.1	0.9	3554	0.00888	0.08265	0.00041	0.00378	0.18773	0.00199	0.70676	0.00022	0.00300	0.00458	-108	0.00933	4.80E-05	0.527	0.145	0.627	1.13
0.2	0.8	3563	0.00881	0.07945	0.00043	0.00400	0.19281	0.00202	0.70453	0.00023	0.00311	0.00461	-104	0.00927	4.81E-05	0.531	0.147	0.624	1.13
0.3	0.7	3574	0.00872	0.07567	0.00046	0.00427	0.19881	0.00205	0.70189	0.00024	0.00325	0.00465	-100	0.00920	4.82E-05	0.536	0.150	0.620	1.14
0.4	0.6	3588	0.00858	0.07114	0.00050	0.00461	0.20602	0.00209	0.69871	0.00025	0.00341	0.00469	-95	0.00911	4.83E-05	0.542	0.153	0.616	1.14
0.5	0.5	3605	0.00837	0.06562	0.00055	0.00505	0.21483	0.00214	0.69480	0.00027	0.00362	0.00474	-89	0.00901	4.85E-05	0.549	0.157	0.610	1.15
0.6	0.4	3626	0.00805	0.05875	0.00062	0.00567	0.22585	0.00219	0.68990	0.00029	0.00390	0.00479	-81	0.00888	4.87E-05	0.558	0.162	0.603	1.16
0.7	0.3	3654	0.00751	0.04997	0.00072	0.00655	0.24001	0.00227	0.68354	0.00032	0.00427	0.00485	-71	0.00872	4.90E-05	0.570	0.169	0.593	1.18
0.8	0.2	3691	0.00654	0.03839	0.00088	0.00792	0.25886	0.00236	0.67500	0.00036	0.00479	0.00489	-56	0.00850	4.93E-05	0.586	0.179	0.580	1.20
0.9	0.1	3744	0.00461	0.02255	0.00117	0.01030	0.28507	0.00249	0.66289	0.00042	0.00560	0.00489	-35	0.00820	4.98E-05	0.610	0.195	0.562	1.23
1.0	0.0	3827	0.00000	0.00000	0.00179	0.01526	0.32366	0.00264	0.64440	0.00054	0.00696	0.00475	0	0.00775	5.06E-05	0.650	0.221	0.536	1.28

TABLE D20a

H$_2$ + CH$_4$ (298K) Stoichiometrically (ϕ = 1.00) Combusted with Air (533K) (Metric Units)

H$_2$	CH$_4$	Exhaust Temp. (K)	CO	CO$_2$	H	H$_2$	H$_2$O(g)	NO	N$_2$	O	OH	O$_2$	Enthalpy, H (kJ/kg)	Density (kg/m^3)	Viscosity (kg/m-s)	Equil. Spec. Heat (kJ/kg-K)	Equil. Therm. Cond. (W/m-K)	Equil. Pr	Le
0.0	1.0	2323	0.01319	0.08077	0.00078	0.00523	0.18001	0.00290	0.70550	0.00046	0.00458	0.00658	-3.02E+01	0.143	7.33E-05	2.46	0.299	0.601	1.17
0.1	0.9	2327	0.01306	0.07812	0.00082	0.00547	0.18425	0.00292	0.70359	0.00047	0.00471	0.00659	-2.32E+01	0.143	7.34E-05	2.47	0.303	0.599	1.17
0.2	0.8	2332	0.01289	0.07503	0.00086	0.00576	0.18921	0.00295	0.70136	0.00048	0.00485	0.00661	-1.50E+01	0.142	7.35E-05	2.49	0.308	0.595	1.18
0.3	0.7	2337	0.01268	0.07138	0.00091	0.00612	0.19506	0.00298	0.69872	0.00050	0.00503	0.00663	-5.26E+00	0.141	7.37E-05	2.51	0.313	0.591	1.19
0.4	0.6	2344	0.01238	0.06703	0.00098	0.00657	0.20208	0.00302	0.69554	0.00052	0.00525	0.00664	6.66E+00	0.140	7.39E-05	2.54	0.320	0.586	1.19
0.5	0.5	2352	0.01198	0.06172	0.00106	0.00716	0.21066	0.00306	0.69164	0.00054	0.00552	0.00665	2.15E+01	0.138	7.41E-05	2.57	0.328	0.580	1.20
0.6	0.4	2363	0.01139	0.05514	0.00118	0.00797	0.22137	0.00311	0.68673	0.00057	0.00587	0.00666	4.04E+01	0.136	7.44E-05	2.61	0.339	0.572	1.22
0.7	0.3	2377	0.01048	0.04677	0.00135	0.00912	0.23511	0.00318	0.68039	0.00062	0.00634	0.00665	6.54E+01	0.134	7.47E-05	2.66	0.354	0.562	1.23
0.8	0.2	2395	0.00897	0.03579	0.00161	0.01088	0.25335	0.00326	0.67185	0.00068	0.00700	0.00660	1.00E+02	0.131	7.52E-05	2.73	0.375	0.549	1.26
0.9	0.1	2422	0.00616	0.02089	0.00206	0.01389	0.27863	0.00336	0.65976	0.00078	0.00799	0.00647	1.51E+02	0.126	7.59E-05	2.84	0.406	0.531	1.29
1.0	0.0	2463	0.00000	0.00000	0.00299	0.01997	0.31560	0.00348	0.64130	0.00094	0.00960	0.00612	2.34E+02	0.120	7.70E-05	3.03	0.461	0.507	1.34

TABLE D20b
H$_2$ + CH$_4$ (298K) Stoichiometrically (ϕ = 1.00) Combusted with Air (500°F) (English Units)

H$_2$	CH$_4$	Exhaust Temp. (°F)	CO	CO$_2$	H	H$_2$	H$_2$O(g)	NO	N$_2$	O	OH	O$_2$	Enthalpy, H (Btu/lb)	Density (lb/ft^2)	Viscosity (lbm/ft-hr)	Equil. Spec. Heat (Btu/lb-°F)	Equil. Therm. Cond. (Btu/ft-hr-°F)	Equil. Pr	Le
0.0	1.0	3722	0.01319	0.08077	0.00078	0.00523	0.18001	0.00290	0.70550	0.00046	0.00458	0.00658	-13	0.00895	4.93E-05	0.587	0.173	0.601	1.17
0.1	0.9	3729	0.01306	0.07812	0.00082	0.00547	0.18425	0.00292	0.70359	0.00047	0.00471	0.00659	-10	0.00890	4.94E-05	0.590	0.175	0.599	1.17
0.2	0.8	3737	0.01289	0.07503	0.00086	0.00576	0.18921	0.00295	0.70136	0.00048	0.00485	0.00661	-6	0.00885	4.95E-05	0.594	0.178	0.595	1.18
0.3	0.7	3747	0.01268	0.07138	0.00091	0.00612	0.19506	0.00298	0.69872	0.00050	0.00503	0.00663	-2	0.00879	4.96E-05	0.599	0.181	0.591	1.19
0.4	0.6	3760	0.01238	0.06703	0.00098	0.00657	0.20208	0.00302	0.69554	0.00052	0.00525	0.00664	3	0.00871	4.97E-05	0.605	0.185	0.586	1.19
0.5	0.5	3774	0.01198	0.06172	0.00106	0.00716	0.21066	0.00306	0.69164	0.00054	0.00552	0.00665	9	0.00862	4.98E-05	0.613	0.190	0.580	1.20
0.6	0.4	3793	0.01139	0.05514	0.00118	0.00797	0.22137	0.00311	0.68673	0.00057	0.00587	0.00666	17	0.00850	5.00E-05	0.623	0.196	0.572	1.22
0.7	0.3	3818	0.01048	0.04677	0.00135	0.00912	0.23511	0.00318	0.68039	0.00062	0.00634	0.00665	28	0.00835	5.03E-05	0.635	0.204	0.562	1.23
0.8	0.2	3851	0.00897	0.03579	0.00161	0.01088	0.25335	0.00326	0.67185	0.00068	0.00700	0.00660	43	0.00816	5.06E-05	0.653	0.216	0.549	1.26
0.9	0.1	3899	0.00616	0.02089	0.00206	0.01389	0.27863	0.00336	0.65976	0.00078	0.00799	0.00647	65	0.00788	5.11E-05	0.679	0.235	0.531	1.29
1.0	0.0	3974	0.00000	0.00000	0.00299	0.01997	0.31560	0.00348	0.64130	0.00094	0.00960	0.00612	101	0.00747	5.18E-05	0.724	0.267	0.507	1.34

TABLE D21a
H$_2$ + CH$_4$ (298K) Stoichiometrically (ϕ = 1.00) Combusted with Air (811K) (Metric Units)

H$_2$	CH$_4$	Exhaust Temp. (K)	CO	CO$_2$	H	H$_2$	H$_2$O(g)	NO	N$_2$	O	OH	O$_2$	Enthalpy, H (kJ/kg)	Density (kg/m^3)	Viscosity (kg/m-s)	Equil. Spec. Heat (kJ/kg-K)	Equil. Therm. Cond. (W/m-K)	Equil. Pr	Le
0.0	1.0	2431	0.01913	0.07429	0.00159	0.00759	0.17489	0.00424	0.70078	0.00098	0.00715	0.00936	2.52E+02	0.136	7.56E-05	2.80	0.374	0.566	1.23
0.1	0.9	2434	0.01887	0.07179	0.00165	0.00791	0.17898	0.00426	0.69888	0.00100	0.00731	0.00934	2.59E+02	0.136	7.57E-05	2.82	0.379	0.563	1.24
0.2	0.8	2438	0.01854	0.06888	0.00172	0.00830	0.18376	0.00428	0.69666	0.00102	0.00751	0.00932	2.68E+02	0.135	7.58E-05	2.84	0.384	0.559	1.24
0.3	0.7	2443	0.01812	0.06545	0.00181	0.00878	0.18941	0.00430	0.69403	0.00104	0.00774	0.00930	2.78E+02	0.134	7.59E-05	2.86	0.391	0.555	1.25
0.4	0.6	2449	0.01759	0.06137	0.00192	0.00939	0.19618	0.00433	0.69087	0.00107	0.00802	0.00926	2.90E+02	0.133	7.61E-05	2.89	0.399	0.550	1.26
0.5	0.5	2456	0.01687	0.05641	0.00207	0.01018	0.20443	0.00436	0.68698	0.00111	0.00837	0.00921	3.06E+02	0.131	7.63E-05	2.92	0.410	0.544	1.27
0.6	0.4	2465	0.01587	0.05028	0.00227	0.01124	0.21473	0.00440	0.68210	0.00116	0.00882	0.00914	3.25E+02	0.130	7.65E-05	2.96	0.423	0.536	1.29
0.7	0.3	2477	0.01441	0.04251	0.00255	0.01274	0.22792	0.00444	0.67579	0.00122	0.00941	0.00902	3.51E+02	0.128	7.68E-05	3.02	0.441	0.526	1.30
0.8	0.2	2493	0.01210	0.03239	0.00297	0.01501	0.24537	0.00450	0.66729	0.00131	0.01022	0.00883	3.86E+02	0.125	7.73E-05	3.10	0.467	0.513	1.33
0.9	0.1	2517	0.00809	0.01880	0.00368	0.01879	0.26946	0.00455	0.65526	0.00144	0.01141	0.00851	4.39E+02	0.121	7.79E-05	3.23	0.506	0.497	1.36
1.0	0.0	2553	0.00000	0.00000	0.00507	0.02615	0.30448	0.00460	0.63687	0.00165	0.01328	0.00790	5.24E+02	0.115	7.89E-05	3.45	0.573	0.476	1.41

TABLE D21b
H_2 + CH_4 (298K) Stoichiometrically (ϕ = 1.00) Combusted with Air (1000°F) (English Units)

H_2	CH_4	Exhaust Temp. (°F)	CO	CO_2	H	H_2	$H_2O(g)$	NO	N_2	O	OH	O_2	Enthalpy, H (Btu/lb)	Density (lb/ft³)	Viscosity (lbm/ft-hr)	Equil. Spec. Heat (Btu/lb-°F)	Equil. Therm. Cond. (Btu/ft-hr-°F)	Equil. Pr	Le
0.0	1.0	3915	0.01913	0.07429	0.00159	0.00759	0.17489	0.00424	0.70078	0.00098	0.00715	0.00936	108	0.00850	5.08E-05	0.670	0.216	0.566	1.23
0.1	0.9	3922	0.01887	0.07179	0.00165	0.00791	0.17898	0.00426	0.69888	0.00100	0.00731	0.00934	112	0.00846	5.09E-05	0.674	0.219	0.563	1.24
0.2	0.8	3929	0.01854	0.06888	0.00172	0.00830	0.18376	0.00428	0.69666	0.00102	0.00751	0.00932	115	0.00841	5.10E-05	0.678	0.222	0.559	1.24
0.3	0.7	3938	0.01812	0.06545	0.00181	0.00878	0.18941	0.00430	0.69403	0.00104	0.00774	0.00930	120	0.00836	5.10E-05	0.683	0.226	0.555	1.25
0.4	0.6	3948	0.01759	0.06137	0.00192	0.00939	0.19618	0.00433	0.69087	0.00107	0.00802	0.00926	125	0.00829	5.11E-05	0.690	0.231	0.550	1.26
0.5	0.5	3961	0.01687	0.05641	0.00207	0.01018	0.20443	0.00436	0.68698	0.00111	0.00837	0.00921	131	0.00821	5.13E-05	0.698	0.237	0.544	1.27
0.6	0.4	3978	0.01587	0.05028	0.00227	0.01124	0.21473	0.00440	0.68210	0.00116	0.00882	0.00914	140	0.00810	5.15E-05	0.708	0.245	0.536	1.29
0.7	0.3	3999	0.01441	0.04251	0.00255	0.01274	0.22792	0.00444	0.67579	0.00122	0.00941	0.00902	151	0.00797	5.17E-05	0.722	0.255	0.526	1.30
0.8	0.2	4028	0.01210	0.03239	0.00297	0.01501	0.24537	0.00450	0.66729	0.00131	0.01022	0.00883	166	0.00779	5.20E-05	0.741	0.270	0.513	1.33
0.9	0.1	4070	0.00809	0.01880	0.00368	0.01879	0.26946	0.00455	0.65526	0.00144	0.01141	0.00851	189	0.00754	5.24E-05	0.771	0.293	0.497	1.36
1.0	0.0	4135	0.00000	0.00000	0.00507	0.02615	0.30448	0.00460	0.63687	0.00165	0.01328	0.00790	225	0.00716	5.30E-05	0.825	0.331	0.476	1.41

TABLE D22a
$H_2 + CH_4$ (298K) Stoichiometrically ($\phi = 1.00$) Combusted with Air (1089K) (Metric Units)

H_2	CH_4	Exhaust Temp. (K)	CO	CO_2	H	H_2	$H_2O(g)$	NO	N_2	O	OH	O_2	Enthalpy, H (kJ/kg)	Density (kg/m³)	Viscosity (kg/m-s)	Equil. Spec. Heat (kJ/kg-K)	Equil. Therm. Cond. (W/m-K)	Equil. Pr	Le
0.0	1.0	2531	0.02572	0.06705	0.00291	0.01041	0.16849	0.00579	0.69505	0.00186	0.01034	0.01238	5.51E+02	0.130	7.76E-05	3.18	0.465	0.531	1.30
0.1	0.9	2534	0.02528	0.06474	0.00301	0.01083	0.17242	0.00580	0.69317	0.00188	0.01054	0.01233	5.59E+02	0.129	7.77E-05	3.20	0.470	0.528	1.31
0.2	0.8	2537	0.02474	0.06206	0.00312	0.01134	0.17701	0.00581	0.69097	0.00191	0.01078	0.01226	5.67E+02	0.129	7.78E-05	3.22	0.477	0.524	1.31
0.3	0.7	2541	0.02408	0.05891	0.00327	0.01196	0.18241	0.00582	0.68836	0.00194	0.01107	0.01218	5.78E+02	0.128	7.79E-05	3.24	0.486	0.520	1.32
0.4	0.6	2546	0.02323	0.05516	0.00344	0.01273	0.18889	0.00583	0.68522	0.00198	0.01141	0.01208	5.90E+02	0.127	7.81E-05	3.27	0.496	0.515	1.33
0.5	0.5	2553	0.02213	0.05063	0.00367	0.01373	0.19679	0.00585	0.68137	0.00203	0.01184	0.01195	6.06E+02	0.126	7.82E-05	3.31	0.509	0.509	1.34
0.6	0.4	2561	0.02065	0.04503	0.00398	0.01507	0.20663	0.00586	0.67653	0.00209	0.01237	0.01178	6.26E+02	0.124	7.85E-05	3.36	0.525	0.501	1.36
0.7	0.3	2571	0.01853	0.03798	0.00441	0.01694	0.21920	0.00587	0.67027	0.00217	0.01308	0.01154	6.53E+02	0.122	7.88E-05	3.42	0.547	0.492	1.38
0.8	0.2	2585	0.01533	0.02885	0.00504	0.01973	0.23581	0.00588	0.66185	0.00228	0.01403	0.01119	6.90E+02	0.119	7.92E-05	3.52	0.579	0.481	1.40
0.9	0.1	2605	0.01002	0.01668	0.00607	0.02429	0.25866	0.00588	0.64990	0.00245	0.01538	0.01066	7.44E+02	0.116	7.97E-05	3.66	0.626	0.467	1.43
1.0	0.0	2636	0.00000	0.00000	0.00799	0.03281	0.29182	0.00583	0.63163	0.00269	0.01746	0.00977	8.32E+02	0.110	8.06E-05	3.93	0.704	0.450	1.47

TABLE D22b

H$_2$ + CH$_4$ (298K) Stoichiometrically (ϕ = 1.00) Combusted with Air (1500°F) (English Units)

H$_2$	CH$_4$	Exhaust Temp. (°F)	CO	CO$_2$	H	H$_2$	H$_2$O(g)	NO	N$_2$	O	OH	O$_2$	Enthalpy, H (Btu/lb)	Density (lb/ft^3)	Viscosity (lbm/ft-hr)	Equil. Spec. Heat (Btu/lb-°F)	Equil. Therm. Cond. (Btu/ft-hr-°F)	Equil. Pr	Le
0.0	1.0	4095	0.02572	0.06705	0.00291	0.01041	0.16849	0.00579	0.69505	0.00186	0.01034	0.01238	237	0.00811	5.22E-05	0.759	0.268	0.531	1.30
0.1	0.9	4101	0.02528	0.06474	0.00301	0.01083	0.17242	0.00580	0.69317	0.00188	0.01054	0.01233	240	0.00807	5.23E-05	0.763	0.272	0.528	1.31
0.2	0.8	4107	0.02474	0.06206	0.00312	0.01134	0.17701	0.00581	0.69097	0.00191	0.01078	0.01226	244	0.00803	5.23E-05	0.768	0.276	0.524	1.31
0.3	0.7	4114	0.02408	0.05891	0.00327	0.01196	0.18241	0.00582	0.68836	0.00194	0.01107	0.01218	248	0.00798	5.24E-05	0.774	0.281	0.520	1.32
0.4	0.6	4124	0.02323	0.05516	0.00344	0.01273	0.18889	0.00583	0.68522	0.00198	0.01141	0.01208	254	0.00792	5.25E-05	0.781	0.286	0.515	1.33
0.5	0.5	4135	0.02213	0.05063	0.00367	0.01373	0.19679	0.00585	0.68137	0.00203	0.01184	0.01195	261	0.00784	5.26E-05	0.790	0.294	0.509	1.34
0.6	0.4	4149	0.02065	0.04503	0.00398	0.01507	0.20663	0.00586	0.67653	0.00209	0.01237	0.01178	269	0.00775	5.28E-05	0.802	0.303	0.501	1.36
0.7	0.3	4168	0.01853	0.03798	0.00441	0.01694	0.21920	0.00587	0.67027	0.00217	0.01308	0.01154	281	0.00762	5.30E-05	0.817	0.316	0.492	1.38
0.8	0.2	4193	0.01533	0.02885	0.00504	0.01973	0.23581	0.00588	0.66185	0.00228	0.01403	0.01119	296	0.00746	5.32E-05	0.840	0.334	0.481	1.40
0.9	0.1	4230	0.01002	0.01668	0.00607	0.02429	0.25866	0.00588	0.64990	0.00245	0.01538	0.01066	320	0.00723	5.36E-05	0.875	0.362	0.467	1.43
1.0	0.0	4285	0.00000	0.00000	0.00799	0.03281	0.29182	0.00583	0.63163	0.00269	0.01746	0.00977	358	0.00688	5.42E-05	0.939	0.407	0.450	1.47

B

Appendix E

Fluid Dynamics Equations

E.1 FLUID DYNAMICS EQUATIONS IN CYLINDRICAL COORDINATES

The unsteady equations of motion for an incompressible Newtonian fluid with constant viscosity in cylindrical coordinates (r,θ,z) are given as follows:

$$\frac{\partial v_r}{\partial \tau} + v_r \frac{\partial v_r}{\partial r} + \frac{v_\theta}{r}\frac{\partial v_r}{\partial \theta} - \frac{v_\theta^2}{r} + w\frac{\partial v_r}{\partial z} =$$

$$f_r - \frac{1}{\rho}\frac{\partial p}{\partial r} + v\left\{ \frac{\partial}{\partial r}\left[\frac{1}{r}\frac{\partial(rv_r)}{\partial r} \right] + \frac{1}{r^2}\frac{\partial^2 v_r}{\partial \theta^2} - \frac{2}{r^2}\frac{\partial v_\theta}{\partial \theta} + \frac{\partial^2 v_r}{\partial z^2} \right\} \tag{E.1}$$

$$\frac{\partial v_\theta}{\partial \tau} + v_r \frac{\partial v_\theta}{\partial r} + \frac{v_\theta}{r}\frac{\partial v_\theta}{\partial \theta} + \frac{v_r v_\theta}{r} + w\frac{\partial v_\theta}{\partial z} =$$

$$f_\theta - \frac{1}{\rho r}\frac{\partial p}{\partial \theta} + v\left\{ \frac{\partial}{\partial r}\left[\frac{1}{r}\frac{\partial(rv_\theta)}{\partial r} \right] + \frac{1}{r^2}\frac{\partial^2 v_\theta}{\partial \theta^2} + \frac{2}{r^2}\frac{\partial v_r}{\partial \theta} + \frac{\partial^2 v_\theta}{\partial z^2} \right\} \tag{E.2}$$

$$\frac{\partial w}{\partial \tau} + v_r \frac{\partial w}{\partial r} + \frac{v_\theta}{r}\frac{\partial w}{\partial \theta} + w\frac{\partial w}{\partial z} = f_z - \frac{1}{\rho}\frac{\partial p}{\partial z} + v\left\{ \frac{1}{r}\frac{\partial}{\partial r}\left(r\frac{\partial w}{\partial r} \right) + \frac{1}{r^2}\frac{\partial^2 w}{\partial \theta^2} + \frac{\partial^2 w}{\partial z^2} \right\} \tag{E.3}$$

where f_i is some type of body force such as buoyancy. The energy equation for an incompressible fluid can be written as:

$$\rho c_p\left(\frac{\partial t}{\partial \tau} + v_r \frac{\partial t}{\partial r} + \frac{v_\theta}{r}\frac{\partial t}{\partial \theta} + w\frac{\partial t}{\partial z} \right) = \frac{1}{r}\frac{\partial}{\partial r}\left(rk\frac{\partial t}{\partial r} \right) + \frac{1}{r^2}\frac{\partial}{\partial \theta}\left(k\frac{\partial t}{\partial \theta} \right) + \frac{\partial}{\partial z}\left(k\frac{\partial t}{\partial z} \right) + \dot{q} + \Phi \tag{E.4}$$

where

$$\Phi = 2\mu \left\{ \begin{array}{l} \left(\dfrac{\partial v_r}{\partial r}\right)^2 + \left[\dfrac{1}{r}\left(\dfrac{\partial v_\theta}{\partial \theta} + v_r\right)\right]^2 + \left(\dfrac{\partial w}{\partial z}\right)^2 + \dfrac{1}{2}\left(\dfrac{\partial v_\theta}{\partial z} + \dfrac{1}{r}\dfrac{\partial w}{\partial \theta}\right)^2 \\[4mm] + \dfrac{1}{2}\left(\dfrac{\partial w}{\partial r} + \dfrac{\partial v_r}{\partial z}\right)^2 + \dfrac{1}{2}\left[\dfrac{1}{r}\dfrac{\partial v_r}{\partial \theta} + r\dfrac{\partial}{\partial r}\left(\dfrac{v_\theta}{r}\right)\right]^2 \end{array} \right\} \qquad \text{(E.5)}$$

E.2   FLUID DYNAMICS EQUATIONS
IN SPHERICAL COORDINATES

The unsteady equations of motion for an incompressible Newtonian fluid with constant viscosity in spherical coordinates (r,θ,ϕ) are given as follows:

$$\frac{\partial v_r}{\partial \tau} + v_r\frac{\partial v_r}{\partial r} + \frac{v_\theta}{r}\frac{\partial v_r}{\partial \theta} + \frac{v_\phi}{r\sin\theta}\frac{\partial v_r}{\partial \phi} - \frac{v_\theta^2 + v_\phi^2}{r} =$$

$$f_r - \frac{1}{\rho}\frac{\partial p}{\partial r} + v\left\{\nabla^2 f_r - \frac{2v_r}{r^2} - \frac{2}{r^2}\frac{\partial v_\theta}{\partial \theta} - \frac{2}{r^2}v_\theta\cot\theta - \frac{2}{r^2\sin\theta}\frac{\partial v_\phi}{\partial \phi}\right\} \qquad \text{(E.6)}$$

$$\frac{\partial v_\theta}{\partial \tau} + v_r\frac{\partial v_\theta}{\partial r} + \frac{v_\theta}{r}\frac{\partial v_\theta}{\partial \theta} + \frac{v_\phi}{r\sin\theta}\frac{\partial v_\theta}{\partial \phi} + \frac{v_r v_\theta}{r} - \frac{v_\phi^2\cot\theta}{r} =$$

$$f_\theta - \frac{1}{\rho r}\frac{\partial p}{\partial \theta} + v\left\{\nabla^2 v_\theta + \frac{2}{r^2}\frac{\partial v_r}{\partial \theta} - \frac{v_\theta}{r^2\sin^2\theta} - \frac{2\cos\theta}{r^2\sin^2\theta}\frac{\partial v_\phi}{\partial \phi}\right\} \qquad \text{(E.7)}$$

$$\frac{\partial v_\phi}{\partial \tau} + v_r\frac{\partial v_\phi}{\partial r} + \frac{v_\theta}{r}\frac{\partial v_\phi}{\partial \theta} + \frac{v_\phi}{r\sin\theta}\frac{\partial v_\phi}{\partial \phi} + \frac{v_r v_\phi}{r} - \frac{v_\theta v_\phi\cot\theta}{r} =$$

$$f_\phi - \frac{1}{\rho r\sin\theta}\frac{\partial p}{\partial \phi} + v\left\{\nabla^2 v_\phi - \frac{v_\phi}{r^2\sin^2\theta} + \frac{2}{r^2\sin\theta}\frac{\partial v_r}{\partial \phi} + \frac{2\cos\theta}{r^2\sin^2\theta}\frac{\partial v_\theta}{\partial \phi}\right\} \qquad \text{(E.8)}$$

where f_i is some type of body force such as buoyancy and

$$\nabla^2 \equiv \frac{1}{r^2}\frac{\partial}{\partial r}\left(r^2\frac{\partial}{\partial r}\right) + \frac{1}{r^2\sin\theta}\frac{\partial}{\partial \theta}\left(\sin\theta\frac{\partial}{\partial \theta}\right) + \frac{1}{r^2\sin^2\theta}\frac{\partial^2}{\partial \phi^2}$$

The energy equation for an incompressible fluid can be written as:

$$\rho c_p\left(\frac{\partial t}{\partial \tau} + v_r\frac{\partial t}{\partial r} + \frac{v_\theta}{r}\frac{\partial t}{\partial \theta} + \frac{v_\phi}{r\sin\theta}\frac{\partial t}{\partial \phi}\right) = \frac{1}{r^2}\frac{\partial}{\partial r}\left(r^2 k\frac{\partial t}{\partial r}\right)$$

$$+ \frac{1}{r^2\sin\theta}\frac{\partial}{\partial \theta}\left(k\sin\theta\frac{\partial t}{\partial \theta}\right) + \frac{1}{r^2\sin^2\theta}\frac{\partial}{\partial \phi}\left(k\frac{\partial t}{\partial \phi}\right) + \dot{q} + \Phi \qquad \text{(E.9)}$$

where

$$
\Phi = 2\mu \left\{
\begin{array}{c}
\left(\dfrac{\partial v_r}{\partial r}\right)^2 + \left[\dfrac{1}{r}\left(\dfrac{\partial v_\theta}{\partial \theta} + v_r\right)\right]^2 \\[2ex]
+ \dfrac{1}{r^2}\left(\dfrac{1}{\sin\theta}\dfrac{\partial v_\phi}{\partial \phi} + v_r + v_\theta \cot\theta\right)^2 + \dfrac{1}{2}\left[r\dfrac{\partial}{\partial r}\left(\dfrac{v_\theta}{r}\right) + \dfrac{1}{r}\dfrac{\partial v_r}{\partial \theta}\right]^2 \\[2ex]
+ \dfrac{1}{2}\left[\dfrac{1}{r\sin\theta}\dfrac{\partial v_r}{\partial \phi} + r\dfrac{\partial}{\partial r}\left(\dfrac{v_\phi}{r}\right)\right]^2 + \dfrac{1}{2}\left[\dfrac{\sin\theta}{r}\dfrac{\partial}{\partial \theta}\left(\dfrac{v_\phi}{\sin\theta}\right) + \dfrac{1}{r\sin\theta}\dfrac{\partial v_\theta}{\partial \phi}\right]^2
\end{array}
\right\} \qquad (E.10)
$$

Appendix F

Material Properties

TABLE F.1a
Properties of Air (English Units)

T (°F)	ρ (lb/ft³)	c_p (Btu/lb-°F)	μ (lb_m/hr-ft)	k (Btu/hr-ft-°F)	Pr
−9	0.08707	0.2403	0.0386	0.0129	0.720
81	0.07251	0.2405	0.0447	0.0152	0.707
171	0.06212	0.2410	0.0504	0.0173	0.700
261	0.05438	0.2422	0.0557	0.0195	0.690
351	0.04832	0.2439	0.0607	0.0216	0.686
441	0.04348	0.2460	0.0654	0.0235	0.684
531	0.03951	0.2484	0.0698	0.0254	0.683
621	0.03623	0.2510	0.0740	0.0271	0.685
711	0.03344	0.2539	0.0781	0.0287	0.690
801	0.03106	0.2567	0.0820	0.0303	0.695
891	0.02899	0.2596	0.0859	0.0317	0.702
981	0.02718	0.2625	0.0895	0.0331	0.709
1071	0.02558	0.2651	0.0931	0.0344	0.716
1161	0.02415	0.2677	0.0964	0.0358	0.720
1251	0.02289	0.2701	0.0996	0.0372	0.723
1341	0.02174	0.2725	0.1028	0.0385	0.726
1521	0.01977	0.2768	0.1087	0.0413	0.728
1701	0.01812	0.2806	0.1145	0.0441	0.728
1881	0.01673	0.2840	0.1201	0.0474	0.719
2061	0.01553	0.2883	0.1283	0.0526	0.703
2241	0.01450	0.2938	0.1349	0.0578	0.685

Source: From F. Kreith, Ed., *The CRC Handbook of Mechanical Engineering,* CRC Press, Boca Raton, FL, 1998, A-32–A-33.

TABLE F.1b
Properties of Air (Metric Units)

T (K)	ρ (kg/m³)	c_p [kJ/(kg · K)]	$\eta \times 10^7$](N · s)/m²]	$\lambda \times 10^3$ [W/(m · K)]	Pr
100	3.5562	1.032	71.1	9.34	0.786
150	2.3364	1.012	103.4	13.8	0.758
200	1.7458	1.007	132.5	18.1	0.737
250	1.3947	1.006	159.6	22.3	0.720
300	1.1614	1.007	184.6	26.3	0.707
350	0.9950	1.009	208.2	30.0	0.700
400	0.8711	1.014	230.1	33.8	0.690
450	0.7740	1.021	250.7	37.3	0.686
500	0.6964	1.030	270.1	40.7	0.684
550	0.6329	1.040	288.4	43.9	0.683
600	0.5804	1.051	305.8	46.9	0.685
650	0.5356	1.063	322.5	49.7	0.690
700	0.4975	1.075	338.8	52.4	0.695
750	0.4643	1.087	354.6	54.9	0.702
800	0.4354	1.099	369.8	57.3	0.709
850	0.4097	1.110	384.3	59.6	0.716
900	0.3868	1.121	398.1	62.0	0.720
950	0.3666	1.131	411.3	64.3	0.723
1000	0.3482	1.141	424.4	66.7	0.726
1100	0.3166	1.159	449.0	71.5	0.728
1200	0.2902	1.175	473.0	76.3	0.728
1300	0.2679	1.189	496.0	82	0.719
1400	0.2488	1.207	530	91	0.703
1500	0.2322	1.230	557	100	0.685

Source: From G.F. Hewitt, G.L. Shires, and T.R. Bott, Eds., *Process Heat Transfer,* CRC Press, Boca Raton, FL, 1994, 1018.

TABLE F.2
Gas Properties

Common name(s)	Acetylene (Ethyne)	Air [mixture]	Ammonia, anhyd.	Argon
Chemical formula	C_2H_2		NH_3	Ar
Refrigerant number	—	729	717	740
CHEMICAL AND PHYSICAL PROPERTIES				
Molecular weight	26.04	28.966	17.02	39.948
Specific gravity, air = 1	0.90	1.00	0.59	1.38
Specific volume, ft^3/lb	14.9	13.5	23.0	9.80
Specific volume, m^3/kg	0.93	0.842	1.43	0.622
Density of liquid (at atm bp), lb/ft^3	43.0	54.6	42.6	87.0
Density of liquid (at atm bp), kg/m^3	693.	879.	686.	1 400.
Vapor pressure at 25 deg C, psia			145.4	
Vapor pressure at 25 deg C, MN/m^2			1.00	
Viscosity (abs), lbm/ft·sec	6.72×10^{-6}	12.1×10^{-6}	6.72×10^{-6}	13.4×10^{-6}
Viscosity (abs), centipoises[a]	0.01	0.018	0.010	0.02
Sound velocity in gas, m/sec	343	346	415	322
THERMAL AND THERMO-DYNAMIC PROPERTIES				
Specific heat, c_p, Btu/lb·deg F or cal/g·deg C	0.40	0.240 3	0.52	0.125
Specific heat, c_p, J/kg·K	1 674.	1 005.	2 175.	523.
Specific heat ratio, c_p/c_v	1.25	1.40	1.3	1.67
Gas constant R, ft-lb/lb·deg R	59.3	53.3	90.8	38.7
Gas constant R, J/kg·deg C	319	286.8	488.	208.
Thermal conductivity, Btu/hr·ft·deg F	0.014	0.015 1	0.015	0.010 2
Thermal conductivity, W/m·deg C	0.024	0.026	0.026	0.017 2
Boiling point (sat 14.7 psia), deg F	− 103	− 320	− 28.	− 303.
Boiling point (sat 760 mm), deg C	− 75	− 195	− 33.3	− 186
Latent heat of evap (at bp), Btu/lb	264	88.2	589.3	70.
Latent heat of evap (at bp), J/kg	614 000	205 000.	1 373 000	163 000
Freezing (melting) point, deg F (1 atm)	− 116	− 357.2	− 107.9	− 308.5
Freezing (melting) point, deg C (1 atm)	− 82.2	− 216.2	− 77.7	− 189.2
Latent heat of fusion, Btu/lb	23.	10.0	143.0	
Latent heat of fusion, J/kg	53 500	23 200	332 300	
Critical temperature, deg F	97.1	− 220.5	271.4	− 187.6
Critical temperature, deg C	36.2	− 140.3	132.5	− 122
Critical pressure, psia	907.	550.	1 650.	707.
Critical pressure, MN/m^2	6.25	3.8	11.4	4.87
Critical volume, ft^3/lb		0.050	0.068	0.029 9
Critical volume, m^3/kg		0.003	0.004 24	0.001 86
Flammable (yes or no)	Yes	No	No	No
Heat of combustion, Btu/ft^3	1 450	—	—	—
Heat of combustion, Btu/lb	21 600	—	—	—
Heat of combustion, kJ/kg	50 200	—	—	—

[a]For N·sec/m^2 divide by 1 000.

TABLE F.2 (continued)
Gas Properties

Common name(s)	Butadiene	n-Butane	Isobutane (2-Methyl propane)	1-Butene (Butylene)
Chemical formula	C_4H_6	C_4H_{10}	C_4H_{10}	C_4H_8
Refrigerant number	—	600	600a	—
CHEMICAL AND PHYSICAL PROPERTIES				
Molecular weight	54.09	58.12	58.12	56.108
Specific gravity, air = 1	1.87	2.07	2.07	1.94
Specific volume, ft³/lb	7.1	6.5	6.5	6.7
Specific volume, m³/kg	0.44	0.405	0.418	0.42
Density of liquid (at atm bp), lb/ft³		37.5	37.2	
Density of liquid (at atm bp), kg/m³		604.	599.	
Vapor pressure at 25 deg C, psia		35.4	50.4	
Vapor pressure at 25 deg C, MN/m²		0.024 4	0.347	
Viscosity (abs), lbm/ft·sec		4.8×10^{-6}		
Viscosity (abs), centipoises[a]		0.007		
Sound velocity in gas, m/sec	226	216	216	222
THERMAL AND THERMO-DYNAMIC PROPERTIES				
Specific heat, c_p, Btu/lb·deg F or cal/g·deg C	0.341	0.39	0.39	0.36
Specific heat, c_p, J/kg·K	1 427.	1 675.	1 630.	1 505.
Specific heat ratio, c_p/c_v	1.12	1.096	1.10	1.112
Gas constant R, ft-lb/lb·deg F	28.55	26.56	26.56	27.52
Gas constant R, J/kg·deg C	154.	143.	143.	148.
Thermal conductivity, Btu/hr·ft·deg F		0.01	0.01	
Thermal conductivity, W/m·deg C		0.017	0.017	
Boiling point (sat 14.7 psia), deg F	24.1	31.2	10.8	20.6
Boiling point (sat 760 mm), deg C	−4.5	−0.4	−11.8	−6.3
Latent heat of evap (at bp), Btu/lb		165.6	157.5	167.9
Latent heat of evap (at bp), J/kg	386 000	366 000	391 000	
Freezing (melting) point, deg F (1 atm)	−164.	−217.	−229	−301.6
Freezing (melting) point, deg C (1 atm)	−109.	−138	−145	−185.3
Latent heat of fusion, Btu/lb		19.2		16.4
Latent heat of fusion, J/kg		44 700		38 100
Critical temperature, deg F		306	273.	291.
Critical temperature, deg C	171.	152.	134.	144.
Critical pressure, psia	652.	550.	537.	621.
Critical pressure, MN/m²		3.8	3.7	4.28
Critical volume, ft³/lb		0.070		0.068
Critical volume, m³/kg		0.004 3		0.004 2
Flammable (yes or no)	Yes	Yes	Yes	Yes
Heat of combustion, Btu/ft³	2 950	3 300	3 300	3 150
Heat of combustion, Btu/lb	20 900	21 400	21 400	21 000
Heat of combustion, kJ/kg	48 600	49 700	49 700	48 800

[a]For N·sec/m² divide by 1 000.

TABLE F.2 (continued)
Gas Properties

Common name(s)	Carbon monoxide	Chlorine	Deuterium	Ethane
Chemical formula	CO	Cl_2	D_2	C_2H_6
Refrigerant number	—	—	—	170
CHEMICAL AND PHYSICAL PROPERTIES				
Molecular weight	28.011	70.906	2.014	30.070
Specific gravity, air = 1	0.967	2.45	0.070	1.04
Specific volume, ft³/lb	14.0	5.52	194.5	13.025
Specific volume, m³/kg	0.874	0.344	12.12	0.815
Density of liquid (at atm bp), lb/ft³		97.3		28.
Density of liquid (at atm bp), kg/m³		1 559.		449.
Vapor pressure at 25 deg C, psia			0.756	
Vapor pressure at 25 deg C, MN/m²			0.005 2	
Viscosity (abs), lbm/ft·sec	12.1×10^{-6}	9.4×10^{-6}	8.75×10^{-6}	$64. \times 10^{-6}$
Viscosity (abs), centipoises[a]	0.018	0.014	0.013	0.095
Sound velocity in gas, m/sec	352.	215.	930.	316.
THERMAL AND THERMO-DYNAMIC PROPERTIES				
Specific heat, c_p, Btu/lb·deg F or cal/g·deg C	0.25	0.114	1.73	0.41
Specific heat, c_p, J/kg·K	1 046.	477.	7 238.	1 715.
Specific heat ratio, c_p/c_v	1.40	1.35	1.40	1.20
Gas constant R, ft-lb/lb·deg F	55.2	21.8	384.	51.4
Gas constant R, J/kg·deg C	297.	117.	2 066.	276.
Thermal conductivity, Btu/hr·ft·deg F	0.014	0.005	0.081	0.010
Thermal conductivity, W/m·deg C	0.024	0.008 7	0.140	0.017
Boiling point (sat 14.7 psia), deg F	−312.7	−29.2		−127.
Boiling point (sat 760 mm), deg C	−191.5	−34.		−88.3
Latent heat of evap (at bp), Btu/lb	92.8	123.7		210.
Latent heat of evap (at bp), J/kg	216 000.	288 000.		488 000.
Freezing (melting) point, deg F (1 atm)	−337.	−150.		−278.
Freezing (melting) point, deg C (1 atm)	−205.	−101.		−172.2
Latent heat of fusion, Btu/lb	12.8	41.0		41.
Latent heat of fusion, J/kg		95 400.		95 300.
Critical temperature, deg F	−220.	291.	−390.6	90.1
Critical temperature, deg C	−140.	144.	−234.8	32.2
Critical pressure, psia	507.	1 120.	241.	709.
Critical pressure, MN/m²	3.49	7.72	1.66	4.89
Critical volume, ft³/lb	0.053	0.028	0.239	0.076
Critical volume, m³/kg	0.003 3	0.001 75	0.014 9	0.004 7
Flammable (yes or no)	Yes	No		Yes
Heat of combustion, Btu/ft³	310.	—		
Heat of combustion, Btu/lb	4 340.	—		22 300.
Heat of combustion, kJ/kg	10 100.	—		51 800.

[a] For N·sec/m² divide by 1 000.

TABLE F.2 (continued)
Gas Properties

Common name(s)	Ethyl chloride	Ethylene (Ethene)	Fluorine
Chemical formula	C_2H_5Cl	C_2H_4	F_2
Refrigerant number	160	1 150	—
CHEMICAL AND PHYSICAL PROPERTIES			
Molecular weight	64.515	28.054	37.996
Specific gravity, air = 1	2.23	0.969	1.31
Specific volume, ft^3/lb	6.07	13.9	10.31
Specific volume, m^3/kg	0.378	0.87	0.706
Density of liquid (at atm bp), lb/ft^3	56.5	35.5	
Density of liquid (at atm bp), kg/m^3	905.	569.	
Vapor pressure at 25 deg C, psia			
Vapor pressure at 25 deg C, MN/m^2			
Viscosity (abs), lbm/ft·sec		6.72×10^{-6}	16.1×10^{-6}
Viscosity (abs), centipoises[a]		0.010	0.024
Sound velocity in gas, m/sec	204.	331.	290.
THERMAL AND THERMODYNAMIC PROPERTIES			
Specific heat, c_p, Btu/lb·deg F or cal/g·deg C	0.27	0.37	0.198
Specific heat, c_p, J/kg·K	1 130.	1 548.	828.
Specific heat ratio, c_p/c_v	1.13	1.24	1.35
Gas constant R, ft-lb/lb·deg F	24.0	55.1	40.7
Gas constant R, J/kg·deg C	129.	296.	219.
Thermal conductivity, Btu/hr·ft·deg F		0.010	0.016
Thermal conductivity, W/m·deg C		0.017	0.028
Boiling point (sat 14.7 psia), deg F	54.	−155.	−306.4
Boiling point (sat 760 mm), deg C	12.2	−103.8	−188.
Latent heat of evap (at bp), Btu/lb	166.	208.	74.
Latent heat of evap (at bp), J/kg	386 000.	484 000.	172 000.
Freezing (melting) point, deg F (1 atm)	−218.	−272.	−364.
Freezing (melting) point, deg C (1 atm)	−138.9	−169.	−220.
Latent heat of fusion, Btu/lb	29.3	51.5	11.
Latent heat of fusion, J/kg	68 100.	120 000.	25 600.
Critical temperature, deg F	368.6	49.	−200
Critical temperature, deg C	187.	9.5	−129.
Critical pressure, psia	764.	741.	810.
Critical pressure, MN/m^2	5.27	5.11	5.58
Critical volume, ft^3/lb	0.049	0.073	
Critical volume, m^3/kg	0.003 06	0.004 6	
Flammable (yes or no)	No	Yes	
Heat of combustion, Btu/ft^3	—	1 480.	
Heat of combustion, Btu/lb	—	20 600.	
Heat of combustion, kJ/kg	—	47 800.	

[a] For N·sec/m^2 divide by 1 000.

TABLE F.2 (continued)
Gas Properties

Common name(s)	Fluorocarbons			
Chemical formula	CCl_3F	CCl_2F_2	$CClF_3$	$CBrF_3$
Refrigerant number	11	12	13	13B1
CHEMICAL AND PHYSICAL PROPERTIES				
Molecular weight	137.37	120.91	104.46	148.91
Specific gravity, air = 1	4.74	4.17	3.61	5.14
Specific volume, ft³/lb	2.74	3.12	3.58	2.50
Specific volume, m³/kg	0.171	0.195	0.224	0.975
Density of liquid (at atm bp), lb/ft³	92.1	93.0	95.0	124.4
Density of liquid (at atm bp), kg/m³	1 475.	1 490.	1 522.	1 993.
Vapor pressure at 25 deg C, psia		94.51	516.	234.8
Vapor pressure at 25 deg C, MN/m²		0.652	3.56	1.619
Viscosity (abs), lbm/ft·sec	7.39×10^{-6}	8.74×10^{-6}		
Viscosity (abs), centipoises^a	0.011	0.013		
Sound velocity in gas, m/sec				
THERMAL AND THERMODYNAMIC PROPERTIES				
Specific heat, c_p, Btu/lb·deg F or cal/g·deg C	0.14	0.146	0.154	
Specific heat, c_p, J/kg·K	586.	611.	644.	
Specific heat ratio, c_p/c_v	1.14	1.14	1.145	
Gas constant R, ft-lb/lb·deg F				
Gas constant R, J/kg·deg C				
Thermal conductivity, Btu/hr·ft·deg F	0.005	0.006		
Thermal conductivity, W/m·deg C	0.008 7	0.010 4		
Boiling point (sat 14.7 psia), deg F	74.9	−21.8	−114.6	−72.
Boiling point (sat 760 mm), deg C	23.8	−29.9	−81.4	−57.8
Latent heat of evap (at bp), Btu/lb	77.5	71.1	63.0	51.1
Latent heat of evap (at bp), J/kg	180 000.	165 000.	147 000.	119 000.
Freezing (melting) point, deg F (1 atm)	−168.	−252.	−294.	−270.
Freezing (melting) point, deg C (1 atm)	−111.	−157.8	−181.1	−167.8
Latent heat of fusion, Btu/lb				
Latent heat of fusion, J/kg				
Critical temperature, deg F	388.4	233.	83.9	152.
Critical temperature, deg C	198.	111.7	28.8	66.7
Critical pressure, psia	635.	582.	559.	573.
Critical pressure, MN/m²	4.38	4.01	3.85	3.95
Critical volume, ft³/lb	0.028 9	0.287	0.027 7	0.021 5
Critical volume, m³/kg	0.001 80	0.018	0.001 73	0.001 34
Flammable (yes or no)	No	No	No	No
Heat of combustion, Btu/ft³	—	—	—	—
Heat of combustion, Btu/lb	—	—	—	—
Heat of combustion, kJ/kg	—	—	—	—

^{a}For N·sec/m² divide by 1 000.

TABLE F.2 (continued)
Gas Properties

Common name(s)	Fluorocarbons			
Chemical formula	CF_4	$CHCl_2F$	$CHClF_2$	$C_2Cl_2F_4$
Refrigerant number	14	21	22	114
CHEMICAL AND PHYSICAL PROPERTIES				
Molecular weight	88.00	102.92	86.468	170.92
Specific gravity, air = 1	3.04	3.55	2.99	5.90
Specific volume, ft³/lb	4.34	3.7	4.35	2.6
Specific volume, m³/kg	0.271	0.231	0.271	0.162
Density of liquid (at atm bp), lb/ft³	102.0	87.7	88.2	94.8
Density of liquid (at atm bp), kg/m³	1 634.	1 405.	1 413.	1 519.
Vapor pressure at 25 deg C, psia		26.4	151.4	30.9
Vapor pressure at 25 deg C, MN/m²		0.182	1.044	0.213
Viscosity (abs), lbm/ft·sec		8.06×10^{-6}	8.74×10^{-6}	8.06×10^{-6}
Viscosity (abs), centipoises[a]		0.012	0.013	0.012
Sound velocity in gas, m/sec				
THERMAL AND THERMO-DYNAMIC PROPERTIES				
Specific heat, c_p, Btu/lb·deg F or cal/g·deg C		0.139	0.157	0.158
Specific heat, c_p, J/kg·K		582.	657.	661.
Specific heat ratio, c_p/c_v		1.18	1.185	1.09
Gas constant R, ft-lb/lb·deg F				
Gas constant R, J/kg·deg C				
Thermal conductivity, Btu/hr·ft·deg F			0.007	0.006
Thermal conductivity, W/m·deg C			0.012	0.010
Boiling point (sat 14.7 psia), deg F	−198.2	48.1	−41.3	38.4
Boiling point (sat 760 mm), deg C	−127.9	9.0	−40.7	3.55
Latent heat of evap (at bp), Btu/lb	58.5	104.1	100.4	58.4
Latent heat of evap (at bp), J/kg	136 000.	242 000.	234 000.	136 000.
Freezing (melting) point, deg F (1 atm)	−299.	−211.	−256.	−137.
Freezing (melting) point, deg C (1 atm)	−183.8	−135.	−160.	−93.8
Latent heat of fusion, Btu/lb	2.53			
Latent heat of fusion, J/kg	5 880.			
Critical temperature, deg F	−49.9	353.3	204.8	294.
Critical temperature, deg C	−45.5	178.5	96.5	
Critical pressure, psia	610.	750.	715.	475.
Critical pressure, MN/m²	4.21	5.17	4.93	3.28
Critical volume, ft³/lb	0.025	0.030 7	0.030 5	0.027 5
Critical volume, m³/kg	0.001 6	0.001 91	0.001 90	0.001 71
Flammable (yes or no)	No	No	No	No
Heat of combustion, Btu/ft³	—	—	—	—
Heat of combustion, Btu/lb	—	—	—	—
Heat of combustion, kJ/kg	—	—	—	—

[a]For N·sec/m² divide by 1 000.

TABLE F.2 (continued)
Gas Properties

Common name(s)	Fluorocarbons			Helium
Chemical formula	C_2ClF_5	$C_2H_3ClF_2$	$C_2H_4F_2$	He
Refrigerant number	115	142b	152a	704
CHEMICAL AND PHYSICAL PROPERTIES				
Molecular weight	154.47	100.50	66.05	4.002 6
Specific gravity, air = 1	5.33	3.47	2.28	0.138
Specific volume, ft³/lb	2.44	3.7	5.9	97.86
Specific volume, m³/kg	0.152	0.231	0.368	6.11
Density of liquid (at atm bp), lb/ft³	96.5	74.6	62.8	7.80
Density of liquid (at atm bp), kg/m³	1 546.	1 195.	1 006.	125.
Vapor pressure at 25 deg C, psia	132.1	49.1	86.8	
Vapor pressure at 25 deg C, MN/m²	0.911	0.338 5	0.596	
Viscosity (abs), lbm/ft·sec				13.4×10^{-6}
Viscosity (abs), centipoises[a]				0.02
Sound velocity in gas, m/sec				1 015.
THERMAL AND THERMO-DYNAMIC PROPERTIES				
Specific heat, c_p, Btu/lb·deg F or cal/g·deg C	0.161			1.24
Specific heat, c_p, J/kg·K	674.			5 188.
Specific heat ratio, c_p/c_v	1.091			1.66
Gas constant R, ft-lb/lb·deg F				386.
Gas constant R, J/kg·deg C				2 077.
Thermal conductivity, Btu/hr·ft·deg F				0.086
Thermal conductivity, W/m·deg C				0.149
Boiling point (sat 14.7 psia), deg F	− 38.0	14.	− 13.	− 452.
Boiling point (sat 760 mm), deg C	− 38.9	− 10.0	− 25.0	4.22 K
Latent heat of evap (at bp), Btu/lb	53.4	92.5	137.1	10.0
Latent heat of evap (at bp), J/kg	124 000.	215 000.	319 000.	23 300.
Freezing (melting) point, deg F (1 atm)	− 149.			[b]
Freezing (melting) point, deg C (1 atm)	− 100.6			−
Latent heat of fusion, Btu/lb				−
Latent heat of fusion, J/kg				−
Critical temperature, deg F	176.		387.	− 450.3
Critical temperature, deg C				5.2 K
Critical pressure, psia	457.6			33.22
Critical pressure, MN/m²	3.155			
Critical volume, ft³/lb	0.026 1			0.231
Critical volume, m³/kg	0.001 63			0.014 4
Flammable (yes or no)	No	No	No	No
Heat of combustion, Btu/ft³	−	−	−	−
Heat of combustion, Btu/lb	−	−	−	−
Heat of combustion, kJ/kg	−	−	−	−

[a] For N·sec/m² divide by 1 000.
[b] Helium cannot be solidified at atmospheric pressure.

TABLE F.2 (continued)
Gas Properties

Common name(s)	Hydrogen	Hydrogen chloride	Hydrogen sulfide	Krypton
Chemical formula	H_2	HCl	H_2S	Kr
Refrigerant number	702	—	—	—
CHEMICAL AND PHYSICAL PROPERTIES				
Molecular weight	2.016	36.461	34.076	83.80
Specific gravity, air = 1	0.070	1.26	1.18	2.89
Specific volume, ft³/lb	194.	10.74	11.5	4.67
Specific volume, m³/kg	12.1	0.670	0.093 0	0.291
Density of liquid (at atm bp), lb/ft³	4.43	74.4	62.	150.6
Density of liquid (at atm bp), kg/m³	71.0	1 192.	993.	2 413.
Vapor pressure at 25 deg C, psia				
Vapor pressure at 25 deg C, MN/m²				
Viscosity (abs), lbm/ft·sec	6.05×10^{-6}	10.1×10^{-6}	8.74×10^{-6}	16.8×10^{-6}
Viscosity (abs), centipoises[a]	0.009	0.015	0.013	0.025
Sound velocity in gas, m/sec	1 315.	310.	302.	223.
THERMAL AND THERMO-DYNAMIC PROPERTIES				
Specific heat, c_p, Btu/lb·deg F or cal/g·deg C	3.42	0.194	0.23	0.059
Specific heat, c_p, J/kg·K	14 310.	812.	962.	247.
Specific heat ratio, c_p/c_v	1.405	1.39	1.33	1.68
Gas constant R, ft-lb/lb·deg F	767.	42.4	45.3	18.4
Gas constant R, J/kg·deg C	4 126.	228.	244.	99.0
Thermal conductivity, Btu/hr·ft·deg F	0.105	0.008	0.008	0.005 4
Thermal conductivity, W/m·deg C	0.018 2	0.014	0.014	0.009 3
Boiling point (sat 14.7 psia), deg F	−423.	−121.	−76.	−244.
Boiling point (sat 760 mm), deg C	20.4 K	−85.	−60.	−153.
Latent heat of evap (at bp), Btu/lb	192.	190.5	234.	46.4
Latent heat of evap (at bp), J/kg	447 000.	443 000.	544 000.	108 000.
Freezing (melting) point, deg F (1 atm)	−434.6	−169.6	−119.2	−272.
Freezing (melting) point, deg C (1 atm)	−259.1	−112.	−84.	−169.
Latent heat of fusion, Btu/lb	25.0	23.4	30.2	4.7
Latent heat of fusion, J/kg	58 000.	54 400.	70 200.	10 900.
Critical temperature, deg F	−399.8	124.	213.	
Critical temperature, deg C	−240.0	51.2	100.4	−63.8
Critical pressure, psia	189.	1 201.	1 309.	800.
Critical pressure, MN/m²	1.30	8.28	9.02	5.52
Critical volume, ft³/lb	0.53	0.038	0.046	0.017 7
Critical volume, m³/kg	0.033	0.002 4	0.002 9	0.001 1
Flammable (yes or no)	Yes	No	Yes	No
Heat of combustion, Btu/ft³	320.	—	700.	—
Heat of combustion, Btu/lb	62 050.	—	8 000.	—
Heat of combustion, kJ/kg	144 000.	—	18 600.	—

[a] For N·sec/m² divide by 1 000.

TABLE F.2 (continued)
Gas Properties

Common name(s)	Methane	Methyl chloride	Neon	Nitric oxide
Chemical formula	CH_4	CH_3Cl	Ne	NO
Refrigerant number	50	40	720	—
CHEMICAL AND PHYSICAL PROPERTIES				
Molecular weight	16.044	50.488	20.179	30.006
Specific gravity, air = 1	0.554	1.74	0.697	1.04
Specific volume, ft^3/lb	24.2	7.4	19.41	13.05
Specific volume, m^3/kg	1.51	0.462	1.211	0.814
Density of liquid (at atm bp), lb/ft^3	26.3	62.7	75.35	
Density of liquid (at atm bp), kg/m^3	421.	1 004.	1 207.	
Vapor pressure at 25 deg C, psia		82.2		
Vapor pressure at 25 deg C, MN/m^2		0.567		
Viscosity (abs), lbm/ft·sec	7.39×10^{-6}	7.39×10^{-6}	21.5×10^{-6}	12.8×10^{-6}
Viscosity (abs), centipoises^a	0.011	0.011	0.032	0.019
Sound velocity in gas, m/sec	446.	251.	454.	341.
THERMAL AND THERMO-DYNAMIC PROPERTIES				
Specific heat, c_p, Btu/lb·deg F or cal/g·deg C	0.54	0.20	0.246	0.235
Specific heat, c_p, J/kg·K	2 260.	837.	1 030.	983.
Specific heat ratio, c_p/c_v	1.31	1.28	1.64	1.40
Gas constant R, ft-lb/lb·deg F	96.	30.6	76.6	51.5
Gas constant R, J/kg·deg C	518.	165.	412.	277.
Thermal conductivity, Btu/hr·ft·deg F	0.02	0.006	0.028	0.015
Thermal conductivity, W/m·deg C	0.035	0.010	0.048	0.026
Boiling point (sat 14.7 psia), deg F	−259.	−10.7	−410.9	−240.
Boiling point (sat 760 mm), deg C	−434.2	−23.7	−246.	−151.5
Latent heat of evap (at bp), Btu/lb	219.2	184.1	37.	
Latent heat of evap (at bp), J/kg	510 000.	428 000.	86 100.	
Freezing (melting) point, deg F (1 atm)	−296.6	−144.	−415.6	−258.
Freezing (melting) point, deg C (1 atm)	−182.6	−97.8	−248.7	−161.
Latent heat of fusion, Btu/lb	14.	56.	6.8	32.9
Latent heat of fusion, J/kg	32 600.	130 000.	15 800.	76 500.
Critical temperature, deg F	−116.	289.4	−379.8	−136.
Critical temperature, deg C	−82.3	143.	−228.8	−93.3
Critical pressure, psia	673.	968.	396.	945.
Critical pressure, MN/m^2	4.64	6.67	2.73	6.52
Critical volume, ft^3/lb	0.099	0.043	0.033	0.033 2
Critical volume, m^3/kg	0.006 2	0.002 7	0.002 0	0.002 07
Flammable (yes or no)	Yes	Yes	No	No
Heat of combustion, Btu/ft^3	985.		—	—
Heat of combustion, Btu/lb	2 290.		—	—
Heat of combustion, kJ/kg			—	—

^{a}For N·sec/m^2 divide by 1 000.

TABLE F.2 (continued)
Gas Properties

Common name(s)	Nitrogen	Nitrous oxide	Oxygen	Ozone
Chemical formula	N_2	N_2O	O_2	O_3
Refrigerant number	728	744A	732	—
CHEMICAL AND PHYSICAL PROPERTIES				
Molecular weight	28.013 4	44.012	31.998 8	47.998
Specific gravity, air = 1	0.967	1.52	1.105	1.66
Specific volume, ft³/lb	13.98	8.90	12.24	8.16
Specific volume, m³/kg	0.872	0.555	0.764	0.509
Density of liquid (at atm bp), lb/ft³	50.46	76.6	71.27	
Density of liquid (at atm bp), kg/m³	808.4	1 227.	1 142.	
Vapor pressure at 25 deg C, psia				
Vapor pressure at 25 deg C, MN/m²				
Viscosity (abs), lbm/ft·sec	12.1×10^{-6}	10.1×10^{-6}	13.4×10^{-6}	8.74×10^{-6}
Viscosity (abs), centipoises[a]	0.018	0.015	0.020	0.013
Sound velocity in gas, m/sec	353.	268.	329.	
THERMAL AND THERMO-DYNAMIC PROPERTIES				
Specific heat, c_p, Btu/lb·deg F or cal/g·deg C	0.249	0.21	0.220	0.196
Specific heat, c_p, J/kg·K	1 040.	879.	920.	820.
Specific heat ratio, c_p/c_v	1.40	1.31	1.40	
Gas constant R, ft-lb/lb·deg F	55.2	35.1	48.3	32.2
Gas constant R, J/kg·deg C	297.	189.	260.	173.
Thermal conductivity, Btu/hr·ft·deg F	0.015	0.010	0.015	0.019
Thermal conductivity, W/m·deg C	0.026	0.017	0.026	0.033
Boiling point (sat 14.7 psia), deg F	− 320.4	− 127.3	− 297.3	− 170.
Boiling point (sat 760 mm), deg C	− 195.8	− 88.5	− 182.97	− 112.
Latent heat of evap (at bp), Btu/lb	85.5	161.8	91.7	
Latent heat of evap (at bp), J/kg	199 000.	376 000.	213 000.	
Freezing (melting) point, deg F (1 atm)	− 346.	− 131.5	− 361.1	− 315.5
Freezing (melting) point, deg C (1 atm)	− 210.	− 90.8	− 218.4	− 193.
Latent heat of fusion, Btu/lb	11.1	63.9	5.9	97.2
Latent heat of fusion, J/kg	25 800.	149 000.	13 700.	226 000.
Critical temperature, deg F	− 232.6	97.7	− 181.5	16.
Critical temperature, deg C	− 147.	36.5	− 118.6	− 9.
Critical pressure, psia	493.	1 052.	726.	800.
Critical pressure, MN/m²	3.40	7.25	5.01	5.52
Critical volume, ft³/lb	0.051	0.036	0.040	0.029 8
Critical volume, m³/kg	0.003 18	0.002 2	0.002 5	0.001 86
Flammable (yes or no)	No	No	No	No
Heat of combustion, Btu/ft³	—	—	—	—
Heat of combustion, Btu/lb	—	—	—	—
Heat of combustion, kJ/kg	—	—	—	—

[a] For N·sec/m² divide by 1 000.

TABLE F.2 (continued)
Gas Properties

	Propane	Propylene (Propene)	Sulfur dioxide	Xenon
Common name(s)				
Chemical formula	C_3H_8	C_3H_6	SO_2	Xe
Refrigerant number	290	1 270	764	—
CHEMICAL AND PHYSICAL PROPERTIES				
Molecular weight	44.097	42.08	64.06	131.30
Specific gravity, air = 1	1.52	1.45	2.21	4.53
Specific volume, ft^3/lb	8.84	9.3	6.11	2.98
Specific volume, m^3/kg	0.552	0.58		
Density of liquid (at atm bp), lb/ft^3	36.2	37.5	42.8	190.8
Density of liquid (at atm bp), kg/m^3	580.	601.	585.	3 060.
Vapor pressure at 25 deg C, psia	135.7	166.4	56.6	
Vapor pressure at 25 deg C, MN/m^2	0.936	1.147	0.390	
Viscosity (abs), lbm/ft·sec	53.8×10^{-6}	57.1×10^{-6}	8.74×10^{-6}	15.5×10^{-6}
Viscosity (abs), centipoises[a]	0.080	0.085	0.013	0.023
Sound velocity in gas, m/sec	253.	261.	220.	177.
THERMAL AND THERMO-DYNAMIC PROPERTIES				
Specific heat, c_p, Btu/lb·deg F or cal/g·deg C	0.39	0.36	0.11	0.115
Specific heat, c_p, J/kg·K	1 630.	1 506.	460.	481.
Specific heat ratio, c_p/c_v	1.2	1.16	1.29	1.67
Gas constant R, ft-lb/lb·deg F	35.0	36.7	24.1	11.8
Gas constant R, J/kg·deg C	188.	197.	130.	63.5
Thermal conductivity, Btu/hr·ft·deg F	0.010	0.010	0.006	0.003
Thermal conductivity, W/m·deg C	0.017	0.017	0.010	0.005 2
Boiling point (sat 14.7 psia), deg F	−44.	−54.	14.0	−162.5
Boiling point (sat 760 mm), deg C	−42.2	−48.3	−10.	−108.
Latent heat of evap (at bp), Btu/lb	184.	188.2	155.5	41.4
Latent heat of evap (at bp), J/kg	428 000.	438 000.	362 000.	96 000.
Freezing (melting) point, deg F (1 atm)	−309.8	−301.	−104.	−220.
Freezing (melting) point, deg C (1 atm)	−189.9	−185.	−75.5	−140.
Latent heat of fusion, Btu/lb	19.1		58.0	10.
Latent heat of fusion, J/kg	44 400.		135 000.	23 300.
Critical temperature, deg F	205.	197.	315.5	61.9
Critical temperature, deg C	96.	91.7	157.6	16.6
Critical pressure, psia	618.	668.	1 141.	852.
Critical pressure, MN/m^2	4.26	4.61	7.87	5.87
Critical volume, ft^3/lb	0.073	0.069	0.03	0.014 5
Critical volume, m^3/kg	0.004 5	0.004 3	0.001 9	0.000 90
Flammable (yes or no)	Yes	Yes	No	No
Heat of combustion, Btu/ft^3	2 450.	2 310.	—	—
Heat of combustion, Btu/lb	21 660.	21 500.	—	—
Heat of combustion, kJ/kg	50 340.	50 000.	—	—

[a]For N·sec/m^2 divide by 1 000.

TABLE F.2 (continued)
Gas Properties

Common name(s) Chemical formula Refrigerant number	cis-2- Butene C_4H_8 -	trans-2- Butene C_4H_8 —	Isobutene C_4H_8 —	Carbon dioxide CO_2 744
CHEMICAL AND PHYSICAL **PROPERTIES**				
Molecular weight	56.108	56.108	56.108	44.01
Specific gravity, air = 1	1.94	1.94	1.94	1.52
Specific volume, ft^3/lb	6.7	6.7	6.7	8.8
Specific volume, m^3/kg	0.42	0.42	0.42	0.55
Density of liquid (at atm bp), lb/ft^3				—
Density of liquid (at atm bp), kg/m^3				—
Vapor pressure at 25 deg C, psia				931.
Vapor pressure at 25 deg C, MN/m^2				6.42
Viscosity (abs), lbm/ft·sec				9.4×10^{-6}
Viscosity (abs), centipoises[a]				0.014
Sound velocity in gas, m/sec	223.	221.	221.	270.
THERMAL AND THERMO- **DYNAMIC PROPERTIES**				
Specific heat, c_p, Btu/lb·deg F or cal/g·deg C	0.327	0.365	0.37	0.205
Specific heat, c_p, J/kg·K	1 368.	1 527.	1 548.	876.
Specific heat ratio, c_p/c_v	1.121	1.107	1.10	1.30
Gas constant R, ft-lb/lb·deg F				35.1
Gas constant R, J/kg·deg C				189.
Thermal conductivity, Btu/hr·ft·deg F				0.01
Thermal conductivity, W/m·deg C				0.017
Boiling point (sat 14.7 psia), deg F	38.6	33.6	19.2	-109.4^b
Boiling point (sat 760 mm), deg C	3.7	0.9	−7.1	−78.5
Latent heat of evap (at bp), Btu/lb	178.9	174.4	169.	246.
Latent heat of evap (at bp), J/kg	416 000.	406 000.	393 000.	572 000.
Freezing (melting) point, deg F (1 atm)	−218.	−158.		
Freezing (melting) point, deg C (1 atm)	−138.9	−105.5		
Latent heat of fusion, Btu/lb	31.2	41.6	25.3	—
Latent heat of fusion, J/kg	72 600.	96 800.	58 800.	—
Critical temperature, deg F				88.
Critical temperature, deg C	160.	155.		31.
Critical pressure, psia	595.	610.		1 072.
Critical pressure, MN/m^2	4.10	4.20		7.4
Critical volume, ft^3/lb				
Critical volume, m^3/kg				
Flammable (yes or no)	Yes	Yes	Yes	No
Heat of combustion, Btu/ft^3	3 150.	3 150.	3 150.	—
Heat of combustion, Btu/lb	21 000.	21 000.	21 000.	—
Heat of combustion, kJ/kg	48 800.	48 800.	48 800.	—

[a] For N·sec/m^2 divide by 1 000.
[b] Sublimes.

Source: From F. Kreith, Ed., *The CRC Handbook of Mechanical Engineering,* CRC Press, Boca Raton, FL, 1998, A-18–A-29.

TABLE F.3
Properties of Metals

Composition	Melting Point (K)	Properties at 300 K — ρ (kg/m³)	cp [J/(kg·K)]	λ [W/(m·K)]	κ × 10⁶ (m²/sec)	\(=κ×10^6\)	Properties at Various Temperatures (K): λ [W/(m·K)] / cp[J/(kg·K)] — 100	200	400	600	800	1000	1200	1500	2000	2500

Composition	Melting Point (K)	ρ (kg/m³)	c_p [J/(kg·K)]	λ [W/(m·K)]	$κ × 10^6$ (m²/sec)	100	200	400	600	800	1000	1200	1500	2000	2500
Aluminum Pure — λ	933	2702	903	237	97.1	302	237	240	231	218					
Pure — c_p						482	798	949	1033	1146					
Alloy 2024-T6 (4.5% Cu, 1.5% Mg, 0.6% Mn) — λ	775	2770	875	177	73.0	65	163	186	186						
— c_p						473	787	925	1042						
Alloy 195, cast (4.5% Cu) — λ		2790	883	168	68.2			174	185						
— c_p								—	—						
Beryllium — λ	1550	1850	1825	200	59.2	990	301	161	126	106	90.8	78.7			
— c_p						203	1114	2191	2604	2823	3018	3227	3519		
Bismuth — λ	545	9780	122	7.86	6.59	16.5	9.69	7.04							
— c_p						112	120	127							
Boron — λ	2573	2500	1107	27.0	9.76	190	55.5	16.8	10.6	9.60	9.85				
— c_p						128	600	1463	1892	2160	2338				
Cadmium — λ	594	8650	231	96.8	48.4	203	99.3	94.7							
— c_p						198	222	242							
Chromium — λ	2118	7160	449	93.7	29.1	159	111	90.9	80.7	71.3	65.4	61.9	57.2	49.4	
— c_p						192	384	484	542	581	616	682	779	937	
Cobalt — λ	1769	8862	421	99.2	26.6	167	122	85.4	67.4	58.2	52.1	49.3	42.5		
— c_p						236	379	450	503	550	628	733	674		
Copper Pure — λ	1358	8933	385	401	117	482	413	393	379	366	352	339			
Pure — c_p						252	356	397	417	433	451	480			
Commercial bronze (90% Cu, 10% Al) — λ	1293	8800	420	52	14		42	52	59						
— c_p							785	460	545						
Phosphor gear bronze (89% Cu, 11% Sn) — λ	1104	8780	355	54	17		41	65	74						
— c_p							—	—	—						
Cartridge brass (70% Cu, 30% Zn) — λ	1188	8530	380	110	33.9	75	95	137	149						
— c_p							360	395	425						

TABLE F.3 (continued)
Properties of Metals

Composition	Melting Point (K)	ρ (kg/m³)	c_p [J/(kg·K)]	λ [W/(m·K)]	$\kappa \times 10^6$ (m²/sec)	100	200	400	600	800	1000	1200	1500	2000	2500
Constantan (55% Cu, 45% Ni)	1493	8920	384	23	6.71	17	19								
						237	362								
Germanium	1211	5360	322	59.9	34.7	232	96.8	43.2	27.3	19.8	17.4	17.4			
						190	290	337	348	357	375	395			
Gold	1336	19,300	129	317	127	327	323	311	298	284	270	255			
						109	124	131	135	140	145	155			
Iridium	2720	22,500	130	147	50.3	172	153	144	138	132	126	120	111		
						90	122	133	138	144	153	161	172		
Iron Pure	1810	7870	447	80.2	23.1	134	94.0	69.5	54.7	43.3	32.8	28.3	32.1		
						216	384	490	574	680	975	609	654		
Armco		7870	447	72.7	20.7	95.6	80.6	65.7	53.1	42.2	32.3	28.7	31.4		
						215	384	490	574	680	975	609	654		
Carbon steels Plain carbon (Mn ≤ 1%, Si ≤ 0.1%)		7854	434	60.5	17.7			56.7	48.0	39.2	30.0				
								487	559	685	1169				
AISI 1010		7832	434	63.9	18.8			58.7	48.8	39.2	31.3				
								487	559	685	1168				
Carbon-silicon (Mn ≤ 1%, 0.1% < Si ≤ 0.6%)		7817	446	51.9	14.9			49.8	44.0	37.4	29.3				
								501	582	699	971				
Carbon-manganese-silicon (1% < Mn ≤ 1.6%, 0.1% < Si ≤ 0.6%)		8131	434	41.0	11.6			42.2	39.7	35.0	27.6				
								487	559	685	1090				
Chromium (low) steels ½Cr-¼Mo-Si (0.18% C, 0.65% Cr, 0.23% Mo, 0.6% Si)		7822	444	37.7	10.9			38.2	36.7	33.3	26.9				
								492	575	688	969				

Properties at 300 K

Properties at Various Temperatures (K)

λ [W/(m·K)]/c_p[J/(kg·K)]

TABLE F.3 (continued)
Properties of Metals

Composition	Melting Point (K)	ρ (kg/m³)	c_p [J/(kg·K)]	λ [W/(m·K)]	κ×10⁶ (m²/sec)	100	200	400	600	800	1000	1200	1500	2000	2500
1 Cr-½ Mo (0.16% C, 1% Cr, 0.54% Mo, 0.39% Si)		7858	442	42.3	12.2			42.0	39.1	34.5	27.4				
								492	575	688	969				
1Cr-V (0.2% C, 1.02% Cr, 0.15% V)		7836	443	48.9	14.1			46.8	42.1	36.3	28.2				
								492	575	688	969				
Stainless steels															
AISI 302		8055	480	15.1	3.91			17.3	20.0	22.8	25.4				
								512	559	585	606				
AISI 304	1670	7900	477	14.9	3.95	9.2	12.6	16.6	19.8	22.6	25.4	28.0	31.7		
						272	402	515	557	582	611	640	682		
AISI 316		8238	468	13.4	3.48			15.2	18.3	21.3	24.2				
								504	550	576	602				
AISI 347		7978	480	14.2	3.71			15.8	18.9	21.9	24.7				
								513	559	585	606				
Lead	601	11,340	129	35.3	24.1	39.7	36.7	34.0	31.4						
						118	125	132	142						
Magnesium	923	1740	1024	156	87.6	169	159	153	149	146					
						649	934	1074	1170	1267					
Molybdenum	2894	10,240	251	138	53.7	179	143	134	126	118	112	105	98	90	86
						141	224	261	275	285	295	308	330	380	459
Nickel Pure	1728	8900	444	90.7	23.0	164	107	80.2	65.6	67.6	71.8	76.2	82.6		
						232	383	485	592	530	562	594	616		
Nichrome (80% Ni, 20% Cr)	1672	8400	420	12	3.4			14	16	21					
								480	525	545					
Inconel X-750 (73% Ni, 15% Cr, 6.7% Fe)	1665	8510	439	11.7	3.1	8.7	10.3	13.5	17.0	20.5	24.0	27.6	33.0		
							372	473	510	546	626	—	—		

Properties at Various Temperatures (K): λ [W/(m·K)]/c_p [J/(kg·K)]

TABLE F.3 (continued)
Properties of Metals

Composition	Melting Point (K)	ρ (kg/m³)	c_p [J/(kg·K)]	λ [W/(m·K)]	κ × 10⁶ (m²/sec)	\[λ [W/(m·K)]/c_p[J/(kg·K)] at Temperatures (K)\] 100	200	400	600	800	1000	1200	1500	2000	2500
Niobium	2741	8570	265	53.7	23.6	55.2	52.6	55.2	58.2	61.3	64.4	67.5	72.1	79.1	
						188	249	274	283	292	301	310	324	347	
Palladium	1827	12,020	244	71.8	24.5	76.5	71.6	73.6	79.7	86.9	94.2	102	110		
						168	227	251	261	271	281	291	307		
Platinum Pure	2045	21,450	133	71.6	25.1	77.5	72.6	71.8	73.2	75.6	78.7	82.6	89.5	99.4	
						100	125	136	141	146	152	157	165	179	
Alloy 60Pt-40Rh (60% Pt, 40% Rh)	1800	16,630	162	47	17.4			52	59	65	69	73	76		
								—	—	—	—	—	—		
Rhenium	3453	21,100	136	47.9	16.7	58.9	51.0	46.1	44.2	44.1	44.6	45.7	47.8	51.9	
						97	127	139	145	151	156	162	171	186	
Rhodium	2236	12,450	243	150	49.6	186	154	146	136	127	121	116	110	112	
						147	220	253	274	293	311	327	349	376	
Silicon	1685	2330	712	148	89.2	884	264	98.9	61.9	42.2	31.2	25.7	22.7		
						259	556	790	867	913	946	967	992		
Silver	1235	10,500	235	429	174	444	430	425	412	396	379	361			
						187	225	239	250	262	277	292			
Tantalum	3269	16,600	140	57.5	24.7	59.2	57.5	57.8	58.6	59.4	60.2	61.0	62.2	64.1	65.6
						110	133	144	146	149	152	155	160	172	189
Thorium	2023	11,700	118	54.0	39.1	59.8	54.6	54.5	55.8	56.9	56.9	58.7			
						99	112	124	134	145	156	167			
Tin	505	7310	227	66.6	40.1	85.2	73.3	62.2							
						188	215	243							
Titanium	1953	4500	522	21.9	9.32	30.5	24.5	20.4	19.4	19.7	20.7	22.0	24.5		
						465	551	591	633	675	620	686			
Tungsten	3660	19,300	132	174	68.3	208	186	159	137	125	118	113	107	100	95
						87	122	137	142	145	148	152	157	167	176

TABLE F.3 (continued)
Properties of Metals

Composition	Melting Point (K)	Properties at 300 K				Properties at Various Temperatures (K) λ [W/(m·K)]/c_p [J/(kg·K)]									
		ρ (kg/m³)	c_p [J/(kg·K)]	λ [W/(m·K)]	$\kappa \times 10^6$ (m²/sec)	100	200	400	600	800	1000	1200	1500	2000	2500
Uranium	1406	19,070	116	27.6	12.5	21.7	25.1	29.6	34.0	38.8	43.9	49.0			
						94	108	125	146	176	180	161			
Vanadium	2192	6100	489	30.7	10.3	35.8	31.3	31.3	33.3	35.7	38.2	40.8	44.6	50.9	
						258	430	515	540	563	597	645	714	867	
Zinc	693	7140	389	116	41.8	117	118	111	103						
						297	367	402	436						
Zircronium	2125	6570	278	22.7	12.4	33.2	25.2	21.6	20.7	21.6	23.7	26.0	28.8	33.0	
						205	264	300	322	342	362	344	344	344	

Source: From G.F. Hewitt, G.L. Shires, and T.R. Bott, Eds., *Process Heat Transfer*, CRC Press, Boca Raton, FL, 1994, 1022-1025.

Author Index

A

Abbasi, H.A., 313, 331, 339–43, 393, 417, 430
Abdullin, A.M., 166, 186
Abou Ellail, M.M.M., 190
Acharya, P., 431
Acharya, S., 26, 430
Adelt, M., 444, 447
Afgan, N.H., 119–21, 181, 185–6, 226, 228, 282, 392, 430
Agarwal, G., 298, 338
Ahluwalia, R.K., 158, 167, 183, 186–7
Ahmady, F., 303, 339
Ahokainen, T., 191
Albert, S., 225
Aldersley, A.E., 25
Ali Ebadian, M., 26
Alliat, I., 340
Almgren, A.S., 183
Altenkirch, R., 26
Altman, D., 253, 284
Alzeta Corp., 302, 339
Amin, E.M., 317, 341
Ammouri, F., 147
Anandan Unni, V.K., 192, 418, 430
Andersen, M., 146
Anderson, J.E., 197, 219, 221, 225, 237, 241–2, 247, 249, 251, 254, 279, 282
Anderson, W.R., 121
Annamalai, K., 26, 189, 192, 229, 282, 339, 431
API, 27, 340
Arai, H., 147
Arai, N., 367
Areaux, L.D., 429
Arnold, G.D., 318–9, 341
Arpaci, V.S., 117, 122, 163, 184, 334, 343
Åsblad, A., 391
Ashgriz, N., 25, 193, 227, 230, 428
Ashurst, W.T., 163, 184
ASM, 392
ASME, 430
ASTM, 147
Aung, W., 117, 147
Auth, G.H., 367, 392
Ayers, J.F., 227

B

Babiy, V.I., 67, 119
Bachmann, R.C., 199, 226
Baer, A.D., 342

Bahadori, M.Y., 225
Bahmed, A., 230
Bai, T., 343
Bai, X.S., 168, 187
Bailey, R.T., 192
Baker, A.J., 181
Baker, R.J., 218, 229
Balakrishnan, A., 89, 120
Balchen, J.G., 145, 192
Baldyga, J., 184
Ball, K.S., 26, 182, 188, 190–2, 339, 431
Ballintijn, C.M., 392
Baltasar, J., 228
Balthasar, M., 187
Banner, D., 147
Baritaud, T., 182
Barnard, J.A., 7, 25
Barnes, A., 226
Barr, P.K., 190, 192, 334, 343
Bartelds, H., 185
Bartok, W., 25, 121, 187–8
Bartz, D.F., 302, 339
Basevich, V.Y., 188
Basov, V.I., 338
Batchelor, G.K., 183
Batten, R., 26, 64
Battles, D.D., 398, 428
Baukal, C.E., 26–7, 64, 91–102, 108–9, 118, 120–1, 124–7, 131–3, 135–41, 146, 191, 196, 201, 204–6, 213–4, 227–8, 232, 237, 239–42, 244–5, 247–50, 252–3, 255–60, 262–3, 269–70, 276, 281–4, 290–6, 318, 320–2, 337–8, 342, 408, 428–31, 446–7
Baulch, D.L., 187
Baum, M., 182
Baxter, L.L., 341
Baxter, W.J., 120, 429
Bayazitoglu, Y., 189, 338
Bayley, S.E., 192, 431, 427
Bazarian, E.R., 408, 429, 446
Bazinet, C., 429
Bean, C.E., 391
Bechara, W., 147
Becker, H.A., 195, 217–9, 224, 229
Becker, J.S., 428, 446
Beckner, V.E., 183
Beér, J.M., 69, 105–6, 118–22, 146, 164, 181, 185–6, 188, 195, 197, 203, 208, 224–6, 228, 238, 240, 243–4, 247, 249–51, 257, 272, 281–2, 285, 289, 337, 392, 430
Bejan, A., 24, 117

Bell, H.S., 379, 392
Bell, J.B., 183
Bell, R.D., 307, 340
Beltagui, S.A., 204, 227
Bender, D., 430
Bendre, A., 430
Benedek, K.R., 446
Benson, S.W., 188
Berg, J.I., 147
Bergman, K., 428
Berman, H.L., 379, 381, 384, 392
Bernstein, S., 189
Berntsson, T., 391
Best, P.E., 225
Bhattacharjee, S., 157, 165, 182, 185
Bhattacharya, S.P., 120
Biede, O., 129, 146
Bigzadeh, E., 26, 189, 192, 228–9, 339, 345, 367, 431
Bilger, R.W., 163–4, 185
Bingham, F.B., 431
Bird, R.B., 64, 183
Bishop, N.G., 447
Bittner, J.D., 104, 121
Bjørge, T., 187
Blackburn, G.F., 209, 228
Blake, T.R., 294–5, 338
Blanco, J.M., 191, 403, 429
Bland, W.F., 392
Blokh, A.G., 7, 24, 105, 121
Blomquist, C.A., 334, 342
Blumenschein, C.D., 367, 428
Boβ, M., 342, 447
Boateng, A.A., 192, 367, 367
Bocherel, P., 441, 447
Bockhorn, H., 105, 121
Boersma, D., 316, 341
Boerstoel, P., 170, 188
Boger, M., 182
Bogstra, A.N., 310, 340
Bohn, J., 429
Bohn, M.S., 24, 147
Booker, P.I., 412, 430
Borghi, R., 188, 190
Boris, J.P., 150, 181–2, 192
Borisov, A., 182
Borkowicz, R., 26, 64
Borman, G., 7, 25
Bortz, S., 189
Bott, T.R., 3, 24, 502, 519
Boughanem, H., 182
Bouma, P.H., 340
Bourke, P.J., 229
Boussuge, M., 340
Bowen, J.R., 182
Bowen, P.J., 343
Bowman, B., 428
Bowman, C.T., 187, 195, 224
Boyd, R., 191
Boyer, H.E., 392
Boyle, J.G., 341

Bradley, D., 211–2, 228
Bradley, J.N., 7, 25
Bradshaw, P., 184
Bramlette, T.T., 281
Braud, Y., 203, 207, 227
Bray, K.N.C., 163, 184
Brebbia, C.A., 190, 192–3
Breithaupt, P.P., 181
Bressloff, N.W., 168, 187
Brewster, B.S., 191
Brewster, M.Q., 26, 119, 192, 340
Brinckman, G.A., 334, 343
Briselden, T., 342
Brogan, R.J., 367
Brookes, S.J., 168, 187
Brooks, D.L., 327, 342, 390, 393
Brookshaw, L., 166, 186
Broughton, F.P., 122
Brown, A.P.G., 217–8, 229
Brown, J.T., 430
Brown, R.A., 195, 224
Brune, M., 342, 447
Brunner, M., 428
Bryant, G.B., 129, 146
Bucci, D., 431
Buck, E., 64
Buckius, R.O., 165, 185, 340
Buhr, E., 118, 146, 126, 197, 212, 220, 225–6, 229, 241, 249, 251, 256–7, 278, 282, 284
Bundy, F.P., 229
Buriko, Y.Y., 290, 338
Burmeister, L.C., 117
Burwell, R.L., 147
Burzynski, J.P., 340
Busson, C., 340
Butler, B.W., 203, 227
Butler, G.W., 177, 189

C

Cain, B.E., 341
Calcote, H.F., 104, 121, 221, 230
Caldwell, F.R., 209, 228
Candler, E.M., 209, 228
Cappelli, M.A., 281
Carangelo, R.M., 225
Card, J.M., 182
Caretto, L.S., 190
Carey, V.P., 184, 338
Carlomagno, G.M., 192
Carnicer, P., 428
Carpenter, N., 140, 147
Carreau, P.J., 447
Carvalho, M.G., 26, 121, 165, 170, 178, 185, 188, 190–3, 228, 414, 419, 430
Cassiano, J., 197, 226, 411, 429
Cavaseno, V., 392
Cess, R.D., 6, 24, 89, 119–20, 164, 185, 212, 228
Chace, A.S., 3, 24, 446
Chaffin, C., 340

Chai, J.C., 186
Chambers, J.T., 226
Champinot, C., 147
Chan, I., 146, 232, 281, 340, 393
Chan, S.H., 181, 184, 188–90
Chandrasekhar, S., 186
Chang, C.H.H., 181
Chang, F.C., 192
Chang, S.L., 191
Chao, C.M., 208, 228
Chaouki, J., 343
Chapman, M.A., 179–80, 403, 446
Chapman, K.S., 193, 428–9
Char, J.-M., 208, 228
Charette, A., 182
Chaussavoine, C., 119
Chawla, T.C., 183, 338
Chedaille, J., 119, 203, 207, 318, 341
Chellaiah, S., 26, 119, 184, 189, 191–2, 230, 367, 446
Chen, C.J., 184
Chen, D.C.C., 253, 258, 283–4
Chen, J.H., 182
Chen, M.M., 228
Chen, Y.-K., 298, 338
Cheremisinoff, N.P., 25, 166, 185, 188
Chien, P.-L., 225
Chigier, N.A., 118, 146, 195, 197–8, 203, 208, 212, 218, 221, 224–6, 229, 238, 240–1, 243–4, 247, 249–51, 256–7, 272, 281–2, 285, 289, 337
Cho, K.W., 191, 429
Cho, P., 26
Cho, S.H., 218, 229
Chopey, N.P., 431
Chorin, A.J., 184
Chou, S.K., 391
Chou, T., 447
Choudhury, D., 26, 189, 192, 229, 282, 339, 431
Chouinard, J.-G., 441, 447
Chua, K.J., 391
Chue, S.H., 218, 229
Chuech, S.G., 119, 147, 183, 227, 338
Chui, E.H., 186
Chung, J.N., 26
Chung, T.J., 150, 181–2
Churchill, S.W., 7, 26, 184
Clay, D.T., 230
Claytor, L.E., 225
Clement, R., 430
Clinch, J.M., 342
Cloutman, L.D., 166, 186
Cobb, L.L., 120, 145, 226, 281
Cobos, C.J., 187
Cobos, L., 428
Cocagne, J.M., 391
Coelho, P.J., 170, 188
Cole, R., 122
Colella, P., 183
Colette, P., 146
Collier, J.G., 122
Collin, R., 190, 192

Coltrin, M.E., 64
Connor, N.E., 25
Conolly, R., 69, 118, 145, 199, 219, 221, 227, 229, 239, 248, 251, 254, 256–7, 259, 267–9, 274, 281, 283
Cookson, R.A., 118, 124, 145–6, 197, 209, 212, 225, 228, 239, 243, 248, 251, 256, 259, 266, 271, 283
Cooper, J.R., 383, 392
Copes, J.S., 390, 393
Copin, C., 222, 230
Coppalle, A., 89, 120
Corio, P., 429
Corliss, J.M., 334, 343
Cornforth, J.R., 25, 391
Costa, M., 212, 228
Costello, F.A., 275, 284
Cox, G., 182
Cox, R.A., 187
Cramer, T.L., 391, 429–31, 447
Crawford, M.E., 117, 284
Crosbie, A.L., 119, 165, 185
Crosley, D.R., 183, 228, 339
Crutchfield, W.Y., 183
CSI Resource Systems, 431
Cuervo, L.A., 446
Cundy, V.A., 192, 430–1
Cunningham, T.S., 338
Cygan, D., 339
Czerniak, D., 26, 64

D

D'Aquino, R., 447
Dahm, W.J.A., 4, 181–2, 338
Dalton, A.I., 64, 446
Dalzell, W.H., 187
Damköhler, G., 289, 338
Daniel, B.R., 342, 343
Davidson, D.F., 187
Davidson, R.L., 392
Davies, D.R., 91, 120–1, 196, 225, 237–8, 246–7, 251, 254, 274, 282
Davies, J.T., 183
Davies, R.M., 69, 118, 196, 199, 225, 227, 237, 239–40, 243, 247–8, 251, 253–4, 256–7, 259, 267–9, 274, 281–2
Davies, T., 23, 27, 327, 342
Davila, J.R., 399, 428
Davis, D.H., 430
Davis, S.R., 429
Day, M., 182
de Goey, L.P.H., 340
de La Faire, A., 24, 367, 392–3
de Lemos, M.J.S., 192
De Lucia, M., 446
De Ris, J., 338
De Riu, L., 147
Dearden, L.M., 341
DeBellis, C.L., 119, 346, 367
Dec, J.E., 334, 343
DeGregorio, M.B., 447

Delichatsios, M.A., 170, 188, 290, 338
Delplanque, J.-P., 121, 189, 192, 367, 341
DeLucia, M., 318, 341
Deng, X.X., 192, 431
Denison, M.K., 165, 186
Desai, T.M., 408, 429
Desgroux, P., 187
Desmond, R.M., 7, 24
DeVincentis, D., 431
DeWerth, D.W., 339
Dewitt, D.P., 24, 229
Dhir, V.K., 122
Dibble, R.W., 183, 228, 339
Ding, M.G., 428, 431, 446
Dittman, H.J., 342, 447
Dixon-Lewis, G., 64
Djavdan, E., 147, 165, 185
Docherty, P., 119, 165–6, 186
Dolenc, D.A., 24, 99–102, 108–9, 131, 146, 182, 189–90, 192, 205, 225, 227, 230, 232, 281, 337–343, 367, 391–3, 428–30, 447
Dolinar, J., 182
Donaldson, L.W., 341, 343, 430
Doran, D.J., 367, 428
Dorresteijn, W.R., 392
Dougherty, R.L., 76, 185
Drake, R.A., 3, 24, 429
Driscoll, J.F., 338
Dryer, F.L., 155, 182, 187–8, 222, 230
Du, Z., 446
Duggan, P.A., 339
Dunduchenko, V.E., 429
Dunham, P.G., 90, 118, 120, 146, 197, 212, 221, 225, 228, 230, 239, 248, 251, 256, 259, 266, 271, 283–4
Duong, H.T., 185
Durão, D.F.G., 195, 224, 227, 430
Dwyer, H.A., 191–2
Dwyer, J.B., 24
Dybbs, A., 229
Dzyuzer, V.Ya., 429

E

Echigo, R., 102, 121
Eckbreth, A.C., 195, 208, 224, 225
Eckert, E.R.G., 284
Edwards, D.K., 89, 119–20, 185, 187
Egolf, C.J., 339
Eguchi, K., 147
Eibeck, P.A., 231, 241, 249, 281
Eleazer, P.B., 430, 446
Elich, J.J., 120, 192
Elliston, D.G., 76, 119
Ellzey, J.L., 189–90, 338
El-Mahallawy, F.M., 25
Endrys, J., 147
Erauskin, I., 428
Erinov, A.E., 391, 393
Ertl, D., 430
Es'kov, A.I., 338
Escalera, R., 285, 337

Esser, C., 187
Eswaran, V., 182
Evans, D.D., 227
Ezekoye, O.A., 189–90
Ezerski, B.J., 391

F

Faeth, G.M., 119, 121, 147, 163, 168, 182, 184, 186, 188, 222, 227, 230, 290, 338
Fairweather, M., 118, 166, 186, 199, 213, 218–9, 226, 228, 238–9, 248, 250, 252–3, 255–6, 259, 268, 270, 275, 282–4
Fang, C.S., 117
Farias, T.L., 105, 121, 165, 185
Farmer, L.K., 146, 232, 281, 446–7
Farouk, B., 26, 119, 184, 189, 191–2, 227, 230, 367, 438, 446
Farrell, L.M., 446
Fay, J.A., 145, 263–4, 266–7, 270–1, 284
Fay, R.H., 197, 225, 242, 246, 249, 254, 282
Federov, A.Y., 188
Feintuch, H.M., 381, 393
Fells, I., 196, 213, 225, 237, 247, 251, 256, 274, 284
Felske, J.D., 26, 228
Feng, J., 190, 341
Ferguson, N.T., 387, 393
Fernandez, R., 145, 192
Ferrell, J.K., 184
Ferron, J.R., 192, 367
Féry, C., 213, 229
Fingerson, L.M., 218, 229
Fioravanti, K.J., 27, 342
Fish, F.F., 340
Fiveland, W.A., 119, 176, 186, 189, 191, 338
Flamme, M., 329, 342, 444, 447
Flanagan, P., 300, 339
Fleshman, J.D., 392
Fogo, D., 431
Fornaciari, N.R., 225
Foster, P.J., 89, 120
Fouhy, K., 431
Franck, P.-A., 391
Frank, P., 187
Franz, H., 147
Frazier, W.F., 341
Freeman, H.M., 431
Freeman, R.A., 326, 341
Freihaut, J.D., 339
Frenklach, M., 170, 182, 187–8
Fridman, A.A., 105, 121, 337, 447
Frisch, U., 184
Fristrom, R.M., 25, 147, 195, 211–2, 224, 228
Frizzell, D.H., 187
Frost, V.A., 188
Fruehan, R.J., 119, 367, 428
Fu, W., 191
Fu, X., 178, 190, 299, 339
Fuchs, L., 187
Fureby, C., 181
Fusaro, D., 431
Futami, H., 392

G

Gaddone, J., 337, 447
Gad-el-Hak, M., 229
Gaffuri, P., 188
Galloway, T.R., 275, 284
Galsworthy, R.A., 182
Ganapathy, V., 7, 24, 392
Ganic, E.N., 24, 122, 184
Garcia, G., 227
Garde, R.J., 184
Gardiner, W.C., 170, 187
Gardon, R., 198, 226, 411, 429
Garg, A., 375–6, 392
Garner, W.E., 82, 120
Garo, A., 188
Garrote, A., 428
Gartling, D.K., 175, 188
Gaydon, A.G., 318, 323, 341
Gebhart, B., 24, 64, 91–7, 113, 118–20, 122, 125–7,
    132–41, 146–7, 196, 198, 201, 206, 226–7, 232,
    237, 239–42, 244–5, 247–50, 252–3, 255–60,
    262–3, 269–70, 276, 279, 281–4, 290–6, 318,
    320–2, 338
Gegelashvili, V.K., 429
George, P.E., 333, 342
Geotti-Bianchini, F., 147
Gershtein, V.Y., 191
Ghoniem, A.F., 181
Ghorashi, B., 229
Ghosh, H., 392
Ghoshdastidar, P.S., 192, 418, 430
Gibbs, B.N., 446
Giedt, W.H., 91, 113, 118, 120, 125, 145, 147, 198, 206,
    226, 238, 244, 248–9, 254, 256–7, 280, 281–2
Gilchrist, J.D., 122
Gildersleeve, R.E., 429
Gill, D.W., 107, 122
Gillis, P.A., 154, 182, 190
Gitman, G.M., 431, 446
Giversen, F., 189
Givi, P., 182
Glassen, L.K., 187
Glassman, I., 25, 54, 103, 121
Glaws, P.C., 428
Glinkov, M.A., 76, 119
Godridge, A.M., 338
Goff, S., 431
Golchert, B., 191
Goldberg, M., 187
Golden, D.M., 170, 187
Goldman, Y., 207, 228
Goldstein, S., 284
Gonzalez, M., 190
Goody, R.M., 169, 187
Goracci, E., 105, 119
Gordon, S., 64, 125, 146, 284
Gore, J.P., 119, 147, 182–3, 189–90, 201, 227, 229, 300,
    338–40
Görner, K., 190
Gorski, L.M., 313, 340

Gosling, T.M., 27, 122
Gosman, A.D., 181, 185, 189–91
Gostowski, V.J., 275, 284
Gottardi, R., 428
Goulard, R., 208, 224, 228
Gouldin, F.C., 195, 224
Gover, M., 183
Graham, M.D., 183
Gray, P., 188
Gray, W.A., 6, 24–5, 102, 118, 121, 159, 183, 200, 227
Green, D., 64
Greeves, G., 170, 188
Gretsinger, K., 339
Grey, J., 208, 228
Grief, R., 87, 120
Griffith, C.R., 431
Grigull, U., 26, 118, 122, 145, 227, 283
Gripenberg, H., 428
Grisham, S., 446
Griswold, J., 7, 25, 68–9, 119
Gritzo, L.A., 121, 157, 183, 192, 341, 367
Grosman, R.E., 335, 343, 430
Grosshandler, W.L., 26, 121, 157, 181–3, 185, 227–8, 230
Guénebaut, H., 318, 323, 341
Gülder, Ö.L., 183
Gulic, M., 120
Gunardson, H., 392
Guo, K., 192, 392
Gupta, A.K., 3, 24, 26, 431
Gupta, R.P., 120
Guruz, K.H., 335, 343
Gushchin, S.N., 191, 429
Gutierrez, A.M., 428
Gutmann, M., 340
Guy, C., 343, 444, 447
Guy, J.J., 25

H

Hackert, C.L., 189–90
Hagiwara, A., 177, 189
Hahne, E., 26, 118, 145, 227, 283
Hale, W., 145, 192, 338
Hall, M.J., 189, 338
Hall, R.E., 342
Hamaker, H.C., 185
Hammond, E.G., 105, 121
Hampartsoumian, E., 183
Hampson, R.F., 187
Hanby, V.I., 334, 342
Handa, K., 402, 428
Hanjalic, K., 184
Hanson, R.K., 187, 225
Harb, J.N., 120, 129, 146, 341
Hardy, W.A., 146
Hargrave, G.K., 118, 197, 199, 212, 218, 222, 226, 228,
    237, 239, 247–8, 250, 252, 256, 259, 262, 265,
    268–9, 271, 273, 276, 283–4
Harker, J.H., 196, 213, 225, 237, 247, 251, 256, 274, 284
Härkönen, E., 391
Harris, A., 227

Hartmann, J.M., 169, 187
Hartnett, J.P., 24, 120, 122, 185–6
Hasegawa, T., 328, 342
Hashimoto, R., 392
Hattori, M., 118, 226, 282
Haupt, G., 118, 225, 284
Haupt, R., 146, 226, 282
Hauser, W.S., 187
Hawlader, M.N.A., 374, 391
Hayasaka, H., 165, 185
Hayhurst, A.N., 211–2, 228
Hayman, G., 187
Haynes, B.S., 104, 121
Hazard, H.R., 24, 446
He, X., 120
Heap, M.P., 185, 190
Heberle, N.H., 183, 228, 339
Heffron, J.F., 408, 429, 446
Hein, K.R.G., 188
Heitor, M.V., 207, 209, 218, 224, 226–7, 229, 429
Hellander, J.C., 129, 146
Hemeson, A.O., 199, 227, 237, 239, 247–8, 252, 259, 263, 265, 271, 283
Hemsath, K.H., 282, 389–90, 393
Henderson, R.D., 181
Henry, V.I., 415, 430
Herron, J.T., 187
Hersch, C.A., 341
Hewitt, G.F., 1, 3, 24, 70–1, 217, 229, 367, 391–2, 428, 502, 519
Heym, J., 342, 447
Hibberd, D.F., 119
Hill, S.C., 191
Hilliard, J.C., 26
Hindman, D.L., 410, 429
Hinze, J.O., 184, 228
Hirata, M., 121
Hirose, Y., 367
Hjertager, B.H., 170, 188
Hnat, J., 430
Ho, M.-D., 428, 431
Ho, T.-Y., 119
Hofmann, D., 208, 228
Hoke, B.C., 430
Holen, J., 187
Holman, J.P., 24, 142, 147
Holmstedt, G., 157, 183
Hoogendoorn, C.J., 91, 118, 120, 157, 183, 188, 190–2, 197, 212, 218, 220, 226, 241, 249, 251, 256–7, 276, 283, 379, 392
Hope, S., 430
Hope, S.R., 430
Horbatch, W.J., 430
Hornbaker, D.R., 198, 200, 226
Horsley, M.E., 118, 145, 199, 212, 220, 222, 227, 241, 247, 249, 251, 256, 259, 265, 283
Hottel, H.C., 6, 24, 105, 119, 122, 147, 183, 185, 187, 345, 367, 375, 392, 446
Howard, J.B., 104, 121
Howarth, C.R., 106, 122
Howell, J.R., 89, 119–20, 185, 189, 297–8, 338–9

Howell, L.H., 183
Hsu, J.C.L., 222, 230
Hsu, P.F., 189, 298–9, 338–9
Hsu, S.T., 6, 24
Hubble, D.H., 119, 428
Huber, A.M., 225
Huebner, S.R., 315, 341, 393
Hughes, D.L., 387, 392
Hughes, P.M.J., 208, 228
Hull, R., 447
Hunn, B.D., 227
Hurd, A.J., 187
Hussaini, M.Y., 182
Hustad, J.E., 67, 107, 119, 199, 203–4, 212, 221, 227, 237, 243, 247, 252, 256–7, 273–5, 281–2
Hutchinson, F.W., 6, 24
Hutchinson, P., 190, 292, 338

I

Ibbotson, A., 293, 338
Ichiraku, Y., 190
IHEA, 25
Iino, H., 102, 121
Il'yashenko, I.S., 338
Im, K.H., 158, 167, 183, 186–7
Imbach, J., 225
Incropera, F.P., 24
Industrial Heating Equip. Association, 446
Institute of Gas Technology, 312, 340, 445–7
Irvine, T.F., 120, 185–6
Ito, S., 190
Ivernel, A.B., 82, 91, 118–9, 146, 199, 208, 227, 238–9, 246–8, 253–4, 257, 259, 268, 276, 282

J

Jackson, E.G., 90, 118, 205, 219, 225, 237, 247–8, 251, 254, 246, 273, 282
Jacob, M., 121
Jacobs, H.R., 186
Jacobsen, M., 227, 281
Jagoda, J.I., 342
Jain, R.C., 24, 337, 341, 447
Jakway, A.L., 192, 431
Jaluria, Y., 119, 338
Jamaluddin, A.S., 186
Janna, W.S., 24
Javorka, D.J., 438, 446
Jeng, S.-M, 119, 147, 183, 227, 338
Jenkins, B.G., 357
Jessee, J.P., 189
Ji, B., 98–102, 108–9, 120, 131, 146, 204–6, 227
Johansson, M., 300, 339
Johnson, A.J., 367, 392
Johnson, R.W., 173–4, 181, 183–4, 189, 229
Johnson, T.R., 185
Jokilaakso, A., 191
Joklik, R.G., 230
Jones, J.A.T., 188, 428
Jones, J.M., 341

Jones, W.P., 163, 184
Jørgensen, K., 146, 339
Joshi, M.L., 26, 337, 343, 430
Joshi, S.V., 446
Joulain, P., 183
Jugjai, S., 304, 339
Just, Th., 187

K

Kabashnikov, V.P., 183
Kagel, R., 430
Kaji, H., 367
Kakaç, S., 117, 147
Kakhi, M., 188
Kaminsky, V.A., 188
Kampp Rasmussen, N.B.
Kandlikar, S.G., 122
Kanury, A.M., 26, 192
Karki, K.C., 174, 181
Karlekar, B.V., 7, 24
Karniadakis, G.E., 181
Kaskan, W.E., 212, 229
Kassemi, M., 186
Kataoka, A., 300, 339
Kataoka, K., 197, 212, 226, 238, 242, 249, 253, 256–7,
        272, 277, 282, 284
Katsuki, M., 328, 342
Kaufman, F., 176, 187
Kaufman, K.C., 189
Kaviany, M., 117
Kays, W.M., 117, 284
Keating, E.L., 7, 25
Kee, R.J., 64, 187
Keey, R.B., 391
Keffer, J.F., 227, 282
Keldar, K.M., 174, 181
Keller, J.G., 24, 341
Keller, J.O., 190, 281, 334, 343
Kenbar, A.M.A., 227
Kendall, R.M., 159, 183, 307, 340
Kent, J.H., 211–2, 228, 338
Kern, D.Q., 6, 378–9, 392
Kerr, J.A., 187, 343
Kessinger, G.F., 24
Kezerle, J.A., 182, 190
Khalil, E.E., 25, 150, 153, 181, 190, 192, 338, 418, 430
Khalil, K.H., 25
Khan, I.M., 170, 188
Khan, J.A., 192
Khavkin, Y.I., 346, 367
Khinkis, M.J., 24, 340, 367, 392–3, 447
Kilham, J.K., 25, 90, 118, 121, 133, 116–7, 126, 138,
        145–6, 196–7, 199, 205, 212–3, 219–20, 225–8,
        237, 239, 242–3, 246–8, 249–52, 254, 256, 259,
        262, 265–71, 273, 282–284
Kimura, Y., 410, 431
King, I.R., 221, 230
Kirilenko, V.I., 293, 338
Kishimoto, S., 118, 226, 282
Kissel, R.R., 446

Kistler, M.D., 428
Kittleson, D.B., 211–2, 228
Klarsfeld, S., 147
Klima, R., 191, 428
Klingensmith, L.K., 446
Kmit, G.I., 183
Knorr, R.E., 27, 122
Knowles, D.F., 1, 24, 398, 428–9
Kobayashi, H., 318, 341, 446
Kocaefe, Y.S., 156, 182
Koch, G.S., 428
Koeroghlian, M., 340
Kohl, R.F., 198, 226
Kollmann, W., 184
Komornicki, W., 165, 185
Kondic, N.N., 208, 228
Kontz, F., 431
Kor, G.J.W., 428
Kovotny, J.L., 89, 120
Köylü, Ü.Ö., 121, 168, 186, 222, 230
Kramnik, B.M., 340
Kreith, F., 24–5, 76, 85–7, 147, 209, 228, 501, 514
Kremer, H., 118, 146, 197, 199, 212, 221, 225–6, 236, 241,
        249, 251, 256, 278, 282, 284
Krichten, D.J., 429
Krill, W.V., 339
Ku, J.C., 105, 121
Kubota, T., 226, 283
Kudo, K., 191, 377, 392
Kuhl, A.L., 182
Kukarkin, A.I., 429
Kulkarni, M.R., 146, 339
Kumar, K., 189
Kunc, W., 24, 367, 392–3, 447
Kunitomo, T., 105, 121
Kuo, J.T., 153, 182, 426, 431
Kurek, H.S., 5, 340, 367, 391–3
Kurosaki, Y., 24
Kut'in, V.B., 191, 429
Kuznetsov, V.R., 184, 290, 338

L

La Course, W.C., 147
Lacelle, R.J., 228
Ladenburg, R.W., 230
Lahaye, J., 104, 121
Lahey, R.T., 122
Lallemant, N., 165, 185
Landahl, M., 184
Landais, T., 24, 340, 367, 392–3
Lange, E., 367
Languir, I., 221, 230
Lanier, W.S., 24
Lankhorst, A.M., 189
Lannutti, J.L., 341
Lanyi, M.D., 447
Lappe, V., 215–6, 229
Larrouturou, B., 150, 181
Larson, D.A., 229
Launder, B.E., 181, 184

Laurendeau, N.M., 208, 212, 228
Law, C.K., 25
Lawn, C.J., 338
Lazzeretti, R., 25
Leah, A.S., 140, 147
Leblanc, B., 72, 105, 119
Leckner, B., 84, 120, 187
Lee, H.S., 186
Lee, J., 189
Lee, K.B., 170, 188
Lee, S.C., 103, 120, 187
Lee, Y.K., 191, 403, 429
Lefebre, M., 429
Lefrank, P.A., 428
Leger, C.B., 192, 430–1
Lehner, M., 122
Lehrman, A., 346, 367, 396, 428
Leipertz, A., 208, 228
Lemieux, P.M., 342
Leon, L., 187
Lepoutre, E., 429
Lesieur, M., 184
Lester, T.W., 430
Leuckel, W., 190, 192
Leung, K.M., 170, 188
Levy, A., 24, 446
Lewis, B., 25, 147, 216, 229–30, 284
Leyens, R.E., 177, 189
Leyer, J.–C., 182
Li, S., 191
Liang, W.W., 388, 340, 393, 429
Libby, P.A., 170, 187
Lide, D.R., 64, 122, 202, 227
Lieberman, N.P., 392
Lievoux, P., 190
Lightfoot, E.N., 64
Lighty, J.S., 192, 431
Lihou, D.A., 72, 119
Liley, P.E., 64
Lilley, D.G., 3, 24–5, 150, 181, 185, 227–8, 230
Lin, K.-C., 188
Lin, W.-Y., 298, 338
Lindstedt, R.P., 163, 170, 185, 188
Linnett, J.W., 146
Lior, N., 7, 26
Lisienko, V.G., 191, 429
Lisin, F., 406, 429
Lissianski, V., 187
Litka, A., 430
Litkouhi, B., 186
Liu, D., 157, 192
Liu, F., 183
Lloyd, J.R., 24
Lockwood, F.C., 26, 185–6, 190–1
Longacre, S., 27
Longwell, J.P., 104, 121
Loo, O., 337, 447
Lopes, J.B., 190
Lottes, S.A., 191
Louis, G., 313, 330, 340, 377
Love, T.J., 6, 24, 119

Loveridge, D.J., 122
Lovett, J.A., 230
Lowe, A., 191
Lowes, T.M., 72, 118–9, 164, 185, 190, 196, 198, 203, 208, 220, 225, 238, 241–3, 246, 247, 249–53, 257, 272, 278, 282–3
Lowy, L., 379, 392
Lucas, D.M., 182
Ludwig, C.B., 83, 120, 164, 169, 185, 227
Lukasiewicz, M.A., 24, 27, 189–90, 192, 339, 341, 343, 393, 428–30, 446
Lukens, J.A., 431
Lund, J.S., 146
Lundgren, E., 181, 192, 335, 343
Lynen, A., 342, 447
Lytle, G.C., 446

M

Maccallum, N.R.L., 227
Machida, M., 147
Machii, N., 189, 192
Madnia, C.K., 182
Madrzykowski, D., 227
Madsen, O.H., 128, 146, 300, 339
Madson, J.M., 212, 229
Maesawa, M., 106, 122, 360, 367
Magel, H.C., 188
Magnussen, B.F., 170, 187–8
Mahajan, R., 119
Mahalingam, S., 182
Majumdar, A.S., 282, 391
Malkmus, W., 120, 185, 227
Mallard, W.G., 187
Manickavasagam, S., 222, 230
Mariasine, J., 225
Marie, B., 147
Marino, J.A., 406, 429
Marinskii, V.S., 429
Markatos, N.C., 191
Markham, J.R., 225
Markstein, G.H., 192, 290, 338
Marksten, U., 343
Marr, R.J., 428
Marsano, S., 335, 343
Martin, G.R., 392
Martin, J.E., 187
Martinez, M.P., 447
Martins, L.-F., 181
Martins, N., 192
Matavosian, R., 89, 120
Matsumura, M., 190
Matsuo, H., 118, 282
Matsuo, M., 82, 198, 226, 240–1, 247, 249, 251, 256–7, 278, 282
Matthews, K.J., 211–2, 228, 338
Matthews, L., 202, 227
Matthews, R.D., 189, 338–9
Mattsson, P., 391
Mauillon, L., 343
Mauss, F., 187

Mawhinney, M.H., 25, 345, 367, 393
Maximuk, D.
Maximuk, A., 447
Mayinger, F., 122
Mazzei, M., 429
McAdams, W.H., 121, 253, 273–4, 283
McBride, B.J., 64, 146, 284
McBride, C., 431
McComb, W.D., 184
McConnell, M.M.B., 340
McDonald, M., 294–5, 338
McGrath, I.A., 253, 283
McIntosh, A.C., 340
McLeod, B.M., 431
McMahon, A., 430
McMurtry, P.A., 182
McQuay, M.C., 26, 189, 192, 225, 229, 339, 343, 431
Megahed, I.E.A., 190–1
Megan, G.P., 391
Mei, F., 189, 315, 341
Meisingset, H.C., 145, 192
Menguc, M.P., 26, 119, 165–6, 183–5, 189, 191–2, 222, 230
Menon, R.K., 218, 229, 447
Merrick, D., 341
Meunier, H., 189–90, 315, 341
Michaud, M., 446
Michelfelder, S., 118, 185, 190, 225, 283
Milewski, E.B., 340
Milewski, J.V., 340
Miller, D.L., 334, 343
Miller, J.A., 64, 187, 188
Mills, A.F., 24
Milson, A., 118, 146, 198, 212, 221, 226, 241, 243, 249, 251, 256, 282
Miquel, P.F., 341
Mish, K.D., 173, 181
Mishra, B., 429
Mital, R., 175, 189, 212, 229, 300, 306–7, 338–40
Miyami, H., 366–7
Mocsari, J.C., 340
Modak, A.T., 185
Modest, M.F., 26, 119, 186
Moffat, R.J., 211, 227–8
Mohebi-Ashtiani, A., 226, 283
Mohr, J.W., 238, 241, 249, 282
Mohr, P., 337, 341, 447
Moilanen, G.L., 26, 64
Moles, F.D., 357
Möller, S.–I., 181, 192, 343
Mongia, H.C., 184
Monrad, C.C., 68, 119
Montestruc, A.N., 430
Moppett, B.E., 182
Moreira, A.L.N., 207, 209, 227, 229
Moreno, F.E., 339
Mori, I., 367
Mori, Y., 430
Morita, M., 328, 341
Morse, J.S., 192, 430
Mortazavi, S., 121

Moss, J.B., 168, 187–8
Most, J.-M., 218, 229
Mott-Smith, H.M., 230
Mourão, M., 228
Moussa, N.A., 341
Mucher, M.B., 431
Mujumdar, A.S., 226
Müller, R., 6, 24–5, 119, 121, 200, 227
Munger, M., 182
Munroe, N.D.H., 191
Murrells, T., 187
Murthy, S.N.B., 25
Myers, G.E., 122
Myers, R.A., 392, 430
Myhr, F.H., 286, 290, 338

N

Nabors, J.K., 337, 430
Nagle, J., 170, 188
Naraghi, M.H.N., 165, 186
Nassaralla, C.L., 428
Nawaz, S., 113, 117–8, 125, 127, 146, 199, 213, 219, 221, 226, 238–9, 248, 253–4, 256–7, 259, 267, 270, 282
Neff, D., 337, 341, 444, 447
Neff, G.C., 26
Negin, K.M., 381, 393
Nelson, R.A., 26, 190–2
Nelson, S.M., 340
Nelson, W.L., 392
Nerad, A.J., 147
Nestor, S., 337, 447
Nestor, S.A., 121
Neumeier, Y., 333, 342
Newall, A.J., 72, 119
Newbold, J., 195, 225
Newby, J.N., 327, 341–2
Nichols, E.L., 212, 228
Nishimura, K., 192
Nishiwaki, N., 121
Nogay, R., 392
Nogueira, M., 178, 190–3, 419, 430
Nowak, A.J., 190, 193
Nutter, G.D., 229

O

O'Doherty, T., 343
Ogisu, Y., 122, 367
Ohta, T., 118, 226, 282
Okamoto, T., 184
Okoh, C.I., 195, 224
Olabin, V., 443, 447
Ondrey, G., 431, 447
Oosthuizen, P.H., 117
Oran, E.S., 150, 181–2, 192
Orloff, L., 170, 188, 338
Orzechowski, N.J., 387, 392
Osuwan, S., 367
Owens, W.D., 192, 426, 431
Özisik, M.N., 6, 24, 122, 173, 188

P

Page, R.H., 282
Paget, M.W., 429
Pai, B.R., 185, 190
Pakulski, D., 429
Pal, D., 192
Palmer, H.B., 121
Palosaari, S., 391
Pam, R., 339
Panahi, S.K., 26
Papadelis, E.P., 431
Parameswaran, T., 228
Parekh, U., 189
Paris, J., 447
Park, H.M., 183
Park, H.S., 191, 429
Parkinson, G., 431, 447
Patankar, S.V., 150, 163, 174, 181, 183, 186, 190
Patrick, D.K., 229
Patton, J.B., 24, 341
Paul, P.H., 281
Pauwels, J.F., 187
Pavlack, T.T., 446
Payne, R., 118, 225, 283
Pearce, K.W., 417, 430
Pease, R.N., 229–30
Pechurkin, V.A., 91, 118, 197, 202, 226, 236, 240, 242,
    249, 251, 254, 256–7, 277, 279–80, 282
Peck, R.E., 146, 189, 339–40
Pember, R.B., 175, 183
Pennati, G., 188
Penner, S.S., 225
Pereira, J.C.F., 184, 430
Perkonen, J., 391
Perlmutter, M., 185
Perrin, M., 177, 187, 190, 225
Perry, R.H., 64
Pershing, D.W., 192, 431
Perthuis, E., 25
Peters, A.A., 181
Peters, J.E., 340
Peters, N., 182, 342
Peterson, R.B., 26
Peterson, R.C., 212, 228
Petersson, M., 428
Petrick, M., 191
Pettersson, M., 296, 338, 371, 391
Petukhov, B.S., 184
Peureux, J., 340
Phillips, B., 429
Picot, J.J.C., 367
Pilling, M.J., 187
Pinar Menguc, M, 367, 446.
Pincus, A.G., 411, 430
Pitz, W.J., 188
Plessing, T., 331, 342
Pletcher, R.H., 147
Pohl, J.H., 1, 24
Poinsot, T., 182
Poitou, S., 187

Polezhaev, Y.V., 217, 229, 367, 391–2, 428
Poling, B.E., 64
Poll, I., 367
Pollock, D.D., 212, 229
Polyakov, A.F., 184
Ponizy, B., 191
Pope, S.B., 163, 182, 184, 188
Popiel, C.O., 118, 197, 212, 218, 220, 222, 226, 241, 249,
    251, 256–7, 276, 283
Posillico, C.J., 196, 208, 218, 221, 225, 241, 248–9, 252,
    257, 283
Post, L., 183, 191
Poulikakos, D., 122
Pourkashanian, M., 293, 338, 341
Povarenkov, V.A., 393
Prado, G., 104, 121
Prasad, A., 392
Prausnitz, J.M., 64
Presser, C., 24–6, 119, 184–5, 189, 191, 192, 227, 228,
    230, 367, 446
Pritchard, R., 25
Prudnikov, A.G., 163, 184
Purvis, M.R.I., 91, 117–8, 120, 126, 138, 145–6, 199,
    219–21, 226–27, 229, 239, 242, 248–9, 254, 256,
    259, 266–7, 270, 283–4
Putnam, A.A., 331, 333–4, 342–3
Pye, L.D., 147

Q

Qingchang, G., 119
Quack, R., 185, 418, 430
Queiroz, M., 228
Quinqueneau, A., 225, 341
Quintiere, J.G., 26, 193, 227, 230, 428

R

Rabhan, A.B., 342
Radulovic, P.T., 191
Ragland, K., 7, 25
Raithby, G.D., 186
Rajani, J.B., 116, 118, 196, 203, 208, 220–1, 225, 241–3,
    247, 249, 251, 254, 257, 283
Rall, D.L., 198, 200, 226
Ramachandran, C.V., 430
Ramachandran, R., 447
Ramadhyani, S., 189, 191, 193, 340, 428–9
Ramamurthy, H., 189, 191, 314, 340
Ramos, J.I., 181
Ranzi, E., 188
Rasmussen, N.B.K., 146, 189, 339
Rathmann, O., 146
Rauenzahn, R.M., 117–8, 198, 226, 242, 249, 254, 279, 281
Reardon, J.E., 120, 185, 227
Reddy, J.N., 175, 188
Reed, R.D., 7, 25
Reed, R.J., 23, 25, 27, 124, 140, 147, 446
Reed, T.B., 197, 219, 225, 242, 246, 249, 254, 283
Reese, J.L., 26, 64
Reese, S.D., 431

Reid, R.C., 64
Reid, R.J., 24, 429
Reid, W.T., 221–2, 230
Reiner, D., 343
Reis, A., 190, 192
Ressent, S., 230
Rezakhanlou, R, 340
Rhine, J.M., 25, 190
Rhodes, C.A., 192
Rich, L., 446
Richards, G., 341
Richards, G.H., 146
Richards, G., 120
Richardson, A.P., 341
Richter, W., 185, 190, 418, 430
Riddell, F.R., 145, 263–4, 266–7, 270–1, 284
Rigby, J.R., 241, 246, 249, 283
Riikonen, A., 391
Riikonen, J., 371, 391
Roache, P.J., 176, 189
Robertson, T.F., 341
Rodi, W., 184
Rohsenow, W.M., 24, 122
Røkke, N.A., 227, 282
Romano, F.J., 431
Rose, J.W., 383, 392
Rosner, D.E., 127, 146, 253, 263–4, 271, 284
Rubini, P.A., 187
Rue, D., 337, 341, 447
Ruiz, R., 26, 182, 192, 193, 308, 339–40, 342–3
Rumminger, M.D., 159, 183, 208, 212, 228, 306, 339
Runchal, A., 26, 189, 192, 229, 282, 339, 431
Rupley, R.M., 187
Russ, E.J., 120, 145, 226, 281
Russell, R.O., 428
Ryan, N.W., 342

S

Sacks, J., 430
Saeki, T., 190
Sage, B.H., 275, 284
Sager, D., 343
Saha, D., 408, 429
Said, R., 170, 188
Sailor, D.J., 281
Sala, J.M., 191, 403, 429
Salama, S.Y., 408, 429
Salooja, A.P., 190
Sammakia, B., 119
Sandner, H., 122
Sanitjai, S., 304, 339
Santoleri, J.J., 335, 343
Santoro, R.J., 26, 225
Sarjeant, M., 338
Sarofim, A.F., 6, 24–5, 119–21, 183, 185, 187–8, 446
Sathe, S.B., 189, 299, 304, 339–40
Sathuvalli, U.B., 338
Saveliev, A.V., 121
Saxena, V., 188
Sayre, A., 185

Schafer, L.L., 431
Schefer, R.W., 225
Schemberg, S., 430
Schmall, R.A., 340, 393
Schmidt, B., 177, 189
Schmidt, H., 119
Schmücker, A., 177, 189
Schnell, U., 188
Schreiber, R.J., 341
Schreiber, W., 26, 189, 192, 229, 282, 339, 431
Schreiner, M.E., 340, 393
Schroeder, D.L., 428
Schulte, E.M., 198, 226, 240–2, 249, 251, 254, 282
Schultz, T.J., 315, 340, 391, 393
Seeker, W.R., 420, 430
Segeler, C.G., 25
Seiyama, T., 147
Selçuk, N., 166, 186, 379, 392
Semerin, O.M., 392
Semerjian, H.G., 26, 121, 181–2, 193, 225, 227, 230, 428
Semernin, A.M., 393
Semernin, O., 24, 340, 367, 393
Serauskas, R.V., 182
Severens, P.F.J., 308, 340
Seyed-Yagoobi, J., 282
Shadlesky, P.S., 284
Shah, N.G., 186, 191
Shah, R.K., 25, 117, 147, 186, 189, 192–3, 227–8, 340, 343, 428, 430
Shahani, G.H., 431
Shamp, D.E., 430
Shibata, K., 192
Shim, K.-H., 105, 121, 367, 391–3, 428
Shires, G.L., 3, 24, 217, 229, 502, 519
Shoji, M., 122
Shorin, S.N., 91, 118, 197, 202, 226, 236, 240, 242, 249, 251, 254, 256–7, 277, 279–80, 282
Shoultz, R.A., 340
Shundoh, H., 226, 282
Shyy, W., 175, 189
Sibulkin, M., 67, 118, 219, 229, 263–4, 282
Sidawi, M.M., 189, 192, 438, 446
Siddall, R.G., 166, 186, 392
Sidlauskas, V., 190
Siegel, R., 119
Sikirica, S.J., 340, 393, 399, 428–9
Silcox, G.D., 192, 431
Silva, T.F., 226, 429
Simon, T.W., 26
Singer, S., 341
Singh, D.K., 192, 367
Singh, S., 154, 189, 229, 282, 329, 339, 342
Singh, S.N., 308, 340, 393
Singh, S.S., 182, 189, 313, 340, 342
Sipowicz, W.W., 342
Sivathanu, Y.R., 157, 182, 206, 227
Skinner, S.M., 227
Skorupska, N., 341
Slater, P.N., 146
Slavejkov, A.G., 27, 122, 287, 337, 430
Slezak, S.E., 193, 227, 230, 428

Sloan, C.E., 372, 391
Sloane, T.M., 170, 187
Smallwood, G.J., 183
Smith, A.M., 119
Smith, C.D., 429
Smith, D.C., 192, 431
Smith, G.P., 187
Smith, P.J., 154, 182, 186, 190, 341
Smith, R.B., 82, 118, 196, 198, 203, 208, 225, 241, 243, 246–7, 249, 251, 257, 272, 278, 283
Smith, T.F., 119
Smith, T.M., 27, 117
Smoot, L.D., 190, 191, 341
Smouse, S.M., 341
Smulyanskii, I.B., 338
Snyder, R.E., 222, 230
So, R.M.C., 184
Sochet, L.R., 187
Soelberg, N.R., 24
Sogaro, A., 188
Solomon, P.R., 195, 225
Solid Waste Association of N.A., 431
Somervell, T.R., 445, 447
Sommer, H.J., 230
Son, S.F., 212, 228
Song, G., 169, 187
Song, T.H., 168–9, 178, 186, 192
Song, W., 191
Sønju, O.K., 119, 227, 281–2
Soufiani, A., 165, 185
Sousa, A., 192
Spalding, B., 150
Spalding, D.B., 151, 163, 181, 183–4, 289, 337
Sparrow, E.M., 6, 24, 119, 212, 228
Spence, G.T., 341
Speyer, R.F., 298, 338
Speziale, C.G., 184
Spiegelhauer, B., 189
Spoon, H., 429
Stala, C., 410, 429
Stambuleanu, A., 7, 25, 105, 121
Stanisic, M.M., 184
Stenström, S., 296, 338, 371, 391
Stephan, K., 26, 118, 122, 227, 283
Sterling, A.M., 430
Stevens, H.J., 147
Steward, F.R., 335, 343, 360, 367
Stewart, C.D., 170, 188
Stewart, I.M., 185
Stewart, S.E., 367, 428
Stewart, W.E., 64
Stralen, S.V., 122
Straub, J., 26, 118, 145, 227, 283
Strauss, W.S., 431
Street, P.J., 338
Strehlow, R.A., 25
Stresino, E.F., 197, 219, 221, 225, 237, 241–2, 247, 249, 251, 254, 279, 282
Strickland, J.H., 157, 183
Strickland-Constable, R.F., 170, 188

Strong, H.M., 229
Stroup, D.W., 227
Su, A., 343
Sugiyama, S., 367
Sullivan, J.D., 159, 183, 307, 340
Sultzbaugh, J., 182
Sunderland, P.B., 121
Suzukawa, Y., 367
Svidzinskii, G.A., 429
Swaminathan, N., 163, 184
Swan, C., 229
Swithenbank, J., 352, 367
Syed, K.J., 188
Sykes, J., 183

T

Tabb, E.S., 393
Tabor, G., 181
Taine, J., 26, 169, 187
Takagi, T., 184
Takamichi, S., 328, 342, 408
Talmor, E., 121, 384, 394
Tamhankar, S., 447
Tamonis, M., 190
Tanaka, Y., 122, 367
Tang, Y.S., 122
Taniguchi, H., 191–2
Taniwaki, K., 410, 431
Tariq, A.S., 118, 145, 219, 227, 229, 241, 249, 251, 283
Tatsumi, N., 442, 447
Tatsumi, T., 184
Taylor, A.M.K.P., 195, 224
Taylor, G., 447
Taylor, H.S., 229–30
Taylor, J.M., 183
Taylor, P.B., 89, 120
Teppo, O., 191
Tessari, D., 430
Tester, M.E., 26
Teyssandier, F., 119
Theby, E.A., 212, 229
Thekdi, A.C., 24, 388, 393, 446
Theologos, K.N., 191
Thiemke, R., 146
Thome, J.R., 122
Thompson, H.A., 146, 183, 190–1, 337, 339–40, 342
Thomson, J.A.L., 120, 185, 227
Thring, M.W., 188
Thurlow, G.G., 122
Tien, C.L., 89, 103, 120, 165, 183–5, 187, 338
Till, M., 147
Tillman, D.A., 341
Tiwari, S.N., 164, 185
Tomazin, C.E., 401, 428
Tomeczek, J., 165, 185, 341
Tomita, Y., 402, 428
Tong, L.S., 122
Tong, T.W., 189, 307, 339, 340
Touzet, A., 24, 367, 392–3

Traub, D., 391
Trinks, W., 25, 346, 367, 393
Troe, J., 187
Trout, H.E., 89, 120
Trouvé, A., 182
Truelove, J.S., 185
Tryggvason, G., 181–2
Tsederberg, N.V., 64
Tsujimoto, Y., 189, 367
Tsukamoto, Y., 122
Tuchscher, J.S., 187
Tucker, R.J., 25, 76, 119, 190
Turbiez, A., 187
Turns, S.R., 7, 41, 222, 230, 286, 290, 337–8
Tyndall, D.W., 64, 446

U

U.S. Dept. of Energy, 1–2, 5–6, 24
U.S. Envir. Prot. Agency, 423, 446
Uchida, H., 392
Ungar, E.W., 24, 446
Upton, E.A., 428
Urban, D.L., 121–2. 230
Ushimaru, K., 189

V

Vafin, D.V., 166, 186
van de Ven, C.J.H., 340
van der Drift, A., 129, 146, 306, 339–40
van der Meer, T.H., 82, 91, 118–20, 188, 198, 212, 218–20,
    226, 229, 241, 249, 252, 256, 259, 262, 265, 283–4
Váos, E.M., 163, 185
Varga, G., 343
Vazquez del Mercado, L., 342
Veldman, C.C., 198, 206, 222, 226, 241, 243, 249, 251,
    256, 283
Velthuis, J.F.M., 189
Vernon, H.L., 428
Vernotte, P., 118, 146, 199, 208, 227, 238–9, 246–8, 253–4,
    257, 259, 268, 276, 282
Vervisch, P., 89, 120
Veynante, D., 182
Villasenor, R., 285, 337
Vincent, M.W., 367
Viskanta, R., 4, 24, 26, 119, 145, 159, 165–6, 168–9, 178–9,
    182–6, 189–93, 229–30, 284, 299, 304, 337,
    339–40, 367, 428–9, 446–7
Visser, B.M.V., 181
Vizioz, J.–P., 69, 82, 118, 196, 203, 208, 220, 225, 238,
    241–3, 246–7, 249–51, 253, 257, 278, 282
von Elbe, G., 25, 284
Vonesh, F.A., 428
Vrable, D.L., 420, 430

W

Wagner, H.G., 102, 121, 338
Wagner, J.C., 335, 343

Waibel, R.T., 338
Walker, R.W., 187
Wall, T.F., 80, 120, 165, 185, 316, 341
Walsh, L.T., 428, 403
Walsh, P.M., 225
Wang, C.P., 225
Wang, H., 170, 188
Wang, J., 191
Wang, Y.B., 69, 119
Ward, J., 190, 192
Ward, M.E., 407–8, 429
Warnatz, J., 64, 187
Warrington, R.O.,
Watson, R., 447
Webb, B.W., 165, 186, 203, 225, 227, 241, 246, 249, 283
Weber, R., 151, 181, 185, 290, 338
Wechsler, T., 431
Weinberg, F.J., 146, 225, 341, 430
Weller, H.G., 181
Wells, M.B., 428
Welty, J.R., 6, 24
Wendt, J.O.L., 420, 430
Wessel, R.A., 191
Westbrook, C.K., 155, 182, 187–8
Westenberg, A.A., 211–2, 228
Westley, F., 187
Westra, L.F., 430
Whitelaw, J.H., 163, 184, 190, 224, 227, 229, 338
Wiebelt, J.A., 119
Wieringa, J.A., 89, 120, 183, 191–2
Wigley, G., 338
Wilcox, D.C., 184
Wilde, J.D., 3, 24
Williams, A., 119, 316, 341, 446
Williams, F.A., 25, 170, 187
Williams, G.C., 187
Williams, J.H., 428
Williams, S.J., 446
Williams-Gardner, A., 391
Willis, R.A., 428
Willmot, A.J., 367
Wills, B.J., 341
Wilson, R.P., 446
Wilson, T.L., 159, 183
Wimpress, N., 384, 392
Winchester, D.C., 447
Winter, E.M., 327, 341
Wise, H., 138, 147, 253, 284
Witze, P.O., 224, 227
Wofrum, J., 187
Wojcicki, S., 191
Wolff, L., 146
Wood, B.J., 138, 147
Wood, C.G., 228
Woodruff, L.W., 91, 118, 147, 198, 206, 226, 244, 249,
    254, 256–7, 280, 282
Wright, D.D., 367
Wrobel, L.C., 190,193
Wunning, J.A., 342, 447
Wunning, J.G., 342, 447

X

Xin, X., 119
Xiong, T.-Y., 304, 308, 339–40
Xu, F., 170, 188
Xu, X., 191
Xu, Z.X., 334, 343

Y

Yagi, S., 102, 121
Yamada, T., 147
Yang, W.-J., 430
Yang, Y., 191
Yao, S.C., 26
Yao, W., 191
Yap, L.T., 293, 338
Yeh, J.-H., 208, 228
Yener, Y., 117
Yerinov, A., 24, 367, 392–3
Yerynov, A.E., 340
Yetman, M.E., 300, 339
Yetter, R.A., 341
Yi, J., 119
Yokosh, S., 342
Yoshimoto, T., 163, 184
You, H.-Z., 69, 82, 107, 118, 198, 202, 212, 218, 221–2,
      226, 241, 243, 249–50, 252, 256–7, 279, 281
Young, A.M., 229

Young, F.K., 431
Young, P.J., 120
Younis, L.B., 159, 183, 299, 339
Youssef, C.F., 343, 430
Yu, B., 191
Yuen, W.W., 26, 182, 188, 338, 339, 431

Z

Zabielski, M.F., 304, 339
Zang, T.A., 182
Zeleznik, F.J., 64, 146, 284
Zelson, L.S., 27, 342
Zhang, B.L., 336, 343
Zhang, D.K., 120
Zhang, Z.M., 26, 190–2
Zhenghua, Y., 157, 183
Zhou, C.Q., 191
Zhou, L., 191
Zhuang, M., 182
Zhukovskii, V.V., 429
Zinn, B.T., 333, 342–3
Zinser, W., 190
Ziolkowski, M., 182
Ziugzda, J., 273, 284
Zukauskas, A., 273, 284
Zukoski, E.E., 226, 283
Zung, J.T., 316, 341
Zwecker, M., 431

Subject Index

A

Absorptance, 80, 128
Absorption, 308, 411
Absorption coefficient, 105, 131, 157–8, 166, 205, 377
Absorptivity, 35, 70–2, 74–5, 80, 88–9, 129, 167, 197, 315
AC discharge, 196
Acetylene, 41, 157, 199–200, 240, 280
Acoustics, 331, 352
Advanced Glass Melter, 416–7
Aerospace, 7, 150, 236, 250, 253
Aerothermochemistry (see thermochemical heat release)
Aging, 2
Agglomeration, 2
Agricultural products, 372
Air, 501–3
    infiltration, 3, 127, 143–5, 292, 346, 403
    preheat, 3, 12–3, 19, 21, 34, 38, 40–1, 43–8, 50–1, 53–4,
        56–8, 60–1, 63, 149, 157, 240, 314, 319, 325–30,
        405, 407–8, 411–4, 419, 438
Airfoil, 238, 280
Air Products, 397, 406, 409–10, 441
AISI Steel Foundation, 398, 400
Albedo, 105
Alumina, 134, 135–6, 138, 248, 298, 311, 335, 439
Aluminum, 113–4, 128, 145, 248, 294, 336, 352–3, 355,
        360, 363, 395–6, 408–10, 438, 440, 442
    anode baking, 2
    dross, 408
    furnace (see furnace, aluminum)
    heat treating, 4
    industry, 2, 5, 293
    melting, 4, 351, 358
    molten, 352, 360
    ovens, 3
    oxide, 360
    production, 408–10
    reverberatory furnace (see furnace, aluminum
        reverberatory)
    scrap, 8, 20–1, 114, 116, 351, 363, 408
    titanite, 300
Alzeta Corp., 302
American Gas Association, 302, 306
American Petroleum Institute, 12, 15, 312
American Society of Mechanical Engineers, 7, 179–80,
        201, 203, 244–5, 304–5, 307, 315, 357, 390, 404–6,
        419, 4, 427
American Society of Metals, 177, 314, 318, 327, 335
Annealing, 2–3, 384–5, 390–1

API (see American Petroleum Institute)
Argon-oxygen decarburization, 328
Artifical intelligence, 149
Ash, 80–1, 158, 168, 178, 316, 420–1, 423, 441
ASME (see American Society of Mechanical Engineers)
Asphalt melting, 4
Automotive, 296
Available heat, 29, 37, 43–7
Avco Research Lab, 416

B

Baking oven, 178
Band models (see radiation, band models)
Bark, 374
Barometric pressure, 12
Basic oxygen furnace, 396
Batch furnace (see furnace, batch)
Black foil, 198, 202
Black liquor, 222
Blackbody, 11, 71–5, 81, 102, 108, 128, 197–8, 203
Blackbody emissive power (see emissive power,
        blackbody)
Blackened surface, 202, 206
Blast furnace, 3, 7, 396
BOF (see basic oxygen furnace)
Boilers, 6, 7, 176, 178, 396, 423, 441
Boiling, 114, 117, 231, 235, 250
    external, 114, 116, 231, 233, 235
    internal, 114, 116, 231, 233, 235
Bonding, 2
Boundary
    condition, 171–3, 198
    layer, 65, 125–7, 140, 220, 232, 238, 250, 261, 263–4,
        280, 334
Boundary-fitted coordinates, 173
Box heater (see heater, box)
Brass, 133–4, 136, 138–9, 141, 206, 248–50, 293, 396, 408,
        438, 440
    forming, 4
    heat treating, 4
    melting, 4
Brazing, 2, 384–5, 391
Bread baking, 4
Brick making, 4
Bricks, 3, 113, 365, 410, 418, 440
Bronze melting, 4
Bunsen burner (see burner, bunsen)
Buoyancy, 162, 231, 279, 290, 292, 326

Buoyancy-induced flow (see natural convection)
Burner, 6, 8, 98, 109, 151, 158, 166, 172, 177, 204, 207, 212, 218, 221–2, 243, 245, 265, 274, 285–343, 349
 air/fuel, 30, 350, 401, 408–9
 air-oxy/fuel, 336, 409
 bunsen, 323
 coal, 7, 158
 combination (see burner, dual fuel)
 controls, 22
 diffusion, 14, 15, 246
 dual fuel, 12, 15
 flat-flame, 293
 forced draft, 16
 hearth, 323–4
 immersion, 410, 442–4
 infrared, 21–2, 129–30, 300, 369, 371–2, 374, 391
 natural draft, 16–7, 325, 384–5
 oil, 7, 285
 open-flame, 285–96
 oxygen/fuel, 14, 287, 293, 331–2, 350, 396–9, 401, 408–9, 441, 444
 partially premixed, 15
 porous radiant, 13, 154, 159–60, 177–8, 215, 297, 302–9
 preheated air, 16
 premixed, 13–5, 245, 297
 pulse, 177
 radiant, 177, 296–316
 radiant tube, 177, 311–6, 329–31, 348–9, 388, 391
 radiant wall, 311–2, 380
 recuperative, 1, 329–30
 regenerative, 27, 327–8, 408
 reticulated ceramic, 159
 ribbon, 286
 roof-fired, 325–6
 staged-air, 16
 staged-fuel, 17
 swirl, 177
 terrace-fired, 324
 tunnel, 245
 wall-fired, 324–5, 380
Butane, 11, 40, 204, 240, 278

C

Cabin heater (see heater, cabin)
Cake baking, 4
Calcining, 1, 3, 124, 346
 alumina, 2
 cement, 2
 gypsum, 2
 lime, 2, 362
 soda ash, 2
Calorimeter, 132–3, 138, 196–7, 199, 204, 290–2
Candy cooking, 4
Carbon dioxide, 13, 29, 30, 34–9, 81, 84–9, 92, 108, 125, 137, 155, 157, 164–5, 167, 169, 209, 221, 231–2, 238, 316, 353, 382, 438
Carbon monoxide, 10–11, 30, 35, 37–9, 41, 81–2, 89–90, 123, 125, 136, 155, 157, 164, 169, 206, 221, 231–2, 266, 273, 286, 322, 348, 376, 385, 423

Carburizing, 384–5, 388
Carpet, 129, 296
CARS, 209
Catalytic, 125, 127, 135, 139–40, 210, 212, 248
 cracker, 178
 surface, 134, 266
CCM (see cellular ceramic material)
Ceiling tiles, 296
Cellular ceramic material, 327
Cement, 20, 178, 345, 355, 362, 417–9, 440, 443
 clinker, 8
 industry, 3
 making, 4, 129, 410
CERAJET, 330, 377
Ceramic, 20, 23, 110–3, 129, 159, 199, 297–8, 302, 311–2, 318, 328, 345, 347, 364, 390–1, 410, 438, 445
 foam, 299–300
 industry, 77, 372, 419
 tube, 1, 312, 348, 377, 388, 442
CFD (see computational fluid dynamics)
CH_4 (see methane)
C_2H_2 (see acetylene)
C_2H_4 (see ethylene)
C_2H_6 (see ethane)
C_3H_6 (see propylene)
C_3H_8 (see propane)
C_4H_{10} (see butane)
$C_{16}H_{10}$ (see pyrene)
Char, 158, 168, 222
Charging, 19
Chemical
 diffusion, 127
 dissociation, 36, 38, 66, 124–6, 135–6, 140, 221, 231, 253, 263–4, 271, 328
 enthalpy (see enthalpy, chemical)
 industry, 2–3, 5, 16, 372
 kinetics, 35, 159, 169
 manufacturing, 41
 reaction, 13, 20, 29, 161, 170, 178–9
 recombination (see thermochemical heat release)
 vapor deposition, 231
Chemical/petroleum feedstock preheating, 2
Chemiluminescence, 82, 101, 105, 205
Chemistry, 34, 150, 154–6, 159, 169–70, 176
 reduced reactions, 155–6
CHEMKIN, 169
Chequer, 364, 412–4
CHERUB, 310
Clay, 3
 burning, 4
 refractories, 2
 structural products, 2
CO, (see carbon monoxide)
CO_2 (see carbon dioxide)
Coal, 1, 7, 11, 67, 80, 105, 107, 123, 128, 167–8, 176–8, 203, 240, 297, 316, 417–8
 boiler, 168
 burners (see burners, coal)
Coating, 77, 128–9, 134, 140, 211, 234, 248–50, 296, 306, 315, 346, 369–70, 390

Coffee roasting, 4
Coke, 20–1, 123–4, 222, 240, 316
  formation, 375, 379
  oven, 4, 396
  oven gas, 106, 238, 240
Cold flow (see flow, nonreacting)
COMBUST, 159
Combustibles, 123
Combustion
  chemistry, 29
  products, 34, 36
Combustion Institute, 103, 107, 167–8, 308–9, 361
Combustor, 18, 21–2, 48
  cylindrical, 20
  rectangular, 20
Composite wall, 110
Computational fluid dynamics, 6, 153–5, 159–80, 223–4,
    414
Computer simulation (see modeling, computer)
Condensation, 114, 117, 231, 236
Conduction heat transfer, 6, 24, 65, 109, 127, 178, 197,
    211, 231, 315, 354, 357, 359, 361–3, 369, 377
  inverse, 279
  steady-state, 110, 113, 231, 233
  transient, 110, 113–4, 231, 233, 418
Contact resistance, 110, 132
Continuous furnace (see furnace, continuous)
Convection heat transfer, 6, 11, 13, 19, 21, 24, 35, 48–51,
    65, 67, 80, 82, 128, 130, 159, 169, 178, 195,
    199–200, 202, 206–7, 211, 272, 290, 297, 301, 317,
    334, 352–3, 359, 362, 369, 378–9, 387, 390, 401,
    408, 419, 422, 440
  coefficient, 49, 65–6, 116, 171, 317, 334, 377, 401,
    418–9, 441
  forced, 65–6, 68–9, 127–8, 141, 231, 233, 246, 253,
    263–81, 361, 377, 395
  natural, 65, 68–9, 142, 231, 233
Convection section, 21, 68, 379, 381, 384
Convection vivre (see thermochemical heat release)
Cookie baking, 4
Copper, 117, 133–4, 137, 170, 196–9, 206, 220, 248–50,
    318, 396, 408, 440
  annealing, 4
  heat treating, 4
Cordierite, 298–9
Creosote, 425–6
Cupola, 3
Curing, 2
Cyanide, 336
Cylinder, 69

D

Damköhler number, 33, 289
Dehydrogenation, 102
Delayed coking, 2
Delta Burner, 177
Density, 49–51, 65, 258
Diagnostics, 3

Diamond coating, 231
Diffusion flame (see flame, diffusion)
Dilute oxygen combustion, 436
Dioxins, 424
Direct heating (see heating, direct)
Direct numerical simulation, 153, 157
Discrete exchange factor method, 165
Discrete ordinate radiation model, 89, 158, 166, 168
Discrete transfer radiation model, 156, 165–7
Discretization, 173
Dissociation (see chemical, dissociation)
Distillation, 2
DNS (see direct numerical simulation)
DOE (see U.S. Dept. of Energy)
DRE, 421, 425
Dryer, 8, 19, 345, 372–4
  convection, 22, 129
  electric, 296
  floater, 374
Drying, 3, 129
  animal food, 2
  coal, 2
  food, 2
  glass products, 2
  plastic, 2
  rubber, 2
  surface film, 2
  wood, 2

E

Eclipse, 306
Eddy, 33
Effective angle model, 165
Efficiency, 5–6, 8, 12–3, 19, 23, 43, 45–6, 128–9, 132, 140,
    149, 178, 291–2, 295–8, 304, 322, 329, 366, 391,
    398, 403, 407, 421, 433, 442, 444–5
Electric arc furnace (see furnace, electric arc)
Electric Furnace, 389
Electromagnetic
  spectrum, 70
  waves, 69
Ellipsoidal radiometer (see radiometer, ellipsoidal)
Emissive power
  blackbody, 74
  flame, 91
  hemispherical, 73
  monochromatic, 73
Emissivity, 6, 71–2, 74–80, 84, 133–4, 141, 157, 179–80,
    198, 213, 216, 307, 314, 360, 377, 418, 427
  coating, 248
  flame, 103–5, 384, 401
  gas, 85–7, 103, 105, 206, 411
  load, 128, 403
  radiant, 128
  solids, 426
  spectral, 76, 83–4
Emittance, 80
  monochromatic, 105

Empirical, 67–8
Enamel baking, 4
Energy
    consumption, 5
    efficiency, 1
Energy Information Administration, 5, 24
Energy optimizing furnace (see furnace, energy optimizing)
Enhancement, 5, 24
Engines, 7
    internal combustion, 7
Ensign Ribbon Burners LLC, 286
Enthalpy, 258, 272, 329, 345
    chemical, 125–6, 136–7
    sensible, 125–6
    total, 125–6
EPA (see U.S. Environmental Protection Agency)
Equivalence ratio, 32, 34, 37, 41–2, 50–5, 57–8, 60–1, 63,
        91–2, 94, 98, 101, 109, 178, 196, 205, 233, 238–43,
        246, 309
Ethane, 41
Ethylene, 41, 222, 286, 376
    glycol, 117, 250
    production, 313, 380
Eulerian, 152
Exothermic, 66
Exothermic displacement of equilibrium (see thermochem-
        ical heat release)
Experimentation, 149, 151, 176, 195–230, 233–57, 259,
        264–8, 444
Explosion, 29
Exponential wide band model, 165
Extinction coefficient, 105, 159
EZ-Fire, 409–10

                            F

Favre-averaging, 163
Film temperature, 142, 258, 272
Finite
    difference, 152, 165, 173–4
    element, 89, 152, 165, 175
    volume, 152, 174–5, 177
Firebrick, 23, 76, 364, 386
Fired-heater, 7
Firing rate, 243, 286–93
Flame, 195
    chemistry, 7
    density, 286
    diffusion, 13, 14, 67, 102–3, 107, 157, 168, 170, 222,
        243, 274, 279, 294
    emissivity (see emissivity, flame)
    forced-draft, 17
    impingement, 14, 21, 66–7, 82, 91, 109–10, 113–4, 116–7,
        126, 132, 134–41, 180, 196–7, 200–1, 203, 216, 218,
        221, 231–84, 311, 320, 372, 375, 390, 415, 440
    length, 96, 286, 292, 294–6, 323, 326, 349, 379, 384, 411
    luminous, 6, 11, 23, 35, 40, 131, 232, 285, 287, 294, 316,
        322, 326, 347, 402, 411
    luminosity, 285–7, 317

    natural draft, 17–8
    nonluminous, 6, 11, 35, 40, 131, 232, 288, 331, 402, 411,
        438
    oxygen/fuel, 66
    premixed, 14, 243
    radiation (see radiation, flame)
    temperature, 34–5, 40–4, 57, 125–6
    thickness, 33
    width, 286
    wrinkled, 33
Flameless oxidation, 331
Flares, 1
Flash smelting, 170, 178
Flashback, 14, 297, 300
FlGR (see recirculation, flue gas)
Floater dryer (see dryer, floater)
Flow
    control, 8, 349
    nonreacting, 155
    visualization, 3
FLOX (see flameless oxidation)
Flue gas
    recirculation (see recirculation, flue gas)
    volume, 34, 46, 48
Fluid
    catalytic cracker, 441
    dynamics, 3, 151–3, 161–4, 331, 497–9
    heating, 2
Fluidized bed, 7
Food, 296
    drying, 374
    industry, 3, 5
    production, 2, 372
Forced draft burner (see burner, forced draft)
Forest products, 2
Forging, 2, 407–8
Forming, 2–3
Foster Wheeler, 377
Fourier transform infrared, 195
Fractal, 168
Frit furnace (see furnace, frit)
Froude number, 289
Fuel, 11, 14, 34
    burn-up ratio, 314
    combustion, 123
    efficiency, 1
    gaseous, 11–2, 29, 316–7
    heating value, 29
    injectors
    -lean, 10, 29, 42, 52, 54, 57, 92, 98, 209, 240, 243, 266,
        321, 333–4, 398
    liquid, 7, 11–2, 316, 335
    low heating value, 10
    preheat, 34, 38–40, 43–4, 317
    -rich, 10–1, 29, 54, 57, 98, 101, 107–8, 126, 240, 266,
        317, 322, 333–4, 398, 408
    solid, 7, 11, 316
    waste liquid, 10
FuGR (see recirculation, furnace gas)

Furnace, 6, 8, 19, 21, 25, 43, 76–7, 82, 113, 129–30, 143–5, 149–51, 154, 158–9, 163–5, 167, 172–3, 178–9, 204, 207–8, 214, 234, 238, 243, 246–7, 278, 280, 290, 317–8, 320, 323, 326–7, 331, 345–67, 375, 395
  aluminum, 3, 353
  aluminum reverberatory, 20, 145, 178, 352, 358, 360–1
  batch, 19, 20, 351, 361
  bell, 386, 388
  car bottom, 386–7
  continuous, 19, 20, 144, 351–2, 361, 386
  conveyor, 386
  cylindrical, 8
  design, 1, 6
  direct-fired, 159, 179–80, 300, 347–9, 387
  electric arc, 325, 396, 441
  energy optimizing, 396
  forge, 407–8
  frit, 411
  fluidized bed, 386
  gas-fired, 159
  glass, 131–2, 157, 178, 353
  hearth, 281
  heat-treating, 145, 387, 391
  indirect-fired, 24, 159, 179, 347–9, 386–91
  liquid bath, 386
  muffle, 20
  operation
  petroleum refinery, 6
  pit, 386
  process, 7
  pusher, 386
  reheat, 77, 109, 145, 178–80, 346, 402–7, 438
  roller hearth, 386, 388–9
  rotary, 20–1, 130, 349, 362–3, 399, 409
  rotary hearth, 386–7
  rotary retort, 386
  round-top, 351
  semicontinuous, 20
  shaft, 20, 130, 442
  soaking pit, 438
  tube, 167, 178
  vacuum, 315, 352–3, 386, 391
  vertical cylindrical, 107, 379
  wall, 293
  wall-fired, 21

G

Gage
  heat flux, 82, 113, 195, 197–9, 307
  radiant heat flux, 200–6
Galvanizing, 442
Gas chromatograph, 221
Gaseous
  fuels, 7
  radiation (see radiation, non-luminous)
Gas infrared radiation factor, 302–3
Gasoline, 378

Gas Research Institute, 160, 169, 176, 298, 301, 309, 311–2, 330, 371, 374, 389, 416, 433–4, 441, 444
Gas turbine, 396
Gauss Elimination Method, 175
Gauss-Seidel Iteration, 175
Gaz de France, 169
GC (see gas chromatograph)
Germanium crystal, 198
Gibbs free energy, 35
GIR (see gas infrared radiation factor)
Glass, 70, 132, 143, 149, 179, 395, 410–7, 440
  annealing, 2, 4
  forming, 2
  industry, 2–3, 24, 131, 293, 411, 438
  making, 4
  melt, 3, 131
  melting, 2, 19, 131, 170, 293, 308, 318
  melting furnace, 3, 327, 350
  molten, 5, 23, 128, 131, 154, 178–9, 204–5, 351, 353, 396
  production, 16, 23, 364
  regenerative furnace, 9, 89, 157, 411–4
  tempering, 2
  torch, 246
Glycol, (see ethylene glycol)
Gold, 198, 202–3
Granular solid (see solid, granular)
Grashoff number, 68, 142–3
Graybody, 23, 72, 75, 128, 215, 285, 316
Grid, 173–5
GRI-Mech, 169
Gypsum production, 2

H

H_2O (see water)
H_2 (see hydrogen)
Hall reduction cell, 336
HCl (see hydrochloric acid)
Heat
  exchanger, 6, 23, 128, 329, 362, 364, 369
  flux, 14, 17, 20, 135–41, 150, 158, 167–9, 171, 179–80, 195–207, 213, 236, 245–6, 250–5, 263–81, 290–4, 301, 317–22, 326, 331–2, 379, 388, 411, 418
  flux gage (see gage, heat flux)
  loss, 19, 128
  recovery, 1, 5–6, 8, 19, 23, 128, 363–6
  recuperation, 3, 21, 46, 326–32
  sink (see load)
  treatment, 2–3, 19, 323, 348, 352, 386
Heat of fusion, 114–6
Heater, 8, 214, 323–4, 375
  box, 379
  cabin, 359, 379
  fired, 27, 375–85
  infrared, 21–2
  process, 19–21, 129, 166, 214, 297, 311, 348, 358, 378–85
  tube, 375
  vertical cylindrical, 20, 360

Heating
  direct, 18
  indirect, 18, 20
Heating value, 124
  higher, 29, 123, 124
  lower, 29, 124
Helmholtz combustor, 333–5
Hemi-nosed cylinder, 69
Hemispherical emissive power (see emissive power,
      hemispherical)
Heptane, 201
Hexamethyldisiloxane, 212
HF (see hydrofluoric acid)
HHV (see heating value, higher)
Higher heating value (see heating value, higher)
Hog fuel, 374
Honeycomb, 140, 177
Hotspot, 14
Hottel's zone method, 82, 159, 165, 179
Hubbard probe, 222
Humidity, 374
Hydraulic diameter, 299
Hydrocarbon fluid, 21
Hydrochloric acid, 164
Hydrofluoric acid, 164
Hydrocracking, 2
Hydrogen, 11, 30, 33, 40–2, 44–5, 47, 49–64, 125, 136,
      206, 231–2, 238–40, 264, 266, 273, 286, 348,
      376–7, 382, 384–5, 387, 433, 457–8, 460–8,
      488–95
Hydrogen sulfide, 177
Hydrotreating, 2
Hypersonic flow, 231, 263

                                    I

IFRF (see International Flame Research Foundation)
Ignition point, 245
Imperial College, 178
Incineration, 11–2, 153, 178, 333, 335, 337, 395–6, 419–27,
      441
Inconel, 213
Indirect heating (see heating, indirect)
In-flame treatment, 335
Infrared, 215
  burner (see burner, infrared)
  detector, 214–6
  radiation (see radiation, infrared)
Initial conditions, 171–3
Ink, 296, 348
Inland Steel Co., 180
Institute of Gas Technology, 312, 333, 442, 445
Insulation, 76, 111–2
Internal combustion engine (see engine, internal combustion)
International Flame Research Foundation, 203, 243, 418,
      434
IR (see infrared)
Iron, 2, 158, 170, 408
  heat treating, 4
  melting, 4, 318

Iron and steel
  industry, 3, 396
  melting, 2
Irradiation, 71, 98

                                    J

John Zink Co., 16–7, 208, 222, 224, 287–8, 325, 380

                                    K

Kerosene, 378
Kiln, 7, 345
  rotary, 8–9, 177–8, 321–2, 352, 355–8, 362–3, 417–20,
      423–6
  shaft, 362
  tunnel, 419
Kilning, 3
Kinematic viscosity, 33
Kirchoff's law, 75
Kolmogorov length, 33, 170

                                    L

Ladle, 20, 353, 358–9
  preheating, 2, 335, 396, 399–401, 438
Lagrangian, 152, 157
Laminar, 126, 142, 150, 161, 170, 243, 264–77, 279–80
  flame, 213, 218, 305
  flame speed, 33
  smoke point, 104
Lampblack, 133–4, 198
Langmuir probe, 221
Laser, 3, 176, 206
  Doppler velocimetry, 218–20
  spectroscopy, 3
  tomography, 222
LDV (see laser Doppler velocimetry)
Lead, 2, 396, 408, 440
Lewis number, 60–3, 258, 264, 266–8
LHV (see heating value, lower)
LIF, 209
Lifted flame, 275
Liftoff distance, 274
Lime, 129, 178, 345, 355, 362, 410, 417–9, 440
Limestone, 19, 417–8
Line reversal, 213, 256
Liquid fuel (see fuel, liquid)
Literature, 6, 238, 449–50
Lithium, 221
Load, 1, 3, 8, 11, 18–21, 29, 35, 73, 81, 113, 116, 123,
      127–8, 151, 179, 285, 290, 318, 333, 335, 345,
      351–2, 353, 355, 369, 403–6, 437
  gas, 18
  liquid, 18
  solid, 18
Lower heating value (see heating value, lower)
LU Decomposition, 175
Luminous radiation (see radiation, luminous)

M

Mach number, 204, 275
Magnesia, 335
MAPP gas, 240
Marsden Inc., 130, 307
MDA (see modified differential approximation)
Mean beam length, 89–90, 166
Meat smoking, 4
Melting, 3, 114–6, 123, 346
Metal, 515–9
   casting industry, 2, 5
   heating, 2, 231
   industry, 372, 395–410
   melting, 2, 19, 358
   molten, 20, 128, 359, 395, 400
   production, 441–2
   rapid heating, 200
   scrap, 116, 158–9, 325, 395, 397–9, 441
   sheet, 129
Methane, 11, 13, 30–64, 67, 89, 101, 108, 117, 125–6, 138,
      155, 157, 170, 201, 204, 209, 218, 238–40, 246,
      266, 268, 274, 277, 286, 302, 317, 319, 323, 326,
      353, 376, 382, 453–4, 469–77, 488–95
   reburn, 170
Methylacetylene, 240
Mie scattering, 168
Mie theory, 158
Minerals, 372, 395, 410–9
Missile, 236, 250
Mixture ratio, 30–4, 37, 40–1, 43, 48
Modeling
   computer, 5, 25–6, 33, 149–93
   physical, 222–4
Modified differential approximation, 158, 168
Moisture removal, 1
Molecular diffusivity, 60
Molybdenum plate, 198
Moment averaging, 151–2
Momentum diffusivity, 65
Monochromatic emissive power (see emissive power,
      monochromatic)
Monochromator, 98, 204
Monte Carlo, 89, 165–7
Mottle, 370–1
Muffle, 348–9
Muffle furnace (see furnace, muffle)
Mullite, 298, 329
Municipal solid waste, 422–3

N

N_2 (see nitrogen)
Naphtha, 378
NASA, 35, 48, 83–4
Natural draft burner (see burner, natural draft)
Natural gas
   fuel, 1, 11–2, 26, 89, 91, 100, 107, 123, 132, 136–9, 167,
      170, 176–7, 204, 214, 218, 220, 238–40, 245, 268,
      285, 290–6, 320–2, 377, 384–5, 388, 402, 416, 438
   production, 2

Navier-Stokes, 153, 175
Nephelometer, 222
Newton-Raphson Iteration, 175
Nickel, 199
Nitrogen, 13, 29, 30–2, 34–9, 42, 48, 50, 52, 55, 57, 81,
      86, 92, 96, 100, 102, 125–6, 155, 221, 266, 348,
      382, 387, 421, 433, 438
Nitrogen oxides, 10–11, 13–4, 26–7, 30, 35, 42, 64, 89,
      101, 155, 164, 170, 221, 294, 307, 315, 317, 320,
      326, 331–3, 366, 379, 391, 417, 421, 423–6,
      435–6
Noise, 379
Nonferrous
   heating, 2
   industry, 3
   melting, 2
Nonluminous radiation (see radiation, nonluminous)
Non-participating medium, 81
Nordsea Gas Technology, Inc., 98
Normalizing, 384, 386
NOx, (see nitrogen oxides)
Numerical modeling (see computer simulation)
Nusselt number, 49, 65, 271, 334
   volumetric, 300

O

O_2, (see oxygen)
OEC (see oxygen-enhanced combustion)
OH, 101, 125, 136, 164, 221, 231, 266
Oil, 436
   burners (see burners, oil)
   field steam generation, 4
   flame, 105, 287, 316, 318
   fuel, 1, 7, 11–2, 89, 106–7, 123, 201, 297, 317, 360, 375,
      378, 382, 402–3, 417, 434
   production, 2
Open hearth, 396
Optical
   density, 89
   filter, 204–5
   techniques, 195, 207, 209
   thickness, 306
Optically thin, 159, 286
Optimization, 149
Ore
   roasting, 2
   smelting, 3
Oven, 369
Oxidized surface, 202
Oxidizer, 10–2, 14, 16, 29–31, 34–6, 40–3, 46–8, 50–1,
      57–8, 60, 62, 90–1, 94–5, 97–8, 117, 123, 196,
      238, 240, 243–6, 290, 292, 295, 297, 315–6, 318–9,
      321–2, 326, 329
Oxygen, 13, 30–2, 34–48, 51–2, 54–5, 57, 60, 62, 92, 96,
      100–1, 117, 125–6, 134, 135–9, 141, 144, 149, 151,
      155, 170, 199–200, 221, 231, 238–40, 245, 264,
      266, 322, 331, 348, 366, 387, 433–8
   enrichment, 19, 26, 30–1, 46, 98–9, 102, 132, 196, 250,
      258, 277–8, 290, 293, 315, 411, 414–5, 417, 434–5

Oxygen-enhanced combustion, 13, 16, 30, 47, 64, 91–2, 98, 132, 136–7, 178, 214, 231, 238–40, 250, 253, 268, 273, 290, 317–8, 335, 411–2, 414–7, 420–6, 433–43, 460, 469, 478
  high-level enrichment, 437
  low-level enrichment, 437
  medium level enrichment, 437
Oxygen-free high conductivity copper, 199

P

Packed bed, 442, 445
Paint, 129, 250
Paint drying, 4
Paper, 21–2, 129, 296–7, 306, 369–72, 374, 445
  industry, 3
Participating medium, 81, 89, 157–8
Particle, 3, 102, 200, 222, 369, 409
PCBs, 424
PDF, 170
Petrochemical, 11, 40, 77
  industry, 16, 20, 67, 123, 323–4, 375
  production, 21
Petroleum, 381
  products, 3, 21, 128, 378, 445
  refining industry, 2, 5
Pharmaceutical industry, 372
Phase change, 26, 114
Pickling, 441–2
Pigment, 372
Pitot-static probe, 212, 216–8, 250–4, 257
Pitot tube, 216–9, 280
Planck mean absorption coefficient, 286
Plasma, 11
Plastics, 70, 129
  fabrication, 2
Plate, flat, 69, 126
Platinum, 126, 134–6, 138–40, 199, 212, 250, 315
Plexiglas, 222–3, 345
Polished surface, 202, 206
Pollutant emissions, 9, 10, 123, 139, 149, 151, 155, 170, 178, 297, 331, 352, 382, 421, 433, 444–5
Polycyclic aromatic hydrocarbons, 104
Polymerization, 102
Porcelain, 4, 248, 296
Porous ceramic fiber, 76
Porous radiant burners (see burners, porous radiant)
Potassium, 222
Potato chip frying, 4
Potential flow theory, 261
Prandtl probe, 216, 222
Prandtl number, 49, 53, 55, 58–61, 65, 143, 281
Precipitation hardening, 2
Predryer, 369–72
Preheated air (see air, preheating)
Preheater
  air, 1
Premixed burner (see burner, premixed)
Pressure drop, 23
Primary combustion chamber, 421, 424

Process
  furnace (see furnace, process)
  heater (see heater, process)
  industry, 1, 231
  tubes, 128–9
Productivity, 1, 5, 13
PROF, 159–60
Propadiene, 240
Propane, 11, 33, 40–1, 44–5, 47, 49–63, 67, 117, 126, 138, 170, 204, 209, 239–40, 253, 274, 279, 286, 360, 376–7, 455–6, 478–87
Propylene, 11, 40, 157
Protective atmosphere, 348, 387–91
Pulp and paper, 5
Pulse combustion, 1, 8, 179, 331–5
Pyrene, 105
Pyrex, 134
Pyrolysis, 3, 11, 104, 421
Pyrometer, 215–6

Q

Quarl, 150, 292–3, 301–2, 311, 324
Quartz, 212

R

RADCAL, 157
Radial jet reattachment, 238, 244–5
Radiance, 98
Radiant
  absorptivity, 129
  efficiency, 73, 76, 300, 306
  fraction, 287–8
  heat flux gage (see gage, radiant heat flux)
  loss, 141, 143, 159
  section, 21, 324, 379, 381, 384
  tube, 27, 390
Radiation, 5, 6, 10, 19, 24–6, 35, 69–70–1, 75–6, 80–2, 96–7, 106, 128–9, 132–3, 140, 143, 161, 168–9, 172, 177–8, 195, 200, 202, 206–7, 211–2, 215, 248, 272–4, 276, 280–1, 285, 287, 290, 293, 301, 305, 317, 323, 353–4, 359, 361, 377–9, 387, 395, 401, 408–9, 426, 440
  band models, 6, 165
  flame, 5–7, 10, 90–2, 96, 98, 100, 102, 106, 205, 437
  infrared, 70–1, 83–4, 89, 105, 165
  luminous, 71, 89, 98, 101–2, 105–6, 108, 131, 134, 157, 164, 168, 170, 201, 204, 234, 321, 347
  models, 89, 156–8, 166, 169
  nonluminous, 6–7, 48, 71, 82, 89–90–6, 98–9, 101–2, 131, 157–8, 164, 200, 206, 234, 243, 317, 321, 347
  penetrating, 109, 132, 204–5
  pyrometer, 71, 300
  spectral, 72, 98–9, 108, 132, 158, 168–9, 300
  surface, 71–2, 200, 232, 234
  thermal, 1, 69
Radiative heat ray, 165
Radiative transport equation, 165–6
Radicals, 127

Radiometer, 91–2, 98–9, 201–3, 360
  ellipsoidal, 72, 203–4
  spectral, 204
Radiosity, 298
Raman, 207, 209, 221
Rapidfire, 406
Raw gas burner (see burner, diffusion)
Rayleigh-Debye-Gans, 168
Rayleigh number, 68–9, 280
Rayleigh scattering, 168, 207, 209
RDF, 421
Recirculation
  flue gas, 12–4, 19, 331, 366
  furnace gas, 13, 19, 331–2, 366
Recombination (see thermochemical heat release)
Recombination coefficient, 134, 138
Recovery factor, 211
Recuperative melter, 412–3
Recuperator, 16, 19, 21, 23, 364, 407, 410
Red and Black Ordering Scheme, 175
Reference temperature, 258, 272
Reflectivity, 70–1
Reformer, 326, 376–7
  steam, 381
  terrace wall, 377
Refractive index, 105
Refractory, 7, 19–20, 76, 80, 90, 109–10, 128, 138, 140,
        145, 178, 196, 202, 205, 235, 245–8, 310, 324,
        326, 336, 346, 358, 365, 372, 376, 395, 399–401,
        418, 421–2, 438–40, 442
  industry, 3
  production, 410, 419
Regenerative melter (see Siemens glass furnace)
Regenerator, 16, 19, 21, 23, 364–6
Reheat, 2
  furnace, 5
Research needs, 5–6
Residence time, 286–8, 421
Resistance temperature detector, 132
Reynolds number, 33, 49, 55–6, 65–6, 141, 233, 243, 281,
        289, 314
Reynolds-Forchheimer equation, 299
Reynolds Stress Model, 315
Richardson number, 68
Rijke combustor, 333–4
Rocket exhaust, 83
Rocket, 7, 236, 250, 264
Rotary kiln (see kiln, rotary)
Rotary hearth calciner, 124, 178
RTD (see resistance temperature detector)

S

Safety, 14, 140, 149, 231, 297
Sand, 336, 395, 425
Sandia Labs, 48, 169, 176
Sankey diagram, 127–8
Scattering albedo, 159
Schleiren photography, 221–2
Schmidt combustor, 333

Schuster-Hamaker model, 165
Schuster-Schwarzschild model, 165
Screw conveyor, 19
Secondary combustion chamber, 424–6
Selective noncatalytic reduction, 170
Semianalytical, 67, 127, 250, 253, 257–71
Sensible enthalpy (see enthalpy, sensible)
Sensible heat, 43
Shadowgraph photography, 219, 221
Shaft furnace (see furnace, shaft)
Siconex, 390
Siemens glass furnace, 3, 412–4
Silica, 114, 140, 212, 250
Silicate, 131
Silicon carbide, 126, 199, 298, 311–2, 329–30, 391
Sillimanite, 248
SIMPLE, 163
Simulation, 3
Sintering, 2–3, 384, 386, 391, 419
Slag, 80, 129, 398–9
Slagging cyclonic combustor, 336
Sludge, 423
Slug calorimeter, 114
Smelting, 399
  bath, 399
  copper, 2
  flash, 399
  iron, 2
  lead, 2
SO_2 (see sulfur dioxide)
Soaking, 3
Sodium, 213, 221–2
Sodium silicate, 443
Softening point, 1
Solaronics, 303
Solid fuel (see fuel, solid)
Solids, 426, 442
  granular, 18, 128–30, 357, 363
Solvents, 420
Solution heat treating, 2
Soot, 11, 29, 39, 72, 89, 102–5, 108–9, 158, 161, 170, 202,
        206, 209, 218, 222, 240, 243, 317, 377
  agglomeration, 104–5
  aggregation, 104
  coagulation, 104
  destruction, 104
  formation, 104, 170
  index, 109
  oxidation, 104, 170
  nucleation, 104
  radiation index, 108
Sooting tendency, 103, 290
Spallation, 110
Species, 5, 7, 29, 35–9, 221, 257
Specific heat, 51–4, 58, 133, 272, 345
Spectral
  element, 152
  emissive power, 75
  emissivity (see emissivity, spectral)
  methods, 152–3

Spectrometer, 109, 213, 300
Spectrophotometer, 303
Spectroradiometric, 105
Spent aluminum potliner, 335–6
Spherical harmonic model, 165–6
SPL (see spent aluminum potliner)
Stability, 14
Staged-air burner (see burner, staged-air)
Staged-fuel burner (see burner, staged-fuel)
Staging, 10, 320–2
  air, 14, 316, 320, 322
  fuel, 14, 316, 320–1
Stagnation
  flow, 258
  point, 67, 132, 199, 202, 219, 222, 236, 246, 258, 261,
    263, 273, 276–8, 290, 293, 321–2
  surface, 69, 220
  target, 246–50, 253
  velocity gradient, 219–20, 257, 261–3
  zone, 220, 250, 261–2, 279
Stainless steel, 70, 133–5, 196–9, 206, 214, 220–1, 243,
    248–50, 291–2, 294, 300, 306, 320–2, 374, 391
Statistical narrow-band radiation model, 157
Steam cylinder, 22, 129, 369, 371, 374, 445
Steam reformer, 178
Steel, 116, 247, 278, 408
  forming, 4
  furnace, 3
  heat treating, 4
  industry, 2–3, 5, 335
  making, 2–4
  melting, 4
  soaking, 2
Stoichiometry, 10, 30–2, 41–2
Stoker, 7
Stone products, 3
Stress relief, 2
Submerged combustion, 441–4
Substrate (see web)
Successive Over-Relaxation, 175
Suction pyrometer, 207–9
Sulfur, 170
Sulfur dioxide, 89, 353
Sulfur recovery unit, 441
Superequilibrium, 267
Superkinetic, 311
Supersonic velocity, 231, 250
Surface
  catalytic efficiency, 266, 271
  combustor-heater, 310
  oxidation, 133
Swirl, 150, 164, 176, 178, 204, 220, 278, 290, 335

T

Target, 66, 113, 116, 132–3, 135–40, 197–8, 200, 203, 231,
    233–4, 238, 291–4, 320–2
Taylor length, 33
TCHR (see thermochemical heat release)
Technology roadmap, 6, 24
Teflon, 296

Tempering, 2, 384, 386
Terrace-fired burner (see burner, terrace-fired)
Textile, 296–7
  manufacturing, 2, 22, 369, 373
  web, 27, 129
Thermal
  capacitance, 114
  conductivity, 49, 53–6, 58, 64, 81, 109–13, 129–30, 159,
    171–2, 178, 197, 199, 214, 248, 258, 260, 272, 300,
    316, 353, 377
  cycling, 20
  diffusivity, 60, 65
  spallation, 231
Thermochemical heat release, 124–5, 132, 134, 136–8, 140,
    221, 231, 235, 238, 264–76, 279–81
  catalytic, 125, 127, 140, 231–2, 235, 263–4, 270–1
  equilibrium, 125–7, 140, 231–3, 235, 263–4, 270–1
  mixed, 125, 127, 140, 231–3, 235
Thermocouple, 138, 140, 149, 196–9, 207–14, 248, 250–6,
    307
Thermophoretic sampling, 222
Thermophysical properties, 258–60, 272
Thomas algorithm, 175
Throughput, 1
Thymol, 345
Tile (see quarl)
Total Schmidt Method, 71
Town gas, 218, 240, 318–9
Trace species, 13, 30, 34–5
Transmission, 23, 205, 411
Transmission electron microscopy, 222
Transmissivity, 70–1, 369–70
Tube heaters, 7
Tungsten filament lamp, 213
Turbulence, 66–7, 102–3, 116–7, 125–7, 150, 152, 157–8,
    161, 163–4, 168, 170, 176, 213, 218, 222, 243,
    262, 264–5, 268–9, 272–81, 286, 294, 314–5, 421
  enhancement factor, 265
  measurements, 3
  -particle interactions, 3
Turbulent chemically reacting flow
Twin sonic orifice probe, 209
Two phase flow, 3, 114

U

UBCs (see used beverage containers)
UHC (see unburned hydrocarbon)
Unburned hydrocarbon, 29, 32, 426
U.S. Dept. of Energy, 1–2, 5–6, 24, 311, 433
U.S. Environmental Protection Agency, 423–4
Used beverage containers, 363
Ultraviolet, 100–1
Unit melter, 412
UV (see ultraviolet)

V

Vacuum, 81, 209, 331, 334, 353, 386–7, 389
Vacuum furnace (see furnace, vacuum)
Validation, 176

Vaporization, 117, 346

Varnish cooking, 4

VC (see vertical cylindrical heater or furnace)

Velocity, 3, 33, 65, 68, 245, 261–2, 286, 289, 352, 366,
    397, 404
  gas, 21, 48–9, 68, 207, 216–9, 250–5, 257

Venturi, 331

Verification, 176

Vertical cylindrical heater or furnace (see heater, vertical
    cylindrical)

View factor, 81, 140, 345, 401

Visbreaking, 2

Viscosity, 55, 57–9, 65, 243, 250, 272

VOCs (see volatile organic compounds)

Volatile organic compounds, 302, 307, 425

Volatiles, 123–4, 151, 345, 363, 369, 374, 421, 440

Volume coefficient of expansion, 68

Volumetric heat transfer coefficient, 299

Vortec Corp., 416

Vortex methods, 152

W

Wall-fired furnace (see furnace, wall-fired)

Wall losses, 140–3

Ware, 419

Waste
  heat recovery
  incineration (see incineration)
  liquids, 12

Water, 13, 22, 29–32, 34–9, 81–2, 85–9, 92, 108, 117, 125,
    128, 132, 137, 155, 157, 164–5, 167, 169, 196,
    205, 209, 221, 231–2, 266, 316, 348, 352, 376,
    387, 400, 438, 445
  cooling, 66, 82, 91, 113, 116, 126, 132–3, 145, 149, 176,
    196, 200, 220–1, 231–3, 236, 243, 274, 278, 290,
    292, 318, 379

Wavelength, 70–1, 73–7, 82, 90, 98–9, 109, 128, 131,
    143–4, 151, 204–5, 216, 316–7, 353, 369–70

Web, 18, 21–2, 129, 139, 348

Weighted-sum-of-gray-gases, 165

Wein's displacement law, 73

Wind tunnel simulator, 224

Wood, 296

Wrinkled flame (see flame, wrinkled)

WS Thermal Process Technology, Inc., 313

X

Xenon arc lamp, 213

Y

Ytterbia, 308–9

Z

Zinc, 2

Zirconia, 298, 439

Milton Keynes UK
Ingram Content Group UK Ltd.
UKHW052027071024
449327UK00027B/2456